EKELAND VARIATIONAL PRINCIPLE

WITH GENERALIZATIONS AND VARIANTS

EKELAND VARIATIONAL PRINCIPLE

WITH GENERALIZATIONS AND VARIANTS

IRINA MEGHEA

édition des archives contemporaines

OLD CITY PUBLISHING

Old City Publishing, Inc.
628 North 2nd Street
Philadelphia, PA 19123
USA

Éditions des Archives Contemporaines
41, rue Barrault
75013 Paris
France

Europe : ISBN 978-2-914610-96-4

Rest of the world : ISBN 978-1-933153-08-7

Technical computing : **Dana Savu**

CONTENTS

INTRODUCTION

In the frame of Variational Calculus, the elementary proposition

"*If* $\varphi : X \to \mathbf{R}$, X *real normed space, has in* x_0 *a local minimum point* (*hence in particular a global minimum point*) *and it is Gâteaux differentiable at* x_0 , *then* $\varphi'_w(x_0) = 0$" (x_0 is *critical point*)

is called **variational principle** ([30], 1; *w* comes from *weak*).

This is the reason why the neighbour propositions, for instance those in which *X* is replaced by a metric space, or in which the statement "$\varphi'_w(x_0) = 0$" is replaced by "$\|\varphi'_w(x_\varepsilon)\| \leq \varepsilon$, $\varepsilon > 0$ any" etc, are called **variational principles (perturbed).** The adjective "perturbed" is imposed by the fact that not the function φ is minimized, but a function of the form $\varphi + \varepsilon\|(\cdot) - x_\varepsilon\|$ (Ekeland), or of more general form $\varphi + \varepsilon \sum_{n=1}^{\infty} \mu_n d(\cdot, v_n)^2$ (Borwein - Preiss), or of even more general form $\varphi + f$ (Deville-Godefroy - Zizler), *f* having some given properties (the second term is the *perturbation function*).

The Ekeland principle is a perturbed variational principle discovered in 1972 ([30], 1) and nowadays, after more than 30 years, it was proved to be, and this monograph will strongly support this, the foundation of the modern Variational Calculus (see, for instance, the minimax theorems in Banach spaces or in the Finsler manifolds, in which the key step of demonstration is made by the application of the Ekeland principle).

As referring the applications, these are numerous and various: the geometry of Banach spaces, nonlinear analysis, differential equations and partial differential equations, global analysis, probabilistic analysis, differential geometry, fixed point theorems, nonlinear semigroups, dynamical systems, optimization, mathematical programming, optimal control.

Finally, we cannot close this INTRODUCTION without the confession of Ekeland ([30], 4):

"The grandfather of these all is the celebrated 1961 theorem of Bishop and Phelps that the set of continuous linear functionals on a Banach space E, which attain their maximum on a prescribed non void closed convex bounded subset $X \subset E$, is norm-dense in E^*".

This Bishop-Phelps theorem can be found in **2.3**$_5$, Chapter I, 2.

*

* *

All the normed spaces that appear in this monograph are real.

Let $f : X \to (-\infty, +\infty]$ be given. The *domain* of f, $\operatorname{dom} f$, is

$$\operatorname{dom} f \overset{\text{def}}{=} \{x \in X : f(x) < +\infty\}.$$

f is *proper* $\overset{\text{def}}{\Longleftrightarrow}$ $\operatorname{dom} f \neq \varnothing$.

N will denote the set of integer numbers ≥ 1.

Both for the functions and for the sequences will be used the terms *increasing* and *strictly increasing*, *decreasing* and *strictly decreasing* instead of *nondecreasing* and *increasing* respectively *nonincreasing* and *decreasing*.

Chapter I

Ekeland variational principle

1

Ekeland principle in complete metric spaces

1.1 Ekeland principle ([30], 1, 2). *Let* (X, d) *be a complete metric space and* $\varphi : X \to \to (-\infty, +\infty]$ *bounded from below, lower semicontinuous and proper. For any* $\varepsilon > 0$ *and* u *of* X *with*

$$\varphi(u) \le \inf \varphi(X) + \varepsilon$$

and for any $\lambda > 0$, *there exists* v_ε *in* X *such that*

(1) $$\varphi(v_\varepsilon) < \varphi(w) + \frac{\varepsilon}{\lambda} d(v_\varepsilon, w)\ \forall\ w \in X \setminus \{v_\varepsilon\}$$

and

(2) $\varphi(v_\varepsilon) \le \varphi(u)$, (3) $d(v_\varepsilon, u) \le \lambda$.

■ Consider the product metric space $X \times \mathbf{R}$, a complete metric space, and, α being a strict positive number, define on the correspondent set the order relation $\le$:

(4) $$(v_1, r_1) \le (v_2, r_2) \Leftrightarrow r_2 - r_1 + \alpha d(v_2, v_1) \le 0$$

(Bishop - Phelps, [4]).

Verify only the antisymmetry:

$$(v_1, r_1) \le (v_2, r_2) \Rightarrow r_1 - r_2 \ge 0,\ (v_2, r_2) \le (v_1, r_1) \Rightarrow r_2 - r_1 \ge 0, \text{ hence } r_1 = r_2 \text{ etc.}$$

Note that, for every (v_0, r_0), the set $A_0 := \{(v, r) \in X \times \mathbf{R}: (v, r) \ge (v_0, r_0)\}$ is closed: $(v', r') \in \overline{A_0} \Rightarrow \exists$ a sequence (v_n, r_n), $n \in \mathbf{N}$ in A_0 with $(v_n, r_n) \to (v', r')$, that is $v_n \to v'$ and $r_n \to r'$, but $r_n - r_0 + \alpha d(v_n, v_0) \le 0\ \ \forall\ n$, take the limit for $n \to \infty$, $(v', r') \in A_0$.

Now we will show

(*) *Let* S *be a closed nonempty set of* $X \times \mathbf{R}$ *for which there is* $\rho > 0$ *so that* $(v, r) \in S \Rightarrow r \ge \rho$. *Then for any* (v_1, r_1) *of* S *there is* (v', r') *in* S, *maximal element relative to* $\le$, *with the property*

$$(v_1, r_1) \le (v', r').$$

▲ Define by induction $(v_n, r_n)_{n\geq 1}$ a sequence of points of S as follows. Let be $S_1 := \{(v, r) \in S : (v, r) \geq (v_1, r_1)\}$ and $\rho_1 := \inf \{r \in \mathbf{R} : (v, r) \in S_1\}$. Suppose (v_n, r_n) defined and let be (5) $S_n := \{(v, r) \in S : (v, r) \geq (v_n, r_n)\}$, (6) $\rho_n := \inf \{r \in \mathbf{R} : (v, r) \in S_n\}$. Then (v_{n+1}, r_{n+1}) is a point taken from S_n, that is (7) $r_{n+1} - r_n + \alpha d(v_{n+1}, v_n) \leq 0$, where r_{n+1} is chosen so that

(8) $$r_n - r_{n+1} \geq \frac{1}{2}(r_n - \rho_n),$$

which is possible as (8) is equivalent to $2r_{n+1} \leq r_n + \rho_n$ and (7) gives us the possibility to diminish r_{n+1} in a convenient manner. The sets S_n are closed (see above A_0), nonempty and $S_n \supset S_{n+1}$ $\forall n \geq 1$. Moreover,

(9) $$|r_{n+1} - \rho_{n+1}| \leq \frac{1}{2^n}|r_1 - \rho| \ \forall n.$$

Indeed, from (8), $2\rho_{n+1} \leq 2r_{n+1} \leq r_n + \rho_n$, hence $0 \leq 2(r_{n+1} - \rho_{n+1}) \leq r_n + \rho_n - 2\rho_{n+1} \leq$ $\leq r_n + \rho_n - 2\rho_n$, hence $|r_{n+1} - \rho_{n+1}| \leq \frac{1}{2}|r_n - \rho_n|$ and hence $|r_n - \rho_n| \leq \frac{1}{2}|r_{n-1} - \rho_{n-1}|$ etc.

Then, from (9) and taking into account the definitions (4) and (5), we have, $\forall$ (v, r) of S_{n+1}, $|r_{n+1} - r| \leq \frac{1}{2^n}|r_1 - \rho|$ and $d(v_{n+1}, v) \leq \frac{1}{2^n}\frac{1}{\alpha}|r_1 - \rho|$ and this implies $\lim_{n\to\infty} \delta(S_n) = 0$. Apply the Cantor theorem, $\bigcap_{n=1}^{\infty} S_n = \{(v', r')\}$. Obviously $(v', r') \geq (v_1, r_1)$ and it remains to show that (v', r') is a maximal element of S. Let be *par absurdum* (v'', r'') of S, $(v'', r'') > (v', r')$. This implies $(v'', r'') > (v_n, r_n)$ $\forall n \geq 1$, hence $(v'', r'') \in S_n$ $\forall n \geq 1$ and hence $(v'', r'') = (v', r')$, contradiction. ▲

Retake the demonstration of the principle. Take $\alpha = \frac{\varepsilon}{\lambda}$ in (4) and then, in (*), $S = \text{epi}\,\varphi$, that is $S = \{(v, r) \in X \times \mathbf{R} : r \geq \varphi(v)\}$ (S is closed as φ is l.s.c.), and therefore $\rho = \inf \varphi(X)$ for instance, $(v_1, r_1) = (u, \varphi(u))$ and let (v_ε, r_0) be a maximal element in S, (10) $(u, \varphi(u)) \leq (v_\varepsilon, r_0)$. Moreover, as $(v_\varepsilon, r_0) \in S$, that is $r_0 \geq \varphi(v_\varepsilon)$, it follows (the definition (4)) (10_1) $(v_\varepsilon, \varphi(v_\varepsilon)) \geq (v_\varepsilon, r_0)$, that implies $r_0 = \varphi(v_\varepsilon)$, $(v_\varepsilon, \varphi(v_\varepsilon))$ a maximal element in S. Then $\forall$ w from $X \setminus \{v_\varepsilon\}$ we have *a fortiori* (*par absurdum*)

$$(w, a) \in S \Rightarrow a - \varphi(v_\varepsilon) + \frac{\varepsilon}{\lambda} d(w, v_\varepsilon) > 0.$$

Take $a = \varphi(w)$ and it gives (1).

From (10) and (10_1) it results (11) $(v_\varepsilon, \varphi(v_\varepsilon)) \geq (u, \varphi(u))$, that implies $\varphi(v_\varepsilon) \leq \varphi(u)$, that is (2). Finally, coming back to (11), that is to $\varphi(v_\varepsilon) - \varphi(u) + \frac{\varepsilon}{\lambda} d(v_\varepsilon, u) \leq 0$, as $\varphi(u) \leq$ $\leq \inf_{x\in X} \varphi(x) + \varepsilon$ and so $\varphi(v_\varepsilon) \geq \varphi(u) - \varepsilon$, it results $\frac{\varepsilon}{\lambda} d(v_\varepsilon, u) \leq \varepsilon$, $d(v_\varepsilon, u) \leq \lambda$. ■

Remark 1. In the frame of the variational calculus, the theorem 1.1 was raised up to the rank of a principle. Nevertheless, it is not justified to assess this title to the other statements. The author could not oppose to the existent situation.

Remark 2. Another demonstration of the Ekeland principle, simple and elementary, follows from the demonstration of 4.1, chapter II.

Remark 3. Here it is another demonstration of the Ekeland principle that does not use even the Cantor theorem (*M. Crandall*). We keep the notations from 1.1 ($\lambda = 1$).

■ Define by recurrence a sequence $(u_n)_{n \geq 0}$. Take $u_0 = u$. Suppose u_n, $n \geq 0$ defined and consider the set

(*) $$S_n := \{w \in X : \varphi(w) \leq \varphi(u_n) - \varepsilon\, d(u_n, w)\}.$$

If $S_n = \varnothing$, one takes $u_{n+1} = u_n$. If $S_n \neq \varnothing$, one takes u_{n+1} from S_n so that

(**) $$\varphi(u_{n+1}) - \inf \varphi(S_n) \leq \frac{1}{2}[\varphi(u_n) - \inf \varphi(S_n)].$$

The definition is still correct in the second case as the second member of (**) is > 0 (attention to the definition of S_n !).

The sequence $(u_n)_{n \geq 0}$ is *fundamental* (*Cauchy sequence*). Indeed, if there is $p \geq 0$ so that $S_p = \varnothing$, we have $u_n = u_p$ $\forall\, n \geq p$. Suppose then $S_n \neq \varnothing$ $\forall\, n \geq 0$. From (*),

(***) $$\varepsilon\, d(u_n, u_{n+1}) \leq \varphi(u_n) - \varphi(u_{n+1}) \;\forall\, n \geq 0$$

and by adding

(****) $$\varepsilon d(u_n, u_p) \leq \varphi(u_n) - \varphi(u_p) \;\forall\, p \geq n \geq 0.$$

The sequence $(\varphi(u_n))_{n \geq 0}$ is decreasing (see (***)) and bounded from below, hence fundamental, this implying, *via* (****), the same property for $(u_n)_{n \geq 0}$. Then let be $v_\varepsilon = \lim\limits_{n \to \infty} u_n$.

v_ε *verifies* (2): $\varphi(u) \geq \varphi(u_n) \geq \underline{\lim}_{n \to \infty} \varphi(u_n) \overset{\text{l.s.c.}}{\geq} \varphi(v_\varepsilon)$.

v_ε *verifies* (3): for $n = 0$ (****) yields

$$\varepsilon d(u, u_p) \leq \varphi(u) - \varphi(u_p) \leq \varphi(u) - \inf \varphi(X) \leq \varepsilon,$$

go to the limit for $p \to \infty$.

v_ε *verifies* (1). Indeed, suppose *par absurdum* $\exists\, w \neq v_\varepsilon$ so that $\varphi(w) \leq \varphi(v_\varepsilon) - \varepsilon d(v_\varepsilon, w)$.

Then $\varphi(w) \leq \varphi(v_\varepsilon) \leq \lim_{p\to\infty} \varphi(u_p) \overset{(****)}{\leq} \varphi(u_n) - \varepsilon d(u_n, v_\varepsilon)\ \forall\, n \geq 0$, that is $w \in S_n\ \forall\, n \geq 0$.

So (**) gives

$$2\varphi(u_{n+1}) - \varphi(u_n) \leq \inf \varphi(S_n) \leq \varphi(w),$$

hence $\lim_{n\to\infty} \varphi(u_n) \leq \varphi(w)$ and hence $\varphi(v_\varepsilon) \leq \varphi(w)$, in contradiction with the assumption. ■

Geometrical interpretation of the fundamental inequality (1) *of the Ekeland principle.* This was given by R. Phelps ([56], 2, p. 45; [2], p. 256).

Let X be a real normed space. In the product vector space $X \times \mathbf{R}$ is considered, for every $\lambda > 0$, the set

$$C_\lambda := \{(x, \alpha) \in X \times \mathbf{R} : \lambda\|x\| \leq -\alpha\}.$$

C_λ is a convex cone with vertex (see §8, preliminaries of affine geometry, 8.1, $\lambda\|x + x'\| \leq \lambda\|x\| + \lambda\|x'\| \leq -(\alpha + \alpha')$), we call it the **cone of Phelps**.

Suppose verified the inequality (1) of 1.1 with $\lambda = 1$. Then we have

(*) $$[(v_\varepsilon, \varphi(v_\varepsilon)) + C_\varepsilon] \cap \operatorname{epi}\varphi = (v_\varepsilon, \varphi(v_\varepsilon)).$$

The inclusion $\supset$ being evident as $(0, 0) \in C_\varepsilon$, pass to $\subset$. Let (x, α) be in C_ε, that is $\varepsilon\|x\| \leq -\alpha$, and $(v_\varepsilon, \varphi(v_\varepsilon)) + (x, \alpha) = (v_\varepsilon + x, \varphi(v_\varepsilon) + \alpha) \in \operatorname{epi}\varphi$, that is

$$\varphi(v_\varepsilon + x) \leq \varphi(v_\varepsilon) + \alpha.$$

But $\varphi(v_\varepsilon) + \alpha \leq \varphi(v_\varepsilon) - \varepsilon\|x\|$, hence $\varphi(v_\varepsilon) \geq \varphi(v_\varepsilon + x) + \varepsilon\|x\|$, therefore, setting $y = v_\varepsilon + x$, $\varphi(v_\varepsilon) \geq \varphi(y) + \varepsilon\|v_\varepsilon - y\|$. This inequality compared with (1) gives $y = v_\varepsilon$, hence $x = 0$, that leads to $\varphi(v_\varepsilon) \leq \varphi(v_\varepsilon) + \alpha \leq \varphi(v_\varepsilon)$, hence $\alpha = 0$ and thus (*).

Reciprocally,

$$(*) \Rightarrow (1)\ (\lambda = 1).$$

Indeed, suppose

$$\varphi(v_\varepsilon) \geq \varphi(y) + \varepsilon\|v_\varepsilon - y\|.$$

It has to demonstrate $y = v_\varepsilon$. For $y = v_\varepsilon + x$ the inequality becomes

$$\varphi(v_\varepsilon) \geq \varphi(v_\varepsilon + x) + \varepsilon\|x\|.$$

But $(x, -\varepsilon\|x\|)\in C_\varepsilon$, hence, according to the assumption, if $(v_\varepsilon, \varphi(v_\varepsilon)) + (x, -\varepsilon\|x\|) = (v_\varepsilon + x, \varphi(v_\varepsilon) - \varepsilon\|x\|) \in \text{epi}\,\varphi$, that is $\varphi(v_\varepsilon + x) \le \varphi(v_\varepsilon) - \varepsilon\|x\|$, then $x = 0$. But the last inequality has been accepted, therefore $x = 0$, $y = v_\varepsilon$.

So

$$(**) \qquad (1)\ (\lambda = 1) \Leftrightarrow (*).$$

Come back to C_λ. This is pointed, that is $C_\lambda \cap (-C_\lambda) = \{0\}$ (see §8, 8.2): (x, α), $(-x, -\alpha) \in C_\lambda \Rightarrow \lambda\|x\| \le -\alpha$ and $\lambda\|x\| \le \alpha$, hence $\lambda\|x\| \le \alpha \le -\lambda\|x\|$ and it results $x = 0$ and $\alpha = 0$.

Consider in $X \times \mathbf{R}$ the binary relation $\overset{C_\lambda}{\le}$: $(x_1, y_1) \overset{C_\lambda}{\le} (x_2, y_2) \overset{\text{def}}{\Leftrightarrow} (x_2, y_2)\in(x_1, y_1) + C_\lambda$. This is an order relation (§8, 8.3). Suppose (*) checked and let (x, α) be whatever element of epi φ, $(x, \alpha) \overset{C_\varepsilon}{\ge} (v_\varepsilon, \varphi(v_\varepsilon))$, that is $(x, \alpha) \in (v_\varepsilon, \varphi(v_\varepsilon)) + C_\varepsilon$. But this implies $(x, \alpha) = (v_\varepsilon, \varphi(v_\varepsilon))$, hence $(v_\varepsilon, \varphi(v_\varepsilon))$ is a maximal element of epi φ. Reciprocally, this property of $(v_\varepsilon, \varphi(v_\varepsilon))$ implies (*): let (x, α) be of epi φ and $(x, \alpha) \in (v_\varepsilon, \varphi(v_\varepsilon)) + C_\varepsilon$, that is $(x, \alpha) \overset{C_\varepsilon}{\ge} (v_\varepsilon, \varphi(v_\varepsilon))$, but then $(x, \alpha) = (v_\varepsilon, \varphi(v_\varepsilon))$, this being maximal, (*) is checked. So, taking also into account for (**), it can say:

The fundamental inequality (1) ($\lambda = 1$) *of the Ekeland principle is equivalent to the property* : $(v_\varepsilon, \varphi(v_\varepsilon))$ *is a maximal element in* epi φ *ordered by the Phelps cone* C_ε.

Now present an improved formulation of the Ekeland principle (Jean-Paul Penot, [55]).

1.1$_0$ *Under the conditions of the theorem* 1.1, *there exists* v_ε *in* X *verifying* (1) *and* (3), *while* (2) *is replaced by*

$$(2_0) \qquad \varphi(v_\varepsilon) \le \varphi(u) - \frac{\varepsilon}{\lambda} d(v_\varepsilon, u).$$

■ Consider the subset

$$M := \{x \in X : \varphi(x) + \frac{\varepsilon}{\lambda} d(x, u) \le \varphi(u)\}.$$

M is nonempty ($u \in M$) and closed (φ is lower semicontinuous), hence M is a complete metric space. Take the function Φ: $M \to (-\infty, +\infty]$,

$$\Phi(x) = \gamma^{-1}\varphi(x),\ \gamma := \frac{\varepsilon}{\lambda}.$$

It is obviously possible to apply 1.1 to Φ with u and $\varepsilon = \lambda$, $\exists\ v_\varepsilon \in M$ so that $d(v_\varepsilon, u) \le \lambda$ and $\Phi(v_\varepsilon) < \Phi(x) + d(v_\varepsilon, x)\ \forall\ x \in M \setminus \{v_\varepsilon\}$, that is

$$\varphi(v_\varepsilon) < \varphi(x) + \gamma\, d(v_\varepsilon, x)\ \forall\ x \in M \setminus \{v_\varepsilon\}.$$

Taking a sequence $(x_n)_{n\ge 1}$, $x_n \in M$, $x_n \to v_\varepsilon$, by going to the lower limit in the inequation defining M it results the inequality (2_0). It remains to show ($v_\varepsilon \in M$!)

$$\varphi(v_\varepsilon) < \varphi(x) + \gamma\, d(v_\varepsilon, x)\ \forall\ x \in X \setminus M.$$

$$x \notin M \Rightarrow \varphi(x) + \gamma\, d(v_\varepsilon, x) = [\varphi(x) + \gamma\, d(u, x)] - \gamma\, d(u, x) + \gamma\, d(v_\varepsilon, x) \overset{x\notin M}{>} \varphi(u) - \gamma\, d(u, x) +$$

$$+ \gamma\, d(v_\varepsilon, x) \overset{(2_0)}{\ge} \varphi(v_\varepsilon) + \gamma\, d(v_\varepsilon, u) - \gamma\, d(u, x) + \gamma\, d(v_\varepsilon, x) \ge \varphi(v_\varepsilon) + \gamma\, d(u, x) - \gamma\, d(u, x) =$$

$= \varphi(v_\varepsilon)$, whence the conclusion. ■

REMARK 1. The fact is that (2_0) is the inequality (11) already obtained during the demonstration of the Ekeland principle 1.1.

REMARK 2. The appearance of (2_0) has a deeper meaning as it is revealed by the equivalence which fololows.

Consider the statement **E**,

E. *Let* X *be a complete metric space and* φ: X→ (−∞, +∞] *bounded from below, lower semicontinuous and proper. For each* u *from* dom φ *and each* $\varepsilon > 0$ *there exists* v_ε *in* X *so that*

(e_1) $$\varphi(v_\varepsilon) \le \varphi(u) - \varepsilon\, d(v_\varepsilon, u),$$

(e_2) $$\varphi(v_\varepsilon) < \varphi(x) + \varepsilon\, d(v_\varepsilon, x)\ \forall\ x \in X \setminus \{v_\varepsilon\}.$$

And now

The Ekeland principle **1.1** *and* **E** *are equivalent.*

■ **E ⇒ 1.1**. Let u be from X and $\varepsilon > 0$ so that

$$\varphi(u) \le \inf \varphi(X) + \varepsilon.$$

According to the assumption, for u and $\eta := \dfrac{\varepsilon}{\lambda}$ there is v_η in X so that (e_1) and (e_2) are verified. Then (1) and (2) from 1.1 are obviously checked, but also (3):

$$d(v_\eta, u) \overset{(e_1)}{\le} \frac{\lambda}{\varepsilon}[\varphi(u) - \varphi(v_\eta)] \le \frac{\lambda}{\varepsilon}[\varphi(u) - \inf \varphi(X)] \le \frac{\lambda}{\varepsilon}\varepsilon = \lambda.$$

1.1 ⇒ E. Let u be from dom φ and $\varepsilon > 0$. Consider the set

$$M := \{x \in X : \varphi(x) - \varphi(u) + \varepsilon d(x, u) \le 0\}.$$

M is nonempty set ($u \in M$) and closed and so a complete metric subspace. Apply 1.1 to φ on M for u, ρ ($> \varphi(u) - \inf \varphi(X)$) and $\lambda = \frac{\rho}{\varepsilon}$, there is v_ε in M so that

$$\varphi(v_\varepsilon) < \varphi(x) + \varepsilon d(v_\varepsilon, x) \ \forall\, x \in M \setminus \{v_\varepsilon\},$$

$$\varphi(v_\varepsilon) \leq \varphi(u),\ d(v_\varepsilon, u) \leq \frac{\rho}{\varepsilon}.$$

Since $v_\varepsilon \in M$, (e_1) is verified. Pass to (e_2). It only remains to show that the last but one inequality is also verified for $x \in X \setminus M$. So

$$\varphi(u) \overset{\text{def}}{<} \varphi(x) + \varepsilon d(x, u).$$

But then

$$\varphi(v_\varepsilon) \overset{(e_1)}{\leq} \varphi(u) - \varepsilon d(v_\varepsilon, u) < \varphi(x) + \varepsilon d(x, u) - \varepsilon d(v_\varepsilon, u) \leq \varphi(x) + \varepsilon d(v_\varepsilon, x). \ \blacksquare$$

We mention here a beautiful remark of F. Sullivan ([67]), who shows that the Ekeland principle, and even a simplified form of this, can operate on a metric space only when it is complete. Precisely,

Let (X, d) *be a metric space so that any function* φ: X$\to$ **R** *bounded from below and continuous has the property*: $\forall\, \varepsilon > 0\ \exists\ v_\varepsilon$ *in* X *so that*

(1) $$\varphi(v_\varepsilon) \leq \varphi(w) + \varepsilon d(v_\varepsilon, w)\ \forall\, w \in X$$

and

(2′) $$\varphi(v_\varepsilon) \leq \inf \varphi(X) + \varepsilon.$$

Then X *is complete* (*F. Sullivan*).

$\blacksquare$ Let $(x_n)_{n \geq 1}$ be a Cauchy sequence in X. Consider the function φ: X$\to$ **R**,

$$\varphi(x) = \lim_{n \to \infty} d(x_n, x).$$

The definition is correct as the sequence $(u_n)_{n\geq 1}$, $u_n := d(x_n, x)$ is fundamental: $|u_n - u_m| \leq$ $\leq d(x_n, x_m)$. φ is uniformly continuous: $|d(x_n, x) - d(x_n, y)| \leq d(x, y)$, take the limit for $n \to \infty$. Moreover, $\inf \varphi(X) = 0$. Indeed, suppose *par absurdum* that there is $\rho > 0$ so that $\varphi(x) > \rho\ \forall\, x \in X$. For $\rho\ \exists N \in \mathbf{N}$ so that $n \geq N \Rightarrow d(x_n, x_N) \leq \rho$, one takes the limit for $n \to \infty$ and the contradiction $\varphi(x_N) \leq \rho$ is obtained.

It will be shown that $\exists\ v_\varepsilon$ in X so that $\varphi(v_\varepsilon) = 0$, that is $v_\varepsilon = \lim\limits_{n\to\infty} x_n$, and so X is complete. Let be ε, $0 < \varepsilon < 1$. The assumption supplies v_ε in X so that $\varphi(v_\varepsilon) = \lim\limits_{n\to\infty} d(x_n,$

$v_\varepsilon) \overset{(2')}{\leq} \varepsilon$ and $\varphi(v_\varepsilon) \leq \varphi(x_n) + \varepsilon d(v_\varepsilon, x_n)$ $\forall\, n \geq 1$. Let m be whatever fixed in **N.** $\exists N \in \mathbf{N}$ so that $n, p \geq N \Rightarrow d(x_n, x_p) \leq \frac{1}{m}$, hence $n \geq N \Rightarrow \varphi(x_n) \leq \frac{1}{m}$. Moreover, as $\lim_{n\to\infty} d(x_n, v_\varepsilon) <$ $< \varepsilon + \frac{1}{m}$, N can be assumed high enough so that $n \geq N \Rightarrow d(x_n, v_\varepsilon) \leq \varepsilon + \frac{1}{m}$. One deduces $\varphi(v_\varepsilon) \leq \frac{1}{m} + \varepsilon(\varepsilon + \frac{1}{m})$. m being whatever, it can take the limit for $m \to \infty$, so $\varphi(v_\varepsilon) \leq \varepsilon^2$. The reasoning is repeated starting from the inequality $\varphi(v_\varepsilon) \leq \varepsilon^2$ and one obtains $\varphi(v_\varepsilon) \leq \varepsilon^3$ and, by induction, $\varphi(v_\varepsilon) \leq \varepsilon^n$ $\forall\, n \geq 1$, whence the conclusion $\varphi(v_\varepsilon) = 0$. ■

Next, a completion is brought to the Ekeland principle, referring to the fundamental inequality (1) ([12]). Firstly,

Let X *be a real normed space and* A *nonempty subset. Then*

$$\text{a) } \overline{\text{conv}}\, A = X$$

and

b) *for each* T *from* $X^* \setminus \{0\}$ *and each* r *from* **R** *there is* $x_r \in A$ *so that* $T(x_r) \leq r$

are equivalent.

Explanation. $\overline{\text{conv}}\, A \equiv$ the closed convex hull of A.

■ a) $\Rightarrow$ b). Suppose *par absurdum*

$$(11_1) \qquad T(x) > r \;\; \forall\, x \in A.$$

$T^{-1}([r, +\infty))$ is closed, convex and includes A ((11_1)), consequently $T^{-1}([r,+\infty)) \supset \overline{\text{conv}}\, A$ and hence $T^{-1}([r, +\infty)) \overset{a)}{=} X$, that is (11_2) $T(x) \geq r$ $\forall\, x \in X$. But $\exists\, x_0$ so that $T(x_0) < 0$ ($T \neq 0 \Rightarrow \exists\, x'$ with $T(x') \neq 0$) and then $nT(x_0) = T(nx_0) \geq r$ $\forall\, n \geq 1$. Pass to the limit for $n \to \infty$ and obtain the contradiction.

b) $\Rightarrow$ a). Suppose *par absurdum* $x_0 \in X \setminus \overline{\text{conv}}\, A$. $\{x_0\}$ being compact, $\overline{\text{conv}}\, A$ closed and both convex, there is a closed hyperplane that separates them very strongly (see hereafter *Memento*), hence $\exists\, T$ in $X^* \setminus \{0\}$ so that $T(x) > T(x_0)$ $\forall\, x \in \overline{\text{conv}}\, A$, in contradiction with b). ■

Memento (separation in real vector spaces)

Let X be real vector space, H hyperplane and $h(x) = \alpha$ its equation, A and B nonempty subsets of X.

H separates A and *B* resp. *H separates very strongly A* and *B*

if

$$h(x) \le \alpha \ \forall\, x \in A \text{ and } h(y) \ge \alpha \ \forall\, y \in B$$

resp.

$$\exists\, \rho > 0 \text{ so that } h(x) \le \alpha - \rho \ \forall\, x \in A \text{ and } h(y) \ge \alpha + \rho \ \forall\, y \in B.$$

Kakutani-Tuckey theorem. Let X be a real vector space and A, B convex disjoint subsets. Then there exist C, D convex subsets such that

$$C \supset A,\ D \supset B,\ C = X \setminus D.$$

Hahn-Banach-Mazur theorem (*first geometric form*). Let X be a real normed space, M open convex nonempty subset and V a linear manifold with

$$M \cap V = \varnothing.$$

Then there is H closed hyperplane so that

$$H \supset V,\ H \cap M = \varnothing.$$

Hahn-Banach theorem (*second geometric form*). Let X be a real normed space, A convex with $\overset{\circ}{A} \ne \varnothing$ and B convex nonempty. If

$$\overset{\circ}{A} \cap B = \varnothing$$

then A and B are separating by a closed hyperplane.

Hahn-Banach theorem (*third geometric form*). In a real normed space two disjoint convex nonempty sets, one of them being closed and the other compact, are separating very strongly by a closed hyperplane.

Dunford-Schwartz theorem. In a real locally convex space any two closed disjoint nonempty convex sets, one of them being compact, are separating very strongly by a continuous linear functional.

For all these theorems see [47], vol. III, pages 706, 707, 708, 709 and 721 respectively.

Now present the announced completion.

Let X *be a real Banach space and* $\varphi : \mathrm{X} \to \mathbf{R}$ *bounded from below, lower semicontinuous. Suppose*

$$(11_3) \qquad \overline{\lim_{\|x\| \to +\infty}} \frac{\varphi(x)}{\|x\|} < +\infty.$$

Whatever $\varepsilon > 0$, *if*

(11_4) $$\varepsilon > \overline{\lim_{\|x\|\to+\infty}} \frac{\varphi(x)}{\|x\|},$$

then

(11_5) $$\overline{\text{conv}}\, E_{\varphi,\varepsilon} = X,$$

$$E_{\varphi,\varepsilon} := \{v \in X: \varphi(v) < \varphi(x) + \varepsilon\|v-x\| \ \forall\, x \in X \setminus \{v\}\}$$

(*F. Cammaroto, A. Chinni*).

■ Let T be from $X^* \setminus \{0\}$ and r from **R** whatever fixed. According to b) from the last proposition, it remains to show

(11_6) $$E_{\varphi,\varepsilon} \cap T^{-1}((-\infty, r]) \neq \varnothing.$$

Consider the function $f: X \to \mathbf{R}$,

(11_7) $$f(x) = \frac{1}{\varepsilon}[\varphi(x) - \inf \varphi(X)].$$

For any x from X, we have

(11_8) $$\sup_{y\in X, \|y\in\|=1} \ \inf_{\lambda\in(0,+\infty)} \frac{f(x-\lambda y)}{\lambda} < 1.$$

Indeed, for $y \in X$, $\|y\| = 1$

$$\overline{\lim_{\lambda\to+\infty}} \frac{f(x-\lambda y)}{\lambda} = \overline{\lim_{\lambda\to+\infty}} \frac{f(x-\lambda y)}{\|x-\lambda y\|}\frac{\|x-\lambda y\|}{\lambda} = \overline{\lim_{\lambda\to+\infty}} \frac{f(x-\lambda y)}{\|x-\lambda y\|} = \overline{\lim_{\|z\|\to+\infty}} \frac{f(z)}{\|z\|}$$ (for the last equality: $F(\lambda) := \dfrac{f(x-\lambda y)}{\|x-\lambda y\|} \overset{x-\lambda y=z}{=} \dfrac{f(z)}{\|z\|} =: G(\|z\|)$ etc.), consequently

$$\sup_{y\in X, \|y\|=1} \ \inf_{\lambda\in(0,+\infty)} \frac{f(x-\lambda y)}{\lambda} \leq \sup_{y\in X, \|y\|=1} \ \overline{\lim_{\lambda\to+\infty}} \frac{f(x-\lambda y)}{\lambda} = \overline{\lim_{\|z\|\to+\infty}} \frac{f(z)}{\|z\|} \overset{(11_7)}{=}$$

$$= \frac{1}{\varepsilon} \overline{\lim_{\|z\|\to+\infty}} \frac{\varphi(z)}{\|z\|} \overset{(11_4)}{<} 1$$ and hence (11_8).

Now let us demonstrate that there exists x_0 in $T^{-1}((-\infty, r])$ so that

(11_9) $$f(x_0) \leq \text{dist}(x_0, T^{-1}(r)).$$

Indeed, let be (11_{10}) $x_1 \in T^{-1}(r)$. We have

(11_{11}) $$\sup_{y\in X, \|y\|=1} \ \inf_{\lambda\in(0,+\infty)} \frac{f(x_1-\lambda y)}{\lambda} \overset{(11_8)}{<} 1 = \sup_{\|x\|=1} \frac{|T(x)|}{\|T\|}.$$

From (11_{11}) it results that there is x_2 with $\|x_2\| = 1$ so that

$$\sup_{y\in X,\|y\|=1}\ \inf_{\lambda\in(0,+\infty)} \frac{f(x_1-\lambda y)}{\lambda} < \frac{T(x_2)}{\|T\|}$$

(in case it is needed x_2 is replaced by $-x_2$) and hence in particular

$$\inf_{\lambda\in(0,+\infty)} \frac{f(x_1-\lambda x_2)}{\lambda} < \frac{T(x_2)}{\|T\|}.$$

Take then $\mu > 0$ so that

$$(1\,1_{12}) \qquad \frac{f(x_1-\mu x_2)}{\mu} < \frac{T(x_2)}{\|T\|}.$$

As $T(x_2) > 0$, $T(x_1 - \mu x_2) \overset{(1\,1_{10})}{=} r - \mu T(x_2) < r$, that is $x_1 - \mu x_2 \in T^{-1}(r)$. Moreover, $f(x_1 - \mu x_2) \overset{(1\,1_{12})}{<} \mu \frac{T(x_2)}{\|T\|} \overset{(1\,1_{10})}{=} \frac{r-T(x_1-\mu x_2)}{\|T\|}$. Nevertheless, according to the lemma of Ascoli, the last term is equal to dist $(x_1 - \mu x_2, T^{-1}(r))$. Take $x_0 = x_1 - \mu x_2$ and yields $(1\,1_9)$.

Note $S := \{x \in X : \|x_0-x\| \le f(x_0)\}$. We have

$$(1\,1_{13}) \qquad S \subset T^{-1}((-\infty, r]).$$

Indeed, if $x_0 \in T^{-1}(r)$, then $f(x_0) \overset{(1\,1_9)}{=} 0$ and $(1\,1_{13})$ is verified as $S = \{x_0\}$ and $x_0 \in$ $\in T^{-1}((-\infty, r])$. Suppose now $(1\,1_{14})$ $x_0 \notin T^{-1}(r)$ and *par absurdum* $\exists\, y \in S$ so that $T(y) > r$. It follows, as $T(x_0) < r$, $\exists\, \lambda \in (0, 1)$ so that $(1\,1_{15})$ $T(\lambda x_0 + (1 - \lambda)y) = r$. Then we have $\|x_0-y\| \le f(x_0) \overset{(1\,1_9)}{\le}$ dist $(x_0, T^{-1}(r)) \overset{(1\,1_{15})}{\le} \|x_0-\lambda x_0 -(1-\lambda)y\| = (1 - \lambda) \times$ $\times \|x_0 - y\| < \|x_0 - y\|$, contradiction and consequently $(1\,1_{13})$.

Now, if $f(x_0) = 0$, then $\varphi(x_0) \overset{(1\,1_7)}{=} \inf \varphi(X)$ and hence $x_0 \in E_{\varphi,\varepsilon}$, $(1\,1_6)$ is verified. Suppose $f(x_0) > 0$. We have $\varphi(x_0) = \inf \varphi(X) + \varepsilon f(x_0)$. Apply 1.1 with $u = x_0$, ε replaced by $\varepsilon f(x_0)$ and $\lambda = f(x_0)$. There is v_ε in S so that $\varphi(v_\varepsilon) < \varphi(x) + \varepsilon\|v_\varepsilon - x\|\ \forall\, x \in X \setminus \{v_\varepsilon\}$, that is $v_\varepsilon \in E_{\varphi,\varepsilon}$ and consequently $(1\,1_6)$ *via* $(1\,1_{13})$. ■

Remark. The statement will be extended over the vector-valued functions (§9, C).

We go on with a variant of Ekeland principle 1.1 ([71], 1).

1.1_1 Theorem. *Let* X *be a complete metric space and* φ: X $\to (-\infty, +\infty]$ *bounded from below, lower semicontinuous and proper. For any* $\varepsilon > 0$ *and* u *from* X *with*

$$\varphi(u) < \inf \varphi(X) + \varepsilon$$

and whatever be $\lambda > 0$, u_0 *from* X *and* h: $[0, +\infty) \to [0, +\infty)$ *increasing with*

$$\int_0^{+\infty} \frac{dr}{1+h(r)} = +\infty, \int_{r_0}^{r_0+r_1} \frac{dr}{1+h(r)} \geq \lambda, r_0 := d(u, u_0), r_0 + r_1 > 0,$$

there is v_ε *in* X *so that*

(12$_1$) $$\varphi(v_\varepsilon) \leq \varphi(w) + \frac{\varepsilon}{\lambda[1+h(d(v_\varepsilon, u_0))]} d(v_\varepsilon, w) \ \forall\, w \in X$$

and

(12$_2$) $\varphi(v_e) \leq \varphi(u)$, (12$_3$) $d(v_\varepsilon, u_0) \leq r_0 + r_1$

(*Zhong Cheng-Kui*).

Elucidation. The claim for r_0, r_1 is correct as $\int_{r_0}^{+\infty} \frac{dr}{1+h(r)} = +\infty$. $\int_{r_0}^{r_0+r_1} \frac{dr}{1+h(r)} \geq \lambda$ imposes $r_1 > 0$ (*par absurdum*).

■ *Building the sequence* $(x_n)_{n \geq 1}$. By recurrence is built a sequence $(x_n)_{n \geq 1}$, $x_n \in X$. Take (12$_5$) $x_1 = u$. Suppose x_n defined. There are two possible cases:

$$\text{I} \quad \varphi(x_n) \leq \varphi(x) + \frac{\varepsilon}{\lambda[1+h(d(x_n, u_0))]} d(x_n, x) \ \forall\, x \in X$$

and

$$\text{II} \quad E_n := \{x \in X : \varphi(x_n) > \varphi(x) + \frac{\varepsilon}{\lambda[1+h(d(x_n, u_0))]} d(x_n, x)\} \neq \varnothing.$$

In the case I take $x_{n+1} = x_n$. In the case II take x_{n+1} of E_n that also verifies

(12$_6$) $$\varphi(x_{n+1}) < \inf \varphi(E_n) + \frac{1}{n+1}.$$

We see that in the case I it follows *a fortiori* (12$_7$) $x_{n+p} = x_n$ $\forall\, p \geq 1$ and $x_i \in E_i$, $i = \overline{1, n-1}$. In the case II it is obviously $x_{n+1} \neq x_n$. Anyway, for the sequence $(x_n)_{n \geq 1}$ is verified

(12$_8$) $$\varphi(x_{n+1}) \leq \varphi(x_n) - \frac{\varepsilon}{\lambda[1+h(d(x_n, u_0))]} d(x_n, x_{n+1}) \ \forall\, n \geq 1.$$

$(x_n)_{n \geq 1}$ *verifies*

(12$_9$) $$d(x_n, u_0) < r_0 + r_1 \ \forall\, n \geq 1.$$

Demonstration by induction. n = 1. This is evident as $r_0 + r_1 > 0$. (12_{10}) < n ⇒ n. Suppose *par absurdum* $\exists\, n_0 > 1$ so that

(12_{11}) $$d(x_k, u_0) < r_0 + r_1,\ 1 \le k < n_0 \text{ and } d(x_{n_0}, u_0) \ge r_0 + r_1.$$

Apply successively (12_8), $\varphi(x_{n_0}) \le \varphi(x_{n_0-1}) - \dfrac{\varepsilon}{\lambda[1+h(d(x_{n_0-1}, u_0))]}\, d(x_{n_0-1}, x_{n_0}) \le \ldots$

$\le \varphi(x_1) - \sum\limits_{k=1}^{n_0-1} \dfrac{\varepsilon}{\lambda[1+h(d(x_k,u_0))]}\, d(x_k, x_{k+1})$, hence $\sum\limits_{k=1}^{n_0-1} \dfrac{\varepsilon}{\lambda[1+h(d(x_k,u_0))]}\, d(x_k, x_{k+1}) \le \varphi(x_1) -$

$- \varphi(x_{n_0}) \le \varphi(u) - \inf \varphi(X) < \varepsilon$ (hypothesis!) and therefore

(12_{12}) $$\sum_{k=1}^{n_0-1} \frac{1}{\lambda[1+h(d(x_k,u_0))]}\, d(x_k, x_{k+1}) < 1.$$

Suppose the sequence $d(x_k, u_0)$, $k = \overline{1, n_0}$ *increasing.*

Then, as h is also increasing, $\sum\limits_{k=1}^{n_0-1} \dfrac{1}{1+h(d(x_k,u_0))}\, d(x_k, x_{k+1}) \ge \sum\limits_{k=1}^{n_0-1} \dfrac{1}{1+h(d(x_k,u_0))} \times$

$\times [d(x_{k+1}, u_0) - d(x_k, u_0)] \ge \sum\limits_{k=1}^{n_0-1} \displaystyle\int_{d(x_k,u_0)}^{d(x_{k+1},u_0)} \frac{dr}{1+h(r)} = \int_{d(x_1,u_0)}^{d(x_{n_0},u_0)} \frac{dr}{1+h(r)} \overset{(12_{11}),(12_5)}{\ge} \int_{r_0}^{r_0+r_1} \frac{dr}{1+h(r)} \ge$

$\ge \lambda$ (hypothesis!) and it gets in contradiction with (12_{12}), consequently (12_{10}), and hence also (12_9), is demonstrated.

Suppose the sequence $d(x_k, u_0)$, $k = \overline{1, n_0}$ *not necessarily increasing.* Even in this situation a contradiction will be found that will demonstrate (12_{10}), that is (12_9). Let k_1 be the lowest index for which (12_{13}) $d(x_{k_1}, u_0) > d(x_{k_1+1}, u_0)$. Obviously $k_1 \ge 1$. As $d(x_{n_0},$ $u_0) \overset{(12_{11})}{\ge} r_0 + r_1 \overset{(12_{11})}{>} d(x_k, u_0)$, $k = \overline{1, n_0 - 1}$, we have *a fortiori* $k_1 \le n_0 - 2$. $\dfrac{1}{1+h(d(x_{k_1}, u_0))} \times$

$\times\, d(x_{k_1}, x_{k_1+2}) \le \dfrac{1}{1+h(d(x_{k_1}, u_0))}\, [d(x_{k_1}, x_{k_1+1}) + d(x_{k_1+1}, x_{k_1+2})] \overset{(12_{13})}{\le} \dfrac{1}{1+h(d(x_{k_1}, u_0))} \times$

$\times\, d(x_{k_1}, x_{k_1+1}) + \dfrac{1}{1+h(d(x_{k_1+1}, u_0))}\, d(x_{k_1+1}, x_{k_1+2})$, so that by replacing in the first

member of (12_{12}) the sum of the two terms where x_{k_1+1} appears by $\dfrac{1}{1+h(d(x_{k_1}, u_0))} \times$

$\times$ d(x_{k_1}, x_{k_1+2}), this is diminished. If d(x_{k_1}, u_0) $\le$ d(x_{k_1+2}, u_0), it is going further with x_{k_1+2} by repeating what was done in the case of x_{k_1}. If d(x_{k_1}, u_0) > d(x_{k_1+2}, u_0), let k_2 be the first index for which d(x_{k_1}, u_0) $\le$ d(x_{k_2}, u_0), d(x_{k_2}, u_0) is compared as above with d(x_{k_2+1}, u_0) etc. In this way a strict increasing sequence of indices is obtained

$$1, k_1, k_2, \ldots, k_p, n_0$$

for which we have

$$d(x_1, u_0) \le d(x_{k_1}, u_0) \le \ldots \le d(x_{k_p}, u_0)$$

and the inequality

$$(12_{14}) \qquad \sum_{j=0}^{p} \frac{1}{\lambda[1+h(d(x_{k_j}, u_0))]} d(x_{k_j}, x_{k_{j+1}}) < 1,\ k_0 := 1,\ k_{p+1} := n_0.$$

But

$$\sum_{j=0}^{p} \frac{1}{1+h(d(x_{k_j}, u_0))} d(x_{k_j}, x_{k_{j+1}}) \ge \sum_{j=0}^{p} \frac{1}{1+h(d(x_{k_j}, u_0))} [d(x_{k_{j+1}}, u_0) - d(x_{k_j}, u_0)] \ge$$

$$\sum_{j=0}^{p} \int_{d(x_{k_j}, u_0)}^{d(x_{k_{j+1}}, u_0)} \frac{dr}{1+h(r)} = \int_{d(x_1, u_0)}^{d(x_{n_0}, u_0)} \frac{dr}{1+h(r)} \overset{(12_{11})}{\ge} \int_{r_0}^{r_0+r_1} \frac{dr}{1+h(r)} \ge \lambda \text{ (hypothesis)}$$ and it gets in contradiction with (12_{14}).

$(x_n)_{n\ge1}$ *is a Cauchy sequence.* As it was built, the sequence $(x_n)_{n\ge1}$ is in one of the following hypostases (see above):

$$(12_{15}) \qquad \exists N \in \mathbf{N} \text{ so that } x_N = x_{N+p}\ \forall p \ge 1\ ((12_7))$$

and

$$(12_{16}) \qquad x_n \ne x_{n+1}\ \forall n \ge 1.$$

We pass obviously to (12_{16}). From (12_8) yields (see above)

$$\sum_{k=1}^{n} \frac{\varepsilon}{\lambda[1+h(d(x_k, u_0))]} d(x_k, x_{k+1}) \le \varphi(x_1) - \varphi(x_{n+1}) \overset{(12_5)}{\le} \varphi(u) - \inf \varphi(X) < \varepsilon,$$

consequently, getting to the limit,

$$(12_{17}) \qquad \sum_{n=1}^{\infty} \frac{1}{1+h(d(x_n, u_0))} d(x_n, x_{n+1}) \le \lambda.$$

Let $\gamma > 0$ be so that $\gamma \ge 1 + h(r_0 + r_1)$. h being increasing, it follows (see (12_9))

$$(12_{18}) \qquad \frac{\gamma}{1+h(d(x_n,u_0))} \ge 1 \ \forall\, n \ge 1.$$

Then, $\forall\, p \ge 1$,

$$d(x_{n+p}, x_n) \le \sum_{k=n}^{n+p-1} d(x_k, x_{k+1}) \overset{(12_{18})}{\le} \gamma \sum_{k=n}^{n+p-1} \frac{1}{1+h(d(x_k,u_0))} d(x_k, x_{k+1}) \le \gamma \sum_{k=n}^{\infty} \frac{1}{1+h(d(x_k,u_0))} \times$$

$\times\ d(x_k, x_{k+1})$ and as the series $\sum_{n=1}^{\infty} \frac{1}{1+h(d(x_n,u_0))} d(x_n, x_{n+1})$ is convergent ((12_{17})), one concludes $(x_n)_{n\ge1}$ is a Cauchy sequence.

Let be (12_{19}) $v_\varepsilon := \lim_{n\to\infty} x_n$.

v_ε *verifies* (12_3). In (12_9) take the limit for $n \to \infty$.

v_ε *verifies* (12_2). From (12_8) it follows $(\varphi(x_n))_{n\ge1}$ decreasing sequence. Let N be whatever fixed in $\mathbf{N}$. $n \ge N \Rightarrow \varphi(x_n) \le \varphi(x_N)$, one takes the lower limit for $n \to \infty$ and yields (12_{20}) $\varphi(v_\varepsilon) \le \varphi(x_N)\ \forall\, N \ge 1$.

In particular $\varphi(v_\varepsilon) \le \varphi(x_1)$, that is (12_2).

v_ε *verifies* (12_1). Let v_1 be, *par absurdum*, from X so that

$$(12_{21}) \qquad \varphi(v_1) < \varphi(v_\varepsilon) - \frac{\varepsilon}{\lambda[1+h(d(v_\varepsilon,u_0))]} d(v_\varepsilon, v_1).$$

So we have (12_{22}) $v_1 \ne v_\varepsilon$. Since $x_n \to v_\varepsilon$ and hence $d(x_n, u_0) \to d(v_\varepsilon, u_0)$, $d(x_n, v_1) \to$ $\to d(v_\varepsilon, v_1)$, $\exists\, N \in \mathbf{N}$ so that

$$(12_{23}) \qquad n \ge N \Rightarrow \varphi(v_1) < \varphi(x_n) - \frac{\varepsilon}{\lambda[1+h(d(x_n,u_0))]} d(x_n, v_1).$$

We read in (12_{23}) that (12_{24}) $v_1 \in E_n\ \forall n \ge N$. Suppose $n \ge N$. Then

$$\inf \varphi(E_n) + \frac{1}{n+1} \overset{(12_6)}{\ge} \varphi(x_{n+1}) \overset{(12_{23})}{>} \varphi(v_1) + \frac{\varepsilon}{\lambda[1+h(d(x_{n+1},u_0))]} d(x_{n+1}, v_1) \overset{(12_{24})}{\ge} \inf \varphi(E_n) +$$

$$+ \frac{\varepsilon}{\lambda[1+h(d(x_{n+1},u_0))]} d(x_{n+1}, v_1),$$

consequently, by combining also with (12_{18}),

$$d(x_{n+1}, v_1) \le \frac{\lambda}{\varepsilon} \frac{\gamma}{n+1} \quad \forall\, n \ge N.$$

Passing to the limit it yields $v_\varepsilon = v_1$, in contradiction with (12_{22}). ■

Remark 1. In the theorem 1.1_1 taking $h = 0$ and $u_0 = u$ one obtains the Ekeland principle 1.1, but with the fundamental inequality not strict [$d(v_\varepsilon, u) \le r_1 \;\; \forall\, r_1 \ge \lambda$ $\left(\int_{r_0}^{r_0+r_1} \frac{dr}{1+h(r)} = r_1\right)$ implies $d(v_\varepsilon, u) \le \lambda$] and with a claim for a strict inequality.

Remark 2. The claim "*h* continuous" of the Theorem 1.1 ([71], 1) is obviously useless, as the monotony of *h* assures its Riemann integrability.

Remark 3. The following statement is clearly inspired by the theorem 1.1_1. It will show, on one side, that 1.1_1 can be demonstrated with the aid of the Ekeland principle 1.1 and, on the other side, that the claim "$\varphi(u) < \inf \varphi(X) + \varepsilon$" may be relaxed to $\le$, while (12_1) may be strenghtened to $<$ with $w \in X \setminus \{v_\varepsilon\}$.

Let X *be a complete metric space,* $\varphi : X \to (-\infty, +\infty]$ *lower bounded, lower semicontinuous, proper and* $u \in \operatorname{dom} \varphi$. *Whatever* u_0 *of* X *and* $H: [0,+\infty) \to [0, +\infty)$ *decreasing with* $\int_0^{+\infty} H(r)dr = +\infty$, $\int_{r_0}^{r_0+r_1} H(r)dr \ge \varphi(u) - \inf \varphi(X)$, $r_0 := d(u, u_0)$, $r_0 + r_1 > 0$, *there is* v^* *in* X *so that*

a) $\varphi(v^*) < \varphi(x) + H(d(v^*, u_0))d(v^*, x) \;\; \forall\, x \in X \setminus \{v^*\}$,

b) $\varphi(v^*) \le \varphi(u)$, c) $d(v^*, u_0) \le r_0 + r_1$

(*T. Q. Bao - P. Q. Khanh*)*).

■ Start with the remark $H(r) > 0 \;\; \forall\, r \ge 0$ (*par absurdum*, *H* is decreasing with divergent integral).

Note, for any *x* of *X*, $\rho(x) := d(x, u_0)$ and consider the set

$$M(x) := \{y \in X : \rho(x) \le \rho(y) \text{ and } \varphi(y) - \varphi(x) + \int_{\rho(x)}^{\rho(y)} H dr \le 0\}.$$

Obviously $x \in M(x)$ and

(α) $$y \in M(x) \Rightarrow M(y) \subset M(x)$$

---

*) *Truong Quang Bao - Phan Quoc Khanh, Are several recent generalizations of Ekeland's variational principle more general than the original principle* ? Acta Mathematica Vietnamica, vol. 28, **3**, 2003, 345 - 350.

(add the inequalities where appear the integrals). Moreover, $M(x)$ is closed: let be $x_n \to x_0$, $x_n \in M(x)\ \forall\ n \geq 1$, thus

$$\rho(x) \leq \rho(x_n) \text{ and } \varphi(x_n) - \varphi(x) + \int_{\rho(x)}^{\rho(x_n)} H(r)\mathrm{d}r \leq 0;$$

take the lower limit for $n \to \infty$ and as ρ is continuous, φ l.s.c. and $s \to \int_0^s H\mathrm{d}r$ continuous (H is Riemann integrable on every compact interval of $[0, +\infty)$ being monotonous), it yields

$$\rho(x) \leq \rho(x_0) \text{ and } \varphi(x_0) - \varphi(x) + \int_{\rho(x)}^{\rho(x_0)} H\mathrm{d}r \leq \varliminf_{n\to\infty} \varphi(x_n) - \varphi(x) + \int_{\rho(x)}^{\rho(x_0)} H\mathrm{d}r \leq 0,\ x_0 \in M(x).$$

So $M(x)$ is complete metric space.

We have

$$(\beta) \qquad y \in M(u) \Rightarrow \rho(y) \leq r_0 + r_1 .$$

Indeed,

$$\int_{\rho(u)}^{\rho(y)} H\mathrm{d}r \overset{\text{def}}{\leq} \varphi(u) - \varphi(y) \leq \varphi(u) - \inf \varphi(X) \leq \int_{r_0}^{r_0+r_1} H\mathrm{d}r ,$$

hence $\int_{r_0}^{\rho(y)} H\mathrm{d}r \leq \int_{r_0}^{r_0+r_1} H\mathrm{d}r$ and hence (β) because $H \geq 0$.

Now the statement **E** (Remark 2 of 1.1_0) enters in action (it was seen there that **E** and Ekeland principle 1.1 are equivalent). Apply **E** to $\varphi|M(u)$ with u initial point and $\varepsilon = H(r_0 + r_1)$, there exists u^* in $M(u)$ so that

$$(\gamma) \qquad \varphi(u^*) - \varphi(u) + H(r_0 + r_1)\mathrm{d}(u^*, u) \leq 0,$$

$$(\delta) \qquad \varphi(x) - \varphi(u^*) + H(r_0 + r_1)\mathrm{d}(u^*, x) > 0\ \forall\ x \in M(u) \setminus \{u^*\}.$$

Apply once again **E** to φ on X with u^* initial point and $\varepsilon = H(\rho(u^*))$, there is v^* in X so that

$$(\nu) \qquad \varphi(v^*) - \varphi(u^*) + H(\rho(u^*))\mathrm{d}(v^*, u^*) \leq 0,$$

$$(\omega) \qquad \varphi(x) - \varphi(v^*) + H(\rho(u^*))\mathrm{d}(v^*, x) > 0\ \forall\ x \in X \setminus \{v^*\}.$$

We show that v^* verifies a), b), and c). By adding (γ) and (ν) it gets $\varphi(u) - \varphi(v^*) \geq 0$, that is b). It remains to check a) and c).

The case $v^* = u^*$. In this situation (ω) is identical to a) and (β) implies c).

The case $v^* \neq u^*$. Suppose *par absurdum* $v^* \in M(u)$. Then, taking $x = v^*$ in (δ) and $y = u^*$ in (β), it follows (*H* is decreasing)

$$\varphi(v^*) - \varphi(u^*) + H(\rho(u^*))d(v^*, u^*) > 0,$$

in contradiction with (ν).

So $v^* \notin M(u)$, hence also, according to (α), $v^* \notin M(u^*)$ as $u^* \in M(u)$. Suppose *par absurdum* $\rho(u^*) \leq \rho(v^*)$. Then

$$\varphi(v^*) - \varphi(u^*) + H(\rho(u^*))d(u^*, v^*) \geq \varphi(v^*) - \varphi(u^*) + \int_{\rho(u^*)}^{\rho(v^*)} H dr > 0$$

$$\left(\int_{\rho(u^*)}^{\rho(v^*)} H dr \leq H(\rho(u^*))[\rho(v^*) - \rho(u^*)] \leq H(\rho(u^*))d(u^*, v^*) \right)$$

and one gets in collision with (ν).

So, $v^* \notin M(u)$ and $\rho(v^*) < \rho(u^*)$. Then $d(v^*, u_0) = \rho(v^*) < \rho(u^*) \overset{(\beta)}{\leq} r_0 + r_1$ and consequently c). Moreover, as H is decreasing we have $H((\rho(v^*)) \geq H(\rho(u^*))$ and combining with (ω) yields a). ■

We are now able to justify all the statements contained in the Remark 3. For this it is enough to take in the statement Bao - Khanh

$$H(r) = \frac{\varepsilon}{\lambda(1 + h(r))}.$$

Moreover, we can also assert (**E** is found in the Remark 2 of 1.1_0)

The Ekeland principle 1.1, *the statement* **E** *and the statement Bao - Khanh are equivalent.*

■ Indeed, taking in B - K $u_0 = u$, $H(r) = \dfrac{\varepsilon}{\lambda}$ and $r_1 = \lambda$ yields 1.1. ■

Now, here it is a statement whence the Ekeland principle 1.1 results, but for functions with values in **R** and which catch only the fundamental inequality (1) ([10])

1.1_2 *Let* $(X, \overset{B}{\leq})$ *be a preordered set, with the property that any decreasing sequence has a minorant, and* φ: X → (−∞, +∞]. *Suppose* φ *increasing and lower bounded. Then* $\forall$ x_0 *from* X *there exists* v *in* X, $v \leq x_0$ *so that*

(12_{25}) $$x \in X, x \overset{B}{\leq} v \Rightarrow \varphi(x) = \varphi(v)$$

(*H. Brezis, F.E. Browder*).

■ A decreasing sequence $(x_n)_{n\geq 0}$ from X is built by recurrence, having x_0 as first term. Suppose x_n is obtained. If x_n verifies (12_{25}), the building is stopped, the demonstration is finished. In the contrary case, $\exists\, x' \overset{B}{<} x_n$ with $\varphi(x') < \varphi(x_n)$ and take x_{n+1} so that

$$(12_{26}) \qquad x_{n+I} \in I_n := \{y \in X : y \overset{B}{\leq} x_n\}$$

and

$$(12_{27}) \qquad \varphi(x_{n+1}) - \inf \varphi(I_n) \leq \frac{1}{2}\,[\varphi(x_n) - \inf \varphi(I_n)].$$

(12_{27}) is indeed possible as $\varphi(x_n) - \inf \varphi(I_n) > 0$ since $x_n \in I_n$. Let y_0 be of X so that $x_n \overset{B}{\geq} y_0$ $\forall n \geq 0$. It follows that the sequence $(\varphi(x_n))_{n\geq 0}$ is decreasing and lower bounded, therefore convergent, and (12_{28}) $\varphi(y_0) \leq \lim\limits_{n\to\infty} \varphi(x_n)$. y_0 verifies (12_{25}). Indeed, suppose *par absurdum* $\exists\, z_0 \in X$, $z_0 \overset{B}{\leq} y_0$ and (12_{29}) $\varphi(z_0) < \varphi(y_0)$ (φ is increasing!). Then $z_0 \in I_n$ and hence, from (12_{27}),

$$\varphi(x_{n+1}) - \frac{1}{2}\,\varphi(x_n) \leq \frac{1}{2}\,\varphi(z_0)\ \forall\, n \geq 0.$$

Taking the limit yields, by using (12_{28}), $\varphi(y_0) \leq \varphi(z_0)$, in contradiction with (12_{29}). ■

We are placed now under the conditions of the Ekeland principle 1.1 with $\varphi\colon X \to \mathbf{R}$, take $\varepsilon > 0$ and, by using 1.1_2, we find v in X checking the fundamental inequality (1):

$$(12_{30}) \qquad \varphi(v) < \varphi(x) + \varepsilon \mathrm{d}(v, x)\ \forall\, x \in X \setminus \{v\}.$$

Consider in the set X the binary relation

$$(12_{31}) \qquad x \overset{B}{\leq} y \overset{def}{\Leftrightarrow} \varphi(x) \leq \varphi(y) - \varepsilon \mathrm{d}(x, y).$$

This is a preorder relation:

$x \overset{B}{\leq} x$ (evidently); $x \overset{B}{\leq} y$ and $y \overset{B}{\leq} z \Rightarrow x \overset{B}{\leq} z$: $\varphi(x) \leq \varphi(y) - \varepsilon \mathrm{d}(x, y)$ and $\varphi(y) \leq \varphi(z) - \varepsilon \mathrm{d}(y, z)$ $\Rightarrow \varphi(x) \leq \varphi(z) - \varepsilon \mathrm{d}(y, z) - \varepsilon \mathrm{d}(x, y) \leq \varphi(z) - \varepsilon \mathrm{d}(x, z)$.

Let us demonstrate that any sequence $(x_n)_{n\geq 1}$ decreasing from X has a minorant. $x_{n+1} \overset{B}{\leq} x_n \Rightarrow (12_{32})$ $\varphi(x_n) - \varphi(x_{n+1}) \geq \varepsilon \mathrm{d}(x_n, x_{n+1})$, hence $(\varphi(x_n))_{n\geq 1}$ is decreasing and thus convergent, being lower bounded by $\inf \varphi(X)$. It follows from (12_{32}), $\forall\, p \geq 1$, $\varepsilon \mathrm{d}(x_n, x_{n+p})$

$\leq \varepsilon d(x_n, x_{n+1}) + \varepsilon d(x_{n+1}, x_{n+2}) + \ldots + \varepsilon d(x_{n+p-1}, x_{n+p}) \leq \varphi(x_n) - \varphi(x_{n+1}) + \varphi(x_{n+1}) - \varphi(x_{n+2}) +$
$+ \ldots + \varphi(x_{n+p-1}) - \varphi(x_{n+p}),$

$$(12_{33}) \qquad \varepsilon d(x_n, x_{n+p}) \leq \varphi(x_n) - \varphi(x_{n+p}).$$

$(\varphi(x_m))_{m\geq 1}$ being Cauchy sequence, (12_{33}) tells that $(x_m)_{m\geq 1}$ has the same property, let be $y_0 = \lim_{m\to\infty} x_m$. It only remains to take the limit for $p \to \infty$ in (12_{33}) and to take into account $\varphi(y_0) \leq \underline{\lim}_{m\to\infty} \varphi(x_m)$ to obtain

$$\varphi(y_0) \leq \varphi(x_n) - \varepsilon d(x_n, y_0)\ \forall\ n \geq 1,$$

that is y_0 is a minorant of the sequence $(x_n)_{n\geq 1}$. Moreover, as φ is increasing ($x \overset{B}{\leq} y \Rightarrow \varphi(x) \leq \varphi(y) - \varepsilon d(x, y) \Rightarrow \varphi(x) \leq \varphi(y)$), apply carelessly 1.1_2, $\exists\ v$ in X so that

$$(12_{34}) \qquad x \overset{B}{\leq} v \Rightarrow \varphi(x) = \varphi(v).$$

Suppose *par absurdum* the contrary of (12_{30}), that is $\exists\ x \neq v$ so that (12_{35}) $\varphi(x) \leq \varphi(v) - \varepsilon d(v, x)$. This means $x \overset{B}{\leq} v$, that implies $\varphi(x) \overset{(12_{34})}{=} \varphi(v)$ and, from (12_{35}), $\varepsilon d(v, x) \leq 0$ and hence $x = v$, contradiction. ■

Remark. In [30], 4, the absence of the claim φ: X $\to$ **R** makes incorrect the statement "Theorem 1 is a special case of the Theorem 4" (page 46).

The following corollary shows the intrinsic value of the Brezis - Browder statement.

Corollary. Let $(X, \overset{B}{\leq})$ *be a preordered set and* φ: X $\to$ **R** *increasing with the properties*:

1) *whatever a decreasing sequence* $(x_n)_{n\geq 1}$, *if* $(\varphi(x_n))_{n\geq 1}$ *is lower bounded, there exists* y_0 *so that* $x_n \geq y_0\ \forall\ n \geq 1$ *and* $\varphi(x_n) \to \varphi(y_0)$;

2) $\forall$ x *of* X *and* $\forall\ \varepsilon > 0$ *there exists* z *in* X *so that* $x \overset{B}{\geq} z$ *and* $\varphi(x) > \varphi(z) \geq \varphi(x) - \varepsilon$.

Then, whatever x *of* X, *the interval*

$$I(x) := \{y \in X: y \overset{B}{\leq} x\}$$

is surjectively applied by φ *on the interval* $(-\infty, \varphi(x)]$ (*H. Brezis, F. E. Browder*).

■ Let us fix x of X and let a be from $(-\infty, \varphi(x))$. Consider the set (nonempty!)

$$(12_{35}) \qquad X_a := \{y \in I(x): \varphi(y) \geq a\}$$

provided with the preorder relation $\overset{B}{\le}$. X_a has the property requested by 1.1_2. Indeed, let $(x_n)_{n\ge 1}$ be decreasing sequence from X_a. Then $\varphi(x_n) \ge a$ $\forall$ $n \ge 1$ and hence, according to 1), $\exists$ y_0 in X so that $x_n \overset{B}{\ge} y_0$ $\forall$ $n \ge 1$ and $\varphi(x_n) \to \varphi(y_0)$. But $y_0 \in \in I(x)$ and $\varphi(y_0) \ge a$ (go to the limit!), consequently $y_0 \in X_a$. Apply 1.1_2 (φ is lower bounded on X_a!), $\exists$ $v \in X_a$ with the property

$$(12_{36}) \qquad z \in X_a,\, z \overset{B}{\le} v \Rightarrow \varphi(z) = \varphi(v).$$

So we have $\varphi(v) \ge a$. Suppose *par absurdum* (12_{37}) $\varphi(v) > a$. According to 2), for $\varepsilon := \varphi(v) - a$ and for v $\exists$ $z \in X$ so that $z \overset{B}{\le} v$ and $\varphi(v) > \varphi(z) \ge \varphi(v) - \varepsilon$, hence $\varphi(z) \in [a, \varphi(v))$, consequently we have $z \in X_a$, as $z \in I(x)$, but also $\varphi(z) < \varphi(v)$, in contradiction with (12_{36}). ■

Remark. In [30], 4, Corollary 5 and also in [10] the condition 2) of 1.1_2 is amputated (see (17)): $\varphi(z) \ge \varphi(x) - \varepsilon$ and in this way the demonstration for the Corollary 5 becomes incorrect. Nevertheless, we underline that the demonstration of the Theorem 6 is however correct, but this happens because, by chance, 2) is verified.

Next we present a variant of the Ekeland principle on real Banach spaces that will provide for the perturbed function a point of strong minimum, but without localizing it. This variant will be obtained with a Deville-Godefroy-Zizler theorem ([27], 1).

Definition. Let $f: X \to \mathbf{R}$ be lower bounded, X real normed space. x_0 of X is a *point of strong minimum* for f, when $f(x_0) = \inf f(X)$ and, whatever $(x_n)_{n\ge 1}$ a sequence of points in X,

$$f(x_n) \to f(x_0) \Rightarrow x_n \to x_0.$$

In this case, x_0 is also a point of strict minimum for f. Indeed, let be $x' \in X$ with $f(x') = \inf f(X)$. Take the sequence (x_n), $x_n = x'$ for $n \ge 1$, then $f(x_n) \to f(x_0)$, hence $x_n \to x_0$ and $x' = x_0$. But if x_0 is a point of strict minimum, then it is not by all means a point of strong minimum. For instance, let X be a real non-reflexive Banach space. Then, according to James theorem ([47], vol. III, Chapter VIII, § 6, 4.33_1), there exists $f \in X^*$ with $\|f\| = 1$ and $\|x\| \le 1 \Rightarrow f(x) < 1$. We have then, for any x of $X \setminus \{0\}$, $f\left(-\frac{x}{\|x\|}\right) < 1$, that is $f(x) + \|x\| > 0$ $\forall$ $x \ne 0$, and then $g := f + \|\cdot\|$ has in 0 a point of strict minimum.

Nevertheless, 0 is not a point of strong minimum for g : take $(x_n)_{n\geq 1}$, $\|x_n\| = 1$ and $f(x_n) \to -1$, hence $g(x_n) \to 0 = g(0)$, even though $x_n \not\to 0$.

Before passing to the announced theorem –

Definition. The function $f: X \to \mathbf{R}$, X real normed space, is called a *bump function* (*fonction bosse*), provided it is bounded and its support is nonempty and bounded ($\operatorname{supp} f \overset{\text{def}}{=} \overline{\{x \in X : f(x) \neq 0\}}$).

Notation. $f: E \to (-\infty, +\infty]$ bounded from below, E set, $\alpha > 0 \Rightarrow S(f; \alpha) := \{x \in E: f(x) \leq \inf f(E) + \alpha\}$.

If $f_1 \leq f_2 + \frac{\alpha}{3}$ and $f_2 \leq f_1 + \frac{\alpha}{3}$, then

(13) $$S(f_1; \frac{\alpha}{3}) \subset S(f_2; \alpha).$$

Indeed, $x \in S(f_1; \frac{\alpha}{3}) \Rightarrow f_2(x) \leq f_1(x) + \frac{\alpha}{3} \leq \inf f_1(X) + \frac{2\alpha}{3} \leq \inf f_2(X) + \alpha$, hence $x \in S(f_2; \alpha)$.

1.2 Theorem. *Let* X *be real Banach space and* $\mathcal{F}$ *a Banach space of bounded continuous real functions defined on* X *having the properties*:

1° $f \in \mathcal{F} \Rightarrow \|f\|_\infty \leq \|f\|$;

2° $f \in \mathcal{F}$ *and* $x \in X \Rightarrow$ *the function* $y \to f_x(y) := f(x + y)$ *found in* $\mathcal{F}$ *and* $\|f_x\| = \|f\|$;

3° $f \in \mathcal{F}$ *and* $\lambda \in \mathbf{R} \Rightarrow$ *the function* $y \to f(\lambda y)$ *found in* $\mathcal{F}$;

4° *There exists a bump function in* $\mathcal{F}$.

Then, if $\varphi: X \to (-\infty, +\infty]$ *is lower bounded, lower semicontinuous and proper, the set* $\mathcal{M}$ *of functions* f *of* $\mathcal{F}$ *so that* $\varphi+f$ *has a point of strong minimum on* X *is a* G_δ *everywhere dense* (*R. Deville, G. Godefroy, V. Zizler*).

■ For every n of **N** take

$$U_n := \{f \in \mathcal{F}: \exists\, \alpha > 0, \operatorname{diam} S(\varphi + f; \alpha) < \frac{1}{n}\}.$$

It will be shown

(14) U_n *is open*, (15) U_n *is everywhere dense*, (16) $\bigcap_{n=1}^{\infty} U_n = \mathcal{M}$.

Demonstration of (14). Let be $f \in U_n$ with corresponding $\alpha > 0$. We have, $\forall g \in \mathscr{F}$ with $\|f-g\| \leq \frac{\alpha}{3}$, $\|f-g\|_\infty \leq \frac{\alpha}{3}$ and hence for $f_1 := \varphi + f$ and $f_2 := \varphi + g$ one finds, according to (13), $S(\varphi + g; \frac{\alpha}{3}) \subset S(\varphi + f; \alpha)$, consequently diam $S(\varphi + g; \frac{\alpha}{3}) < \frac{1}{n}$, $g \in U_n$, U_n is open.

Demonstration of (15). Let f be whatever in $\mathscr{F}$ and $\varepsilon > 0$. It has to find $g \in \mathscr{F}$ and $\alpha > 0$ so that $\|g\| < \varepsilon$ and (17) diam $S(\varphi + f + g; \alpha) < \frac{1}{n}$. Let b be a bump function from $\mathscr{F}$. It can admit $\|b\| < \varepsilon$. The assumption 2° allows to suppose $b(0) \neq 0$ and therefore even $b(0) > 0$. Moreover, it could be admitted, *via* the assumption 3°, that (18) supp $b \subset S(0, \frac{1}{2n})$. Take $\alpha = \frac{1}{2} b(0)$ and let be $x_0 \in X$ so that (19) $(\varphi + f)(x_0) < \inf(\varphi + f)(X) + \alpha$. Consider $g: X \to \mathbf{R}$, $g(x) = -b(x - x_0)$. According to 2°, $g \in \mathscr{F}$, $\|g\| = \|b\| < \varepsilon$ and $g(x_0) = -b(0)$. In order to check (17) it is enough to show that the corresponding set is included in the open ball $S(x_0, \frac{1}{2n})$, that is $\|x - x_0\| \geq \frac{1}{2n} \Rightarrow x \notin S(\varphi + f + g; \alpha)$, this being equivalent to

(20) $$(\varphi + f + g)(x) > \inf(\varphi + f + g)(X) + \alpha.$$

But supp $g \overset{(18)}{\subset} S(x_0, \frac{1}{2n})$, hence $\|x - x_0\| \geq \frac{1}{2n} \Rightarrow g(x) = 0$, consequently $(\varphi + f + g)(x) =$

$= (\varphi + f)(x) \geq \inf(\varphi + f)(X) \overset{(19)}{>} (\varphi + f)(x_0) - \alpha = (\varphi + f + g)(x_0) + b(0) - \frac{1}{2} b(0) \geq$

$\geq \inf(\varphi + f + g)(X) + \alpha$ and hence (20).

Demonstration of (16). $\bigcap_{n=1}^{\infty} U_n \subset \mathscr{M}$. Let be $f \in \bigcap_{n=1}^{\infty} U_n$. It is going to show that $\varphi + f$ attains on X a strong minimum. Consider, for every n of $\mathbf{N}$, $\beta_n > 0$ so that diam $S(\varphi + f; \beta_n) < \frac{1}{n}$ (definition of U_n!) and a decreasing sequence $(\alpha_n)_{n \geq 1}$ is built by induction with the property diam $S(\varphi + f; \alpha_n) < \frac{1}{n}$ $\forall n \geq 1$: take $\alpha_1 = \beta_1$; if $\beta_2 \leq \beta_1$ take $\alpha_2 = \beta_2$, in a contrary case take $\alpha_2 = \beta_1$ and notice that in both cases $S(\varphi + f; \alpha_2) \subset S(\varphi + f; \beta_2)$ etc. The set

$S(\varphi + f;\, \alpha_n)$ is closed as $\varphi + f$ is l.s.c., apply Cantor theorem, $\bigcap_{n=1}^{\infty} S(\varphi + f;\, \alpha_n) = \{x_0\}$. x_0 is a point of strong minimum for $\varphi + f$. Indeed, let be $(x_n)_{n\geq 1}$, $x_n \in X$, $\lim_{n\to\infty} (\varphi + f)(x_n) = $ $= \inf (\varphi + f)(X)$. Therefore, for every α_n $\exists\, p_n \in \mathbf{N}$ so that $m \geq p_n \Rightarrow (\varphi + f)(x_m) \leq \inf (\varphi + $ $+ f)(X) + \alpha_n$, that is $x_m \in S(\varphi + f;\, \alpha_n)$, that imposes, as x_0 belongs to the intersection, $m \geq p_n \Rightarrow \|x_m - x_0\| \leq \operatorname{diam} S(\varphi + f;\, \alpha_n) < \dfrac{1}{n}$, consequently $\lim_{n\to\infty} x_n = x_0$.

$\mathcal{M} \subset \bigcap_{n=1}^{\infty} U_n$ – justification *par absurdum*.

Finally, the Baire theorem assures us that $\mathcal{M}$ is everywhere dense. ■

The theorem D - G - Z will interfere in a decisive manner in the proof of several statements. For the moment, here it is the promised variant of the Ekeland principle ([56], 2, section 4, corollary 4.6).

1.3 *Let* X *be a real Banach space and* φ: X $\to (-\infty, +\infty]$ *bounded from below, lower semicontinuous and proper. Then whatever* $\varepsilon > 0$ *there exists* x_0 *in* X *so that*

$$\varphi(x_0) \leq \inf \varphi(X) + 2\varepsilon$$

and x_0 *is a point of strong minimum for the perturbed function* $\varphi + \varepsilon\|\cdot - x_0\|$.

■ Let $\mathcal{F}$ be the vector space of Lipschitz functions bounded on X provided with the norm

$$(21) \qquad \|f\| = \|f\|_{\infty} + \|f\|_{\mathrm{Lip}},\ \|f\|_{\mathrm{Lip}} := \sup_{\substack{x,y\in X\\ x\neq y}} \frac{|f(x) - f(y)|}{\|x - y\|}.$$

$\mathcal{F}$ is a Banach space and it verifies the requests 1°, 2°, 3° of 1.2. In order to verify 4° take $F: \mathbf{R} \to \mathbf{R}$ of C^1 class with a nonempty support included in the interval [1,2], b: $b(x) = F(\|x\|)$ is a bump function on X and it is found in $\mathcal{F}$. Apply 1.2, $\exists\, f \in \mathcal{F}$ so that (22) $\|f\| < \varepsilon$ and $\exists\, x_0 \in X$ point of strong minimum for $\varphi + f$. So, $\forall\, x \in X$

$$|f(x)| \overset{(21),(22)}{<} \varepsilon,\ |f(x) - f(x_0)| \overset{(21),(22)}{\leq} \varepsilon\|x - x_0\|,\ (\varphi + f)(x) \geq (\varphi + f)(x_0).$$

Then, $\forall\, x \in X$

$$\varphi(x) \geq \varphi(x_0) + f(x_0) - f(x) \geq \varphi(x_0) - \varepsilon\|x - x_0\|,$$

that is x_0 is a minimum point for the perturbed function. We have also (see (22))

$\varphi(x) \geq \varphi(x_0) + f(x_0) - f(x) \geq f(x_0) - 2\varepsilon$, that implies $\inf \varphi(X) \geq \varphi(x_0) - 2\varepsilon$.

Now let $(x_n)_{n\geq 1}$ be with $\varphi(x_n) + \varepsilon\|x_n - x_0\| - \varphi(x_o) \to 0$. Then, since

$$\varphi(x_n) - \varphi(x_0) + \varepsilon\|x_n - x_0\| \geq (\varphi + f)(x_n) - (\varphi + f)(x_0),$$

we have

$$(\varphi + f)(x_n) - (\varphi + f)(x_0) \to 0$$

and consequently $x_n \to x_0$, as x_0 is a point of strong minimum for $\varphi + f$. ■

Finally, here it is a *local variant* of the Ekeland principle 1.1 ([30], 7).

But firstly

Definition. Let X be a real normed space, $\varphi: X \to (-\infty, +\infty]$ proper, $\varepsilon > 0$ and v from X. φ is ε - *supported in* v if there exists ξ in X^* and $\delta > 0$ so that

(23) $$\|w - v\| \leq \delta \Rightarrow \varphi(v) \leq \varphi(w) + \varepsilon\|v - w\| + \xi(v - w).$$

We see that, for $\xi = 0$ and δ whatever positive, (23) becomes the fundamental inequality (1) of the Ekeland principle.

And now, the announced variant –

1.4 *Let* X *be a real Banach space and* φ: X $\to (-\infty, +\infty]$ *lower semicontinuous and proper. Suppose there is* F: X $\to$ **R** *with Fréchet derivative*, F(0) > 0 *and* F(x) $\leq$ 0 *on* $\{x \in X: \|x\| \geq 1\}$. *Then, whatever* $\varepsilon > 0$, *the set of points where* φ *is* ε - *supported is dense in* dom φ (*I. Ekeland - G. Lebourg*).

■ Let be $u \in$ dom φ and $r > 0$. It follows to find in $S := S(u, r)$ (open ball) a point $v \in$ dom φ in which φ is ε - supported. φ being l.s.c. in u, r can be chosen small enough so that φ is bounded from below on S. Consider the function $\Phi: X \to (0, +\infty]$,

(24) $$\Phi(x) = \frac{1}{\max\left(0, F\left(\frac{x-u}{r}\right)\right)}.$$

Φ is l.s.c.: let x_0 be of X and $a < \dfrac{1}{\max\left(0, F\left(\frac{x_0-u}{r}\right)\right)}$; when $a \leq 0$, the inequality is valid on X; when $a > 0$, then $\dfrac{1}{a} > \max\left(0, F\left(\frac{x_0-u}{r}\right)\right)$ and the inequality remains valid on a neighborhood of x_0, as F is continuous. We have (25) dom $\Phi \subset S$, since $\Phi(x) = +\infty$ on

$X \setminus S$. Obviously (26) $u \in \text{dom } \Phi$. Moreover, dom Φ is open, as $x \in \text{dom } \Phi \Rightarrow F\left(\frac{x-u}{r}\right) > 0$ and F is continuous, Φ is Fréchet differentiable on dom Φ.

Therefore, the function $\Phi + \varphi: X \to (-\infty, +\infty]$ is bounded from below ($\Phi(x) + \varphi(x) \overset{(26)}{=} +\infty$ on $X \setminus S$, $\Phi + \varphi$ is lower bounded on S), l.s.c. (Φ is l.s.c.) and proper ($u \in \text{dom}\,\Phi \cap \cap \text{dom } \varphi$).

According to 1.1, $\exists\, v \in X$ so that

(27) $$\Phi(v) + \varphi(v) \le \Phi(w) + \varphi(w) + \frac{\varepsilon}{2}\|v - w\| \;\forall w \in X.$$

Taking $w = u$ in (27) we find $v \in \text{dom } \Phi \cap \text{dom } \varphi$ (*par absurdum*, use (26)). So, Φ being Fréchet differentiable in v, for $\xi := -\Phi'(v)$ and for $\delta > 0$ we have

(28) $$\|w - v\| \le \delta \Rightarrow \Phi(w) - \xi(v - w) - \frac{\varepsilon}{2}\|v - w\| \le \Phi(v).$$

By adding (27) to (28) it results (23). ■

Remark 1. The demand "φ bounded from below" does not exists in the statement 1.4.

Remark 2. The Hilbert spaces have got the functions F like in 1.4 ($F(x) = 1 - x \cdot x$), also the spaces $\mathbf{L}^p$, $p \in (1, +\infty)$ ($F(x) = 1 - \|x\|^p$). $\mathbf{L}^1$ and $\mathbf{L}^\infty$ have not this property ([30], 7).

SEVERAL APPLICATIONS OF EKELAND PRINCIPLE

Sometimes the justifications are summary, we then appeal to the references.

Nonlinear semigroups ([10])

Let (X, d) metric space and S *nonlinear semigroup* on X [for every t of $[0, +\infty)$ $S(t)$ is a map $X \to X$ having the properties:

(29) $$S(0) = \text{id}_X \text{ (identical map of } X),$$

(30) $$S(t_1 + t_2) = S(t_1) \circ S(t_2) \;\forall\, t_1, t_2 \in [0, +\infty)].$$

Suppose S a *continuous semigroup of contractions*, that is

(31) $\forall\, u \in X$ the map $t \to S(t)u$ is continuous on $[0, +\infty)$,

(32) $\forall\, t \geq 0,\ \forall\, u, v \in X\ \ d(S(t)u, S(t)v) \leq d(u, v)$.

With every S and every $\rho \geq 0$ a binary relation $\underset{\approx}{<}$ in $X \times [0, +\infty)$ is associated ([10]):

(33) $$(u, a) \underset{\approx}{>} (v, b) \overset{\text{def}}{\Leftrightarrow} b \geq a \text{ and } d(S(b-a)u, v) \leq \rho(b-a).$$

$\underset{\approx}{<}$ is an order relation. Indeed, $(u, a) \underset{\approx}{>} (u, a)$, $(u, a) \underset{\approx}{>} (v, b)$ and $(v, b) \underset{\approx}{>} (u, a) \Leftrightarrow (u, a) = (v, b)$ (use in both cases (29) and (32)) and $(u, a) \underset{\approx}{>} (v, b)$ and $(v, b) \underset{\approx}{>} (w, c) \Rightarrow (u, a) \underset{\approx}{>} (w, c)$: $d(S(c-a)u, w) \leq d(S(c-a)u, S(c-b)v) + d(S(c-b)v, w) \overset{(30)}{=} d(S(c-b) \circ S(b-a)u, S(c-b)v) + d(S(c-b)v, w) \overset{(32)}{\leq} d(S(b-a)u, v) + S((c-b)v, w)) \overset{(33)}{\leq} \rho(b-a) + \rho(c-b) = \rho(c-a)$.

Retain:

(33_1) $$(u, a) \underset{\approx}{>} (v, b) \text{ and } (u, a) \neq (v, b) \Rightarrow b > a,$$

as if *par absurdum* is admitted $b = a$, then $d(S(0)u, v) = 0$, $S(0)u \overset{(29)}{=} u$, hence $d(u, v) = 0$, $u = v$, contradiction.

Let V be a complete metric space, S a continuous semigroup of contractions on V and $\rho \geq 0$. Consider the function $\varphi\colon V \times [0, +\infty) \to \mathbf{R}$,

(34) $$\varphi(u, a) = -a.$$

φ is increasing referring to $\underset{\approx}{>}$. Moreover, φ verifies 1) from the corollary to 1.1_2, where $X = F \times [0, +\infty)$, $F \subset V$ closed and $\overset{B}{\leq}$ is $\underset{\approx}{<}$ (see Lemma 1 in [10]).

1.5 *Let* V *be a complete metric space,* S *a continuous semigroup of contractions on* V, F *a closed subset and* $\rho \geq 0$ *so that*

(35) $$\varliminf_{t \to 0} \frac{1}{t} \operatorname{dist}(S(t)u, F) \leq \rho \ \forall\, u \in F.$$

Then

$$\operatorname{dist}(S(t)u, F) \leq \rho t \ \forall\, u \in F,\ \forall\, t \geq 0$$

(*H. Brezis, F.E. Browder*).

■ Let α be whatever, (36) $\alpha > \rho$. Apply the corollary of 1.1_2 with $X = F \times [0, +\infty)$ and $\underset{\approx}{<}$ with ρ replaced by α (see (33)) as $\overset{B}{\leq}$, φ like in (34). As it was mentioned, φ is increasing and 1) is checked. Go to 2) and let (u, a) be whatever from X and $\varepsilon > 0$. (35) provides $t \in (0, \varepsilon)$ so that

(37) $$\operatorname{dist}(S(t)u, F) < \alpha t.$$

Then, let $v \in F$ be so that $d(S(t)u, v) < \alpha t$. Taking $b := a + t$ it yields $(u, a) \overset{(33)}{\underset{\approx}{>}} (v, b)$ and $\varphi(u, a) > \varphi(v, b) \geq \varphi(u, a) - \varepsilon$, 2) is checked.

Therefore, let u be of F and $t \geq 0$ whatever. $\varphi(u, t) = -t$, $\varphi(u, 0) = 0$ and then the surjectivity of φ relative to the interval $(-\infty, \varphi(u, 0))$ gives $v \in F$ so that $(v, t) \underset{\approx}{<} (u, 0)$, that is ((33))

(38) $$d(S(t)u, v) \leq \alpha t.$$

From (38) it results $\operatorname{dist}(S(t)u, v) \leq \alpha t$ and taking into account (36) one obtains the conclusion. ■

Remark. Taking $\rho = 0$ in 1.5 a statement own to R. Martin is obtained: F is invariant relative to the semigroup S (*Invariant sets for evolution systems*, Proc. International Conference on Differential Equations, Academic Press, New York, 1975).

1.6 *Let* V, S, F *be like in* 1.5 *and*

(39) $$c := \underline{\lim}_{t \to +\infty} \frac{1}{t} \operatorname{dist}(S(t)u, F).$$

Then c *remains constant while* u *describes* F *and*

$$c = \sup_{u \in F} \inf_{t > 0} \frac{1}{t} \operatorname{dist}(S(t)u, F) \text{ ([10]).}$$

■ Let u, v be whatever in F and

$$c_u = \underline{\lim}_{t \to +\infty} \operatorname{dist}(S(t)u, F),\ c_v = \underline{\lim}_{t \to +\infty} \operatorname{dist}(S(t)v, F).$$

Suppose *par absurdum* $c_u \neq c_v$, for instance $c_u < c_v$.

$$\left|\operatorname{dist}(S(t)u, F) - \operatorname{dist}(S(t)v, F)\right| \leq d(S(t)u, S(t)v) \overset{(32)}{\leq} d(u, v),$$

hence

(40) $$\lim_{t \to +\infty} \frac{1}{t} [\operatorname{dist}(S(t)u, F) - \operatorname{dist}(S(t)v, F)] = 0.$$

Let $r' > 0$ be so that (41) $c_u < \inf\limits_{t\in[r',+\infty)} \frac{1}{t}\operatorname{dist}(S(t)v, F) =: \alpha$. Take $\beta \in (c_u, \alpha)$ and a sequence $(r_n)_{n\geq 1}$, $r_n \geq r'$, $r_n \to +\infty$. Since

$$\inf_{t\in[r_n,+\infty)} \frac{1}{t}\operatorname{dist}(S(t)u, F) \leq c_u < \beta < \alpha \ \forall\, n \geq 1,$$

it can find a sequence $(t_n)_{n\geq 1}$, $t_n \in [r_n, +\infty)$, $\frac{1}{t_n}\operatorname{dist}(S(t_n)u, F) \leq \beta < \alpha\ \forall\, n \geq 1$, hence

$$\frac{1}{t_n}\operatorname{dist}(S(t_n)u, F) \leq \beta < \alpha \leq \frac{1}{t_n}\operatorname{dist}(S(t_n)v, F),\ \frac{1}{t_n}[\operatorname{dist}(S(t_n)v, F) - \operatorname{dist}(S(t_n)u, F)] \geq \alpha - \beta$$

$\forall\, n \geq 1$, in contradiction with (40) since $t_n \to +\infty$. One concludes $c_u = c_v$.

We have also

$$\sup_{u\in F}\inf_{t>0}\frac{1}{t}\operatorname{dist}(S(t)u, F) \leq c. \tag{42}$$

Indeed, supposing *par absurdum* the contrary, we find $u_1 \in F$ so that $c < \inf\limits_{t>0}\frac{1}{t}\operatorname{dist}(S(t)u_1, F)$, contradiction, since $c = \varliminf\limits_{t\to+\infty} \frac{1}{t}\operatorname{dist}(S(t)u_1, F)$.

Note l the first member of (42) and let ρ be whatever, (43) $\rho > l$. Put on the set $X := F \times [0, +\infty)$ the order relation $\underset{\approx}{<}$ ((33)) and let φ be the function of (34). Demonstrate that the condition (12_{25}) of 1.1_2 is not verified. Indeed, let (u, a) be whatever from X. By using (43) yields $t > 0$ so that dist $(S(t)u, F) < \rho t$, hence $\exists\ v \in F$ with $d(S(t)u, v) < \rho t$ and taking $b := a + t$ is found $(v, b) \underset{\approx}{<} (u, a)$ and $\varphi(v, b) = -b = \varphi(u, a) - t < \varphi(u, a)$.

Consider for any (u, a) of X the interval in X

$$S(u, a) := (\leftarrow, (u, a)).$$

φ is not lower bounded on $S(u, a)$. Indeed, suppose *par absurdum* the contrary. This imposes that any decreasing sequence $(u_n, a_n)_{n\geq 1}$ from $S(u, a)$ has a minorant in $S(u, a)$, otherwise with (33_1) a subsequence $(u_{k_n}, a_{k_n})_{n\geq 1}$ could be found with $a_{k_n} \to +\infty$, that yields $\lim\limits_{n\to\infty} \varphi(u_{k_n}, a_{k_n}) = -\infty$, contradiction. Therefore, it can apply 1.1_2 to $S(u, a)$, $\exists\ (v, b)$ in $S(u, a)$, and hence in X, having the property (12_{25}), that has been already proven as being impossible (see above), φ is not bounded from below on any $S(u, a)$.

Then, let u be whatever from F. It can find, by using the last claim, a sequence $(u_n, t_n)_{n\geq 1}$ in $S(u, 0)$ with $t_n \to +\infty$ and $(u, 0) \underset{\approx}{>} (u_n, t_n)$ $\forall\, n \geq 1$, that is $d(S(t_n)u, u_n) \leq \rho t_n$ $\forall\, n \geq 1$, this imposes ([47], vol. II, page 651, 11.9_1) $\overline{\lim_{t\to+\infty}} \frac{1}{t}\, \mathrm{dist}(S(t)u, F) \leq \rho$ and in confrontation to (43) one obtains, since ρ is any,

$$\overline{\lim_{t\to+\infty}} \frac{1}{t}\, \mathrm{dist}(S(t)u, F) \leq l.$$

By comparing with (42) the demonstration is finished (see also (39)). ■

Optimization with regular constraints ([30], 2)

Let X be a real normed space, $\varphi : X \to \mathbf{R}$ with Fréchet derivative and $G_i: X \to \mathbf{R}$, $i = \overline{1, m}$ of C^1 class. Consider *the constrained optimization problem*

$$(*)\begin{cases} \inf \varphi(v) \\ G_i(v) = 0,\ i = \overline{1, p} \\ G_i(v) \geq 0,\ i = \overline{p+1, m}. \end{cases}$$

Let $\mathscr{C}$ (the first letter of the word *constraint*) be *the feasible set*

(44) $$\mathscr{C} = \{v \in X : G_i(v) = 0,\ i = \overline{1, p},\ G_i(v) \geq 0,\ i = \overline{p+1, m}\}.$$

$\mathscr{C}$ is closed. Consider for each $v \in \mathscr{C}$ $I(v)$, *the set of saturated constraints,*

(45) $$i \in I(v) \Leftrightarrow G_i(v) = 0.$$

It is obviously $\{1, \ldots, p\} \subset I(v)$. Moreover, it is accepted the *regularity assumption*:

(46) whatever v of $\mathscr{C}$, the functionals $G_i'(v)$, $i \in I(v)$ are linear independent.

The following theorem provides for the problem (*) some points that are *almost optimal* and *almost* satisfy the necessary conditions for optimality.

But firstly, we present a variant of Farkas - Minkovski lemma.

1.7 *Lemma. Let* X *be a real reflexive space,* ξ_i, $i = \overline{1, p}$, ζ_j, $j = \overline{1, q}$, ξ *in* X^* *and* $\varepsilon > 0$. *Suppose, for any* h *of* X,

(47) $$\begin{cases} \xi_i(h) = 0, i = \overline{1, p} \\ \zeta_j(h) \geq 0, j = \overline{1, q} \end{cases} \Rightarrow \xi(h) \geq -\varepsilon \|h\|.$$

Then $\exists\lambda_i \in \mathbf{R}$, $i = \overline{1, p}$, $\exists\mu_j \geq 0$, $j = \overline{1, q}$ *so that*

$$(48)\qquad \left\| \xi - \sum_{i=1}^{p} \lambda_i \xi_i - \sum_{j=1}^{q} \mu_j \zeta_j \right\| \leq \varepsilon$$

(*I. Ekeland*).

■ Consider in X^* the convex cone with vertex

$$(49)\qquad C := \{ \sum_{i=1}^{p} \lambda_i \xi_i + \sum_{j=1}^{q} \mu_j \zeta_j : \lambda_i \in \mathbf{R}, \mu_j \geq 0 \}.$$

Note $S^* := \{\eta \in X^* : \|\eta\| \leq 1\}$ and let be

$$(50)\qquad \Omega := C + \varepsilon S^*.$$

Ω is obviously convex, but also closed, since it is $\sigma(X^*, X)$ - closed, as C has this property and S^* is $\sigma(X^*, X)$ - compact (theorem of Alaoglu - Bourbaki - Kakutani).

Firstly, suppose *par absurdum* the contrary of (48). Then $\xi \notin \Omega$ (see (50)), hence ξ and Ω are separating very strongly (Hahn - Banach theorem, the third geometric form, see *Memento*, after 1.1_0), $\exists\, h \in X$ and $\alpha \in \mathbf{R}$ so that

$$(51)\qquad \xi(h) < \alpha,$$

$$(52)\qquad \eta(h) \geq \alpha \ \forall\, \eta \in \Omega.$$

From (52) and (50) it follows

$$\eta_1(h) + \varepsilon\eta_2(h) \geq \alpha \ \forall\, \eta_1 \in C, \forall\, \eta_2 \in S^*,$$

hence

$$\eta_1(h) \geq \alpha + \varepsilon \sup_{\eta_2 \in S^*} \eta_2(-h) \ \forall\, \eta_1 \in C$$

and hence

$$(53)\qquad \eta_1(h) \geq \alpha + \varepsilon\|h\| \ \forall\, \eta_1 \in C.$$

Take $\eta_1 = t\xi_i$, $1 \leq i \leq p$ and $t \in \mathbf{R}$, then (53) gives $t\xi_i(h) \geq \alpha + \varepsilon\|h\| \ \forall\, t \in \mathbf{R}$, that imposes

$$(54)\qquad \xi_i(h) = 0,\ i = \overline{1, p}.$$

Now, take $\eta_1 = t\zeta_j$, $1 \leq j \leq q$ and $t \geq 0$. Then (see (53)) $t\zeta_j(h) \geq \alpha + \varepsilon\|h\| \ \forall\, t \geq 0$, hence

$$(55)\qquad \zeta_j(h) \geq 0,\ j = \overline{1, q}.$$

Finally, take $\eta_1 = 0$ in (53),

$$(56)\qquad \alpha \leq -\varepsilon\|h\|.$$

By combining (56) with (51) gives

(57) $$\xi(h) < -\varepsilon\|h\|.$$

But (54) and (55) impose, according to (47), $\xi(h) \geq -\varepsilon\|h\|$, in contradiction with (57). ■

Remark. The lemma 1.7 is the Lemma 3.3 of [30], 2 with $V = X$. But the author did not notice the fact that V must be *reflexive* in order to apply correctly the Hahn - Banach theorem cited above, since in this case we have $J(V) = V^{**}$, J the Hahn immersion. Consequently, both the Theorem 3.1 and Corollary 3.4 have to be completed by the same assumption "V reflexive".

1.8 Theorem. *Suppose* X *real reflexive space. If*

(58) $$\inf_{v\in\mathscr{C}} \varphi(v) > -\infty,$$

then for any $\varepsilon > 0$ *there is* v_ε *in* $\mathscr{C}$ *and* $\lambda_1,\ldots,\lambda_m$ in $\mathbf{R}$ *so that*

(59) $$\varphi(v_\varepsilon) \leq \inf_{v\in\mathscr{C}} \varphi(v) + \varepsilon^2$$

and

(60) $$\begin{cases} \lambda_i \geq 0, i = \overline{p+1, m}, \\ \lambda_i = 0 \textit{ when } G_i(v_\varepsilon) > 0, \\ \left\| \varphi'(v_\varepsilon) - \sum_{i=1}^{m} \lambda_i G_i'(v_\varepsilon) \right\| \leq \varepsilon \end{cases}$$

(*I. Ekeland*).

Consider the function $\Phi: X \to (-\infty, +\infty]$,

(61) $$\Phi(v) = \begin{cases} \varphi(v), v \in \mathscr{C} \\ +\infty, v \notin \mathscr{C}. \end{cases}$$

Φ is lower semicontinuous: $\mathscr{C}$ is closed, $v_0 \notin \mathscr{C} \overset{(61)}{\Rightarrow} \Phi(v) = +\infty$ on a neighborhood of v_0; $v_0 \in \mathscr{C} \Rightarrow \Phi(v_0) = \varphi(v_0)$, hence $a < \Phi(v_0) \Rightarrow a < \varphi(v)$ on a neighborhood of v_0 and consequently $a < \Phi(v)$ on it. Moreover, Φ is bounded from below (take into account (58) and (61)), apply to Φ the Ekeland principle 1.1 with ε^2 and $\lambda = \varepsilon$, $\exists\, v_\varepsilon \in X$ so that

(62) $$\Phi(v_\varepsilon) \leq \inf \Phi(X) + \varepsilon^2,$$

(63) $$\Phi(v_e) < \Phi(w) + \varepsilon\|v_\varepsilon - w\| \ \forall\, w \in X \setminus \{v_\varepsilon\}.$$

From (62) and (63) it follows

(64) $$v_\varepsilon \in \mathscr{C}, \varphi(v_\varepsilon) \leq \inf_{v\in C} \varphi(v) + \varepsilon^2,$$

(65) $$\varphi(v_\varepsilon) \le \varphi(w) + \varepsilon\| v_\varepsilon - w\| \ \forall\, w \in \mathscr{C}.$$

Now demonstrate

(66) If $G_i'(v_\varepsilon)(h) = 0,\ i = \overline{1, p}$ and $G_i'(v_\varepsilon)(h) \ge 0,\ i \in \{p + 1, \ldots, m\} \cap I(v_\varepsilon)$, then $\varphi'(v_\varepsilon)(h) \ge -\varepsilon\| h \|$.

▲ By using the theorem of implicit functions and (46) one finds $u : [0, r] \to \mathscr{C}$ of C^1 class so that

(66_1) $$u(0) = v_\varepsilon,\ u'(0) = h.$$

By replacing in (65) with $u(t)$,

$$\varphi(u(t)) - \varphi(u(0)) \ge -\varepsilon\|u(t) - u(0)\| \ \forall\, t \in [0, r].$$

Dividing by $t > 0$ and taking $t \to 0+$ gives (see (66_1))

$$\varphi'(v_\varepsilon)(h) \ge -\varepsilon\| h \|. \ ▲$$

In the following, applying (66) combined with 1.7 where $\xi_i = G_i'(v_\varepsilon),\ i = \overline{1, p},\ \zeta_j = G_j'(v_\varepsilon),\ j \in \{p + 1, \ldots, m\} \cap I(v_\varepsilon),\ \xi = \varphi'(v_\varepsilon)$ and taking $\lambda_j = 0$ for $j \in \{p + 1, \ldots, m\} \setminus I(v_\varepsilon)$, one finds (60). ■

For $m = 1$, from 1.8 is obtained

1.9 *Corollary. Let* φ: X → **R** *be with Fréchet derivative and* G: X → **R** *of* C^1 *class so that*

$$G(v) = 0 \Rightarrow G'(v) \ne 0.$$

Suppose that there is α *in* **R** *having the property*

$$G(v) = 0 \Rightarrow \varphi(v) \ge \alpha.$$

Then $\forall\, \varepsilon > 0\ \exists\, v_\varepsilon$ *in* X *and* λ *in* **R** *so that*

$$G(v_\varepsilon) = 0,$$

$$\|\varphi'(v_\varepsilon) - \lambda G'(v_\varepsilon)\| \le \varepsilon.$$

Control theory ([30], 2; for some concepts and statements – *Pallu de la Barrière*, *Cours d'automatique théorique*, Dunod, 1965)

Consider a *system* governed by the differential equation (μ the Lebesgue measure in **R**)

$$(67)\qquad \begin{cases} \dfrac{dx}{dt}(t) = f(t, x(t), u(t)) & \mu - \text{a.e.} \\ x(0) = x_0 \in \mathrm{R}^n \end{cases}$$

on the prescribed time interval $[0, T]$. The vector $x(t)$ describes the *state* of the system and $u(t)$ is the *control at the time* t and belongs to a compact metric space K. The function

$$f: [0, T] \times \mathbf{R}^n \times K \to \mathbf{R}^n$$

is continuous and

$$(68)\qquad \frac{\partial f}{\partial x_i}, i = \overline{1, n} \text{ are continuous,}$$

$$(69)\qquad x \cdot f(t, x, u) \leq c(1 + \|x\|^2), c > 0.$$

Let $u : [0, T] \to K$ be a measurable *control*. The condition (68) assures the existence of a unique solution of (67) on an interval $[0, \tau]$ small enough. The condition (69) becomes, *via* the Gronwall lemma,

$$(70)\ \|x(t)\|^2 \leq (\|x_0\|^2 + 2cT)e^{2cT}$$

and consequently it assures the existence of a unique solution even on $[0, T]$. Therefore, the measurable control u once being chosen, the final point $x(T)$ is well defined.

So, let g: $\mathbf{R}^n \to \mathbf{R}$ be of C^1 class. The next statement provides a measurable control u_ε so that the corresponding trajectory x_ε minimizes $g(x(T))$ among all solutions x of (67), u a measurable control.

1.10 Theorem. *For each* $\varepsilon > 0$ *there exists* u_ε *measurable control, let* x_ε *be the corresponding trajectory, so that*

$$(71)\qquad g(x_\varepsilon(T)) \leq \inf g(x(T)) + \varepsilon$$

and

$$(72)\qquad f(t, x_\varepsilon(t), u_\varepsilon(t)) \cdot p(t) \leq \inf_{u \in K} [f(t, x_\varepsilon(t), u) \cdot p(t)] + \varepsilon \quad \mu\text{ - a. e. } on\ [0, T],$$

where $p = (p_1, \ldots, p_n)$ *is the solution of the system of linear differential equations*

$$(73)\qquad \begin{cases} \dfrac{dp_i}{dt}(t) = -\displaystyle\sum_{j=1}^{n} \frac{\partial f_j}{\partial x_i}(t, x_\varepsilon(t), u_\varepsilon(t))p_j(t), \\ p(T) = g'(x_\varepsilon(T)) \end{cases}$$

(*I. Ekeland*).

■ Let X be the set of measurable controls $u : [0, T] \to K$ and, on this, the distance

$$(74)\qquad d(u_1, u_2) = \text{mes}\ \{t \in [0, T]: u_1(t) \neq u_2(t)\}$$

(the Lebesgue measure on $\mathbf{R}$).

d is indeed a distance on X : $d(u_1, u_1) = \text{mes } \varnothing = 0$, $d(u_1, u_2) = 0 \Rightarrow u_1 = u_2$, since if *par absurdum* $\exists\, t_0 \in [0, T]$ so that $u_1(t_0) \neq u_2(t_0)$, take disjoint neighbourhoods U_1, U_2 of $u_1(t_0)$ respectively $u_2(t_0)$ and then, by using the continuity, an interval J in which there is t_0 so that $u_i(J) \subset U_i$, $i = 1,2$ and a contradiction is obtained as mes $J > 0$; $d(u_1, u_3) \leq d(u_1, u_2) + d(u_2, u_3)$, since obviously $\{t : u_1(t) \neq u_3(t)\} \subset \{t : u_1(t) \neq u_2(t)\} \cup \{t : u_2(t) \neq u_3(t)\}$.

X *is complete*. Indeed, let $(u_n)_{n\geq 1}$ be Cauchy sequence in X. A subsequence $(u_{N_n})_{n\geq 1}$ is extracted from it with the property

(75) $$d(u_{N_n}, u_{N_{n+1}}) \leq 2^{-n}.$$

For each n of **N** consider the set

$$A_n = \bigcup_{p\geq n} \{t \in [0, T] : u_{N_p}(t) \neq u_{N_{p+1}}(t)\}.$$

We have mes $A_n \overset{(75)}{\leq} \sum_{p=n}^{\infty} 2^{-p} = 2^{1-n}$ and $A_n \supset A_{n+1}$. Define u_0 in X by

$$t \notin A_n \Rightarrow u_0(t) = u_{N_n}(t).$$

Then $u_{N_n} \to u_0$ (use (74)) and this implies, since $(u_n)_{n\geq 1}$ is fundamental, X complete.

Consider the map $\varphi: X \to \mathbf{R}$,

(76) $$\varphi(u) = g(x(T)),$$

where $t \to x(t)$ is the solution of (67) that corresponds to u. φ is bounded from below and continuous ([30], 2), apply the Ekeland principle 1.1 with ε and $\lambda = 1$, there is u_ε in X so that

(77) $$\varphi(u_\varepsilon) \leq \inf \varphi(X) + \varepsilon,$$

(78) $$\varphi(u_\varepsilon) \leq \varphi(u) + \varepsilon d(u_\varepsilon, u) \ \forall\, u \in X.$$

(71) has been already obtained (see (76)). Pass to prove (72). Let x_ε the corresponding trajectory of u_ε ((67)),

(79) $$\begin{cases} \dfrac{dx_\varepsilon}{dt}(t) = f(t, x_\varepsilon(t), u_\varepsilon(t)) \ \ \mu-\text{a.e. on } [0, T], \\ x_\varepsilon(0) = x_0. \end{cases}$$

Take t_0 in $(0, T)$ that verifies (79) and let u_0 be whatever from K. Consider, for each $\tau > 0$, $v_\tau : [0, T] \to K$,

(80) $$\begin{cases} v_\tau(t) = u_0 \text{ for } t \in [0, T] \cap (t_0 - \tau, t_0), \\ v_\tau(t) = u_\varepsilon(t) \text{ for } t \notin [0, T] \cap (t_0 - \tau, t_0). \end{cases}$$

$v_\tau \in X$ and for τ small enough we have (81) $d(v_\tau, u_\varepsilon) = \tau$. Let x_τ be the trajectory corresponding to v_τ ((67)). Then (see, for instance, *Cours d'automatique théorique* cited above)

(82)
$$\frac{d}{d\tau} g(x_\tau(T))\bigg|_{\tau=0} = p(t_0) \cdot [f(t_0, x_\varepsilon(t_0), u_0) - f(t_0, x_\varepsilon(t_0), u_\varepsilon(t_0))].$$

But from (78), (76) and (81) it follows

$$g(x_\tau(T)) - g(x_\varepsilon(T)) \geq -\varepsilon\tau, \ \tau \text{ small.}$$

Divide by τ and take the limit for $\tau \to 0+$. By combining also with (82) yields (72), since u_0 is whatever in K. ■

Remark. Referring to the theorem 1.10, Ekeland confessed ([30], 4, page 452):

"This (optimization and control theorem) was really the motivation which led me to prove Theorem 1" (the Ekeland principle 1.1 in this monograph).

The geometry of Banach spaces ([30], 4)

Definitions. Let V be a complete metric space. A subset G of V is called *residual* if $G = \bigcap_{n=1}^{\infty} U_n$, U_n open and everywhere dense. Obviously, G is also everywhere dense (theorem of Baire). A property $\pi(u)$, that has significance for every u of V, is called *generic*, when $\pi(u)$ is true for every point of a residual set. Obviously, if $\pi_1(u), \pi_2(u), \ldots, \pi_n(u), \ldots$ are generic, then $\bigwedge_{n=1}^{\infty} \pi(u_n)$ has the same property.

Let X be a set, V a real Banach space, $f: X \times V \to \mathbf{R}$ and $F: V \to [-\infty, +\infty)$,

(83)
$$F(v) = \inf_{x \in X} f(x, v).$$

Suppose

(84)
$$F(v) > -\infty \ \forall\, v \in V,$$

(85)
$$v \to f(x, v) \text{ is of } C^1 \text{ class } \forall\, x \in X.$$

1.11 *Suppose there exists* Φ: $V \to \mathbf{R}$ *with Fréchet derivative and* $\Phi(0) > 0$, $\Phi(x) \leq 0$ *on the set* $\{x \in V: \|x\| \geq 1\}$. *Moreover,* $\forall\, u \in V\ \exists\, \delta, \theta > 0$ *so that*

(86)
$$\{ f'_v(x, v): \|v - u\| \leq \delta, f(x, v) \leq F(u) + \theta\} \text{ is bounded in } V^*$$

and

(87)
$$\{v \to f'_v(x, v): f(x, u) \leq F(u) + \theta\} \text{ is equicontinuous in } u.$$

Then F *is Fréchet differentiable on a residual set* G *and*, $\forall$ u *in* G,

$$f(x_n, u) \to F(u) \Rightarrow f'_v(x_n, u) \to F'(u)$$

(*I. Ekeland - G. Lebourg*).

The conditions (86) and (87) are coherent, both sets are nonempty: F being upper semicontinuous in u (see (83)), it can suppose δ small enough to have $\|v-u\| \le \delta \Rightarrow F(v) < F(u) + \theta$ and then for v in $S_\delta(u)$ (closed ball centred in u of radius δ) there exists, according to (83), x in X so that $f(x, v) < F(u) + \theta$; similar justification for (87).

■ ([30], 7). Consider for every $\varepsilon > 0$ the set G_ε of points u of V for which there is $\delta > 0$ and $\theta > 0$ so that

(88) $$\|v_i - u\| \le \delta, f(x_i, u) \le F(u) + \theta,\ i = 1, 2 \Rightarrow \|f'_v(x_1, v_1) - f'_v(x_2, v_2)\| \le \varepsilon.$$

G_ε *is open.* Indeed, let u be *whatever fixed* in V. According to (86), $\exists\ \delta_0, \theta_0, \rho > 0$ so that

(89) $$\|v-u\| \le \delta_0,\ f(x, v) \le F(u) + \theta_0 \Rightarrow \|f'_v(x, v)\| \le \rho.$$

By using (89) and the upper semicontinuity of F we find, applying the first formula of the finite increments, δ_1 and θ_1, $\delta_1 < \delta_0$ and $\theta_1 < \theta_0$, so that

$\|v-u\| \le \delta_1$, $f(x, v) \le F(v) + \theta_1$ and $\|w-u\| \le \delta_1$, $f(x, w) \le F(w) + \theta_1 \Rightarrow$

(90) $$|f(x, v) - f(x, w)| \le \rho\|v-w\|,$$

(91) $$|F(v) - F(w)| \le \rho\|v-w\|.$$

Now, let δ, θ be relative to u given by (88). If need, we diminish them to have $\delta < \delta_1$, $\theta < \theta_1$. Take (92) $\delta' := \min\left(\frac{\delta}{2}, \frac{\theta}{3\rho}\right)$, $\theta' = \frac{\theta}{3}$ and let v be whatever in V with $\|v-u\| \le \delta'$.

Show

(93) $$v \in G_\varepsilon$$

and hence G_ε is open.

Suppose (94) $\|v_i - v\| \le \delta'$, $i = 1, 2$, (95) $f(x_i, v) \le F(v) + \theta'$, $i = 1, 2$. (93) will be validated if (see (88))

(96) $$\|f'_v(x_1, v_1) - f'_v(x_2, v_2)\| \le \varepsilon.$$

But (94) implies $\|v_i - u\| \le \delta$, $i = 1, 2$ and (95) implies, *via* (90) and (91) where one takes $w = u$,

$$f(x_i, u) \le F(u) + \theta,\ i = 1, 2:\ f(x_i, u) \le f(x_i, v) + \rho\|v-u\| \le F(v) + \theta' + \rho\|v-u\| \le$$
$$\le F(u) + \theta' + 2\rho\|v-u\| \le F(u) + \theta' + 2\rho\delta' \le F(u) + \theta.$$

But $u \in G_\varepsilon$ and hence, according to (88), it follows (96).

G_ε *is everywhere dense.* Indeed, let S_ε be the set of points of V in which F is $\frac{\varepsilon}{8}$ - supported (see (23)). Since F is lower semicontinuous (F is locally Lipschitz, see (91)) and has values in **R** and V possesses the function Φ from the statement, the application of 1.4 allows to affirm that S_ε is everywhere dense. So it remains to show

(97) $$S_\varepsilon \subset G_\varepsilon.$$

Let u be whatever fixed in S_ε. According to the definition $\exists\ \xi$ in V^* and $\delta_0 > 0$ so that

(98) $$\|v-u\| \le \delta_0 \Rightarrow F(v) \ge F(u) + \xi(v-u) - \frac{\varepsilon}{8}\|v-u\|.$$

On the other side, the equicontinuity in u present at (87) gives $\delta_1, \theta > 0$ with (99) $\delta_1 < \delta_0$, $\theta < \frac{\varepsilon}{8}\delta_1$ so that

(100) $$\|v-u\| \le \delta_1,\ f(x, u) \le F(u) + \theta \Rightarrow \|f'_v(x,v) - f'_v(x,u)\| \le \frac{\varepsilon}{8}.$$

Applying the second formula of finite increments gives, *via* (100),

(101) $\|v-u\| \le \delta_1$ and $f(x, u) \le F(u) + \theta \Rightarrow |f(x, v) - f(x, u) - f'_v(x, u)(v-u)| \le$ $\le \frac{\varepsilon}{8}\|v-u\|.$

Combining (101), (98) and (99) ($F(v) \le f(x, v)$!), yields

(102) $$\|v-u\| \le \delta_1 \text{ and } f(x, u) \le F(u) + \theta \Rightarrow (\xi - f'_v(x, u))(v-u)) \le \frac{3\varepsilon}{8}\delta_1.$$

But (102) is also true for $u - v$ (set $u - v$ as $v' - u$), hence actually we have

$$\|v-u\| \le \delta_1 \text{ and } f(x, u) \le F(u) + \theta \Rightarrow |(\xi - f'_v(x, u))(v-u)| \le \frac{3\varepsilon}{8}\delta_1$$

and consequently

$$\|v-u\| \le \delta_1 \text{ and } f(x, u) \le F(u) + \theta \Rightarrow \|\xi - f'_v(x,u)\| \le \frac{3\varepsilon}{8},$$

which combined with (100) gives, by using the triangle inequality,

$$\|v-u\| \le \delta_1 \text{ and } f(x, u) \le F(u) + \theta \Rightarrow \|\xi - f'_v(x,v)\| \le \frac{\varepsilon}{2}$$

and also

$$\|v_i - u\| \le \delta_1,\ i = 1, 2 \text{ and } f(x, u) \le F(u) + \theta \Rightarrow \|f'_v(x, v_1) - f'_v(x, v_2)\| \le \varepsilon,$$

that is (88) is checked, $u \in G_\varepsilon$ and consequently (97).

$G := \bigcap_{n=1}^{\infty} G_{\frac{1}{n}}$ *has the property claimed by the statement.* Indeed, let u be whatever fixed in G. Obviously (103) $u \in G_\varepsilon\ \forall\ \varepsilon > 0$. Let $(x_n, v_n)_{n\ge1}$ be a sequence in $X \times V$ with $v_n \to u$ and $f(x_n, u) \to F(u)$. Then $(f'_v(x_n, v_n))_{n\ge1}$ is a Cauchy sequence (use (103) and (88)), let be $\xi = \lim_{n\to\infty} f'_v(x_n, v_n)$. By using (88) yields

(104) $$\xi = \lim_{n\to\infty} f'_v(y_n, w_n), \text{ where } w_n \to u \text{ and } f(y_n, u) \to F(u).$$

Prove

(105) $$\xi = F'(u).$$

Let $\varepsilon > 0$ be whatever. For $\frac{\varepsilon}{2}$ $\exists\ \delta_1, \theta > 0$ so that

(106) $$\|v - u\| \le \delta_1 \text{ and } f(x, u) \le F(u) + \theta \Rightarrow \|f'_v(x, v) - \xi\| \le \frac{\varepsilon}{2}$$

(*par absurdum*, use (104)). (106) gives us, by integration over a segment of extremities u and v,

(107) $$\|v - u\| \le \delta_1 \text{ and } f(x, u) \le F(u) + \theta \Rightarrow |f(x, v) - f(x, u) - \xi(v - u)| \le \varepsilon\|v - u\|.$$

In particular for $x = x_n$ where x_n is so chosen as $f(x_n, u) \le F(u) + \frac{1}{n}$ and $\frac{1}{n} < \theta$,

$$F(v) \le f(x_n, v) \le f(x_n, u) + \xi(v - u) + \varepsilon\|v - u\|$$

and taking the limit

(108) $$\|v - u\| \le \delta_1 \Rightarrow F(v) \le F(u) + \xi(v - u) + \varepsilon\|v - u\|.$$

On the other hand, from (107),

$$\|v - u\| \le \delta_1 \text{ and } f(x, u) \le F(u) + \theta \Rightarrow$$

(109) $$f(x, v) \ge f(x, u) + \xi(v - u) - \varepsilon\|v - u\|.$$

Take δ_2, $0 < \delta_2 < \min\left(\delta_1, \frac{\theta}{3L}\right)$, L the Lipschitz constant of F around u. Suppose $\|u - v\| \le \delta_2$ and $f(x, v) \le F(v) + \frac{\theta}{3}$. Then $f(x, u) \le F(u) + \theta$ [$f(x, v) < F(v) + \frac{2\theta}{3}$, for δ_2

small enough we also have $f(x, u) < F(v) + \frac{2\theta}{3}$ ((85)) and $F(v) \le F(u) + L\|v - u\| \le F(u) +$

$+ L\delta_2 < F(u) + \frac{\theta}{3}$] and, using (109),

(110) $$\|v - u\| \le \delta_2 \text{ and } f(x, v) \le F(v) + \frac{\theta}{3} \Rightarrow f(x, v) \ge F(u) + \xi(v - u) - \varepsilon\|v - u\|.$$

Take a sequence $(x_n)_{n \ge N}$ where $\frac{1}{N} < \frac{\theta}{3}$ and $f(x_n, v) \le F(v) + \frac{1}{n}$. By using (110) and passing to limit,

$$\|v - u\| \le \delta_2 \Rightarrow F(v) \ge F(u) + \xi(v - u) - \varepsilon\|v - u\|$$

and this added to (108) gives (105).

Finally, referring to the last claim of the statement, take (see above) $v_n = u$ and then $\xi = \lim_{n \to \infty} f'_v(x_n, u)$. ■

1.12 *Corollary. If the norm of Banach space* V *is Fréchet differentiable* (*off the origin*), *then* V *is an Asplund space*
(*I. Ekeland - G. Lebourg*).

Elucidation. V is by definition an *Asplund space* when any real convex continuous function on *V* is Fréchet differentiable on a residual set.

■ Let $F : V \to \mathbf{R}$ be convex continuous. Then $\exists\ \Phi: V^* \to (-\infty, +\infty]$ convex lower semicontinuous so that

$$F(v) = \sup\{\xi(v) - \Phi(\xi) : \xi \in V^*\} \quad \forall\, v \in V$$

(Hahn - Banach theorem in $V \times \mathbf{R}$, [30], 7). Then, whatever u in $V\ \exists\ U$ a neighborhood of u and M a bounded subset of V^* so that

(111) $$F(v) = \sup\{\xi(v) - \Phi(\xi) : \xi \in M\} \ \forall\, v \in U \text{ ([30], 7).}$$

Apply 1.11 with $\Phi(\xi) = 1 - \|\xi\|$, $X = M$, $f(\xi, v) = \Phi(\xi) - \xi(v)$. $f'_v(\xi, v) = -\xi$, (86) and (87) are checked (use (111)), $-F$ is generic Fréchet differentiable. ■

Remark. The reciprocal statement of the corollary 1.12 is not true. R. Haydon gave the example of a compact space X with the property that $C(X\,;\mathbf{R})$ is an Asplund space, but without admitting an equivalent norm, nor even one Gâteaux differentiable off the origin.

Variant in infinite dimension of Hopf - Rinow theorem ([30], 3; previously, see ANNEX, *Riemann manifolds*). The Hopf - Rinow theorem claims that, in a complete Riemann manifold (the connection was included in the definition of a Riemann manifold) of finite dimension, any two points may be joint by a minimal geodesic. This property is not longer true for infinite dimension (*N. Grossman, Hilbert manifolds without epiconjugate points,* Proc. Amer. Math. Soc. **16** (1965), 1365 - 1371; *J. Mac Alpin, Thesis,* Columbia University, 1965).

In this subsection a version in the infinite dimension of Hopf - Rinow theorem is presented. This was obtained by I. Ekeland by using an abbreviated form of his principle 1.1. The demonstration will devolve from a succession of statements as it is presented below ([30], 3).

Definition. Let M be a Riemann manifold, $f: M \to \mathbf{R}$ and $\varepsilon > 0$. f is *locally* ε *- supported* at the point p of M if there is an open neighborhood U of p and $g: U \to \mathbf{R}$ of C^∞ class so that

$$g(p) = 0 \text{ and } f(m) - f(p) \geq g(m) - \varepsilon d(m, p)\ \forall\ m \in U. \tag{112}$$

1.13 *If* f *is locally* ε *- supported at the point* p *of* M, M *complete, then* $\forall\ \varepsilon' > \varepsilon\ \exists\ \eta' > 0$ and $\zeta' \in T_pM$ *so that*

$$f(\exp_p\xi) - f(p) - \langle \xi, \zeta' \rangle_p \geq -\varepsilon' \|\xi\|_p\ \forall\ \xi \in S_p^{\eta'}. \tag{113}$$

Reciprocally, if (113) *is verified with* $\varepsilon' = \varepsilon$, *then* f *is locally* ε *- supported at* p.

■ *First claim.* Let U, g be like in (112) and $\varepsilon' > \varepsilon$. Take η, $0 < \eta < \Delta(p)$ so that $\mathscr{S}_p^\eta \subset U$. (112) is rewritten (ANNEX, (49))

$$f(\exp_p\xi) - f(p) \geq g(\exp_p\xi) - \varepsilon \|\xi\|_p\ \forall\ \xi \in S_p^\eta. \tag{114}$$

But $g \circ \exp_p$ is Fréchet differentiable in 0, hence $\exists\ \eta'$, $0 < \eta' < \eta$ and $\exists\ \zeta'$ in T_pM so that

$$\|\xi\|_p < \eta' \Rightarrow \left| g(\exp_p\xi) - \langle \xi, \zeta' \rangle_p \right| \leq (\varepsilon' - \varepsilon)\|\xi\|_p \tag{115}$$

($g(p) = 0$, Riesz representation theorem). Combining (114) with (115) yields (113).

Second claim. Let be $\eta > 0$ and $\zeta \in T_pM$ that together with ε check (113). It may obviously suppose $\eta < \Delta(p)$ and then $\exp_p: S_p^\eta \to \mathscr{S}_p^\eta$ is a C^∞ - diffeomorphism and $\|\xi\|_p = d(p, \exp_p \xi)$ $\forall\ \xi \in S_p^\eta$ (ANNEX, (49)). The assumption (113) becomes $f(m) - f(p) - \langle \exp_p^{-1} m, \zeta\rangle_p \geq -\varepsilon d(p, m)$ $\forall\ m \in \mathscr{S}_p^\eta$ and it remains only to take $g : \mathscr{S}_p^\eta \to \mathbf{R}$, $g(m) = \langle \exp_p^{-1} m, \zeta\rangle_p$ (ANNEX, (44)). ■

1.14 *Let* M *be a complete Riemann manifold and* f : M → **R** *lower semicontinuous. Whatever* $\varepsilon > 0$ *the set of points of* M, *in which* f *is locally* ε - *supported, is everywhere dense.*

■ Let q be a point of M and V a neighborhood of q. It follows to find in V a point p where f is locally ε - supported.

Take δ, $0 < \delta < \Delta(q)$ so that $\mathscr{S}_q^\delta \subset V$ and, moreover, f is bounded from below on $\mathscr{S}_q^\delta$ (f is l.s.c. in q!). Consider the function $\psi: \overline{\mathscr{S}_q^\delta} \to \mathbf{R}$ (see ANNEX, (49))

$$\psi(m) = \begin{cases} \left(\delta^2 - \left\|\exp_q^{-1} m\right\|_q^2\right)^{-1}, m \in \mathscr{S}_q^\delta \\ +\infty, m \in \Sigma_q^\delta. \end{cases}$$

ψ is bounded from below, l.s.c. on $\overline{\mathscr{S}_q^\delta}$ and of C^∞ class on $\mathscr{S}_q^\delta$. Take the function $\varphi := \psi + f$. φ is l.s.c. and bounded from below on the complete metric space $\overline{\mathscr{S}_q^\delta}$. $\exists\ p$ in $\overline{\mathscr{S}_q^\delta}$ so that

$$\varphi(p) \leq \inf \varphi(\overline{\mathscr{S}_q^\delta}) + \varepsilon, \tag{116}$$

$$\varphi(m) \geq \varphi(p) - \varepsilon d(m, p)\ \forall\ m \in \overline{\mathscr{S}_q^\delta} \tag{117}$$

(see 1.1).

(116) provides us $\varphi(p) \in \mathbf{R}$, consequently $p \in \mathscr{S}_q^\delta$, hence $p \in V$. Replacing φ by $\psi + f$ in (117), it follows

$$f(m) - f(p) \geq g(m) - \varepsilon d(m, p)\ \forall\ m \in \mathscr{S}_q^\delta,$$

where $g : \mathscr{S}_q^\delta \to \mathbf{R}$, $g(m) = \psi(p) - \psi(m)$, that finishes the demonstration according to (112). ■

The following five statements will interfere straightforwardly in the demonstration of the Ekeland theorem. M is a *complete* Riemann manifold and *q* a *fixed* point in *M.*

Notations. Let p be whatever in $M \setminus \{q\}$ and $\rho > 0$.

(118) $\mathscr{C}_p^q = \{\gamma : [0, 1] \to M$ a piecewise C^∞ path $|\ \gamma(0) = p, \gamma(1) = q\}$

(see ANNEX, *Finsler manifolds*),

(119) $D(p) = \min\{\Delta(p), d(p, q)\}$ (hence $D(p) > 0$; for $d(p, q)$

see ANNEX, (42)),

(120) $\delta \in (0, D(p)), \gamma \in \mathscr{C}_p^q \Rightarrow T_p^\delta(\gamma) = \{t \in [0, 1]: d(\gamma(t), p) = \delta\}$

(nonempty, Cauchy - Bolzano theorem),

(121) $C_p^\delta(\rho) = \{\gamma(t): \gamma \in \mathscr{C}_p^q, L(\gamma) \leq d(p, q) + \rho, t \in T_p^\delta(\gamma)\}$

($C_p^\delta(\rho)$ is closed nonempty included in Σ_p^δ; for $L(\gamma)$ see ANNEX, (41)).

(122) d_q is the function $M \to \mathbf{R}$, $d_q(m) = d(q, m)$.

1.15 *Suppose* d_q *is locally* ε*- supported,* $\varepsilon < 1$, *at the point* p *of* M $\setminus$ {q}. *Then* $\forall\ \theta > 4\sqrt{\varepsilon}$ $\exists\ \eta$ *in* (0, D(p)) *having the property:*

(123) $$\forall\, \delta \textit{ from } (0, \eta)\ \exists\, \rho > 0 \textit{ so that } \operatorname{diam} C_p^\delta(\rho) \le \theta\delta.$$

■ Take $\varepsilon' > \varepsilon$ and $\beta > 0$ so small that

(124) $$\theta > 4\sqrt{\varepsilon' + \frac{\beta}{2}},\ (124_1)\ \varepsilon' + \beta - 1 < 0.$$

Apply 1.13 to d_q, $\exists\, \eta$ in $(0, D(p))$ and $\exists\, \zeta$ in $T_p M$ so that

(125) $$\|\xi\|_p \le \eta \Rightarrow d(q, \exp_p \xi) \ge d(q, p) + \langle \xi, \zeta\rangle_p - \varepsilon' \|\xi\|_p .$$

$d(q, \exp_p \xi) \le d(q, p) + d(p, \exp_p \xi)$ and combining (125) with (49), ANNEX yields

$$\|\xi\|_p \le \eta \Rightarrow (1 + \varepsilon')\|\xi\|_p \ge \langle \xi, \zeta\rangle_p,$$

consequently

(126) $$\|\zeta\|_p \le 1 + \varepsilon'.$$

Now, let δ be whatever fixed in $(0, \eta)$ and $\gamma(s)$ any in $C_p^\delta(\beta\delta)$, that is $\gamma \in \mathscr{C}_p^q$, (127) $L(\gamma) \le d(q, p) + \beta\delta$ and (128) $d(\gamma(s), p) = \delta$, $s \in [0, 1]$. We have $d(q, \gamma(s)) + d(\gamma(s), p) \le L(\gamma) \overset{(127)}{\le} d(q, p) + \beta\delta$, hence (see (128))

(129) $$d(q, \gamma(s)) \le d(q, p) + \beta\delta - \delta.$$

Let $\xi(s)$ be in σ_p^δ so that (130) $\xi(s) = \exp_p^{-1}(\gamma(s))$. By inserting into (125) (correctly, $\delta < < \eta$) and taking into account for (129) and (130), yields

$$-(1 - \varepsilon')\|\xi(s)\|_p + \beta\delta \ge \langle \xi(s), \zeta\rangle_p,$$

that dividing by δ $(= \|\xi(s)\|_p)$ becomes

(131) $$\left\langle \frac{\xi(s)}{\delta}, \zeta \right\rangle_p \le -1 + \varepsilon' + \beta$$

and using (126) combined with (124_1) it follows

(132) $$\left\langle \frac{\xi(s)}{\delta}, \frac{\zeta}{\|\zeta\|_p} \right\rangle_p \le \frac{-1+\varepsilon'+\beta}{1+\varepsilon'}.$$

Then

$$\left\| \frac{\xi(s)}{\delta} + \frac{\zeta}{\|\zeta\|_p} \right\|_p^2 = 2 + 2\left\langle \frac{\xi(s)}{\delta}, \frac{\zeta}{\|\zeta\|_p} \right\rangle_p \overset{(132)}{\le} 2\frac{2\varepsilon'+\beta}{1+\varepsilon'},$$

(133) $$\left\| \frac{\xi(s)}{\delta} + \frac{\zeta}{\|\zeta\|_p} \right\|_p \le \sqrt{2}\sqrt{\frac{2\varepsilon'+\beta}{1+\varepsilon'}}.$$

So for $\gamma'(s')$ from $C_p^\delta(\beta\delta)$ we have also ($\xi'(s') = \exp_p^{-1}\gamma'(s')$)

(134) $$\left\| \frac{\xi'(s')}{\delta} + \frac{\zeta}{\|\zeta\|_p} \right\|_p \le \sqrt{2}\sqrt{\frac{2\varepsilon'+\beta}{1+\varepsilon'}}.$$

(133) and (134) give

$$\left\| \frac{\xi(s)}{\delta} - \frac{\xi'(s')}{\delta} \right\|_p \le 2\sqrt{2}\sqrt{\frac{2\varepsilon'+\beta}{1+\varepsilon'}} \le 2\sqrt{2}\sqrt{2\varepsilon'+\beta} \overset{(124)}{\le} \theta$$

and consequently diam $C_p^\delta(\beta\delta) \le \theta\delta$. ■

Remark. In [30], 3 the relation (2.12) is not generally true. It should add here the condition $-1 + \varepsilon' + \beta < 0$ (see above (124_1)). Consequently, the demonstration for the Lemma 2.1 is vicious. Fortunately, the imposed condition will not affect, as it will see later, the demonstration of the Ekeland theorem.

Now consider, for every $\theta > 0$, the subset A_θ of M (see (119), (121))

(135) $$A_\theta = \{p \in M \setminus \{q\} : \exists\, \delta \in (0, D(p)), \exists\, \rho > 0 \text{ so that diam } C_p^\delta(\rho) \le \theta\delta\}.$$

If d_q is locally ε - supported, $\varepsilon < 1$, at the point p of $M\setminus\{q\}$, then, according to 1.15, $p \in A_\theta\ \forall\, \theta > 4\sqrt{\varepsilon}$. But it can affirm even more.

1.16 *Let* d_q *be locally* ε *- supported,* $\varepsilon < 1$, *at the point* p *of* $M \setminus \{q\}$. *If* $\theta > 4\sqrt{\varepsilon}$, *then*

$$p \in \operatorname{int} A_\theta.$$

■ Choose λ from (1, 2) and $\theta' > 4\sqrt{\varepsilon}$ so that

$$\frac{\lambda^2}{2-\lambda}\theta' = \theta, \tag{136}$$

this being possible as $\lim\limits_{\lambda \to 1+} \dfrac{\lambda^2}{2-\lambda} = 1$ and $\theta > 4\sqrt{\varepsilon}$. Let δ_1 be from (0, $D(p)$). For every point m of $\mathscr{S}_p^{\delta_1}$ consider the map

$$\varphi_m\colon S_p^{\delta_1} \to T_m M,\ \varphi_m = \exp_m^{-1} \circ \exp_p. \tag{137}$$

Obviously, φ_p is the identity on $S_p^{\delta_1}$. Moreover, the class of $(m, \xi) \to \varphi_m(\xi)$ is C^∞. Hence, it can find δ_2 in (0, δ_1) so that

$$\|\xi\|_p \le \delta_2 \text{ and } d(p, m) \le \delta_2 \Rightarrow \frac{1}{\lambda} \le \|T_\xi \varphi_m\| \le \lambda. \tag{138}$$

Set $\delta_3 := \dfrac{\delta_2}{3}$. Whatever m of $\mathscr{S}_p^{\delta_3}$ and whatever ξ, ξ' of $S_p^{\delta_3}$, the reciprocal image by φ_m of the segment of extremities $\varphi_m(\xi)$, $\varphi_m(\xi')$ is found in $S_p^{3\delta_3}$, consequently (138) is working and then it results

$$m, \exp_p\xi, \exp_p\xi' \in \mathscr{S}_p^{\delta_3} \Rightarrow \frac{1}{\lambda}\|\xi - \xi'\|_p \le \|\varphi_m(\xi) - \varphi_m(\xi')\|_m \le \lambda\|\xi - \xi'\|_p. \tag{139}$$

1.15 provides δ in (0, δ_3) and $\rho > 0$ so that ($\theta' > 4\sqrt{\varepsilon}$!)

$$\operatorname{diam} C_p^{\frac{3\delta}{4}}(\rho) \le \frac{3}{4}\delta\theta'. \tag{140}$$

Take $\eta := \min\left\{\dfrac{\delta}{4}, \dfrac{\rho}{3}, \dfrac{3\delta}{4}\left(1 - \dfrac{1}{\lambda}\right)\right\}$ and it will be shown

$$\mathscr{S}_p^{\eta} \subset A_\theta, \tag{142}$$

that closes the demonstration.

Firstly, remark

$$m \in \mathscr{S}_p^{\frac{\delta}{4}} \Rightarrow \mathscr{S}_p^{\frac{3\delta}{4}} \subset \mathscr{S}_m^{\delta} \subset \mathscr{S}_p^{\frac{5\delta}{4}} \text{ (the triangle inequality).} \tag{143}$$

Since $\frac{5\delta}{4} < 3\delta < 3\delta_3 = \delta_2 < \delta_1$ and $\delta_1 < D(p)$, according to (143) any two points in $\mathscr{S}^{\delta}_{m}$ can be joint by a minimal geodesic depending smoothly on the end points (see ANNEX, (48)), and then *a fortiori* ($\eta < \delta$!)

(144) $$m \in \mathscr{S}^{\frac{\delta}{4}}_{p} \Rightarrow \eta < D(m).$$

Let m be of $\mathscr{S}^{\eta}_{p}$, γ of $\mathscr{C}^{q}_{m}$ with (145) $L(\gamma) \le \mathrm{d}(m, q) + \frac{\rho}{3}$ and s of $T^{\delta}_{m}(\gamma)$. Note $\xi_m := \exp^{-1}_{m}\gamma(s)$, hence $\xi_m \in \sigma^{\delta}_{m}$. Consider the path γ_1 from $\mathscr{C}^{q}_{m}$,

$$\gamma_1(t) = \begin{cases} \exp_m 2t\xi_m, t \in [0, \frac{1}{2}] \\ \gamma(2t(1-s) + 2s - 1), t \in [\frac{1}{2}, 1]. \end{cases}$$

We have (146) $L(\gamma_1) \le L(\gamma)$. Set (147) $\mu := \exp^{-1}_{p} m$ and consider a new path γ_2 of $\mathscr{C}^{q}_{p}$,

$$\gamma_2(t) = \begin{cases} \exp_p 2t\mu, t \in [0, \frac{1}{2}] \\ \gamma_1(2t-1), t \in [\frac{1}{2}, 1]. \end{cases}$$

We have (see (145), (146))

$L(\gamma_2) = L(\gamma_1) + \mathrm{d}(m, p) \le \mathrm{d}(m, q) + \frac{\rho}{3} + \mathrm{d}(m, p) \le \mathrm{d}(p, q) + \mathrm{d}(m, p) + \frac{\rho}{3} + \mathrm{d}(m, p) \le$ $\le \mathrm{d}(p, q) + \rho$, since $m \in \mathscr{S}^{\eta}_{p}$ (see also (141)).

Take τ from $T^{\frac{3\delta}{4}}_{p}(\gamma_2)$ and let be $\zeta_p := \exp^{-1}_{p}(\gamma(\tau))$, hence $\zeta_p \in S^{\frac{3\delta}{4}}_{p}$. If γ' is another path from $\mathscr{C}^{q}_{m}$ like in (145), in a similar manner are defined ξ'_m of σ^{δ}_{m}, γ'_1 of $\mathscr{C}^{q}_{m}$, γ'_2 of $\mathscr{C}^{q}_{p}$, ζ'_p of $S^{\frac{3\delta}{4}}_{p}$ and we still have $L(\gamma'_2) \le \mathrm{d}(p, q) + \rho$. Taking into account for (140), gives

$$\|\zeta_p - \zeta'_p\|_p \le \frac{3}{4}\delta\theta',$$

which combined with (139) gives

(148) $$\|\varphi_m(\zeta_p) - \varphi_m(\zeta'_p)\|_m \le \frac{3}{4}\delta\lambda\theta'.$$

The same estimation applied to $\|\zeta_p\|_p \left(=\frac{3\delta}{4}\right)$ yields

$$\|\varphi_m(\zeta_p) - \exp_m^{-1}(p)\|_m \geq \frac{3}{4}\frac{\delta}{\lambda}. \tag{149}$$

But $\|\exp_m^{-1}(p)\|_m = d(m, p)$ (ANNEX, (49)) and then, using (149),

$$\|\varphi_m(\zeta_p)\|_m \geq \frac{3}{4}\delta\left(\frac{2}{\lambda}-1\right) \text{ and similarly} \tag{150}$$

$$\|\varphi_m(\zeta'_p)\|_m \geq \frac{3}{4}\delta\left(\frac{2}{\lambda}-1\right). \tag{151}$$

But, according to the definitions, $\xi_m = \dfrac{\delta\varphi_m(\zeta_p)}{\|\varphi_m(\zeta_p)\|_m}$, $\xi'_m = \dfrac{\delta\varphi_m(\zeta'_p)}{\|\varphi_m(\zeta'_p)\|_m}$, hence $\|\xi_m - \xi'_m\|_m \leq$

$$\leq \delta\frac{\|\varphi_m(\zeta_p) - \varphi_m(\zeta'_p)\|_m}{\min\{\|\varphi_m(\zeta_p)\|_m, \|\varphi_m(\zeta'_p)\|_m\}} \overset{(148),(150),(151)}{\leq} \delta\frac{\lambda^2}{2-\lambda}\theta' \overset{(136)}{=} \delta\theta, \|\xi_m - \xi'_m\|_m \leq \delta\theta, \text{ that}$$

imposes $\operatorname{diam} C_m^\delta\left(\frac{\rho}{3}\right) \leq \delta\theta$ and, *via* (135), $m \in A_\theta$. m being whatever in $\mathscr{S}_p^\eta$, (142) is demonstrated. ■

Definition. Let p be from $M \setminus \{q\}$ and δ from $(0, D(p))$. The path γ from $\mathscr{C}_p^q$ is *geodesic inside* $\mathscr{S}_p^\delta$, if there is ξ in $T_pM \setminus \{0\}$ so that

$$\gamma(t) = \exp_p t\xi,\ 0 \leq t \leq \frac{\delta}{\|\xi\|_p}. \tag{152}$$

Notation. A_∞ denotes the set of points p of $M \setminus \{q\}$ with the property that there exists an increasing sequence $(r_n)_{n\geq 1}$ with $\lim_{n\to\infty} r_n = D(p)$ so that, for any sequence $(\gamma_n)_{n\geq 1}$ from $\mathscr{C}_p^q$ with $L(\gamma_n) \to d(p, q)$ and σ_n geodesic inside $\mathscr{S}_p^{r_n}$, $n \geq 1$, the sequence $\left(\frac{\dot\gamma_n(0)}{\|\dot\gamma_n(0)\|_p}\right)_{n\geq 1}$ converges in σ_p^1.

1.17 $p \in A_\theta\ \forall\ \theta > 0 \Rightarrow p \in A_\infty$.

■ Since $p \in A_{\frac{1}{n}}\ \forall\ n \geq 1$, there is a sequence $(\delta_n)_{n\geq 1}$, $0 < \delta_n < D(p)$ and also a decreasing sequence $(\rho_n)_{n\geq 1}$ with the limit 0 ($\rho' < \rho'' \Rightarrow C_p^\delta(\rho') \subset C_p^\delta(\rho'')$!) so that

(153) $$\operatorname{diam} C_{\mathrm{p}}^{\delta_{\mathrm{n}}}(\rho_{\mathrm{n}}) \leq \frac{\delta_{\mathrm{n}}}{n}.$$

Take an increasing sequence $(r_{\mathrm{n}})_{\mathrm{n}\geq 1}$ with $\delta_{\mathrm{n}} \leq r_{\mathrm{n}} < D(p)$, $r_{\mathrm{n}} \to D(p)$ and let be now $(\gamma_{\mathrm{n}})_{\mathrm{n}\geq 1}$ a sequence from $\mathscr{C}_{\mathrm{p}}^{\mathrm{q}}$ with $L(\gamma_{\mathrm{n}}) \to \mathrm{d}(p, q)$ and γ_{n}, $n \geq 1$, geodesic inside $\mathscr{S}_{\mathrm{p}}^{r_{\mathrm{n}}}$. A suitable change of the parameter t allows to suppose $\exists\, \xi_{\mathrm{n}}$ in σ_{p}^{1} with $\gamma_{\mathrm{n}}(t) = \exp_{\mathrm{p}} t\xi_{\mathrm{n}}$ for $0 \leq t \leq r_{\mathrm{n}}$ and $\xi_{\mathrm{n}} = \dfrac{\dot{\gamma}_{\mathrm{n}}(0)}{\|\dot{\gamma}_{\mathrm{n}}(0)\|_{\mathrm{p}}}$. We keep: $k \geq n \Rightarrow \delta_{\mathrm{n}} \leq r_{\mathrm{n}} \leq r_{\mathrm{k}}$, consequently γ_{k} cuts $\Sigma_{\mathrm{p}}^{\delta_{\mathrm{n}}}$ in $\exp_{\mathrm{p}}\delta_{\mathrm{n}}\xi_{\mathrm{k}}$.

It will be shown that $(\xi_{\mathrm{n}})_{\mathrm{n}\geq 1}$ is a Cauchy sequence. Let be $\varepsilon > 0$. Take n in **N** so that $\dfrac{1}{n} \leq \varepsilon$. There is N in **N**, $N > n$ so that $k \geq N \Rightarrow L(\gamma_{\mathrm{k}}) \leq \mathrm{d}(p, q) + \rho_{\mathrm{n}}$. When $l \geq k \geq N$, $\delta_{\mathrm{n}}\xi_{l}$ and $\delta_{\mathrm{n}}\xi_{\mathrm{k}}$ are found in $C_{\mathrm{p}}^{\delta_{\mathrm{n}}}(\rho_{\mathrm{n}})$ (see the last remark) and consequently $\|\delta_{\mathrm{n}}\xi_{l} - \delta_{\mathrm{n}}\xi_{\mathrm{k}}\|_{\mathrm{p}} \overset{(153)}{\leq} \dfrac{\delta_{\mathrm{n}}}{n}$, $l, k \geq N \Rightarrow \|\xi_{l} - \xi_{\mathrm{k}}\|_{\mathrm{p}} \leq \varepsilon$. ■

Let p be from $M \setminus \{q\}$ and δ from $(0, D(p))$, hence in particular $\delta < \mathrm{d}(q, p)$ ((119)). We have

(154) $$\operatorname{dist}(q, \Sigma_{\mathrm{p}}^{\delta}) = \mathrm{d}(q, p) - \delta.$$

Indeed, we have $\mathrm{d}(q, p) \leq \mathrm{d}(p, m) + \mathrm{d}(q, m)$, in particular $\forall\, m$ of $\Sigma_{\mathrm{p}}^{\delta}$, hence $\mathrm{d}(p, q) \leq \delta + \operatorname{dist}(q, \Sigma_{\mathrm{p}}^{\delta})$. Suppose *par absurdum* $\mathrm{d}(q, p) < \delta + \operatorname{dist}(q, \Sigma_{\mathrm{p}}^{\delta})$. There is γ in $\mathscr{C}_{\mathrm{q}}^{\mathrm{p}}$ ((118)) so that

(155) $$L(\gamma) < \delta + \operatorname{dist}(q, \Sigma_{\mathrm{p}}^{\delta}).$$

$L(\gamma) = \int_0^1 \|\dot\gamma(t)\|_{\gamma(t)} dt$ and since $L(\gamma) > \delta$ because $d(q, p) > \delta$, there is t_0 in $(0, 1)$ so that (156)

$\int_0^{t_0} \|\dot\gamma(t)\|_{\gamma(t)} dt = \delta$ (γ being C^∞ - piecewise, it can apply, except for a finite number of cases, the Cauchy - Bolzano theorem). Therefore, for $\gamma_1 := \gamma \,|\, [t_0, 1]$ we have ((155), (156))

(157) $$L(\gamma_1) < \mathrm{dist}\,(q, \Sigma_p^\delta).$$

But $\gamma_1(t_0) \overset{(156)}{\in} \Sigma_p^\delta$ and $\gamma_1(1) = q$, hence, according to the definition, $L(\gamma_1) \geq \mathrm{dist}\,(q, \Sigma_p^\delta)$, in contradiction with (157).

Definition. A *nearest point* to q from Σ_p^δ is a point m so that

(158) $$d(q, m) = \mathrm{dist}(q, \Sigma_p^\delta).$$

In the infinite dimensional case such points do not always exist, and in the finite dimensional case their uniqueness is not assured. This specification reveals the importance of the following statement.

1.18 *Let* p *be from* A_∞ *and* $(r_n)_{n\geq 1}$ *the increasing sequence with* $r_n \to D(p)$ *from the definition of* A_∞. *Then all the sequences* $\left(\frac{\dot\gamma_n(0)}{\|\dot\gamma_n(0)\|_p} \right)_{n\geq 1}$, *where* $\gamma_n \in \mathscr{C}_p^q$, $L(\gamma_n) \to d(p, q)$ *and* γ_n *is geodesic inside* $\mathscr{S}_p^{r_n}$, *have the same limit* μ_p *and, whatever* δ *from* $(0, D(p))$, $\exp_p \delta\mu_p$ *is a nearest point to* q *in* Σ_p^δ.

■ *First claim.* Let $(\gamma_n)_{n\geq 1}$, $(\gamma_n')_{n\geq 1}$ be sequences like in the statement and $\xi = \lim_{n\to\infty} \frac{\dot\gamma_n(0)}{\|\dot\gamma_n(0)\|_p}$, $\xi' = \lim_{n\to\infty} \frac{\dot\gamma_n'(0)}{\|\dot\gamma_n'(0)\|_p}$. Consider the sequence $(\gamma_n'')_{n\geq 1}$, n odd $\Rightarrow \gamma_n'' = \gamma_n$, n even $\Rightarrow \gamma_n'' = \gamma_n'$. Since, $\forall\, n \geq 1$, γ_n'' is geodesic inside $\mathscr{S}_p^{r_n}$ and $L(\gamma_n'') \to d(p, q)$, the sequence $\left(\dfrac{\dot{\gamma}''_n(0)}{\left\|\dot{\gamma}''_n(0)\right\|_p}\right)_{n\geq 1}$ converges, let ξ'' be its limit. Then $\xi'' = \xi$, $\xi'' = \xi'$, hence $\xi = \xi'$, we have a common limit. This has been denoted by μ_p. Obviously $\left\|\mu_p\right\|_p = 1$.

Second claim. Let $(\gamma_n)_{n\geq 1}$ be a sequence from $\mathscr{C}^{\,q}_{\,p}$ like in the statement. For every n of **N** take s_n from $T_p^{r_n}(\gamma_n)$ (see (120)) and let ξ_n be from σ_p^1 so that

(159) $$\gamma_n(s_n) = \exp_p r_n \xi_n .$$

γ_n is shorted to γ_n^1,

(160) $$\gamma_n^1(t) = \begin{cases} \exp_p 2tr_n\xi_n, t \in [0, \frac{1}{2}] \\ \gamma_n(2t(1-s_n) + 2s_n - 1), t \in [\frac{1}{2}, 1]. \end{cases}$$

As γ_n^1 is geodesic inside of $\mathscr{S}^{\,r_n}_{\,p}$ ((159)!) we have (the first claim),

(161) $$\mu_p = \lim_{n\to\infty} \zeta_n,\ \zeta_n := \frac{\dot{\gamma}_n^1(0)}{\left\|\dot{\gamma}_n^1(0)\right\|_p}, \text{ as moreover}$$

(162) $$L(\gamma_n^1) \to \mathrm{d}(p, q) \text{ since } L(\gamma_n) \geq L(\gamma_n^1) \geq \mathrm{d}(p, q).$$

Now, let δ be whatever in $(0, D(p))$ and N of **N** so that $n \geq N \Rightarrow r_n \geq \delta$. Then $n \geq N \Rightarrow$ $\Rightarrow \gamma_n^1$ cuts Σ_p^δ in $\exp_p\delta\zeta_n$. But $\exp_p\delta\zeta_n \to \exp_p\delta\mu_p$ since $\zeta_n \to \mu_p$ (ANNEX, (49)) and as $n \geq N \Rightarrow \mathrm{d}(q, \exp_p\delta\zeta_n) \leq L(\gamma_n^1) - \delta$, we have passing to the limit ((162))

$$\mathrm{d}(q, \exp_p\delta\mu_p) \leq \mathrm{d}(p, q) - \delta,$$

which, confronted with (154), gives $\mathrm{d}(q, \exp_p\delta\mu_p) \leq \mathrm{dist}(q, \Sigma_p^\delta)$ and consequently $\mathrm{d}(q, \exp_p\delta\mu_p) = \mathrm{dist}(q, \Sigma_p^\delta)$, as $\exp_p\delta\mu_p \in \Sigma_p^\delta$ since $\delta = \left\|\delta\mu_p\right\|_p = \mathrm{d}(p, \exp_p\delta\mu_p)$ (ANNEX, (49)). ■

1.19 *If* $p \in A_\infty$, *then* $\exp_p \delta\mu_p \in A_\infty\ \forall\ \delta$ *from* $(0, D(p))$.

■ Note $m := \exp_p \delta\mu_p$. We have (169) $d(m, p) = \delta$ as $\|\mu_p\|_p = 1$ (ANNEX, (49)). (163) shows that the demand in the statement is coherent since $m \neq q$, given that $\delta \overset{(119)}{<} d(p, q)$. Show

(164) $$D(m) \geq D(p) - \delta.$$

(164) will be validated if $d(m, q) \geq d(p, q) - \delta$, which is true according to (163), and moreover (165) $\Delta(m) \geq \Delta(p) - \delta$. Let η be whatever, $\eta < \Delta(p) - \delta$, $\exists\ \delta' < \Delta(p)$ so that $\eta < \delta' - \delta$. Then $\mathscr{S}_m^\eta \subset \mathscr{S}_p^{\delta'}$: $x \in \mathscr{S}_m^\eta \Rightarrow d(m, x) < \eta \Rightarrow d(p, x) \leq d(m, x) + d(m, p) \overset{(163)}{=} d(m, x) + \delta < \eta + \delta < \delta' - \delta + \delta = \delta'$. But $\mathscr{S}_p^{\delta'}$ has the property mentioned in ANNEX, (48) since $\delta' < \Delta(p)$, consequently $\mathscr{S}_m^\eta$ has the same property that assures (165), since the definition gives $\eta \leq \Delta(m)$.

Now, let $(r_n)_{n\geq 1}$ be the increasing sequence afferent by definition to p, $0 < r_n < D(p)$, $r_n \to D(p)$. (164) allows to choose an increasing sequence $(\rho_n)_{n\geq 1}$, $0 < \rho_n < D(m)$, $\rho_n \to D(m)$ so that

(166) $$\rho_n \geq r_n - \delta\ \forall\ n \geq 1.$$

Let $(\gamma_n)_{n\geq 1}$ be a sequence from $\mathscr{C}_m^q$ with $L(\gamma_n) \to d(m, q)$ and γ_n geodesic inside $\mathscr{S}_m^{\rho_n}$. One may suppose when need, by using a change of variable, (167) $0 \leq t \leq \rho_n \Rightarrow \gamma_n(t) = \exp_m t\zeta_n$, $\|\zeta_n\|_p = 1$. So it remains to show, as $\gamma_n'(0) = \zeta_n$,

(168) $$(\zeta_n)_{n\geq 1} \text{ convergent in } \sigma_m^1.$$

But (168) will result from

(169) $$\lim_{n\to\infty} \gamma_n(r_n - \delta) = m_1 := \exp_p D(p)\mu_p.$$

Indeed, according to (166) and (167) we have $\gamma_n(r_n - \delta) = \exp_m(r_n - \delta)\zeta_n$ and then (169) yields $m_1 = \lim_{n\to\infty} \exp_m(r_n - \delta)\zeta_n$, that imposes the convergence of the sequence $((r_n - \delta)\zeta_n)_{n\geq 1}$, property equivalent to (168) as $r_n - \delta \to D(p) - \delta$.

Go to the justification of (169). Let N be of $\mathbf{N}$ so that $n \geq N \Rightarrow r_n \geq \delta$. Then $(\gamma_n(r_n - \delta))_{n\geq N}$ is correctly considered. We have

$$d(q, m) = \text{dist}(q, \Sigma_p^\delta) \overset{(154)}{=} d(p, q) - \delta,$$

and since $L(\gamma_n) \to d(m, q)$ it follows

(170) $$L(\gamma_n) + \delta \to d(p, q).$$

Take (171) s_n from $T_p^{r_n}(\gamma_n)$ ((120)) and let be (172) $\xi_n := r_n^{-1} \exp_p^{-1} \gamma_n(s_n)$. Consider also the sequence $(\gamma_n^1)_{n\geq 1}$, γ_n^1 from $\mathscr{C}_p^q$,

(173) $$\gamma_n^1(t) = \begin{cases} \exp_p 2tr_n\xi_n, t \in [0, \frac{1}{2}] \\ \gamma_n(2t(1 - s_n) + 2s_n - 1), t \in [\frac{1}{2}, 1] \end{cases}$$

(correctly, $\gamma_n(0) = \exp_p 0 = p$ (ANNEX, (44)), $\gamma_n^1(1) = \gamma_n(1) = q$, for $t = \frac{1}{2}$ one finds, on both lines, $\gamma_n(s_n)$ ((172)).

The shaded path is γ_n^1, $\gamma_n^1 = \gamma_n^1 \mid [0, \frac{1}{2}] \vee \gamma_n^1 \mid [\frac{1}{2}, 1]$ and $\gamma_n^1 \mid [\frac{1}{2}, 1] = \gamma_n \mid [s_n, 1]$. So we have $L(\gamma_n^1) = L(\gamma_n^1 \mid [0, \frac{1}{2}]) + L(\gamma_n^1 \mid [\frac{1}{2}, 1]) \overset{\text{fig.1}}{\leq} \delta + d(m, \gamma_n(s_n)) + L(\gamma_n \mid [s_n, 1]) \leq \delta +$ $+ \int_0^{s_n} \|\dot{\gamma}_n(t)\|_{\gamma_n(t)} dt + \int_{s_n}^1 \|\dot{\gamma}_n(t)\|_{\gamma_n(t)} dt = \delta + L(\gamma_n)$, $L(\gamma_n^1) \leq \delta + L(\gamma_n)$, hence $0 \leq L(\gamma_n^1) - d(p, q) \leq \delta + L(\gamma_n) - d(p, q)$, go to limit and (170) yields

$$\lim_{n\to\infty} L(\gamma_n^1) = d(p, q).$$

Since $p \in A_\infty$ and $(r_n)_{n\geq 1}$ is the afferent sequence whereas γ_n^1 is clearly geodesic inside

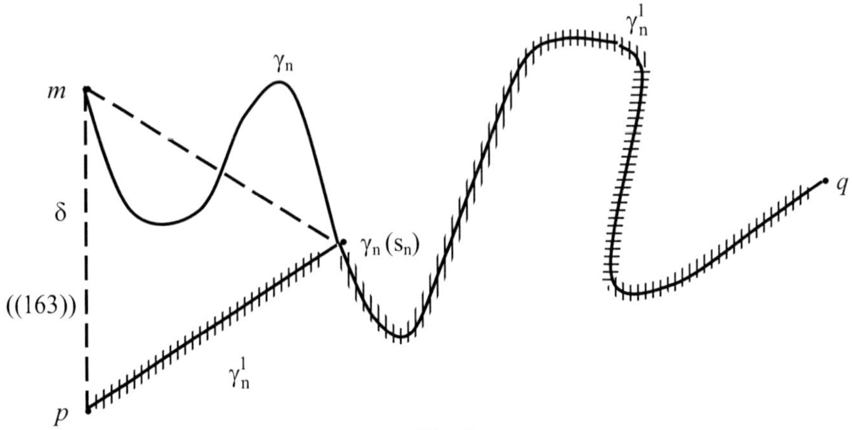

Fig. 1

$\mathscr{S}_{r_n}^{p}$, 1.18 gives (see also (173))

(174) $$\lim_{n\to\infty} \xi_n = \mu_p.$$

But $\gamma_n(s_n) \overset{(172)}{=} \exp_p r_n \xi_n$ and in addition $r_n \to D(p)$, this imposes, *via* (174),

(175) $$\lim_{n\to\infty} \gamma_n(s_n) = m_1 \text{ ((169), notation).}$$

On the other side, $d(m, \gamma_n(s_n)) \overset{\text{fig.1}}{\geq} d(p, \gamma_n(s_n)) - \delta \overset{(171)}{=} r_n - \delta$, $d(m, \gamma_n(s_n)) \geq r_n - \delta$, consequently (176) $s_n \geq r_n - \delta$.

(176) allows to write, for $n \geq N$,

$$d(\gamma_n(s_n), \gamma_n(r_n - \delta)) \leq \int_{r_n-\delta}^{s_n} \|\dot{\gamma}_n(t)\|_{\gamma_n(t)} dt = L(\gamma_n) - \int_0^{r_n-\delta} \|\dot{\gamma}_n(t)\|_{\gamma_n(t)} dt - \int_{s_n}^{1} \|\dot{\gamma}_n(t)\|_{\gamma_n(t)} dt .$$

But γ_n is geodesic inside of $\mathscr{S}_m^{r_n}$ and taking into account for $\rho_n \overset{(166)}{\geq} r_n - \delta$, the last inequality gives

(177) $$d(\gamma_n(s_n), \gamma_n(r_n - \delta)) \leq L(\gamma_n) - (r_n - \delta) - d(q, \gamma_n(s_n)).$$

But $\gamma_n(s_n) \overset{(171)}{\in} \Sigma_p^{r_n}$ and then we find with (154)

$$d(q, \gamma_n(s_n)) \geq d(q, \Sigma_p^{r_n}) = d(q, p) - r_n,$$

that inserted into (177) yields

$$d(\gamma_n(s_n), \gamma_n(r_n - \delta)) \leq L(\gamma_n) + \delta - d(p, q).$$

Go to the limit for $n \to \infty$ and, *via* (170), one obtains

(178) $$\lim_{n\to\infty} d(\gamma_n(s_n), \gamma_n(r_n - \delta)) = 0.$$

It remains only to combine (178) with (175) to obtain (169). ■

Now we can go on to

1.20 Ekeland theorem. *Let* M *be a complete Riemann manifold and* q *a point of* M. *The set of points* p *of* M, *for that there is a unique minimal geodesic from* p *to* q, *contains a subset everywhere dense and of* G_δ *type.*

■ Consider, for every n of **N**, $\Omega_n := \operatorname{int} A_{\frac{1}{n}}$ ((135)). Ω_n is obviously open, but it is also everywhere dense. Indeed, let ε be from $\left(0, \dfrac{1}{4n^2}\right)$. According to 1.14 the set D of points of M, in which d_q is locally ε - supported, is everywhere dense. But $D \subset \operatorname{int} A_{\frac{1}{n}}$ (see 1.16), hence Ω_n is everywhere dense. Let be $\Omega := \bigcap_{n=1}^{\infty} \Omega_n$. M being complete, Ω is everywhere dense. But as $0 < \theta_1 < \theta_2 \Rightarrow A_{\theta_1} \subset A_{\theta_2}$, we have $\Omega \subset \bigcap_{\theta>0} A_\theta$ and taking into account for 1.17 it gives

(179) $$\Omega \subset A_\infty.$$

If it will be shown that for every point p of A_∞ there exists a unique minimal geodesic from p to q, then (179) allows to affirm that the demonstration is finished.

But let p be whatever from A_∞ and (180) $r := d(p, q)$. For every δ of $(0, D(p))$ there is in Σ_p^δ (according to 1.18) a nearest point $\exp_p \delta\mu_p$ to q. Consider the path γ of C^∞ class,

(181) $$\gamma(t) = \exp_p tr\mu_p,\ t \in [0, 1].$$

Suppose demonstrated

(182) $$\gamma(t) \in A_\infty \text{ and } d(q, \gamma(t)) = r(1-t)\ \forall\, t \in [0, 1).$$

$d(q, \gamma(1)) = \lim_{t\to 1} d(q, \gamma(t)) \overset{(182)}{=} 0$, hence $\gamma(1) = q$ and then γ is a geodesic from p to q. Moreover, γ is minimal as $L(\gamma) = r \overset{(180)}{=} d(p, q)$, the existence claimed in the statement is demonstrated. Go to the uniqueness. Let γ' be from $\mathscr{C}_p^q$ minimal geodesic. Take the sequence $(\gamma_n)_{n\geq 1}$ from $\mathscr{C}_p^q$, $\gamma_n = \gamma'$. We have $L(\gamma_n) \to d(p, q)$, γ_n is geodesic inside $\mathscr{S}_p^{r_n}$, 1.18 yields $\frac{\dot{\gamma}'(0)}{r} = \mu_p$ and consequently $\gamma' = \gamma$.

So it remains to demonstrate (182). This is true on $[0, D(p))$ as assure 1.18 and 1.19 corroborated with (154). Let t_0 be the supremum of all numbers s from $(0, 1)$ with the property that (182) is true on $[0, s)$. It remains to show $t_0 = 1$. Suppose *par absurdum*

(183) $$t_0 < 1.$$

The contradiction that will be obtained signifies the end of the demonstration. Obviously, $t_0 \geq D(p) > 0$. Let be $p_0 := \gamma(t_0)$ and δ of $(0,D(p)) \cap (0,D(p_0))$. Take s of $(0, t_0)$ so that (184) $d(p, \gamma(s)) < \frac{\delta}{4}$ (the continuity of γ) and let be (185) $m := \gamma(s)$ ($\overset{(182)}{=} \exp_p sr\mu_p$). Since $\delta < D(p)$, two whatever points of $\mathscr{S}_p^{\delta}$ can be joint by a unique minimal geodesic depending smoothly on the endpoints (ANNEX, (48), the definition of $\Delta(p)$). But $\mathscr{S}_m^{\frac{\delta}{2}} \subset \subset \mathscr{S}_p^{\delta}$ (triangle inequality and (184)), consequently (186) $\frac{\delta}{2} < D(m)$. (182) is checked on $[0, t_0)$, hence (187) $d(q, m) = r(1 - s)$ and $m \in A_\infty$. Apply 1.18 combined with (186), $\exp_m \frac{\delta}{2} \mu_m$ is a nearest point of $\Sigma_m^{\frac{\delta}{2}}$ to q and we have ((187), (154))

(188) $$d(q, \exp_m \frac{\delta}{2} \mu_m) = r(1 - s) - \frac{\delta}{2} > 0.$$

Then

(189) $$d(p, \exp_m \frac{\delta}{2} \mu_m) \geq d(p, q) - d(q, \exp_m \frac{\delta}{2} \mu_m) \overset{(188)}{=} rs + \frac{\delta}{2}.$$

Set (190) $\rho := \frac{rs}{rs + \frac{\delta}{2}}$ and consider the C^∞ - piecewise path γ^1

(191) $$\gamma^1(t) = \begin{cases} \exp_p \frac{trs}{\rho} \mu_p, t \in [0, \rho] \\ \exp_m \frac{(t-\rho)\delta}{2(1-\rho)} \mu_m, t \in [\rho, 1]. \end{cases}$$

The definition is correct, as the first expression is equal to $\exp_p sr\mu_p$ for $t = \rho$, that is to m ((185)), while the second expression is equal to 0 for $t = \rho$, hence also to m (ANNEX,

(44)). γ^1 is a path from p to $\exp_m \frac{\delta}{2}\mu_m$. We have $L(\gamma^1) = rs + \frac{\delta}{2}$ and $\|\dot{\gamma}^1(t)\|_{\gamma(t)}$ = constant, consequently, as (189) imposes, γ^1 is a minimal geodesic. But $\gamma^1(t) = \gamma\left(\frac{ts}{\rho}\right)$ for $t \in [0, \rho]$ and since γ is also geodesic it follows

$$\gamma^1(t) = \gamma\left(\frac{ts}{\rho}\right) \text{ for } t \in [0, 1]$$

and consequently

(192) $$\gamma(t) = \exp_m \frac{\rho(t-s)\delta}{2s(1-\rho)}\mu_m \text{ for } t \in [s, \frac{s}{\rho}].$$

(192) reveals, *via* 1.18, 1.19 and (154), that (182) is true for $t \in [s, \frac{s}{\rho}]$. But (193) $\frac{s}{\rho} \overset{(190)}{=} s + \frac{\delta}{2r}$ and on the other hand, since $d(\gamma(t_0), \gamma(s)) < \frac{\delta}{4}$ and the speed along γ has constant magnitude r, *a fortiori* $t_0 - s < \frac{\delta}{4r}$, that leads, *via* (193), to

$$t_0 < \frac{s}{\rho},$$

contradiction, (183) is false and $t_0 = 1$. ■

2

Ekeland principle versus equivalent theorems

A. EKELAND PRINCIPLE VERSUS THE DROP THEOREM AND THE FLOWER PETAL THEOREM

Definitions. Let (X, d) be a metric space, a, b points in X and γ from $(0, +\infty)$. The "*petal*" $P_\gamma(a, b)$ is

$$(1)\qquad P_\gamma(a, b) \stackrel{\text{def}}{=} \{x \in X : \gamma \mathrm{d}(x, a) + \mathrm{d}(x, b) \le \mathrm{d}(a, b)\}.$$

A graphic representation in the Euclidean plane of elementary geometry yields the intuitive support of this concept and justifies its name. Obviously, $a \in P_\gamma(a, b)$, $\gamma \le 1 \Rightarrow b \in P_\gamma(a, b)$. Moreover, $\delta \le \gamma \Rightarrow P_\gamma(a, b) \subset P_\delta(a, b)$.

Let X be a real normed space, $a \in X$ and M a convex nonempty subset of X.

The "*drop*" $D(a, M)$ is

$$(2)\qquad D(a, M) \stackrel{\text{def}}{=} \{a + t(x - a) : x \in M, t \in [0, 1]\},$$

i. e. $D(a, M)$ is the convex hull of $\{a\} \cup M$.

If M *is closed and bounded, then* D(a, M) *is closed.*

■ Indeed, let $(x_n)_{n\geq 1}$ be a sequence from $D(a, M)$ with $x_n \to x'$. We have, $\forall\, n \geq 1$, $x_n =$ $= (1 - t_n)a + t_n y_n$, $y_n \in M$ and $t_n \in [0, 1]$. If from a rank N we have $t_n = 0$, then $n \geq N \Rightarrow$ $x_n = a$, hence $x' = a \in D(a, M)$. In the contrary case, $\exists$ a subsequence $(t_{k_n})_{n\geq 1}$, $t_{k_n} > 0$ $\forall\, n$. It has a convergent subsequence, use the same notation, let be $t_{k_n} \to t'$. Assume that $t' = 0$. Then, as $x_{k_n} = (1 - t_{k_n})a + t_{k_n} y_{k_n}$, by passing to the limit one obtains $x' = a \in$ $D(a, M)$, because, M being bounded, $t_{k_n} y_{k_n} \to 0$. Let then be $t' > 0$. From $y_{k_n} = \frac{x_{k_n}}{t_{k_n}} +$ $+ \frac{t_{k_n} - 1}{t_{k_n}} a$ one obtains, by passing to the limit, $y' := \lim_{n\to\infty} y_{k_n} = \frac{x'}{t'} + \frac{t'-1}{t'} a$, hence $x' =$ $= (1 - t')a + t'y'$. As $y' \in M$, because M is closed, and $t' \in (0, 1]$ results $x' \in D(a, M)$. ■

Moreover, retain that if $\gamma \leq \frac{t-r}{t+r}$, where $t := d(a, b) > r$, and $M = S_r(b)$ (the closed ball with center b and radius r), then the convexity imposes

$$D(a, M) \subset P_\gamma(a, b). \tag{3}$$

The drop theorem (*Theorem* D, *J. Daneš*). *Let* X *be a real normed space,* H *a complete subset,* $x_0 \in H$ *and* $S := S_r(b)$ *with* $r < \text{dist}(b, H)$. *Then an element* a *exists in* $H \cap D(x_0, S)$ *such that*

$$\{a\} = H \cap D(a, S).$$

The flower petal theorem (*Theorem* F, *J. P. Penot*). *Let* (X, d) *be a metric space,* M *a nonempty complete subset,* $x_0 \in M$ *and* $b \in X \setminus M$. *Then for any* $\gamma > 0$ *there is* a *in* $P_\gamma(x_0, b) \cap M$, *such that*

$$\{a\} = P_\gamma(a, b) \cap M.$$

The Ekeland principle 1.1$_0$ (*Theorem* E).

2.1 Theorem. *Theorem* E, *Theorem* F *and Theorem* D *are equivalent* (*J. P. Penot*).

In order to obtain the proof of this theorem, we need

2.2 *Let* E *be a real normed space,* $E \times \mathbf{R}$ *the product space normed by* $\|(x, \lambda)\| =$ $= \max(\|x\|, |\lambda|)$ *and* $S := S_r((0, h))$ *the closed ball in* $E \times \mathbf{R}$, $0 < r < h$. *Then for the cone* $C := \mathbf{R}_+ S$ *we have*

$$C = \{(x, t) \in E \times \mathbf{R} : t \geq \frac{h-r}{r}\|x\|\}. \tag{4}$$

■ We have

(5) $$(x, t) \in S \Leftrightarrow \max(\|x\|, |t-h|) \le r \Leftrightarrow \|x\| \le r \text{ and } t \in [h-r, h+r].$$

Any (x, t) in $E \times \mathbf{R}$ can be written as $(x, t) = (\frac{h-r}{t}x, h-r)$, hence in particular, according to (5), any (x, t) from S can be written as (6) $(x, t) = (\frac{h-r}{t}x, h-r)$ with $\left\|\frac{h-r}{t}x\right\| \le$ $\le r$. But $\lambda(x, t) \in C \setminus \{0\} \Leftrightarrow \lambda > 0$ and $(x, t) \in S$. From (6) and (7) one obtains the conclusion (4). ■

Pass to the proof of 2.1.

■ *Theorem* E $\Rightarrow$ *Theorem* F. Take the function $\varphi : M \to \mathbf{R}$, $\varphi(x) = d(x, b)$, function continuous and lower bounded. Apply 1.1_0 with $\frac{\varepsilon}{\lambda} = \gamma$ and $u = x_0$, $\exists\, a := v_\varepsilon$ in M so that

(8) $$\varphi(a) < \varphi(x) + \gamma d(a, x)\ \forall x \in M \setminus \{a\},$$

(9) $$\varphi(a) \le \varphi(x_0) - \gamma d(a, x_0).$$

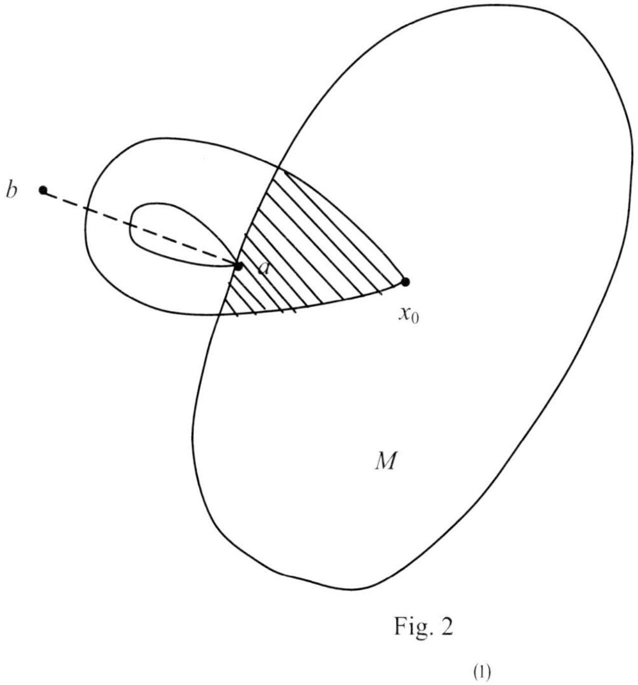

Fig. 2

(9) gives $\gamma\, d(a, x_0) \le d(b, x_0) - d(a, b)$, hence $a \overset{(1)}{\in} P_\gamma(x_0, b)$. Moreover, from (8), $x \in M \setminus \{a\} \Rightarrow x \notin P_\gamma(a, b)$.

Theorem F $\Rightarrow$ *Theorem* D. Take $H_1 := H \cap D(x_0, S)$, $\gamma := \dfrac{\delta - r}{\delta + r} (< 1)$, where $\delta =$ $= \mathrm{dist}(b, H)$, and apply Theorem *F* (correctly, $D(x_0 , S)$ being closed (the debut!) H_1 is complete), let be a the corresponding point from H_1. Since $t := \mathrm{d}(a, b) \geq \delta$, we have $\dfrac{\delta - r}{\delta + r} \leq \dfrac{t - r}{t + r}$, that is $\gamma \leq \dfrac{t - r}{t + r}$, and because $t > r$, we get (10) $D(a, S) \overset{(3)}{\subset} P_\gamma (a, b)$. Moreover, as $a \in D (x_0, S)$ we have (see Fig. 3 or see §9, demonstration for 9.48, 9.45 $\Rightarrow$

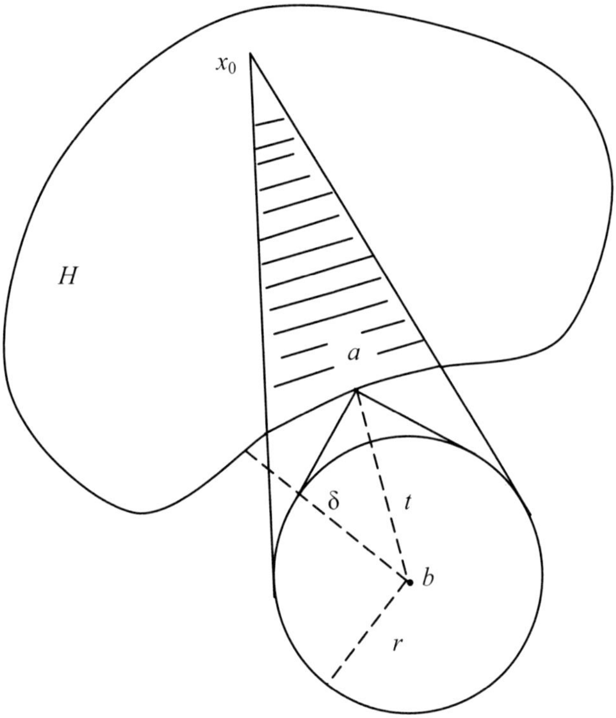

Fig. 3

$\Rightarrow$ 9.40, (*)) (11) $D(a, S) \subset D (x_0 , S)$. According to (10) and (11), one finds $D(a, S) \cap$ $\cap H \subset D(a , S) \cap (D(x_0 , S) \cap H) \subset P_\gamma (a , b) \cap H_1 = \{a\}$.

Theorem D $\Rightarrow$ *Theorem* E. We are in the hypotheses of 1.1_0, that is in those of 1.1. Note $\gamma := \dfrac{\varepsilon}{\lambda}$, $x_0 = u$, $a := v_\varepsilon$. It can suppose that the distance d is bounded. Indeed, replacing d by (12) $\mathrm{d}_1 := \min(\delta, \mathrm{d})$ whith (13) $\delta = \gamma^{-1}[\varphi(x_0) - \inf \varphi(X) + 1]$, when (1) is true for d_1, then (1) is also true for d, and if (2_0) is true for d_1, then (2_0) is true also for d, as $\mathrm{d}_1 (a, x_0) \leq \gamma^{-1}[\varphi(x_0) - \varphi(a)] \overset{(13)}{<} \delta$, that imposes, according to (12), $\mathrm{d}_1(a, x_0) = \mathrm{d}(a,$

x_0). Concerning (3), this follows from (2_0) (the end of the proof for 1.1 + Remark to 1.1_0). Therefore, (*X*, d) can be isometrically embedded into a real normed space, for instance in the vector space of continuous bounded real functions on *X*, endowed with the norm sup, through the mapping $x \to d_x$, where $d_x(y) = d(x, y)$.

So, *X* can be supposed to be a complete metric subspace of a real normed space *E*. In the drop theorem *D* take $X = E \times \mathbf{R}$ with the norm $\|(x,t)\| = \max(\|x\|, |t|)$. We can suppose $x_0 = 0$ and $\varphi(x_0) = 0$, otherwise replace φ by $x \to \varphi(x_0 + x) - \varphi(x_0)$. Consider the function

(14) $$f := -\varphi.$$

f is upper semicontinuous, upper bounded and proper. Let be (15) $\mu := \sup f(X)$. Take (16) $r > \gamma^{-1}\mu$, (17) $h = \gamma r + r$, hence (18) $h > \mu + r$, and consider the closed ball (19) $S := S_r((0, h))$ ($= S_r(0) \times [h - r, h + r]$) and also $C := \mathbf{R}_+ S$. Since $(x, t) \in S \overset{(19)}{\Rightarrow} t \geq h - r$, we have $(x, t) \notin H$,

(20) $$H := \{(x, t) \in X \times \mathbf{R} : t \leq f(x)\}$$

(the hypograph of *f* in $X \times \mathbf{R}$; see (18)). *H* is nonempty (*f* is proper!) and closed: $(x_n, t_n) \in H$, $x_n \to x$ and $t_n \to t \Rightarrow t_n \leq f(x_n)$, $\forall n \geq 1$, take the upper limit for $n \to \infty$, $t \leq$ $\leq f(x)$, i.e. $(x, t) \in H$. So, *H* is complete, apply Theorem *D* with $x_0 = (0, 0)$ ($\in H$!) and $S = S_r((0, h))$, (21) $\exists\, (a, \alpha) \in H \cap D((0, 0), S)$ such that (22) $H \cap D((a, \alpha), S) = \{(a, \alpha)\}$. As $(a, \alpha) \overset{(21)}{\in} D((0,0), S) \overset{(2)}{=} [0, 1]\, S \overset{(19)}{=} [0, 1](S_r(0) \times [h - r, h + r])$, we have $a \in S_r(0)$ and hence $(a, h) \overset{(19)}{\in} S$, but because $S \subset D((a, \alpha), S)$ we get [the convexity, $\lambda(a, \alpha) + (1 - \lambda)(a, h) = (a, \lambda\alpha + (1-\lambda)h)$] $(a, t) \in D((a, \alpha), S)$ $\forall\, t$ of $[\alpha, h]$, in particular, since $\alpha \overset{(21)}{\leq} f(a) \overset{(18)}{\leq} h$, we have $(a, f(a)) \in H \cap D((a, \alpha), S)$, which imposes, *via* (22), $f(a) = \alpha$. Apply 2.2 to the cone *C*, as $(a, \alpha) \in [0, 1]S$ it results $f(a) \geq \dfrac{h-r}{r}\|a\| \overset{(17)}{=} \gamma d(a, x_0)$, that means $f(a) - f(x_0) \geq \gamma\, d(a, x_0)$ ($x_0 = 0$, $f(x_0) = 0$), hence (see (14)) $\varphi(a) \leq \varphi(x_0) - \gamma d(a, x_0)$, the inequality ($2_0$). Pass to the demonstration of (1). Let be (23) (x, t) in $(a, \alpha) + C$, $x \in X \setminus \{a\}$, $t \leq \mu$ ((15)). $(x - a, t - \alpha) \in C$, so $t - \alpha \overset{2.2}{\geq} \dfrac{h-r}{r}\|x - a\| = \gamma\|x - a\| > 0$. Hence we can write

(24) $$(x-a, t-\alpha) = \lambda(y, h-r-\alpha)$$

with $y = \dfrac{x-a}{\lambda}$ and (25) $\lambda = \dfrac{t-\alpha}{h-r-\alpha} \in (0, 1)$, as $h-r-\alpha \geq h-r-\mu \overset{(18)}{>} 0$ ($(a, \alpha) \in H$, $\mu \geq f(a)$) and $t-\alpha \leq \mu-\alpha \overset{(18)}{<} h-r-\alpha$. Since C is a convex cone, we have (see §8, 8.1)

$$(a+y, h-r) = (a, \alpha) + (y, h-r-\alpha) \in C,$$

since (a, α), $(y, h-r-\alpha) \in C$ ((23), 2.2), therefore $(a+y, h-r) \in C \cap (E \times \{h-r\}) =$ $=: A$, but $A \subset S$ (use 2.2), hence (26) $(a+y, h-r) \in S$. Then, according to (25) as $D((a, \alpha), S)$ is convex,

(27) $$(x, t) \overset{(24)}{=} (a, \alpha) + \lambda[(a+y, h-r) - (a, \alpha)] \in D((a, \alpha), S).$$

Since $x \neq a$, (27) and (22) impose

(28) $$(x, t) \notin H, x \in X \setminus \{a\}, t \leq \mu.$$

One infer that, as $x \in X \Rightarrow f(x) \leq \mu$,

(29) $$(x, f(x)) \notin (a, \alpha) + C, \forall x \in X \setminus \{a\}.$$

Indeed, if one supposes *par absurdum* the contrary, we are in the situation of (23) and *via* (28) we derive $(x, f(x)) \notin H$, a contradiction. But (29) gives, *via* 2.2 and (17),

$$x \in X \setminus \{a\} \Rightarrow f(x) - f(a) \leq f(x) - \alpha < \gamma - \|x-a\|,$$

that transformed through (14) yields the fundamental inequality (1). ■

In the case of a Banach space, the ball $S_r(b)$ from the drop theorem may be replaced by a convex closed bounded set. Indeed, we have the variant ([25])

2.2$_1$ *Let* X *be a real Banach space,* A *nonempty closed subset and* B *convex nonempty closed bounded subset and* d(A, B) > 0. *For each* x_0 *from* A *there exists* a *in* A ∩

$\cap$ D(x_0, B) *such that*

$$\{a\} = A \cap D(a, B)$$

(*J. Daneš*).

■ A locally convex version of this statement will be demonstrated in §9, section G (see 9.39$_3$). ■

It follows another variant of the drop theorem ([25]).

2.2$_2$ *Let* X *be a real normed space,* H *a complete nonempty subset and* S : = S_r(b) *with* r < dist (b, H). *For any* ρ > dist (b, H) *there exists* a *in* H *having the property*

$$\{a\} = D(a, S) \cap H \text{ and } \|a - b\| \leq \rho.$$

■ A translation permits to assume that $b = 0$. $F := H \cap S_\rho(0)$ is not empty. Let be $\varphi : F \to \mathbf{R}$,

$$\varphi(x) = \frac{\rho + r}{\alpha - r}\,\|x\|.$$

φ is bounded from below, apply Ekeland principle 1.1 with ε = λ =1, ∃ a in F, hence $\|a\| \leq \rho$, so that

$$(29_1) \qquad \varphi(a) < \varphi(x) + \|x - a\|,\ x \in F \setminus \{a\}.$$

Obviously, $a \in D(a, S) \cap H$. Suppose *par absurdum* that $x \in D(a, S) \cap H \setminus \{a\}$, that is

$$(29_2) \qquad x \in H,\ x = (1 - t)a + t\,v,\ v \in S_r(0),\ t \in (0, 1)$$

($t \neq 1$ as if *par absurdum* $t = 1$, it results $x = v \in H$ and hence $r < \alpha \leq \|v\|$, a contradiction).

Then (29$_1$) gives

$$\frac{\rho+r}{\alpha-r}\,||a|| < \frac{\rho+r}{\alpha-r}\,||x|| + ||x - a|| \overset{(29_2)}{=} \frac{\rho+r}{\alpha-r}\,||x|| + t\,||a - v|| \leq \frac{\rho+r}{\alpha-r}\,||x|| + t\,(\rho+r),$$

$$(29_3) \qquad ||a|| < ||x|| + t(\alpha - r).$$

But, according to (29_2),

$$||x|| \leq (1 - t)||a|| + t\,||v||,$$

hence

$$(29_4) \qquad t\,(\alpha - r) \leq t(||a|| - ||v||) \leq ||a|| - ||x||.$$

Combine (29_4) with (29_3) and find

$$||a|| < ||x|| + ||a|| - ||x||,$$

a contradiction. ■

Remark 1. 2.2_2 remains also true for ρ = dist (b, H) if this distance is attained, i.e. $\exists\ h \in H$ so that $\rho = \mathrm{d}\,(b, h)$, because then $H \cap S_\rho(0) \neq \varnothing$ and the demonstration continues as above.

Remark 2. Section G from §9 presents other confrontations of the Ekeland principle with equivalent theorems.

Here is a simple application of the Daneš drop theorem (J. Borwein, A tangent cone separation principle, preprint, Dalhousie University, N. S., Canada).

2.3 *Let* X *be a real normed space,* M *a complete subset,* x_0 *from* M *and* x_1 *from* X *such that the segment* $[x_0, x_1] \not\subset$ M. *Then for any* $\rho > 0$ *there is an element* a *in* M *with* $\|x_0 - a\| \leq \|x_0 - x_1\| + \rho$ *and*

$$x_1 \notin a + C_M(a).$$

Explanation. $C_M(a)$, *the contingent cone to* M *at* a, is

$$C_M(a) = \{v \in X : \underline{\lim}_{t \to 0+} \frac{1}{t} \operatorname{dist}(a + tv, M) = 0\}.$$

■ It can suppose $x_0 = 0$. Let $\lambda \geq 1$ be such that $x_1 = \lambda u$ with $u \in X \setminus M$ (this is possible, M is closed, $[0, x_1] \not\subset M$), r from $(0, \rho)$ with $r <$ dist (u, M) and the closed ball $S := S_r(u)$. Apply theorem D, $\exists\, a \in M \cap D(0, S)$ such that (29_5) $M \cap D(a, S) = \{a\}$. Set a in the form $a = \mu(u + b)$ with $\mu \in (0, 1)$ and $\|b\| \leq r$ and let be $t := \dfrac{\lambda - 1}{\lambda - \mu}$ $(\in [0, 1))$ and $v :=$ $= t\,a + (1 - t)x_1$. Then

$$\|v - u\| = \|t\mu b + [t\mu + (1 - t)\lambda - 1]u\| = t\mu\,\|b\| < r,$$

that is $v \in \overset{\circ}{S}$. Moreover,

$$x_1 - a = \frac{v - a}{1 - t} \in \mathbf{R}_+(\overset{\circ}{S} - a),$$

but as $[a + (0, 1)(S - a)] \cap M = \varnothing$ (use (29_5)), we have $\mathbf{R}_+(\overset{\circ}{S} - a) \subset X \setminus C_M(a)$, hence $x_1 \notin$ $\notin a + C_M(a)$. Finally,

$$\|a\| < \operatorname{diam} D(0, S) \leq \|u\| + r \leq \|x_1\| + \rho. \ ■$$

Go on with

SUPPORT FUNCTIONAL AND SUPPORT POINT FOR CLOSED CONVEX SETS

Definitions. Let X be a real normed space and C a closed, convex and proper subset (i.e. $\varnothing \neq C \neq X$).

The functional f from $X^* \setminus \{0\}$ is *support functional* for C if there is x_0 in C so that

$$(29_6) \qquad f(x_0) = \sup f(C).$$

The point x_0 of C is a *support point* if there exists f in $X^* \setminus \{0\}$ such that (29_6) is verified. The name "support" comes from the geometric interpretation of these two cases: the hyperplane from X of equation $f(x) = f(x_0)$ goes through x_0 and leaves C in one of the semi-spaces determined by it.

The concept "support functional", modified, appears also in §5, section G. The aim pursued there is specific for the paragraph in which it is found.

The goal of this subsection is to present two theorems of Bishop-Phelps about support functionals and support points. Their proofs will be preceded by three propositions. $S(x_0, r) := \{x \in X : \|x - x_0\| < r\}$. For the notion "cone in a vector space", see §8, "Preliminaries of affine geometry".

2.3_1 *Let* X *be a real Banach space and* F *a closed subset. For any* $\varepsilon > 0$ *and* y_0 *from* Fr F, *there exists* $\delta > 0$, *a closed convex cone with vertex* C, int C $\neq \emptyset$, *and* x_0 *from* Fr F *such that*

$$(29') \qquad \|x_0 - y_0\| < \varepsilon \text{ and } F \cap (x_0 + C) \cap S(x_0, \delta) = \{x_0\}$$

(*F. E. Browder*).

∎ Let b be in $X \setminus F$ with $(29_7)\ \|y_0 - b\| \le \frac{\varepsilon}{3}$. Then $(29_8)\ \alpha := \text{dist}(b, F) \le \frac{\varepsilon}{3}$. Take $\rho = \frac{\varepsilon}{2}$, r from $(0, \alpha)$ and let be the closed ball $S := S_r(b)$. Apply the drop theorem 2.2_2, $\exists\, x_0$ in F so that

$$(29_9) \qquad \|x_0 - b\| \le \frac{\varepsilon}{2},\ D(x_0, S) \cap F = \{x_0\}.$$

We have $\|x_0 - y_0\| \le \|x_0 - b\| + \|b - y_0\| \le \frac{\varepsilon}{2} + \frac{\varepsilon}{3} < \varepsilon$ ((29_7), (29_9)), $\|x_0 - y_0\| < \varepsilon$ and pass to the second assertion of the statement.

Take δ, $(29_{10})\ 0 < \delta < \|x_0 - b\| - r$ and C the convex cone with vertex defined by

$$(29_{11}) \qquad x_0 + C = \{x_0 + t(v - x_0) : v \in S_r(b), t \ge 0\}.$$

If we prove

$$(29_{12}) \qquad (x_0 + C) \cap S(x_0, \delta) \subset D(x_0, S),$$

then the drop theorem will justify the second assertion.

Let be, according to (29_{12}) and (29_{11}) (x_0 is in the first member of (29_{12})),

(29_{13}) $$x = x_0 + t(v - x_0),\ t \geq 0,\ \|v - b\| \leq r,\ \|x - x_0\| < \delta.$$

Thus it remains to prove that in (29_{13}) *a fortiori* $t \leq 1$. But

$$x = x_0 + t(b - x_0) + t(v - b),\ t(b - x_0) + t(v - b) = x - x_0,$$

hence

$$t\|b - x_0\| - t\|v - b\| \leq \|x - x_0\| < \delta \text{ and therefore } ((29_{13}))$$

$$t(\|b - x_0\| - r) < \delta \overset{(29_{10})}{<} \|x_0 - b\| - r,$$

that yields $t < 1$.

Finally, indeed $x_0 \in \operatorname{Fr} F$, this is imposed by the second relation in (29′). ■

2.3₂ *Let* X *be a real Banach space,* F *a closed nonempty subset,* f *an element of* X^* *with* $\sup f(F) < +\infty$ *and* λ *from* (0, 1). *Then the set*

(29_{14}) $$C := \{x \in X : \lambda\|x\| \leq f(x)\}$$

is a closed convex cone with vertex and for any y_0 *from* F *there is* x_0 *in* F *such that*

(29_{15}) $$x_0 \in y_0 + C \text{ and } F \cap (x_0 + C) = \{x_0\}$$

(*R.. Phelps*).

■ *First assertion.* The set C is clearly closed because the function $x \to f(x) - \lambda\|x\|$ is continuous. As $\mu C \subset C\ \forall \mu \geq 0$, C is a cone with vertex. C is also convex, since $C + C \subset$ $\subset C$ (§8, 8.1) : $x', x'' \in C, \lambda\|x' + x''\| \leq \lambda(\|x'\| + \|x''\|) \leq \lambda f(x') + \lambda f(x'') = \lambda f(x' + x'')$.

Second assertion. $M := (y_0 + C) \cap F$ is nonempty closed and therefore complete and $\varphi: M \to \mathbf{R}$, $\varphi(x) = -f(x)$ is continuous lower bounded. Apply Ekeland principle 1.1 with ε from $(0, \lambda)$ and $\lambda = 1$ (the notation in 1.1!), $\exists\, x_0$ in M such that

(29_{16}) $$-f(x_0) < -f(x) + \varepsilon\|x - x_0\|\ \ \forall x \in M \setminus \{x_0\}.$$

Let x' be arbitrary in (29_{17}) $F \cap (x_0 + C)$. x_0 is in this set because $x_0 \in F$ and $0 \in C$. It remains only to show that $x' = x_0$. We have (29_{18}) $x' \in M$: $x' - x_0 \overset{(29_{17})}{\in} C$ and $x_0 - y_0 \in C$ since $x_0 \in M$, hence $x' - y_0 \in C$ (§8, 8.1) and $x' \in M$. Suppose *par absurdum* $x' \neq x_0$. Then (29_{18}) and (29_{16}) yield

$$-f(x_0) < -f(x') + \varepsilon\|x' - x_0\|.$$

Because $x' - x_0 \overset{(29_{17})}{\in} C$, we get $\lambda\|x' - x_0\| \overset{(29_{14})}{\leq} f(x' - x_0)$, this contradicts the last inequality since $\varepsilon < \lambda$. ■

Remark 1. The demand "$\|f\|_{X^*} = 1$" in Lemma 7.6 from [32] is useless.

Remark 2. The convex cone with vertex C is in fact pointed (§8, 8.2): $\lambda\,\|x\| \le f(x)$ and $\lambda\,\|-x\| \le f(-x)$ imply $x = 0$.

2.3$_3$ *Let* X *be a real Banach space,* F *a closed convex subset,* f *from* X^* *and* λ *from* (0, 1). *Suppose there is* x_0 *in* F *such that*

$$F \cap (x_0 + C) = \{x_0\},\ C \text{ the cone from } 2.3_2. \tag{29_{19}}$$

Then a functional g *exists in* X^* *with the properties*

$$\sup g(F) = g(x_0) \text{ and } \|g - f\| \le \lambda \tag{29_{20}}$$

(*R. Phelps*).

■ Consider the function $\varphi\colon X \to \mathbf{R}$,

$$\varphi(x) = \lambda\,\|x\| - f(x). \tag{29_{21}}$$

φ is continuous and convex. Take the nonempty subsets M_1, M_2 of $X \times \mathbf{R}$,

$$M_1 = \{(x, r) \in X \times \mathbf{R}\colon \varphi(x) < r\}, \tag{29_{22}}$$

$$M_2 = \{(x, r) \in X \times \mathbf{R}\colon x \in F - \{x_0\},\ r = 0\}. \tag{29_{23}}$$

M_1 is open (the function $(x, r) \to \varphi(x) - r$ is continuous on $X \times \mathbf{R}$) and both M_1 and M_2 are convex. Moreover (29_{24}) $M_1 \cap M_2 = \varnothing$. Indeed, suppose *par absurdum* there is x in $F - \{x_0\}$ with $\varphi(x) < 0$, that is $x \in C$ (see (29_{21}), (29_{14})). Then $x + x_0 \in F$ and $x + x_0 \in$ $\in x_0 + C$, which, *via* (29_{19}), imposes $x = 0$, a contradiction since $(0, 0) \notin M_1$.

So it can apply Hahn-Banach theorem, the second geometrical form (see *Memento*, §1, after 1.1$_0$), M_1 and M_2 are separating, thus $\exists\,\Phi$ in $(X \times \mathbf{R})^* \setminus \{0\}$ so that

$$\sup \Phi(M_2) \le \inf \Phi(M_1). \tag{29_{25}}$$

Φ is of the form

$$\Phi(x, r) = g(x) + tr\ \ \forall\,(x, r) \in X \times \mathbf{R}, \tag{29_{26}}$$

where $t \in \mathbf{R}$ and $g \in X^*$.

The point (0, 0) belongs to M_2, hence $\sup \Phi(M_2) \ge 0$, and also $(0, 0) \in \overline{M_1}$ ($U\times(-a, a)$ an arbitrary neighbourhood of (0, 0) in $X \times \mathbf{R}$, φ continuous in 0, $\varphi(0) = 0$, consequently we can suppose U sufficiently small such as to have $\varphi(U) \subset (-a, a)$), hence $\inf \Phi(M_1) \le$ ≤ 0 (take a sequence $(x_n, r_n)_{n\ge 1}$ in M_1 with $(x_n, r_n) \to (0, 0)$, $\inf \Phi(M_1) \le \Phi(x_n, r_n)$) and consequently

$$\sup \Phi(M_2) = \inf \Phi(M_1) = 0. \tag{29_{27}}$$

As $(x, \varphi(x)) \in \overline{M_1}$ $\forall\, x \in X$, using (29_{27}) we get (29_{28}) $g(x) + t\,\varphi(x) \geq 0$ $\forall\, x \in X$. The last inequality implies $t \neq 0$, because $t = 0$ yields forcibly $g = 0$ (g is linear!) and hence, *via* (29_{26}), $\Phi = 0$, a contradiction. Moreover, since $(0, 1) \in M_1$, (29_{27}) yields $\Phi(0, 1) \geq 0$ and hence $t > 0$ ((29_{26})). We can then suppose $t = 1$, replacing, if is need, g by the continuous linear functional $x \to g\left(\frac{x}{t}\right)$.

So (29_{28}) can be rewritten (see (29_{21})) $g(x) + \lambda\|x\| - f(x) \geq 0$ $\forall\, x \in X$, hence $f(x) - g(x) \leq \lambda\|x\|$ $\forall\, x \in X$, but also $g(x) - f(x) \leq \lambda\|x\|$ (replace x by $-x$), hence $|g(x) - f(x)| \leq \lambda\|x\|$ $\forall\, x \in X$ and consequently the second inequality of (29_{20}).

Moreover, for every x from F we have $(x - x_0, 0) \in M_2$ ((29_{23})), hence $\Phi(x - x_0, 0) = g(x - x_0) \leq 0$ ((29_{27})), $g(x) \leq g(x_0)$ $\forall\, x \in F$ and consequently the first assertion of (29_{20}). ■

Remark. The demand "$\|f\|_{X^*} = 1$" in Lemma 7.7 ([32]) is useless.

We can pass now to the Bishop-Phelps theorems on support points and support functionals.

2.3_4 Bishop-Phelps theorem (support point). *Let* X *be a real Banach space and* C *a proper convex closed subset.The set of support points of* C *is dense in* Fr C.

■ Use 2.3_1. Let y_0 be in Fr C and $\varepsilon > 0$. There is x_0 in Fr C, $\delta > 0$ and a closed convex cone K with vertex and int $K \neq 0$ such that

(29_{29}) $$C \cap (x_0 + K) \cap S(x_0, \delta) = \{x_0\} \text{ and } \|x_0 - y_0\| < \varepsilon.$$

It remains to prove that x_0 is a support point of C. In fact we have

(29_{30}) $$C \cap (x_0 + K) = \{x_0\}.$$

Indeed, suppose *par absurdum* x' in $C \cap (x_0 + K) \setminus \{x_0\}$. Then $x' = x_0 + t\,(x - x_0)$ is, for sufficiently small t, in $S(x_0, \delta)$ and also in C ($x, x_0 \in C$, C is convex) and in $x_0 + K$ because $t\,(x - x_0) \in K$ (cone!), contradiction with (29_{29}).

So (29_{30}) allows to apply the Hahn-Banach theorem, the second geometrical form (see before), to C and $U := \text{int}(x_0 + K)$, $\exists\, f$ in X^* and α in $\mathbf{R}$ such that

$$f(x) \leq \alpha \leq f(y)\ \forall\, x \in C,\ \forall\, y \in U.$$

The second inequality gives $\alpha \leq f(x_0)$ ($x_0 \in \text{Fr}U$ and f is continuous), consequently $f(x) \leq f(x_0)$ $\forall\, x \in C$ and hence $\sup f(C) = f(x_0)$, x_0 is a support point. ■

2.3_5 Bishop-Phelps theorem (support functional). *Let* X *be a real Banach space and* C *be a proper bounded convex closed subset. The set of support functionals for* C *is dense in* X^*.

■ Let f be in X^* and $\varepsilon > 0$. We will find g in X^* a support functional for C with $\|f - g\| \leq \varepsilon$. It can suppose $\varepsilon < 1$. We have, $\forall\, x \in C$, $|f(x)| \leq \|f\|\ \|x\| \leq a\|f\|$, $a \in \mathbf{R}$, as C is bounded, hence in particular f is upper bounded on C and so it can apply 2.3_2 with $\lambda = \varepsilon$: $\exists\, x_0$ in C so that $C \cap (x_0 + K) = \{x_0\}$. But then we can continue with application of 2.3_3 with $\lambda = \varepsilon$, $\exists\, g$ in X^* so that $\|f - g\| \leq \varepsilon$ and $\sup g(C) = g(x_0)$. ■

Remark. Let X be a real Banach space and $S_X = \{x \in X : \|x\| \leq 1\}$. If X is reflexive, the James theorem tells us that any f from X^* attains its norm on S_X and if X in not reflexive, there exists f in X^* which does not attain its norm on S_X. The Bishop-Phelps theorem specifies the situation : if X is not reflexive, the set of continuous linear functionals which attain their norm on S_X is dense in X^*. Indeed, S_X being closed, convex, bounded and proper, the set $\mathscr{S}$ of support functionals for S_X is dense in X^*, let f be from $\mathscr{S}$ and x_0 from S_X such that $f(x_0) = \sup f(S_X)$. Then $f(x)$, $f(-x) \leq f(x_0)$ $\forall\, x \in S_X$, hence $|f(x)| \leq f(x_0)$ $\forall\, x \in S_X$ and consequently $\|f\| = \sup_{x \in S_X} |f(x)| = f(x_0)$. ■

We finish the section A with

THE SURJECTIVITY OF NONLINEAR OPERATORS

Definitions and notations (we keep the notations from the theory of compact operators). Let X, Y be real normed spaces and $T : X \to Y$.

The kernel $N(T)$ of T is

$$N(T) = \{x \in X : T(x) = 0\}.$$

The range $R(T)$ of T is

$$R(T) = T(X).$$

The graph $G(T)$ of T is

$$G(T) = \{(x, Tx) : x \in X\},$$

a subset of $X \times Y$. T is called *closed* when $G(T)$ is closed.

The left polar $A^\perp$ of a vector subspace A of X is

$$A^{\perp} = \{\xi \in X^*: \xi(x) = 0 \ \forall \ x \in A\}.$$

The right polar $B^{\perp}$ of a vector subspace B of X^* is

$$B^{\perp} = \{x \in X : \xi(x) = 0 \ \forall \ \xi \in B\}.$$

$A^{\perp}$ and $B^{\perp}$ are closed vector subspaces.

Suppose $T : X \to Y$ linear. The *adjoint* T^* of T, $T^*: Y^* \to X^*$ is defined by :

$$\mu \in Y^* \Rightarrow T^*\mu(x) = \mu(Tx) \ \forall \ x \in X.$$

T^* is obviously linear. When T is also continuous, T^* has the same property.

We have the formula (see for example [8], 1, Corollary II. 17)

$$(29_{31}) \qquad N(T^*)^{\perp} = \overline{R(T)},$$

where T is a closed linear operator, hence in particular a continuous linear operator.

Moreover, when $T : X \to Y$ is also a closed linear operator, X and Y Banach spaces, the following statements are equivalent:

1) T is surjectiv;

2) $N(T^*) = 0$ and $R(T^*)$ is closed;

3) There is $\rho > 0$ so that

$$\|\mu\| \le \rho \|T^*\mu\| \ \forall \ \mu \in Y^*$$

(*the surjectivity theorem*; see for instance [8], 1, theorem II.19).

In the following we present some variants of this surjectivity theorem for nonlinear operators. $S_r(a) = \{y \in Y : \|y - a\| \le r\}$. We will use the variant 2.2_2 of the drop theorem.

2.3_6 Theorem. *Let* X, Y *be real Banach spaces and* $f : X \to Y$ *Gâteaux differentiable with* f (X) *closed. Suppose that for the point* y *of* Y *there is* $\rho_0 > 0$ *and* α_0 *in* [0, 1) *such that*

$$(29_{32}) \qquad f^{-1}(S_{\rho_0}(y)) \neq \varnothing,$$

$$(29_{33}) \qquad \inf \{\|y - f(x) - z\| : z \in \overline{R(f'_w(x))}\} \le \alpha_0 \|y - f(x)\| \ \ \forall \ x \in f^{-1}(S_{\rho_0}(y)).$$

Then

$$y \in f(X)$$

(*F. E. Browder*).

■ Suppose *par absurdum* $y \notin F := f(X)$ and let be $d :=$ dist (y, F), hence $d > 0$ as F is closed.

The hypothesis $\alpha_0 > 0$.

(29_{32}) yields $d \le \rho_0$ and so we distinguish the cases

$$1)\ d < \rho_0\ ,\ 2)\ d = \rho_0\ .$$

The case 1). One takes $\varepsilon > 0$ sufficiently small for having $\alpha_0\,(d+\varepsilon) < d$ and $d+\varepsilon < \rho_0$, then r so that

(29_{34}) $$\alpha_0\,(d+\varepsilon) < r < d.$$

Apply 2.2_2 with $\rho = d+\varepsilon$, $\exists\ u_0$ in F, $u_0 = f(x_0)$ so that

(29_{35}) $$\|y-u_0\| \le d+\varepsilon\ (\le \rho_0) \text{ and } F \cap D(u_0\,, S) = \{u_0\},\ S{:}= S_r(y).$$

(29_{33}) gives, with (29_{35}) and (29_{34}),

$$(*)\ \inf\{\|y - f(x_0) - z\| : z \in \overline{R(f'_w(x_0))}\} < r,$$

consequently there is h in X so that

(29_{36}) $$\|y - f(x_0) - f'_w\,(x_0)(h)\| < r.$$

Taking into account the definition of the Gâteaux derivative one finds

$$\left\| y - f(x_0) - \frac{f(x_0+th)-f(x_0)}{t} \right\| < r,\ t \in (0,\,\delta).$$

This inequality expresses that

$$v(t) := f(x_0) + \frac{f(x_0+th)-f(x_0)}{t} \in S\ \forall\ t \in (0,\,\delta)$$

and then, supposing $\delta < 1$,

$$(1-t)u_0 + tv(t) \in D(u_0\,, S)\ \forall\ t\in (0,\,\delta),$$

i.e. $f(x_0+th) \in D(u_0\,, s)\ \forall\ t \in (0,\,\delta)$ and according to (29_{35}) one finds $f(x_0+th) = f(x_0)$ $\forall\ t \in (0,\,\delta)$, this implies $f'_w\,(x_0)(h) = 0$ and consequently, *via* (29_{36}), $\|y - f(x_0)\| < r$, a contradiction because $r < \text{dist}\,(y,\ f(X))$ ((29_{34})).

The case 2. Take r so that $\alpha_0\,d < r < d$ and apply 2.2_2 with $\rho = d$ (see Remark 1), $\exists\ u_0 := f(x_0)$ so that $\|y - u_0\| < d\ (= \rho_0)$ and $F \cap D(u_0\,, S) = \{u_0\}$, $S := S_r(y)$. Then (29_{33}) leads us again to the inequality (*) and the demonstration continues as in the case 1).

The hypothesis $\alpha_0 = 0$.

Take r, $0 < r < d$, (29_{33}) supplies the inequality (*) where $f(x_0) \in S_{\rho_0}\,(y)$ and the proof continues from here as in the case 1). ∎

REMARK. The correct proof for 2.3_6 was obtained with difficulty by the author.

And now –

2.3_7 The surjectivity theorem. *Let* X, Y *be real Banach spaces and* f : X → Y *Gâteaux differentiable with* f (X) *closed. If*

$$\mathrm{N}(\mathrm{f}_{\mathrm{w}}^{\prime *}(\mathrm{x})) = \{0\} \ \forall\ \mathrm{x}\ \textit{of}\ \mathrm{X},$$

then f *is surjective* (*S. I. Pohožaev, F. E. Browder*).

■ Let y be arbitrary in Y. Show that $y \in f(X)$. Because $N(f_{\mathrm{w}}^{\prime *}(x)) = \{0\}$ (the left member is a vector subspace of Y^*), we have $N(f_{\mathrm{w}}^{\prime *}(x))^{\perp} = Y$ (the right polar) and hence, according to (29_{31}), $\overline{\mathrm{R}(f_{\mathrm{w}}'(x))} = Y \ \forall x \in X$. Consequently, the greatest lower bound of the first member of (29_{33}) is equal to 0 (take $z = y - f(x)$ and hence (29_{33}) is satisfied with $\alpha_0 = 0 \ \forall\ x \in X$). Remains only to take $\rho_0 > 0$ so that dist $(y, f(X)) < \rho_0$ for (29_{32}) to be verified and then to apply 2.3_6 in order to obtain $y \in f(X)$. ■

In 2.3_6 and 2.3_7 is formulated the global demand "$f(X)$ closed", which however proves to be unconfortable in the applications of these surjectivity theorems. In the next theorem appears a local demand. Before this proposition – some notions about generalized sequences.

Let X be a topological space and $(x_\alpha)_{\alpha\in A}$ a generalized sequence of points in X (i. e. A is a preordered by $\leq$ rightly filtrable set). This sequence *converges* to the point x_0 of X (in which case we say that it is *convergent* and x_0 is its *limit*), $x_\alpha \xrightarrow[\alpha\in A]{} x_0$, if $\forall\ V$, neighbourhood of x_0, $\exists\ \alpha_0$ in A such that $\alpha \geq \alpha_0 \Rightarrow x_\alpha \in V$ (the *Moore-Smith convergence*). If X is separated, the limit is unique whenever it exists.

Let B be a *cofinal* subset of A, i.e. $\forall\ \alpha \in A\ \exists\ \beta$ in B so that $\beta \geq \alpha$. By definition, $(x_\beta)_{\beta\in B}$ is a *subsequence* of $(x_\alpha)_{\alpha\in A}$. Every subsequence of a generalized sequence, which tends to x_0, also tends to x_0.

The topological space X *is quasi-compact* (i.e. it satisfies the Borel - Lebesgue axiom), *iff every generalized sequence in* X *has a convergent subsequence.*

Moreover, since when X has the initial topology relating to $(f_i)_{i\in I}$, then

$$x_\alpha \xrightarrow[\alpha\in A]{} x_0 \Leftrightarrow f_i(x_\alpha) \xrightarrow[\alpha\in A]{} f_i(x_0)\ \forall\ i \in I,$$

we have, in the case of the $*$ - weak topology on X^* with X complete:

$$\xi_\alpha \xrightarrow[\alpha\in A]{*-\text{weak}} \xi_0 \Leftrightarrow \xi_\alpha(x) \xrightarrow[\alpha\in A]{in\ \mathrm{R}} \xi_0(x)\ \forall\ x \in X;$$

$$\xi_\alpha \xrightarrow[\alpha\in A]{*-\text{weak}} \xi_0 \Rightarrow \text{the family } (\|\xi_\alpha\|)_{\alpha\in A} \text{ is bounded;}$$

$$\xi_\alpha \xrightarrow[\alpha\in A]{*-\text{weak}} \xi_0 \text{ and } x_\alpha \xrightarrow[\alpha\in A]{\text{strong}} x_0 \Rightarrow \xi_\alpha(x_\alpha) \xrightarrow[\alpha\in A]{in\ \mathrm{R}} \xi_0(x_0).$$

And now –

2.3₈ The surjectivity theorem. *Let* X, Y *be real Banach spaces and* f : X → Y *Gâteaux differentiable and continuous. Suppose*

1° $R(f'_w(x)) = Y \ \forall\, x \in X$,

2° $\exists\, \rho > 0$ *such that* $\forall\, x \in X, \forall\, y \in Y\ \exists\, z$ *in* $(f'_w(x))^{-1}(y)$ *having the property*

$$\|z\| \leq \rho\|y\|.$$

Then f *is surjective*

(*I. Ekeland*).

■ It is enough to prove

$$0 \in f(X). \tag{29$_{37}$}$$

Indeed, suppose that this is true and let y_0 be from Y. The function $g := f - y_0$ is continuous and Gâteaux differentiable. Since $g'_w(x) = f'_w(x)\ \forall\, x \in X$, 1^0 and 2^0 are satisfied, consequently $\exists\, x$ in X such that $g(x) = 0$, that is $y_0 = f(x)$.

Pass now to prove (29$_{37}$). Consider the functional $\varphi : X \to \mathbf{R}$, $\varphi(x) = \|f(x)\|$. Apply to φ Ekeland principle 1.1 with $\varepsilon > 0$ anyone and $\lambda = 1$, $\exists\, x_\varepsilon$ in X so that

$$\|f(x_\varepsilon)\| \leq \inf_{x \in X} \|f(x)\| + \varepsilon, \tag{29$_{38}$}$$

$$\|f(x_\varepsilon)\| \leq \|f(x)\| + \varepsilon\|x - x_\varepsilon\| \ \ \forall\, x \in X. \tag{29$_{39}$}$$

In (29$_{39}$) take $x = x_\varepsilon + th$, $t > 0$ and h in X arbitrary and then let μ_t in Y^* be with

$$\|\mu_t\| = 1,\ \mu_t(f(x_\varepsilon + th)) = \|f(x_\varepsilon + th)\| \tag{29$_{40}$}$$

(Hahn corollary). Then we have, taking into account that $\mu_t(f(x_\varepsilon)) \leq \|f(x_\varepsilon)\|$,

$$\frac{1}{t}[\mu_t(f(x_\varepsilon + th)) - \mu_t(f(x_e))] \geq -\varepsilon\|h\|. \tag{29$_{41}$}$$

Add to this the formula

$$\lim_{t \to 0} \frac{f(x_\varepsilon + th) - f(x_\varepsilon)}{t} = f'_w(x_\varepsilon)(h). \tag{29$_{42}$}$$

Set, for every $t > 0$, $J(t) := (0, t]$ and $\mu_{J(t)} := \mu_t$. The set $\mathcal{F} := \{J(t) : t > 0\}$ is rightly filtrable ordered with respect to the relation "⊃" and since $\|\mu_{J(t)}\| = 1$ ((29$_{40}$)), $(\mu_{J(t)})_{J(t) \in \mathcal{F}}$ is a generalized sequence in $S_{Y^*} := \{\mu \in Y^* : \|\mu\| \leq 1\}$. But S_{Y^*} is ∗-weak compact (Alaoglu - Bourbaki - Kakutani theorem), hence $(\mu_{J(t)})_{J(t) \in \mathcal{F}}$ has a ∗-weakly convergent subsequence, denote it identically, and let be

(29_{43}) $\mu_{J(t)} \xrightarrow[J(t)\in\mathcal{F}]{*\text{-weak}} \mu_0$. By (29_{40}) we get, because $\lim\limits_{t\to 0} f(x_\varepsilon + th) = f(x_\varepsilon)$ (use the third relation from the debut with $G_{J(t)} := f(x_\varepsilon + th)$),

(29_{44})
$$\|\mu_0\| = 1,\ \mu_0(f(x_\varepsilon)) = \|f(x_\varepsilon)\|.$$

Moreover, if we denote $F_{J(t)} := \dfrac{f(x_\varepsilon + th) - f(x_\varepsilon)}{t}$, (29_{42}) yields

(29_{45}) $F_{J(t)} \xrightarrow[J(t)\in\mathcal{F}]{\text{strong}} f'_w(x_\varepsilon)(h)$. From (29_{43}) and (29_{45}) one obtains (as before)

$$\mu_{J(t)}(F_{J(t)}) \xrightarrow[J(t)\in\mathcal{F}]{\text{in } \mathbf{R}} \mu_0(f'_w(x_\varepsilon)(h))$$

and combining with (29_{41}),

(29_{46})
$$\mu_0(f'_w(x_\varepsilon)(h)) \geq -\varepsilon\|h\| \ \forall\, h \in X.$$

Let h be in X with $f'_w(x_\varepsilon)(h) = -f(x_\varepsilon)$ and $\|h\| \leq \rho\|f(x_\varepsilon)\|$ (the hypothesis 2^0). By (29_{46}) we get $\mu_0(f(x_\varepsilon)) \leq \varepsilon\rho\|f(x_\varepsilon)\|$. Take ε such that $\varepsilon\rho < 1$, one obtains *a fortiori* $f(x_\varepsilon) = 0$, that is (29_{37}), otherwise the last inequality leads to a contradiction with (29_{44}). ■

B. EKELAND PRINCIPLE VERSUS CARISTI-KIRK THEOREM

Caristi-Kirk theorem. *Let* X *be a complete metric space and* $T: X \to 2^X_*$ *a multivalued function. If there exists a lower bounded and lower semicontinuous function* $\varphi: X \to \mathbf{R}$ *with the property* :

for every x *from* X $\exists$ y *in* T(x) *such that*

$$d(x, y) \leq \varphi(x) - \varphi(y),$$

then $\exists\, x_0$ *in* X *with*

$$x_0 \in T(x_0).$$

x_0 is called a *fixed point* of T.

Remark 1. If T is single valued, we are in the case of the Caristi theorem in the form (2.1)′, [14], form suggested by F. E. Browder.

Remark 2. A generalization of C – K theorem is found at §9, I, 9.59.

Note. In a paper of W. A. Kirk (*Caristi's fixed point theorem and metric convexity,*

Colloq. Math., **36** (1976), 81 – 86), Caristi - Kirk theorem is rightly named "Carsti's theorem".

2.4 *Abbreviated Ekeland principle and Caristi-Kirk theorem are equivalent.*

Explanation. Abbreviated Ekeland principle signifies Ekeland principle 1.1 with $\varepsilon = \lambda = 1$ and reduced to the fundamental inequality (1).

■ *E. P.* $\Rightarrow$ *T. C – K.* We can apply 1.1 and consequently for $\varepsilon = \lambda = 1$ $\exists$ x_0 in X such that (30) $\varphi(x_0) < \varphi(x) + d(x_0, x)$ $\forall$ $x \in X \setminus \{x_0\}$. Suppose *par absurdum* $x_0 \notin T(x_0)$. Since $\forall$ y from $T(x_0)$ we have $y \neq x_0$, it results $y \in T(x_0) \Rightarrow \varphi(x_0) \overset{(30)}{<} \varphi(y) + d(x_0, y)$. On the other hand, the hypothesis gives a point y in $T(x_0)$ so that $\varphi(y) \leq \varphi(x_0) - d(x_0, y)$, a contradiction, hence $x_0 \in T(x_0)$.

T. C – K $\Rightarrow$ *E. P.* Suppose *par absurdum* $\forall$ x of X the set

$$T(x) := \{y \in X \setminus \{x\} : \varphi(x) \geq \varphi(y) + d(x, y)\}$$

is nonempty. According to the hypothesis, the multivalued mapping $T : X \to 2_*^X$ must have a fixed point x_0, that is $x_0 \in T(x_0)$, $x_0 \in X \setminus \{x_0\}$, a contradiction. ■

Remark. In section D of § 9, two generalizations are presented, one for Ekeland principle and the other for Caristi-Kirk theorem. These are also equivalent (9.12).

The same remark can be made relative to §10 (see 10.7, 10.3, 10.8_1 and 10.9).

Complete this section with a proposition due to F. H. Clarke ([20], 1, 7.6.2), the "punctual" version of the Banach fixed point theorem.

Definitions. Let x, y be distinct points in a metric space X. The *open segment* $]x, y[$ is the set of points t of X, $t \neq x, y$ with the property

(31) $$d(x, t) + d(t, y) = d(x, y).$$

The continuous mapping $f : X \to X$, X a metric space, is a *directional contraction* if there exists ρ in $(0, 1)$ having the property:

$$x \neq f(x) \Rightarrow \exists\, y \text{ in }]x, f(x)[\text{ so that}$$

(32) $$d(f(x), f(y)) \leq \rho\, d(x, y).$$

And now –

2.5 *In a complete metric space, every directional contraction has a fixed point.*

■ Let $f : X \to X$ be a directional contraction, X complete. Consider the function $\varphi : X \to \mathbf{R}$, $\varphi(x) = d(x, f(x))$. φ is continuous and lower bounded. Apply 1.1 to it with

$\varepsilon = \frac{1-\rho}{2}$ (ρ the constant from (32)) and $\lambda = 1$, $\exists\, x_0$ in X so that

$$\varphi(x_0) \leq \varphi(y) + \frac{1-\rho}{2}\,\mathrm{d}(x_0, y)\ \forall\, y \in X,$$

that is

$$\text{(33)}\qquad \mathrm{d}(x_0, f(x_0)) \leq \mathrm{d}(y, f(y)) + \frac{1-\rho}{2}\,\mathrm{d}(x_0, y)\ \forall\, y \in X.$$

We have $x_0 = f(x_0)$. Indeed, suppose *par absurdum* $x_0 \neq f(x_0)$.Then, taking into account the definition, for an y of $]x_0, f(x_0)[$, i. e.

$$\text{(34)}\qquad \mathrm{d}(x_0, y) + \mathrm{d}(y, f(x_0)) = \mathrm{d}(x_0, f(x_0)),$$

we have

$$\text{(35)}\qquad \mathrm{d}(f(x_0), f(y)) \leq \rho\, \mathrm{d}(x_0, y).$$

But

$0 \overset{(35)}{\leq} \rho\,\mathrm{d}(x_0, y) - \mathrm{d}(f(y), f(x_0)) \leq \rho\,\mathrm{d}(x_0, y) - \mathrm{d}(f(y), y) + \mathrm{d}(f(x_0), y) \overset{(34)}{=} (\rho - 1)\mathrm{d}(x_0, y) - \mathrm{d}(f(y), y) + \mathrm{d}(x_0, f(x_0)) \overset{(33)}{\leq} \frac{\rho-1}{2}\,\mathrm{d}(x_0, y)$, $0 \leq \frac{\rho-1}{2}\,\mathrm{d}(x_0, y)$, which implies $\mathrm{d}(x_0, y) \leq 0$, that is $x_0 = y$, contradiction and $x_0 = f(x_0)$. ■

C. EKELAND PRINCIPLE VERSUS TAKAHASHI THEOREM, CARISTI-KIRK THEOREM AND OETTLI - THÉRA THEOREM

In this section we present some propositions equivalent to the version 1.1_0 of Ekeland principle, all in a unified form ((52)).

In the following, (X, d) is a complete metric space. Consider $f : X\times X \to (-\infty, +\infty]$ lower semicontinuous with respect to the second argument and with the properties:

$$\text{(36)}\qquad f(v, v) = 0\ \forall\, v \in X,$$

$$\text{(37)}\qquad f(u, v) \leq f(u, w) + f(w, v)\ \forall\, u, v, w \in X.$$

Suppose that an element v_0 exists in X such that

$$\text{(38)}\qquad \inf_{v\in X} f(v_0, v) > -\infty$$

and let be the set

(39) $$S_0 := \{v \in X : f(v_0, v) + d(v_0, v) \le 0\}.$$

S_0 is closed and nonempty ($v_0 \overset{(36)}{\in} S_0$).

Recall and complete one definition (see B).

Let $T : M \to 2_*^M$ be a multivalued function. x_0 is a fixed point resp. *invariant point* of T, if

$$x_0 \in T(x_0) \text{ resp. } \{x_0\} = T(x_0).$$

We prove that the following propositions are equivalent.

Theorem 1. *There is* v^* *in* S_0 *such that*

$$f(v^*, v) + d(v^*, v) > 0 \ \forall\, v \in X \setminus \{v^*\}.$$

Theorem 2 (*W.Takahashi*, [69]). *Suppose*

$$\text{(40) } \forall\ \bar{v} \textit{ from } S_0 \textit{ with } \inf_{v \in X} f(\bar{v}, v) < 0 \textit{ there is } v \textit{ in } X,$$

$$v \ne \bar{v} \textit{ and } f(\bar{v}, v) + d(\bar{v}, v) \le 0.$$

Then there is v^* *in* S_0 *so that*

$$f(v^*, v) \ge 0 \ \forall\, v \in X.$$

Theorem 3. *Let* $T : X \to 2_*^X$ *be a multivalued function with the property*

(41) $$\textit{for every } \bar{v} \textit{ from } S_0 \textit{ there is } v \textit{ in } T\bar{v} \textit{ which verifies}$$

$$f(\bar{v}, v) + d(\bar{v}, v) \le 0.$$

Then T *has a fixed point in* S_0.

Theorem 4. *Let* M *be a subset of* X *with the property*

(42) $$\textit{for every } \bar{v} \textit{ from } S_0 \setminus M \textit{ there is } v \textit{ in } X \textit{ such that}$$

$$v \ne \bar{v} \textit{ and } f(\bar{v}, v) + d(\bar{v}, v) \le 0.$$

Then $S_0 \cap M \ne \varnothing$ (*W. Oettli - M. Théra*).

Proof of Theorem 4. Construct by recurrence a sequence $(v_n)_{n \ge 0}$, $v_n \in X$, so that v_0 is the point from (38) and with every v_n, $n \ge 0$ associate

$$S_n := \{v \in X : f(v_n, v) + d(v_n, v) \le 0\}$$

with

$$\gamma_n := \inf_{v \in S_n} f(v_n, v) > -\infty,$$

which is correct since $S_n \overset{(36)}{\ne} \varnothing$.

Remark that S_0 is the set from relationship (39) and, according to (38), $\gamma_0 > -\infty$. Suppose

v_{n-1}, $n \geq 1$ defined and with $\gamma_{n-1} > -\infty$. Take v_n of S_{n-1} such that

(43) $$f(v_{n-1}, v_n) \leq \gamma_{n-1} + \frac{1}{n}.$$

We have (44) $S_n \subset S_{n-1}$: $v \in S_n \Rightarrow f(v_n, v) + d(v_n, v) \leq 0$, $v_n \in S_{n-1} \Rightarrow f(v_{n-1}, v_n) +$ $+ d(v_{n-1}, v_n) \leq 0$, add and take into account (37). Then

$$\gamma_n = \inf_{v \in S_n} f(v_n, v) \overset{(37)}{\geq} \inf_{v \in S_n} [f(v_{n-1}, v) - f(v_{n-1}, v_n)] \overset{(44)}{\geq} \inf_{v \in S_{n-1}} [f(v_{n-1}, v) - f(v_{n-1}, v_n)] =$$

$$= \inf_{v \in S_{n-1}} f(v_{n-1}, v) - f(v_{n-1}, v_n) \overset{(43)}{\geq} -\frac{1}{n},$$

(45) $$\gamma_n \geq -\frac{1}{n}, \; n \geq 1.$$

In particular, $\gamma_n > -\infty$. Let v be in S_n, $n \geq 1$. (46) $d(v_n, v) \leq -f(v_n, v) \leq -\gamma_n \overset{(45)}{\leq} \frac{1}{n}$, consequently $n \geq 1 \Rightarrow \operatorname{diam} S_n \leq \frac{2}{n}$ (the triangle inequality), $\lim_{n \to \infty} \operatorname{diam} S_n = 0$. Moreover, $p \geq n \overset{(44)}{\Rightarrow} S_p \subset S_n \overset{(46)}{\Rightarrow} d(v_n, v_p) \leq \frac{1}{n}$, consequently $(v_n)_{n \geq 0}$ is a Cauchy sequence, let be $v^* = \lim_{n \to \infty} v_n$. We have

(47) $$\bigcap_{n=0}^{\infty} S_n = \{v^*\},$$

since on the one hand $v^* \in S_n \; \forall \, n \geq 1$ ($p \geq n \Rightarrow d(v_n, v_p) \leq \frac{1}{n}$, $\lim_{p \to \infty} d(v_n, v_p) = d(v_n, v^*) =$ $= 0$, S_n is closed) and, on the other hand, $\operatorname{diam} (\bigcap_{n=1}^{\infty} S_n) \leq \operatorname{diam} S_n \leq \frac{2}{n} \; \forall \, n \geq 1$, $\operatorname{diam} (\bigcap_{n=0}^{\infty} S_n) = 0$.

It remains to prove

(48) $$v^* \in M,$$

since $v^* \in S_0$ ((47)). Suppose *par absurdum* $v^* \notin M$. Then there is v in X, $v \neq v^*$, such that $f(v^*, v) + d(v^*, v) \leq 0$ ((42)). With (47) one finds $f(v_n, v^*) + d(v_n, v^*) \leq 0 \; \forall \, n \geq 0$. We add and with (47) one obtains $f(v_n, v) + d(v_n, v) \le 0 \ \forall\ n \ge 0$, that is $v \in \bigcap_{n=0}^{\infty} S_n$, consequently $v \overset{(47)}{=} v^*$, contradiction and hence (48). ■

Corollary. Every subset of X, *with the property* (42), *is nonempty.*

Remark. A uniform variant of Theorem 4 is to 9.57, §9, I.

2.6 THEOREM. *Theorems* 1, 2, 3 *and* 4 *are equivalent* (*W. Oettli - M. Théra*).

■ *Theorem* 4 ⇒ *Theorem* 1. For every $\bar{v}$ of X consider the set

$$\Gamma(\bar{v}) = \{v \in X : v \ne \bar{v},\ f(\bar{v}, v) + d(\bar{v}, v) \le 0\} \tag{49}$$

and let be $M := \{\bar{v} \in X : \Gamma(\bar{v}) = \varnothing\}$. Let $\bar{v}$ be arbitrary in $S_0 \setminus M$. We have $\Gamma(\bar{v}) \ne \varnothing$, i. e. (42) is verified. The hypothesis imposes $M \ne \varnothing$, let v^* be in M, hence $\Gamma(v^*) = \varnothing$ and (49) yields $f(v^*, v) + d(v^*, v) > 0 \ \forall\ v \in X \setminus \{v^*\}$.

Theorem 4 ⇒ *Theorem* 2. Take

$$M = \{\bar{v} \in X : \inf_{v \in X} f(\bar{v}, v) \ge 0\}. \tag{50}$$

For every $\bar{v}$ in $S_0 \setminus M$ there is, according to (40), v in $X \setminus \{\bar{v}\}$ for which $f(\bar{v}, v) + d(\bar{v}, v) \le 0$, that is (42) is verified, let then be v^* from $S_0 \cap M$. (50) imposes $f(v^*, v) \ge 0 \ \forall\ v \in X$.

Theorem 4 ⇒ *Theorem* 3. Take

$$M = \{\bar{v} \in X : \bar{v} \in T\bar{v}\}. \tag{51}$$

Let $\bar{v}$ be from $S_0 \setminus M$. There is, according to (41), $v \in T\bar{v}$ which verifies $f(\bar{v}, v) + d(\bar{v}, v) \le 0$. Moreover, because $\bar{v} \notin M$, (51) gives $\bar{v} \ne v$, hence (42) is verified and consequently we can take a point v^* in $S_0 \cap M$, that is $v^* \in S_0$ and $v^* \in T v^*$.

Theorem 1 ⇒ *Theorem* 4. Let M be a subset of X with the property (42). According to the hypothesis, $\exists\ v^*$ in S_0 such that $f(v^*, v) + d(v^*, v) > 0 \ \forall\ v \in X \setminus \{v^*\}$. Then (42) imposes $v^* \in M$ and so $v^* \in S_0 \cap M$.

Theorem 2 ⇒ *Theorem* 4. Let M be a subset of X satisfying (42). Suppose *par absurdum* $S_0 \cap M = \varnothing$, that is $\bar{v} \in S_0 \Rightarrow \bar{v} \notin M$. Then, according to (42),

$$\bar{v} \in S_0 \Rightarrow \exists\ v \ne \bar{v},\ f(\bar{v}, v) + d(\bar{v}, v) \le 0. \tag{52}$$

(52) imposes that (40) is verified. Then the hypothesis yields an element v^* in S_0 with the property $f(v^*, v) \ge 0 \ \forall\ v \in X$ and we come in collision with (52), since $v \ne v^* \Rightarrow$

$\Rightarrow \mathrm{d}(v^*, v) > 0$. So $S_0 \cap M \neq \varnothing$.

Theorem 3 $\Rightarrow$ *Theorem* 4. We can suppose that X has at least two points. Let M be a subset of X with the property (42). Suppose, *par absurdum*, (53) $S_0 \cap M = \varnothing$. Consider the mapping $T: X \to 2^X_*$,

$$T(\bar{v}) = \{v \in X : v \neq \bar{v}\}. \tag{54}$$

Then $\bar{v} \in S_0 \overset{(53),(42)}{\Rightarrow} \exists\ v \in T\bar{v}$ such that $f(\bar{v}, v) + \mathrm{d}(\bar{v}, v) \leq 0$, (41) is verified and consequently $\exists\ v^* \in S_0$ such that $v^* \in T v^*$, which contradicts (54), and so $S_0 \cap M \neq \varnothing$. ■

Remark. Because Theorem 4 was proved, one concludes, *via* 2.6, that Theorem 1, Theorem 2 and Theorem 3 are true.

Comments

I. From Theorem 1 it can obtain Ekeland principle 1.1_0, but for the function φ with values in **R** (which justifies the title of this section). Indeed, take f: $f(u, v) = \varphi(v) - \varphi(u)$ and replace d by $\frac{\varepsilon}{\lambda}\mathrm{d}$. From (55) $\varphi(v_0) \leq \inf \varphi(X) + \varepsilon$ it results $\inf\limits_{v \in X} f(v_0, v) > -\infty$, consequently $\exists\ v^*$ in S_0 such that

$$\varphi(v) - \varphi(v^*) + \frac{\varepsilon}{\lambda}\mathrm{d}(v^*, v) > 0\ \forall\ v \in X \setminus \{v^*\},$$

that is the fundamental inequality (1). $v^* \in S_0$ means (56) $\varphi(v^*) - \varphi(v_0) + \frac{\varepsilon}{\lambda}\mathrm{d}(v^*, v_0) \leq 0$, which is the inequality (2_0). Moreover, by (55) and (56), we get $-\varepsilon + \frac{\varepsilon}{\lambda}\mathrm{d}(v^*, v_0) \leq$ $\leq \inf \varphi(X) - \varphi(v_0) + \frac{\varepsilon}{\lambda}\mathrm{d}(v^*, v_0) \leq \varphi(v^*) - \varphi(v_0) + \frac{\varepsilon}{\lambda}\mathrm{d}(v^*, v_0) \leq 0$, hence $\mathrm{d}(v^*, v_0) \leq \lambda$ – the inequality (3).

II. With the aid of Theorem 3 it can obtain Caristi - Kirk theorem (Section B), but with φ taking values in **R** (which justifies once again the title of this section). Indeed, take f: $f(u, v) = \varphi(v) - \varphi(u)$. Clearly $\inf\limits_{v \in X} f(v_0, v) > -\infty$ and since $\forall\ \bar{v} \in X\ \ \exists\ v \in T\bar{v}$ such that $\mathrm{d}(\bar{v}, v) \leq \varphi(\bar{v}) - \varphi(v)$, clearly $f(\bar{v}, v) + \mathrm{d}(\bar{v}, v) \leq 0$, consequently T has a fixed point situated exactly in S_0.

Remark. The statements of this section are generalized in the section J of §9.

D. SIEGEL THEOREM

In this section , following [54], 1, we give a generalization of the Caristi - Kirk theorem, obtaining in addition that the Downing - Kirk theorem is not, as the authors believe, a generalization of Caristi - Kirk theorem, but even is equivalent to this (2.17).

Let (X, d), (Y, δ) be complete metric spaces, $f: X \to Y$ be *closed* ($x_n, x \in X$, $x_n \to x$ and $f(x_n) \to y \Rightarrow y = f(x)$) and $\varphi : f(X) \to \mathbf{R}_+$ lower semicontinuous.

2.7 *Let* $(x_n)_{n\geq 1}$ *be a sequence in* X *with the property*

(57) $$\max \{d(x_n, x_{n+1}), \delta(f(x_n), f(x_{n+1}))\} \leq \varphi(f(x_n)) - \varphi(f(x_{n+1})) \ \forall\, n \geq 1.$$

Then there is $a := \lim_{n\to\infty} x_n$ *and*

(58) $$\max \{d(x_n, a), \delta(f(x_n), f(a))\} \leq \varphi(f(x_n)) - \varphi(f(a)) \ \forall\, n \geq 1.$$

■ From (57) it results that the sequence $(\varphi(f(x_n))_{n\geq 1}$ is decreasing and hence convergent. Also, again from (57),

(59) $$m \geq n \Rightarrow \max \{d(x_n, x_m), \delta(f(x_n), f(x_m))\} \leq \varphi(f(x_n)) - \varphi(f(x_m)),$$

consequently both sequences $(x_n)_{n\geq 1}$, $(f(x_n))_{n\geq 1}$ are Cauchy and therefore convergent, let be $a := \lim_{n\to\infty} x_n$, $y := \lim_{n\to\infty} f(x_n)$. f being closed, it yields $\lim_{n\to\infty} f(x_n) = f(a)$. Moreover, for $n \geq 1$ fixed, $\max \{d(x_n, a), \delta(f(x_n), f(a))\} = \max \{\lim_{m\to\infty} d(x_n, x_m), \lim_{m\to\infty} \delta(f(x_n), f(x_m))\} \overset{(59)}{\leq} \varphi(f(x_n)) - \underline{\lim}_{m\to\infty} \varphi(f(x_m)) \leq \varphi(f(x_n)) - \varphi(f(a))$, because φ is l. s. c. in $f(a)$. ■

Definition. Let X be as before and $(h_n)_{n\geq 1}$ a sequence, $h_n : X \to X$. The *countable composition* of $(h_n)_{n\geq 1}$, $\prod_{n=1}^{\infty} h_n$, a mapping of X in X, is

$$\prod_{n=1}^{\infty} h_n(x) = \lim_{n\to\infty} (h_n \circ h_{n-1} \circ \ldots \circ h_1)(x),$$

in the hypothesis that the limit exists for every x of X.

Notation. Let X, Y, f, φ be as before. $\mathscr{H}^*$ will designate the set of mappings $h : X \to$ $\to X$ with the property

(60) $$\max \{d(x, h(x)), \delta(f(x), f(h(x)))\} \leq \varphi(f(x)) - \varphi(f(h(x))) \ \forall\, x \in X.$$

$\mathcal{H}^*$ is nonempty (the identity map of X!).

2.8 *$\mathcal{H}^*$ is closed with respect to composition and countable composition.*

■ For the convenience of the writing, the composition of the functions will be denoted through juxtaposition. Let h_1, h_2 be in $\mathcal{H}^*$ and x arbitrary in X. $\max\{d(x, h_2 h_1(x)), \delta(f(x), f h_2 h_1(x))\} \le \max\{d(x, h_1(x)), \delta(f(x), f h_1(x))\} + \max\{d(h_1(x), h_2 h_1(x)), \delta(fh_1(x), fh_2 h_1(x))\} \overset{(60)}{\le} \varphi(f(x)) - \varphi(fh_1(x)) + \varphi(fh_1(x)) - \varphi(fh_2 h_1(x)) = \varphi(f(x)) - \varphi(fh_2 h_1(x))$, $h_2 h_1 \in \mathcal{H}^*$.

It remains to show that for every sequence $(h_n)_{m\ge1}$ from $\mathcal{H}^*$, $\prod_{n=1}^{\infty} h_n$ exists and belongs to $\mathcal{H}^*$. Let x be arbitrary in X and $x_n := h_n h_{n-1} \dots h_1(x)$, $n \ge 1$. The sequence $(x_n)_{n\ge1}$ verifies (57) (take $x = x_n$ in (60), $h = h_{n+1} \in \mathcal{H}^*$) and hence, according to 2.7, there exists $\prod_{n=1}^{\infty} h_n(x)$. Finally, for every x from X, $\pi(x) := \prod_{n=1}^{\infty} h_n(x)$ verifies (60). Indeed, as we saw, (61) $\lim_{n\to\infty} f(x_n) = f(\pi(x))$. But $h_n \dots h_1 \in \mathcal{H}^*$, therefore (see (60)) $\max\{d(x, x_n), \delta(f(x), f(x_n))\} \le \varphi(f(x)) - \varphi(f(x_n))$. Take the upper limit for $n \to \infty$ and according to (61) we get

$\max\{d(x, \pi(x)), \delta(f(x), f(\pi(x)))\} \le \varphi(f(x)) - \underline{\lim}_{n\to\infty} \varphi(f(x_n)) \le \varphi(f(x)) - \varphi(f(\pi(x)))$. ■

Notations. Let X, Y, f, φ be as before, $A \subset X$ nonempty, $\mathcal{F} (\ne \emptyset) \subset \mathcal{H}^*$ and $x \in X$.

(62) $$r(A) := \inf\{\varphi(f(x)) : x \in A\},$$

(63) $$\mathcal{F}(x) := \{h(x) : h \in \mathcal{F}\}.$$

Evidently $A \subset B \Rightarrow r(A) \ge r(B)$. We have

(64) $$\operatorname{diam} \mathcal{F}(x) \le 2[\varphi(f(x)) - r(\mathcal{F}(x))].$$

Indeed, let h_1, h_2 be arbitrary in $\mathcal{F}$. $d(h_1(x), h_2(x)) \le d(x, h_1(x)) + d(x, h_2(x)) \overset{(60)}{\le} \varphi(f(x)) - \varphi(fh_1(x)) + \varphi(f(x)) - \varphi(fh_2(x))$ and $r(\mathcal{F}(x)) \le \varphi(fh_i(x))$, $i = 1, 2$.

And now –

2.9 Siegel theorem. *Le* (X, d), (Y, δ) *be complete metric spaces*, $f: X \to Y$ *closed*, $\varphi: f(X) \to \mathbf{R}_+$ *lower semicontinuous*, $\mathcal{H}^*$ *the set of the mappings* $h: X \to X$ *satisfying*

$$(*)\ \max\{d(x, h(x)), \delta(f(x), fh(x))\} \le \varphi(f(x)) - \varphi(fh(x))\ \forall x \in X,$$

$\mathscr{F}$ a nonempty subset of $\mathscr{H}^*$ closed under composition and x_0 *in* X.

1° *If $\mathscr{F}$ is closed under countable composition, there exists* H *in $\mathscr{F}$ such that*

$$h(H(x_0)) = H(x_0) \ \forall \ h \in \mathscr{F}.$$

2° *If all the mappings from $\mathscr{F}$ are continuous, there exists a sequence* $(h_n)_{n\geq 1}$ *in $\mathscr{F}$ and* $a := \lim_{n\to\infty} h_n h_{n-1} \dots h_1(x_0)$ *such that*

$$h(a) = a \ \forall \ h \in \mathscr{F}.$$

■ Take a sequence $(\varepsilon_n)_{n\geq 1}$, $\varepsilon_n > 0$, $\varepsilon_n \to 0$. $\exists\ h_1$ in $\mathscr{F}$ such that (65) $\varphi(f h_1(x_0)) - r(\mathscr{F}(x_0)) < \frac{\varepsilon_1}{2}$ (see (62), (63)). Set $x_1 := h_1(x_0)$. We have (66) $\mathscr{F}(x_1) \subset \mathscr{F}(x_0)$, because $\mathscr{F}$ is closed under composition, and

$$\operatorname{diam}\mathscr{F}(x_1) < \varepsilon_1, \text{ since}$$

$$\operatorname{diam}\mathscr{F}(x_1) \overset{(64)}{\leq} 2[\varphi(f(x_1)) - r(\mathscr{F}(x_1))] \overset{(66)}{\leq} 2[\varphi(f h_1(x_0)) - r(\mathscr{F}(x_0))] \overset{(65)}{<} \varepsilon_1.$$

The relationship (64) was correctly applied because the condition (*) of the statement is condition (60). $\exists\ h_2$ in $\mathscr{F}$ so that $\varphi(f h_2(x_1)) - r(\mathscr{F}(x_1)) < \frac{\varepsilon_2}{2}$. Set $x_2 := h_2(x_1)$. We have $\mathscr{F}(x_2) \subset \mathscr{F}(x_1)$ and $\operatorname{diam} \mathscr{F}(x_2) \leq 2[\varphi(f(x_2)) - r(\mathscr{F}(x_2))] \leq 2[\varphi(f h_2(x_1)) - r(\mathscr{F}(x_1))] < \varepsilon_2$ etc. Thus, by recurrence, we get the sequences $(x_n)_{n\geq 1}$ from X and $(h_n)_{n\geq 1}$ from $\mathscr{F}$ with the properties (67) $x_n = h_n(x_{n-1})$, $\mathscr{F}(x_n) \subset \mathscr{F}(x_{n-1})$, $\operatorname{diam}\mathscr{F}(x_n) < \varepsilon_n\ \forall n \geq 1$.

1° Let $H := \prod_{n=1}^{\infty} h_n$ ($\mathscr{F}$ is closed under countable composition) and $a := H(x_0)$. We have $a = \lim_{n\to\infty} h_n \dots h_1(x_0) = \prod_{m=n+1}^{\infty} h_m(x_n)\ \forall n \geq 0$, therefore, since $\mathscr{F}$ is closed under countable composition, $a \in \mathscr{F}(x_n)\ \forall\ n \geq 0$, i. e. $a \in \bigcap_{n=0}^{\infty} \mathscr{F}(x_n)$. But $\lim_{n\to\infty} \operatorname{diam}\mathscr{F}(x_n) \overset{(67)}{=} 0$, we get (68) $\{a\} = \bigcap_{n=0}^{\infty} \mathscr{F}(x_n)$. On the other hand, for any h in $\mathscr{F}$ we have, $\forall n \geq 0$,

$$h(a) = h\left(\prod_{m=n+1}^{\infty} h_m(x_n)\right) \in \mathscr{F}(x_n)$$ because $h \circ \bigcap_{m=n+1}^{\infty} h_m \in \mathscr{F}$. A confrontation with (68) yields $h(a) = a$.

2° Return to the sequences $(x_n)_{n\geq 1}$ and $(h_n)_{n\geq 1}$ constructed in the beginning of the proof. $(x_n)_{n\geq 1}$ is fundamental: let be $\varepsilon > 0$, $\exists\, N$ in **N** such that $\varepsilon_N < \varepsilon$, $m, n > N \overset{(67)}{\Rightarrow} x_m, x_n \in \mathscr{F}(x_N)$, hence $d(x_m, x_n) \overset{(67)}{<} \varepsilon$. Denote then $a := \lim_{n\to\infty} x_n$. We will show that (69) $h(a) = a$ $\forall\, h \in \mathscr{F}$.

As $x_n \overset{(67)}{\in} \mathscr{F}(x_{n-1})$ $\forall\, n \geq 1$, we get $a \in \overline{\mathscr{F}(x_{n-1})}$ $\forall\, n \geq 1$, that is (70) $a \in \bigcap_{n=0}^{\infty} \overline{\mathscr{F}(x_n)}$.

Let $\varepsilon > 0$ be arbitrary. Since $x_n \to a$, we get $h(x_n) \to h(a)$ (h is continuous), hence $\exists\, N$ in **N** so that $n \geq N \Rightarrow h(x_n) \in S(h(a), \varepsilon)$ (the open ball!). Obviously, $h(x_n) \in \mathscr{F}(x_n)$ $\forall\, n \geq 0$. Moreover, according to (70), $S(a, \varepsilon_n)$ contains an element y_n of $\mathscr{F}(x_n)$ and then for $n \geq N$ we have

$$d(h(a), a) \leq d(h(a), h(x_n)) + d(h(x_n), y_n) + d(y_n, a) \overset{(67)}{<} \varepsilon + \varepsilon_n + \varepsilon_n .$$

Taking the limit, one finds $d(h(a), a) \leq \varepsilon$ and this imposes (69) since ε is arbitrary. ■

Remark. 2.9 is S. Park's version for Siegel theorem ([54], 1).

2.10 *Corollary. Let* X, Y, f, φ, $\mathscr{H}^*$ *be as in* 2.9. *Then all the mappings of* $\mathscr{H}^*$ *have a common fixed point.*

Moreover, if h *from* $\mathscr{H}^*$ *is continuous, then, for every* x_0 *in* X, $a := \lim_{n\to\infty} h^n(x_0)$ *is a fixed point for* h.

Explanation. $h^n = h \circ h \circ \ldots \circ h$, n "factors", $h^0 = h$.

■ *First statement.* $\mathscr{H}^*$ is closed under composition and countable composition (2.8), apply 1° from 2.9. *Second statement.* $\mathscr{F} := \{h^n : n \geq 0\}$ is closed under composition and h^n is continuous $\forall\, n \geq 0$. Apply to $\mathscr{F}$ 2° from 2.9. ■

2.11 Theorem. *Let* (X, d) *be a complete metric space* , φ: X → $\mathbf{R}_+$ *lower semicontinuous,* $\mathscr{H}^*$ *the set of mappings* h: X → X *which satisfy*

(71) $$d(x, h(x)) \leq \varphi(x) - \varphi(h(x)) \ \forall\, x \in X,$$

$\mathscr{F}$ *a nonempty subset of* $\mathscr{H}^*$, *closed under composition, and* x_0 *from* X. *Then the conclusions* 1° *and* 2° *from* 2.9 *hold true* (*J. Siegel*, [66]).

■ Apply 2.9 with $Y = X$ and $f = \mathrm{id}_X$. ■

REMARK. In [54], 1 the false assertion Theorem 2.11 ⇒ Theorem 2.9 is made (in the original Theorem 2 $\Rightarrow$ Theorem 1): the metric space (M, d) (original notation) is replaced by the metric space (M, ρ), $\rho(x, y) := \max\{\mathrm{d}(x, y), \mathrm{d}(f(x), f(y))\}$, but, for instance, when $h: M \to M$ is continuous with respect to d, it is not necessarily continuous with respect to ρ, that blocks up the application of Theorem 2 from which the Theorem 1, b) is derived.

2.12 *Let* X, Y, f, φ *be as in theorem* 2.9 *and* h: X → X. *If*

$$\max\{\mathrm{d}(\mathrm{x}, \mathrm{h}(\mathrm{x})), \delta(\mathrm{f}(\mathrm{x}), \mathrm{fh}(\mathrm{x}))\} \leq \varphi(\mathrm{f}(\mathrm{x})) - \varphi(\mathrm{fh}(\mathrm{x}))\ \forall\, \mathrm{x} \in \mathrm{X},$$

then h *has a fixed point.*

■ $h \in \mathscr{H}^*$, apply 2.10. ■

2.13 Downing - Kirk theorem.[*] *Let* X, Y, f, φ *be as in theorem* 2.9 *and* h: X → X. *If*

$$\max\{\mathrm{d}(\mathrm{x}, \mathrm{h}(\mathrm{x})), \mathrm{c}\,\delta(\mathrm{f}(\mathrm{x}), \mathrm{fh}(\mathrm{x}))\} \leq \varphi(\mathrm{f}(\mathrm{x})) - \varphi(\mathrm{fh}(\mathrm{x})),\ \mathrm{c} > 0,\ \forall\, \mathrm{x} \in \mathrm{X},$$

then h *has a fixed point.*

■ Let be $\rho := \max\left\{1, \frac{1}{c}\right\}$. $\psi := \rho\varphi$ is lower semicontinuous. But $\forall\, x \in X$

$$\frac{1}{\rho}\max\{\mathrm{d}(x, h(x)), \delta(f(x), fh(x))\} = \max\{\frac{1}{\rho}\mathrm{d}(x, h(x)), \frac{1}{\rho}\delta(f(x), fh(x))\} \leq$$

$$\leq \max\{\mathrm{d}(x, h(x)), c\,\delta(f(x), fh(x))\} \overset{\text{hypothesis}}{\leq} \varphi(f(x)) - \varphi(fh(x)),$$

therefore ψ verifies the hypothesis of 2.12 and consequently h has a fixed point. ■

In fact

2.14 2.12 *and* 2.13 *are equivalent.*

■ Indeed, for $c = 1$ one recovers the condition from 2.12. ■

2.15 Caristi - Kirk theorem (*single valued form*). *Let* (X, d) *be a complete metric space*, φ: X → $\mathbf{R}_+$ *lower semicontinuous and* h: X → X. *If* h *verifies*

$$\mathrm{d}(\mathrm{x}, \mathrm{h}(\mathrm{x})) \leq \varphi(\mathrm{x}) - \varphi(\mathrm{h}(\mathrm{x}))\ \forall\, \mathrm{x} \in \mathrm{X},$$

then h *has a fixed point.*

■ Apply 2.12 with $Y = X$, $f = \mathrm{id}_X$. ■

In fact

2.15_1 *Caristi - Kirk theorem for multivalued functions* (Section B) *is equivalent to Caristi - Kirk theorem for single valued functions.*

---

[*] *D. Downing, W. Kirk, A generalization of Caristi's theorem with applications to nonlinear mapping theory*, Pacific J. Math., **69** (1977), 339-346.

■ Keep the hypothesis and notations of the section B. Consider the function $\psi : X \to \mathbf{R}_+$, $\psi(x) = \varphi(x) - \inf \varphi(X)$. ψ is still lower semicontinuous and has the property: $\forall\, x \in X\, \exists\, y \in T(x)$, $y := h(x)$, such that $d(x, y) \leq \psi(x) - \psi(y)$. The function $h : X \to X$ verifies $\forall\, x \in X$ the inequality from 2.15 with φ instead of ψ, hence $\exists\, x_0 \in X$ so that $h(x_0) = x_0$ and therefore $x_0 \in T(x_0)$. ■

Show now

2.16 *Propositions* 2.12 *and* 2.15 *are equivalent.*

■ We have only to prove that 2.15 ⇒ 2.12. Consider the distance on X

(72) $$\rho(x, y) := \max\{d(x, y), \delta(f(x), f(y))\}.$$

(X, ρ) is a complete metric space and $\varphi \circ f$ is lower semicontinuous with respect to ρ. Indeed, let be $x_n \xrightarrow{\rho} f(x_0)$. Then (72) imposes $f(x_n) \xrightarrow{\delta} f(x_0)$. But φ being lower semicontinuous in $f(x_0)$, we get $\underline{\lim}_{n\to\infty} \varphi(f(x_n)) \geq \varphi(f(x_0))$. Or, according to the hypothesis from 2.12, we have

$$\rho(x, h(x)) \leq \varphi \circ f(x) - \varphi \circ f(h(x)) \ \forall\, x \in X,$$

consequently 2.15 supplies for h a fixed point. ■

Confronting 2.14 to 2.16 one finds

2.17 *The Caristi - Kirk and Downing - Kirk theorems are equivalent.*

Retain still (see the section B)

FINAL REMARK. The abbreviated Ekeland principle can be obtained from Siegel theorem.

The following diagram points out graphically the implications and the equivalences established before.

Th. 2.11 Th. Caristi - Kirk ⇔ The abbreviated Ekeland principle

⇑ ⇕ ⇕

Th. Siegel ⇒ Corollary 2.10 ⇒ 2.12 ⇔ Th. Downing - Kirk

3

Ekeland principle in Banach spaces

We use the differential calculus in the normed spaces. This fact also facilitates a better understanding of the force of inequality (1) from Ekeland principle (1.1). As it was already agreed in the INTRODUCTION, all Banach spaces that appear are real.

3.1 *Let* X *be Banach space and* φ: X$\rightarrow$ **R** *bounded from below, lower semicontinuous and Gâteaux differentiable. Then,* $\forall\, \varepsilon > 0$ *and* $\forall$ u *of* X *with* $\varphi(u) \leq \inf \varphi(X) + \varepsilon$, *there exists* v_ε *in* X *so that* :

(1) $$\|\varphi'_w(v_\varepsilon)\| \leq \sqrt{\varepsilon}\,,$$

(2) $$\varphi(v_\varepsilon) \leq \varphi(u) - \sqrt{\varepsilon}\,\|v_\varepsilon - u\|,$$ (3) $$\|v_\varepsilon - u\| \leq \sqrt{\varepsilon}$$

(*I. Ekeland*, [30], 1).

Definition. v_ε is an *almost critical point* for φ.

■ Apply 1.1_0 with $\lambda = \sqrt{\varepsilon}$, there is v_ε in X verifying (2) and (3) of the statement and

(4) $$\varphi(v_\varepsilon) \leq \varphi(v) + \sqrt{\varepsilon}\;\|v_\varepsilon - v\| \quad \forall\, v \in X.$$

If we put, in (4), $v = v_\varepsilon + t\,h$, $t > 0$, $h \in X$, $\|h\| = 1$ one obtains:

$$\varphi(v_\varepsilon + th) - \varphi(v_\varepsilon) \geq -\sqrt{\varepsilon}\, t,$$

$$\frac{\varphi(v_\varepsilon + th) - \varphi(v_\varepsilon)}{t} \geq -\sqrt{\varepsilon}$$

Take the limit for $t \to 0+$ and get

(5) $$\varphi'_w(v_\varepsilon)(h) \geq -\sqrt{\varepsilon} \quad \forall\, h \text{ with } \|h\|=1.$$

But $\|-h\|=\|h\|=1$, hence replacing in (5) h by $-h$ one gets $\varphi'_w(v_\varepsilon)(h) \leq \sqrt{\varepsilon}$ and hence

(6) $$|\varphi'_w(v_\varepsilon)(h)| \leq \sqrt{\varepsilon} \quad \forall\, h,\ \|h\|=1.$$

By (6) one obtains $\|\varphi'_w(v_\varepsilon)\| \leq \sqrt{\varepsilon}$. ■

Show now a variant of 3.1 obtained with Theorem 1.1_1 ([71], 1).

3.1_0 *Let* X *be a Banach space and* $\varphi : X \to \mathbf{R}$ *a lower bounded, lower semicontinuous and Gâteaux differentiable. For any* $\varepsilon > 0$ *and* u *from* X *with* $\varphi(u) \leq \inf \varphi(X) + \varepsilon$ *and for every* $\lambda > 0$ *and* $h : [0, +\infty) \to [0, +\infty)$ *increasing with*

$$\int_0^{+\infty} \frac{dr}{1+h(r)} = +\infty, \quad \int_{\|u\|}^{\|u\|+r_1} \frac{dr}{1+h(r)} \geq \lambda \textit{ where } r_1 > 0,$$

there exists v_ε *in* X *such that*

$$\|\varphi'_w(v_\varepsilon)\| \leq \frac{\varepsilon}{\lambda[1+h(\|v_\varepsilon\|)]},$$

$$\varphi(v_\varepsilon) \leq \varphi(u), \quad \|v_\varepsilon\| \leq \|u\| + r_1$$

(*Zhong Cheng-Kui*).

■ Apply 1.1_1 + Remark 3 with $u_0 = 0$. For every $t > 0$ and x from X with $\|x\|=1$ we have

$$\frac{\varphi(v_\varepsilon + tx) - \varphi(v_\varepsilon)}{t} \geq -\frac{\varepsilon}{\lambda[1+h(\|v_\varepsilon\|)}.$$

Take the limit for $t \to 0+$,

$$\varphi'_w(v_\varepsilon)(x) \geq -\frac{\varepsilon}{\lambda[1+h(\|v_\varepsilon\|)]}.$$

Replace x by $-x$, hence

$$|\varphi'_w(v_\varepsilon)(x)| \leq \frac{\varepsilon}{\lambda[1+h(\|v_\varepsilon\|)} \quad \text{etc. ■}$$

Definition. The sequence $(x_n)_{n\geq 1}$ is called *minimizing* for the lower bounded function $f : X \to \mathbf{R}$ if

$$\lim_{n\to\infty} f(x_n) = \inf_{x\in X} f(x).$$

If we take in 3.1 $\varepsilon = \frac{1}{n}$, $n \in \mathbf{N}$ one obtains $\varphi(v_n) \leq \inf f(X) + \frac{1}{n}$, hence $\lim_{n\to\infty} \varphi(v_n) =$ $= \inf f(X)$, $(v_n)_{n\geq 1}$ is minimizing sequence for φ. Moreover, $\lim_{n\to\infty} \|\varphi'_w(v_n)\| = 0$. This result will be reinforced in 3.1_1, even in a generalization of this. Previously

Definitions. Let X be a real normed space and γ from $C([0,1]; X)$. The *geodesic length* $l(\gamma)$ of γ is:

$$l(\gamma) = \int_0^1 \frac{\|\gamma'(t)\|}{1+\|\gamma(t)\|} dt.$$

The *geodesic distance* $\delta(x, y)$ between the points x, y of X is

$$(6_1) \qquad \delta(x, y) = \inf \{l(\gamma) : \gamma \in C^1([0,1]; X), \gamma(0) = x, \gamma(1) = y\}.$$

δ is a distance on the set X. We have $\delta(x, y) \leq \|x - y\|$. Conversely, let M be a bounded subset of X. Then $\exists\, \rho > 0$ such that $x, y \in M \Rightarrow \|x - y\| \leq \rho\delta(x, y)$.

When $y = 0$, the infimum from (6_1) is obtained on the segment with the ends 0 and x, hence

$$(6_2) \qquad \delta(0, x) = \int_0^1 \frac{\|x\|}{1+t\|x\|} dt = \ln(1+\|x\|).$$

(6_2) shows that every bounded in norm set is also bounded with respect to δ and conversely. When X is a Banach space, the metric space (X, δ) is complete (use the relations between $\delta(x, y)$ and $\|x - y\|$, every Cauchy sequence with respect to δ is bounded with respect to δ and hence bounded in norm).

Pass to

3.1_1 *Let* X *be a Banach space and* $\varphi : X \to \mathbf{R}$ *lower bounded, lower semicontinuous and Gâteaux differentiable. There is a minimizing sequence* $(x_n)_{n\geq 1}$ *for* φ *such that*

$$\lim_{n\to\infty} (1 + \|x_n\|)\|\varphi'_w(x_n)\| = 0$$

(*I. Ekeland*, [30], 5).

■ Apply Ekeland principle 1.1 to the metric space (X, δ) and to φ with $\lambda = \sqrt{\varepsilon}$ and $\varepsilon =$ $= \frac{1}{n^2}$, $n \in \mathbf{N}$. One finds x_n in X so that

$$(6_3) \qquad \varphi(x_n) \leq \inf \varphi(X) + \frac{1}{n^2},$$

(6₄) $$\varphi(x) - \varphi(x_n) \geq -\frac{1}{n}\,\delta(x, x_n)\ \forall\, x \in X.$$

(6₃) tells that $(x_n)_{n\geq 1}$ is a minimizing sequence for φ. In (6₄) take $x = x_n + tu$, $t > 0$, $u \in X$, $\|u\| = 1$ and dividing by t, we get:

$$\frac{\varphi(x_n + tu) - \varphi(x_n)}{t} \geq -\frac{1}{n}\,\frac{1}{t}\int_0^1 \frac{ds}{1 + \|x_n + su\|}$$

and, taking the limit for $t \to 0+$,

(6₅) $$\varphi'_w(x_n)(u) \geq -\frac{1}{n}\,\frac{\|u\|}{1 + \|x_n\|}.$$

In (6₅) one replaces u by $-u$ and proceeding as for (5) one deduces

$$\|\varphi'_w(x_n)\| \leq \frac{1}{n(1 + \|x_n\|)},$$

hence the conclusion. ■

Generalization of 3.1₁. *Let* X *be a Banach space and* $\varphi : X \to \mathbf{R}$ *lower bounded, lower semicontinuous and Gâteaux differentiable. For every increasing function* $h : [0, +\infty) \to [0, +\infty)$, *a minimizing sequence* $(x_n)_{n\geq 1}$ *for* φ *exists such that*

$$\lim_{n\to\infty} [1 + h(\|x_n\|)]\,\|\varphi'_w(x_n)\| = 0$$

(*Zhong Cheng - Kui*, [71], 1).

■ In 3.1₀ take $\varepsilon = \frac{1}{n}$, $\lambda = \frac{1}{\sqrt{n}}$, $n \in \mathbf{N}$, one finds $(x_n)_{n\geq 1}$ such that $\inf \varphi(X) \leq \varphi(x_n) \leq$ $\leq \inf \varphi(X) + \frac{1}{n}$ and $[1 + h(\|x_n\|)]\,\|\varphi'_w(x_n)\| \leq \frac{1}{\sqrt{n}}$, whence the conclusion. ■

Remark. Taking $h : [0, +\infty) \to [0, +\infty)$, $h(r) = r$ we fall indeed over 3.1₁.

To the above generalization we add the next proposition.

3.1₂ Theorem. *Let* X *be a Banach space,* $\varphi : X \to \mathbf{R}$ *lower semicontinuous Gâteaux differentiable and* $\alpha := \lim_{r\to+\infty} (\inf_{\|x\|\geq r} \varphi(x))$. *If* $\alpha \in \mathbf{R}$, *there exists a sequence* $(y_n)_{n\geq 4}$ *of points in* X *such that* $\|y_n\| \to +\infty$, $\varphi(y_n) \to \alpha$ *and* $[1 + h(\|y_n\|)]\,\|\varphi'_w(y_n)\| \to 0$, *where* $h : [0, +\infty) \to [0, +\infty)$ *is increasing and* $\int_0^{+\infty} \frac{dr}{1 + h(r)} = +\infty$

(*Zhong Cheng-Kui*, [71], 2; *H. Brezis - L. Nirenberg* for $h = 0$, [9]).

■ Denote, for every $r \geq 0$, $\mu(r) := \inf_{\|x\|\geq r} \varphi(x)$. μ is increasing and (6_6) $\alpha = \lim_{r\to+\infty} \mu(r)$. Let ε be arbitrary from $\left(0, \frac{1}{3}\right)$. There is $\bar{r}$, (6_7) $\bar{r} \geq \frac{1}{\varepsilon}$ such that $((6_6))$

$$(6_8) \qquad \alpha - \varepsilon^2 \leq \mu(r) \quad \forall\, r \geq \bar{r}.$$

Because $\int_0^{+\infty} \frac{dr}{1+h(r)} = +\infty$, we may take r^*, $r^* > \bar{r}$ such that

$$(6_9) \qquad \int_{\bar{r}}^{r^*} \frac{dr}{1+h(r)} \geq 1$$

(being increasing, h is Riemann integrable on $[\bar{r}, r^*]$).

A partition

$$(6_{10}) \qquad \bar{r} =: r_N < r_{N-1} < \ldots < r_1 < r_0 := r^*$$

of the interval $[\bar{r}, r^*]$ exists with the property

$$(6_{11}) \qquad \left| \int_{\bar{r}}^{r^*} \frac{dr}{1+h(r)} - \sum_{k=0}^{N-1} \frac{1}{1+h(r_k)}(r_k - r_{k+1}) \right| \leq \varepsilon.$$

Consider the complete metric subspaces

$$M_k := \{x \in X : \|x\| \geq r_k\},\ k = \overline{0, N}.$$

Apply 1.1_1 to M_0 and φ (correctly, since $\inf_{\|x\|\geq r_0} \varphi(x) > -\infty$, see (6_8) which imposes this) with ε^2, $\lambda = \varepsilon$ and $u_0 = 0$, there is $z_0 := z_0(\varepsilon)$ in M_0 such that

$$(6_{12}) \qquad \mu(r_0) \leq \varphi(z_0) < \mu(r_0) + \varepsilon^2 \leq \alpha + \varepsilon^2$$

and

$$(6_{13}) \qquad \varphi(x) \geq \varphi(z_0) - \frac{\varepsilon}{1+h(\|z_0\|)}\|x - z_0\| \quad \forall\, x \in M_0.$$

Suppose $a_0)$ $\|z_0\| > r_0$. Take $x = z_0 + ty$ with $t > 0$ and $\|y\| = 1$. For sufficiently small t, from (6_{13}),

$$\frac{\varphi(z_0 + ty) - \varphi(z_0)}{t} > \frac{-\varepsilon}{1+h(\|z_0\|)}$$

and with a usual argument (see the proof of 3.1_0) one finds

$$(6_{14}) \qquad [1 + h(\|z_0\|)]\,\|\varphi'_w(z_0)\| \leq \varepsilon.$$

To the same conclusion (6_{14}) we are leading by the hypothesis

$$b_0)\ (6_{13})\ \textit{is verified}\ \forall\, x \in M_1.$$

Suppose now the situation

$$\textit{neither}\ a_0),\ \textit{nor}\ b_0)\ \textit{are satisfied}$$

that is

$$c_0)\ \|z_0\| = r_0$$

and

$$d_0)\ \exists\, x_1' \in M_1 \setminus M_0 \text{ such that } \varphi(x_1') < \varphi(z_0) - \frac{\varepsilon}{1 + h(\|z_0\|)}\|x_1' - z_0\|.$$

Apply 1.1_1 to M_1 and φ with ε^2, $\lambda = \varepsilon$, $u_0 = 0$ and $u = z_0$, there is $z_1 := z_1(\varepsilon)$ in M_1 such that

$$(6_{15}) \qquad \mu(r_1) \le \varphi(z_1) \le \varphi(z_0) \overset{(6_{12})}{\le} \alpha + \varepsilon^2 \overset{(6_8)}{\le} \mu(r_1) + 2\varepsilon^2$$

and

$$(6_{16}) \qquad \varphi(x) \ge \varphi(z_1) - \frac{\varepsilon}{1 + h(\|z_1\|)}\|x - z_1\| \ \forall\, x \in M_1.$$

When

$$\textit{either}\ a_1)\ \|z_1\| > r_1,\ \textit{or}\ b_1)\ (6_{16})\ \textit{is verified}\ \forall x \in M_2$$

then, as before, from (6_{16}),

$$(6_{17}) \qquad [1 + h(\|z_1\|)]\,\|\varphi_w'(z_1)\| \le \varepsilon.$$

Suppose now the situation

$$\textit{neither}\ a_1),\ \textit{nor}\ b_1)\ \textit{are verified},$$

i.e.

$$c_1)\ \|z_1\| = r_1$$

and

$$d_1)\ \exists\ x_2' \in M_2 \setminus M_1 \text{ such that } \varphi(x_2') < \varphi(z_1) - \frac{\varepsilon}{1 + h(\|z_1\|)}\|x_2' - z_1\|.$$

Prove

$$(6_{18}) \qquad \frac{\varepsilon}{1 + h(\|z_0\|)}\|z_0 - z_1\| \le \varphi(z_0) - \varphi(z_1).$$

Let $(x_n)_{n \ge 1}$ be the sequence defined in 1.1_1 with respect to M_1 and φ. We know that $\forall\, n \ge$ ≥ 1 ($x_1 = z_0$, ε^2, $\lambda = \varepsilon$)

$$\sum_{k=1}^{n} \frac{\varepsilon}{1+h(\|x_k\|)} \|x_k - x_{k+1}\| \le \varphi(z_0) - \varphi(x_{n+1}),$$

take the limit for $n \to \infty$ and because $\varliminf\limits_{n\to\infty} \varphi(x_n) \ge \varphi(z_1)$, one gets

$$(6_{19}) \qquad \sum_{n=1}^{\infty} \frac{\varepsilon}{1+h(\|x_n\|)} \|x_n - x_{n+1}\| \le \varphi(z_0) - \varphi(z_1).$$

Suppose (6_{20}) $\|x_n\| \le \|z_0\| (= r_0)$ $\forall\ n \ge 1$ and prove (6_{18}). h being increasing, we have

$$\frac{\varepsilon}{1+h(\|z_0\|)} \|z_0 - z_1\| \le \frac{\varepsilon}{1+h(\|x_1\|)} \|x_1 - x_2\| + \frac{\varepsilon}{1+h(\|z_0\|)} \|x_2 - z_1\| \overset{(6_{20})}{\le} \frac{\varepsilon}{1+h(\|x_1\|)} \|x_1 - x_2\| +$$

$$+ \frac{\varepsilon}{1+h(\|x_2\|)} \|x_2 - x_3\| + \frac{\varepsilon}{1+h(\|z_0\|)} \|x_3 - z_1\| \le \ . \ . \ . \le \sum_{k=1}^{n-1} \frac{\varepsilon}{1+h(\|x_k\|)} \|x_k - x_{k+1}\| +$$

$+ \frac{\varepsilon}{1+h(\|z_0\|)} \|x_n - z_1\|$, take the limit for $n \to \infty$, $\frac{\varepsilon}{1+h(\|z_0\|)} \|z_0 - z_1\| \le \sum\limits_{n=1}^{\infty} \frac{\varepsilon}{1+h(\|x_n\|)} \times$

$\times \|x_n - x_{n+1}\|$, because $x_n \to z_1$ and applying (6_{19}) one obtains (6_{18}).

Pass to (6_{20}), in fact to

$$(6_{21}) \qquad \|x_n\| \le \|z_0\| \quad \forall\ n \ge 2.$$

d_0) tells that $\exists\ n \ge 1$ such that $E_n \ne \varnothing$ (follow attentively the construction of the sequence (x_n) from 1.1_1). *A fortiori* $x_2 \in E_1$ $(= \{x \in M_1 : \varphi(x) < \varphi(z_0) - \frac{\varepsilon}{1+h(\|z_0\|)} \|x - z_0\|)$. (6_{13}) imposes then $x_2 \in M_1 \setminus M_0$, hence then $\|x_2\| < r_0$. Suppose now, *par absurdum*, that there exists $p > 2$ such that $\|x_k\| < r_0$ for $k = \overline{2, p-1}$ and $\|x_p\| \ge r_0$, that is $x_p \in M_0$. Then $x_p \ne$ $\ne x_{p-1}$, whence $x_2, \ldots, x_p$ are defined as in case II, consequently

$$(6_{22}) \qquad \varphi(x_p) < \varphi(x_{p-1}) - \frac{\varepsilon}{1+h(\|x_{p-1}\|)} \|x_{p-1} - x_p\| < \varphi(x_1) - \sum_{k=1}^{p-1} \frac{\varepsilon}{1+h(\|x_k\|)} \|x_k - x_{k+1}\|.$$

On the other hand, since $x_p \in M_0$ we have $((6_{13}))$

$$(6_{23}) \qquad \varphi(x_p) \ge \varphi(x_1) - \frac{\varepsilon}{1+h(\|x_1\|)} \|x_p - x_1\|.$$

By (6_{22}) and (6_{23}) we get

$$\sum_{k=1}^{p-1} \frac{\varepsilon}{1+h(\|x_k\|)} \|x_k - x_{k+1}\| < \frac{\varepsilon}{1+h(\|x_1\|)} \|x_p - x_1\| \le \frac{\varepsilon}{1+h(\|x_1\|)} \sum_{k=1}^{p-1} \|x_k - x_{k+1}\| \le \sum_{k=1}^{p-1} \frac{\varepsilon}{1+h(\|x_k\|)}$$

$\times \|x_k - x_{k+1}\|$ ($\|x_k\| < r_0 = \| z_0 \| = \| x_1\|$) and the contradiction is clear, hence (6_{21}), and therefore (6_{18}), is proved.

Apply 1.1_1 to M_2 and φ with ε^2, $\lambda = \varepsilon$, $u_0 = 0$ et $u = z_1$, there exists $z_2 := z_2(\varepsilon)$ in M_2 such that

$$(6_{24}) \qquad \mu(r_2) \leq \varphi(z_2) \leq \varphi(z_1) \overset{(6_{15})}{\leq} \alpha + \varepsilon^2 \leq \mu(r_2) + 2\varepsilon^2$$

and

$$(6_{25}) \qquad \varphi(x) \geq \varphi(z_2) - \frac{\varepsilon}{1+h(\|z_2\|)}\|x - z_2\| \ \forall\, x \in M_2.$$

Two cases are possible :

either a_2) $\| z_2 \| > r_2$, or b_2) (6_{25}) is verified $\forall x \in M_3$

or

c_2) $\| z_2 \| = r_2$ and d_2) $\exists\, x_3' \in M_3 \setminus M_2$ such that

$$\varphi(x_3') < \varphi(z_2) - \frac{\varepsilon}{1+h(\|z_2\|)}\|x_2' - z_2\|.$$

The first case supplies as before

$$[1 + h(\|z_2\|\,]\,\|\varphi_w'(z_2)\| \leq \varepsilon.$$

Repeat step by step the process and one gets the points $z_0, z_1,\ldots, z_N$ (look at the partition of $\left[\bar{r}, r^*\right]$, the begining of the proof) in the complete metric spaces $M_0, M_1, \ldots, M_N$ respectively. Prove that there is k, $0 \leq k \leq N-1$, such that the hypostasis *either* a_k, *or* b_k is true. Suppose *par absurdum* the contrary, that is $c_0) + d_0)$, $c_1) + d_1)$,…, $c_{N-1}) + b_{N-1})$ are true. Thus

$$(6_{26}) \qquad \|z_k\| = r_k,\ k = \overline{0, N-1}$$

and (see (6_{18}))

$$(6_{27}) \qquad \frac{\varepsilon}{1+h(\|z_k\|)}\|z_k - z_{k+1}\| \leq \varphi(z_k) - \varphi(z_{k+1}),\ k = \overline{0, N-1}.$$

Because (6_{27}),

$$(6_{28}) \qquad \sum_{k=0}^{N-1} \frac{\varepsilon}{1+h(\|z_k\|)}\|z_k - z_{k+1}\| \leq \varphi(z_0) - \varphi(z_N) \leq \alpha + \varepsilon^2 - \varphi(z_N) \overset{(6_8)}{\leq} 2\varepsilon^2.$$

But

$$1-\varepsilon \overset{(6_9)}{\leq} \int_{\bar{r}}^{r^*} \frac{dr}{1+h(r)} - \varepsilon \overset{(6_{11})}{\leq} \sum_{k=0}^{N-1} \frac{1}{1+h(r_k)}(r_k - r_{k+1}) = \sum_{k=0}^{N-1} \frac{1}{1+h(\|z_k\|)}(\|z_k\| - \|z_{k+1}\|) \leq$$

$$\leq \sum_{k=0}^{N-1} \frac{1}{1+h(\|z_k\|)} \|z_k - z_{k+1}\| \overset{(6_{28})}{\leq} 2\varepsilon^2,$$

hence $1-\varepsilon \leq 2\varepsilon^2$, a contradiction, since $\varepsilon < \frac{1}{3}$.

The proof of theorem 3.1_2 is now rapidely finished. The hypostasis *either* a_k), *or* b_k) means

$$\textit{either } \| z_k(\varepsilon) \| > r_k, \textit{ or } \varphi(x) \geq \varphi(z_k(\varepsilon)) - \frac{\varepsilon}{1+h(\|z_k(\varepsilon)\|)} \|x - z_k(\varepsilon)\| \ \forall x \in M_{k+1}.$$

But from this situation we get (see before)

$$(6_{29}) \qquad [1+h(\| z_k(\varepsilon) \|)] \|\varphi'_w(z_k(\varepsilon))\| \leq \varepsilon.$$

Moreover, we have (see for instance (6_{24})) $\mu(r_k) \leq \varphi(z_k(\varepsilon)) \leq \alpha + \varepsilon^2$, hence (see (6_8))

$$(6_{30}) \qquad \alpha - \varepsilon^2 \leq \varphi(z_k(\varepsilon)) \leq \alpha + \varepsilon^2.$$

Also (see (6_7))

$$(6_{31}) \qquad \| z_k(\varepsilon) \| > \frac{1}{\varepsilon}.$$

Take $\varepsilon = \frac{1}{n}$, $n \in \mathbf{N}$, $n \geq 4$ (in order to have $\varepsilon < \frac{1}{3}$) and $y_n := z_k\left(\frac{1}{n}\right)$. The sequence $(y_n)_{n\geq 4}$ verifies the demands of the statement as can be seen from (6_{29}), (6_{30}) and (6_{31}). ■

Continue with some corollaries of 3.1.

3.2 *Corollary. Let* X *be a Banach space and* $\varphi : X \to \mathbf{R}$ *lower bounded, lower semi-continuous and Gâteaux differentiable. Then for every minimizing sequence* $(u_n)_{n\geq 1}$ *for* φ, *there is* $(v_n)_{n\geq 1}$ *a minimizing sequence for* φ *such that*

$$(7) \qquad \lim_{n\to\infty} \| \varphi'_w(v_n) \| = 0,$$

$$(8) \quad \varphi(v_n) \leq \varphi(u_n) \ \forall\, n \geq 1, \qquad (9) \quad \lim_{n\to\infty} \| v_n - u_n \| = 0$$

(*I. Ekeland*).

■ Let be $\alpha := \inf \varphi(X)$. For every n of $\mathbf{N}$, take

$$\varepsilon_n := \begin{cases} \varphi(u_n) - \alpha, & \text{if } \varphi(u_n) - \alpha > 0, \\ \frac{1}{n}, & \text{if } \varphi(u_n) - \alpha = 0. \end{cases}$$

Apply 3.1. In the first case one finds v_n in X such that $\| \varphi_w'(v_n) \| \leq \sqrt{\varepsilon_n}$, $\varphi(v_n) \leq \varphi(u_n)$, $\|v_n - u_n\| \leq \sqrt{\varepsilon_n}$. In the second case, one finds v_n in X with $\| \varphi_w'(v_n) \| \leq \frac{1}{\sqrt{n}}$, $\varphi(v_n) \leq \varphi(u_n)$, $\|v_n - u_n\| \leq \frac{1}{\sqrt{n}}$. Clearly, the sequence $(v_n)_{n\geq 1}$ has the properties of the statement ($\varepsilon_n \to 0$ because (u_n) is minimizing for φ). ■

3.3 *Corollary. In the conditions of* 3.1, *if we add*

$$\exists\, \rho, k \in \mathbf{R}, \rho > 0 \textit{ such that } \varphi(x) \geq \rho\|x\| + k \ \forall\, x \in X,$$

then the set $\varphi_w'(X)$ *is dense in* ρS^*, $S^* := \{\xi \in X^* : \|\xi\| \leq 1\}$

(*I. Ekeland*, [30], 1).

■ Let be $\varepsilon > 0$ and ξ arbitrary in ρS^*. Show that u_ε exists in X such that $\| \varphi_w'(u_\varepsilon) - \xi\| \leq \leq \varepsilon$. Take the functional $\Phi: X \to \mathbf{R}$, $\Phi(x) = \varphi(x) - \xi(x)$. Φ is lower semicontinuous and Gâteaux differentiable. Φ is also lower bounded: $\|\xi\| \leq \rho \Rightarrow \xi(x) \leq \rho\|x\|$ and hence $\Phi(x) \geq \geq k\ \forall\, x$. Apply 3.1, $\exists\, u_\varepsilon$ such that $\| \Phi_w'(u_\varepsilon)\| \leq \varepsilon$ and $\Phi_w'(x) = \varphi_w'(x) - \xi$. ■

3.4 *Corollary. In the conditions of* 3.1, *if in addition* $\exists\, f : [0, +\infty) \to \mathbf{R}$ *continuous with* $\lim\limits_{t\to+\infty} \frac{f(t)}{t} = +\infty$ *and* $\varphi(x) \geq f(\|x\|)\ \forall\, x \in X$, *then*

$$\varphi'_w(X) \textit{ is everywhere dense}$$

(*I. Ekeland*, [30], 2).

■ It follows to prove that for every $\rho > 0$ $\varphi'_w(X)$ is dense in ρS^*. But $\exists\, t_0 > 0$ such that $t \geq t_0 \Rightarrow \frac{f(t)}{t} \geq \rho$, hence $\varphi(x) \geq \rho\|x\|$ for $\|x\| \geq t_0$. f being continuous on $[0, t_0)$, there exists k such that $f(t) \geq k$ on $[0, t_0)$, hence $\varphi(x) \geq \rho\|x\| + k\ \forall\, x \in X$, apply 3.3. ■

The proposition 3.1 may be relativized.

3.5 *Let* H *be a Hilbert space,* $\varphi: H \to \mathbf{R}$ *lower semicontinuous Gâteaux differentiable and* C *a nonempty, convex, closed subset on which* φ *is lower bounded and* $\Phi(C) \subset C$, $\Phi(u) := u - \varphi_w'(u)$. *Then* $\forall \varepsilon > 0$ *and* $\forall$ u *of* C *with* $\varphi(u) \leq \inf \varphi(C) + \varepsilon$, *there is* v_ε *in* C *such that*

$$\| \varphi_w'(v_\varepsilon) \| \leq \sqrt{\varepsilon}, \tag{10}$$

$$\varphi(v_\varepsilon) \leq \varphi(u) - \sqrt{\varepsilon}\, \|v_\varepsilon - u\|, \tag{11}$$

$$\|v_\varepsilon - u\| \leq \sqrt{\varepsilon} \tag{12}$$

(*H. Hofer*).

Attention to Riesz theorem which identifies H with H^* through an isometry!

■ C is a complete metric space, apply 1.1_0 with $\lambda = \sqrt{\varepsilon}$, $\exists\ v_\varepsilon$ in C which satisfies (11), (12) and moreover

(13) $$\varphi(v_\varepsilon) \leq \varphi(v) + \sqrt{\varepsilon}\ \| v_\varepsilon - v \| \quad \forall\ v \in C.$$

Take $v = (1-\lambda)v_\varepsilon + \lambda\Phi(v_\varepsilon)$, λ arbitrary from (0, 1). As $v = v_\varepsilon - \lambda\varphi'_w(v_\varepsilon)$, according to (13),

$$\frac{\varphi(v_\varepsilon - \lambda\varphi'_w(v_\varepsilon)) - \varphi(v_\varepsilon)}{\lambda} \geq -\sqrt{\varepsilon}\ \|\varphi'_w(v_\varepsilon)\|.$$

Taking the limit for $t \to 0+$ we get

$$\varphi'_w(v_\varepsilon)(-\varphi'_w(v_\varepsilon)) \geq -\sqrt{\varepsilon}\ \| \varphi'_w(v_\varepsilon) \|,$$

i.e.

$$-\| \varphi'_w(v_\varepsilon) \|^2 \geq -\sqrt{\varepsilon}\ \| \varphi'_w(v_\varepsilon) \|$$

and consequently (10). ■

3.2 may also be relativized with the same proof.

3.6 *In the hypotheses of* 3.5, *for every minimizing sequence* $(u_n)_{n\geq 1}$ *of* φ *in* C, *there is* $(v_n)_{n\geq 1}$ *minimizing sequence of* φ *in* C *such that*

$$\lim_{n\to\infty} \| \varphi'_w(v_n) \| = 0,$$

$$\varphi(v_n) \leq \varphi(u_n) \quad \forall\ n \geq 1, \quad \lim_{n\to\infty} \| v_n - u_n \| = 0.$$

Remark. Proposition 3.5 remains true when the Hilbert space H is replaced by a reflexive, strictly convex Banach space – one uses the concepts *metric gradient* and *duality mapping*, the proof remains the same (*Yang Li*, *Almost critical points for functionals defined on convex sets*, Annales de l'Université Bucarest, Mathématique, 1998, n°1, p. 121-126).

β - DERIVATIVE

The concept ([6]) is often used in the variational calculus. Here it is a suitable place for a rapid exposition.

Let X be a real normed space.

A nonempty set β of bounded parts of X, with the properties:

1° $\bigcup_{A\in\beta} A = X$,

2° $A \in \beta \Rightarrow -A \in \beta$ and $\lambda A \in \beta$ $(\lambda > 0)$,

3° for every A, B in β there exists C in β such that $A \subset C$ and $B \subset C$,

is named *bornology on* X.

For instance, the set β_F of bounded symmetrical ($M = -M$) parts of X and the set β_G of finite symmetrical parts of X are bornologies on X (for the justification in the case β_F use the balls).

β_1, β_2 being bornologies on X, the property

$$\forall \text{ A } \textit{from } \beta_1 \ \exists \text{ B } \textit{in } \beta_2 \textit{ such that } \text{A} \subset \text{B}$$

will be designated by

$$\beta_1 \prec \beta_2.$$

3.7 *We have*

$$\beta_G \prec \beta \prec \beta_F$$

for every bornology β *on* X.

■ *First inclusion.* Let $S := \{a_1 \ldots, a_N\}$ be in β_G. Take $A_1, \ldots, A_N$ in β such that $a_i \in A_i$, $i = \overline{1, N}$ (see 1°). For A_1, A_2 $\exists$ B_1 in β such that $A_1 \cup A_2 \subset B_1$ (see 3°). For B_1 and A_3 $\exists$ B_2 in β such that $B_1 \cup A_3 \subset B_2$ etc., for B_{N-2} and $A_N \,\exists\, B_{N-1}$ in β such that $B_{N-2} \cup A_N \subset B_{N-1}$. Clearly, $S \subset B_{N-1}$.

Second inclusion. S being arbitrary in β, take a ball including S. ■

Example 1. The set β_K of the compact, symmetrical parts of X is a bornology on X:

1° $\forall$ x of X the set $\{x\}$ is compact;

2° the homothety $x \to \lambda x$, $\lambda > 0$ is continuous;

3° $A, B \in \beta_K \Rightarrow A \cup B \in \beta_K$.

Thus, we have (3.7)

$$\beta_G \prec \beta_K \prec \beta_F.$$

Let β be a bornology on X and $f: X \to \overline{\mathbf{R}}$ locally finite in the point a (there is a neighborhood of a on which f is finite). By definition f is β - *differentiable in* a, if there exists φ in the dual X^* such that for every S in β we have

$$\text{(14)} \qquad \lim_{\substack{t\to 0 \\ h\in S}} \text{u} \frac{f(a+th) - f(a)}{t} = \varphi(h)$$

(uniform limit on S for $t \to 0$).

φ is the β - *derivative of* f *in* a, and is denoted

$$\nabla_\beta f(a).$$

The definition is correct because f is finite on a neighbourhood V of a and we have, $\forall\, h$ of S, $a + th \in V$ for t sufficiently small, because S is bounded. Underline that the property 2° of bornology allows to replace in (14) "$t \to 0$" by "$t \to 0+$" or by "$t \to 0-$" (change t by $-t$).

f β - *differentiable on a subset of* X has the usual meaning.

3.8 *Formula of finite increments. Let* β *be a bornology on* X *and* $f : X \to \overline{\mathbf{R}}$ β - *differentiable on the segment* [a, b]. *Then there exists* θ *in* (0, 1) *so that*

$$f(b) - f(a) = \nabla_\beta f(a + \theta(b - a))(b - a) \quad ([46]).$$

■ Take $F : [0, 1] \to \mathbf{R}$, $F(t) = f(a + t(b - a))$. Fix t from (0, 1). We have for δ, with $|\delta|$ sufficiently small for not getting out of $[a, b]$,

$$\frac{F(t+\delta) - F(t)}{\delta} = \frac{1}{\delta}[f(a + t(b-a) + \delta(b-a)) - f(a + t(b-a))]. \tag{15}$$

Let S be in β with $b - a \in S$ (see 1°). If in (15) we take the limit for $\delta \to 0$, we get, according to (14),

$$F'(t) = \nabla_\beta f(a + t(b-a))(b-a). \tag{16}$$

Similary, $F'(0) = \nabla_\beta f(a)(b-a)$, $F'(1) = \nabla_\beta f(b)(b-a)$. Thus, $\exists\, \theta$ in (0, 1) such that $F(1) - F(0) = F'(\theta)$ and one takes into account (16). ■

And now we disclose ∇_{β_F} and ∇_{β_G}, what will justify the use of the letters F and G.

3.9 *Let* $f : X \to \overline{\mathbf{R}}$ *be locally finite in a point* a. f *is* β_F - *differentiable in* a respectively β_G - *differentiable in* a, *iff* f *is Fréchet differentiable* respectively *Gâteaux differentiable in* a *and in this case*

$$\nabla_{\beta_F} f(a) = f'(a) \text{ respectively } \nabla_{\beta_G} f(a) = f'_w(a) \quad ([46]).$$

■ *First statement. Necessary.* Let be $\varphi := \nabla_{\beta_F} f(a)$. $\varepsilon > 0$ being arbitrary, we must find, to have $\varphi = f'(a)$, $\eta > 0$ such that

$$0 < \|h\| \le \eta \Rightarrow \frac{|f(a+h) - f(a) - \varphi(h)|}{\|h\|} \le \varepsilon. \tag{17}$$

Fix a closed ball $S_r(0)$ (this is contained in β_F !). According to (14), for εr $\exists\, \delta > 0$ such that

(18) $$0 < |t| \le \delta \Rightarrow \frac{|f(a+t\,h') - f(a) - t\,\varphi(h')|}{|t|} \le \varepsilon\, r \ \ \forall\, h' \in S_r(0).$$

Take $\eta = r\,\delta$ and verify (17). Let $h \neq 0$, $\|h\| \le r\,\delta$. Then $\frac{\|h\|}{r} \le \delta$ and since for $h' :=$ $= h\frac{r}{\|h\|}$ we have $\|h'\| = r$ and consequently $h' \in S_r(0)$, $t := \frac{\|h\|}{r}$ and h' verify (18). Replace this and obtain (17).

First statement. Sufficient. Let $\varphi := f'(a)$. Whatever being A in β_F, we must prove $\varphi(h) = \operatorname*{limu}_{\substack{t\to 0\\ h\in A}} \frac{f(a+t\,h) - f(a)}{t}$, of course this hold if, $S_R(0)$ being a closed ball which contains A, we have

(19) $$\varphi(h) = \operatorname*{limu}_{\substack{t\to 0\\ h\in S_R(0)}} \frac{f(a+t\,h) - f(a)}{t}.$$

Let be $\varepsilon > 0$. For $\frac{\varepsilon}{R}$ $\exists\, S$ a ball centred in 0 such that

(20) $$h \in S \Rightarrow |f(a+h) - f(a) - \varphi(h)| \le \frac{\varepsilon}{R}\,\|h\|.$$

Take $\delta > 0$ so that (21) $\delta\, S_R(0) \subset S$ and let t be in $\mathbf{R} \setminus \{0\}$ with $|t| \le \delta$ and h any from $S_R(0)$. Then $th \overset{(21)}{\in} S$, hence $|f(a+th) - f(a) - \varphi(th)| \le \frac{\varepsilon}{R}|t|\,\|h\| \le \varepsilon\,|t|$, $0 < |t| \le \delta \Rightarrow$ $\Rightarrow \left|\frac{f(a+t\,h) - f(a)}{t} - \varphi(h)\right| \le \varepsilon \ \ \forall\, h \in S_R(0)$, i.e. (19).

Second statement. Necessary. Let $\varphi := \nabla_{\beta_G} f(a)$. h_0 being arbitrary and fixed from X, we must prove

(22) $$\lim_{t\to 0} \frac{f(a+th_0) - f(a)}{t} = \varphi(h_0).$$

But for S from β_G with $h_0 \in S$ we have $\operatorname*{limu}_{\substack{t\to 0\\ h\in S}} \frac{f(a+t\,h) - f(a)}{t} = \varphi(h)$ and so, in particular, (22).

Second statement. Sufficient. Let be $\varphi := f'_w(a)$ and $S := \{h_1, \ldots, h_N\}$ arbitrary in β_G. Whatever $\varepsilon > 0$, $\exists\, \delta_i > 0$, $i = \overline{1,N}$ such that $0 < |t| \le \delta_i \Rightarrow \left|\frac{f(a+th_i) - f(a)}{t} - \varphi(h_i)\right| \le \varepsilon$, hence if $\delta := \min_{1\le i\le N} \delta_i$, then $0 < |t| \le \delta \Rightarrow \left| \frac{f(a+th)-f(a)}{t} - \varphi(h) \right| \le \varepsilon \;\; \forall\, h \in S$,

$$\underset{\substack{t\to 0\\ h\in S}}{\text{limu}} \frac{f(a+th)-f(a)}{t} = \varphi(h),\ \varphi = \nabla_{\beta_G} f(a). \ ■$$

Continue with other properties of the β - derivative.

3.10 *Let* β_1, β_2 *be bornologies on* X, $\beta_1 \prec \beta_2$. *If* $f: X \to \overline{\mathbf{R}}$ *is* β_2*- differentiable in* a, *then* f *is also* β_1*- differentiable in* a *and*

$$\nabla_{\beta_1} f(a) = \nabla_{\beta_2} f(a).$$

In particular, if f *is* β *- differentiable in* a, *then* f *is Gâteaux differentiable in* a *and*

$$\nabla_\beta f(a) = f'_w(a).$$

■ Let S be arbitrary in β_1. ∃ T in β_2 with $T \supset S$. But $\nabla_{\beta_2} f(a) = \underset{\substack{t\to 0\\ h\in T}}{\text{limu}} \frac{f(a+th)-f(a)}{t}$,

hence in particular $\nabla_{\beta_2} f(a) = \underset{\substack{t\to 0\\ h\in S}}{\text{limu}} \frac{f(a+th)-f(a)}{t}$ and consequently $\nabla_{\beta_1} f(a) =$

$= \nabla_{\beta_2} f(a)$.

For the second statement – take into account 3.7 and 3.9. ■

3.11 β *- derivative of the composed function. Let be* $f: X \to \mathbf{R}$ β*-differentiable in* x_0 *and* $F: \mathbf{R} \to \mathbf{R}$ *of* C^1 *class on* J *open interval with* $f(x_0) \in J$. *Then* $g := F \circ f$ *is* β*-differentiable in* x_0 *and*

$$\nabla_\beta\, g(x_0) = F'(f(x_0))\, \nabla_\beta f(x_0) \;\; ([46]).$$

■ Let S be fixed in β and h arbitrary in S. For t sufficiently near to 0, we have (for $t \to 0$ $f(x_0 + th) \to f(x_0)$, f being hemicontinuous in x_0), puting $\alpha := F'(f(x_0))$ and $\xi := \nabla_\beta f(x_0)$ $(\theta(t) \in (0, 1))$,

$$\frac{g(x_0+th)-g(x_0)}{t} - \alpha\xi(h) = \frac{F(f(x_0+th))-F(f(x_0))}{t} - \alpha\xi(h) = F'(f(x_0) + \theta(t)[f(x_0 +$$

$$+ th) - f(x_0)]\frac{f(x_0+th)-f(x_0)}{t} - \alpha\xi(h) = \{F'(f(x_0) + \theta(t)[f(x_0 + th) - f(x_0)]) \times$$

$$\times \frac{f(x_0+th)-f(x_0)}{t} - \alpha\frac{f(x_0+th)-f(x_0)}{t}\Big\} + \left[\alpha\frac{f(x_0+th)-f(x_0)}{t} - \alpha\xi(h)\right] = \{F'(f(x_0) +$$

$$+\,\theta(t)[f(x_0+th)-f(x_0)])-\alpha\}\,\frac{f(x_0+th)-f(x_0)}{t}+\alpha\left[\frac{f(x_0+th)-f(x_0)}{t}-\xi(h)\right].$$

The uniform limit on S of the second term for $t \to 0$ is equal to 0. Concerning the first term, this has the same property because, for $t \in (0, \delta)$, $\frac{f(x_0+t\,h)-f(x_0)}{t}$ is uniformly bounded in norm on S (the definition of $\nabla_\beta f(x_0)$, S is bounded) and the brace has the uniform limit on S for $t \to 0$ equal to 0 – use the continuity of F' on J and also $\operatorname{limu}_{\substack{t\to 0\\ h\in S}} [f(x_0 + th) - f(x_0)] = 0$: $f(x_0 + th) - f(x_0) = t\xi(h) + t\psi(t)$, $\operatorname{limu}_{\substack{t\to 0\\ h\in S}} t\xi(h) = 0$ (S is bounded!), $\operatorname{limu}_{\substack{t\to 0\\ h\in S}} t\psi(t) = 0$ (the definition of $\nabla_\beta f(x_0)$!). ■

Finally,

3.12 β - *differentiability of function series. Let be* $\sum_{n=1}^{\infty} f_n$, $f_n : U \to \mathbf{R}$, U *an open convex bounded set in a real normed space. If*

1° f_n *is*, $\forall\, n \in \mathbf{N}$, β-*differentiable on* U,

2° $\sum_{n=1}^{\infty} \nabla_\beta f_n$ *is uniformly convergent on* U,

3° *there is* x_0 *in* U *such that* $\sum_{n=1}^{\infty} f_n(x_0)$ *is convergent,*

then the given series is uniformly convergent on U, *its sum is* β-*differentiable on* U *and*

$$\nabla_\beta\Big(\sum_{n=1}^{\infty} f_n\Big)(x) = \sum_{n=1}^{\infty} \nabla_\beta f_n(x)\ \forall\, x \in U \text{ ([46]).}$$

It is enough to prove the corresponding proposition for sequences of functions.

3.12$_1$ *Let be* $(f_n)_{n\ge 1}$, f_n: U $\to$ **R**, U *an open convex bounded set in a real normed space. If*

1° $\exists\ \nabla_\beta f_n(x)\ \forall\, n \ge 1, \forall\, x \in U$,

2° $\exists\ \operatorname{limu}_{\substack{n\to\infty\\ x\in U}} \nabla_\beta f_n(x)$,

3° $\exists\ x_0$ *in* U *such that* $\lim_{n\to\infty} f_n(x_0)$ *exists, then* $(\mathrm{f_n})_{n\geq 1}$ *has a uniform limit* F *on* U, F *is* β*-differentiable on* U *and*

$$\nabla_\beta \mathrm{F(x)} = \underset{\substack{n\to\infty \\ x\in U}}{\mathrm{limu}}\ \nabla_\beta \mathrm{f_n(x)}\ \ ([46]).$$

■ *First statement*. Let be $\varepsilon > 0$. For $\frac{\varepsilon}{2r}$, $r := \operatorname{diam} U$, $\exists\, P$ in **N** such that

$$(23)\qquad m, n \geq P \Rightarrow \|\nabla_\beta f_n(x) - \nabla_\beta f_m(x)\| \leq \frac{\varepsilon}{2r}\quad \forall\, x \in U$$

(see 2°, apply 1.1 from V, § 2, [47], vol. II). For $\frac{\varepsilon}{2}$ $\exists\, N$ in **N**, $N \geq P$ such that (see 3°)

$$(24)\qquad m, n \geq N \Rightarrow |\, f_n(x_0) - f_m(x_0)\,| \leq \frac{\varepsilon}{2}.$$

We have, x being arbitrary in U,

$$(25)\ |\, f_n(x) - f_m(x)\,| \leq |\,[f_n(x) - f_m(x)] - [f_n(x_0) - f_m(x_0)]\,| + |\, f_n(x_0) - f_m(x_0)\,| =$$

$$= |\, g(x) - g(x_0)\,| + |\, f_n(x_0) - f_m(x_0)\,|,\ g := f_n - f_m.$$

But $g(x) - g(x_0) \overset{3.8}{=} \nabla_\beta\, g\,(x_0 + \theta(x - x_0))(x - x_0)$, $\theta \in (0, 1)$, hence the first term in the third member of (25) is majorized by $\|\nabla_\beta f_n(x_0 + \theta(x - x_0)) - \nabla_\beta f_m(x_0 + \theta(x - x_0))\| \times$ $\times \|x - x_0\|$ and hence, if $m, n \geq N$, by $\frac{\varepsilon}{2r} r$ (see (23)), therefore, by (25) and (24),

$$m, n \geq N \Rightarrow |\, f_n(x) - f_m(x)\,| \leq \frac{\varepsilon}{2} + \frac{\varepsilon}{2} = \varepsilon\ \forall\, x \in U,$$

$(f_n)_{n\geq 1}$ is uniformly convergent on U.

The second and the third statements. Let be F : $F(x) = \underset{\substack{n\to\infty \\ x\in U}}{\mathrm{limu}}\ f_n(x)$, a an arbitrary fixed point in U and (26) $g(a) := \lim_{n\to\infty} \nabla_\beta f_n(a)$ (see 2°). We must prove

$$(27)\qquad \nabla_\beta F(a) = g(a).$$

Let S be from β and $\varepsilon > 0$. S is bounded, let be $\|h\| \leq \rho$, $\rho > 0$, $\forall\, h$ of S. For $\frac{\varepsilon}{3\rho}$ $\exists\, N$ and n_1 in **N** so that

$$(28)\qquad n \geq N \Rightarrow \|\nabla_\beta f_n(x) - \nabla_\beta f_{n_1}(x)\| \leq \frac{\varepsilon}{3\rho}\quad \forall\, x \in U$$

and

(29) $$\|g(a) - \nabla_\beta f_{n_1}(a)\| \leq \frac{\varepsilon}{3\rho}$$

(see 2° and (26)).

But $\forall\, n \geq N$ and $\forall\, x$ of U, we have ($h := f_n - f_{n_1}$)

(30) $$|[f_n(x) - f_{n_1}(x)] - [f_n(a) - f_{n_1}(a)]| = |h(x) - h(a)| \overset{3.8}{=} |\nabla_\beta h(a + \theta(x-a))(x -$$

$$- a)| \leq \|\nabla_\beta f_n(a+\theta(x-a)) - \nabla_\beta f_{n_1}(a+\theta(x-a))\| \|x-a\| \overset{(28)}{\leq} \frac{\varepsilon}{3\rho}\|x-a\|$$

and taking in (30) the limit for $n \to \infty$ we get

(31) $$|[F(x) - f_{n_1}(x)] - [F(a) - f_{n_1}(a)]| \leq \|x-a\| \frac{\varepsilon}{3\rho} \quad \forall\, x \in U.$$

For $\frac{\varepsilon}{3}$ by the definition of the β-derivative we get $\delta > 0$ such that

(32) $$0 < t < \delta \Rightarrow | f_{n_1}(a + th) - f_{n_1}(a) - t\nabla_\beta f_{n_1}(a)(h) | \leq t\frac{\varepsilon}{3}, \forall\, h \in S.$$

Suppose δ sufficiently small such that $a + th \in U$, then, by (29), (31) and (32),

$$0 < t < \delta,\ h \in S \Rightarrow |F(a+th) - F(a) - tg(a)(h)| \leq |[F(a+th) - F(a)] - [f_{n_1}(a+th) -$$

$$- f_{n_1}(a)]| + | f_{n_1}(a+th) - f_{n_1}(a) - t\nabla_\beta f_{n_1}(a)(h)| + | t\nabla_\beta f_{n_1}(a)(h) - tg(a)(h)| \leq t\frac{\varepsilon}{3} +$$

$$+ t\frac{\varepsilon}{3} + t\frac{\varepsilon}{3} = t\varepsilon$$

and, consequently, we have obtained:

$$\forall\, \varepsilon > 0\ \exists\, \delta > 0 \text{ such that } 0 < t < \delta \Rightarrow \left|\frac{F(a+th) - F(a)}{t} - g(a)(h)\right| \leq \varepsilon\ \forall\, h \in S, \text{ i.e. (27).}$$ ■

Remark. The propositions 3.1 – 3.6 holds true if one replaces Gâteaux derivative by β-derivative, where β is a bornology on X – 3.10 provides us this statement. In the case of Fréchet derivative, the demand "φ lower semicontinuous" becomes, clearly, useless.

4

Palais - Smale type conditions

Let X be a real normed space, β a bornology on X (see §3) and $\varphi: X \to \mathbf{R}$.

Definitions. φ verifies the *Palais - Smale condition*, PS, with respect to β when, $\forall\ (u_n)_{n\geq 1}$ a sequence of points in X for which

(1) $$(\varphi(u_n))_{n\geq 1} \text{ is bounded and } \lim_{n\to\infty} \|\nabla_\beta \varphi(u_n)\| = 0,$$

this sequence has a convergent subsequence.

Let c be in $\mathbf{R}$. φ verifies the *Palais - Smale condition* (respectively the *weak Palais - Smale condition*) *on the level* c, $(PS)_c$, with respect to β, if $\forall\ (u_n)_{n\geq 1}$ a sequence of points in X such that

(2) $$\lim_{n\to\infty} \varphi(u_n) = c \text{ and } \lim_{n\to\infty} \|\nabla_\beta \varphi(u_n)\| = 0,$$

this sequence has a convergent subsequence (respectively c is a critical value for φ, that is $c = \varphi(u_0)$ and $\nabla_\beta \varphi(u_0) = 0$).

Let c be in $\mathbf{R}$ and F a nonempty subset of X. φ verifies the *Palais - Smale condition on the level* c *around* F (or *relative to* F), $(PS)_{c,F}$, with respect to β, when $\forall\ (u_n)_{n\geq 1}$ a sequence of points in X for which

(3) $$\lim_{n\to\infty} \varphi(u_n) = c,\ \lim_{n\to\infty} \|\nabla_\beta \varphi(u_n)\| = 0 \text{ and } \lim_{n\to\infty} \text{dist}(u_n, F) = 0,$$

this sequence has a convergent subsequence ([34], p.5)

Later, other Palais - Smale type conditions will appear.

Note now

4.1 φ *verifies* PS $\Leftrightarrow \varphi$ *verifies* $(\mathrm{PS})_c$ *for every* c *from* **R**.

■$\Rightarrow$. Let $(u_n)_{n\geq 1}$ be such that $\lim_{n\to\infty} \varphi(u_n) = c$ and $\lim_{n\to\infty}\|\nabla_\beta\varphi(u_n)\| = 0$. The sequence $(u_n)_{n\geq 1}$ has a convergent subsequence, since the sequence $(\varphi(u_n))_{n\geq 1}$ is bounded. $\Leftarrow$. Let $(u_n)_{n\geq 1}$ be such that $(\varphi(u_n))_{n\geq 1}$ is bounded and $\lim_{n\to\infty}\|\nabla_\beta\varphi(u_n)\| = 0$. $(\varphi(u_n))_{n\geq 1}$ has a convergent subsequence $(\varphi(u_{k_n}))_{n\geq 1}$, $\varphi(u_{k_n}) \to c$. According to the hypothesis, $(u_{k_n})_{n\geq 1}$ has a convergent subsequence $(u_{l_{k_n}}))_{n\geq 1}$, clearly also a subsequence of $(u_n)_{n\geq 1}$. ■

$(\mathrm{PS})_c$

4.2 *Let* X *be a Banach space and* φ: X $\to$ **R** *lower bounded, lower semicontinuous and Gâteaux differentiable. If* φ *verifies* $(\mathrm{PS})_c$, *where* c = inf f (X), *then* φ *has a global minimum point* u_0 *and* $\varphi'_w(u_0) = 0$ ([44]).

■ Let $(v_n)_{n\geq 1}$ be a minimizing sequence for φ, i.e. $\varphi(v_n) \to c$. A minimizing sequence $(u_n)_{n\geq 1}$ for φ exists, such that $\|\varphi'_w(u_n)\| \to 0$ (§3, 3.2). $(\mathrm{PS})_c$ supplies a convergent subsequence $(u_{k_n})_{n\geq 1}$, $u_{k_n} \to u_0$. Then $\underline{\lim}_{n\to\infty} \varphi(u_{k_n}) \geq \varphi(u_0)$ and because $\varphi(u_{k_n}) \to c$, we get $c \geq \varphi(u_0)$, $c = \varphi(u_0)$.

We also have $\varphi'_w(u_0) = 0$. Indeed, let h be arbitrary fixed, $\|h\| = 1$. For every $\varepsilon > 0\ \exists\ \delta > 0$ such that

$$t \in (0, \delta) \Rightarrow 0 \leq \varphi(u_0 + th) - \varphi(u_0) \leq t\varphi'_w(u_0)(h) + t\varepsilon,$$

hence $\varphi'_w(u_0)(h) \geq -\varepsilon$, take the limit for $\varepsilon \to 0$, $\varphi'_w(u_0)(h) \geq 0\ \ \forall h, \|h\| = 1$. Then we also, clearly, have $\varphi'_w(u_0)(h) \leq 0\ \ \forall h, \|h\| = 1$, hence $\varphi'_w(u_0) = 0$. ■

4.3 *In the conditions of* 3.5, *if* φ *also verifies a* $(\mathrm{PS})_c$ *condition on* C, *then* φ *attains its infimum on* C *in a point* u_0 *and* $\varphi'_w(u_0) = 0$ ([44]).

■ 3.6 supplies a sequence $(u_n)_{n\geq 1}$ of points in C, such that $\lim_{n\to\infty} \varphi(u_n) = \inf \varphi(C)$ and

$\lim\limits_{n\to\infty}\|\varphi'_w(u_n)\| = 0$. Consequently, by $(PS)_c$, we get a subsequence $(u_{k_n})_{n\geq 1}$ convergent in C, $u_{k_n} \to u_0$. Then $\underline{\lim}_{n\to\infty} \varphi(u_{k_n}) \geq \varphi(u_0)$, which imposes $\varphi(u_0) = \inf \varphi(C)$.

Pass to the second statement. Take $v := (1 - t)u_0 + t\,\Phi(u_0)$ (see the proof of 3.5). $v = u_0 - t\varphi'_w(u_0)$, hence $\varphi(u_0 - t\varphi'_w(u_0)) \geq \varphi(u_0)$ $\forall\, t$ from (0, 1). But $\varphi'_w(u_0)(h) =$

$= \lim\limits_{t\to 0+} \dfrac{\varphi(u_0 + th) - \varphi(u_0)}{t}$ $\forall\, h$, therefore taking $h = -\varphi'_w(u_0)$ one finds $\varphi'_w(u_0)(-\varphi'_w(u_0)) \geq$

≥ 0, hence $-\|\varphi'_w(u_0)\|^2 \geq 0$ and hence $\varphi'_w(u_0) = 0$. ■

Pass now to the critique of PS. The following two propositions, first due to H. Brezis - L. Nirenberg and second to N. Ghoussoub, show that PS is sufficiently restrictive.

4.4 *Let* X *be a Banach space and* $\varphi : X \to \mathbf{R}$ *lower bounded, lower semicontinuous and Gâteaux differentiable. If* φ *verifies* PS, *then* φ *is coercive* ([9]).

Explanation. φ coercive $\overset{\text{def.}}{\Leftrightarrow} \lim\limits_{\|u\|\to+\infty} \varphi(u) = +\infty$.

■ Suppose the contrary *par absurdum*, that is $c_1 := \underline{\lim}_{\|u\|\to+\infty} \varphi(u) \in \mathbf{R}$ (φ is lower bounded!). Show that φ does not verify $(PS)_{c_1}$, consequently also not PS (4.1) and here it is the contradiction.

We have $c_1 = \lim\limits_{r\to+\infty} \mu(r), \mu(r) := \inf\limits_{\|u\|\geq r} \varphi(u)$. Obviously, μ is increasing. There is, for each n in $\mathbf{N}$, $\rho_n, \rho_n \geq n$ such that (4) $c_1 - \dfrac{1}{n} \leq \mu(r)$ $\forall\, r \geq \rho_n$. Denote $F_n := \{u \in X : \|u\| \geq \rho_n\}$. F_n being closed, it is a complete metric subspace. Let u_n be, $\|u_n\| \geq 2\rho_n$ such that

$$(5) \qquad \varphi(u_n) < \mu(2\rho_n) + \frac{1}{n}\left(\leq c_1 + \frac{1}{n}\right).$$

Apply to F_{2n} Ekeland principle 1.1 $\left(u_n, \varepsilon = \dfrac{1}{n}, \lambda = \dfrac{1}{\sqrt{n}}\right)$, $\exists v_n$, (6) $\|v_n\| \geq 2\rho_n$ with the properties

$$(7) \qquad \varphi(v_n) \leq \varphi(u) + \frac{1}{\sqrt{n}}\|v_n - u\|\ \forall u \in F_{2n}, \|v_n - u_n\| \leq \frac{1}{\sqrt{n}}.$$

Then

$$c_1 - \frac{1}{n} \overset{(4)}{\leq} \mu(\rho_n) \leq \varphi(v_n) \overset{(7)}{\leq} \varphi(u_n) + \frac{1}{\sqrt{n}}\|v_n - u_n\| \overset{(5),(7)}{\leq} c_1 + \frac{1}{n} + \frac{1}{\sqrt{n}}\frac{1}{\sqrt{n}},$$

consequently $\lim\limits_{n\to\infty} \varphi(v_n) = c_1$. Moreover, since $v_n \overset{(6)\ o}{\in} F_n$, by (7) we get, with the same proof as in 3.1, $\|\varphi'_w(v_n)\| \leq \frac{1}{\sqrt{n}}$, hence $\lim\limits_{n\to\infty}\|\varphi'_w(v_n)\| = 0$. However $(v_n)_{n\geq 1}$ has no convergent subsequence, since $\lim\limits_{n\to\infty}\|v_n\| \overset{(6)}{=} +\infty$ as $\lim\limits_{n\to\infty} \rho_n = +\infty$ $(\rho_n \geq n)$, $(PS)_{c_1}$ is not satisfied, a contradiction. ■

Remark 1. The proof of 4.4 in [34] (Corollary 1.5) is incorrect.

Remark 2. The last proposition can be reformulated as follows (see 4.1):

In the assumptions of 4.4, *if* φ *verifies* $(PS)_c$ *for every* $c \in \mathbf{R}$, *then* φ *is coercive.*

4.5 *Let* X *be a Banach space and* $\varphi: X \to \mathbf{R}$ *lower semicontinuous, Gâteaux differentiable and verifying* PS. *If* u_0 *is a local minimum point for* φ, *then there is* $\varepsilon > 0$, *which has one of following properties*:

1° $\exists\ \alpha$ *in* $(0, \varepsilon)$ *such that* $\varphi(u_0) < \inf\{\varphi(u): \|u - u_0\| = \alpha\}$

or

2° $\forall\ \alpha$ *from* $(0, \varepsilon)$ $\exists\ u_\alpha$ *a local minimum point for* φ *with* $\|u_\alpha - u_0\| = \alpha$ *and* $\varphi(u_\alpha) = \varphi(u_0)$ ([34], Corollary 1.6).

■ Let $\varepsilon > 0$ be such that $\|u - u_0\| \leq \varepsilon \Rightarrow \varphi(u_0) \leq \varphi(u)$. If 1° is not verified, then $\forall\ \alpha$ from $(0, \varepsilon)$ we have

(8) $$\varphi(u_0) = \inf\{\varphi(u) : \|u - u_0\| = \alpha\}.$$

Fix α in $(0, \varepsilon)$, take δ such that $0 < \alpha - \delta < \alpha + \delta < \varepsilon$ and consider the closed set $F := \{u \in X : \alpha - \delta \leq \|u - u_0\| \leq \alpha + \delta\}$. Take a sequence $(\varepsilon_n)_{n\geq 1}$, $\varepsilon_n > 0$, $\varepsilon_n \to 0$ and draw out, according to (8), a sequence $(u_n)_{n\geq 1}$,

(9) $$\|u_n - u_0\| = \alpha,\ \varphi(u_n) < \varphi(u_0) + \varepsilon_n^2\ \forall n.$$

Because $\varphi(u_n) \geq \varphi(u_0)$, we have also

(10) $$\lim_{n\to\infty} \varphi(u_n) = \varphi(u_0).$$

Apply to $\varphi|F$, for every $n \in \mathbf{N}$, Ekeland principle 1.1 ($u_n \overset{(9)\ o}{\in} F\ \ \forall n$) with $\varepsilon = \varepsilon_n^2$ and $\lambda = \varepsilon_n$, $\exists\ v_n \in F$ such that

(11) $\varphi(v_n) \le \varphi(u) + \varepsilon_n \|v_n - u\| \quad \forall u \in F, \quad \varphi(v_n) \le \varphi(u_n), \quad \|v_n - u_n\| \le \varepsilon_n$.

$v_n \in \overset{o}{F}$ for sufficiently big n, since $\|v_n - u_n\| \le \varepsilon_n, \varepsilon_n \to 0$ and one finds, using (11), $\|\varphi'_w(v_n)\| \le \varepsilon_n$ (see the proof of 3.1), hence $\lim_{n\to\infty} \|\varphi'_w(v_n)\| = 0$. Moreover, $(\varphi(v_n))_{n\ge1}$ is bounded because $(\varphi(u_n))_{n\ge1}$ is bounded being convergent ((10)). Apply PS, $(v_n)_{n\ge1}$ has a convergent subsequence $(v_{k_n})_{n\ge1}$, let $u_\alpha := \lim_{n\to\infty} v_{k_n}$. We have $u_\alpha \in \overset{o}{F}$ because $\|u_\alpha - u_0\| = \alpha : \|u_{k_n} - v_{k_n}\| \overset{(11)}{\le} \varepsilon_{k_n}$, hence $u_{k_n} \to u_\alpha$ whereas $\|u_{k_n} - u_0\| \overset{(9)}{=} \alpha$. Thus we take in (11) the lower limit for $n \to \infty$, $\varphi(u_\alpha) \le \underline{\lim}_{n\to\infty} \varphi(v_{k_n}) \le \varphi(u)\ \forall\ u \in F$, hence u_α is a local minimum point for φ. Moreover, $\varphi(u_\alpha) \overset{(11)}{\le} \underline{\lim}_{n\to\infty} \varphi(u_{k_n}) = \lim_{n\to\infty} \varphi(u_{k_n}) \overset{(10)}{=} \varphi(u_0)$, $\varphi(u_\alpha) \le$ $\le \varphi(u_0)$ and on the other hand $\varphi(u_0) \overset{(8)}{\le} \varphi(u_\alpha)$ because $\|u_\alpha - u_0\| = \alpha$, therefore $\varphi(u_\alpha) =$ $= \varphi(u_0)$. ■

Remark concerning 4.5. If u_0 is a strictly local minimum point for φ, then 2° is prohibit and hence $\exists\ \alpha$ in $(0, \varepsilon)$ such that $\varphi(u_0) < \inf\{\varphi(u) : \|u - u_0\| = \alpha\}$.

Specification. The proof shows that ε can be arbitrary in an interval $(0, \varepsilon_0]$, ε_0 such that $\|u_0 - u\| \le \varepsilon_0 \Rightarrow \varphi(u_0) \le \varphi(u)$.

$(PS)_{c,F}$

This condition $(PS)_{c,F}$ attenuates somewhat deficiencies of the PS condition. We will see that through the minimization on F of a functional (*minimization with constraints*) it can obtain global critical points of this.

As a preliminary –

4.6 *Let* H *be a Hilbert space,* φ: H $\to$ **R** *of* C^1*- Fréchet class and* F *a closed subset of* H *such that*:

for every u *from* F *with* $\varphi'(u) \ne 0$, *there is, for sufficiently small* $r > 0$, $f_u : S_r(0) \to$ **R**

Fréchet differentiable such that, denoting $g_u(\delta) = f_u\left(\delta \frac{\varphi'(u)}{\|\varphi'(u)\|}\right)$, $\delta \in [0, r]$, *we have*

(12) $$g_u(0) = 1 \text{ and } g_u(\delta)\left(u - \delta \frac{\varphi'(u)}{\|\varphi'(u)\|}\right) \in F.$$

Then, if φ *is lower bounded on* F, *for every* $(v_n)_{n\geqslant 1}$ *a minimizing sequence for* φ *on* F *there exists a sequence* $(u_n)_{n\geqslant 1}$ *in* F *such that*

(13) $$\|\varphi'(u_n)\| \le \sqrt{\varepsilon_n}\left[1 + \|u_n\| \left|g'_{u_n}(0)\right|\right] + \left|g'_{u_n}(0)\right| \; |\varphi'(u_n)(u_n)|,$$

(14) $$\varphi(u_n) \le \varphi(v_n) \; \forall n,$$

(15) $$\lim_{n\to\infty} \|u_n - v_n\| = 0,$$

where $\varepsilon_n > 0$ *and* $\varepsilon_n \to 0$ ([34], Lemma 1.9).

■ Denote $c := \inf \varphi(F)$ and let n be from **N**. For $\varepsilon_n := \varphi(v_n) - c + \frac{1}{n}$, hence $\varepsilon_n > 0$, we have $\varphi(v_n) < c + \varepsilon_n$. Apply Ekeland principle 1.1 with $\lambda = \sqrt{\varepsilon_n}$, $\exists\, u_n$ in F with known properties. Thus we get the sequence $(u_n)_{n\geqslant 1}$ satisfying (14), (15) ($\|u_n - v_n\| \le \sqrt{\varepsilon_n}$, $\varepsilon_n \to$ $\to 0$) and

(16) $$\varphi(v) \ge \varphi(u_n) - \sqrt{\varepsilon_n}\|v - u_n\| \quad \forall\, v \in F.$$

Verify (13). It is enough to work in the assumption $\|\varphi'(u_n)\| > 0 \;\forall\; n$. Thus apply the hypothesis made in the statement with respect to F with $u = u_n$ and denoting, for $\delta \in (0, r]$, $v_\delta := g_{u_n}(\delta)\left(u_n - \delta \frac{\varphi'(u_n)}{\|\varphi'(u_n)\|}\right)$ ($\in F$), replace v_δ in (16) and find, taking into account the Fréchet derivative definition,

$$\sqrt{\varepsilon_n}\|v_\delta - u_n\| \ge \varphi(u_n) - \varphi(v_\delta) = \varphi'(v_\delta)(u_n - v_\delta) + o(\delta),$$

where $\lim_{\delta\to 0} \frac{o(\delta)}{\delta} = 0$, hence

(17) $$\sqrt{\varepsilon_n}\|v_\delta - u_n\| \ge \left[1 - g_{u_n}(\delta)\right]\varphi'(v_\delta)(u_n) + \delta g_{u_n}(\delta)\varphi'(v_\delta)\left(\frac{\varphi'(u_n)}{\|\varphi'(u_n)\|}\right) + o(\delta).$$

But $\lim_{\delta\to 0} \frac{1 - g_{u_n}(\delta)}{\delta} = \lim_{\delta\to 0} \frac{g_{u_n}(0) - g_{u_n}(\delta)}{\delta} = -g'_{u_n}(0)$, $\lim_{\delta\to 0} v_\delta = u_n$, $\lim_{\delta\to 0} \frac{1}{\delta}\|v_\delta - u_n\| =$ $= \left\|g'_{u_n}(0)u_n - \frac{\varphi'(u_n)}{\|\varphi'(u_n)\|}\right\| \le 1 + \left|g'_{u_n}(0)\right|\|u_n\|$, consequently if in (17) it divides by δ and takes the limit for $\delta \to 0$ one obtains

$$\sqrt{\varepsilon_n}\left(1+\left|g'_{u_n}(0)\right|\|u_n\|\right) \ge -g'_{u_n}(0)\varphi'(u_n)(u_n)+\varphi'(u_n)\left(\frac{\varphi'(u_n)}{\|\varphi'(u_n)\|}\right)$$

and it remains only to remark that $\varphi'(u_n)\left(\frac{\varphi'(u_n)}{\|\varphi'(u_n)\|}\right) = \|\varphi'(u_n)\|$ in order to get *a fortiori* (13).

And now the promised proposition.

Notation. $\varphi: X \to \mathbf{R}$ β-differentiable, $c \in \mathbf{R} \Rightarrow$

$$K_c(\varphi) := \{x \in X : \varphi(x) = c, \nabla_\beta\varphi(x) = 0\}.$$

4.7 *Let* H *be a Hilbert space,* φ: H $\to$ **R** *of* C^1*-Fréchet class and* F *a nonempty convex closed subset such that* (I – φ')(F) $\subset$ F, I *the identity map. If* φ *is lower bounded on* F, *then for every* $(v_n)_{n\geq 1}$ *a minimizing sequence for* φ *on* F, *there is a sequence* $(u_n)_{n\geq 1}$ *in* F *such that*

$$\varphi(u_n) \le \varphi(v_n)\ \forall\ n,\ \lim_{n\to\infty}\|u_n - v_n\| = 0,\ \lim_{n\to\infty}\|\varphi'(u_n)\| = 0\,.$$

Moreover, if φ *satisfies* $(PS)_{c,F}$ *where* c = inf φ(F), *then*

$$F \cap K_c(\varphi) \ne \Phi\,.$$

■ Apply 4.6 with g_u = 1.(12) is indeed satisfied: if $u \in F$ and $\varphi'(u) \ne 0$, then, F being convex,

$$u-\delta\frac{\varphi'(u)}{\|\varphi'(u)\|} = \left(1-\frac{\delta}{\|\varphi'(u)\|}\right)u+\frac{\delta}{\|\varphi'(u)\|}(I-\varphi')(u) \in F\,.$$

Let $(u_n)_{n\geq 1}$ be the sequence given by the statement. $c \le \varphi(u_n) \le \varphi(v_n)\ \forall\ n$, hence $\varphi(u_n) \to$ $\to c$. Since $g'_{u_n}(0) = 0$, $\|\varphi'(u_n)\| \overset{(13)}{\le} \sqrt{\varepsilon_n}$, hence $\|g'(u_n\| \to 0$, clearly dist $(u_n, F) = 0$, consequently $(u_n)_{n\geq 1}$ has a convergent subsequence $(u_{k_n})_{n\geq 1}$, $u_{k_n} \to u_0 \in F$. This implies $\|\varphi'(u_{k_n})\| \to \|\varphi'(u_0)\| = 0$, u_0 is a global critical point of φ contained in F. ■

CRITICAL POINTS AND WEAK SOLUTIONS FOR ELLIPTIC TYPE EQUATION

Open set of C^1 class in $\mathbf{R}^N$. We use the notations (the norm is the Euclidean norm from $\mathbf{R}^{N-1}$):

$$\mathbf{R}_+^N = \{x = (x', x_N): x_N > 0\},$$

$$Q = \{x = (x', x_N) : \|x'\| < 1, |x_N| < 1\},$$

$$Q_+ = Q \cap \mathbf{R}_+^N$$

$$Q_0 = \{x = (x', x_N) : \|x'\| < 1, x_N = 0\}.$$

Let Ω be an open nonempty set of $\mathbf{R}^N$, $\Omega \neq \mathbf{R}^N$ and $\partial\Omega$ its boundary. By definition, Ω is of C^1 *class* if $\forall\, x$ from $\partial\Omega\ \exists\ U$ a neighbourhood of x in $\mathbf{R}^N$ and $f : Q \to U$ one-to-one such that $f \in C^1(\overline{Q})$, $f^{-1} \in C^1(\overline{U})$, $f(Q_+) = U \cap \Omega$, $f(Q_0) = U \cap \partial\Omega$.

Weak solution. Let Ω be an open bounded nonempty set in $\mathbf{R}^N$, $N > 1$, $f : \Omega \times \mathbf{R}^N \to$ $\to \mathbf{R}$ and $u_0 \in H_0^1(\Omega)\left(= W_0^{1,2}(\Omega)\right)$. Consider the problem

$$(*) \quad \begin{cases} -\Delta u = f(x,u) & \text{in } \Omega, \\ u = u_0 & \text{on } \partial\Omega. \end{cases}$$

Actually $u = u_0$ means $u\,|_{\partial\Omega} = u_0$, where γ: $u \to u\,|_{\partial\Omega}$ is the *trace operator*, a continuous linear mapping from $W^{1,p}(\Omega)$ in $\mathbf{L}^p(\partial\Omega)$, $p \in [1, +\infty)$. We have $\gamma^{-1}(0) = W_0^{1,p}(\Omega)$ $\left(= \overline{C_c^1(\Omega}^{W^{1,p}(\Omega)} \right)$ and $u \in W^{1,p}(\Omega) \cap C(\overline{\Omega}) \Rightarrow \gamma(u) = u\,|\,\partial\Omega$.

$\overline{u}$ from $H_0^1(\Omega)$ is by definition a *weak solution* for (*) if $\overline{u} = u_0$ on $\partial\Omega$ and

$$(18) \qquad \int_\Omega \nabla\overline{u} \cdot \nabla v dx - \int_\Omega f(x, \overline{u}(x)) v dx = 0 \quad \forall\, v \in C_c^\infty(\Omega)$$

(∇w, the weak gradient, is equal to $\left(\dfrac{\partial w}{\partial x_1}, \ldots, \dfrac{\partial w}{\partial x_N}\right)$, $\dfrac{\partial w}{\partial x_i}$ the weak derivatives).

Nemytski operator. Let be $\mathbf{R}^N$, $N \geq 1$, μ the Lebesgue measure in $\mathbf{R}$, Ω open nonempty Lebesgue measurable (L. m.) and $\mathcal{M}(\Omega) := \{u : \Omega \to \mathbf{R} |\ u$ Lebesgue measurable (L. m.)$\}$.

By definition $f : \Omega \times \mathbf{R} \to \mathbf{R}$ is a *Carathéodory function* if

1° $f(\cdot, s)$ is L.m. $\forall\, s \in \mathbf{R}$,

2° $f(x, \cdot)$ is continuous $\forall\, x \in \Omega \setminus A$, $\mu(A) = 0$.

In this case, for every u from $\mathcal{M}(\Omega)$ it can consider the function $N_f : \mathcal{M}(\Omega) \to \mathcal{M}(\Omega)$,

$$N_f u : N_f u(x) = f(x, u(x)),$$

Nemytskii operator.

Suppose that $\mu(\Omega) < +\infty$. Then $u_n(x) \underset{x\in\Omega}{\overset{\mu}{\to}} u_0(x) \Rightarrow N_f u_n(x) \underset{x\in\Omega}{\overset{\mu}{\to}} N_f u_0(x)$.

Suppose that f satisfies the *growth condition* :

(19) $|f(x,s)| \le c|s|^{p-1} + \beta(x)$, $\forall\, x \in \Omega \setminus A$ with $\mu(A) = 0$, $\forall\, s \in \mathbf{R}$, where $c \ge 0$, $p > 1$ and $\beta \in \mathbf{L}^q(\Omega)$, $q \in [1, +\infty]$.

Then

(20) $N_f(\mathbf{L}^{(p-1)\,q}(\Omega)) \subset \mathbf{L}^q(\Omega)$;

(21) N_f is continuous ($q < +\infty$) and bounded on $\mathbf{L}^{(p-1)q}(\Omega)$;

(22) If Ω is bounded and $\dfrac{1}{p} + \dfrac{1}{q} = 1$, then $N_f(\mathbf{L}^p(\Omega)) \subset \mathbf{L}^q(\Omega)$ with N_f continuous;

moreover, $N_F(\mathbf{L}^p(\Omega)) \subset \mathbf{L}^1(\Omega)$ with N_F continuous, where $F(x,s) = \int_0^s f(x,t)dt$, and Φ: $\mathbf{L}^p(\Omega) \to \mathbf{R}$, $\Phi(u) = \int_\Omega F(x,u(x))dx$ is of C^1-Frechét class and $\Phi' = N_f$.

4.8 Theorem. *Let Ω be an open bounded nonempty set in $\mathbf{R}^N$ and $f : \Omega \times \mathbf{R} \to \mathbf{R}$ a Carathéodory function with the growth condition*

(23)
$$|f(x, s)| \le c|s|^{p-1} + b(x),$$

where $c > 0$, $2 \le p \le \dfrac{2N}{N-2}$ *when* $N \ge 3$ *and* $2 \le p < +\infty$ *when* $N = 1, 2$, *and where*

$$b \in \mathbf{L}^q(\Omega), \frac{1}{p} + \frac{1}{q} = 1.$$

Then the energy functional $\varphi : H_0^1(\Omega) \to \mathbf{R}$,

(24)
$$\varphi(u) = \frac{1}{2}\int_\Omega |\nabla u(x)|^2 dx - \int_\Omega F(x, u(x))dx,$$

where $F(x,s) = \int_0^s f(x,t)dt$, *is of C^1 - Fréchet class and*

$$\varphi'(u)(v) = \int_\Omega \nabla u \cdot \nabla v\, dx - \int_\Omega f(x, u(x))v dx \quad \forall u, v \in H_0^1(\Omega).$$

Explanation. $|\nabla u|^2 = \sum_{i=1}^{N} \left|\frac{\partial u}{\partial x_i}\right|^2$, $\frac{\partial u}{\partial x_i}$ the weak derivatives.

4.9 *Corollary. Let* Ω *and* f *be as in* 4.8. *Then the weak solutions of* (∗) *are precisely the critical points of the functional* $\varphi : H_0^1(\Omega) \to \mathbf{R}$,

$$\varphi(u) = \frac{1}{2}\int_\Omega |\nabla u(x)|^2 dx - \int_\Omega F(x, u(x))dx, \quad F(x,s) := \int_0^s f(x,t)dt.$$

■ Indeed, if $\bar{u}$ is a weak solution of (∗), then $\varphi'(\bar{u})(v) = 0$ ((18), 4.8) $\forall\, v \in C_c^\infty(\Omega)$, hence $\varphi'(\bar{u}) = 0$, because φ' is continuous and $C_c^\infty(\Omega) = H_0^1(\Omega)$. The converse is obviously. ■

Weak subsolutions and weak supersolutions of (∗). Let Ω be an open bounded set of C^1 class in $\mathbf{R}^N$, $N \geq 3$, $f : \Omega \times \mathbf{R} \to \mathbf{R}$ a Carathéodory function and $\bar{u} \in H_0^1(\Omega)$.

Definition. $\bar{u}$ is a *weak subsolution* respectively *weak supersolution* of (∗) if

(25) $\bar{u} \leq u_0$ on $\partial\Omega$ respectively $\bar{u} \geq u_0$ on $\partial\Omega$

and

(26)
$$\begin{cases} \int_\Omega \nabla\bar{u} \cdot \nabla v dx \leq \int_\Omega f(x, \bar{u}(x)) v dx \;\forall\, v \in C_c^\infty(\Omega),\, v \geq 0 \\ \text{resp.} \\ \int_\Omega \nabla\bar{u} \cdot \nabla v dx \geq \int_\Omega f(x, \bar{u}(x)) v dx \;\forall v \in C_c^\infty(\Omega),\, v \leq 0. \end{cases}$$

4.10 *Let* Ω *be an open bounded of* C^1 *class set in* $\mathbf{R}^N$, $N \geq 3$, $f : \Omega \times \mathbf{R} \to \mathbf{R}$ *a Carathéodory function and* u_1 , u_2 *from* $H_0^1(\Omega)$ *bounded weak subsolution* resp. *weak supersolution of* (∗) *with* $u_1(x) \leq u_2(x)$ *a. e. on* Ω.

Suppose that f *verifies* (23) *and there is* $\rho > 0$ *such that the function* g: $g(x,s) = f(x,s) + \rho s$ *is strictly increasing in* s *on* [inf $u_1(\Omega)$, sup $u_2(\Omega)$]. *Then a weak solution* $\bar{u}$ *of* (∗) *exists in* $H_0^1(\Omega)$ *with the property*

$$u_1(x) \leq \bar{u}(x) \leq u_2(x) \text{ a. e. on } \Omega$$

([34], *Theorem* 1.16).

■ It can suppose $u_0 = 0$. Take the equivalent norm on $H_0^1(\Omega)$

$$\|u\|^2 = \rho\|u\|_2^2 + \|\nabla u\|_2^2$$

and let $<\cdot,\cdot>$ the corresponding scalar product. Consider the functional $\varphi : H_0^1(\Omega) \to \mathbf{R}$,

(27) $$\varphi(u) = \frac{1}{2} < u, u > - \int_\Omega G(x, u(x))dx ,$$

$$G(x, s) := \int_0^s g(x, t)dt .$$

φ is of C^1 class and its critical points are the weak solutions of (∗). Use 4.7. Let be

$$\mathcal{F} := \{u \in H_0^1(\Omega): u_1(x) \le u(x) \le u_2(x) \text{ a. e. on } \Omega\}.$$

$\mathcal{F}$ is closed convex. We also get

(28) $$(I - \varphi')\mathcal{F} \subset \mathcal{F}.$$

Indeed, let u be in $\mathcal{F}$ and $v := (I - \varphi')(u)$. For every w in $C_c^\infty(\Omega)$, $w \ge 0$ we have

$$< v, w > \overset{(27)}{=} < u, w > - < u, w > + \int_\Omega g(x, u(x))\, wdx,$$

$$< v - u_1, w > \overset{(26)}{\ge} \int_\Omega |g(x, u(x)) - g(x, u_1(x))|\, wdx \ge 0 ,$$

apply the maximum principle, we get $u_1(x) \le v(x)$ a. e. on Ω. In the same way, we get $v(x) \le u_2(x)$ a.e. on Ω and hence (28). φ is lower bounded on $\mathcal{F}$: $\forall\ u$ in $\mathcal{F}$ we have $\varphi(u) \overset{(27)}{\ge} \frac{1}{2} < u, u > - c_1$, $c_1 \in \mathbf{R}$ and moreover $(PS)_{c,F}$ is verified, $c := \inf \varphi(F)$ ([34], Theorem 1.16, proof). Finish the proof applying 4.7. ■

Example. Consider the problem (Ω open bounded of C^1 class in $\mathbf{R}^N$, $N \ge 3$)

(29) $$\begin{cases} -\Delta u = \alpha(x)\, u\, |u|^{p-2} \\ u = u_0 \quad \text{on } \partial\Omega, \end{cases}$$

where $p = \dfrac{2N}{N-2}$, α is continuous with $1 \le \alpha(x) \le a < +\infty$ on Ω and $u_0 \in C^1(\overline{\Omega})$, $u_0(x) = 1$ on $\partial\Omega$. Then $u_1 := 1$ is a weak subsolution, $u_2 := M$, $M > 1$ sufficiently big, is a weak supersolution, $|f(x, s)| \le a|s|^{p-1}$ (condition (23)) and $s \to \alpha(x)s|s|^{p-2} + s$ is increasing in s on $[1, M]$, consequently, according to 4.10, (29) has a weak solution $\overline{u}$ with $1 \le$ $\le \overline{u}(x) \le M$ a.e. on Ω.

5

Subdifferential calculus

A. SUBDERIVATIVE

In this section appear some elements of the convex functions theory. These are due to the papers of W. Fenchel, J. J. Moreau and R. T. Rockafellar. Most of the results about subderivatives are also due to Rockafellar.

Let X be a real normed space, $f: E \to (-\infty, +\infty]$, $E \subset X$. The domain of f, dom f, is equal to $\{x \in E: f(x) < +\infty\}$. f is proper if dom $f \neq \varnothing$ (see INTRODUCTION).

Let x_0 be a point of dom f. The functional ξ from X^* is by definition a *subderivative of* f *in* x_0, if

(1) $$f(x) - f(x_0) \geq \xi(x - x_0) \quad \forall\, x \in E.$$

Sometimes is accepted the name ξ *subgradient of* f *in* x_0. The set of subderivatives of f in x_0 is denoted by

$$\partial f(x_0).$$

Someone names even $\partial f(x_0)$ the subderivative in x_0. When $\partial f(x_0) \neq \varnothing$, f is by definition *subdifferentiable in* x_0.

Example 1. *Let* X *be a real normed space and* $f : X \to \mathbf{R}$, $f(x) = \|x\|$. ξ *is a subderivative of* f *in* $x_0 \neq 0$ *iff*

$$\|\xi\| = 1 \text{ and } \xi(x_0) = \|x_0\|.$$

Necessity. According to the definition, $\xi(x - x_0) \le \|x\| - \|x_0\|\ \forall\ x \in X$, hence $\xi(x - x_0) \le$ $\le \|x - x_0\|\ \ \forall\ x \in X$ and so $\|\xi\| \le 1$. On the other hand, for $x = 0$ we get $\xi(x_0) \ge \|x_0\|$, that imposes $\|\xi\| \ge 1$, hence $||\xi|| = 1$, but also $\xi(x_0) = \|x_0\|$ since $\xi(x_0) \le \|x_0\|$.

Sufficiency. Let x be arbitrary in X. $\xi(x - x_0) = \xi(x) - \xi(x_0) = \xi(x) - \|x_0\| \le \|x\| - \|x_0\|$. ■

The Gâteaux derivative provides another example of subderivative. Precisely,

5.0 *Let* X *be a real normed space and* $\mathrm{f} : \mathrm{X} \to (-\infty, +\infty]$ *convex. If* f *is Gâteaux differentiable in* x_0 *from* dom f, *then*

$$\partial \mathrm{f}(\mathrm{x}_0) = \{\mathrm{f}'_{\mathrm{w}}(\mathrm{x}_0)\}.$$

Specification. The concept "f β-differentiable in x_0" (end of §3) was defined, in order to obtain a unitary exposition, for "f locally finite in x_0". Here we weaken this demand to "$x_0 \in \operatorname{dom} f$".

■ Show firstly that $\xi := f'_{\mathrm{w}}(x_0) \in \partial f(x_0)$, i.e.

$$f(x) \ge f(x_0) + \xi(x - x_0)\ \forall\, x \in X\ .$$

When $f(x) = +\infty$, there is nothing to prove. Suppose $f(x) < +\infty$. Then, f being convex, *a fortiori* is finite on the segment with ends x, x_0 and we have

$$f(x_0 + t(x - x_0)) \le (1 - t) f(x_0) + t f(x)\ \forall\, t \in [0,1]\,.$$

f is Gâteaux differentiable at x_0, i.e.

$$f(x_0 + t(x - x_0)) = f(x_0) + t\xi(x - x_0) + \ o(t)\ ,\ t \text{ small},\ \lim_{t\to 0} \frac{o(t)}{t} = 0\,.$$

Combining with the last but one relation, one finds $f(x_0) + t\,\xi(x - x_0) + o(t) \le (1 - t) f(x_0) +$ $+ t f(x)$, $f(x) \ge f(x_0) + \xi(x - x_0) + \dfrac{o(t)}{t}$, t positive and small. Taking the limit for $t \to 0+$ one finds $\xi \in \partial f(x_0)$.

Conversely, let us prove $\mu \in \partial f(x_0) \Rightarrow \mu = \xi$. Let v be arbitrary in X. We must prove that $\mu(v) = \xi(v)$. By assumption, $f(x_0 + tv) - f(x_0) \ge t\,\mu(v)$, $\forall\ t \in \mathbf{R}$. For $t > 0$ we have $\dfrac{1}{t}[f(x_0 + tv) - f(x_0)] \ge \mu(v)$, taking the limit for $t \to 0+$ we get $\xi(v) \ge \mu(v)$. For $t < 0$ and then $t \to 0-$ we find $\xi(x) \le \mu(v)$, hence $\xi(v) = \mu(v)$. ■

Remark 1. The same proposition is in 5.7 but for f with values in **R**.

Remark 2. Proposition 5.0 has a converse: "If the convex function $f : X \to (-\infty, +\infty]$ is continuous and has in x_0 from dom f a unique subderivative, then f is Gâteaux differentiable at x_0" ([6], Ch.I, Proposition 5.3).

Corollary to 5.0. *If* $f : X \to (-\infty, +\infty]$ *is convex Gâteaux differentiable at* x_0 *from* dom f *and* $f'_w(x_0) = 0$, *then* x_0 *is a global minimum point for* f.

Shortly,

If $f'_w(x_0) = 0$ *and* f *is convex, then* x_0 *is a global minimum point.*

■ The formula of 5.0 gives $0 \in \partial f(x_0)$, whence the conclusion. ■

In the next proposition, f is as in (1).

5.1 $\partial f(x_0)$ *is convex and *-weak closed. When* f *is upper semicontinuous at* x_0 *and* $x_0 \in \overset{o}{E}$, *then* $\partial f(x_0)$ *is bounded.*

■ *First statement.* Let ξ_1, ξ_2 be in $\partial f(x_0)$ and $\lambda_1, \lambda_2 > 0$, $\lambda_1 + \lambda_2 = 1$. $f(x) - f(x_0) \geq$ $\geq \xi_i(x - x_0)\ \forall x \in E$, $i = 1, 2$. Multiply with λ_i and add.

Second statement. Let be $\mu \in \overline{\partial f(x_0)}^{*-\text{weak}}$. We must show that $\mu \in \partial f(x_0)$, that is

$$(2) \qquad f(x) - f(x_0) \geq \mu(x - x_0)\ \forall x \in E.$$

Fix x from E and let $\varepsilon > 0$ be arbitrary. $\exists\ \xi$ in $\partial f(x_0)$ such that

$$|(\mu - \xi)(x)| < \varepsilon \quad \text{and} \quad |(\mu - \xi)(x_0)| < \varepsilon,$$

hence

$$\xi(x) > \mu(x) - \varepsilon, \quad -\xi(x_0) > -\mu(x_0) - \varepsilon.$$

Because $f(x) - f(x_0) \geq \xi(x - x_0)$, by addition we find $f(x) - f(x_0) \geq \mu(x - x_0) - 2\varepsilon$ and we get (2) by passing to limit for $\varepsilon \to 0$.

Third statement. Fix $\varepsilon > 0$. $\exists\ \delta > 0$ such that

$$(3) \qquad \|x - x_0\| < \delta \Rightarrow f(x) < f(x_0) + \varepsilon \quad (x_0 \in \overset{o}{E}\,!).$$

Fix t_0, $0 < t_0 < \delta$. We show

$$(4) \qquad \mu \in \partial f(x_0) \Rightarrow \|\mu\| \leq \frac{\varepsilon}{t_0}.$$

Let v be arbitrary, $\|v\| = 1$. Take in (1) $x = x_0 + t_0 v$, we get $f(x_0 + tv) - f(x_0) \geq \mu(t_0 v)$ and combining with (3), $\mu(v) \leq \dfrac{\varepsilon}{t_0}$, that imposes (4) (change v in $-v$). ■

5.2 *Let* X *be a real normed space and* $f : X \to (-\infty, +\infty]$ *convex. If* f *is upper semicontinuous on a neighborhood of* x_0, $x_0 \in$ dom f, *then* $\partial f(x_0) \neq \emptyset$.

■ *A fortiori* f is finite and upper semicontinuous on an open ball $S := S(x_0, \delta)$. In the product normed space $X \times \mathbf{R}$ consider the subset

(5) $$E := \{(x, a) \in X \times \mathbf{R} : x \in S, a > f(x)\}.$$

E is nonempty (f is finite on the ball!), open [let (x', a') be from E, for a_1 of thc interval $(f(x'), a')$ we have $f(x) < a_1$ on a neighbourhood V of x' with $V \subset S$, take $V \times (a_1, +\infty)$] and convex (f is convex!). Since $(x_0, f(x_0)) \notin E$, by the Hahn-Banach separation theorem (the second geometrical form, *Memento*, §1, after 1.1_0), there exists F in $(X \times \mathbf{R})^* \setminus \{0\}$ such that $F(x_0, f(x_0)) \leq F(x, a)\ \forall\ (x, a) \in E$. But $\exists!\ \mu$ in X^* and $\exists!\ r$ in $\mathbf{R}$ so that $F(x, \lambda) = \mu(x) + r\lambda$ on $X \times \mathbf{R}$ and thus

(6) $$\mu(x_0) + r f(x_0) \leq \mu(x) + ra \quad \forall\ (x, a) \in E.$$

Suppose $\mu \neq 0$. (6) imposes $r \neq 0$, in the oposite case we would have $\mu(x_0) \leq \mu(x)\ \forall x \in \in S$, hence $\|h\| \leq R \Rightarrow \mu(h) = 0$ and consequently $\mu = 0$, a contradiction. Moreover, taking in (6) $x = x_0$ one finds $s > 0$ (*par absurdum*, take $a \to +\infty$). Let $\xi := -\dfrac{\mu}{r}$. From (6),

(7) $$-\xi(x_0) + f(x_0) \leq -\xi(x) + a\ \forall\ (x, a) \in E.$$

Let x be arbitrary fixed in S. (7) being verified $\forall\ a > f(x)$, it is also verified by $f(x)$ (*par absurdum*), hence

(8) $$f(x) - f(x_0) \geq \xi(x - x_0) \quad \forall x \in S .$$

Suppose $\mu = 0$. Then *a fortiori* $r \neq 0$ since $F \neq 0$ and (6) becomes

(6_1) $$r f(x_0) \leq ra\ \forall\ (x, a) \in E.$$

If, *par absurdum*, $r < 0$, we would have $r f(x_0) = -\infty$ (take the limit for $a \to +\infty$), a contradiction and hence $r > 0$. From (6_1), $f(x_0) \leq a\ \forall\ (x, a)$ in E, consequently $f(x_0) \leq \leq f(x)\ \forall\ x \in S$. Taking $\xi = 0$ we get again (8).

Let now y be arbitrary in $X \setminus S$. Since $[y, x_0] \cap S \neq \varnothing$, $\exists\ t$ in $(0,1)$ such that $x := ty + (1 - t)x_0 \in S$, hence x satisfies (8). But $f(x) \leq t f(y) + (1 - t) f(x_0)$ because f is convex and then, from (8), we get $t[f(y) - f(x_0)] \geq t\xi(y - x_0)$, therefore $f(x) - f(x_0) \geq \xi(x - x_0)$ $\forall\ x \in X$, $\xi \overset{(1)}{\in} \partial f(x_0)$. ■

5.3 *Corollary. Let* X *be a real Banach space and* f : X → (− ∞, + ∞] *convex. If* f *is upper semicontinuous on a neighbourhood of* x_0 , $x_0 \in$ dom f, *then* $\partial f(x_0)$ *is nonempty, convex and* ∗*-weak compact.*

■ Combine 5.2 with 5.1, in X^* any bounded and ∗-weak closed set is ∗-weak compact (corollary to Alaoglu-Bourbaki-Kakutani theorem). ■

5.4 *Corollary. If* f : X → **R**, X *a real Banach space, is continuous and convex, then for every* x *of* X $\partial f(x)$ *is nonempty, convex and* ∗*-weak compact.*

Explanation regarding the obtaining of (6). X being a real normed space, $X^* \times \mathbf{R}$ can be identified with $(X \times \mathbf{R})^*$ by

$$(\xi, \alpha)(x, t) = \xi(x) + \alpha t.$$

In this place we make some specifications.

Let X be a real normed space and $f: X \to (-\infty, +\infty]$ proper. The *epigraph*, epi f, of f is

$$\operatorname{epi} f = \{(x, t) \in X \times \mathbf{R} : t \geq f(x)\}.$$

epi f is closed $\Leftrightarrow f$ is lower semicontinuous. Also, epi f is convex $\Leftrightarrow f$ is convex (f convex $\Rightarrow$ dom f convex). Moreover, f continuous in x_0 from dom $f \Rightarrow \overbrace{\operatorname{epi} f}^{\mathrm{o}} \neq \varnothing$ (let V be a neighborhood of x_0 on which $f(x) < f(x_0) + 1$, we have $V \subset$ dom f and $V \times (f(x_0) + + 1, +\infty) \subset \operatorname{epi} f$).

In the following we insist on the *continuity of convex functions*. These functions take values in $(-\infty, +\infty]$ and, by convention, $0\,(+\infty) = (+\infty)\,0 = 0$,

5.4$_1$ *Lemma. Let* X *be a real normed space and* f : X → (– ∞, + ∞] *convex. If* f *is upper bounded on a neighbourhood of* x_0, *then* f *is continuous at* x_0.

■ A suitable translation allows to suppose $x_0 = 0$ and $f(x_0) = 0$. According to the hypothesis, there is $S := \{x \in X: \|x\| < r\}$ and $a \in \mathbf{R}$ such that $f(x) \leq a$ on S. Let ε be arbitrary from (0, 1). Take $\delta := \varepsilon r$. $\|x\| < \delta \Rightarrow f(x) \leq (1-\varepsilon) f(0) + \varepsilon f\left(\frac{x}{\varepsilon}\right) = \varepsilon f\left(\frac{x}{\varepsilon}\right) \leq$ $\leq \varepsilon a$. Also, $\|x\| < \delta \Rightarrow f(0) \leq \frac{1}{1+\varepsilon} f(x) + \frac{\varepsilon}{1+\varepsilon} f\left(-\frac{x}{\varepsilon}\right)$, hence $f(x) \geq -\varepsilon a$ and hence $|f(x)| \leq \varepsilon a$ for $\|x\| < \delta$, f is continuous in 0. ■

5.4$_2$ *Let* X *be a real normed space and* f : X → (– ∞, + ∞] *convex. Then the following statements are equivalent*:

1° *There exists an open nonempty set* U *on which* f *is upper bounded*;

2° f *is proper*, $G := \overbrace{\operatorname{dom} f}^{\mathrm{o}} \neq \varnothing$ *and* f *is continuous on* G.

■ 2° ⇒ 1°. Clearly. 1° ⇒ 2°. Clearly, f is proper and $U \subset G$. Moreover, according to 5.4_1 f is continuous at every point of U. Let v be arbitrary in G, we show that f is continuous at v. Fix a point u in U and take $\rho > 1$ such that

$$w := u + \rho(v - u) \subset G \tag{8_1}$$

(such a number exists because for $\rho \to 1$ we have $w \to u$ and $u \in G$). Consider the homothety $h : X \to X$,

$$h(x) = w + \frac{\rho - 1}{\rho}(x - w)\,. \tag{8_2}$$

$h(u) \overset{(8_1)}{=} u + \rho(v - u) + (\rho - 1)(u - v) = v$, hence v is in $h(U)$ which is open. Let y be arbitrary from $h(U)$. We have $y = \frac{\rho - 1}{\rho} z + \frac{1}{\rho} w$, where $z = h^{-1}(y)$, since $y = h(z) \overset{(8_2)}{=} w + \frac{\rho - 1}{\rho}(z - w)$, consequently, f being convex,

$$f(y) \le \frac{\rho - 1}{\rho} f(z) + \frac{1}{\rho} f(w)$$

and since $z \in U$ we get

$$f(y) \le \frac{\rho - 1}{\rho} a + \frac{1}{\rho} f(w)\,,$$

where $f(x) \le a$ on U. We deduce that f is upper bounded on $h(U)$, consequently f is continuous at v (5.4_1). ■

5.4_3 *Corollary. Let* $\mathrm{f} : \mathrm{X} \to (-\infty, +\infty]$ *be convex and proper,* X *a real normed space. Then the following statements are equivalent*:

1° f *is upper bounded on an open nonempty set*;

2° $\mathrm{G} := \overbrace{\mathrm{dom\,f}}^{\circ} \ne \varnothing$ *and* f *is locally Lipschitz on* G.

■ 2° ⇒ 1°. Obviously. 1° ⇒ 2°. By assumption we get f continuous on G (5.4_2). Let u be arbitrary fixed in G. We find a neighbourhood of u on which f will be Lipschitz (notation: $S_r(a) := \{x \in X : \|x - a\| \le r\}$).

f is continuous at u, hence $\exists\, r_0 > 0$ and m, M in **R** such that

$$x \in S_{r_0}(u) \Rightarrow m \le f(x) \le M. \tag{8_3}$$

Take now r from $(0, r_0)$ and let v be arbitrary fixed in $S_r(u)$. Consider the function

(8_4) $$\Phi(x) = f(x+v) - f(v),$$

a correct definition for x with $\|x\|$ small, and let be $V := \{x \in X : \|x\| \le r_0 - r\}$. We have $\Phi(x) \le M - m$ on V, because $x \in V \Rightarrow \|(x+v) - u\| \le (r_0 - r) + r = r_0$ and we take into account (8_3). Then, with a reasoning made in 5.4_1, for any ε from $[0,1]$ ($\Phi(0) = 0!$) we have

(8_5) $$x \in \varepsilon V \Rightarrow |\Phi(x)| \le \varepsilon(M - m).$$

Let x be from $S_{r_0 - r}(v)$, i.e. $\|x - v\| \le r_0 - r$. Then $y := x - v \in \varepsilon V, \varepsilon := \dfrac{\|x - v\|}{r_0 - r}$, hence, according to (8_5), $|\Phi(y)| \overset{(8_4)}{=} |f(x) - f(v)| \le \varepsilon(M - m)$,

(8_6) $$x \in S_{r_0 - r}(v) \Rightarrow |f(x) - f(v)| \le \frac{M - m}{r_0 - r}\|x - v\|, \ v \text{ from } S_r(u).$$

Let now x_1, x_2 be in $S_r(u)$. Consider on the segment $[x_1, x_2]$ (this is included in $S_r(u)$) the equidistant points $z_1 := x_1, z_2, \ldots, z_{n-1}, z_n := x_2$, sufficiently numerous in order to have $\|z_k - z_{k+1}\| \le r_0 - r$, $k = \overline{1, n-1}$. By (8_6) we get successively

$$|f(z_k) - f(z_{k+1})| \le \frac{M - m}{r_0 - r}\|z_k - z_{k+1}\|, \ k = \overline{1, n-1}.$$

By addition we get

$$|f(x_1) - f(x_2)| \le \frac{M - m}{r_0 - r}\|x_1 - x_2\|,$$

f is Lipschitz on $S_r(u)$. ■

5.4_4 *Corollary. Let* f : X → (− ∞, + ∞] *be convex and proper*, X *a real normed space with finite dimension. Then, if* $G := \overbrace{\operatorname{dom} f}^{o} \ne \emptyset$, f *is locally Lipschitz, hence in particular continuous, on* G.

■ Let $n := \dim X$ and $x_0, x_1, \ldots, x_n$ be affinely independent points. The interior of the simplex with n dimensions and vertices $x_0, x_1, \ldots, x_n$ is the set

$$U := \left\{ \sum_{i=0}^{n} \lambda_i x_i : \sum_{i=0}^{n} \lambda_i = 1, \lambda_i > 0 \right\}.$$

U is open and for every x from U we have, because f is convex,

$$f(x) = f\left(\sum_{i=0}^{n} \lambda_i x_i \right) \le \sum_{i=0}^{n} \lambda_i f(x_i) \le \max_{0 \le i \le n} f(x_i).$$

Apply 5.4_3 and get the conclusion. ■

Finally,

5.4_5 *Let* $f : X \to (-\infty, +\infty]$ *be convex lower semicontinuous proper,* X *a real Banach space. Then supposing that* $G := \overset{\circ}{\overbrace{\operatorname{dom} f}}$ *is nonempty,* f *is continuous on* G.

■ Let x_0 be from G. By a translation we can suppose $x_0 = 0$. Take $a > f(0)$. $F: = \{x \in \in X: f(x) \le a\}$ is convex and closed. Moreover F is absorbing because the restriction of f to every line which passes through 0 is continuous in the neighbourhood of 0 (5.4_4). Consequently, $V := F \cap (-F)$ is a cask. But X being a Banach space, it is "tonnelé" and consequently V is a neighbourhood of 0. As f is upper bounded on V, f is continuous in 0 (5.4_1). ■

Remark. Combine 5.4_5 with 5.3 and find $\partial f(x) \neq \varnothing\ \forall\ x$ in $\overset{\circ}{\overbrace{\operatorname{dom} f}}$ supposed nonempty.

Resume the study of the subderivative.

Begin with a variant of 5.2.

5.4_6 *Let* $f : X \to (-\infty, +\infty]$ *be convex,* X *a real normed space. If* f *is continuous in* x_0 *from* dom f, *then* $\partial f(x) \neq \varnothing\ \forall x \in \overset{\circ}{\overbrace{\operatorname{dom} f}}$ *and also* $\partial f(x_0) \neq \varnothing$.

■ f being continuous and finite at x_0, it is upper bounded on an open neighborhood U of x_0, consequently f is continuous on $\overset{\circ}{\overbrace{\operatorname{dom} f}}$, a nonempty set ($5.4_3$). Thus, it is enough to prove that $\partial f(x_0) \neq \varnothing$, because in other points the continuity is also verified.

Because f is convex, epi f has the same property. We have $\overset{\circ}{\overbrace{\operatorname{epi} f}} \neq \varnothing$ because, as $f(x) \le a$, $a \in \mathbf{R}\ \forall\ x \in U$, the open set $U \times (a, +\infty)$ is included in epi f. But $(x_0\,,\ f(x_0)) \in \in$ epi f as boundary point of this set, consequently $(x_0, f(x_0))$ and $\overset{\circ}{\overbrace{\operatorname{epi} f}}$ are separating by a closed hyperplane (Hahn-Banach theorem, the second geometrical form, *Memento*, §1, after 1.1_0). Thus we obtain H a support closed hyperplane of epi f which contains $(x_0\,,\ f(x_0))$, that is $\exists\ \xi \in X^*$, $\exists\ \alpha, \beta \in \mathbf{R}$, $\alpha \ge 0$ such that

$$H = \{(x, t) \in X \times \mathbf{R}\colon \xi(x) + \alpha t = \beta\}$$

and

$$(8_7)\qquad \begin{cases}\xi(x)+\alpha t \geq \beta \quad \forall (x,t)\in \operatorname{epi} f,\\ \xi(x_0)+\alpha f(x_0)=\beta.\end{cases}$$

We have $\alpha > 0$. Indeed, if we suppose *par absurdum* $\alpha = 0$, combining in (8_7) we get $\xi(x - x_0) \geq 0\ \forall\ x \in \operatorname{dom} f$ and since $\operatorname{dom} f$ is a neighborhood of x_0, we yield $\xi = 0$, contradiction. Thus, dividing by α in (8_7) where $t = f(x)$, we get $\frac{\beta}{\alpha} - \frac{1}{\alpha}\xi(x) \leq f(x)\ \forall x \in \operatorname{dom} f$ and $\frac{\beta}{\alpha} - \frac{1}{\alpha}\xi(x_0) = f(x_0)$, hence $-\frac{1}{\alpha}\xi(x - x_0) \leq f(x) - f(x_0)\ \forall\ x \in X$, i.e. $-\frac{1}{\alpha}\xi \in \partial f(x_0)$. ■

5.4$_7$ *Rules of subdifferential calculus. Let* X *be a real normed space,* $f, f_1, f_2 : X \to (-\infty, +\infty]$ *and* $\lambda > 0$.

1° $\partial(\lambda f)(x) = \lambda\partial f(x)$, *in the hypothesis that one of the two members is nonempty.*

2° $\partial(f_1 + f_2)(x) \supset \partial f_1(x) + \partial f_2(x)$, *in the hypothesis that the terms of the second member are nonempty.*

3° *Suppose* f_1, f_2 *convex lower semicontinuous and* $\exists\ x_0$ *in* $\operatorname{dom} f_1 \cap \operatorname{dom} f_2$ *in which* f_1 *is continuous. Then*

$$\partial(f_1 + f_2)(x) \subset \partial f_1(x) + \partial f_2(x),$$

in the hypothesis that the first member is nonempty.

■ 1° Let for instance $\xi \in \partial(\lambda f)(x)$, i.e.

$$\lambda f(y) - \lambda f(x) \geq \xi(y - x)\ \forall\ y \in X.$$

Division by λ preserves the sense of the inequality, hence $\frac{\xi}{\lambda} \in \partial f(x)$, $\xi \in \lambda\partial f(x)$. The reasoning continues in the same way.

2° Let $\xi_i \in \partial f_i(x)$ be, $i = 1, 2$, i. e.

$$f_i(y) - f_i(x) \geq \xi_i(y - x)\ \forall\ y \in X,\ i = 1, 2.$$

We add and get $\xi_1 + \xi_2 \in \partial(f_1 + f_2)(x)$:

$$[f_1(y) + f_2(y)] - [f_1(x) + f_2(x)] \geq (\xi_1 + \xi_2)(y - x)\ \forall\ y \in X.$$

The consideration of the sum $f_1(y) + f_2(y)$ is correct, because both terms are $> -\infty$ and when we add up, $f_1(x)$ and $f_2(x)$, being finite, do not produce nuisances.

3° Let ξ from $\partial(f_1 + f_2)(x)$. We need to prove the decomposition $\xi = \xi_1 + \xi_2$, $\xi_i \in \partial f_i(x)$, $i = 1, 2$. By assumption we get $f_i(x) \in \mathbf{R}$, $i = 1, 2$ and

(8₈) $$f_1(y) + f_2(y) \geq \xi(y - x) + f_1(x) + f_2(x) \;\forall\, y \in X.$$

In $X \times \mathbf{R}$ consider the convex sets

$$C_1 := \{(y, t) \in X \times \mathbf{R} : f_1(y) - \xi(y - x) - f_1(x) \leq t\},$$

$$C_2 := \{(y, t) \in X \times \mathbf{R} : t \leq f_2(x) - f_2(y)\}$$

A common point of C_1 and C_2 establishes everywhere, *via* (8_8), the sign =, hence the common points of C_1 and C_2 can be only boundary points. Remark that C_1 is the epigraph of the function g,

$$y \to g(y) = f_1(y) - \xi(y) - f_1(x) + \xi(x),$$

which is continuous in x_0, that imposes $\overset{\circ}{C_1} \neq \varnothing$. But, as we have remarked, $\overset{\circ}{C_1} \cap C_2 = \varnothing$, hence C_1 and C_2 are separating by a closed hyperplane H of $X \times \mathbf{R}$ (Hahn-Banach theorem, the second geometrical form, *Memento*, §1, after 1.1_0). Because H is the graph of the affine continuous functional

$$y \to \mu(y) + \alpha,\; \mu \in X^*,\; \alpha \in \mathbf{R},$$

the separation can be written

$$f_2(x) - f_2(y) \leq \mu(y) + \alpha \leq f_1(y) - \xi(y - x) - f_1(x) \;\forall\, y \in X.$$

Taking $y = x$ we get $\alpha = -\mu(x)$ and then the double inequality gives

$$-\mu(y - x) \leq f_2(y) - f_2(x) \;\forall\, y \in X, \text{ i. e.} -\mu \in \partial f_2(x),$$

$$(\xi + \mu)(y - x) \leq f_1(y) - f_1(x) \;\forall\, y \in X, \text{ i. e. } \xi + \mu \in \partial f_1(x),$$

consequently, $\xi = \xi_1 + \xi_2$, $\xi_1 := \xi + \mu$, $\xi_2 := -\mu$, the searched decomposition. ■

It can pass now to

5.4₈ Theorem. *Let* X *be a Banach space and* $f : X \to (-\infty, +\infty]$ *convex, lower semi-continuous and proper. For any* x_0 *from* dom f *there exists a sequence* $(x_n)_{n\geq 1}$ *of points in* X *such that*

(8₉) $$\|x_n - x_0\| \leq \frac{1}{n},$$

(8₁₀) $$f(x_n) \to f(x_0),$$

(8₁₁) $$\partial f(x_n) \neq \varnothing \;\forall\, n \geq 1$$

(*A. Brøndsted, R. T. Rockafellar*).

■ The set $E := \operatorname{epi} f = \{(x, t) \in X \times \mathbf{R} : f(x) \leq t\}$ is nonempty (f is proper), convex (f is convex) and closed (f is l.s.c.). Choose a point (x', t') which *is not in* E (this is possible: let x' be arbitrary, since $f(x') > -\infty \;\exists\, t'$ in $\mathbf{R}$ with $t' < f(x')$). The set $\{(x', t')\}$

being convex and compact, one very strongly separates from E through a closed hyperplane, let be $\nu(x, t) = \alpha$ its equation ($\nu \in (X \times \mathbf{R})^*$, Hahn-Banach theorem, the third geometrical form, *Memento,* §1, after 1.1_0), i.e. $\exists\ \rho > 0$ such that

(8_{12}) $$\nu(x', t') \le \alpha - \rho < \alpha + \rho \le \nu(x, t)\ \forall\ (x, t) \in E.$$

But ν has the form $\nu(x, t) = \mu(x) + t r$, where $\mu \in X^*$ and $r \in \mathbf{R}$. Then taking $t = f(x)$ in (8_{12}) we get

$$\mu(x) + r f(x) - (\alpha - \rho) > 0\ \forall\ x \in X,$$

or still, for $\xi := -\dfrac{\mu}{r}$,

(8_{13}) $$f(x) - \xi(x) - \alpha_1 > 0,\ \alpha_1 \in \mathbf{R},\ \forall\ x \in X.$$

Take the function $\varphi: X \to (-\infty, +\infty]$,

(8_{14}) $$\varphi(x) = f(x) - \xi(x) - \alpha_1 .$$

φ is l. s. c., lower bounded ((8_{13})) and proper, apply the Ekeland principle 1.1 with $\varepsilon =$ $= \varphi(x_0) - \inf \varphi(X)$ and $\lambda = \dfrac{1}{n}$, $n \ge 1$, $\exists\ x_n$ in X such that (8_9) and

(8_{15}) $$\varphi(x_n) \le \varphi(x_0),$$

(8_{16}) $$\varphi(x_n) < \varphi(x) + n\varepsilon\|x - x_n\|\ \ \forall\ x \ne x_n .$$

Consider the function $\psi : X \to (-\infty, +\infty]$,

$$\psi(x) = \varphi(x) + \varepsilon n \|x - x_n\| .$$

This is convex l.s.c. and x_n is a global minimum point for f (see (8_{16})), hence $0 \in \partial\psi(x_n)$. But the function $\chi: X \to \mathbf{R}$,

$$\chi(x) = \|x - x_n\| ,$$

is convex continuous and consequently, according to 5.4_7, 3°, there is ζ_1 in $\partial\varphi(x_n)$ and ζ_2 in $\partial\chi(x_n)$ such that $0 = \zeta_1 + \varepsilon n \zeta_2$. But then, as $\partial(f - \alpha_1)(x_n) = \partial f(x_n)$, (8_{14}) and 5.4_7, 2° give $\zeta_1 + \xi \in \partial f(x_n)$, that is (8_{11}).

Finally, if we re-write (8_{16}) in terms of f, we find

$$f(x_n) \le f(x_0) + \xi(x_n - x_0)\ \forall\ n \ge 1.$$

Since $x_n \to x_0$ it results $\overline{\lim_{n\to\infty}} f(x_n) \le f(x_0)$. On the other hand, $f(x_0) \le \underline{\lim_{n\to\infty}} f(x_n)$ because f is l.s.c. in x_0, consequently $\overline{\lim_{n\to\infty}} f(x_n) = \underline{\lim_{n\to\infty}} f(x_n)$, $f(x_n) \to f(x_0)$, i.e. (8_{10}). ■

Remark. The statements of the Brøndsted-Rockafellar theorem can be abbreviated in

$$\operatorname{dom} \partial f \textit{ is dense in } \operatorname{dom} f$$

(see the definitions after 5.4_{10}).

Next let us present a variant of the Brøndsted-Rockafellar theorem. But firstly, let us say some words about ε - subderivative.

Let X be a real normed space, $f: X \to (-\infty, +\infty]$, x_0 in $\operatorname{dom} f$ and $\varepsilon > 0$. The ε-*subderivative of* f *in* x_0 is

$$(8_{17})\ \partial_\varepsilon f(x_0) := \{\xi \in X^*: \xi(x) \le f(x + x_0) - f(x_0) + \varepsilon\ \forall\, x \in X\}.$$

5.4_9 *If* f *is convex lower semicontinuous, then for every* x_0 *in* dom f *and* $\varepsilon > 0$ *we have*

$$\partial_\varepsilon f(x_0) \ne \varnothing.$$

■ Since epi f is convex and closed (f is l.s.c.) and $\{(x_0, f(x_0) - \varepsilon)\}$ is convex, compact and disjoined of epi f, the two sets are very strongly separating (see before): $\exists\ \mu \in X^*$ and $\exists\ \alpha \in \mathbf{R}$ such that

$$(8_{18})\ \mu(x_0) + \alpha(f(x) - \varepsilon) < \mu(y) + \alpha t\ \forall\, y \in \operatorname{dom} f,\ \forall\, t \ge f(y).$$

For $y = x_0$ and $t = f(x_0)$ we find $-\alpha\varepsilon < 0$, hence $\alpha > 0$. Taking $\xi = -\dfrac{\mu}{\alpha}$ and $t = f(y)$, (8_{18}) becomes

$$-\xi(x_0) + f(x_0) - \varepsilon < -\xi(y) + f(y)\ \forall\, y \in X$$

(for $y \notin \operatorname{dom} f$ the inequality remains true) and consequently taking $x := y - x_0$ one finds (8_{17}). ■

And now present the variant announced of 5.4_8.

5.4_{10} Theorem. *Let* $f: X \to (-\infty, +\infty]$ *be convex lower semicontinuous,* X *a real Banach space. For every* x_0 *in* dom f, $\varepsilon > 0$, $\lambda > 0$ *and every* ξ_0 *from* $\partial_\varepsilon f(x_0)$, *there exists* $x \in \operatorname{dom} f$ *and* $\xi \in \operatorname{dom} X^*$ *such that*

$$(8_{19})\qquad \xi \in \partial f(x),\ \|x - x_0\| \le \frac{\varepsilon}{\lambda},\ \|\xi - \xi_0\| \le \lambda$$

(*A. Brøndsted - R. T. Rockafellar*).

■ The statement is coherent, $\partial_\varepsilon f(x_0) \ne \varnothing$ (5.4_9). We have, according to (8_{17}),

$$\xi_0(y - x_0) \le f(x) - f(x_0) + \varepsilon\ \forall\, y \in X,$$

hence the function

$$(8_{20})\qquad g: g(y) = f(y) - \xi_0(y)$$

has the values in $(-\infty, +\infty]$, is proper and l.s.c. Moreover, $g(x_0) \le \inf g(X) + \varepsilon$. Apply

Ekeland principle 1.1 with $\frac{\varepsilon}{\lambda}$ instead of λ, there is z in dom g (= dom f) so that,

$$(8_{21}) \qquad \|z - x_0\| \le \frac{\varepsilon}{\lambda}, \quad g(z) \le g(y) + \lambda\|y - z\| \quad \forall y \in X.$$

Consider the function $h : X \to \mathbf{R}$, $h(y) = \lambda\|y - z\|$. By (8_{21}) we get $0 \in \partial(g + h)(z)$. But g and h are convex, g l. s. c. and h continuous, hence 0 can be decomposed (5.4_7, 3°), (8_{22}) $0 = \mu + \eta$, $\mu \in \partial g(z)$, $\eta \in \partial h(z)$. But (8_{23}) $\partial g(z) = \partial f(z) - \xi_0$ ((8_{20}), the definition!) and

$\partial h(z) = \{\zeta \in X^*: ||\zeta|| \le \lambda\}$: $\zeta \in \partial h(z) \Rightarrow \forall\, y \in X\ \zeta(y - z) \le h(y) - h(z) = \lambda\,\|y - z\| \Rightarrow ||\zeta|| \le$

$\le \lambda$; $||\zeta|| \le \lambda \Rightarrow \forall\, y \in X\ \ \zeta(y - z) \le ||\zeta||\ ||y - z|| \le \lambda\ ||y - z|| = h(y) - h(z)$, $\zeta \in \partial h(z)$.

Thus $\mu \overset{(8_{23})}{=} \xi - \xi_0$ with $\xi \in \partial f(z)$ and $\mu \overset{(8_{22})}{=} -\eta$, hence $\|\xi - \xi_0\| \le \lambda$ and we only have to take $z = x$ in order to all the statements of (8_{19}) be verified. ∎

The subderivative in the theory of the monotonous operators will end the section A. Our aim is to prove that the monotonous operator $x \to \partial f(x)$ is maximal (theorem of Rockafellar).

Definitions. Let X be a real normed space, $E \subset X$ nonempty and T: $E \to 2^{X^*}$ a multi-valued application.

The *domain* of T, dom T, is

$$\text{dom } T = \{x \in E: T(x) \ne \varnothing\}.$$

The *graph* $\Gamma(T)$ of T is

$$\Gamma(T) = \{(x, \xi) \in E \times X^*: \xi \in T(x)\}.$$

T is *monotonous* when $\forall\, x, y \in E$ and $\forall\, \xi \in T(x)$, $\forall\, \mu \in T(y)$ we have

$$(8_{24}) \qquad (\xi - \mu)(x - y) \ge 0.$$

Example. Let $f : X \to (-\infty, +\infty]$ be proper, X a real normed space. Consider the operator T: dom $f \to 2^{X^*}$, $T(x) = \partial f(x)$, designated ∂f. For instance, when X is complete and f convex continuous with values in $\mathbf{R}$, then dom ∂f = dom $f = X$ (5.4).

The operator ∂f is monotonous:

$$x, y \in \text{dom} f \text{ and } \xi \in \partial f(x),\ \mu \in \partial f(y) \Rightarrow \xi(y - x) \le f(y) - f(x),$$

$$-\mu(y - x) = \mu(x - y) \le f(x) - f(y),$$

add and change the sign, $(\xi - \mu)(x - y) \ge 0$.

The subset M of $E \times X^*$, X a real normed space and $E \subset X$ nonempty, is called *monotonous* (in $E \times X^*$) if

(8₂₅) $\forall\ (x, \xi), (y, \mu) \in M$ we have $(\xi - \mu)(x - y) \geq 0$

and *maximal monotonous* if M is a maximal element in the set of the monotonous subsets of $E \times X^*$ordered by inclusion. The definition (8_{25}) is coherent, since then we also have $(\mu - \xi)(y - x) \geq 0$.

5.4₁₁ *The operator* T *is monotonous iff the graph* $\Gamma(T)$ *of* T *is a monotonous set* (*in* $E \times X^*$).

■ It is obvious. ■

For instance, the operator ∂f being monotonous, $\Gamma(\partial f)$ is monotonous in $\mathrm{dom} f \times X^*$.

The monotonous operator T is *maximal monotonous* when its graph is a maximal monotonous set.

5.4₁₂ 1°. T *is maximal monotonous iff* $y \in E$, $\mu \in X^*$ *and* $(\mu - \xi)(y - x) \geq 0\ \forall\ x \in$ $\in \mathrm{dom}\, T$, $\forall\ \xi \in T(x) \Rightarrow \mu \in T(y)$.

2°. *If* T *is maximal monotonous, then* $T(x)$, *for every* $x \in E$, *is convex.*

■ 1° *Necessity.* Suppose T maximal and, *par absurdum*, $\exists\ y \in E$ and $\exists\ \mu \in X^*$ satisfying the inequality of the statement but however $\mu \notin T(y)$, that is $(y, \mu) \notin \Gamma(T)$. But $\Gamma(T) \cup \{(y, \mu)\}$ is monotonous in $E \times X^*$ (the relation of the hypothesis!), a contradiction since $\Gamma(T)$ is maximal. *Sufficiency.* Analogous.

2° Let ξ, μ from $T(x)$ and $\alpha, \beta > 0$, $\alpha + \beta = 1$. Suppose, *par absurdum*, that $\alpha\xi + \beta\mu \notin$ $\notin T(X)$, i.e. (8_{26}) $(x, \alpha\xi + \beta\mu) \notin \Gamma(T)$. But $\{(x, \alpha\xi + \beta\mu)\} \cup \Gamma(T)$ is monotonous: let (y, ζ) be arbitrary in $\Gamma(T)$, $\Gamma(T)$ being monotonous we have $(\xi - \zeta)(x - y) \geq 0$, $(\mu - \zeta)(x -$ $- y) \geq 0$, one multiplies respectively by α, β and one adds, we obtain $(\alpha\xi + \beta\mu - \zeta)(x -$ $- y) \geq 0$, a contradiction *via* (8_{26}), because T is maximal. ■

Continue with two propositions, due to S. Simons, that will be used for the proof of the Rockaffelar theorem.

5.4₁₃ *Let* X *be a real Banach space and* $f\colon X \to (-\infty, +\infty]$ *convex, lower semicontinuous and proper. If*

$$f(x_0) < \inf f(X) + \alpha\beta,$$

where $\alpha > 0$, $\beta > 0$, *there exists* $x \in \mathrm{dom}\, f$ *and* $\xi \in \partial f(x)$ *such that*

$$\|x - x_0\| < \beta \ \ and\ \ \|\xi\| < \alpha.$$

■ We use 5.4_{10}, *a fortiori* $x_0 \in \mathrm{dom} f$. Take $\varepsilon > 0$ such that $f(x_0) - \inf f(X) < \varepsilon < \alpha\beta$

(the condition of the statement implies $\inf f(X) > -\infty$ and also $x_0 \in \operatorname{dom} f$) and then $\lambda >$ > 0 such that $\frac{\varepsilon}{\beta} < \lambda < \alpha$. Since $0 \in \partial_{\varepsilon} f(x_0)$ (see (8_{17}), $f(x_0) \le f(x) + \varepsilon \ \forall\, x \in X$), there is x in $\operatorname{dom} f$ and ξ in $\partial f(x)$ such that $\|x - x_0\| \le \frac{\varepsilon}{\lambda} < \beta$ and $\|\xi\| \le \lambda < \alpha$ ($\xi_0 = 0!$). ■

5.4$_{14}$ *Let* f *be as in* 5.4$_{13}$ *and* x_0 *from* dom f *with*

$$\inf f(X) < f(x_0).$$

Then there is $z \in \operatorname{dom} f$ *and* $\zeta \in \partial f(z)$ *such that*

$$f(z) < f(x_0) \text{ and } \zeta(x_0 - z) > 0.$$

Attention! We did not ask $\inf f(X) > -\infty$.

■ Take λ in **R**, (8_{27}) $\inf f(X) < \lambda < f(x_0)$ and let be

$$(8_{28}) \qquad M := \sup_{\substack{x \in X \\ x \neq x_0}} \frac{\lambda - f(x)}{\|x - x_0\|}.$$

Show

$$(8_{29}) \qquad 0 < M < +\infty.$$

epi f is nonempty, convex (f is convex) and closed (f is l.s.c.). Since $(x_0, \lambda) \notin \operatorname{epi} f$ and $\{(x_0, \lambda)\}$ is convex compact, the two sets are strongly separating in $X \times \mathbf{R}$ (see before): there is on the hand $\xi \in X^*$ and $r \in \mathbf{R}$ and on the other hand $\alpha_1, \alpha_2 \in \mathbf{R}$ such that

$$(8_{30}) \qquad \xi(x_0) + \lambda r \le \alpha_1 < \alpha_2 \le \xi(x) + tr \ \forall\, x \in \operatorname{dom} f, \ \forall\, t \ge f(x).$$

Take in (8_{30}) $x = x_0$, $t = f(x_0)$ and obtain

$$r(f(x_0) - \lambda) \ge \alpha_2 - \alpha_1 > 0,$$

hence $r > 0$ and then, according to (8_{30}) where $t = f(x)$ and $\frac{\xi}{r} = \xi_1$,

$$-f(x) \le \xi_1(x) - \frac{\alpha_2}{r} \ \forall\, x \in X$$

($x \notin \operatorname{dom} f \Rightarrow -f(x) = -\infty$).

Let be $x \in F \setminus \{x_0\}$, where $F = \{x \in X\colon f(x) \le \lambda\}$. Then

$$\frac{\lambda - f(x)}{\|x - x_0\|} \le \frac{\lambda + \xi_1(x) - \frac{\alpha_2}{r}}{\|x - x_0\|} = \frac{\left|\lambda + \xi_1(x_0) - \frac{\alpha_2}{r} + \xi_1(x) - \xi_1(x_0)\right|}{\|x - x_0\|} \le$$

$$\leq \frac{\left|\lambda + \xi_1(x_0) - \frac{\alpha_2}{r}\right|}{\operatorname{dist}(x_0, F)} + \frac{|\xi_1(x) - \xi_1(x_0)|}{\|x - x_0\|} \leq \frac{\left|\lambda + \xi_1(x_0) - \frac{\alpha_2}{r}\right|}{\operatorname{dist}(x_0, F)} + \|\xi_1\|,$$

consequently $M < +\infty$ when $x \in F \setminus \{x_0\}$. When $x \notin F$ and $x \neq x_0$ we have $\frac{\lambda - f(x)}{\|x - x_0\|} < 0$, hence $M < +\infty$. We also have $M > 0$. Indeed, take x'', $f(x'') < \lambda$ ((8_{27})). Since $\lambda < f(x_0)$, *a fortiori* $x'' \neq x_0$ and then $\frac{\lambda - f(x'')}{\|x'' - x_0\|} > 0$, hence $M > 0$ and (8_{29}) is proved.

Now, let ε be from $(0,1)$, $(1 - \varepsilon)M < M$, which is correct – see (8_{29}). $\exists\, x' \in X \setminus \{x_0\}$ such that

(8_{31})
$$\frac{\lambda - f(x')}{\|x' - x_0\|} > (1 - \varepsilon)M.$$

We underline that (8_{31}) imposes $x' \in \operatorname{dom} f$. Consider the functional Φ, $\Phi(x) = M\|x - x_0\|$.

(8_{31}) amounts to $(1 - \varepsilon)\Phi(x') + f(x') < \lambda$,

(8_{32})
$$(\Phi + f)(x') < \lambda + \varepsilon\Phi(x').$$

But

(8_{33})
$$\lambda \leq \inf\,(\Phi + f)(X),$$

because for $x = x_0$ we have $\lambda < f(x_0) = (\Phi + f)(x_0)$ and for $x \neq x_0$ we have $\frac{\lambda - f(x)}{\|x - x_0\|} \leq M$, $\lambda \leq (\Phi + f)(x)$. Now combine ($8_{32}$) with ($8_{33}$) and get

$$(\Phi + f)(x') < \inf(\Phi + f)(X) + \varepsilon M\|x' - x_0\|.$$

Apply 5.4_{13} with $\Phi + f$ instead of f, x' instead of x_0 and $\alpha = \varepsilon M, \beta = \|x' - x_0\|$, there is $z \in \operatorname{dom}(\Phi + f)$ ($= \operatorname{dom} f$) and $\xi \in \partial(\Phi + f)(z)$ such that $\|z - x'\| < \|x_0 - x'\|$, hence *a fortiori* $z \neq x_0$, that is $\|z - x_0\| > 0$, and $\|\xi\| < \varepsilon M$.

Apply 5.4_7, 3°, correctly since Φ is continuous, $\exists\, \mu \in \partial\Phi(z)$, $\zeta \in \partial f(z)$ such that $\xi = \mu + \zeta$. Then $\mu(z - x_0) \geq \Phi(z) - \Phi(x_0) = M\|z - x_0\|$ and consequently

$$\zeta(x_0 - z) = \xi(x_0 - z) + \mu(z - x_0) \geq M\|z - x_0\| - \|\xi\|\,\|x_0 - z\| > (1 - \varepsilon)M\|z - x_0\| > 0,$$

$\zeta(x_0 - z) > 0$ (we take $\|\xi\| < \varepsilon M$ into account).

Moreover, since $\zeta \in \partial f(z)$ we have

$$f(x_0) \geq f(z) + \zeta(x_0 - z) > f(z),\ f(z) < f(x_0). \blacksquare$$

And now

5.4$_{15}$ Theorem. *If* f: X $\to (-\infty, +\infty]$ *is convex lower semicontinuous proper*, X *a real Banach space, then the monotonous operator* ∂f *is maximal* (*R. T. Rockafellar*).

■ Let be $x \in \mathrm{dom} f$ and $\xi \notin \partial f(x)$. Then $0 \notin \partial(f-\xi)(x)$ (the definition!), this implies

$$\inf(f-\xi)(X) < (f-\xi)(x).$$

Apply 5.4$_{14}$, there is $z \in \mathrm{dom}(f-\xi)$ ($= \mathrm{dom} f$) and $\zeta \in \partial(f-\xi)(z)$ such that $\zeta(x-z) >$ > 0. Let $\mu \in \partial f(z)$ be (hence $(z, \mu) \in \Gamma(\partial f)$) such that $\zeta = \mu - \xi$ (5.4$_7$, 3°). Then $(\mu - \xi)(z - x) < 0$, consequently the set $\Gamma(\partial f) \cup \{(x, \xi)\}$ is not monotonous, in other words $\Gamma(\partial f)$ does not support any monotonous set which contains it, except itself, ∂f is maximal. ■

B. SUPERDERIVATIVE

Let X be a real normed space, $f: E \to [-\infty, +\infty)$, $E \subset X$ and x_0 a point in dom f: $= \{x \in$ $\in E$: $f(x) > -\infty$. The functional ξ from X^* is by definition a *superderivative of* f *in* x_0 if

(8$_{34}$) $$f(x) - f(x_0) \le \xi(x - x_0) \ \forall\, x \in E.$$

Sometimes is accepted denomination *supergradient of* f *in* x_0. When there is at least one of this, f is by definition *superdifferentiable in* x_0.

The passage of the properties of the subderivative to the superderivative can be made by a simple change of sign (convex $\to$ concave, concave $\to$ convex, l.s.c. $\to$ u.s.c., u.s.c. $\to$ l.s.c., f proper $\to$ $f \not\equiv -\infty$). For instance

5.4$_{16}$ *If* f: X $\to$ **R**, X *a Banach space, is continuous and concave, then, for every* x *from* X, *the set of superderivatives of* f *in* x *is nonempty, concave and *-weak compact.*

C. DIRECTIONAL DERIVATIVE

Let X be a real normed space, $E \subset X$ and $x_0 \in \overline{E}$. The vector v of X is by definition a *possible direction for* E *according to* x_0, if there is $\rho > 0$ such that $x_0 + tv \in E\ \forall\, t$ in

$(0, \rho)$. In addition, let be $f : E \to (-\infty, +\infty]$ and $x_0 \in \operatorname{dom} f$. If

$$\lim_{t\to 0+} \frac{f(x_0+tv)-f(x_0)}{t}$$

exists and is finite, it is designated by

$$f'(x_0;v),$$

the derivative of f *at* x_0 *according to the vector* v or *according to the direction* v (*directional derivative*). Thus

$$f'(x_0;v) = \lim_{t\to 0+} \frac{f(x_0+tv)-f(x_0)}{t}. \tag{9}$$

Suppose E *convex*, f *convex* and *finite* on a neighborhood V of x_0. Show

(∗) *there exists* $f'(x_0; v)$.

Obviously, it can suppose V an open ball with the center in x_0, let δ be its radius, $\delta \le \rho\|v\|$. Consider the function $F : \left[0, \frac{\delta}{\|v\|}\right) \to \mathbf{R}$, $F(t) = f(x_0 + tv)$ (we supposed $v \neq 0$, $f'(x_0; 0) = 0$). F is convex: $t_1, t_2 \in \left[0, \frac{\delta}{\|v\|}\right)$ and $\lambda_1, \lambda_2 \ge 0$, $\lambda_1 + \lambda_2 = 1 \Rightarrow F(\lambda_1 t_1 + \lambda_2 t_2) = f((\lambda_1 + \lambda_2)x_0 + (\lambda_1 t_1 + \lambda_2 t_2)v) = f(\lambda_1(x_0 + t_1 v) + \lambda_2(x_0 + t_2 v)) \le \lambda_1 f(x_0 + t_1 v) + \lambda_2 f(x_0 + t_2 v) = \lambda_1 F(t_1) + \lambda_2 F(t_2)$. Thus the function $\Phi : \left(0, \frac{\delta}{\|v\|}\right) \to \mathbf{R}$, $\Phi(t) = \frac{F(t)-F(0)}{t}$ is increasing, that is the function $t \to \frac{f(x_0+tv)-f(x_0)}{t}$ has this property on $\left(0, \frac{\delta}{\|v\|}\right)$, consequently there is $\lim_{t\to 0+} \frac{f(x_0+tv)-f(x_0)}{t}$ finite. We underline

$$f'(x_0;v) \le \frac{f(x_0+tv)-f(x_0)}{t} \quad \forall\, t \in (0, \rho_1),\ \rho_1 > 0. \tag{10}$$

Continue in the same assumptions.

(10_1) *The function* $v \to f'(x_0; v)$ *is positive homogenous and subadditive, hence a forti-ori convex.*

■ Indeed, let be $\lambda > 0$.

$$\lambda f'(x_0;v) = \lambda \lim_{t\to 0+} \frac{f(x_0+t\lambda v)-f(x_0)}{\lambda t} = \lim_{t\to 0+} \frac{f(x_0+t\lambda v)-f(x_0)}{t} = f'(x_0;\lambda v).$$

$$f'(x_0;v_1+v_2)=\lim_{t\to 0+}\frac{f(x_0+\frac{t}{2}(v_1+v_2))-f(x_0)}{\frac{t}{2}}=\lim_{t\to 0+}\frac{2f\left(\frac{x_0+tv_1}{2}+\frac{x_0+tv_2}{2}\right)-2f(x_0)}{t}\le$$

$$\le\lim_{t\to 0+}\frac{f(x_0+tv_1)-f(x_0)}{t}+\lim_{t\to 0+}\frac{f(x_0+tv_2)-f(x_0)}{t},$$

since $2f\left(\frac{x_0+tv_1}{2}+\frac{x_0+tv_2}{2}\right)\le f(x_0+tv_1)+f(x_0+tv_2)$. ∎

Continue in the assumptions "E open convex, f convex finite".

(10_2) *If* f *is continuous at* x_0, *then the function* Φ: $v\to f'(x_0; v)$ *is continuous on* X.

∎ It is enough to show that Φ is upper bounded on a neighborhood of 0 (see *Explanation* below). Consider $S_\delta(x_0)$ a closed ball included in E on which $|f(x)|\le M$. Let v be arbitrary fixed from $X\setminus\{0\}$, $\|v\|<1$. Take the convex function $F:\left[0,\frac{\delta}{\|v\|}\right]\to\mathbf{R}$, $F(t)=$ $=f(x_0+tv)$. We have

$$F'_+(0)\le\frac{F\left(\frac{\delta}{\|v\|}\right)-F(0)}{\frac{\delta}{\|v\|}},$$

that is

$$f'(x_0;v)\le\frac{f\left(x_0+\frac{\delta}{\|v\|}v\right)-f(x_0)}{\frac{\delta}{\|v\|}}$$

and hence

$$f'(x_0;v)\le\frac{2M}{\delta}\|v\|\le\frac{2M}{\delta},$$

Φ is upper bounded on the open ball of X with the center at 0 and with radius 1 and consequently Φ is continuous on X. ∎

Explanation. We used the statement: "U open convex set of a real normed space, $g:U\to\mathbf{R}$ convex; if g is upper bounded on an open nonempty subset of U, then g is continuous on U". This proposition follows from 5.4_2.

Place in the hypotheses of (*) and suppose $\partial f(x_0)\ne\varnothing$. Let ξ be from $\partial f(x_0)$. Obviously we have $\xi(v)\le\frac{f(x_0+tv)-f(x_0)}{t}$ $\forall\, t>0$ (see (1)), by passing to the limit for

$t \to 0+$ one finds $\xi(v) \le f'(x_0\,; v)$. ξ being arbitrary, in fact we have obtained

$$(11) \qquad f'(x_0; v) \ge \sup_{\xi \in \partial f(x_0)} \xi(v)\,.$$

If f is, in addition, upper semicontinuous, then in (11) we have an equality. Precisely,

5.5 *Let* X *be a real normed space and* $f : X \to (-\infty, +\infty]$ *convex. If* f *is upper semicontinuous on a neighborhood of* $x_0 \in \operatorname{dom} f$, *then, for every* v *of* X,

$$f'(x_0; v) = \sup_{\xi \in \partial f(x_0)} \xi(v)\,.$$

■ $f'(x_0\,; v)$ exists (see (*)), because f is convex and moreover finite on a ball with the center at x_0 (being u.s.c. at x_0). We also have $\partial f(x_0) \ne \varnothing$ (5.2). Fix v in X. It remains to find, taking (11) into account, ξ in $\partial f(x_0)$ such that (12) $f'(x_0\,; v) \le \xi(v)$. The proof is analogous with that for 5.2. Let $S(x_0, \delta)$ be an open ball on which f is finite and u.s.c. Consider the sets

$$A := \{(y, a) \in X \times \mathbf{R} : y \in S(x_0, \delta),\ a > f(y)\},$$
$$B : \{(x_0 + tv,\ f(x_0) + t f'(x_0\,; v)) : t \in [0, \rho_1)\}$$

(ρ_1 that from (10)). A and B are nonempty and convex, A open. Moreover, $A \cap B = \varnothing$ since $f(x_0) + t f'(x_0\,; v) \le f(x_0 + tv)$ $\forall\, t \in [0, \rho_1)$ (it is obviously for $t = 0$; for $t > 0$ see (10)). Apply the Hahn-Banach theorem, the second geometrical form (*Memento*, §1, after 1.1_0), there exists (μ, r), $\mu \in X^*$ and $r \in \mathbf{R}$, such that (the necessary explanations are found in the proof of 5.2)

$$(13)\ \mu(y) + ra \ge \mu(x_0 + tv) + r[f(x_0) + t f'(x_0\,; v)],\ \forall\, (y, a) \in A,\ \forall\, t \in [0, \rho_1)\,.$$

Suppose $\mu \ne 0$. Taking in (13) $y = x_0$ and $t = 0$ we find $r > 0$. Let be $\xi := -\dfrac{\mu}{r}$. It can replace in (13) a by $f(y)$ and for $\mu = -r\xi$ we get

$$(14) \qquad -\xi(y) + f(y) \ge -\xi(x_0 + tv) + f(x_0) + t f'(x_0\,; v)\ \forall\, y \in S(x_0, \delta),\ \forall\, t \in [0, \rho_1).$$

For $t = 0$ in (14) one finds

$$(15)\ f(y) - f(x_0) \ge \xi(y - y_0)\ \forall\, y \in S(x_0, \delta).$$

Because (15) is also true for $y \in X \setminus S(x_0, \delta)$ (the same justification as in 5.2), we get $\xi \in$ $\in \partial f(x_0)$.

It only remains to take $y = x_0$ and $t > 0$ in (14) in order to fall on (12).

Suppose $\mu = 0$. Then we have *a fortiori* $r \ne 0$ and by (13) we get $r > 0$ and then

$$(15_1)\ f(y) - f(x_0) \ge t f'(x_0\,; v)\ \forall\, y \in S(x_0, \delta),\ \forall\, t \in [0, \rho_1).$$

In particular $f(y) - f(x_0) \ge \xi(y - x_0)$, where $\xi = 0$, $\forall\, y \in S(x_0, \delta)$). From this we find

$f(x) - f(x_0) \ge \xi(x - x_0)\ \forall\, x \in X$, i.e. $\xi \in \partial f(x_0)$. Now take $y = x_0$ and $t > 0$ in (15_1), we find $f'(x_0\,;\, v) \le 0$, i.e. (12) is satisfied with $\xi = 0$. ■

Continue with the connexion between the directional derivative and the subderivative. But first the definition of $f'_-(x_0; v)$.

Let X be a real normed space, $E \subset X$, $f : E \to (-\infty, +\infty]$, $x_0 \in \overset{\circ}{E}$, $f(x_0) < +\infty$ and $v \in X$. If there exists

$$(15_1) \qquad \lim_{t\to 0-} \frac{f(x_0 + tv) - f(x_0)}{t}$$

and is finite, then it is designated by

$$f'_-(x_0; v)\,.$$

Suppose E convex, f convex, f finite on an open ball $S(x_0\,, \delta)$ included in E. Consider, as for $f'(x_0\,;\, v)$, the function (we suppose $v \ne 0$, $f'_-(x_0; 0) = 0$)

$$F : \left(-\frac{\delta}{\|v\|}, \frac{\delta}{\|v\|}\right) \to \mathbf{R},\ F(t) = f(x_0 + tv).$$

This is convex, hence the lateral limits exist and are finite and $F'_-(0) \le F'_+(0)$, i.e.

$$(15_2) \qquad f'_-(x_0; v) \le f'(x_0; v)\,.$$

Remark

$$(15_3) \qquad f'_-(x_0; v) = -f'(x_0; -v)\,.$$

5.6 *Let* E *be an open convex set,* $\mathrm{f} : \mathrm{E} \to \mathbf{R}$ *convex,* $\mathrm{x}_0 \in \mathrm{E}$ *and* $\xi \in \mathrm{X}^*$. *Then*

$$\xi \in \partial \mathrm{f}(\mathrm{x}_0) \Leftrightarrow \mathrm{f}'_-(\mathrm{x}_0; \mathrm{v}) \le \xi(\mathrm{v}) \le \mathrm{f}'(\mathrm{x}_0; \mathrm{v})\quad \forall\ \mathrm{v} \in \mathrm{E}.$$

■ The statement is coherent, see (15_2).

$\Rightarrow$. (15_4) $f(x) - f(x_0) \ge \xi(x - x_0)\ \forall\, x \in E$. Let $S(x_0; \delta)$ be an open ball included in E and $I := \left(-\frac{\delta}{\|v\|}, \frac{\delta}{\|v\|}\right)$ (we suppose $v \ne 0$, otherwise…). Then, by (15_4), $t \in I \Rightarrow f(x_0 + tv) - f(x_0) \ge t\xi(v)$, hence $t > 0 \Rightarrow \frac{f(x_0 + tv) - f(x_0)}{t} \ge \xi(v)$, $t < 0 \Rightarrow \frac{f(x_0 + tv) - f(x_0)}{t} \le \xi(v)$, pass to the limit for $t \to 0+$ and $t \to 0-$.

$\Leftarrow$. Let x be arbitrary in E. For $\varepsilon > 0$ sufficiently small it can consider the function $F : J := (-\varepsilon, 1 + \varepsilon) \to \mathbf{R}$, $F(t) = f(x_0 + tv)$, where $v = x - x_0$. F being convex we have

$$F'_+(0) \le F(1) - F(0)\,,$$

consequently

$$f(x) - f(x_0) \geq f'(x_0; v) \geq \xi(v) = \xi(x - x_0). \ \blacksquare$$

A direct consequence of 5.6 is

5.7 *If* $f: X \to \mathbf{R}$ *convex,* X *a real normed space, is Gâteaux differentiable at* x_0*, then*

$$\partial f(x_0) = \{f'_w(x_0)\}.$$

5.7₁ *Corollary. The norm of the Gâteaux derivative of the norm on a real vector space, when it exists, is equal to* 1. ■

■ Combine 5.7 with the example 1. ■

The following variant of 5.2 separates us from the "total" space.

5.8 *Let* E *be an open convex set of a real normed space* X *and* $f: E \to \mathbf{R}$ *convex continuous. Then for every* x_0 *from* E

$$\partial f(x_0) \neq \varnothing.$$

■ According to 5.6, it is enough to find ξ in X^* for which

$$f'_-(x_0; v) \leq \xi(v) \leq f'(x_0; v) \ \forall\, v \in E. \tag{15_5}$$

Fix v_0 in X. Let be $V := \{\lambda v_0 : \lambda \in \mathbf{R}\}$ and $\mu: V \to \mathbf{R}$, $\mu(\lambda v_0) = \lambda f'(x_0; v_0)$, the definition being correct as $f'(x_0; v_0)$ exists (see before (*)). μ is linear and continuous. We prove

$$\mu(y) \leq f'(x_0; y) \ \forall\, y \in V. \tag{15_6}$$

Indeed, for $\lambda \geq 0$ we have

$$\mu(\lambda v_0) = \lambda f'(x_0; v_0) \overset{(10_1)}{=} f'(x_0; \lambda v_0)$$

and for $\lambda < 0$ we have

$$\mu(\lambda v_0) = \lambda\, f'(x_0; v_0) = -|\lambda|\, f'(x_0; v_0) \overset{(15_3)}{=} |\lambda| f'_-(x_0; -v_0) \overset{(15_2)}{\leq} |\lambda|\, f'(x_0; -v_0) =$$
$$= f'(x_0; \lambda v_0).$$

Return to (15_6). $y \to f'(x_0; y)$ is not only positive homogeneous and subadditive ((10_1)), but also continuous ((10_2), consequently there exists ξ in X^* such that $\xi(v) \leq$ $\leq f'(x_0; v) \ \forall\, v \in X$. It remains to verify the first inequality of (15_5). But, $\forall\, v \in X$, $\xi(v) =$

$$= -\xi(-v) \geq -f'(x_0; -v)) \overset{(15_3)}{=} f'_-(x_0; v). \ \blacksquare$$

D. β - SUBDERIVATIVE AND β - SUPERDERIVATIVE

Let X be a real normed space, β bornology on X (§3) and $f: X \to \overline{\mathbf{R}}$ locally finite in x_0. ξ from X^* is by definition β-*subderivative of* f *in* x_0 if $\forall\ \varepsilon > 0$ and $\forall\ S$ from $\beta\ \exists\ \delta > 0$ such that

$$(16)\ t \in (0, \delta) \Rightarrow -\varepsilon \leq \frac{f(x_0 + tv) - f(x_0)}{t} - \xi(v)\ \forall\ v \in S.$$

The definition is correct since the sets of β are bounded. The set of β-subderivatives of f at x_0 is denoted

$$\partial_\beta f(x_0).$$

By definition f is β-*subdifferentiable in* x_0 if $\partial_\beta f(x_0) \neq \emptyset$. A *variant* of these definitions is with

$$\text{“}f: X \to (-\infty, +\infty],\ x_0 \in \operatorname{dom} f\text{”}.$$

Obviously

$$(17)\qquad \beta_1 \prec \beta_2 \Rightarrow \partial_{\beta_2} f(x_0) \subset \partial_{\beta_1} f(x_0).$$

5.9 *Let* X *be a real normed space,* β *a bornology on* X *and* $f: X \to (-\infty, +\infty]$ *convex. If* f *is upper semicontinuous on a neighbourhood of* x_0 *from* dom f, *then* $\partial_\beta f(x_0) \neq \emptyset$ *and*

$$\partial_\beta f(x_0) = \partial_{\beta_G} f(x_0) = \partial f(x_0).$$

■ f is, *a fortiori*, locally finite at x_0. $\partial f(x_0) \neq \emptyset$ (5.2) and we have

$$(18)\qquad \partial f(x_0) \subset \partial_\beta f(x_0).$$

Indeed, let $\xi \in \partial f(x_0)$, $\varepsilon > 0$ and $S \in \beta$. $\forall\ v$ in S and $\forall\ t > 0$ we have $f(x_0 + tv) - f(x_0) \geq$ $\geq \xi(tv)$, hence $\dfrac{f(x_0 + tv) - f(x_0)}{t} \geq \xi(v) - \varepsilon$ and therefore (18). Because $\partial_\beta f(x_0) \overset{(17),3.7}{\subset}$ $\subset \partial_{\beta_G} f(x_0)$, it remains to prove

$$\partial_{\beta_G} f(x_0) \subset \partial f(x_0),$$

i.e. $\xi \in \partial_{\beta_G} f(x_0) \Rightarrow \xi \in \partial f(x_0)$, that is

$$(19)\ f(x) - f(x_0) \geq \xi(x - x_0)\ \forall\ x \in X.$$

Fix x and let be $v := x - x_0$ and $\varepsilon > 0$ arbitrary. According to the hypothesis there is $\delta > 0$ such that $t \in (0, \delta) \Rightarrow \xi(v) - \varepsilon \leq \dfrac{f(x_0 + tv) - f(x_0)}{t}$. Take successively the limit for $t \to$

$\to 0+$ and then the limit for $\varepsilon \to 0+$, one finds (see above (∗)) $\xi(v) \le f'(x_0; v)$, i.e. $\xi(v) \le \le \sup_{\mu \in \partial f(x_0)} \mu(v)$ (see 5.5). Thus there are two possibilities:

$$(20) \qquad \xi(v) < \sup_{\mu \in \partial f(x_0)} \mu(v), \qquad (21) \qquad \xi(v) = \sup_{\mu \in \partial f(x_0)} \mu(v).$$

In the case (20) $\exists\ \mu \in \partial f(x_0)$ such that $\xi(v) < \mu(v)$, but $f(x) - f(x_0) \ge \mu(x - x_0) = \mu(v) > > \xi(v) = \xi(x - x_0)$ and consequently (19). In the case (21), since $f(x) - f(x_0) \ge \mu(v)\ \forall \mu \in \in \partial f(x_0)$, *a fortiori* $f(x) - f(x_0) \ge \xi(v)$ and consequently again (19). ■

The simple statement which follows connects, as we shall see, the β-subderivative with a smooth variational principle ([56], 2, 4.12, c)).

5.10 *Let* X *be a real normed space and* $\mathrm{f}, \mathrm{g} : \mathrm{X} \to \overline{\mathbf{R}}$ *locally finite at* x_0. *If* g *is* β-*differentiable at* x_0 *and* f + g *has in* x_0 *a minimum point, then* f *is* β-*subdifferentiable at* x_0.

■ According to the hypothesis, we have

$$(22)\ (-g)(x) - (-g)(x_0) \le f(x) - f(x_0)\ \forall\ x \in X.$$

But, setting $\xi := \nabla_\beta(-g)$, $\forall\ \varepsilon > 0$ and $\forall\ S \in \beta\ \exists\ \delta > 0$ such that $t \in (0, \delta)$, $v \in S \Rightarrow -\varepsilon \le$
$\le \dfrac{(-g)(x_0 + tv) - (-g)(x_0)}{t} - \xi(v) \overset{(22)}{\le} \dfrac{f(x_0 + tv) - f(x_0)}{t} - \xi(v)$, whence the conclusion. ■

Remark. In 5.10 it can take $f : X \to (\infty, +\infty]$ and $x_0 \in \operatorname{dom} f$.

Let X be a real normed space, β bornology on X, $f : X \to \overline{\mathrm{R}}$ locally finite at x_0. ξ from X^* is by definition a β-*superderivative of* f *in* x_0, if $\forall\ \varepsilon > 0$ and $\forall\ S$ from $\beta\ \exists\ \delta > 0$ such that

$$(22_1) \qquad t \in (0, \delta) \Rightarrow \frac{f(x_0 + tv) - f(x_0)}{t} - \xi(v) \le \varepsilon\ \forall\ v \in S.$$

The set of the β-superderivatives of f at x_0 is designated

$$\partial^\beta f(x_0).$$

By definition f is β-*superdifferentiable at* x_0 if $\partial^\beta f(x_0) \ne \varnothing$. A *variant* of this definitions is with

$$\text{“}\ f : X \to [-\infty, +\infty),\ x_0 \in \operatorname{dom} f\ \text{”}.$$

Obviously

$$(22_2) \qquad \beta_1 \prec \beta_2 \Rightarrow \partial^{\beta_2} f(x_0) \subset \partial^{\beta_1} f(x_0),$$

$$(22_3) \qquad \partial^\beta f(x_0) = -\partial_\beta(-f)(x_0).$$

5.10$_1$ *Let* $f: X \to \overline{\mathbf{R}}$ *be locally finite at* x_0. f *is* β-*differentiable at* $x_0 \Leftrightarrow$ f *is* β-*super-differentiable and* β-*subdifferentiable at* x_0. *In this case*

$$\{\nabla_\beta f(x_0)\} = \partial^\beta f(x_0) = \partial_\beta f(x_0).$$

■ *Necessity.* Obvious. *Sufficiency.* Take ξ_1 in $\partial_\beta f(x_0)$, ξ_2 in $\partial^\beta f(x_0)$ and let be $\varepsilon > 0$ and S in β arbitrary. $\exists\ \delta > 0$ such that $t \in (0, \delta) \Rightarrow (22_4)\ -\varepsilon \le \dfrac{f(x_0+tv)-f(x_0)}{t} - \xi_1(v)$ $\forall\ v \in S$ and $\dfrac{f(x_0+tv)-f(x_0)}{t} - \xi_2(v) \le \varepsilon\ \forall\ v \in S$. In particular $-\varepsilon + \xi_1(v) \le \varepsilon + \xi_2(v)$,

$$(22_5) \qquad \xi_1(v) - \xi_2(v) \le 2\varepsilon\ \forall\ v \in S.$$

In (22_5) v can be replaced by $-v$, because for the same ε and for $-S\ \exists\ \delta_1 > 0$ with a certain property, consequently, taking (22_5) into account, $\xi_1(v) - \xi_2(v) \ge -2\varepsilon\ \forall\ v \in S$, hence $\xi_1|S = \xi_2|S$ and therefore $\xi_1 = \xi_2$ since $\bigcup_{S\in\beta} S = X$. It only remains to confront once more with (22_4) and we get the conclusion. ■

5.10$_2$ *Let* $f: X \to \overline{\mathbf{R}}$ *be concave and moreover finite lower semicontinuous on a neighborhood of* x_0. *If* $\partial_\beta f(x_0) \ne \varnothing$, *then* f *is* β-*differentiable at* x_0 *and*

$$\partial_\beta f(x_0) = \{\nabla_\beta f(x_0)\}.$$

■ $-f$ being convex, we have $\partial_\beta(-f)(x_0) \ne \varnothing$ (5.9, consider the restriction $-f\,|\,(-\infty, +\infty]$). But $\partial_\beta(-f)(x_0) \overset{(22_3)}{=} -\partial^\beta f(x_0)$, hence $\partial^\beta f(x_0) \ne \varnothing$ and since also $\partial_\beta f(x_0) \ne \varnothing$, apply 5.10$_1$ and get the conclusion. ■

For the purpose of presenting a substantial property of the β-subderivative pass now to

E. THE PROPERTY (H_β) AND THE BANACH SPACE $D_\beta(X)$

Definition. Let X be a real normed space and β a bornology on X. X has *the property* (H_β) if there is a bump function on X (§1, before 1.2) with a bounded β-derivative.

We underline that f β-differentiable is Lipschitz iff the function $x \to \nabla_\beta f(x)$ is bounded on X (use the finite increment formula, 3.8).

It is interesting to mention

5.11 *If the real normed space* X *admits an equivalent norm* β*-differentiable, then it has the property* (H_β).

■ Let Φ: **R** → **R** be of C^1 class with bounded derivative and nonempty support included into the interval $\left[\frac{1}{2},\frac{3}{2}\right]$. The function $F: X \to \mathbf{R}$, $F(x) = \Phi(\|x\|)$ is β-differentiable and equal with 0 for $\|x\| < \frac{1}{2}$ and $\|x\| > \frac{3}{2}$. Moreover, for $x \neq 0$ we have $F'(x) =$ $= \Phi'(\|x\|)\nabla_\beta \|x\|$. But $\nabla_\beta\|x\| \overset{3.7}{=} \|x\|'_w$ and the norm of the functional $\|x\|'_w$ is equal to 1 ((5.7₁)) and so X has the property (H_β). ■

Definition. Let X be a real normed space and β a bornology on X. $D_\beta(X)$ will denote the real vector space of the bounded functions $f: X \to \mathbf{R}$ with bounded β-derivative. *A fortiori* f is Lipschitz. Endow $D_\beta(X)$ with the norm

$$(22_6) \qquad \|f\|_\beta = |||f||| := \|f\|_\infty + \|\nabla_\beta f\|_\infty .$$

We underline that when X has the property (H_β), the set $D_\beta(X)$ is nonempty.

5.12 *The normed space* $D_\beta(X)$ *is complete.*

■ Let $(f_n)_{n\geq 1}$ be a Cauchy sequence with respect to the norm $|||\cdot|||$. Then the sequences $(f_n)_{n\geq 1}$ and $(g_n)_{n\geq 1}$, $g_n := \nabla_\beta f_n$ are also Cauchy in the norm $\|\cdot\|_\infty$ ((22₆)), let $f: X \to \mathbf{R}$, $g: X \to X^*$ be the uniform limits of these. f is Lipschitz: $|f_n(x) - f_n(y)| \overset{3.8}{\leq} \|\nabla_\beta f_n\|_\infty \|x -$ $- y\|$, $\|\nabla_\beta f_n\| \leq \rho$ $\forall n \geq 1$ (Cauchy sequence), pass to the limit for $n \to \infty$. Thus as f is evidently bounded, it remains to show

$$(23) \qquad f \ \beta\text{-differentiable and } \nabla_\beta f = g,$$

i.e. $\forall x \in X$ and $\forall S \in \beta$ we have

$$(23_1) \qquad \lim_{\substack{t\to 0 \\ h\in S}} u \frac{f(x+th) - f(x)}{t} = g(x)(h).$$

(because then $|||f - f_n||| = \|f - f_n\|_\infty + \|\nabla_\beta f - \nabla_\beta f_n\|_\infty \to 0$).

Fix x and S and let be $M := \sup\{\|h\|: h \in S\}$ (the sets of β are bounded). For every n in **N** consider the function $F_n: \mathbf{R} \times S \to \mathbf{R}$,

$$F_n(t,h) = \begin{cases} \dfrac{f_n(x+th) - f_n(x)}{t}, & t \neq 0 \\ g_n(x)(h), & t = 0. \end{cases}$$

Let be $\varepsilon > 0$. $\exists\ N$ in **N** such that (24) $n \geq N \Rightarrow \|g_n - g_N\|_\infty \leq \varepsilon$ (Cauchy sequence). By (24) we get

(25) $$n \geq N, h \in S \Rightarrow |F_n(0,h) - F_N(0,h)| \leq \|g_n - g_N\|_\infty \|h\| \leq \varepsilon M .$$

Now suppose $n \geq N$, $t \neq 0$, $h \in S$.

$$F_n(t,h) - F_N(t,h) = \frac{1}{t}[(f_n - f_N)(x+th) - (f_n - f_N)(x)]$$

and with the aid of 3.8 and (24), $|F_n(t, h) - F_N(t, h)| \leq \varepsilon M$, consequently, in combination with (25),

(25_1) $$n \geq N, t \in \mathbf{R}, h \in S \Rightarrow | F_n(t, h) - F_N(t, h)| \leq \varepsilon M.$$

f_N being β-differentiable, $\exists\ \delta > 0$ such that $t \in (0, \delta)$, $h \in S \Rightarrow |F_N(t, h) - F_N(0, h)| \leq \varepsilon$ and then using (25_1) we get

(26) $$n \geq N, t \in (0, \delta), h \in S \Rightarrow |F_n(t, h) - F_N(0, h)| \leq (1 + M)\, \varepsilon.$$

But setting, for $t \neq 0$ $F(t,h) := \dfrac{f(x+th) - f(x)}{t}$, we have $\lim_{n\to\infty} F_n(t,h) = F(t,\ h)$. Of course, $\lim_{n\to\infty} g_n(x)(h) = g(x)(h)$. Taking in (26) the limit for $n \to \infty$ one obtains $|F(t, h) - F_N(0, h)| \leq \varepsilon(1 + M)$ and doing the same in (25) we get $|g(x)(h) - F_N(0, h)| \leq \varepsilon M$. From the last two inequalities one infers $t \in (0, \delta)$, $h \in S \Rightarrow \left|\dfrac{f(x+th) - f(x)}{t} - g(x)(h)\right| \leq \varepsilon(1 + 2M)$, that imposes ($23_1$). ■

As an application of 5.12

5.13 *Corollary. Let* X *be a Banach space with the property* (H_β) *and for every* n *from* **N** $\varphi_n : X \to (-\infty, +\infty]$ *lower bounded lower semicontinuous and proper. For every* $\varepsilon > 0$ *there exists* f *in* $D_\beta(X)$ *such that*

$$\| f\|_\infty \leq \varepsilon,\ \|\nabla_\beta f\|_\infty \leq \varepsilon$$

and every function $\varphi_n + f$ *attains a strong minimum on* X.

■ Take in 1.2 $\mathscr{F} = D_\beta(X)$, it is allowed by 5.12, and let be $\mathscr{M}_n := \bigcap_{m=1}^{\infty} U_m^n$ (see (16)) the set corresponding to φ_n. U_m^n is open and dense everywhere (see (14), (15)). Then $\bigcap_{n=1}^{\infty} \mathscr{M}_n$, being a countable intersection of open everywhere dense sets, it is everywhere dense (Baire theorem) and hence the conclusion in virtue of 1.2, taking the open ball with the center at 0 and the radius ε. ■

Now it can return to the β-subderivative.

5.14 *Let* X *be a real Banach space having the property* (H_β), $f : X \to (-\infty, +\infty]$ *lower semicontinuous proper and*

$$A := \{x \in X : f \textit{ is locally finite in } x\}$$

nonempty. Then the set

$$\{x \in A : f \textit{ is } \beta\textit{-subdifferentiable in } x\}$$

is dense in A ([56], 2, Corollary 4.13).

■ Let x_0 be in A and $\varepsilon > 0$. We show that in the open ball $S(x_0, \varepsilon)$ there is a point of A in which f is β-subdifferentiable. Since $f(x_0) > f(x_0) - 1$ $\exists\, \eta$ in $(0, \varepsilon)$ such that (27) $+\infty > f(x) > f(x_0) - 1$ on the closed ball $S_\eta(x_0)$. Consider the function $g: X \to (-\infty, +\infty]$, $g(x) = f(x)$ on $S_\eta(x_0)$, $g(x) = +\infty$ on $X \setminus S_\eta(x_0)$. g is lower bounded proper and l.s.c. [let x' be with $\|x' - x\| = \eta$, if $a < g(x')(= f(x'))$, $\exists\, V$ a neighborhood of x' on which $f(x) > a$, *a fortiori* $g(x) > a$ because $g(x) = +\infty$ on $X \setminus S_\eta(x_0)$ etc.].

Take φ in $D_\beta(X)$ such that (28) $\varphi(x_0) = 0$, $\varphi(x) = 2$ on $X \setminus S(x_0, \frac{\eta}{2})$ (the property (H_β)!). For $g + \varphi$, a lower bounded, l.s.c. (φ is continuous being Lipschitz) and proper function, we have (29) $(g + \varphi)(x_0) \overset{(28)}{=} f(x_0)$, $(g + \varphi)(x) > f(x_0) + 1$ on $X \setminus S(x_0, \frac{\eta}{2})$ (for $x \in S_\eta(x_0) \setminus S(x_0, \frac{\eta}{2})$ use (28) and (27)). Apply 5.13, $\exists\, h$ in $D_\beta(X)$ such that (30) $\|h\|_\infty < \frac{1}{4}$ and $(g + \varphi) + h$ has a point of minimum x_1. We have $(g + \varphi + h)(x) > f(x_0) + \frac{3}{4}$ on $X \setminus S(x_0, \frac{\eta}{2})$ ((29), $h(x) > -\frac{1}{4}$ on X) and $(g + \varphi + h)(x_0) \overset{(29)}{<} f(x_0) + \frac{1}{4}$, consequently $x_1 \in S(x_0, \frac{\eta}{2})$ (*par absurdum*) and since $g(x) = f(x)$ on $S\left(x_0, \frac{\eta}{2}\right)$, it results (see (27)) that g, and also f, is locally finite at x_1. Apply 5.10 to $g + (\varphi + h)$, we get g β-subdifferentiable in x_1 and it only remains to remark again that $f(x) = g(x)$ on $S\left(x_0, \frac{\eta}{2}\right)$ to end the proof ($\eta < \varepsilon$!). ■

5.14₁ *Corollary. Let* X *be a real Banach space having the property* (H_β), D *an open nonempty convex subset and* $f : D \to \mathbf{R}$ *continuous convex. Then* f *is* β*-differentiable on a dense subset of* D.

■ Let S be an arbitrary open ball with $\overline{S} \subset D$. Consider the function g: $X \to (-\infty, +\infty]$, $g(x) = -f(x)$ on $\overline{S}$, $g(x) = +\infty$ on $X \setminus \overline{S}$. g is l.s.c. and proper, one applies 5.14 ($A = S$!), g is β-subdifferentiable on a dense subset of S, let x_0 be from S in which g has this property. g being concave (on X) and continuous on S, it can apply 5.10_2, g is β-differentiable at x_0 and consequently f has the same property. ■

F. CLARKE DERIVATIVE

Let X be a real normed space, $E \subset X$, $f: E \to \mathbf{R}$, $x_0 \in \overset{o}{E}$ and $v \in X$. We set

$$f^0(x_0; v) := \overline{\lim_{\substack{x \to x_0 \\ t \to 0+}}} \frac{f(x+tv) - f(x)}{t}.$$

The upper limit exists obviously. $f^0(x_0; v)$ is by definition *Clarke derivative* (or the *generalized directional derivative*) of the function *f at* x_0 *in the direction v*. Looking back to the definition (9), it becomes clear the generalization, due to F. H. Clarke, of the concept.

Thus by definition

$$f^0(x_0; v) = \inf_{\substack{V \in \mathcal{V}(x_0) \\ r \in (0,+\infty)}} \sup_{\substack{x \in V \\ t \in (0,r)}} \frac{f(x+tv) - f(x)}{t}. \tag{31}$$

Remark. Peano proposed, in 1892, as the definition of the derivative at x_0 the limit, when it exists,

$$\lim_{\substack{x \to x_0 \\ y \to x_0}} \frac{f(x) - f(y)}{x - y},$$

considering it more suitable for the physical sciences than the usual definition. Thus Peano is the first who made x_0 to disappear in the expression of the fraction whose limit is taken to obtain the derivative.

Let f and x_0 be as before. By definition f is *Lipschitz around* x_0 *with the constant* L, $L \in (0, +\infty)$, if there is a neighborhood V of x_0, $V \subset E$ such that

$$(32)\ |f(x)-f(x')| \le L\|x-x'\| \ \forall\, x, x' \in V.$$

f is *locally Llipschitz on an open set* U, if it is Lipschitz around every point of U.

Those are the first properties.

5.15 *Let* f *be Lipschitz around* x_0 *with the constant* L. *Then*

1° *the function* $v \to f^0(x_0; v)$ *is with values in* **R**, *positive homogeneous, subadditive on* X *and*

$$|f^0(x_0; v)| \le L\|v\| \ \forall\, v \in X;$$

2° $f^0(x_0; -v) = (-f)^0(x_0; v)\ \forall\, v \in X$, $\lambda \ge 0 \Rightarrow (\lambda f)^0(x_0; v) = \lambda f^0(x_0; v)\ \forall\, v \in X$;

3° $v \to f^0(x_0; v)$ *is Lipschitz on* X *with the constant* L ([20], 1).

■ 1° $f^0(x_0; v) \in$ **R**. For x near x_0 and t strictly positive near 0 we have

$$(33)\qquad \left|\frac{f(x+tv)-f(x)}{t}\right| \le \frac{1}{t} L\|tv\| = L\|v\|.$$

By (33) we get (34) $|f^0(x_0; v)| \le L\|v\|$ and so $f^0(x_0; v) \in$ **R**. Indeed, suppose *par absurdum* $f^0(x_0; v) > L\|v\|$ for instance. Then $\forall\, V$ from $\mathcal{V}(x_0)$ and $\forall\, r$ from $(0, +\infty)$ we have

$$\sup_{\substack{x\in V\\ t\in(0,r)}} \frac{f(x+tv)-f(x)}{t} > L\|v\|,$$

in contradiction with (33).

$v \to f^0(x_0; v)$ *is positive homogeneous.* Since $f^0(x_0; 0) = 0$, let be $\lambda > 0$. Then $f^0(x_0; \lambda v) = \lambda \overline{\lim_{\substack{x\to x_0\\ \lambda t\to 0+}}} \frac{f(x+t\lambda v)-f(x)}{\lambda t}$ etc.

$v \to f^0(x_0; v)$ *is subadditive.*

$$(35)\qquad f(x_0; v_1+v_2) = \overline{\lim_{\substack{x\to x_0\\ t\to 0+}}} \left[\frac{f((x+tv_1)+tv_2)-f(x+tv_1)}{t} + \frac{f(x+tv_1)-f(x)}{t}\right] \le$$

$$\le \overline{\lim_{\substack{x\to x_0\\ t\to 0+}}} \frac{f((x+tv_1)+tv_2)-f(x+tv_1)}{t} + \overline{\lim_{\substack{x\to x_0\\ t\to 0+}}} \frac{f(x+tv_1)-f(x)}{t}.$$

As $\lim_{\substack{x\to x_0\\ t\to 0+}} (x+tv_1) = x$, in the third member of (35) the first term is equal to $f^0(x_0; v_2)$

and the second to $f^{\circ}(x_0\,;\,v_1)$.

$$2^{\circ}\ f^0(x_0;-v) = \overline{\lim_{\substack{x\to x_0\\ t\to 0+}}}\ \frac{f(x-tv)-f(x)}{t} \overset{u=x-tv}{=} \overline{\lim_{\substack{u\to x_0\\ t\to 0+}}}\ \frac{(-f)(u+tv)-(-f)(u)}{t} = (-f)^0\,(x_0\,;\,v),$$

since $\lim\limits_{\substack{x\to x_0\\ t\to 0+}}(x-tv)=x_0$.

For the second statement use the formula (31).

3° Let v, w be arbitrary in X. For x near to x_0 and t strictly positive near 0 we have

$$f(x_0+tv)-f(x_0) \le f(x_0+tw)-f(x_0)+Lt\|v-w\|.$$

Divide by t and take the upper limit for $x\to x_0$ and $t\to 0+$, one finds

$$f^0(x_0;v) \le f^0(x_0;w)+L\|v-w\|.$$

Changing v by w we get the desired conclusion. ■

5.16 *Let* f *be locally Lipschitz on* X. *The function* Φ: $X\times X\to\mathbf{R}$,

$$\Phi(x;\,v) = f^0\,(x_0\,;\,v)$$

is upper semicontinuous ([20], 1).

■ Let $(x_0\,,\,v_0)$ be a point in $X\times X$ and $(x_n\,,\,v_n)_{n\ge 1}$ an arbitrary sequence such that $(x_n\,,\,v_n)\to(x_0\,,\,v_0)$. We show (36) $\overline{\lim\limits_{n\to\infty}} f^0(x_n,v_n)\le f^0(x_0;v_0)$. According to (31) we have $\forall\, m,p\in$ $\in\mathbf{N}$

$$f^0(x_n;v_n)-\frac{1}{n} < \sup_{\substack{x\in S(x_n,\frac{1}{m})\\ t\in(0,\frac{1}{p})}} \frac{f(x+tv_n)-f(x)}{t}. \tag{37}$$

Fix m, p in $\mathbf{N}$ and let k_n be in $\mathbf{N}$, $k_n>n$ such that (38) $\frac{1}{m}+\frac{1}{p}<\frac{1}{k_n}$. We find with (37) y_{k_n} and t_{k_n} such that $\|y_{k_n}-x_n\|<\frac{1}{m}$ and $t_{k_n}\in\left(0,\frac{1}{p}\right)$, hence (39) $\|y_{k_n}-x_n\|+t_{k_n}\overset{(38)}{<}\frac{1}{k_n}$,

and moreover (see (37))

$$f^0(x_n;v_n)-\frac{1}{k_n} < \frac{f(y_{k_n}+t_{k_n}v_n)-f(y_{k_n})}{t_{k_n}}. \tag{40}$$

The second member of (40) is equal to

$$\frac{f(y_{k_n}+t_{k_n}v_0)-f(y_{k_n})}{t_{k_n}} + \frac{f(y_{k_n}+t_{k_n}v_n)-f(y_{k_n}+t_{k_n}v_0)}{t_{k_n}},$$

consequently passing in (40) to the upper limit for $n \to \infty$ and taking into account $y_{k_n} \to$ $\to x_0$, $t_{k_n} \to 0$ (see (39)) and $|f(y_{k_n} + t_{k_n} v_n) - f(y_{k_n} + t_{k_n} v_0)| \le L t_{k_n} \|v_n - v_0\|$, we find even (36), since $\overline{\lim_{n\to\infty}} \frac{f(y_{k_n} + t_{k_n} v_0) - f(y_{k_n})}{t_{k_n}} \le f^0(x_0; v_0)$. ■

G. CLARKE SUBDERIVATIVE

CLARKE SUBDERIVATIVE

Let X be a real normed space, $E \subset X$, $f: E \to \mathbf{R}$ and $x_0 \in \overset{o}{E}$. The functional ξ from X^* is by definition *Clarke subderivative* (or *generalized gradient*) *of* f *in* x_0 if

$$(41)\ f^0(x_0; v) \ge \xi(v)\ \forall\ v \in X.$$

The comparison of the definitions (41) and (1) justifies the name "Clarke subderivative". The set of these generalized gradients is designated by

$$\partial f(x_0).$$

Some call even $\partial f(x_0)$ the Clarke subderivative at x_0.

The notation coincides with that of the set of the subderivatives of f at x_0. The explanation will be given in 5.18: if in addition f is convex, then Clarke subderivative and the subderivative coincide. As a preliminary, here is an example.

Example 2. Calculate $\partial f(x_0)$ for the function $f: \mathbf{R} \to \mathbf{R}$, $f(x) = |x|$.

$x_0 > 0$. $f^0(x_0; v) = \overline{\lim_{\substack{x\to x_0\\ t\to 0+}}} \frac{x_0 + tv - x_0}{t} = v$. $\xi \in \partial f(x_0) \Leftrightarrow v \ge \xi\, v\ \forall\ v \in \mathbf{R}$, hence $\xi = 1$, $\partial f(x_0) = \{1\}$.

$x_0 < 0$. $\partial f(x_0) = \{-1\}$.

$x_0 = 0$. $f^0(0; v) = \overline{\lim_{\substack{x\to 0\\ t\to 0+}}} \frac{|tv|}{t} = |v|$, so $\partial f(0) = [-1, 1]$.

5.17 *If* f *is Lipschitz around* x_0, L *the constant, then*

1° $\partial f(x_0)$ *is nonempty, convex, *-weak compact* (*for* X *complete*) *and*

$$\|\xi\| \leq L \ \forall \ \xi \in \partial f(x_0);$$

2° $f^0(x_0; v) = \sup_{\xi \in \partial f(x_0)} \xi(v) \ \forall \ v \in X$ ([20], 1).

■ 1° The function $v \to f^0(x_0; v)$ being positive homogeneous and subadditive (5.15), there is ξ a linear functional such that

$$-f^0(x_0; -v) \leq \xi(v) \leq f^0(x_0; v) \ \forall \ v \in X.$$

But

$$f^0(x_0; v) \overset{5.15}{\leq} L\|v\|, \ -f^0(x_0; -v) \geq -L\|-v\|,$$

consequently $|\xi(v)| \leq L\|v\| \ \forall \ v \in X$, hence $\xi \in X^*$, $\xi \in \partial f(x_0)$ and $\|\xi\| \leq L$.

The convexity is obvious according to the definition (41).

For the remaining statement, $\partial f(x_0)$ is $*$-weak compact, it is enough to show that $\partial f(x_0)$ is *-weak closed (corollary of Alaoglu-Bourbaki-Kakutani theorem, [47], vol. III, p. 744, 3.37_1). Let be $\xi \in \overline{\partial f(x_0)}^{*-\text{weak}}$. Prove

$$(42) \ f^0(x_0; v) \geq \xi(v) \ \forall \ v \in X.$$

Fix v in X and let $\varepsilon > 0$ be arbitrary. There is μ in $\partial f(x_0)$such that $|(\xi - \mu)(v)| < \varepsilon$. Thus $\mu(v) > \xi(v) - \varepsilon$ and we have

$$f^0(x_0; v) \geq \mu(v) \geq \xi(v) - \varepsilon.$$

Pass to the limit for $\varepsilon \to 0$ and get (42).

2° Suppose *par absurdum* $\exists \ v_0 \in X$ such that $f^0(x_0; v_0) \neq \sup_{\xi \in \partial f(x_0)} \xi(v_0)$. Since $f^0(x_0; v_0) \geq \xi(v_0) \ \forall \ \xi \in \partial f(x_0)$, this implies (43) $f(x_0; v_0) > \xi(v_0) \ \forall \ \xi \in \partial f(x_0)$. Take μ a linear functional on X with (44) $\mu(v_0) = f^0(x_0; v_0)$ and $-f^0(x_0; -v) \leq \mu(v) \leq f^0(x_0; v) \ \forall \ v \in X$ ([47], vol. III, p. 693, 2.2). But $\mu \in \partial f(x_0)$ (see the beginning of the proof), consequently (44) knoks (43). ■

Continue with two propositions of [16].

5.17_1 *Let* X *be a reflexive or separable Banach space and* $f : X \to \mathbf{R}$ *locally Lipschitz. For every* x_0 *from* X *and* $\varepsilon > 0$, *there is* $\delta > 0$ *such that, for every* ξ *in* $\partial f(x)$ *with* $\|x - x_0\| \leq \delta$, *there exists* ξ' *in* $\partial f(x_0)$ *having the property*

$$|(\xi - \xi')(v)| \leq \varepsilon \ \forall \ v \in X \ \ ([16], \text{p. } 105).$$

■ Suppose *par absurdum* the contrary, $\exists \ x_0, \ v_0 \in X$, $\varepsilon > 0$ and also the sequences $(x_n)_{n\geq 1}$ in X, $(\xi_n)_{n\geq 1}$, ξ_n in $\partial f(x_n)$ such that $\forall \ n \in \mathbf{N}$

$$(44_1)\ \|x_n - x_0\| \le \frac{1}{n} \text{ and } |(\xi_n - \xi)(v_0)| > \varepsilon_0\ \forall\ \xi \in \partial f(x_0).$$

According to (44_1), x_n is, for $n \ge N$, in a neighborhood of x_0 as in 5.17, hence $n \ge N \Rightarrow$ $\Rightarrow \|\xi_n\| \le L$, so there is a subsequence $(\xi_{k_n})_{n\ge1}$ weakly convergent (Šmulian corollary, [47], vol. III, p. 770, 4.29) and consequently $*$-weak convergent (X reflexive $\Leftrightarrow X^*$ reflexive (Pettis, [47], vol. III, p. 750, 4.7), let be $\xi_{k_n} \xrightarrow{*-\text{weak}} \xi_0$, i.e. (44_2) $\xi_{k_n}(v) \to \xi_0(v)$ $\forall\ v \in X$ ([47], vol. III, p. 745, 3.39 or p.746, ex.9 or p. 763, 4.25). But $\xi_0 \in \partial f(x_0)$. Indeed, u being arbitrary fixed in X, $\exists\ (h_n)_{n\ge1}$, $h_n \to 0$ and $(t_n)_{n\ge1}$, $t_n \downarrow 0$ such that (use (31))

$$\forall\ n\ \frac{1}{t_n}\left[f(x_{k_n} + h_n + t_n u) - f(x_{k_n} + h_n)\right] \ge \xi_{k_n}(u) - \frac{1}{n},$$

pass to the limit for $n \to \infty$ taking (44_2) into account, $f^0(x_0; u) \ge \xi_0(u)$, i.e. $\xi_0 \in \partial f(x_0)$. Thus it can take $\xi = \xi_0$ in (44_1) and one gets in collision with (44_2). ■

Remark. In [16] 5.17_1, alias 1, (6), has an erroneus formulation. Moreover, X must be reflexive or separable.

5.17_2 *Let* X *be a real reflexive space and* f : X→**R** *locally Lipschitz.*

1° *For every* x_0 *at* X, *there is* ξ_0 *in* $\partial f(x_0)$ *such that*

$$\|\xi_0\| = \inf\{\|\xi\| : \xi \in \partial f(x_0)\}.$$

2° *The function* μ: X → **R**

$$\mu(x) = \inf\{\|\xi\| : \xi \in \partial f(x)\}$$

is lower semicontinous ([16], p.105).

■ 1° The application $\xi \to \|\xi\|$ of X^* in **R** is weak lower semicontinous and hence $*$-weak lower semicontinous ([47], vol III, p.746, (0)), since X^* is reflexive. On the other hand, $\partial f(x_0)$ is $*$-weak compact (5.17), hence the conclusion.

2° Suppose *par absurdum* the contrary and let be x_0 and x_n, $n \in$ **N** from X, $x_n \to x_0$ with

$$(44_3)\qquad \underline{\lim}_{n\to\infty}\ \mu(x_n) < \mu(x_0).$$

Take ξ_n in $\partial f(x_n)$ with $\mu(x_n) = \|\xi_n\|$ (see 1°). 5.17_1 yields, for every n from **N**, ξ_n in $\partial f(x_n)$ and ξ_n^0 in $\partial f(x_0)$ such that

$$(44_4)\qquad |(\xi_n - \xi_n^0)(v)| \le \frac{1}{n}\ \forall\ v \in X.$$

$\partial f(x_0)$ being $*$-weak compact (5.17), it is bounded, hence $(\xi_n^0)_{n\geq 1}$ has a weakly convergent subsequence and hence $*$-weakly convergent, $\left(\xi_{k_n}^0\right)_{n\geq 1}$,

(44₅) $$\xi_{k_n}^0 \xrightarrow{*-\text{weak}} \xi^0 \in \partial f(x_0).$$

Since

$$|(\xi_{k_n} - \xi^0)(v)| \leq |(\xi_{k_n} - \xi_{k_n}^0)(v)| + |(\xi_{k_n}^0 - \xi^0)(v)|,$$

from (44₄) and (44₅) we get $\xi_{k_n} \xrightarrow{*-\text{weak}} \xi^0$, and this attracts $\varliminf_{n\to\infty} \|\xi_{k_n}\| \geq \|\xi^0\|$ ([47], p. 745, 3.39, 3°), i. e. $\varliminf_{n\to\infty} \mu(x_{k_n}) \geq \|\xi^0\| \geq \mu(x_0)$ and we come in collision with (44₃). ■

Remark. In [16], 1, (7) (alias 5.17₂) the demand „ X reflexive" is lacking, and this is an error.

5.18 *Let* E *be a convex subset of a real Banach space and* $f : E \to \mathbf{R}$ *convex. If* f *is Lipschitz around* $x_0 \in \overset{\circ}{E}$, *then for every* v *at* X *we have*

$$f^0(x_0; v) = f'(x_0; v)$$

and the set of the subderivatives in x_0 *coincides with the set of Clarke subderivatives in* x_0 ([20], (1)).

■ It is enough to prove the first statement because the second is obtained, *via* 5.19, 2° (see below), by the formulae which are found in 5.17 and 5.5. We suppose obviously $v \neq$ $\neq 0$ and let $\varepsilon > 0$ be arbitrary fixed. Set

$$\alpha := \inf_{\substack{V\in\mathcal{V}(x_0)\\ r\in(0,+\infty)}} \sup_{\substack{x\in V\\ t\in(0,r)}} \frac{f(x+tv)-f(x)}{t} \overset{(31)}{(=} f^0(x_0; v)),$$

(44₆) $$\beta := \lim_{r\to 0+} \sup_{\|x-x_0\|<r\varepsilon} \sup_{t\in(0,r)} \frac{f(x+tv)-f(x)}{t}.$$

The suppositions $\alpha < \beta$ and $\beta < \alpha$ give a contradiction (attention to the definitions: limit, least upper bound, greatest lower bound), therefore $\alpha = \beta$. But the function $t \to$ $\to \frac{f(x+tv)-f(x)}{t}$, x near x_0, being increasing on an interval $(0, \eta)$ (see section C, the proof of the statement (*)), we get, via (44₆),

(44₇) $$f^0(x_0; v) = \lim_{r\to 0+} \sup_{\|x-x_0\|<r\varepsilon} \frac{f(x+rv)-f(x)}{r}.$$

But we have

$\dfrac{f(x+rv)-f(x)}{r} = \left[\dfrac{f(x+rv)-f(x)}{r} - \dfrac{f(x_0+rv)-f(x_0)}{r}\right] + \dfrac{f(x_0+rv)-f(x_0)}{r}$ and, *via*

Lipschitz condition verified on the open ball $S(x_0\,;\,r\varepsilon)$ for r near 0, $x \in S(x_0\,;\,r\varepsilon) \Rightarrow$

$\left|\dfrac{f(x+rv)-f(x)}{r} - \dfrac{f(x_0+rv)-f(x_0)}{r}\right| \le 2\varepsilon L$, hence, using ($44_7$), $f^0(x_0\,;\,v) \le$

$\le \lim\limits_{r\to 0+} \dfrac{f(x_0+rv)-f(x_0)}{r} + 2\varepsilon L = f'(x_0\,;\,v) + 2\varepsilon L$ and hence $f^0(x_0\,;\,v) \le f'(x_0\,;\,v)$.

Pass to the opposite inequality. We have, $\forall\ t$ from $(0, r)$, $\dfrac{f(x_0+tv)-f(x_0)}{t} \le$

$\le \sup\limits_{\substack{x\in V\\ t\in(0,r)}} \dfrac{f(x+tv)-f(x)}{t}$, where $V \in \mathscr{V}(x_0)$, hence $\lim\limits_{t\to 0+} \dfrac{f(x_0+tv)-f(x_0)}{t} \le$

$\le \sup\limits_{\substack{x\in V\\ t\in(0,r)}} \dfrac{f(x+tv)-f(x)}{t}$ and consequently, according to (31), $f'(x_0\,;\,v) \le f^0(x_0\,;\,v)$. ■

Remark. The second part of the proof uses only the hypothesis „ there is $f'(x_0\,;\,v)$".

Example 3. As an application of 5.18 calculate $\partial f(x)$, $f: \mathbf{R}^n \to \mathbf{R}$, $f(x_1, ..., x_n) =$ $= \max\{x_i : i = \overline{1,n}\}$. f is convex (f_i convex $\forall\ i \in I \Rightarrow \sup\limits_{i\in I} f_i$ convex), but also Lipschitz on $\mathbf{R}^n$: $|f(x_1', ..., x_n') - f(x_1'', ..., x_n'')| = |x_i' - x_j''| =$ (for example) $(x_i' - x_j'') = (x_i' - x_i'') +$ $+ (x_i'' - x_j'') \le x_i' - x_i'' = |x_i' - x_i''| \le \|(x_1', ..., x_n') - (x_1'', ..., x_n'')\|_2$ (the Euclidean norm). Set

$J(x) := \{i\colon 1 \le i \le n,\ f(x) = x_i\}$. $f'(x; v) = \lim\limits_{t\to 0+} \dfrac{f(x+tv)-f(x)}{t} =$

$= \lim\limits_{t\to 0+} \dfrac{\max\{x_i + tv_i : i = \overline{1,n}\} - f(x)}{t} = \lim\limits_{t\to 0+}\max\limits_{1\le i\le n} \dfrac{x_i + tv_i - f(x)}{t} = \lim\limits_{t\to 0+}\max\limits_{i\in J(x)} \dfrac{x_i + tv_i - x_i}{t} =$

$= \max\limits_{i\in J(x)} v_i$.

Thus, according to 5.18, $f^0(x; v) = \max\limits_{i\in J(x)} v_i$, consequently $\partial f(x) = \{\xi \in \mathbf{R}^n : \max\limits_{i\in J(x)} v_i \ge$ $\ge \xi \cdot v\ \ \forall\ v \in \mathbf{R}^n\}$, hence $\partial f(x) = \{\xi \in \mathbf{R}^n : \xi_i \ge 0,\ \sum\limits_{i=1}^{n} \xi_i = 1,\ i \ne J(x) \Rightarrow \xi_i = 0\}$.

Here we should add a few words about the support functional (a change as compared to the definition (29_6), §2).

Let X be a real normed space and E a nonempty subset of X. The function $\sigma_E : X^* \to$

$\to (-\infty, +\infty]$,

(45) $$\sigma_E(\xi) = \sup_{x \in E} \xi(x)$$

is by definition the *support functional* of *E*.

Suppose *E* a nonempty subset of the normed space X^*. Then σ_E is defined on the bidual X^{**}. But the Hahn embedding $x \to Jx$ from *X* into X^{**} is linear, injective and preserves the norms, one identifies *X* and *J*(*X*) and then, $\forall\, x$ at *X*,

(46) $$\sigma_E(x) = \sup_{\xi \in E} \xi(x)$$

makes sense.

(46) makes the connection with 5.5 and with 2° of 5.17. It can state

5.19 *Let* X *be a real Banach space,* A *and* B *nonempty convex closed subsets,* S *and* T *nonempty convex *-weak closed subsets of* X^*. *Then*

1° $A \subset B \Leftrightarrow \sigma_A(\xi) \le \sigma_B(\xi)\ \ \forall\, \xi \in X^*$;

2° $S \subset T \Leftrightarrow \sigma_S(x) \le \sigma_T(x)\ \forall\, x \in X$;

3° S *is *-weak compact* $\Leftrightarrow$ σ_S *takes on* X *finite values*;

4° *Let* $f : X \to (-\infty, +\infty]$. *If there is* S *a nonempty convex *-weak closed subset of* X^* *such that* $f = \sigma_S$, *then* f *is a proper positive homogeneous subadditive and upper semicontinous. This representation is unique.*

Return to Clark subderivative. But as a preliminary we present two definitions.

Let *X*, *Y* be nonempty subsets and $f : X \to 2^Y$ a multivalued function. The *graph* of *f*, Gr *f*, is by definition the set

$$\{(x, y) \in X \times Y : x \in X, y \in f(x)\}.$$

Suppose *X*, *Y* be normed spaces and let be $x_0 \in X$. *f* is by definition *upper semicontinuous at* x_0, if $\forall\, \varepsilon > 0\ \exists\, \delta > 0$ such that (open ball)

$$(47)\ x \in S(x_0, \delta) \Rightarrow f(x) \subset f(x_0) + S(0, \varepsilon).$$

5.20 *Let* X *be a real normed space and* $f : X \to \mathbf{R}$ *Lipschitz around* x_0, L *the constant.*

1° *Let be* $(x_n)_{n\ge1}$, $x_n \in X$, $x_n \to x_0$ *and* $(\xi_n)_{n\ge1}$, $\xi_n \in \partial f(x_n)\ \forall\, n \ge 1$ *sequences. If* ξ_0 *is a *-weak adherence value of* $(\xi_n)_{n\ge1}$, *then* $\xi_0 \in \partial f(x_0)$;

2° $\partial f(x_0) = \bigcap_{r\in(0,+\infty)} \bigcup_{x\in S(x_0,r)} \partial f(x)$;

3° *If* $\dim X < +\infty$, *the multivalued function* $x \to \partial f(x)$ *is upper semicontinous at* x_0

([20], 1)

■ 1° Let v be arbitrary fixed in X. We must prove

(48) $$f^0(x_0; v) \geq \xi_0(v).$$

ξ_0 being a *-weak adherence value, we can find a subsequence $(\xi_{k_n})_{n\geq 1}$ such that $\xi_{k_n}(v) \to$

$\to \xi_0(v)$ ([47], p.738, 3.29). But $f^0(x_{k_n}; v) \geq \xi_{k_n}(v)\ \forall\ n \geq 1$, take the upper limit for

$n \to \infty$ and because $\overline{\lim_{n\to\infty}} f^0(x_{k_n}; v) \overset{5.16}{\leq} f^0(x_0; v)$ we get even (48).

2° Use 1°.

3° Suppose the contrary. Then $\exists\ \varepsilon_0 > 0$ and $\exists\ (x_n)_{n\geq 1}$, $(\xi_n)_{n\geq 1}$, $\xi_n \in \partial f(x_n)$ such that $\forall\ n \geq 1$

(49) $$\|x_n - x_0\| < \frac{1}{n},\ \|\xi_n - w\| > \varepsilon_0\ \forall\ w \in \partial f(x_0).$$

But for $n \geq N$ we have $\|\xi_n\| \leq L$ (5.17), hence there is a convergent subsequence $(\xi_{k_n})_{n\geq 1}$, $\xi_{k_n} \to \xi_0$ in the *-weak topology. But $\xi_0 \in \partial f(x_0)$ (see 2°) and according to (49) one finds $\|\xi_{k_n} - \xi_0\| > \varepsilon_0\ \forall\ n$, consequently $\xi_{k_n} \not\to \xi_0$ and hence, since $\dim V < +\infty$, $\xi_{k_n} \not\to \xi_0$ in the *-weak topology ([47], vol III, p.740, 3.31; p.745, 3.39, 1°), a contradiction. ■

STRICT DERIVATIVE OF HADAMARD TYPE

Let X, Y be normed spaces on the same field of scalars, $E \subset X$, $f: E \to Y$, $x_0 \in \overset{\circ}{E}$. ξ from $L(X; Y)$ is by definition the *strict derivative* (of *Hadamard type*) of f *at* x_0, set $\xi = D_s f(x_0)$, if for every K a compact subset of X we have

(50) $$\operatorname*{limu}_{\substack{x\to x_0 \\ t\to 0+ \\ v\in K}} \frac{f(x+tv) - f(x)}{t} = \xi(v)\ (Bourbaki)$$

(uniform limit on K for $x \to x_0$, $t \to 0+$). ■

Remark. $D_s f(x_0)$ is what we could call *generalized* β_H*-derivative* of f *at* x_0, β_H the bornology of the compact subsets of X (*Hadamard bornology*), see §3.

5.21 *Let* ξ *be from* L(X; Y). $\xi = D_s f(x_0)$ *iff* f *is Lipschitz around* x_0 *and, for every* v *at* X,

$$\xi(v) = \lim_{\substack{x \to x_0 \\ t \to 0+}} \frac{f(x+tv) - f(x)}{t}. \text{ ([20], 1).}$$

■ *Necessity*. The second condition is obviously verified since $\{v\}$ is compact. Suppose *par absurdum* that f is not Lipschitz around x_0. Then there is, for every $n \in \mathbf{N}$, $n \geq N$, x_n and x_n' from $S(x_0, \frac{1}{n})$ (open ball) such that

(51) $$\|f(x_n) - f(x_n')\| \geq n\,\|x_n - x_n'\|.$$

Take v_n from X with $\|v_n\| = \frac{1}{\sqrt{n}}$ and then $t_n > 0$ such that (52) $x_n' = x_n + t_n v_n$. Thus $t_n \to 0$.

The set $K := \{v_n : n \geq N\} \cup \{0\}$ is compact since $v_n \to 0$, consequently, according to the definition (50), $\forall\, \varepsilon > 0\ \exists\, n_0 \in \mathbf{N}$, $n_0 \geq N$ such that

$$n \geq n_0 \Rightarrow \left\| \frac{f(x_n + t_n v) - f(x_n)}{t_n} - \xi(v) \right\| \leq \varepsilon\ \forall\, v \in K,$$

hence

(53) $$\left\| \frac{f(x_n + t_n v) - f(x_n)}{t_n} \right\| \leq \varepsilon + \|\xi(v)\|,\ \forall\, n \geq n_0,\ \forall\, v \in K.$$

Take $v = v_n$ in (53) and taking (51) and (52) into account we get $\sqrt{n} \leq \varepsilon + \|\xi(v_n)\|\ \forall\, n \geq n_0$, contradiction because $\xi(v_n) \to 0$.

Sufficiency. Let K be a compact part of X and $\varepsilon > 0$ arbitrary. According to the hypothesis, in particular for every v at K $\exists\, \delta_v > 0$ such that

(54) $$x \in S(x_0, \delta_v),\ t \in \left(0, \frac{\delta_v}{\delta_v + \|v\|}\right) \Rightarrow \left\| \frac{f(x+tv) - f(x)}{t} - \xi(v) \right\| \leq \varepsilon.$$

Suppose δ_v sufficiently small such that $(L + \|\xi\|)\delta_v \leq \varepsilon$, L Lipschitz constant, and in addition f verifies on $S(x_0, 2\delta_v)$ the Lipschitz condition. Then

$$x \in S(x_0, \delta_v),\ w \in S(v, \delta_v),\ t \in \left(0, \frac{\delta_v}{\delta_v + \|v\|}\right) \Rightarrow \left\| \frac{f(x+tw) - f(x)}{t} - \xi(w) \right\| \leq \left\| \frac{f(x+tw) - f(x)}{t} - \frac{f(x+tv) - f(x)}{t} \right\| + \left\| \frac{f(x+tv) - f(x)}{t} - \xi(v) \right\| + \|\xi(v-w)\| \overset{(54)}{\leq} L\|v - w\| + \varepsilon + \|\xi\|\,\|v - w\|$$

(in the considered situation $x + tv$ and $x + tw$ are found in $S(x_0, 2\delta_v)$),

(55) $$\left\| \frac{f(x+tw) - f(x)}{t} - \xi(w) \right\| \leq 2\varepsilon,\ \forall\, x \in S(x_0, \delta_v),\ \forall\, w \in S(v, \delta_v),\ \forall\, t \in \left(0, \frac{\delta_v}{\delta_v + \|v\|}\right).$$

The open balls $S(v, \delta_v)$, $v \in K$ cover K, let $S(v_1, \delta_{v_1}), \ldots, S(v_N, \delta_{v_N})$ which also cover K.

Take $\delta = \min_{1\le i\le N} \delta_{v_i}$ and for this (55) shows that $x \in S(x_0, \delta)$, $t \in (0, \delta) \Rightarrow$

$$\Rightarrow \left\| \frac{f(x+tw)-f(x)}{t} - \xi(w) \right\| \le 2\varepsilon, \ \forall\, w \in K, \xi = D_s f(x_0). \ \blacksquare$$

The connection between the strict derivative and the Gâteaux derivative is now at hand through 5.21.

5.22 *If* f *is Gâteaux differentiable on an open neighbourhood of* x_0 *included in* E *and* f'_w *is continuous at* x_0, *then* f *is strictly differentiable at* x_0 *and*

$$D_s f(x_0) = f'_w(x_0).$$

■ Use 5.21. Let $S := S(x_0, \delta)$ an open ball included in E on which f is Gâteaux differentiable. $x \to f'_w(x)$ being continuous at x_0, it is bounded on a neighbourhood of x_0, we diminish δ for having $\| f'_w(x)\| \le L$ on S. For every x, y from S we have

$$\|f(x) - f(y)\| \le \|x-y\| \sup_{z\in[y,x]} \|f'_w(z)\|$$

([47], vol I, p.403,1.10), hence $\|f(x) - f(y)\| \le L\|x-y\|$, f is Lipschitz around x_0. It remains to prove

(56) $$\lim_{\substack{x\to x_0\\ t\to 0+}} \frac{f(x+tv)-f(x)}{t} = f'_w(x_0)(v) \ \forall\, v \in X.$$

Fix v in X. We have, for x near x_0 and t strictly positive near 0 (we go into S),

(57) $$\|f(x+tv) - f(x) - f'_w(x_0)(tv)\| \le t\,\|v\| \sup_{z\in[x,x+tv]} \|f'_w(z) - f'_w(x_0)\|$$

([47], vol I, p.403, 1.11).

Let $\varepsilon > 0$ be arbitrary, $\exists\, \eta > 0$ such that (we suppose $v \ne 0$)

(58) $$\|x - x_0\| \le \eta \Rightarrow \|f'_w(x) - f'_w(x_0)\| \le \frac{\varepsilon}{\|v\|}.$$

Divide by t in (57) taking (58) into account and get (56). ■

Remark relating to 5.22. It is obvious that $D_s f(x_0) = f'(x_0)$, the Fréchet derivative at x_0.

Make now the link with the directional derivative.

5.23 *If* f *is strictly differentiable in* x_0 *then, for every* v *at* X,

$$D_s f(x_0)(v) = f'(x_0; v)$$

and hence $D_s f(x_0) = f'_w(x_0)$.

■ Fix v in X, put $\alpha := D_s f(x_0)(v)$, $F(x,t) := \dfrac{f(x+tv)-f(x)}{t}$ and let be $\varepsilon > 0$. Since

$\alpha \overset{5.21}{=} \lim_{\substack{x\to x_0\\ t\to 0+}} F(x,t)$, $\exists\, \delta > 0$ such that

$$0 < \|x - x_0\| \le \delta,\ 0 < t \le \delta \Rightarrow |\alpha - F(x,t)| \le \varepsilon.$$

Pass to the limit for $x \to x_0$, it follows, because f being continous in x_0 (since f is Lipschitz around x_0, 5.21) we have $\lim_{x\to x_0} F(x,t) = \dfrac{f(x_0+tv)-f(x_0)}{t}$, $0 < t \le \delta \Rightarrow \left|\alpha - \dfrac{f(x_0+tv)-f(x_0)}{t}\right| \le \varepsilon$, i.e. the first formula of the statement. The second formula is obvious since $D_s f(x_0) \in X^*$. ■

Pass to the connection with Clarke subderivative.

Let X be a real normed space, $E \subset X$, $f : E \to \mathbf{R}$, $x_0 \in \overset{o}{E}$.

5.24 *If* f *is Gâteaux differentiable in* x_0, *then*

$$f'_w(x_0) \in \partial f(x_0).$$

■ Let v be arbitrary in X. We must prove ((41))

$$f^0(x_0; v) \ge f'_w(x_0)(v). \tag{59}$$

But

$$f'_w(x_0)(v) = \lim_{t\to 0}\frac{f(x_0+tv)-f(x_0)}{t} = \lim_{t\to 0+}\frac{f(x_0+tv)-f(x_0)}{t} = f'(x_0; v),\ f'(x_0; v) \le$$
$$\le f^0(x_0; v)$$

(the justification of the end of the proof for 5.18 remains true) and consequently (59). ■

5.25 *If* f *is strictly differentiable in* x_0, *then* $\partial f(x_0) = \{D_s f(x_0)\}$. *Reciprocally, if* f *is Lipschitz around* x_0 *and* $\partial f(x_0) = \{\xi\}$, *then* $\xi = D_s f(x_0)$ ([20], 1).

■ *First statement.* We have, because f is Lipschitz around x_0 (5.21), $f^0(x_0; v) \overset{5.17}{=} \sup_{\xi\in\partial f(x_0)} \xi(v)\ \forall\, v \in X$. But $D_s f(x_0)(v) = \sup_{\xi\in\{D_s f(x_0)\}} \xi(v)\ \forall\, v \in X$ and the subsets of X^* $\partial f(x_0)$ and $\{D_s f(x_0)\}$ are nonempty convex *-weak closed (5.17), consequently, as $f^0(x_0; \cdot) = D_s f(x_0)$ (attention to definitions), we have $\partial f(x_0) = \{D_s f(x_0)\}$ (5.19, 2°).

Second statement. It remains to prove, according to 5.21,

$$(60)\qquad \lim_{\substack{x\to x_0\\ t\to 0+}} \frac{f(x+tv)-f(x)}{t} = \xi(v)\ \ \forall\ v \in X.$$

$f^{\,0}(x_0\,;\,v) \overset{5.17}{=} \sup\limits_{\mu\in\partial f(x_0)} \mu(v)$, $\partial f(x_0) = \{\xi\}$, consequently

$$(61)\qquad f^{\,0}(x_0\,;\,v) = \xi(v)\ \forall\ v \in X.$$

Fix v in X.

$$\underline{\lim}_{\substack{x\to x_0\\ t\to 0+}} \frac{f(x+tv)-f(x)}{t} = -\ \overline{\lim}_{\substack{x\to x_0\\ t\to 0+}} \frac{f(x)-f(x+tv)}{t} = -\ \overline{\lim}_{\substack{x\to x_0\\ t\to 0+}} \frac{f(x+tv-tv)-f(x+tv)}{t} =$$

$$= -f^{0}(x_0\,;\,-v) \overset{(61)}{=} -\ \xi(-\,v) = \xi(v) = f^{0}(x_0\,;\,v) = \overline{\lim}_{x\to x_0} \frac{f(x+tv)-f(x)}{t}$$ and consequently

(60). ■

RULES OF SUBDIFFERENTIAL CALCULUS

Calculus with the generalized gradients was mostly established by F. H. Clarke ([20], 1).

In the following X is real normed space, $E \subset X$, $f: E \to \mathbf{R}$, $x_0 \in \overset{o}{E}$.

5.26 Scalar factor. *Let* f *be Lipschitz around* x_0. *For every* λ *from* **R** *we have*

$$\partial(\lambda\, f)(x_0) = \lambda\, \partial\, f(x_0).$$

■ $\lambda > 0$. $\xi \in \partial(\lambda f)(x_0) \Rightarrow \forall\ v \in X\ \ \xi(v) \le (\lambda f)^{0}(x_0\,;\,v) \overset{5.15}{=} \lambda\, f^{\,0}(x_0\,;\,v)$, $\frac{\xi}{\lambda}(v) \le f^{0}(x_0\,;\,v)$, $\frac{\xi}{\lambda} \in \partial f(x_0)$, $\partial(\lambda f)(x_0) \subset \lambda\, \partial f(x_0)$. $\xi \in \lambda\, \partial f(x_0) \Rightarrow \forall\ v \in X\ \ \frac{\xi}{\lambda}(v) \le f^{0}(x_0\,;\,v)$, $\xi(v) \le \lambda f^{0}(x_0\,;\,v) \overset{5.15}{=} (\lambda f)^{0}(x_0\,;\,v)$, $\xi \in \partial(\lambda f)(x_0)$, $\lambda\, \partial f(x_0) \subset \partial(\lambda f)(x_0)$, these two sets are equal.

$\lambda = 0$. Obviously.

$\lambda = -1$. $\xi \in \partial(-f)(x_0) \Rightarrow \forall\ v \in X\ \ \xi(v) \le (-f)^{0}(x_0\,;\,v) \overset{5.15}{=} f^{0}(x_0\,;\,-v)$, $-\,\xi(-\,v) \le f^{\,0}(x_0\,;\,-v)$, hence $-\,\xi \in \partial f(x_0)$, $\xi \in -\,\partial f(x_0)$. $\xi \in -\,\partial f(x_0) \Rightarrow \forall\ v \in X\ \ -\xi(v) \le f^{\,0}(x_0\,;\,v) = (-f)^{0}(x_0\,;\,-v)$, $\xi(-\,v) \le (-\,f)^{0}(x_0\,;\,-v)$, hence $\xi \in \partial(-\,f)(x_0)$. ■

5.27 Local extremum. *Let* f *be Lipschitz around* x_0. *If* x_0 *is a point of local extremum for* f *we have*

$$0 \in \partial f(x_0).$$

■ x_0 *local minimum point*. Let v be arbitrary in X. We must prove that $f^0(x_0; v) \geq 0$. Suppose *par absurdum* that $f^0(x_0; v) < 0$. There is, according to (31), V in $\mathcal{V}(x_0)$ and $r >$ > 0 such that $\sup\limits_{x \in V} \dfrac{f(x+tv)-f(x)}{t} < 0$, hence in particular we have $t \in (0, r) \Rightarrow f(x_0 + tv) -$ $- f(x_0) < 0$, which hinders x_0 to be a local minimum point for f, a contradiction.

x_0 *local maximum point*. In this case x_0 is a local minimum point for $-f$, hence $0 \in$ $\in \partial(-f)(x_0) \overset{5.26}{=} -\partial f(x_0)$, $0 \in \partial f(x_0)$. ■

5.28 Finite sums. *Let* X *be a Banach space. If* $f_1, \ldots, f_n$ *are Lipschitz around* x_0, *then*

$$\partial\left(\sum_{i=1}^{n} f_i\right)(x_0) \subset \sum_{i=1}^{n} \partial f_i(x_0).$$

■ $\sum\limits_{i=1}^{n} f_i$ is Lipschitz around x_0. The recurrence allows to prove the statement only for the case $n = 2$. Use the support functional ((45)) and 5.19, 2°. Setting $A := \partial(f_1 + f_2)(x_0)$ and $B := \partial f_1(x_0) + \partial f_2(x_0)$, we have, with v arbitrary in X, $\sigma_A(v) = (f_1 + f_2)^0(x_0; v)$, $\sigma_B(v) = f_1^0(x_0; v) + f_2^0(x_0; v)$. But $(f_1 + f_2)^0(x_0; v) \leq f_1^0(x_0; v) + f_2^0(x_0; v)$, because

$$\overline{\lim_{\substack{x \to x_0 \\ t \to 0+}}} \left[\frac{f_1(x+tv)-f_1(x)}{t} + \frac{f_2(x+tv)-f_2(x)}{t}\right] \leq \overline{\lim_{\substack{x \to x_0 \\ t \to 0+}}} \frac{f_1(x+tv)-f_1(x)}{t} +$$

$$+ \overline{\lim_{\substack{x \to x_0 \\ t \to 0+}}} \frac{f_2(x+tv)-f_2(x)}{t},$$

consequently $A \subset B$. ■

5.28 can be strengthened in an equality.

Precisely

5.29 *Let* X *be a Banach space. If* $f_1, \ldots, f_n$ *Lipschitz around* x_0 *are, except at most one, strictly differentiable at* x_0, *then*

$$\partial\Big(\sum_{i=1}^{n} f_i\Big)(x_0) = \sum_{i=1}^{n} \partial f_i(x_0).$$

■ Since every finite sum of strictly differentiable functions is strictly differentiable (use (50)), it is sufficient to consider only the case $f_1 + f_2, f_1$ strictly differentiable.

(62) $$\partial(f_1 + f_2)(x_0) \overset{5.28}{\subset} \partial f_1(x_0) + \partial f_2(x_0) \overset{5.25}{=} D_s f_1(x_0) + \partial f_2(x_0).$$

Let ξ be arbitrary from $\partial(f_1 + f_2)(x_0)$. Then, according to (62), (63) $\xi = D_s f(x_0) + \xi_2$, $\xi_2 \in \partial f_2(x_0)$. Let v be fixed arbitrary in X. Taking (63) into account, $\xi(v) = D_s f_1(x_0)(v) + \xi_2(v)$, consequently

$$\sup_{\xi \in \partial(f_1+f_2)(x_0)} \xi(v) = D_s f_1(x_0)(v) + \sup_{\xi_2 \in \partial f_2(x_0)} \xi_2(v),$$

hence (see 5.17)

$$(f_1 + f_2)^0(x_0; v) = D_s f_1(x_0)(v) + f_2^0(x_0; v)$$

and one applies 5.19, 2°. ■

5.30 *Corollary. Let* $f_1, \dots, f_n$ *be Lipschitz around* x_0, *among which at least one is not strictly differentiable. Then for every* $\lambda_1, \dots, \lambda_n$ *from* **R**,

$$\partial\Big(\sum_{i=1}^{n} \lambda_i f_i\Big)(x_0) = \sum_{i=1}^{n} \lambda_i \partial f_i(x_0).$$

■ Combine 5.29 and 5.26. ■

The following concept proved to be useful in questions concerning equalities where generalized gradients appear.

Regular function ([20], 1). Let X be a real normed space, $E \subset X, f: E \to \mathbf{R}$, $x_0 \in \overset{o}{E}$. f is by definition *regular at* x_0, if, for every v of X, there is $f'(x_0; v)$ and

(64) $$f'(x_0; v) = f^0(x_0; v).$$

Here are the first statements on the concept.

5.31 *Let* f *be Lipschitz around* x_0.

1° *If* f *is strictly differentiable at* x_0, *then* f *is regular at* x_0;

2° *If,* X *being complete,* f *is convex, then* f *is regular at* x_0;

3° *Every linear combination with positive coefficients of regular functions at* x_0 *is regular at* x_0;

4° *If,* X *being complete,* f *is regular and Gâteaux differentiable at* x_0, *then*

$$\partial f(x_0) = \{ f'_w(x_0) \}.$$

■ 1° Let v be arbitrary in X. $f^0(x_0; v) \overset{5.17}{=} \sup_{\xi \in \partial f(x_0)} \xi(v)$, but $\partial f(x_0) \overset{5.25}{=} \{D_s f(x_0)\}$, hence $f^0(x_0; v) = D_s f(x_0)(v)$. On the other hand $D_s f(x_0)(v) \overset{5.23}{=} f'(x_0; v)$.

2° Apply 5.18.

3° We must show

$$(65)\qquad f \text{ regular at } x_0,\ \lambda \geq 0 \Rightarrow \lambda f \text{ regular at } x_0,$$

and

$$(66)\qquad f, g \text{ regular at } x_0 \Rightarrow f+g \text{ regular at } x_0.$$

Let v be arbitrary in X. It can suppose $\lambda > 0$. $f^0(x_0;v) \overset{5.15}{=} f'(x_0;v)$, $(\lambda f)^0(x_0;v) = \lambda f^0(x_0;v)$, $(\lambda f)'(x_0;v) = \lambda f'(x_0;v)$ (the definition!) and consequently (65).

$$(f+g)'(x_0;v) \overset{\text{the definition}}{=} f'(x_0;v) + g'(x_0;v) = f^0(x_0;v) + g^0(x_0;v) \geq (f+g)^0(x_0;v)$$

(see the proof for 5.28), hence $(f+g)'(x_0;v) \geq (f+g)^0(x_0;v)$. The opposite inequality is supplied by the remark to 5.18 and so one gets (66).

4° Let v be arbitrary in X. $f'_w(x_0)(v) = f'(x_0;v) = f^0(x_0;v)$, hence $f^0(x_0;v) = \sup_{\xi\in\{f'_w(x_0)\}} \xi(v)$. On the other hand, $f^0(x_0;v) = \sup_{\xi\in\partial f(x_0)} \xi(v)$, apply 5.19, 2°. ■

5.31 leads to a variant of 5.29.

5.32 *Let* X *be a Banach space. If* $f_1, \ldots, f_n$ *Lipschitz around* x_0 *are regular at* x_0, *then*

$$\partial\Big(\sum_{i=1}^{n} f_i\Big)(x_0) = \sum_{i=1}^{n} \partial f_i(x_0).$$

■ The recurrence and 3° of 3.51 allow us to prove it only in the case $n = 2$. Let A, B be the sets from the statement. We have $\sigma_A = (f_1+f_2)^0(x_0;\cdot)$, $\sigma_B = f_1^0(x_0;\cdot) + f_2^0(x_0;\cdot)$. But $\sigma_A = (f_1+f_2)'(x_0;\cdot) = f_1'(x_0;\cdot) + f_2'(x_0;\cdot) = f_1^0(x_0;\cdot) + f_2^0(x_0;\cdot) = \sigma_B$, $\sigma_A = \sigma_B$ and consequently $A = B$ (5.19, 2°). ■

Remark 1. The inclusion of 5.29 can be strict : if $\partial f(x_0)$ has only an element, then for $g = -f$ we have $\partial(f+g)(x_0) = \{0\}$, while $\partial f(x_0) + \partial g(x_0) = \partial f(x_0) - \partial f(x_0)$ (5.26).

Remark 2. The formula of 5.32 remains true for the assumption that not $f_1, \ldots, f_n$ are regular at x_0 but $-f_1, \ldots, -f_n$, because

$$\partial\Big(-\sum_{i=1}^{n} f_i\Big)(x_0) = \sum_{i=1}^{n} \partial(-f_i)(x_0)$$

and $\partial(-g)(x_0) \overset{5.26}{=} -\partial g(x_0)$.

5.33 Formula of the finite increments (Mean - Value Theorem). *Let* f *be locally Lipschitz on an open set* U *of* X *Banach space and* a,b *distinct points in* U. *If* [a,b] ⊂ U, *then there exists* c *in* (a, b) *such that*

$$f(b) - f(a) \in \partial f(c)(b-a) \text{ (*G. Lebourg*)}.$$

Explanation. $\partial f(c)(b-a) = \{\xi(b-a) : \xi \in \partial f(c)\}$.

■ We use two auxiliary functions. The first is $g : [0, 1] \to \mathbf{R}$,

$$g(t) = f(c_t),\ c_t := a + t(b-a).$$

g is locally Lipschitz on [0, 1] [t, for instance, is in (0, 1), on a ball with the center at $a + t(b - a)$ the Lipschitz condition is verified, the intersections of the ball with (a, b) give t', t'' with $t \in (t', t'')$ etc.] and we have

(67) $$\partial g(t) \subset \partial f(c_t)(b-a).$$

Justify (67), $\mathbf{R}$ and $\mathbf{R}^*$ are identified by Riesz theorem. The two sets in (67) are convex (see 5.17), hence they are intervals of $\mathbf{R}$, and $*$-weak compact, hence compact ([47], vol III, p.740, 3.31). Consequently, (67) reverts to

(68) $$\sup\{\partial g(t)v\} \le \sup\{\partial f(c_t)(b-a)v\},\ v = \pm 1.$$

$$\sup\{\partial g(t)v\} \overset{5.17}{=} g^0(t; v) = \overline{\lim_{\substack{\tau\to t\\ \lambda\to 0}}} \frac{g(\tau+\lambda v) - g(\tau)}{\lambda} =$$

$$\overline{\lim_{\substack{\tau\to t\\ \lambda\to 0+}}} \frac{f(a+(\tau+\lambda v)(b-a)) - f(a+\tau(b-a))}{\lambda} \overset{\text{def}}{\le} \overline{\lim_{\substack{x\to c_t\\ \lambda\to 0+}}} \frac{f(x+\lambda v(b-a)) - f(x)}{\lambda} = f^0(c_t; v(b-$$

$$-a)) \overset{5.17}{=} \sup\{\partial f(c_t)(b-a)v\}$$ and therefore (68).

Pass to the second auxiliary function $h : [0, 1] \to \mathbf{R}$,

$$h(t) = g(t) + t[f(a) - f(b)].$$

h is continous, because g is continous (being locally Lipschitz), and $h(0) = h(1)$ $(= f(a))$.

If h is constant it has in (0, 1) a point of local extremum t' and then $0 \overset{5.27}{\in} \partial h(t')$. Suppose h not constant and let $M := \sup_{t\in[0,1]} h(t)$, $m := \inf_{t\in[0,1]} h(t)$. Thus $m \ne M$ and let be $m = f(t_1)$, $M = f(t_2)$, hence $t_1 \ne t_2$. Since $h(0) = h(1)$, either t_1 or t_2 are *a fortiori* in (0, 1) (*par absurdum*), i.e. there is t' in (0, 1) a point of local extremum of h and consequently again (69) $0 \in \partial h(t')$. But, setting $\varphi(t) = t(f(a) - f(b))$, we have

$$\partial h(t') \overset{5.28}{\subset} \partial g(t') + \partial\varphi(t') \overset{(67)}{\subset} \partial f(c_{t'})(b-a) + \partial\varphi(t'),\ \partial\varphi(t') = f(a) - f(b)$$

(5.22,5.25), consequently, by (69), $0 \in \partial f(c_{t'})(b-a) + [f(a) - f(b)]$, i. e. $f(b) - f(a) \in$ $\in \partial f(c_{t'})(b-a)$. Take $c = c_{t'}$. ■

Let X be a real Banach space, $E \subset X$, $x_0 \in \overset{o}{E}$ and $h := (h_1, \dots, h_n) : E \to \mathbf{R}^n$, $g: \mathbf{R}^n \to$ $\to \mathbf{R}$.

5.34 Clarke subderivative of the composed function. *Let be* $f := g \circ h$. *If* $h_1, \dots, h_n$ *are Lipschitz around* x_0 *and* g *is Lipschitz around* $h(x_0)$, *then* f *is Lipschitz around* x_0 *and*

$$\partial f(x_0) \subset \overline{co}\left\{\sum_{i=1}^{n} \alpha_i \xi_i : \xi_i \in \partial h_i(x_0), (\alpha_1, \dots, \alpha_n) \in \partial g(h(x_0))\right\}$$

(*F. H. Clarke*).

Explanation. $\overline{co}\{\ \}$ is the ∗-weak closed convex hull of the set in braces. $\mathbf{R}^n$ and $(\mathbf{R}^n)^*$ are identified by Riesz theorem.

■ f is indeed Lipschitz around x_0 : take V a neighbourhood of $h(x_0)$ on which g is Lipschitz, $\exists\ U$ a neighbourhood of x_0 such that $h(U) \subset V$ (the continuity in x_0) and we diminish U such that h is Lipschitz on U, then f is lipschitz on U. Let be

$$S := \left\{\sum_{i=1}^{n} \alpha_i \xi_i : \xi_i \in \partial h_i(x_0), (\alpha_1, \dots, \alpha_n) \in \partial g(h(x_0))\right\}.$$

Since $\partial h_1(x_0) \times \dots \times \partial h_n(x_0)$ is *-weak compact (5.17 and Tihonov theorem), and $\partial g(h(x_0))$ as well, S is *-weak compact and then we have for the support functionals, for every v from X,

(70) $$\sigma_S(v) = \sigma_{\overline{co}\, S}(v).$$

Consequently, if we prove

(71) $$\sigma_{\partial f(x_0)}(v) \le \sigma_S(v)$$

we get the wanted inclusion (5.19, 2°).

But $\sigma_{\partial f(x_0)}(v) \overset{5.17}{=} f^0(x_0; v)$ and

$$\sigma_S(v) = \sup\left\{\sum_{i=1}^{n} \alpha_i \xi_i(v) : \xi_i \in \partial h_i(x_0), (\alpha_1, \dots, \alpha_n) \in \partial g(h(x_0))\right\} =: \varphi_0(v),$$

hence (71) becomes

(72) $$f^0(x_0; v) \le \varphi_0(v).$$

Let $\varepsilon > 0$ be arbitrary and, v being that of (72), $\varphi(\varepsilon) := \sup\{\sum_{i=1}^{n} \alpha_i\, \xi_i(v) : \xi_i \in \partial h_i(x_i)$, $(\alpha_1, \dots, \alpha_n) \in \partial g(y)$, $x_i \in S(x_0, \varepsilon)$, $y \in S(h(x_0), \varepsilon)\}$.

Obviously $\varphi(\varepsilon) \geq \varphi_0(v)$. Prove

(73)
$$\lim_{\varepsilon \to 0+} \varphi(\varepsilon) = \varphi_0(v).$$

Let $\eta > 0$ be arbitrary. Take $\varepsilon_0 > 0$ such that $h_1, \ldots, h_n$ are Lipschitz on $S(x_0, \varepsilon)$ ($\subset U$), L the constant as for g with respect to V, and moreover, for every i, $1 \leq i \leq n$ and every x_i from $S(x_0, \varepsilon_0)$,

(74)
$$h_i^0(x_i; \pm v) \leq h_i^0(x_0; \pm v) + \frac{\eta}{L + \eta}$$

(take (5.16) into account). Let $\alpha := (\alpha_1, \ldots, \alpha_n)$ be in $\partial g(h(x_0)) + S(0, \eta)$, $\alpha = \alpha' + \alpha''$. Then $\|\alpha'\| \overset{5.17}{\leq} L$, $\|\alpha''\| < \eta$, hence $|\alpha_i| \leq \|\alpha\| \leq L + \eta$, $i = \overline{1, n}$. Mulptiply by $|\alpha_i|$ in (74), we get (see 5.15)

(75)
$$x_i \in S(x_0, \varepsilon_0) \Rightarrow h_i^0(x_i; \alpha_i v) \leq h_i^0(x_0; \alpha_i v) + \eta, \; i = \overline{1, n}.$$

In the following, we diminish ε_0 for having also

(76)
$$y \in h(x_0) + S(0, \varepsilon_0) \Rightarrow \partial g(y) \subset \partial g(h(x_0)) + S(0, \eta)$$

(see 5.20, 3°). And now, taking (76) into account,

$$\varphi(\varepsilon_0) \leq \sup\{\sum_{i=1}^{n} \sup\{\alpha_i \xi_i(v): \xi_i \in \partial h_i(x_i), x_i \in S(x_0; \varepsilon_0)\}: (\alpha_1, \ldots, \alpha_n) \in \partial g(h(x_0)) + S(0,$$

$$\eta)\} \overset{5.17}{\leq} \sup\{\sum_{i=1}^{n} \sup\{h_i^0(x_i; \alpha_i v): x_i \in S(x_0; \varepsilon_0)\}: (\alpha_1, \ldots, \alpha_n) \in \partial g(h(x_0)) + S(0, \eta)\}$$

$$\overset{(75)}{\leq} \sup\{\sum_{i=1}^{n} [h_i^0(x_0; \alpha_i v + \eta] : (\alpha_1, \ldots, \alpha_n) \in \partial g(h(x_0)) + S(0, \eta)\} = \sup\{\sum_{i=1}^{n} h_i^0(x_0; \alpha_i v):$$

$$(\alpha_1, \ldots, \alpha_n) \in \partial g(h(x_0)) + S(0, \eta)\} + n\eta \overset{5.17}{=} \sup\{\sum_{i=1}^{n} \sup\{(\alpha_i' + \alpha_i'')\xi_i(v): \xi_i \in \partial h_i(x_0)\}:$$

$$(\alpha_1', \ldots, \alpha_n') \in \partial g(h(x_0)), (\alpha_1'', \ldots, \alpha_n'') \in S(0, \eta)\} + n\eta \leq \varphi_0(v) + \sup\{\sum_{i=1}^{n} \sup\{\xi_i(\alpha_i'' v):$$

$\xi_i \in \partial h_i(x_0)\}: (\alpha_1'', \ldots, \alpha_n'') \in S(0, \eta)\} + n\eta \leq \varphi_0(v) + n\eta\, \|v\| L + n\eta$, since $|\xi_i(\alpha_i'' v)| \leq$ $\leq |\alpha_i''|\, \|v\|\, \|\xi_i\| \leq \|\alpha''\|\, \|v\| L \leq \eta \|v\| L$,

consequently, since $\varepsilon \to \varphi(\varepsilon)$ is decreasing, $0 < \varepsilon \leq \varepsilon_0 \Rightarrow \varphi(\varepsilon) \leq \varphi_0(v) + n\eta(\|v\|L + 1)$. As $\varphi_0(v) \leq \varphi(\varepsilon)$, (73) is proved.

Pass now to the proof of (72). According to the definition, $\exists\, x$ in a neighbourhood of

x_0 and $\exists\, t$ in an interval $(0, r)$ such that

(77) $$f^0(x_0; v) \le \frac{f(x+tv)-f(x)}{t} + \varepsilon.$$

Take x and t so near of x_0 respectively 0 such that we have

(78) $$x,\, x+tv \in S(x_0, \varepsilon),\ h(x),\, h(x+tv) \in S(h(x_0), \varepsilon).$$

Apply the finite increment formula (5.33), there is y in $(h(x), h(x+tv))$, hence $y \overset{(78)}{\in} S(h(x_0), \varepsilon)$, and $(\alpha_1, \dots, \alpha_n)$ in $\partial g(y)$ such that

(79) $$f(x+tv) - f(x) = g(h(x+tv)) - g(h(x)) = \sum_{i=1}^{n} \alpha_i\, [h_i(x+tv) - h_i(x)].$$

Apply again 5.33, $\exists\, x_i \in (x, x+tv)$, hence $x_i \in S(x_0, \varepsilon)$, and $\exists\, \xi_i \in \partial h_i(x_i)$ such that $h_i(x + tv) - h_i(x) = \xi_i(tv)$ and (79) becomes $\dfrac{f(x+tv)-f(x)}{t} = \sum_{i=1}^{n} \alpha_i\, \xi_i(v)$, consequently (the definition!) $\dfrac{f(x+tv)-f(x)}{t} \le \varphi(\varepsilon)$ and combining with (77) we find $f^0(x_0; v) \le \varphi(\varepsilon) + \varepsilon$. Take the limit for $\varepsilon \to 0+$, we get (72) *via* (73). ■

5.35 *Completion to* 5.34. *The inclusion of the formula becomes equality in each of the following cases*:

1° $h_1, \dots, h_n$ *are regular at* x_0, g *is regular at* $h(x_0)$ *and for every* $(\alpha_1, \dots, \alpha_n)$ *from* $\partial g(h(x_0))$ *we have* $\alpha_i \ge 0$, $i = \overline{1, n}$ (*in this case* f *is regular at* x_0);

2° n = 1 *and* g *is strictly differentiable at* $h(x_0)$ (*in this case* $\overline{\text{co}}$ *disappears from the formula*);

3° h *is strictly differentiable at* x_0 *and* g *is regular at* $h(x_0)$ (*in this case* f *is regular at* x_0 *and* $\overline{\text{co}}$ *dissapears from the formula*).

■ 1° $\varphi_0(v) \le \sup\{\sum_{i=1}^{n} \sup\{\alpha_i \xi_i(v) : \xi_i \in \partial h_i(x_0)\} : (\alpha_1, \dots, \alpha_n) \in \partial g(h(x_0))\} \overset{\alpha_i \ge 0}{=}$

$= \sup\{\sum_{i=1}^{n} \alpha_i \sup\{\xi_i(v) : \xi_i \in \partial h_i(x_0)\} : (\alpha_1, \dots, \alpha_n) \in \partial g(h(x_0))\} \overset{5.17}{=} \sup\{\sum_{i=1}^{n} \alpha_i h_i^0(x_0; v) :$

$(\alpha_1, \dots, \alpha_n) \in \partial g(h(x_0))\} \overset{\text{the hypothesis}}{=} \sup\{\sum_{i=1}^{n} \alpha_i h_i'(x_0; v) : (\alpha_1, \dots, \alpha_n) \in \partial g(h(x_0))\} \overset{5.17}{=}$

$$= g^0(h(x_0); w),\ w := (h_1'(x_0; v), \dots, h_n'(x_0; v)) \overset{\text{the hypothesis}}{=} g'(h(x_0); w) \overset{C,(9)}{=}$$

$$= \lim_{t\to 0+} \frac{g(h(x_0)+tw) - g(h(x_0))}{t} = \lim_{t\to 0+}\left[\frac{g(h(x_0+tv)) - g(h(x_0))}{t} + \right.$$

$$\left. + \frac{g(h(x_0)+tw) - g(h(x_0+tv))}{t}\right],$$

the limit of the second term of the sum is equal to 0:

$$\left\|\frac{g(h(x_0)+tw) - g(h(x_0+tv))}{t}\right\| \le L\left\|w - \frac{h(x_0+tv)-h(x_0)}{t}\right\| \text{ and } \lim_{t\to 0+}\frac{h(x_0+tv)-h(x_0)}{t} \overset{(9)}{=}$$

$= (h_1'(x_0; v), \dots) = w$, consequently

$$\varphi_0(v) \le \lim_{t\to 0+}\frac{g(h(x_0+tv)) - g(h(x_0))}{t} \overset{(9)}{=} f'(x_0; v).$$

So, taking into account (72) and the remark to 5.18, we have

$$f^0(x_0; v) \le \varphi_0(v) \le f'(x_0; v) \le f^0(x_0; v),$$

hence $f^0(x_0; v) = f'(x_0; v)$ (f is regular at x_0) and

$$f^0(x_0; v) = \varphi_0(v),\ \sigma_{\partial f(x_0)}(v) = \sigma_S(v) \overset{(70)}{=} \sigma_{\overline{\text{co}}S}(v),$$

hence $\partial f(x_0) = \overline{\text{co}}\, S$ (5.19, 2°).

2° $\partial g(h(x_0)) \overset{5.25}{=} \{D_s g(h(x_0))\}$, set $\alpha := D_s g(h(x_0))$, hence $\alpha \in \mathbf{R}$. It can suppose $\alpha > 0$ and thus we have

$$\text{(80)} \qquad \varphi_0(v) = \sup\{\alpha\,\xi(v) : \xi \in \partial h(x_0)\} \overset{5.17}{=} \alpha\, h^0(x_0; v) \overset{\text{def.}}{=} \overline{\lim_{\substack{x\to x_0\\ t\to 0+}}} \frac{\alpha[h(x+tv) - h(x)]}{t}.$$

But

$$\frac{h(x+tv)-h(x)}{t} = \frac{g(h(x+tv)) - g(h(x))}{t}\ \frac{h(x+tv)-h(x)}{g(h(x+tv)) - g(h(x))} \text{ and}$$

$$\lim_{\substack{x\to x_0\\ t\to 0+}} \frac{h(x+tv)-h(x)}{g(h(x+tv)) - g(h(x))} = \frac{1}{\alpha},$$

consequently, taking (80) into account,

$$\varphi_0(v) = \overline{\lim_{\substack{x\to x_0\\ t\to 0+}}} \frac{g(h(x+tv)) - g(h(x))}{t} = f^0(x_0; v),$$

$\varphi_0(v) = f^0(x_0; v)$ and the reasoning follows as in the case 1° (the support functionals!).

The second statement is obvious, the set $\alpha\ \{\xi : \xi \in \partial h(x_0)\}$ being convex and *-weak closed (5.17).

3° The justification is similar to those for 2° (again taking 5.31, 1° into account). ■

Pass to a variant of the formula for the generalized gradient of the composed function (5.34).

5.36 Clarke subderivative of the composed function. *Let be* $f := g \circ h$, $h : X \to Y$, X *and* Y *real Banach spaces*, $g : Y \to \mathbf{R}$. *If* h *is strictly differentiable at* x_0 *and* g *Lipschitz around* $h(x_0)$, *then* f *is Lipschitz around* x_0 *and*

$$\partial f(x_0) \subset \partial g(h(x_0)) \circ D_s h(x_0)$$

(*J. P. Aubin, F. H. Clarke*).

Explanation. $\partial g(h(x_0)) \circ D_s h(x_0) = \{\xi \circ D_s h(x_0) : \xi \in \partial g(h(x_0))\}$.

■ For the first statement – the usual justification, h is also Lipschitz around x_0 (5.21). For the second statement – the same proof as for 5.34 : the sets of the statement are convex and *-weak closed (5.17) and one passes to the support functionals $v \to f^0(x_0; v)$, $v \to \sup\{\xi(\mu_0(v)) : \xi \in \partial g(h(x_0))\}$, where $\mu_0 = D_s h(x_0)$. ■

5.37 *Completions to* 5.36.

1° *If moreover* g (resp. – g) *is regular in* $h(x_0)$, *then* f (resp. – f) *is regular in* x_0 *and the inclusion of the formula becomes an equality.*

2° *If moreover the image of every neighbourhood of* x_0 *through* h *is dense in a neighbourhood of* $h(x_0)$, *then the inclusion of the formula becomes an equality.*

■ Suppose g regular at $h(x_0)$ (the case $-g$ regular at $h(x_0)$ is treated as in 5.26) and let v be arbitrary in X.

$$f^0(x_0; v) \overset{5.36}{\leq} \sup\{\xi(\mu_0(v)) : \xi \in \partial g(h(x_0))\} = g^0(h(x_0); \mu_0(v)) \overset{(9)}{=} g'(h(x_0); \mu_0(v)) =$$

$$= \lim_{t \to 0+} \frac{g(h(x_0) + t\mu_0(v)) - g(h(x_0))}{t} = \lim_{t \to 0+} \Big[\frac{g(h(x_0 + tv) - g(h(x_0)))}{t} +$$

$$+ \frac{g(h(x_0) + t\mu_0(v)) - g(h(x_0 + tv))}{t}\Big],$$ in brackets the second term has the limit 0:

$$\left\| \frac{g(h(x_0) + t\mu_0(v)) - g(h(x_0 + tv))}{t} \right\| \leq L \left\| \frac{h(x_0 + tv) - h(x_0)}{t} - \mu_0(v) \right\|$$

and

$$\lim_{t \to 0+} \frac{h(x_0 + tv) - h(x_0)}{t} = \mu_0(v) \ (5.23),$$

consequently

$$f^0(x_0; v) \le g^0(h(x_0); \mu_0(v)) = \lim_{t\to 0+} \frac{g(h(x_0+tv)) - g(h(x_0))}{t} = f'(x_0; v) \le f^0(x_0; v)$$

(remark to 5.18), hence $f^0(x_0; v) = f'(x_0; v)$, i. e. f is regular in x_0, and $f^0(x_0; v) =$ $= g^0(h(x_0); \mu_0(v))$, i. e. the sets in the statement are equal (5.19, 2°).

2° The hypothesis allows to write, v arbitrary in X (in a certain place one repeats the calculus in the brackets [ ] of 1°):

$$f^0(x_0; v) \overset{5.36}{\le} g^0(h(x_0); \mu_0(v)) = \overline{\lim_{\substack{y\to h(x_0)\\ t\to 0+}}} \frac{g(y+t\mu_0(v)) - g(y)}{t} =$$

$$= \overline{\lim_{\substack{x\to x_0\\ t\to 0+}}} \frac{g(h(x)+t\mu_0(v)) - g(h(x))}{t} = \overline{\lim_{\substack{x\to x_0\\ t\to 0+}}} \frac{g(h(x+tv)) - g(h(x))}{t} = f^0(x_0; v),$$

hence $f^0(x_0; v) = g^0(h(x_0); \mu_0(v))$ and hence, v being arbitrary, the corresponding sets are equal. ■

5.38 Clarke subderivative of the product. *Let be* $f_1, f_2 : X \to \mathbf{R}$, X *a real Banach space. If* f_1, f_2 *are Lipschitz around* x_0, *then* $f_1 f_2$ *has the same property and*

$$\partial(f_1 f_2)(x_0) \subset f_2(x_0)\, \partial f_1(x_0) + f_1(x_0)\, \partial f_2(x_0).$$

If moreover f_1, f_2 *are regular at* x_0 *and* $f_1(x_0) \ge 0$, $f_2(x_0) \ge 0$, *then* $f_1 f_2$ *is regular at* x_0 *and the inclusion becomes an equality* (*J. B. Hiriart - Urruty*).

■ *First statement.* Consider the functions $h : X \to \mathbf{R}^2$, $h = (f_1, f_2)$ and $g : \mathbf{R}^2 \to \mathbf{R}$, $g(u_1, u_2) = u_1 u_2$. Obviously $f_1 f_2 = g \circ h$. g being of C^∞-Fréchet class, it is strictly differentiable (5.22) and hence regular (5.31), in particular Lipschitz (5.21). Moreover $\partial g(u_1, u_2) \overset{5.25}{=} \{g'(u_1, u_2)\}$ (Fréchet derivative), $g'(u_1, u_2) = u_2 p_1 + u_1 p_2$, p_1 and p_2 the projection functions of $\mathbf{R}^2$, consequently $\partial g(u_1, u_2) = \{(u_1, u_2)\}$ (identification by Riesz theorem).

Apply 5.34 and get the inclusion of the statement (the second member is convex and *-weak closed, 5.17).

Second statement. Apply 5.35, 1°. ■

5.39 Clarke suderivative of the quotient. *Let be* $f_1, f_2 : X \to \mathbf{R}$, X *a real Banach space. If* f_1, f_2 *are Lipschitz around* x_0 *and* $f_2(x_0) \ne 0$, *then* $\frac{f_1}{f_2}$ *is Lispchitz around* x_0 *and*

$$\partial\left(\frac{f_1}{f_2}\right)(x_0) \subset \frac{1}{f_2^2(x_0)}\left[f_2(x_0)\partial f_1(x_0) - f_1(x_0)\partial f_2(x_0)\right].$$

If in addition f_1 *and* $-f_2$ *are regular at* x_0 *and* $f_1(x_0) \geq 0$, $f_2(x_0) > 0$, *then* $\frac{f_1}{f_2}$ *is regular at* x_0 *and the inclusion becomes an equality.* (*J. B. Hiriart - Urruty*).

■ The same proof as for 5.38. ■

We give in the following two statements useful in applications.

5.40 *Let* X *be a real Banach space and* $f_1, \dots, f_n : X \to \mathbf{R}$ *Lipschitz around* x_0. *Then* $f : X \to \mathbf{R}$,

$$f(x) = \max(f_1(x), \dots, f_n(x))$$

is Lipschitz around x_0 *and*

$$\partial f(x_0) \subset \overline{\text{co}}\,\{\partial f_i(x_0) : i \in J(x_0)\},$$

where $J(x_0) = \{j : 1 \leq j \leq n, f(x_0) = f_j(x_0)\}$.

Moreover, if, for every i *from* $J(x_0)$, f_i *is regular at* x_0, *then* f *is regular at* x_0 *and the inclusion becomes an equality* ([20], 1).

■ *First statement.* The same justification as for the function f from example 3.

Second and third statement. Consider the functions $g : \mathbf{R}^n \to \mathbf{R}$, $g(u_1, \dots, u_n) = \max\limits_{1 \leq i \leq n} u_i$ (see ex.3), $h : X \to \mathbf{R}^n$, $h(x) = (f_1(x), \dots, f_n(x))$. Obviously $f = g \circ f$. g being Lipschitz around $h(x_0)$ and convex, it is regular at $h(x_0)$ (5.31). Apply 5.34 taking into account the expression of $\partial f(x)$ in the example 3 and get the inclusion, apply 5.35, 1° and get an equality. ■

5.41 *Let* X *be a real normed space,* $\varphi : [0, 1] \to X$ *of* C^1 *class and* $f : X \to \mathbf{R}$ *locally Lipschitz. Then* $F := f \circ \varphi$ *is differentiable* μ *- a.e. on* [0, 1], μ *the Lebesgue measure on* **R**, *and*

$$F'(t) \leq \sup_{\xi \in \partial f(\varphi(t))} \xi(\varphi'(t)) \quad \mu - \text{a. e.}$$

■ Use Rademacher theorem. F is locally Lipschitz (use the definition, the continuity of f and a formula of finite increments), hence differentiable μ – a. e. Let t_0 be a point in which we have derivative.

$$F'(t_0) = \lim_{t \to 0+} \frac{f(\varphi(t_0 + \lambda)) - f(\varphi(t_0))}{\lambda} = \lim_{\lambda \to 0} \frac{f(\varphi(t_0) + \lambda\varphi'(t_0) + o(\lambda)) - f(\varphi(t_0))}{\lambda} =$$

$$= \lim_{\lambda \to 0} \frac{f(\varphi(t_0) + \lambda\varphi'(t_0)) - f(\varphi(t_0))}{\lambda}, \text{ as}$$

$\lim\limits_{\lambda \to 0} \dfrac{f(\varphi(t_0) + \lambda\varphi'(t_0) + o(\lambda)) - f(\varphi(t_0) + \lambda\varphi'(t_0))}{\lambda} = 0$, since $\| f(\varphi(t_0) + \lambda\varphi'(t_0) + o(\lambda)) -$

$- f(\varphi(t_0) + \lambda\varphi'(t_0))\| \leq L\,|(o(\lambda))|$. Thus $F'(t_0) = f'(\varphi(t_0); \varphi'(t_0))$ and $f'(\varphi(t_0); \varphi'(t_0)) \leq$ $\leq f^0(\varphi(t_0); \varphi'(t_0))$ (remark to 5.18), therefore the conclusion (5.17). ■

In the final part of this paragraph we connect to the subject of the chapter within we are by the sections H and I.

H. CRITICAL POINTS FOR NONDIFFERENTIABLE FUNCTIONALS

The title want to say „not obligatory differentiable".

Definitions. x_0 is a *critical point* (in the sense of Clarke subderivative) for the real function f if $0 \in \partial f(x_0)$. In this case $f(x_0)$ is a *critical value* (in the sense of Clarke subderivative) for f.

Here is an application of the concept ([22]).

Let Ω be a bounded domain of $\mathbf{R}^N$ with the smooth boundary $\partial\Omega$ (topological boundary). Consider the nonlinear boundary value problem

$$(*)\ \begin{cases} -\Delta u = f(x,u), & x \in \Omega \\ Bu = 0 \text{ on } \partial\Omega, \end{cases}$$

where B designates the boundary condition Dirichlet or Neumann and $f: \Omega \times \mathbf{R} \to \mathbf{R}$ is a measurable function with *subcritical growth*, i.e.

$$\text{(I)}\ \ |f(x, s)| \leq a + b|s|^{\sigma}\ \forall\, s \in \mathbf{R}, x \in \Omega \text{ a.e.}$$

where $a, b > 0$, $0 \leq \sigma < \dfrac{N+2}{N-2}$ for $N > 2$ and $\sigma \in [0, +\infty)$ for $N = 1$ or $N = 2$.

Set ([16])

$$\underline{f}\,(x, t) = \varliminf_{s\to t} f(x, s),\ \ \overline{f}\,(x, t) = \varlimsup_{s\to t} f(x, s).$$

Suppose

(II) $\underline{f}$, $\overline{f}: \Omega \times \mathbf{R} \to \mathbf{R}$ are N-measurable (i.e. with respect to x).

We emphasize that (II) is verified in the following two cases:

1° f is indepedent of x;

2° f is Baire measurable and $s \to f(x, s)$ is decreasing $\forall\, x \in \Omega$, in which case we have

$$\overline{f}(x,t)=\max\{f(x,t+),\ f(x,t-)\},\ \underline{f}(x,t)=\min\{f(x,t+),\ f(x,t-)\}.$$

Definition. u from $W^{2,p}$, $p>1$ is *solution* of (∗) if $Bu=0$ on $\partial\Omega$ in the sense of trace and

(81) $$-\Delta u(x)\in[\underline{f}(x,u(x)),\ \overline{f}(x,u(x))] \text{ in } \Omega \text{ a.e.}$$

Let be $X:=H_0^1(\Omega)$ resp. $X:=H^1(\Omega)$ as the boundary condition is Dirichlet resp. Neumann. Associate to (∗) the locally Lipschitz functional $\Phi: X\to\mathbf{R}$,

(82) $$\Phi(u)=\frac{1}{2}\int_\Omega|\nabla u|^2\,dx-\int_\Omega F(x,u)\,dx,\ u\in X,$$

where $F(x,s)=\int_0^s f(x,t)\,dx$. Set

(83) $$Q(u):=\frac{1}{2}\int_\Omega|\nabla u|^2\,dx,\ u\in X,\ \Psi_1(u):=\int_\Omega F(x,u)\,dx,\ u\in X.$$

F, an application of $\Omega\times\mathbf{R}$ in $\mathbf{R}$, is locally Lipschitz (use (I)). The functional Ψ: $\mathbf{L}^{\sigma+1}(\Omega)\to$ $\to\mathbf{R}$, $\Psi(u)=\int_\Omega F(x,u)\,dx$ is also locally Lipschitz (again (I)). Using Sobolev embedding $X\subset\mathbf{L}^{\sigma+1}(\Omega)$, we get that $\Psi_1:=\Psi\,|\,X$ is locally Lipschitz on X, what implies Φ locally Lipschitz on X, and consequently, according to 5.27, the critical points of Φ for Clarke subderivative can be taken into account. It can state

5.42 *If* (I) *and* (II) *are verified, every critical point of* Φ *is solution for* (∗).

To prove this use the proposition (the proof in [16], theorems 2.1,2.2).

5.43 *Suppose* (I) *and* (II) *satisfied. Then* Ψ *is locally Lipschitz on* $\mathbf{L}^{\sigma+1}(\Omega)$ *and*

1° $\partial\Psi(u)\subset[\underline{f}(x,u(x)),\overline{f}(x,u(x))]$ *in* Ω *a.e.*

2° *If* $\Psi_1=\Psi\,|\,X$, *where* $X=H_0^1(\Omega)$ *or* $X=H^1(\Omega)$,

then

$$\partial\Psi_1(u)\subset\partial\Psi(u)\ \forall\ u\in X$$

(*J.* P. *Aubin - F. H. Clarke, K. C. Chang*).

Pass to the the proof of 5.42.

■ Let u_0 be a critical point for Φ. We have

(84) $$0\in\partial\Psi(u_0)\subset\partial Q(u_0)+\partial(-\Psi_1)(u_0),$$

since $\Phi\overset{(82)}{=}Q-\Psi_1$ and one applies 5.28. $\partial Q(u_0)=\{Q'(u_0)\}$, where $Q'(u_0)(v)=$

$$= \int_\Omega \nabla u_0 \cdot \nabla v \, dx.$$

By (84) and 5.17, 2° we find

$$0 \le \int_\Omega \nabla u_0 \cdot \nabla v \, dx + (-\Psi_1)^0 (u_0 ; v).$$

But $(-\Psi_1)^0 (u_0 ; v) = \Psi_1^0 \; (u_0 ; -v)$ (5.12,2°) and thus

$$\int_\Omega \nabla u_0 \cdot \nabla(-v) \, dx \le \Psi^0 (u_0 ; -v) \; \forall \; v \in X,$$

i.e.

$$\mu_0 (v) : = \int_\Omega \nabla u_0 \cdot \nabla v \, dx \le \Psi_1^0 (u_0 , v) \; \forall \; v \in X,$$

$\mu_0 = -\Delta u_0 \in \partial \Psi_1 (u_0)$ and , using 5.43, $-\Delta u_0 \in \partial \Psi(u_0)$. Since $\partial \Psi(u_0) \subset (\mathbf{L}^{\sigma+1}(\Omega))^* =$ $= \mathbf{L}^{(\sigma+1)/\sigma}(\Omega)$, it results $u_0 \in W^{2,(\sigma+1)/\sigma}(\Omega)$ and (5.43)

$$-\Delta u_0 (x) \in [\, \underline{f} (x, u_0(x)), \overline{f} (x, u_0(x))] \text{ in } \Omega \text{ a.e. } \blacksquare$$

We present in the following other results of [22] , paper affected in fact by serious errors. These results contain conditions of Palais-Smale type suggested by Ekeland principle.

Let X be a complete metric space, φ: X → **R** and $c \in \mathbf{R}$.

φ satisfies $(PS)^*_{c,+}$ *condition* when, for every sequences $(u_n)_{n\ge1}$, $u_n \in X$, $(\varepsilon_n)_{n\ge1}$ and $(\delta_n)_{n\ge1}$, $\varepsilon_n , \delta_n \in \mathbf{R}_+$, $\varepsilon_n \to 0$ and $\delta_n \to 0$, if

(85) $$\varphi(u_n) \to c$$

and

(86) $$\forall \; u \in X \;\; d(u_n , u) \le \delta_n \Rightarrow \varphi(u_n) \le \varphi(u) + \varepsilon_n d(u_n , u),$$

then

$(u_n)_{n\ge1}$ *has a convergent subsequence.*

We change in (86) u_n and u to each other, one obtains $(PS)^*_{c,-}$ *condition*. Finally, $(PS)^*_c$ *condition* means $(PS)^*_{c,+} + (PS)^*_{c,-}$.

When, X being a real Banach space, the conclusion required by the hypothesis

„$(u_n)_{n\ge1}$ has a convergent subsequence"

is replaced by

$(u_n)_{n\ge1}$ *has a weak convergent subsequence*,

we get respectively the conditions

$$(PS)^*_{c,\,w,\,+},\ (PS)^*_{c,\,w,\,-},\ (PS)^*_{c,\,w}\,.$$

Suppose X a real Banach space and φ locally Lipschitz. φ satisfies $[PS]^*_{c,+}$ *condition* (obvious definition for $[PS]^*_{c,-}$, $[PS]^*_{c}$) when the properties (85) and (86) imply

c *is critical value of* φ (for Clarke subderivative).

The definition is, according to 5.17, coherent. We have

$$(PS)^*_{c,+} \Rightarrow [PS]^*_{c,+}\,,$$

reciprocally it is not true (consider φ: $\mathbf{R} \to \mathbf{R}$ Lipschitz, periodic and $c = \inf \varphi(\mathbf{R})$ or $c = \sup \varphi(\mathbf{R})$).

Finally, a last version of Palais-Smale condition due to Chang ([16]). Let X be a real Banach space, φ: $X \to \mathbf{R}$ locally Lipschitz and $c \in \mathbf{R}$. φ satisfies $(PS)_c^{ch}$ *condition* when, for every sequence $(u_n)_{n\ge 1}$ from X, if

(87) $$\varphi(u_n) \to c$$

and

(88) $$\mu(u_n) := \inf\{\|\xi_n\| : \xi_n \in \partial\,\varphi(u_n)\} \to 0,$$

then

$(u_n)_{n\ge 1}$ *has a convergent subsequence.*

The definition is correct, $\partial\,\varphi(u_n) \ne \varnothing\ \forall\ n$ (5.17, see also 5.17_2). It can state

Let X *be a real Banach space and* φ: X $\to$ **R** *locally Lipschitz and convex. Then*

$$\varphi \textit{ verifies } (PS)_c^{ch} \Rightarrow \varphi \textit{ verifies } (PS)_{c,-}^{*}\,.$$

■ Let (u_n) be a sequence from X and (ε_n), (δ_n) sequences from $\mathbf{R}_+$, $\varepsilon_n \to 0$, $\delta_n \to 0$ such that

$$\varphi(u_n) \to c$$

and

(88_1) $$\|u_n - u\| \le \delta_n \Rightarrow \varphi(u) \le \varphi(u_n) + \varepsilon_n\,\|u_n - u\|\ \forall\ u \in X.$$

We must prove, for finding a convergent subsequence of (u_n),

(88_2) $$\mu(u_n) := \inf\{\,\|\xi_n\| : \xi_n \in \partial\,\varphi(u_n)\} \to 0.$$

Take $u := u_n + tv$, $\|v\| = 1$, $0 < t \le \delta_n$. (88_1) gives, since $\|u_n - u\| \le \delta_n$,

$$\frac{\varphi(u_n + tv) - \varphi(u_n)}{t} \le \varepsilon_n$$

and passing to the limit for $t \to 0+$ we get (5.18)

$$\varphi^0(u_n\,;v) = \varphi'(u_n\,;v) \le \varepsilon_n\,,$$

consequently $\xi(v) \le \varepsilon_n\ \forall\ \xi \in \partial\,\varphi(u_n)$ (5.17) and hence, changing v in $-v$,

(88_3) $$\|\xi\| \leq \varepsilon_n \ \forall\ \xi \in \partial\, \varphi(u_n).$$

Let ξ_n be in $\partial\varphi(u_n)$ such that $\|\xi_n\| = \mu(u_n)$ (see 5.17). Then, taking (88_3) into account,

$$\mu(u_n) \leq \varepsilon_n\ ,$$

which yields (88_2) by passing to the limit. ∎

Remark. The last statement represents all that the author could recover from Proposition 3, [22], p.475. The proof of this ([22], p.483) contains among other things the implicit statement that $v \to \Phi^0\,(u_0\,;\,v)$ is not a subnorm, but even a linear functional.

Pass now to some propositions of [22].

5.44 *Let* X *be a complete metric space,* $\varphi : \mathrm{X} \to \mathbf{R}$ *lower bounded, lower semicontinuous and* $\mathrm{c} := \inf \varphi\,(\mathrm{X})$. c *is attained when* φ *verifies* $(\mathrm{PS})^*_{\mathrm{c},+}$.

∎ Let $(v_n)_{n\geq 1}$, $v_n \in X$ be a minimizing sequence for φ such that, for every n, $\varepsilon_n := \varphi(v_n) - c > 0$, hence $\varepsilon_n \to 0$. Apply Ekeland principle 1.1 with $\varepsilon = \varepsilon_n$, $\lambda = 1$, one finds $(u_n)_{n\geq 1}$ a sequence in X with the properties

$$\varphi(u_n) \leq \varphi(v_n),$$

$$\varphi(u_n) \leq \varphi(u) + \varepsilon_n d(u_n\,,\,u)\ \forall\ u \in X.$$

Since $c \leq \varphi(u_n)$, we have $\varphi(u_n) \to c$, apply $(\mathrm{PS})^*_{c,+}$, let $(u_{k_n})_{n\geq 1}$ be a convergent subsequence, $u_{k_n} \to u_0$. But $\varphi(u_0) \leq \underline{\lim}_{n\to\infty}\ \varphi(u_{k_n}) = c$, this imposes $\varphi(u_0) = c$. ∎

5.45 *Let* X *be a real Banach space,* $\varphi : \mathrm{X} \to \mathbf{R}$ *lower bounded, locally Lipschitz and* $\mathrm{c} := \inf \varphi(\mathrm{X})$.

If φ *satisfies* $(\mathrm{PS})^*_{\mathrm{c},+}$, *then* φ *has critical points* (for Clarke subderivative).

∎ Apply 5.44 combined with 5.27. ∎

5.46 *Let* X *be a real Banach space,* $\varphi : \mathrm{X} \to \mathbf{R}$ *lower bounded, locally Lipschitz and* $\mathrm{c} := \inf \varphi(\mathrm{X})$.

If φ *verifies* $[\mathrm{PS}]^*_{\mathrm{c},+}$, *then* c *is a critical value of* φ (for Clarke subderivative).

∎ Let be $(\varepsilon_n)_{n\geq 1}$, $\varepsilon_n > 0$, $\varepsilon_n \to 0$. For every ε_n take v_n such that $\varphi(v_n) \leq c + \varepsilon_n$ and apply Ekeland principle 1.1 with $\lambda = 1$. $\exists\ u_n$ such that $\varphi(u_n) \leq \varphi(v_n)$, (89) $\varphi(u_n) \leq \varphi(u) + \varepsilon_n \|u_n - u\|\ \forall\ u \in X$. Since $\varphi(u_n) \to c$, (89) allows to apply the hypothesis $[\mathrm{PS}]^*_{c,+}$, c is a critical value. ∎

As an application we continue with the problem (∗). But at first

5.47 *Let be* $\mathrm{X} := \mathrm{H}^1(\Omega)$ *and* $\Phi : \mathrm{X} \to \mathbf{R}$, $\Phi(\mathrm{u}) = \frac{1}{2}\int_\Omega |\nabla \mathrm{u}|^2 \mathrm{dx} - \int_\Omega \mathrm{G(u)dx} - \int_\Omega \mathrm{hu\,dx}$, *where* G : **R** → **R** is *with the period* T *and Lipschitz*, $h \in \mathbf{L}^2(\Omega)$ *and* $\int_\Omega h\,dx = 0$.

Then, for every c *from* **R**, Φ *verifies* $[PS]^*_c$.

■ It is enough to prove it for $[PS]^*_{c,+}$. Let $(u_n)_{n\geq 1}$ be a sequence from X, $(\varepsilon_n)_{n\geq 1}$ and $(\delta_n)_{n\geq 1}$ sequences from $\mathbf{R}_+$ convergent to 0. Suppose $\Phi(u_n) \to c$ and

(90)
$$\|u_n - u\| \leq \delta_n \Rightarrow \Phi(u_n) \leq \Phi(u) + \varepsilon_n \|u_n - u\|.$$

Decompose X in direct sum

(91)
$$X = X_0 \oplus X_1,$$

X_1 the vector space of the constant functions, $X_0 = X_1^{\perp}$, the vector space of H^1 - functions having the mean value equal to 0. Let $u_n = v_n + c_n$, $v_n \in X_0$, $c_n \in \mathbf{R}$ and $|G(s)| \leq M$ on **R**.

Since $\int_\Omega c_n h\,dx = 0$, we have

(92)
$$\Phi(u_n) = \frac{1}{2}\int_\Omega |\nabla v_n|^2\,dx - \int_\Omega G(v_n + c_n)dx - \int_\Omega h v_n\,dx \geq \frac{1}{2}\int_\Omega |\nabla v_n|^2\,dx - M\,\text{mes}\,(\Omega) -$$

$$-\rho^{\frac{1}{2}}\left(\int_\Omega h^2 dx\right)^{\frac{1}{2}}\left(\int_\Omega |\nabla v_n|^2 dx\right)^{\frac{1}{2}},$$ since $\int_\Omega |hv_n|\,dx \leq \left(\int_\Omega h^2 dx\right)^{\frac{1}{2}}\left(\int_\Omega v_n^2 dx\right)^{\frac{1}{2}}$ and $\left(\int_\Omega v_n^2 dx\right) \leq$

$\leq \rho \int_\Omega |\nabla v_n|^2\,dx$ (Poincaré inequality). As $(\Phi(u_n))_{n\geq 1}$ is bounded, (92) imposes

$(\int_\Omega |\nabla v_n|^2\,dx)_{n\geq 1}$ bounded and hence also $(\|v_n\|)_{n\geq 1}$ bounded.

Consider the sequence $(\tilde{u}_n)_{n\geq 1}$, $\tilde{u}_n = v_n + \tilde{c}_n$, where $\tilde{c}_n \equiv c_n$ (modulo T) and $\tilde{c}_n \in [0, T]$. Since Φ has the period T, (90) gives

$$\|\tilde{u}_n - (u + \tilde{c}_n - c_n)\| \leq \delta_n \Rightarrow \Phi(\tilde{u}_n) \leq \Phi(u) + \varepsilon_n \|(\tilde{u}_n - u) + (c_n - \tilde{c}_n)\|,$$

i.e.

(93)
$$\|\tilde{u}_n - w\| \leq \delta_n \Rightarrow \Phi(\tilde{u}_n) \leq \Phi(w) + \varepsilon_n \|\tilde{u}_n - w\|.$$

But (v_n) and $(\tilde{c}_n)$ are bounded, hence $(\tilde{u}_n)$ is bounded, consequently it has a weak convergent subsequence $(\tilde{u}_{k_n})_{n\geq 1}$, $\tilde{u}_{k_n} \xrightarrow{\text{weak}} \tilde{u}$ (Eberlein-Šmulian), whence the existence of a convergent subsequence of $(\tilde{u}_{k_n})_{n\geq 1}$, the same notation, (94) $\tilde{u}_{k_n} \to \tilde{u}$ (the same proof as in Proposition 4, [22], p.484). In (93) take $w = \tilde{u}_{k_n} + \delta_{k_n} v$, $\|v\| = 1$, we get

$\Phi(\tilde{u}_{k_n} + \delta_{k_n} v) - \Phi(\tilde{u}_{k_n}) \geq -\varepsilon_{k_n} \delta_{k_n}$, $-\varepsilon_{k_n} \leq \frac{1}{\delta_{k_n}}[\Phi(\tilde{u}_{k_n} + \delta_{k_n} v) - \Phi(\tilde{u}_{k_n})]$ and passing to the limit we find, since $(\tilde{u}_{k_n}, \delta_{k_n}) \xrightarrow{(94)} (\tilde{u}, 0)$,

$0 \leq \overline{\lim_{n\to\infty}} \frac{1}{\delta_{k_n}}[\Phi(\tilde{u}_{k_n} + \delta_{k_n} v) - \Phi(\tilde{u}_{k_n})] \leq \Phi^0(\tilde{u}; v)$, $0 \leq \Phi^0(\tilde{u}; v)$, $\|v\| = 1$, whence $0 \leq$ $\leq \Phi^0(\tilde{u}; v)\ \forall\ v \in X$ $(0 \leq \Phi^0(\tilde{u}; \frac{v}{\|v\|}) \overset{5.15}{=} \frac{1}{\|v\|}\Phi^0(\tilde{u}; v))$, i.e. $0 \in \partial\Phi(\tilde{u})$. Moreover, $c = \Phi(\tilde{u})$, since $\Phi(u_{k_n}) = \Phi(\tilde{u}_{k_n} - \tilde{c}_{k_n} + c_{k_n}) = \Phi(\tilde{u}_{k_n}) \to c$ and also $\Phi(\tilde{u}_{k_n}) \xrightarrow{(94)}$ $\to \Phi(\tilde{u})$ (Φ is continuous being locally Lipschitz), c is a critical value for Φ. ■

And now

5.48 *Nonlinear Neumann problem*

$$(\mathrm{N}) \qquad \begin{aligned} -\Delta u &= g(u) + h(x),\ x \in \Omega \\ \frac{\partial u}{\partial n} &= 0 \quad \text{on} \quad \partial\Omega, \end{aligned}$$

with the conditions

III $g : \mathbf{R} \to \mathbf{R}$ *bounded measurable* T-*periodic*, $\int_0^T g(s)\,ds = 0$

and

IV h *bounded measurable*, $\int_\Omega h\,dx = 0$,

has a solution in $H^1(\Omega)$ *in the sense of* (81).

■ We are in the presence of a problem of type (*) with $f(x, u) = g(u) + h(x)$. The conditions III and IV imply (I) ($\sigma = 0$) and (II). The associate functional is

$$\Phi(u) = \frac{1}{2}\int_\Omega |\nabla u|^2\,dx - \int_\Omega G(u)\,dx - \int_\Omega hu\,dx,\ u \in H^1(\Omega),$$

where $G(u(x)) = \int_0^{u(x)} g(t)\,dt$. G is Lipschitz and has the period T (use (III)). Since (I) and (II) are satisfied, any critical point of Φ is, according to 5.42, a solution for the problem (N). But Φ verifies $[PS]^*_c$ for every c at $\mathbf{R}$ (5.47), in particular for $c = \inf_{u\in H^1(\Omega)} \Phi(u)$, this is correct since Φ is lower bounded (justification as for (92)). It only remains to apply 5.46,

c is a critical value, $c = \Phi(u_0)$, u_0 a critical point, u_0 is a solution for (N). ■

I. MATHEMATICAL PROGRAMMING ([30], 4)

Let f, g_i, $i=\overline{1,r}$, h_j, $j=\overline{1,s}$ be real locally Lipschitz functions on $\mathbf{R}^n$. Consider the problem of minimizing f subject to the following constraints (equalities and inequalities):

(95) $$g_i(x) = 0,\ i=\overline{1,r},$$

(96) $$h_j(x) \le 0, j=\overline{1,s}.$$

5.49 Theorem. *If* x_0 *minimizes* f *subject to the constraints* (95) + (96), *there exist real numbers* $\lambda \ge 0$, μ_i, $\nu_j \ge 0$ *not all zero such that*

$$0 \in \lambda\, \partial f(x_0) + \sum_{i=1}^{r} \mu_i\, \partial g_i(x_0) + \sum_{j=1}^{s} \nu_j\, \partial h_j(x_0)$$

(*F. H. Clarke*).

■ Consider, for every $\varepsilon > 0$, the function $F_\varepsilon : \mathbf{R}^n \to \mathbf{R}$,

(97) $$F_\varepsilon(x) = \max_{\substack{1\le i\le r\\ 1\le j\le s}} \{f(x) - f(x_0) + \varepsilon, |g_i(x)|, h_j(x)\}.$$

F_ε is locally Lipschitz (5.40). We have $F_\varepsilon(x_0) = \varepsilon$ (see (95), (96)), hence

(98) $$F_\varepsilon(x_0) \le \inf_{x\in R^n} F_\varepsilon(x) + \varepsilon,$$

since $F_\varepsilon(x) \ge 0$ on $\mathbf{R}^n$. Apply to F_ε Ekeland principle 1.1 with $\lambda = \sqrt{\varepsilon}$, $\exists\, x_\varepsilon$ in $\mathbf{R}^n$ such that

(99) $$\|x_0 - x_\varepsilon\| \le \sqrt{\varepsilon}$$

and x_ε minimizes on $\mathbf{R}^n$ the function $G_\varepsilon : \mathbf{R}^n \to \mathbf{R}$,

(100) $$G_\varepsilon(x) = F_\varepsilon(x) + \sqrt{\varepsilon}\, \|x - x_\varepsilon\|.$$

(100) gives

$$\partial G_\varepsilon(x_\varepsilon) \subset \partial F_\varepsilon(x_\varepsilon) + \sqrt{\varepsilon}\, S,\ S := \{x \in \mathbf{R}^n : \|x\| \le 1\}\ (5.28)$$

and since $0 \in \partial G_\varepsilon(x_\varepsilon)$ (5.27),

(101) $$0 \in \partial F_\varepsilon(x_\varepsilon) + \sqrt{\varepsilon}\, S.$$

For $\partial F_\varepsilon(x_\varepsilon)$ we make use of the formula from 5.40. We have (102) $F_\varepsilon(x_\varepsilon) > 0$, since if *par absurdum* $F_\varepsilon(x_\varepsilon) = 0$, then $g_i(x_\varepsilon) = 0$, $h_j(x_\varepsilon) \leq 0$ and $f(x_\varepsilon) \leq f(x_0) - \varepsilon$, a contradiction as x_0 is a minimum point with the constraints (95,(96). According to 5.40, we have to consider the Clarke subderivatives in x_ε for those functions f, $|g_i|$, h_j whose value in x_ε is equal to $F_\varepsilon(x_\varepsilon)$, which imposes in particular, *via* (102), the indices i for which $g_i(x_\varepsilon) \neq 0$. Then, since $t \to |t|$ is of C^1 class on $\mathbf{R} \setminus \{0\}$ and its derivative is sgn t, we have $\partial |g_i(x_\varepsilon)| =$ $= [\operatorname{sgn} g_i(x_\varepsilon)]\, \partial g_i(x_\varepsilon)$ (combine 5.35, 2° with 5.22).

Thus, using (101) and 5.40, we find

$$\lambda_\varepsilon \geq 0,\ \mu_{\varepsilon,i} \geq 0,\ \nu_{\varepsilon,j} \geq 0,\ \lambda_\varepsilon + \sum \mu_{\varepsilon,i} + \sum \nu_{\varepsilon,j} = 1$$

such that

(103) $$0 \in \lambda_\varepsilon \partial f(x_\varepsilon) + \sum \mu_{\varepsilon,i} [\operatorname{sgn} g_i(x_\varepsilon)]\, \partial g_i(x_\varepsilon) + \sum \nu_{\varepsilon,j} \partial h_j(x_\varepsilon).$$

(103) yields ξ_ε in $\partial f(x_\varepsilon)$, $\eta_{\varepsilon,i}$ in $\partial g_i(x_\varepsilon)$ and $\zeta_{\varepsilon,j}$ in $\partial h_j(x_\varepsilon)$ such that

(104) $$\lambda_\varepsilon \xi_\varepsilon + \sum (\pm \mu_{\varepsilon,i})\, \eta_{\varepsilon,i} + \sum \nu_{\varepsilon,j} \zeta_{\varepsilon,j} = 0.$$

Take $\varepsilon = \frac{1}{n}$, $n \in \mathbf{N}$ and pass to the limit, we find on one hand ξ from $\partial f(x_0)$, η_i from $\partial g_i(x_0)$ and ζ_j from $\partial h_j(x_0)$ (use 5.20,1°, the strong and *-weak topologies coincide in finite dimension) and on the other hand $\lambda \geq 0$, $\mu_i \geq 0$, $\nu_j \geq 0$ with $\lambda + \sum \mu_i + \sum \nu_j = 1$ (use a $r + s$ – simplex) such that (see (104))

$$\lambda \xi + \sum (\pm \mu_i) \eta_i + \sum \nu_j \zeta_j = 0$$

and the proof is finished. ■

Remark. When the functions f, g_i, h_j are of C^1 class, the relation in the statement becomes (take into account 5.25 combined with the remark to 5.22)

$$\lambda f'(x_0) + \sum \mu_i g_i'(x_0) + \sum \nu_j h_j'(x_0) = 0.$$

With certain conditions of regularity we have $\lambda \neq 0$ and dividing by λ *Lagrange multipliers* appear.

6

Variational theorems of minimax type in Banach spaces

In this paragraph, Ekeland principle 1.1 is used to prove three variational theorems of minimax type. We deduce from these some variants of the famous Mountain Pass and Saddle Point theorems. Versions of these for locally Lipschitz functions will be given in the second section.

All the Banach spaces will be real.

MINIMAX THEOREM I

6.1 Minimax theorem. *Let* X *be a Banach space,* K *a compact metric space,* K_0 *a nonempty compact part of* K, θ *from* $C(K_0 ; X)$ *and*

$$T := \{g \in C(K; X) : g \mid K_0 = \theta\}$$

the metric space with usual distance.

Let $\varphi : X \to \mathbf{R}$ *be continuous and Gâteaux differentiable with* $\varphi'_w(\cdot)(v)$ *upper semi-continuous* $\forall$ v *of* X *and*

$$c := \inf_{g \in T} \sup_{K} (\varphi \circ g), \; c_0 := \sup_{K_0} (\varphi \circ \theta).$$

If

$$c > c_0$$

then, for every $\varepsilon > 0$ *and* f *from* T *with* $\sup_K (\varphi \circ f) < c + \varepsilon$, *there is* v_ε *in* X *such that*

$$c - \varepsilon \le \varphi(v_\varepsilon) \le \sup_K (\varphi \circ f)$$

and

$$\| \varphi'_w (v_\varepsilon) \| \le \sqrt{\varepsilon}$$

(*Shi Shuzhong, J. Mawhin - M. Willem*).

■ The statement is correct, we have $T \neq \varnothing$ (Dugundji theorem), c is *a fortiori* finite. We may suppose

(1) $$0 < \varepsilon < c - c_0 .$$

Indeed, suppose the proof done in the situation (1). If $c < \sup_K (\varphi \circ f)$, take ε_1, with $0 < \varepsilon_1 < \min(\varepsilon, c - c_0)$ and $c + \varepsilon_1 < \sup_K (\varphi \circ f)$, and f_1 in T with $\sup_K (\varphi \circ f_1) < c + \varepsilon_1$ and apply to them what has been accepted. If $c = \sup_K (\varphi \circ f)$ take ε_1 with $0 < \varepsilon_1 < \min(\varepsilon, c - c_0)$ and apply to f and ε_1 what has been accepted.

Consider the function

$$\Phi: T \to \mathbf{R}, \; \Phi(g) = \sup_K (\varphi \circ g).$$

Φ is lower bounded ($c = \inf_T \Phi > c_0$), but also l.s.c. on T : let g_0 be arbitrary fixed from T and $g_n \to g_0$, $g_n \in T$; if $\Phi(g_0) = \varphi(g_0(t_0))$, then $\Phi(g_n) - \Phi(g_0) \ge \varphi(g_n(t_0)) - \varphi(g_0(t_0))$, take the lower limit for $n \to \infty$. T being complete (as closed subset of $C(K; X)$), apply Ekeland principle 1.1 with respect to Φ and ε with $\lambda = \sqrt{\varepsilon}$, there is h in T such that

(2) $$\Phi(h) \le \Phi(f) < c + \varepsilon,$$

(3) $$\Phi(h) - \sqrt{\varepsilon}\, d(h, g) < \Phi(g) \; \forall \, g \in T \setminus \{h\}.$$

We prove that there exists t_0 in K such that

(4) $$c - \varepsilon \le \varphi(h(t_0))$$

and

(5) $$\|\varphi'_w(h(t_0))\| \le \sqrt{\varepsilon} .$$

Suppose *par absurdum* the contrary, i.e.

(6) $$A \cap B = \emptyset,$$

$A := \{t \in K : c - \varepsilon \le \varphi(h(t))\}$, $B := \{t \in K : \|\varphi'_w(h(t))\| \le \sqrt{\varepsilon}\}$.

Thus we have, for every t in A, $\|\varphi'_w(h(t))\| > \sqrt{\varepsilon}$, consequently there is v_t in X with $\|v_t\| =$ $= 1$ such that

(7) $$\varphi'_w(h(t))(v_t) < -\sqrt{\varepsilon}$$

(if $\varphi'_w(h(t))(v_t) > \sqrt{\varepsilon}$, replace v_t by $-v_t$, $\|-v_t\| = \|v_t\| = 1$).

By (7) we get

There are $\rho_t > 0$ *and* S_t an open ball centred in t such that

(8) $$\varphi'_w(h(\tau) + u)(v_t) < -\sqrt{\varepsilon} \quad \forall \tau \in S_t, \forall u \in X, \|u\| \le \rho_t .$$

Prove (8). There is $S(h(t), R)$, an open ball centred in $h(t)$, on which $\varphi'_w(u)(v_t) < -\sqrt{\varepsilon}$ (the hypothesis „ $\varphi'_w(\cdot)(v_t)$ u.s.c.") and for $S(h(t), \frac{R}{2})$ there is an open ball S_t centred in t with the property $\tau \in S_t \Rightarrow h(\tau) \in S(h(t), \frac{R}{2})$, take $\rho_t = \frac{R}{2}$.

$(S_t)_{t\in A}$ is an open covering of the compact set A, let be $S_{t_1}, S_{t_2}, \dots, S_{t_N}$ a finite covering and $\rho := \min_{1\le k\le N} \rho_{t_k}$. Consider the continous function $\psi : K \to [0, 1]$ with the property

$$\psi(t) = \begin{cases} 1, c \le \varphi(h(t)) \\ 0, \varphi(h(t)) \le c - \varepsilon \end{cases}$$

and let be $g : K \to X$,

(9) $$g(t) = h(t) + \frac{1}{N}\rho\psi(t)\sum_{k=1}^{N} v_{t_k} .$$

g, obviously continuous, will furnish the contradiction – consequence of supposition (6).

We have $g \in T$: $t \in K_0 \Rightarrow \varphi(h(t)) \overset{(1)}{=} \varphi(\theta(t)) \le c_0 < c - \varepsilon$, hence $\psi(t) = 0$, $g(t) \overset{(9)}{=} h(t) =$ $= \theta(t)$. Estimate $\varphi(g(t)) - \varphi(h(t))$ for $t \in A$, hence t belongs to one of the balls $S_{t_1}, \dots,$ S_{t_N}. Apply the finite increment formula 3.8, there is α in $(0, 1)$ such that

$$(10)\ \varphi(g(t)) - \varphi(h(t)) = \varphi'_w(h(t)+\frac{\alpha\rho}{N}\psi(t)\sum_{k=1}^{N} v_{t_k})(\frac{\rho}{N}\psi(t)\sum_{k=1}^{N} v_{t_k}) = \frac{\rho}{N}\psi(t)\sum_{i=1}^{N}\psi'_w(h(t)+$$

$$+\frac{\alpha\rho}{N}\psi(t)\sum_{k=1}^{N} v_{t_k})(v_{t_i}) \overset{(8)}{\leq} -\sqrt{\varepsilon}\rho\psi(t)$$ (the inequality is strict when $\varphi(t) > 0$!). If $t \notin A$ we

have $\psi(t) = 0$, hence ((9))

$$(11) \qquad \varphi(g(t)) = \varphi(h(t)).$$

Let now t_0 be from K such that $\varphi(g(t_0)) = \Phi(g)$. Then $\varphi(h(t_0)) \overset{(10),(11)}{\geq} \varphi(g(t_0)) \geq c$, consequently $t_0 \in A$ and $\psi(t_0) = 1$, (10) works,

$$\varphi(g(t_0)) - \varphi(h(t_0)) < -\sqrt{\varepsilon}\rho,$$

hence

$$(12) \qquad \Phi(g) + \sqrt{\varepsilon}\rho < \Phi(h),$$

and this implies $g \neq h$. Moreover, since $d(g, h) \leq \rho$ (use (9), $\psi(t) \leq 1$, $\| v_{t_k} \| = 1$), by (12) we get

$$\Phi(g) + \sqrt{\varepsilon}\, d(g, h)) < \Phi(h)$$

and we come in collision with (3). ■

REMARK. Minimax theorem I (6.1) was obtained by Shi Shuzhong and Mawhin-Willem in the hypothesis **φ of C[1]-Fréchet class**. The passage to the **Gâteaux derivative** was done by the author ([46]).

Of course, in the statement 6.1 Gâteaux derivative can be replaced by any β-derivative, β bornology on X (see 3.10).

6.2 *Corollary. In the conditions of* 6.1, *for every sequence* $(g_n)_{n\geq 1}$, $g_n \in T$, *if*

$$\lim_{n\to\infty} \sup_K (\varphi \circ g_n) = c,$$

a sequence $(v_n)_{n\geq 1}$ *exists in* X *such that*

$$\lim_{n\to\infty} \varphi(v_n) = c$$

and

$$\lim_{n\to\infty} \|\varphi'_w(v_n)\| = 0.$$

■ Take, for every n from **N**,

$$\varepsilon_n := \begin{cases} \sup_K(\varphi \circ g_n) - c, & \text{when } \sup_K(\varphi \circ g_n) > c, \\ \frac{1}{n}, & \text{otherwise.} \end{cases}$$

Apply 6.1 to ε_n and g_n with the remark $\varepsilon_n \to 0$. ■

It is useful to isolate in a statement

6.3 Let K , K_0, T , φ , c , c_0 *be as in the theorem* 6.1. *If there is a subset* E *of* X *with the property*

$$g(K) \cap E \neq \emptyset \ \forall\ g \in T$$

and in addition $c_0 < \inf_E \varphi$, *then*

$$c > c_0.$$

■ Let u be from $g(K) \cap E$. $\sup \varphi(g(K)) \geq \varphi(u) \geq \inf_E \varphi$, hence, since g is arbitrary, $c \geq$

$\geq \inf_E \varphi$, $c > c_0$. ■

We also add (see §4)

6.4 *In the conditions of* 6.1, *if* φ *satisfies the weak* $(PS)_c$ *condition, then* c *is a critical value for* φ.

■ The statement is correct, *a fortiori* $c \in \mathbf{R}$. Consider, for every n from **N**, g_n in T such that

$$c \leq \sup_K (\varphi \text{ o } g_n) < c + \frac{1}{n}.$$

Pass to the limit for $n \to \infty$, apply 6.2 and then $(PS)_c$ weak. ■

Pass to the variants corresponding to 6.1 announced at the beginning.

6.5 Mountain Pass theorem. *Let* X *be a Banach space,* $\varphi : X \to \mathbf{R}$ *continuous and Gâteaux differentiable with* $\varphi'_w(\cdot)(v)$ *upper semicontinuous* $\forall\ v \in X$. *Suppose that for* u_0, u_1 *from* X *and* Ω *an open neighbourhood of* u_0 *such that* $u_1 \in X \setminus \overline{\Omega}$ *we have*

$$\inf_{Fr\Omega} \varphi > \max(\varphi(u_0),\varphi(u_1)).$$

Let be

$$\Gamma := \{g \in C([0, 1]; X) : g(0) = u_0,\ g(1) = u_1\}$$

and

$$c := \inf_{g\in\Gamma} \sup_{[0,1]} (\varphi \text{ o } g).$$

If φ *verifies the weak* $(PS)_c$ *condition, then* c *is a critical value of* φ *and*

$$c > \max(\varphi(u_0), \varphi(u_1))$$

(*A. Ambrosetti - P. H. Rabinowitz*).

■ Apply 6.4 taking into account 6.3, where $K = [0,1]$, $K_0 = \{0,1\}$, $\theta(0) = u_0$, $\theta(1) = u_1$, $T = \Gamma$, $E = \text{Fr}\,\Omega$. We have $c_0 = \max(\varphi(u_0), \varphi(u_1))$. The application of 6.3 is correct since

$$g([0, 1]) \cap \text{Fr}\,\Omega \neq \varnothing \ \forall\, g \in \Gamma,$$

as $g([0, 1]) \cap \Omega \neq \varnothing$, $g([0, 1]) \cap (X \setminus \overline{\Omega}) \neq \varnothing$ and $g([0, 1])$ is connected. ■

Remark 1. The intuitive name „ Mountain Pass " given to the theorem 6.5 comes from a topographical situation which takes its source in the case $X = \mathbf{R}^2$ with the representation in the Euclidean space $\mathscr{E}_3$ of the elementary geometry. The conditions for u_0, u_1 and φ show that the points $M_0 := (u_0, \varphi(u_0))$ and $M_1 := (u_1, \varphi(u_1))$ are lower then the mountain wall

$$\{(x, y, \varphi(x, y)) : (x, y) \in \text{Fr}\,\Omega\}$$

and are situated on both sides of it. The path

$$t \to \varphi(g(t)),\ t \in [0, 1]$$

starts from M_0 and arrives at M_1 passing the mountain wall, since, as we have seen above,

$$g([0, 1]) \cap \text{Fr}\,\Omega \neq \varnothing.$$

But the highest point of the path has the quota

$$\sup_{t\in[0,1]} \varphi(g(t)),$$

an attained supremum. So, taking into account the significance that c receives now, the „Mountain pass" intuitive denomination is seen to be well chosen.

6.5$_1$ *Corollary. Let* X *be a Banach space and* $\varphi : X \to \mathbf{R}$ *lower unbounded, continuous and Gâteaux differentiable with* $\varphi'_w(\,\cdot\,)(v)$ *u.s.c. for every* v *at* X. *If* φ *has a point of locally strict minimum* u_0 *and verifies the* PS *condition, then* φ *has also a critical point distinct from* u_0.

■ Take, according to the remark at 4.5, $\varepsilon_0 > 0$ such that $\|u_0 - u\| \le \varepsilon_0 \Rightarrow \varphi(u_0) \le \varphi(u)$ and α from $(0, \varepsilon_0)$ such that $\varphi(u_0) < \inf\{\varphi(u) : \|u_0 - u\| = \alpha\}$. Finally, let u_1 be with $\varphi(u_1) < \varphi(u_0)$ (φ is not lower bounded). Apply 6.5 with $\Omega = \{u \in X : \|u_0 - u\| < \alpha\}$, which is correct since $u_1 \notin \overline{\Omega}$ and $\inf \varphi(\text{Fr}\,\Omega) > \max(\varphi(u_0), \varphi(u_1))$. c is a critical value and if $c = \varphi(u_2)$, *a fortiori* $u_2 \neq u_0$ since $\varphi(u_2) > \varphi(u_0)$. ■

Remark 2. If in 6.5$_1$ φ is lower bounded, we add the condition $\inf \varphi(X) < \varphi(u_0)$ and the conclusion still holds: take u' with $\varphi(u') < \varphi(u_0)$ and apply 4.5 combined with 6.5.

6.5_2 *Corollary. Let* X *be a Banach space and* $\varphi : X \to \mathbf{R}$ *continuous Gâteaux differentiable with* $\varphi'_w(\cdot)(v)$ *u.s.c.* $\forall\, v \in X$. *If* φ *verifies* PS *and*

$$(12_1) \qquad \inf\{\varphi(u): \|u\| = r\} \geq \max(\varphi(0), \varphi(u_0)),\ 0 < r < \|u_0\|,$$

then φ *has a critical point different from* 0 (*P. H. Rabinowitz*).

■ (*D. G. Figueiredo - S. Solimini*). If the inequality of (12_1) is strict, the answer is given by 6.5 with $\Omega = \{u \in X : \|u\| < r\}$. Let be then

$$(12_2) \qquad \inf\{\varphi(u): \|u\| = r\} = \max(\varphi(0), \varphi(u_0)).$$

If u_0 is a local minimum point, the discussion is closed, otherwise, there is u_0' near u_0, $r < \| u_0' \|$ such that $\varphi(u_0') < \varphi(u_0)$. Then, taking into account (12_2), we have

$$(12_3) \qquad \inf\{\varphi(u): \|u\| = r\} \geq \max(\varphi(0), \varphi(u_0')).$$

In the case of an inequality, apply 6.5 with $u_0 = 0$ and $u_1 = u_0'$, this yields a critical point different from 0 since $c > \varphi(0)$. In the case of an equality, applying (12_3) we get

$$(12_4) \qquad \inf\{\varphi(u) : \|u\| = r\} = \varphi(0) > \varphi(u_0').$$

Introduce the ball $S_r := \{u \in X : \|u\| \leq r\}$. Obviously $\inf\varphi(S_r) \leq \varphi(0)$. If $\inf\varphi(S_r) < \varphi(0)$, then $\inf \varphi(\overset{\circ}{S}_r) < \varphi(0)$ (*par absurdum*, take into account (12_4)) and the proof is finished by applying 6.5. Finally, the last posibility is $\inf \varphi(S_r) = \varphi(0)$. This expresses the fact that 0 is a point of local minimum. Then comes into action 4.5 with $u_0 = 0$ and $\varepsilon = r$: in the case 1°, 6.5 comes also into action with respect to the points 0 and u_0' taking (12_4) into account; in the case 2°, 4.5 is sufficient for obtaining the conclusion. ■

6.5_3 *Corollary. Let* X *be a Banach space and* $\varphi : X \to \mathbf{R}$ *continuous Gâteaux differentiable with* $\varphi'_w(\cdot)(v)$ *u.s.c.* $\forall\, v \in X$ *and verifying* PS. *If* φ *has two local minimum points, then it still has a critical point.*

■ *Let* u_1, u_2 be two distinct local minimum points and, for instance, $\varphi(u_1) > \varphi(u_2)$. Take $\varepsilon_0 > 0$ such that $\|u_1 - u\| \leq \varepsilon_0 \Rightarrow \varphi(u_1) \leq \varphi(u)$ and in addition $\|u_1 - u_2\| > \varepsilon_0$. Apply 4.5. In the case 1°, 6.5 comes into action with $\Omega = \{u \in X : \|u_1 - u\| < \alpha\}$. In the case 2° – nothing to add. ■

We put here some words about

Brouwer degree (elucidation for the following demonstration of 6.6).

Let Ω be open bounded subset of $\mathbf{R}^n$ and $\mathcal{M}$ the set of the triplets (f, Ω, y), $f : \overline{\Omega} \to \mathbf{R}^n$ continuous and $y \in \mathbf{R}^n \setminus \mathrm{Fr}\, \Omega$.

There is a function and it is unique

$$\deg : \mathcal{M} \to \mathbf{Z} \text{ (Brouwer degree)}$$

with the properties (Id is identical application)

(d1) $\deg(\mathrm{Id}, \Omega, y) = 1 \ \forall\, y \in \Omega$;

(d2) $\deg(f, \Omega, y) = \deg(f, \Omega_1, y) + \deg(f, \Omega_2, y)$, where Ω_1, Ω_2 are disjoint open subsets of Ω and $y \notin f(\overline{\Omega} \setminus (\Omega_1 \cup \Omega_2))$;

(d3) $\deg(h(t, \cdot), \Omega, y(t))$ remains constant for $t \in [0, 1]$, if $h : [0, 1] \times \overline{\Omega} \to \mathbf{R}^n$ and $y : [0, 1] \to \mathbf{R}^n$ are continuous and $y(t) \in \mathbf{R}^n \setminus h(t, \mathrm{Fr}\,\Omega) \ \forall\, t \in [0, 1]$ ([26], th. 1.1, def. 2.3).

Moreover, Brouwer degree has the properties:

(d4) $\deg(f, \Omega, y) \neq 0 \Rightarrow f^{-1}(y) \neq \emptyset$;

(d5) $f \mid \mathrm{Fr}\,\Omega = g \mid \mathrm{Fr}\,\Omega \Rightarrow \deg(f, \Omega, y) = \deg(g, \Omega, y)$

([26], th. 3.1).

Brouwer degree is generalized at finite-dimensional topological vector space (t.v.s.).

Let X be a real n - dimensional t.v.s. and $\mathcal{M}$ the set of triplets (f, Ω, y), Ω open bounded subset of X, $f : \overline{\Omega} \to X$ continuous and $y \in X \setminus f(\mathrm{Fr}\,\Omega)$.

$\deg : \mathcal{M} \to \mathbf{Z}$ is defined by

$$\deg(f, \Omega, y) \overset{\text{def}}{=} \deg(F f F^{-1}, F(\Omega), F(y)),$$

where $F : X \to \mathbf{R}^n$ is the linear homeomorphism defined by

$$F(x_i) = e_i,\ i = \overline{1, n},$$

where $(x_1, \dots, x_n)$ is a basis of X and $(e_1, \dots, e_n)$ the natural basis of $\mathbf{R}^n$.

6.6 Saddle Point theorem. *Let* X *be a Banach space and* $\varphi : X \to \mathbf{R}$ *continuous and Gâteaux differentiable with* $\varphi'_w(\cdot)(v)$ *upper semicontinuous* $\forall\, v \in X$. *Suppose* $X = V \oplus$ $\oplus\, W$, *direct sum, with* V, W *closed subspaces and*

$$\dim V < +\infty,\ \sup_{\sigma_R} \varphi < \inf_{W} \varphi,\ \sigma_R := \{v \in V : \|v\| = R\}.$$

Let be

$$S_R := \{v \in V : \|v\| \le R\},\ T := \{g \in C(S_R; X) : g(t) = t \text{ on } \sigma_R\},$$

$$c := \inf_{g\in T} \sup_{S_R} (\varphi \circ g).$$

If φ *verifies the weak* $(PS)_c$ *condition, then* c *is a critical value for* φ *(P. H. Rabinowitz).*

■ Take $K = S_R$ (which is correct since dim $V < +\infty$), $K_0 = \sigma_R$, $\theta : K_0 \to X$, $\theta(t) = t$, $c_0 = \sup_{\sigma_R} \varphi$. If we prove

(13) $$c > c_0,$$

then, *via* 6.4, we obtain c a critical value for φ. For (13) use 6.3 with $E = W$, it remains to show

$$g(S_R) \cap W \neq \varnothing \ \forall\, g \in T,$$

which is equivalent to

$\pi \circ g : S_R \to V$ (π the projection of X on V along W) has at least a zero.

We have $\pi \circ g = \mathrm{Id}$ on $\sigma_R(= \mathrm{Fr}\, S_R)$, hence $\deg(\pi \circ g, \overset{o}{S}_R, 0) \overset{(d5)}{=} \deg(\mathrm{Id}, \overset{o}{S}_R, 0)$, but $\deg(\mathrm{Id}, \overset{o}{S}_R, 0) \overset{(d1)}{=} 1$, hence $\deg(\pi \circ g, \overset{o}{S}_R, 0) = 1$ and hence, in accordance with (d4), $(\pi \circ g)^{-1}(0) \neq \varnothing$. ■

REMARK. Theorems 6.5 and 6.6 were obtained by Ambrosetti-Rabinowitz and Rabinowitz in the assumption **φ of C^1-Fréchet class**. The passing to **φ continuous with $\varphi'_w(\cdot)(v)$ u.s.c. $\forall\, v \in X$** was done by the author ([46]). The same remark is obviously true also for 6.5_1, 6.5_2, 6.5_3.

Of course, the Gâteaux derivative in the statements 6.5 and 6.6 can be replaced by any β-derivative, β bornology on X (3.10).

In the following we present a generalization of the minimax theorem I ([71], 1, Theorem 3.1). This will be deduced, naturally, with Zhong Cheng - Kui theorem (1.1_1).

Generalization of 6.1. *Let* K *be a compact metric space,* K_0 *a closed subset of* K, X *a Banach space,* θ *from* $C(K_0; X)$ *and* T *the metric space, with the usual distance,*

$$T = \{g \in C(K; X) : g \mid K_0 = \theta\}.$$

Let $\varphi : X \to \mathbf{R}$ *be continuous and Gâteaux differentiable with* $\varphi'_w(\cdot)(v)$ *upper semi-continuous* $\forall\, v \in X$. *If*

$$c := \inf_{g\in T} \sup_{K} (\varphi \circ g) > c_0 := \sup_{K_0} (\varphi \circ \theta),$$

then for every $\varepsilon > 0$ *and* f *from* T *with* $\sup_K (\varphi \circ f) < c + \varepsilon$ *and for every* $h : [0, +\infty) \to [0, +\infty)$ *increasing with* $\int_0^{+\infty} \frac{dr}{1 + h(r)} = +\infty$, *there is* v_ε *in* X *such that*

$$c - \varepsilon \le \varphi(v_\varepsilon) \le \sup_K (\varphi \circ f)$$

and

$$\|\varphi'_w(v_\varepsilon)\| \le \frac{\sqrt{\varepsilon}}{1 + h(\|v_\varepsilon\|)}.$$

■ We follow closely the proof of 6.1. It can suppose $0 < \varepsilon < c - c_0$. Let be $\Phi : T \to \mathbf{R}$,

$$\Phi(g) = \sup_K (\varphi \circ g).$$

Φ is lower bounded ($c = \inf_{g \in T} \Phi(g)$, $c > c_0$) and lower semicontinuous. Apply to Φ 1.1_1 with $u_0 = 0$ and $\lambda = \sqrt{\varepsilon}$, there is g_ε in T such that

$$\Phi(g_\varepsilon) \le \sup_K (\varphi \circ f)$$

and

$$(13_1) \qquad \Phi(g_\varepsilon) - \frac{\sqrt{\varepsilon}}{1 + h(\sup_{t \in K} \|g_\varepsilon(t)\|)} d(g_\varepsilon, g) \le \Phi(g) \ \forall\, g \in T.$$

So we must prove the existence of one t_0 at K such that

$$(13_2) \qquad c - \varepsilon \le \varphi(g_\varepsilon(t_0))$$

and

$$(13_3) \qquad \|\varphi'_w(g_\varepsilon(t_0))\| \le \frac{\sqrt{\varepsilon}}{1 + h(\|g_\varepsilon(t_0)\|)}.$$

Since h is increasing , (13_3) is obviously a consequence of

$$(13_4) \qquad \|\varphi'_w(g_\varepsilon(t_0))\| \le \frac{\sqrt{\varepsilon}}{1 + h(\sup_{t \in K} \|g_\varepsilon(t)\|)}.$$

Note by A and B respectively the set of points t from K satisfying (13_2) respectively (13_4) and suppose *par absurdum*

$$A \cap B = \varnothing.$$

Reasoning like for (8) (the proof of 6.1), we find, for every t in A , v_t in X , $\|v_t\| = 1$ and it can state

There is $\rho_t > 0$ *and an open ball* S_t *centred at* t, *such that*

$$\varphi'_w (g_\varepsilon(\tau) + u) (v_t) < -\frac{\sqrt{\varepsilon}}{1 + h(\sup_{t\in K}\|g_\varepsilon(t)\|)}, \ \forall\, \tau \in S_t, \forall\, u \in X, \|u\| \leq \rho_t .$$

Draw out from the covering $(S_t)_{t\in A}$ a finite covering S_{t_1} , ... , S_{t_N} , let be $\rho := \min_{1\leq k\leq N} \rho_{t_k}$.

Consider the functions $\psi : K \to [0, 1]$ continuous with $\psi(t) = 1$ for $c \leq \varphi(g_\varepsilon(t))$, $\psi(t) = 0$ for $\varphi(g_\varepsilon(t)) \leq c - \varepsilon$, and $g : K \to X$,

$$g(t) = g_\varepsilon(t) + \frac{1}{N}\rho\psi(t)\sum_{k=1}^{N} v_{t_k} .$$

The proof continues as in 6.1 and we come in collision with (13_1). ■

REMARK. The generalization of 6.1 was obtained in the assumption **φ of C¹-Fréchet class** ([71],1). The passage to **Gâteaux derivative** was carried out by the author ([46]). Certainly these derivative can be replaced by any β-derivative, β bornology on *X* (3.10).

MINIMAX THEOREM II

Let X be a real normed space and F a closed nonempty subset of it. By definition F *separates* the distinct points y_0 , y_1 from X when these points are in distinct connected components of $X \setminus F$. Since X is locally connected and $X \setminus F$ is open, the connected components of $X \setminus F$ are all open, let Ω_0 be the component which contains y_0, and Ω_1 the union of all other components. Obviously, $y_1 \in \Omega_1$ and $\{\Omega_0 , \Omega_1\}$ is an open partition of $X \setminus F$.

6.7 Minimax theorem. *Let* X *be a Banach space and* φ: X → **R** *continuous Gâteaux differentiable with* $\varphi'_w : X \to X^*$ *continuous from the norm topology to the *-weak topology. Take the distinct points* y_0, y_1 *in* X *and let be*

$$\Gamma := \{\gamma \in C([0, 1]; X): \gamma(0) = y_0 , \gamma(1) = y_1\}$$

and

$$c: = \inf_{\gamma\in\Gamma} \sup_{[0,1]} (\varphi \circ \gamma).$$

Suppose there is a closed subset F *having the property* $F \cap A_{\varphi,c}$ *nonempty and separates* y_0, y_1, $A_{\varphi, c} := \{x \in X: \varphi(x) \geq c\}$.

Then there is a sequence $(x_n)_{n\geq 1}$ *in* X *such that*

(14) $$\lim_{n\to\infty} d(x_n, F) = 0,$$

(15) $$\lim_{n\to\infty} \varphi(x_n) = c,$$

(16) $$\lim_{n\to\infty} (1 + \|x_n\|)^{-1} \|\varphi'_w(x_n)\| = 0$$

(*N. Ghoussoub - D. Preiss*).

■ Set $F_c := F \cap A_{\varphi, c}$, a closed set, and let $\{\Omega_0, \Omega_1\}$ be an open partition of $X \setminus F_c$, $y_i \in \Omega_i$, $i = 0, 1$. Take ε such that

(17) $$0 < \varepsilon < \frac{1}{2} \min(1, d(y_0, F_c), d(y_1, F_c))$$

and let γ be from Γ with

(18) $$\sup_{[0,1]} (\varphi \circ \gamma) < c + \frac{\varepsilon^2}{4}.$$

Define, *via* γ, the numbers t_0, t_1, $0 \leq t_0 < t_1 < 1$ (nonempty sets appear, see (17)),

(19) $$t_0 = \sup\{t \in [0, 1]: \gamma(t) \in \Omega_0, d(\gamma(t), F_c) \geq \varepsilon\},$$

(20) $$t_1 = \inf\{t \in [t_0, 1]: \gamma(t) \in \Omega_1, d(\gamma(t), F_c) \geq \varepsilon\}.$$

Remark ((19)) $t_0 < t_1 < 1$ (Ω_0, Ω_1 are disjoint open sets). Obviously

(21) $$d(\gamma(t), F_c) \leq \varepsilon \ \forall\, t \in [t_0, t_1].$$

Consider the nonempty set (see $\gamma \mid [t_0, t_1]$)

(22) $$\Gamma(t_0, t_1) = \{f \in C([t_0, t_1]; X): f(t_0) = \gamma(t_0), f(t_1) = \gamma(t_1)\},$$

endow it with the usual distance (the same notation)

(23) $$d(f_1, f_2) = \sup_{t\in[0,1]} \|f_1(t) - f_2(t)\|$$

and it obtains a complete metric space. Consider the functions

(24) $$\psi_\varepsilon: X \to \mathbf{R},\ \psi_\varepsilon(x) = \max(0, \varepsilon^2 - \varepsilon d(x. F_c)),$$

(25) $$\Phi: \Gamma(t_0, t_1) \to \mathbf{R},\ \Phi(f) = \sup_{[t_0,t_1]} (\varphi \circ f + \psi_\varepsilon \circ f).$$

Φ is lower semicontinous.

Let f be arbitrary in $\Gamma(t_0, t_1)$. Since $f(t_0) \in \Omega_0$, $f(t_1) \in \Omega_1$ and $f([t_0, t_1])$ is connected, $\exists\, t_f$ in (t_0, t_1) such that $f(t_f) \in \mathrm{Fr}\,\Omega_0$, but $\mathrm{Fr}\,\Omega_0 \subset F_c$, we get

(26) $$\Phi(f) \geq \varphi(f(t_f)) + \psi_\varepsilon(f(t_f)) \overset{(24)}{\geq} c + \varepsilon^2$$

as $d(f(t_f), F_c) = 0$.

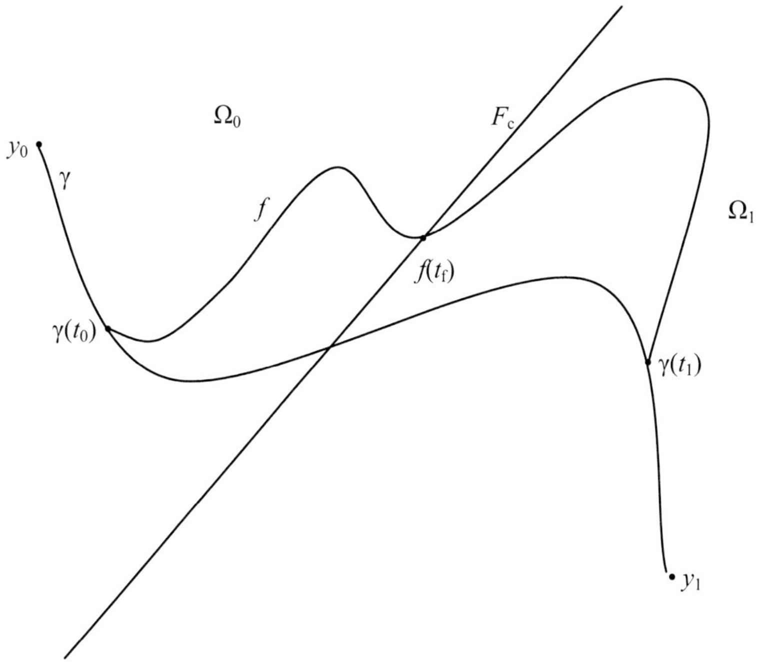

Fig. 4

Moreover, let be $\gamma_\varepsilon := \gamma \mid [t_0, t_1]$. Then

$$\Phi(\gamma_\varepsilon) \le \sup_{t\in[0,1]} [\varphi(\gamma(t)) + \psi_\varepsilon(\gamma(t))],$$

hence (see (18) and (24))

$$\Phi(\gamma_\varepsilon) < (c + \frac{\varepsilon^2}{4}) + \varepsilon^2 \overset{(26)}{\le} \inf_{\Gamma(t_0,t_1)} \Phi + \frac{\varepsilon^2}{4}. \tag{27}$$

Apply to Φ Ekeland principle 1.1 with $\frac{\varepsilon^2}{4}$, $\lambda = \frac{\varepsilon}{2}$ and $u = \gamma_\varepsilon$ (see (27)), $\exists f_\varepsilon$ in $\Gamma(t_0, t_1)$ such that

$$\Phi(f_\varepsilon) \le \Phi(f) + \frac{\varepsilon}{2} \mathrm{d}(f, f_\varepsilon) \ \forall f \in \Gamma(t_0, t_1), \tag{28}$$

$$\Phi(f_\varepsilon) \le \Phi(\gamma_\varepsilon),\ \mathrm{d}(f_\varepsilon, \gamma_\varepsilon) \le \frac{\varepsilon}{2}. \tag{29}$$

Consider the set

$$M := \{t \in [t_0, t_1]: \varphi(f_\varepsilon(t)) + \psi_\varepsilon(f_\varepsilon(t)) = \Phi(f_\varepsilon)\}. \tag{30}$$

M is nonempty (Weierstrass theorem, $(\varphi + \psi_\varepsilon) \circ f_\varepsilon$ is continuous) and compact being closed. Moreover, $t_0, t_1 \notin M$. Indeed, since $d(\gamma(t_i), F_c) \overset{(19),(20)}{\geq} \varepsilon$ we have $\psi_\varepsilon(\gamma(t_i)) \overset{(24)}{=} 0$ and then $\varphi(f_\varepsilon(t_i)) + \psi_\varepsilon(f_\varepsilon(t_i)) \overset{(22)}{=} \varphi(\gamma_c(t_i)) \overset{(18)}{<} c + \frac{\varepsilon^2}{4} < c + \varepsilon^2 \overset{(26)}{\leq} \Phi(f_\varepsilon)$.

Prove now (a more difficult part of the proof)

(31) $$\exists\, t \text{ in } M \text{ such that } \|\varphi'_w(f_\varepsilon(t))\|\,(1 + \|f_\varepsilon(t)\|)^{-1} \leq \frac{3\varepsilon}{2}.$$

Suppose *par absurdum*

(32) $$t \in M \Rightarrow \|\varphi'_w(f_\varepsilon(t))\|\,(1 + \|f_\varepsilon(t)\|)^{-1} > \frac{3\varepsilon}{2}.$$

Fix t arbitrary in M. By (32) we get

$$\|\varphi'_w(f_\varepsilon(t))\| > \frac{3\varepsilon}{2}(1 + \|f_\varepsilon(t)\|),$$

hence $\exists\, u \in X$, $\|u\| = 1$ such that $|\varphi'_w(f_\varepsilon(t))(u)| > \frac{3\varepsilon}{2}(1 + \|f_\varepsilon(t)\|)$, which is equivalent with

$$\varphi'_w(f_\varepsilon(t))(u) < -\frac{3\varepsilon}{2}(1 + \|f_\varepsilon(t)\|)$$

or

$$\varphi'_w(f_\varepsilon(t))(u) > \frac{3\varepsilon}{2}(1 + \|f_\varepsilon(t)\|).$$

If eventually the second posibility holds true, by replacing u by $-u$ the norm of this does not change and hence we find u at X, $\|u\| = 1$ such that

$$\varphi'_w(f_\varepsilon(t))(u) < -\frac{3\varepsilon}{2}(1 + \|f_\varepsilon(t)\|).$$

By division we get

(33) $$\begin{cases} \exists u_t \in X, \|u_t\| = (1 + \|f_\varepsilon(t)\|)^{-1} \text{ such that} \\ \qquad \varphi'_w(f_\varepsilon(t))(u_t) < -\frac{3\varepsilon}{2}. \end{cases}$$

φ'_w being continuous as in the statement, $\exists\, J_t$ an open neighbourhood of t in $[t_0, t_1]$ such that

(34) $$s \in J_t \Rightarrow \varphi'_w(f_\varepsilon(s))(u_t) < -\frac{3\varepsilon}{2}.$$

And now, since M is compact, there exists $J_{t_1}, \ldots, J_{t_N}$ a finite covering of M. Associate to these $\pi_1, \ldots, \pi_N : M \to [0,1]$ a continuous partition of the unit on M with (35) $\operatorname{supp} \pi_i \subset \subset J_{t_i}$, $i = \overline{1, N}$ and consider the continuous function $v_1 : M \to X$,

$$v_1(t) = \sum_{i=1}^{N} \pi_i(t)\, u_{t_i} .$$

We have

(36) $$\varphi'_w(f_\varepsilon(t))(v_1(t)) < -\frac{3\varepsilon}{2} \;\; \forall\, t \in M,$$

(37) $$\|v_1(t)\| \le (1 + \|f_\varepsilon(t)\|)^{-1} \;\forall\, t \in M.$$

Indeed, by (33) we get (37) and as for (36) proceed in the following way: let I be the set, obviously nonempty, of the indices i for which $\pi_i(t) > 0$

$$(\pi_i(t) \ge 0, \sum_{i=1}^{N} \pi_i(t) = 1),\ \varphi'_w(f_\varepsilon(t))(v_1(t)) = \sum_{i=1}^{N} \pi_i(t)\varphi'_w(f_\varepsilon(t))(u_{t_i}) = \sum_{i\in I} \pi_i(t) \times$$

$$\times\, \varphi'_w(f_\varepsilon(t))(u_{t_i}) \overset{(34)}{<} -\frac{3\varepsilon}{2} \sum_{i\in I} \pi_i(t) = -\frac{3\varepsilon}{2} .$$

Let be $t' := \inf M$, $t'' := \sup M$. Obviously $t', t'' \in M$ and since $t_0, t_1 \notin M$ we have the situation $t_0 < t' \le t'' < t_1$. Dugundji theorem allows us to extend v_1 to $v_2 : [t', t''] \to X$

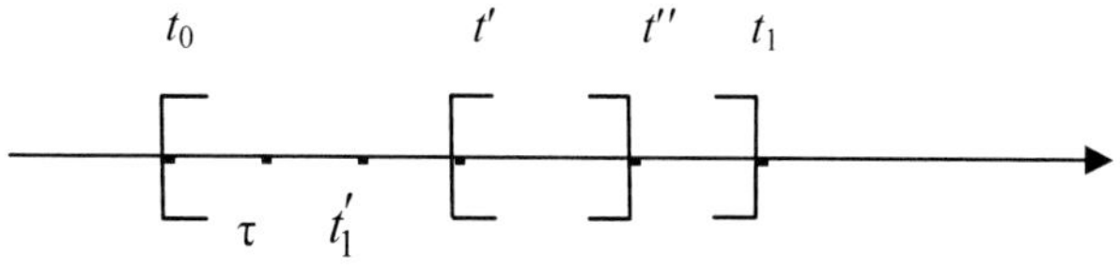

Fig. 5

continuous. Since every $v_2(t)$ is a convex combination of elements from $v_1(M)$, (37) allows to write

(38) $$\|v_2(t)\| \le (1 + \|f_\varepsilon(t)\|)^{-1} \;\; \forall\, t \in [t', t''].$$

Take two small disjoint intervals $[t_0, \tau]$, $[t_1', t']$ and the continuous function $w : [t_0, \tau] \cup \cup [t_1', t'] \to X$, $t \in [t_0, \tau] \Rightarrow w(t) = 0$, $t \in [t_1', t'] \Rightarrow w(t) = \frac{t}{t'} v_2(t')$. Prolong w according to Dugundji theorem to a continuous function $v_3 : [t_0, t'] \to X$, emphasize that $v_3(t_0) = 0$, for this we have again (pay attention to $w(t)$ on the two intervals!)

(39) $$\|v_3(t)\| \le (1 + \|f_\varepsilon(t)\|)^{-1} \;\forall\, t \in [t_0, t'].$$

Let now be $v_4: [t_0, t''] \to X$, $v_4(t) = v_3(t)$, $t \in [t_0, t']$, $v_4(t) = v_2(t)$, $t \in [t', t'']$. v_4 is continuous, since $\lim\limits_{t\to t'+} v_4(t) = v_2(t') = \lim\limits_{t\to t'-} v_3(t) = v_3(t') = v_4(t')$, $\lim\limits_{t\to t'-} v_4(t) = \lim\limits_{t\to t'-} v_3(t) = v_3(t') =$ $= v_4(t')$. Emphasize that $v_4(t_0) = 0$. Moreover, taking into account (38) and (39),

$$\|v_4(t)\| \le (1 + \|f_\varepsilon(t)\|)^{-1} \ \forall\, t \in [t_0, t''].$$

Continuing in the same way on the interval $[t'', t_1]$, we finally find a continuous function $v: [t_0, t_1] \to X$ with the properties (see also (36))

$$\varphi'_w(f_\varepsilon(t))(v(t)) < -\frac{3\varepsilon}{2} \ \forall\, t \in M, \tag{40}$$

$$v(t_0) = v(t_1) = 0, \tag{41}$$

$$\|v(t)\| \le (1 + \|f_\varepsilon(t)\|)^{-1} \ \forall\, t \in [t_0, t_1]. \tag{42}$$

Let $\lambda > 0$ be arbitrary. $f_\varepsilon + \lambda v \overset{(41)}{\in} \Gamma(t_0, t_1)$, replace it in (28), we get

$$\Phi(f_\varepsilon + \lambda v) \ge \Phi(f_\varepsilon) - \frac{\varepsilon}{2}\, d(f_\varepsilon + \lambda v, f_\varepsilon). \tag{43}$$

Let t_λ be from $[t_0, t_1]$ such that

$$\Phi(f_\varepsilon + \lambda v) \overset{(25)}{=} (\varphi + \psi_\varepsilon)(f_\varepsilon(t_\lambda) + \lambda v(t_\lambda)). \tag{44}$$

(43) and (44) yields

$$(\varphi + \psi_\varepsilon)(f_\varepsilon(t_\lambda) + \lambda v(t_\lambda)) \ge (\varphi + \psi_\varepsilon)(f_\varepsilon(t_\lambda) - \frac{\varepsilon}{2}\, d(f_\varepsilon + \lambda v, f_\varepsilon),$$

i.e.

$$\varphi(f_\varepsilon(t_\lambda) + \lambda v(t_\lambda)) - \varphi(f_\varepsilon(t_\lambda)) \ge -[\psi_\varepsilon(f_\varepsilon(t_\lambda) + \lambda v(t_\lambda)) - \psi_\varepsilon(f_\varepsilon(t_\lambda))] - \tag{45}$$

$$-\frac{\varepsilon}{2}\, d(f_\varepsilon + \lambda v, f_\varepsilon) \ \forall\, \lambda > 0.$$

But ψ_ε is Lipschitz with the constant ε, it is sufficient to verify $\psi_\varepsilon(x) - \psi_\varepsilon(y) \le \varepsilon\|x - y\|$. The case $d(x, F_c), d(y, F_c) \le \varepsilon$: the first member $= \varepsilon(d(y, F_c) - d(x, F_c)) \le \varepsilon\, \|x - y\|$. The case $d(x, F_c) \le \varepsilon$, $d(y, F_c) > \varepsilon$: the first member $= \varepsilon^2 - \varepsilon d(x, F_c) = \varepsilon[\varepsilon - d(x, F_c)] \le$ $\le \varepsilon[d(y, F_c) - d(x, F_c)] \le \varepsilon\|x - y\|$. The case $d(x, F_c) > \varepsilon$, $d(y, F_c) \le \varepsilon$: the first member $=$ $= -\varepsilon^2 + \varepsilon d(y, F_c) \le 0 \le \varepsilon\, \|x - y\|$. The case $d(x, F_c), d(y, F_c) > 0$: the first member $= 0 \le$ $\le \varepsilon\|x - y\|$.

In this way, the great bracket of the second member of (45) is majorised by $\varepsilon\| f_\varepsilon(t_\lambda) +$ $+ \lambda v(t_\lambda) - f_\varepsilon(t_\lambda)\|$, which is majorised in its turn by $\varepsilon d(f_\varepsilon + \lambda v, f_\varepsilon)$, consequently, taking into account (45),

(46) $$\varphi(f_\varepsilon(t_\lambda) + \lambda v(t_\lambda)) - \varphi(f_\varepsilon(t_\lambda)) \geq -\frac{\varepsilon}{2}\, d(f_\varepsilon + \lambda v, f_\varepsilon)\ \forall\, \lambda > 0.$$

Apply to (46) the finite increment formula 3.8 and divide by λ, $\forall\, \lambda > 0\ \exists\, \theta_\lambda \in (0, 1)$ such that

(47) $$\varphi'_w(f_\varepsilon(t_\lambda) + \theta_\lambda \lambda v(t_\lambda))(v(t_\lambda)) \geq -\frac{3\varepsilon}{2}\frac{1}{\lambda}\, d(f_\varepsilon + \lambda v, f_\varepsilon).$$

For the second member of (47) we have

$$\lim_{\lambda\to 0+} \frac{1}{\lambda}\, d(f_\varepsilon + \lambda v,\ f_\varepsilon) = \lim_{\lambda\to 0+} \frac{1}{\lambda} \sup_{t\in[t_0,t_1]} \|\lambda v(t)\| = \sup_{t\in[t_0,t_1]} \|v(t)\|.$$

Pass to the first member of (47). Look at (44). For $\lambda = \frac{1}{n}$ we have ($t_n := \frac{t_1}{n}$)

(48) $$\Phi(f_\varepsilon + \frac{1}{n} v) = (\varphi + \psi_\varepsilon)(f_\varepsilon(t_n) + \frac{1}{n} v(t_n)).$$

$(t_n)_{n\geq 1}$ has a convergent subsequence $(t_{k_n})_{n\geq 1}$, $t_{k_n} \to \tau \in [t_0, t_1]$.

Then as $f_\varepsilon(t_{k_n}) + \frac{1}{k_n} v(t_{k_n}) \to f_\varepsilon(\tau)$, one finds, taking in (48) the lower limit for $n \to \infty$,

$\Phi(f_\varepsilon) \leq (\varphi + \psi_\varepsilon)(f_\varepsilon(\tau))$, and this imposes (see (25)) $\Phi(f_\varepsilon) = (\varphi + \psi_\varepsilon)(f_\varepsilon(\tau))$, i.e. (49) $\tau \in$ $\in M((30))$.

Come back to (47), where we replace t_λ by t_{k_n}, pass to the limit and one obtains (the same justification as for (59), see below)

(50) $$\varphi'_w(f_\varepsilon(\tau))(v(\tau)) \geq -\frac{3\varepsilon}{2} \sup_{t\in[t_0,t_1]} \|v(t)\| \overset{(42)}{\geq} -\frac{3\varepsilon}{2}.$$

Comparing (50) and (49) with (40) we find a contradiction and this imposes (31).

This is the end of the proof. Work in the following with the same t of (31).

Since $d(f_\varepsilon, \gamma_\varepsilon) \overset{(29)}{\leq} \frac{\varepsilon}{2}$, we have $d(f_\varepsilon(t), F_c) \leq \frac{\varepsilon}{2} + d(\gamma_\varepsilon(t), F_c) \overset{(21)}{\leq} \frac{\varepsilon}{2} + \varepsilon$,

(51) $$d(f_\varepsilon(t), F_c) \leq \frac{3\varepsilon}{2}.$$

But $\Phi(f_\varepsilon) \overset{(29)}{\leq} \Phi(\gamma_\varepsilon)$, we get

(52) $$c + \varepsilon^2 \overset{(26),(30)}{\leq} \varphi(f_\varepsilon(t)) + \psi_\varepsilon(f_\varepsilon(t)) \overset{(27)}{\leq} c + \frac{5}{4}\varepsilon^2.$$

Take $\varepsilon = \frac{1}{n}$, $n \in \mathbf{N}$, set $f_{\frac{1}{n}}(t) = x_n$ and taking into account (51),(52) and (31) one finds the relations (14),(15) and (16) from the statement. ■

REMARK *relative to Ghoussoub-Preiss theorem* 6.7.

In the statement of [30], 5, p.140, theorem 6, the property (16) of 6.7 is replaced by

$$\lim_{n\to\infty} (1 + \|x_n\|)\, \|\varphi'_w(x_n)\| = 0.$$

Its proof is erroneus. The author could recover the statement by replacing the quoted relation above by

$$\lim_{n\to\infty} (1 + \|x_n\|)^{-1}\, \|\varphi'_w(x_n)\| = 0,$$

obviously a weaker one.

The propositions that follow (Corollary 9, Corollary 10, Theorem 12, Corollary 13) remain unchanged.

Pass to the corresponding variant for Mountain Pass theorem. Firstly, give the following

Definition. Let X be a real normed space. The Gâteaux differentiable map $\varphi: X \to \mathbf{R}$ verifies the *condition* (C) *at the level* c, $c \in \mathbf{R}$, *with respect to a nonempty closed subset* F of X, if every sequence $(x_n)_{n\geq 1}$ in X with the properties

(53) $$\lim_{n\to\infty} d(x_n, F) = 0,$$

(54) $$\lim_{n\to\infty} \varphi(x_n) = c,$$

(55) $$\lim_{n\to\infty} (1 + \|x_n\|)^{-1}\, \|\varphi'_w(x_n)\| = 0,$$

has a convergent subsequence.

Let $(x_{k_n})_{n\geq 1}$ be such a convergent subsequence. Remark that $x_0 := \lim_{n\to\infty} x_{k_n} \in F$: $d(x_0, F) \leq d(x_0, x_{k_n}) + d(x_{k_n}, F)$, pass to the limit for $n \to \infty$. Remark also : the condition (C) at the level c for F is implied by $(PS)_{c,F}$ (see §4), since if $\|\varphi'_w(x_n)\| \to 0$ then $(1+ \|x_n\|)^{-1} \times \times \|\varphi'_w(x_n)\| \to 0$.

Definition. x_0 is a *critical point at the level* c, $c \in \mathbf{R}$, for $\varphi : X \to \mathbf{R}$, X a real normed space, if

$$\varphi(x_0) = c,\ \varphi'_w(x_0) = 0.$$

Come back to the first definition and suppose in addition $\varphi'_w : X \to X^*$ continuous from the norm topology on X to the *-weak topology on X^*. Then, if $x_0 = \lim_{n\to\infty} x_{k_n}$,

(56) $\qquad x_0$ *is a critical point at the level* c *for* φ.

■ Obviously $\varphi(x_{k_n}) \to \varphi(x_0)$, hence, taking into account (54), $\varphi(x_0) = c$. Pass to the second condition. Since $x_{k_n} \to x_0$, $(1 + \|x_{k_n}\|)^{-1} \to (1 + \|x_0\|)^{-1}$, whence

(57) $\lim_{n\to\infty} \|\varphi'_w(x_{k_n})\| = 0$. Let u be arbitrary in X. $|\varphi'_w(x_{k_n})(u)| \le \|\varphi'_w(x_{k_n})\|\,\|u\|$ and (57) implies $\lim_{n\to\infty} \varphi'_w(x_{k_n})(u) = 0$, consequently

(58) $$\varphi'_w(x_{k_n}) \xrightarrow{*-\text{weak}} 0.$$

On the other hand, since $x_{k_n} \to x_0$ we get

(59) $$\varphi'_w(x_{k_n}) \xrightarrow{*-\text{weak}} \varphi'_w(x_0).$$

Indeed, let u be arbitrary fixed in X. We must show

(60) $$\varphi'_w(x_{k_n})(u) \to \varphi'_w(x_0)(u).$$

([47], vol III, p.745, 3.31).

Let $\varepsilon > 0$ be arbitrary. Since φ'_w is continuous at x_0, for the neighbourhood $V :=$ $= V(\varphi'_w(x_0); \varepsilon; u)$ of $\varphi'_w(x_0)$ in the *-weak topology ([47], vol III, p.738, 3.29) there is a neighbourhood U of x_0 in the strong topology with the corresponding property. From a rank N on, we have $x_{k_n} \in U$, hence $\varphi'_w(x_{k_n}) \in V$, i. e. $|\varphi'_w(x_{k_n})(u) - \varphi'_w(x_0)(u)| \le \varepsilon$ and consequently (60).

As the *-weak topology is separated, (58) and (59) give $\varphi'_w(x_0) = 0$ and consequently (56). ■

The following version ([30], 5, 145, Corollary 9) is an immediate consequence of Ghoussoub-Preiss theorem (6.7).

6.8 Mountain Pass theorem. *Let* X *be a Banach space,* φ: X $\to$ **R** *continuous Gâteaux differentiable and* $\varphi'_w : X \to X^*$ *continuous from the norm topology to the *-weak topology. Take the distinct points* y_0, y_1 *in* X *and let be*

$$\Gamma := \{\gamma \in C([0, 1]; X) : \gamma(0) = y_0,\ \gamma(1) = y_1\}$$

and

$$c := \inf_{\gamma\in\Gamma} \sup_{[0,1]} (\varphi \circ \gamma).$$

If φ *verifies the condition* (C) *at the level* c *with respect to* X *and*

$$c > \max(\varphi(y_0), \varphi(y_1)),$$

then there is a critical point of φ *at the level* c

(*A. Ambrosetti - P. H. Rabinowitz*).

■ Apply 6.7 with $F = X$, $A_{\varphi, c}$ separates the points y_0, y_1. Indeed, $y_0, y_1 \notin A_{\varphi, c}$ (the condition in the statement) and if *par absurdum* we suppose that y_0, y_1 belong to the same connected component U of $X \setminus A_{\varphi, c}$, then, since $X \setminus A_{\varphi, c}$ is locally path connected, the component U is path connected, let γ_1 be a continuous path in U which joints y_0 to y_1. We have (the supremum is attained!)

$$\varphi(\gamma_1(t)) < c \ \forall\, t \in [0, 1], \text{ hence } \sup_{[0,1]} (\varphi \circ \gamma_1) < c,$$

in contradiction with the definition of c. Thus, since φ verifies the condition (C) at the level c, (56) finishes the proof. ■

REMARK. The demand of the variant 6.8 for the Mountain Pass theorem ([30], 5, p.145, Corollary 9):

$\varphi'_w : X \to X^*$ *continuous from the norm topology to the *-weak topology*,

a variant proved using Ghoussoub-Preiss theorem, requires more than the demand of the variant 6.5 for the same Mountain Pass theorem obtained by the author ([46]):

$\varphi'_w(\cdot)(v)$ *is upper semicontinuous* $\forall\, v \in X$,

since the first requirement imply $x \to \varphi'_w(x)(v)$ continuous $\forall\, v$ at X: let $x_0 \in X$, $\varepsilon > 0$ and $V := V(\varphi'_w(x_0); \varepsilon; v)$ a neighbourhood of $\varphi'_w(x_0)$ in the *-weak topology; for V there is a neighbourhood U of x_0 in wiew of the continuity in x_0 of φ'_w, $x \in U \Rightarrow \varphi'_w(x) \in V$, i.e. $|\varphi'_w(x)(v) - \varphi'_w(x_0)(v)| < \varepsilon$.

Otherwise, conversely $x \to \varphi'_w(x)(v)$ continuous $\forall\, v$ at X implies the first requirement (every finite intersection of neighbourhoods is a neighbourhood).

There is still another variant for the Mountain Pass theorem ([30], 5, p.145, Corollary 10; the requirement (87) is superfluous).

6.9 *Let* X, φ, Γ, c *be as to* 6.8 *and* F *a nonempty closed subset of* X *included in* $A_{\varphi, c}$ *that separates* y_0, y_1. *If* φ *verifies* (C) *at the level* c *with respect to* F, *then* φ *has a critical point at the level* c *belonging to* F.

■ Apply successively 6.7, the condition (C) and (56). ■

MOUNTAIN PASS POINTS

Let X, φ, Γ, c, F be as to 6.7. Suppose φ verifies the condition (C) at the fixed level c with respect to F. Consider the following sets $K_c(\varphi)$ and $M_c(\varphi)$.

(61) $$K_c(\varphi) = K_c := \{x \in X : \varphi(x) = c,\ \varphi'_w(x) = 0\},$$

the set of critical points of φ at the level c (see also above at 4.7). This is closed (the *-weak topology being separate, every point is a closed set).

(62) $$M_c(\varphi) = M_c := \{x \in K_c : x \text{ point of local minimum for } \varphi\}.$$

We have, according to (56), $K_c \cap F \neq \varnothing$. $K_c \cap F$ is even compact: let $(x_n)_{n\geq 1}$ be an arbitrary sequence from $K_c \cap F$, it has a convergent subsequence in $K_c \cap F$ since the conditions (53), (54), (55) are satisfied ($d(x_n, F) = 0$, $\varphi(x_n) = c$, $(1 + \|x_n\|)^{-1}\,\|\varphi'_w(x_n)\| = 0$), and φ verifies (C) at the c level.

Definition (*Hofer*). x_0 from K_c is a *mountain pass point* (*m. p. point*) for φ if, for every open neighbourhood U of x_0, the set

$$\varphi_U^c := \{x \in U : \varphi(x) < c\}$$

is nonempty and disconnected. In this case x_0 is obviously not a point of local minimum for φ.

Consider also the set

(63) $$P_c(\varphi) = P_c := \{x \in K_c(\varphi) : x \text{ is a m.p. point for } \varphi\}.$$

It can state

6.10 Theorem. *Suppose that* φ *verifies the condition* (C) *at the* c *level relative to* X. *Then*

$$\textit{either } F \cap \overline{M}_c \neq \varnothing \textit{ or } F \cap P_c \neq \varnothing$$

(*H. Hofer, I. Ekeland*).

■ Suppose (64) $F \cap P_c = \varnothing$. Prove

(65) $$F \cap \overline{M}_c \neq \varnothing,$$

which is enough, since the assumption $F \cap \overline{M}_c = \varnothing$ implies $F \cap P_c \neq \varnothing$ (*par absurdum*, see (65)).

The assumption (64) allows us to state:

(66) $$\exists\, \varepsilon_1 > 0 \text{ such that the set } \{x \in X : d(x, F_c \cap K_c) < \varepsilon_1\}$$

cuts only a finite number of connected components of φ_X^c, let then be $U_1, \ldots, U_N$.

Remark (67) $F_c \cap K_c = F \cap K_c$, a nonempty compact set (see above), and (68) $K_c \cap \cap \varphi_X^c = \varnothing$ and pass to prove (66). Suppose *par absurdum* the contrary, i.e. $\forall\ \varepsilon > 0$ the set $\{x \in X : d(x, F \cap K_c) < \varepsilon\}$ cuts an infinite number of connected components of φ_X^c. Let U_1 be a connected component of φ_X^c splitted by $\{x \in X : d(x, F \cap K_c) < 1\}$, let U_2 be a connected component of φ_X^c, $U_2 \neq U_1$, splitted by $\{x \in X : d(x, F \cap K_c) < \frac{1}{2}\}$etc. This yields a sequence of two by two distinct connected components $(U_n)_{n\geq 1}$ of φ_X^c, a sequence $(z_n)_{n\geq 1}$, $z_n \in U_n$ such that $d(z_n, F \cap K_c) < \frac{1}{n}$ and a sequence $(x_n)_{n\geq 1}$ in $F \cap K_c$ such that $d(z_n, x_n) < \frac{1}{n}$, which implies (69) $d(x_n, U_n) < \frac{1}{n}$. Let $(x_{k_n})_{n\geq 1}$ be a subsequence of $(x_n)_{n\geq 1}$ convergent to x_0 from $F \cap K_c$ (a compact set !), hence (70) $d(x_0, U_{k_n}) \to 0$ (see (69)). But x_0 is a m.p. point for φ, in contradiction with (64): by (70), for every open neighbourhood V of x_0, $V \cap \varphi_X^c$ is a nonempty and disconnected set, since the sets U_n are disjoint two by two.

Consider now the nonempty compact sets

$$(71) \qquad Q_i := F \cap K_c \cap \overline{U_i} \ (\overset{(68)}{=} F \cap K_c \cap \mathrm{Fr}\, U_i),\ i = \overline{1, N}.$$

Reasoning as above *par absurdum*, we find ε_2, $\varepsilon_2 < \varepsilon_1$ such that

$$(72) \qquad \{x \in X : d(x, Q_i) < \varepsilon_2\} \cap (\bigcup_{j\neq i} \overline{U_j}) = \varnothing.$$

If necessary we diminish ε_2 to have also (73) $d(y_1, F_c) > \varepsilon_2$. Let be (74) $X \setminus F_c = \Omega_0 \cup \cup \Omega_1$, open disjoint sets, $y_i \in \Omega_i$, $i = 0, 1$. We have for every U_i, $1 \leq i \leq N$,

(75) either $U_i \subset \Omega_0$ or $U_i \subset \Omega_1$ (which is equivalent to $U_i \cap \overline{\Omega}_0 = \varnothing$).

Indeed, if *par absurdum* we suppose the contrary, then, since U_i is a connected set included in $\Omega_0 \cup \Omega_1$ since φ_X^c is in this situation ((74)), we have $U_i \cap \Omega_0 \neq \emptyset$, $U_i \cap \Omega_1 \neq \neq \varnothing$, a contradiction, since Ω_0, Ω_1 are open disjoint sets.

Consider for every ε, $0 < \varepsilon < \varepsilon_2$, the set

$$(76) \qquad R(\varepsilon) := \Omega_0 \cup \bigcup_{i=1}^{N} \{x \in X : d(x, Q_i) < \varepsilon\},$$

hence, putting $J_1 := \{i : 1 \le i \le N,\ U_i \subset \Omega_0\}$, $J_2 := \{i : 1 \le i \le N,\ U_i \cap \Omega_0 = \varnothing\}$, taking into account (75), (76) becomes

$$R(\varepsilon) = \Omega_0 \cup \bigcup_{i \in J_1} \{x \in X : d(x, Q_i) < \varepsilon\} \cup \bigcup_{i \in J_2} \{x \in X : d(x, Q_i) < \varepsilon\}. \tag{77}$$

$R(\varepsilon)$ is open and $y_0 \in R(\varepsilon)$ since $y_0 \in \Omega_0$. Moreover, (78) $y_1 \notin \overline{R(\varepsilon)}$. Indeed, $\overline{R(\varepsilon)} = \overline{\Omega}_0 \cup \bigcup_{i=1}^{N} \{x \in X : d(x, Q_i) \le \varepsilon\}$ and $y_1 \notin \overline{\Omega}_0$ since $\overline{\Omega}_0 \cap \Omega_1 = \varnothing$ and $y_1 \notin \{x \in X : d(x, Q_i) \le \varepsilon\}$, $i = \overline{1, N}$ since $d(y_1, Q_i) \overset{(71)}{\ge} d(y_1, F_c) \overset{(73)}{>} \varepsilon_2 > \varepsilon$. Therefore Fr $R(\varepsilon)$ separates y_0 and y_1.

The definition (77) gives

$$\bigcup_{i \in J_1} U_i \subset R(\varepsilon), \tag{79}$$

$$\bigcup_{i \in J_2} U_i \subset X \setminus \overline{R(\varepsilon)} \tag{80}$$

(for (80) use (75) and (72), $\varepsilon < \varepsilon_2$), consequently

$$\mathrm{Fr}\, R(\varepsilon) \cap \Big(\bigcup_{i=1}^{N} U_i\Big) = \varnothing. \tag{81}$$

We also have

$$\mathrm{Fr}\, R(\varepsilon) \cap \varphi_X^c \subset M \cap \varphi_X^c, \tag{82}$$

$M := \{x \in X : d(x, F_c \cap K_c) < \varepsilon_1\}$. Indeed,

$\mathrm{Fr}\, R(\varepsilon) \overset{(76)}{\subset} \mathrm{Fr}\, \Omega_0 \cup \bigcup_{i=1}^{N} \{x \in X : d(x, Q_i) = \varepsilon\}$, $\mathrm{Fr}\, \Omega_0 \cap \varphi_X^c = \varnothing$ and $\{x \in X : d(x, Q_i) = \varepsilon\} \subset M$ since $d(x, F_c \cap K_c) \overset{(71)}{\le} d(x, Q_i) = \varepsilon < \varepsilon_1$.

Since the only connected components of φ_X^c intersected by M are $U_1, \ldots, U_N$ ((66)), according to (81) and (82) one concludes

$$\mathrm{Fr}\, R(\varepsilon) \cap \varphi_X^c = \varnothing, \text{ i.e. } \varphi(x) \ge c \text{ on } \mathrm{Fr}\, R(\varepsilon).$$

So it can apply 6.9 with $F = \mathrm{Fr}\, R(\varepsilon)$ (φ verifies (C) at the level c on X !), φ has a critical point x_ε,

$$x_\varepsilon \in K_c \cap \mathrm{Fr}\, R(\varepsilon). \tag{83}$$

x_ε is even a local minimum point. Indeed, otherwise $\exists\, i$, $1 \le i \le N$, such that $x_\varepsilon \in U_i$. If $i \in J_1$ we have $U_i \overset{(79)}{\subset} R(\varepsilon)$, hence $x_\varepsilon \in R(\varepsilon)$ in contradiction with (83) and if $i \in J_2$ we have $U_i \cap \mathrm{Fr}\, R(\varepsilon) \overset{(80)}{=} \varnothing$, but we also have $x_\varepsilon \in U_i \cap \mathrm{Fr}\, R(\varepsilon)$, again a contradiction.

Take $\varepsilon = \frac{1}{n}$, $n \in \mathbf{N}$. We have (84) $\lim_{n\to\infty} d(x_{\frac{1}{n}}, F_c) = 0$ since $d(x_{\frac{1}{n}}, F_c) \le d(x_{\frac{1}{n}}, Q_i) \overset{(83)}{=} \frac{1}{n}$ (see also(67), (71)). The sequence $(x_{\frac{1}{n}})_{n\ge1}$ has a convergent subsequence since K_c is compact (φ verifies (C) at the level c relative to X), one notes the same, let $x_0 = \lim_{n\to\infty} x_{\frac{1}{n}}$.

We have $x_0 \in F_c$ since $d(x_0, F_c) \le d(x_0, x_{\frac{1}{n}}) + d(x_{\frac{1}{n}}, F_c)$ and we take into account (84).

Consequently, $F_c \cap \overline{M}_c \ne \varnothing$ and hence (65). ■

Remark. By theorem 6.10, it results in particular either $M_c \ne \varnothing$ or $P_c \ne \varnothing$, i.e. there is in K_c or a local minimum point or a m.p. point.

6.11 *Corollary. Let* X, φ, Γ, c *be as in* 6.8. *Suppose that* φ *verifies* (C) *at the level* c *relative to* X *and*

$$c > \max(\varphi(y_0), \varphi(y_1)).$$

Then we have

$$\textit{either } \overline{M_c} \setminus M_c \ne \varnothing \textit{ or } P_c \ne \varnothing$$

(*H. Hofer*).

In other words :

„There is, at the level *c*, either a m.p. point, or a point which is not of local minimum, but it is the limit of a sequence of local minimum points".

■ Let F be the boundary of the set $A_{\varphi, c}$ (6.7), i.e. $F = \{x \in X : \varphi(x) = c\}$. Apply 6.10, which is correct since $F \cap A_{\varphi, c} = F$ and F separates y_0, y_1 (the same justification as for 6.8). So we have either $F \cap M_c \ne \varnothing$ or $F \cap P_c \ne \varnothing$. Moreover, $F \cap M_c = \varnothing$, since every local minimum point on the level c is an inner point of $A_{\varphi, c}$. It only remains to remark the inclusion

$$P_c, \overline{M_c} \subset K_c \text{ (a closed set, (61))} \subset F. \ \blacksquare$$

Remark. The propositions 7.11 and 7.12 will also give information about the structure of the set K_c of critical points in the Mountain Pass theorem.

MOUNTAIN PASS THEOREM AND SADDLE POINT THEOREM FOR LOCALLY LIPSCHITZ FUNCTIONS

In the following, we present some versions (a beautiful illustration of subdifferential calculus), with the needed corrections in the statements and proofs, of the theorems quoted in the title, all owned to Shi Shuzhong ([65]). You should see §5, Clarke derivative and subderivative.

Definition. Palais-Smale $(PS)_*$ *condition.* The locally Lipschitz function $f: X \to \mathbf{R}$, X real normed space, verifies $(PS)_*$ when every sequence $(x_n)_{n\geq 1}$ in X, for which $(f(x_n))_{n\geq 1}$ is convergent and

$$\lim_{n\to\infty} \inf \{\|\xi\|: \xi \in \partial f(x_n)\} = 0,$$

has a convergent subsequence.

This definition is correct since $\forall\, n \geq 1$ $\partial f(x_n)$ is nonempty (5.17, 1°; see also 5.17_2).

The following propositions, together with Ekeland principle, will help to prove the theorems named in the title.

6.12 *Let* X *be a Banach space,* $\varphi : X \to \mathbf{R}$ *locally Lipschitz and* C(K; X), K *a compact metric space, with the usual distance. Then the function* $F: C(K; X) \to \mathbf{R}$,

$$F(g) = \sup_K (\varphi \circ g)$$

in locally Lipschitz.

■ Let g be arbitrary fixed in $C(K; X)$. We must find $\delta_g > 0$ and $\rho_g > 0$ such that (E a normed space, $S(a, r) := \{x \in E : \|x - a\| < r\}$)

(85) $$h_1, h_2 \in S(g, \delta_g) \Rightarrow |\varphi(h_1) - \varphi(h_2)| \leq \rho_g \|h_1 - h_2\|.$$

Since φ is locally Lipschitz, there exists, for every t in K, $\delta_t > 0$ and $\rho_t > 0$ such that

(86) $$x_1, x_2 \in S(g(t), \delta_t) \Rightarrow |\varphi(x_1) - \varphi(x_2)| \leq \rho_t \|x_1 - x_2\|.$$

But $(S(g(t), \delta_t))_{t\in K}$ is an open covering of the compact set $g(K)$, hence $\exists\, t_1, \dots, t_N$ such that

(87) $$g(K) \subset \bigcup_{i=1}^{N} S(g(t_i), \delta_{t_i}).$$

Let be δ_g a Lebesgue coefficient of the covering indicated by (87) ($x \in g(K) \Rightarrow \exists\, i$ such that (88) $S(x, \delta_g) \subset S(g(t_i), \delta_{t_i})$) and $\rho_g := \max_{1\leq i\leq N} \rho_{t_i}$. Then, according to (86) and (88),

(89) $$t \in K, x_1, x_2 \in S(g(t), \delta_g) \Rightarrow |\varphi(x_1) - \varphi(x_2) \leq \rho_g \|x_1 - x_2\|.$$

So

$$h_1, h_2 \in C(K; X), \sup_{t\in K} \|h_i(t) - g(t)\| < \delta_g, i = 1, 2 \Rightarrow |F(h_1) - F(h_2)| = |\sup_{t\in K} \varphi(h_1(t)) -$$

$$- \sup_{t\in K} \varphi(h_2(t))| \leq \sup_{t\in K} |\varphi(h_1(t)) - \varphi(h_2(t))| \overset{(89)}{\leq} \rho_g \sup_{t\in K} \|h_1(t) - h_2(t)\| = \rho_g \|h_1 - h_2\|$$

and consequently (85) holds. ■

6.13 *Under the hypotheses of* 6.12 *but with* X *reflexive, for every* g, h *in* C(K; X) *we have*

$$F^0(g; h) \leq \sup_{s\in\mu(g)} \varphi^0(g(s); h(s)),$$

where $\mu(g) = \{s \in K: F(g) = \varphi(g(s))\}$.

■ $\mu(g)$ is nonempty since $\varphi o g$ attains its supremum on K. Choose the sequences $(u_n)_{n\geq 1}$ in $C(K; X)$, $u_n \to g$ and $(\lambda_n)_{n\geq 1}$ strictly decreasing in $\mathbf{R}_+$, $\lambda_n \to 0$ such that

(90) $$F^0(g; h) = \lim_{n\to\infty} \frac{F(u_n + \lambda_n h) - F(u_n)}{\lambda_n}.$$

For every n in $\mathbf{N}$ take s_n in $\mu(u_n + \lambda_n h)$.

We have

(91) $$\frac{F(u_n + \lambda_n h) - F(u_n)}{\lambda_n} \leq \frac{\varphi(u_n(s_n) + \lambda_n h(s_n)) - \varphi(u_n(s_n))}{\lambda_n}.$$

By G. Lebourg formula (5.33), there is θ_n in (0, 1) and ξ_n in $\partial\varphi(u_n(s_n) + \theta_n \lambda_n h(s_n))$ such that

$$\frac{\varphi(u_n(s_n) + \lambda_n h(s_n)) - \varphi(u_n(s_n))}{\lambda_n} = \xi_n(h(s_n)),$$

hence, taking into account (91),

(92) $$\frac{F(u_n + \lambda_n h) - F(u_n)}{\lambda_n} = \xi_n(h(s_n)) \ \forall\, n \geq 1.$$

K being compact, the sequence $(s_n)_{n\geq 1}$ has a convergent subsequence, which we denote identically, let be $\lim_{n\to\infty} s_n = s_0$. Then we have

$\theta_n \lambda_n h(s_n) \to 0$, $(\|\theta_n \lambda_n h(s_n)\| \leq \lambda_n \|h(s_n)\|$, $h(s_n) \to h(s_0)$, $\lambda_n \to 0)$ and $u_n(s_n) \to g(s_0)$ $(u_n(s_n) - g(s_0) = u_n(s_n) - g(s_n) + g(s_n) - g(s_0)$, $g(s_n) \to g(s_0)$ since g is continuous, $\|u_n(s_n) -$ $- g(s_n)\| \leq \sup_{t\in K} \|u_n(t) - g(t)\| \to 0)$, consequently.

(93) $$u_n(s_n) + \theta_n \lambda_n h(s_n) \to g(s_0).$$

The sequence $(\xi_n)_{n\geq 1}$ is bounded. Indeed, let V be a neighbourhood of $g(s_0)$ on which φ has the Lipschitz constant L. For $n \geq N$ $g(s_n) \in V$, hence $n \geq N \Rightarrow ||\xi_n|| \leq L$ (5.17, 1°). But X^* is reflexive since X is reflexive ([47], vol III, p.750), consequently there is a weakly convergent subsequence $(\xi_{k_n})_{n\geq 1}$ (Šmulian corollary, [47], vol III, p.770), hence also *-weak convergent ([47], vol III, p.738, 3.28), let be $\xi_{k_n} \xrightarrow{\text{*-weak}} \xi_0$. ξ_0 being obviously also a *-weak adherent value, apply (93) combined with 1° of 5.20, one gets

(94) $$\xi_0 \in \partial\varphi(g(s_0)).$$

Moreover, (95) $\xi_n(h(s_0)) \to \xi_0(h(s_0))$ ([47], vol III, p.475, 3.39). Thus, taking into account (90), (92) and (95),

$F^0(g; h) \leq \lim\limits_{n\to\infty} \xi_n(h(s_n)) = \lim\limits_{n\to\infty} \xi_n(h(s_n) - h(s_0)) + \lim\limits_{n\to\infty} \xi_n(h(s_0)) = \xi_0(h(s_0))$ ($x_n \to 0$ $\Rightarrow \xi_n(x_n) \to 0$: $\xi_n(x_n) = \xi_n(x_n) - \xi_0(x_n) + \xi_0(x_n)$, $(\xi_n - \xi_0)(x_n) \to 0$),

hence, using also (94),

$$F^0(g; h) \leq \xi_0(h(s_0)) \overset{5.17,2^{\circ}}{\leq} \varphi^0(g(s_0); h(s_0)).$$

It remains to prove (96) $s_0 \in \mu(g)$. But $s_n \in \mu(u_n + \lambda_n h) \Rightarrow \varphi(u_n(s_n) + \lambda_n h(s_n)) \geq \varphi(u_n(t) + \lambda_n h_n(t))$ $\forall\, t \in K$, we pass to the limit for $n \to \infty$ and taking into account (93) and also $u_n \to g$ (in $C(K; X)$), we get (96). ■

Note. The additional requirement „ X reflexive" was introduced in the statement 6.13 by the author. It could be found the adherent value ξ_0, the clue element of the proof. In [65] ξ_0 (x^* in original) appears without justification. With this all was relieved. The requirement „X reflexive" can be replaced by „X separable" ((ξ_n) has also in this case a *-weak convergent subsequence, [47], vol III, p.763, 4.25).

6.14 *Under the assumptions of* 6.13, *let* K_0 *be a nonempty closed subset of* K, $C_0(K; X) := \{h \in C(K; X): h\,|\,K_0 = 0\}$ *and* g *in* C(K; X) *with*

(97) $$\mu(g) \subset K \setminus K_0.$$

If there is $\varepsilon > 0$ *such that*

(98) $$F^0(g; h) \geq -\varepsilon\,||h||\ \forall\, h \in C_0(K; X),$$

then there is s_0 *in* $\mu(g)$ *such that*

(99) $$\varphi^0(g(s_0); v)\,) \geq -\varepsilon\,||v||\ \forall\, v \in X.$$

■ Suppose *par absurdum* the contrary. Then for every t in $\mu(g)$ there is v_t in X, $||v_t|| = 1$, such that

(100) $$\varphi^0(g(t); v_t) < -\varepsilon.$$

Since g is continuous and φ^0 u.s.c. (5.16), there is $\delta_t > 0$ such that (E normed space, $S(a, r) := \{x \in E: \|x - a\| < r\}$)

(101) $$s \in S(t, \delta_t) \Rightarrow \varphi^0(g(s); v_t) < \quad \varepsilon.$$

If necessary we diminish δ_t ((97) allows this) in order to have also:

(102) $$K_0 \cap S(t, \delta_t) = \varnothing.$$

$(S(t, \delta_t))_{t\in\mu(g)}$ is an open covering of the compact set $\mu(g)$, hence there is $S(t_i, \delta_{t_i})$, $i = \overline{1, N}$ such that

(103) $$\mu(g) \subset \bigcup_{i=1}^{N} S(t_i, \delta_{t_i}).$$

Retain also (see (103))

(104) $$K = \bigcup_{i=1}^{N} S(t_i, \delta_{t_i}) \cup (K \setminus \mu(g)).$$

Now consider the functions

$$\rho_0 : K \to \mathbf{R},\ \rho_0(t) = \text{dist}(t, \mu(g)),$$

$$\rho_i : K \to \mathbf{R},\ \rho_i(t) = \text{dist}(t, K \setminus S(t_i, \delta_{t_i})),\ i = \overline{1, N}.$$

The sets $\mu(g)$ and $K \setminus S(t_i, \delta_{t_i})$ being closed, we have

(105) $$\rho_0(t) = 0 \Leftrightarrow t \in \mu(g),\ \rho_i(t) = 0 \Leftrightarrow t \notin S(t_i, \delta_{t_i}),$$

consequently (see (104))

(106) $$\sum_{i=0}^{N} \rho_i(t) > 0\ \forall\, t \in K.$$

Now, let $h : K \to X$,

(107) $$h(t) = \frac{\sum_{i=1}^{N} v_{t_i} \rho_i(t)}{\sum_{i=0}^{N} \rho_i(t)},$$

which is correct according to (106).

h is continuous, $||h|| \le 1$ since $||h(t)|| \le \dfrac{\sum_{i=1}^{N} \rho_i(t)}{\sum_{i=0}^{N} \rho_i(t)}$, and $h(t) = 0$ on K_0 (see (102), attention to the numerator), hence (108) $h \in C_0(K; X)$, $||h|| \le 1$. We have for t arbitrary in K

$$\varphi^0(g(t); h(t)) \overset{5.15}{\le} \frac{1}{\sum_{i=0}^{N} \rho_i(t)} \sum_{i=1}^{N} \rho_i(t)\, \varphi^0(g(t); v_{t_i}),$$

hence, taking into account (101), (103), (105) and (106),

$$(109) \qquad t \in \mu(g) \Rightarrow \varphi^0(g(t); h(t)) < -\varepsilon \overset{(108)}{\le} -\varepsilon\, ||h||.$$

But we have

$$F^0(g; h) \overset{(6.13)}{\le} \sup_{t \in \mu(g)} \varphi^0(g(t); h(t)) \overset{(109)}{<} -\varepsilon\, ||h||$$

(the supremum is attained, $t \to \sup \varphi^0(g(t); h(t))$ is u.s.c. on $\mu(g)$ compact, 5.16), one gets in collision with (98), and so (99) is proved. ■

And now

6.15 Mountain Pass Theorem. *Let* X *be a Banach reflexive or separable space and* $\varphi : X \to \mathbf{R}$ *locally Lipschitz which verifies the* $(PS)_*$ *condition. Suppose that for the distinct points* u_0, u_1 *in* X *and for an open neighbourhood* Ω *of* u_0 *with* $u_1 \in X \setminus \overline{\Omega}$ *we have*

$$(110) \qquad \inf_{Fr\Omega} \varphi > \max(\varphi(u_0), \varphi(u_1)).$$

Then

$$c := \inf_{g \in \Gamma} \sup_{[0,1]} (\varphi \circ g),$$

where $\Gamma = \{g \in C([0,1]; X) : g(0) = u_0,\ g(1) = u_1\}$, *is a critical value of* φ *and*

$$c \ge \inf_{Fr\Omega} \varphi$$

(*Shi Shuzhong* version, [65]).

Remark. The requirement „X reflexive or separable" has been added by the author (see *Note* to 6.13).

Explanation. „c a critical value for φ" signifies „$\exists\ x_0$ in X such that $\varphi(x_0) = c$ and $0 \in \partial\varphi(x_0)$".

■ Use without notice the notations of 6.12 ($K = [0, 1]$), 6.12 and 6.13 ($K_0 = \{0,1\}$).

Let g be arbitrary in Γ. Since $g([0, 1])$ is connex and $g([0, 1]) \cap \Omega$, $g([0, 1]) \cap (X \setminus \overline{\Omega})$ are nonempty, we also have

(111) $$g([0, 1]) \cap \mathrm{Fr}\,\Omega \neq \emptyset,$$

let $g(t')$ be in this intersection. Then, since $\sup\limits_{t\in[0,1]} \varphi(g(t)) \geq \varphi(g(t'))$, we have

(112) $$\sup_{t\in[0,1]} \varphi(g(t)) \geq \inf_{\mathrm{Fr}\Omega} \varphi \overset{(110)}{>} \max(\varphi(u_0), \varphi(u_1))\ \forall\, g \in \Gamma,$$

consequently, according to (111),

(113) $$g \in \Gamma \Rightarrow \mu(g) \subset (0, 1) = [0, 1] \setminus K_0 .$$

Γ is, as a closed part, a complete metric subspace of $C([0, 1]; K)$. Consider the function

(114) $$F_1\colon \Gamma \to \mathbf{R},\ F_1(g) = \sup_{[0,1]} (\varphi \circ g).$$

Since F from 6.12 is locally Lipschitz and $F_1 = F \mid \Gamma$, F_1 is continuous. Moreover, F_1 is lower bounded ((112)). Take $(\varepsilon_n)_{n\geq 1}$ a strictly decreasing sequence in $\mathbf{R}_+$ with $\varepsilon_n \to 0$ and apply to F_1 Ekeland principle 1.1. We find a sequence $(g_n)_{n\geq 1}$ in Γ such that, $\forall\, n \geq 1$,

(115) $$F_1(g_n) \leq c + \varepsilon_n \ (c = \inf_{g\in\Gamma} F_1(g)!)$$

and

(116) $$F_1(u) > F_1(g_n) - \varepsilon_n \|u - g_n\|\ \forall\, u \in \Gamma \setminus \{g_n\}.$$

Now let h be arbitrary in $C_0([0, 1]; X)$ (6.14). Since $g_n + \lambda_n h$, $\lambda_n > 0$ and $\lambda_n \to 0$, is in Γ, (116) gives ([47], vol.II, p. 651, 11.9$_1$) $F_1^0(g_n; h) \geq \overline{\lim\limits_{m\to\infty}} \dfrac{F_1(g_m + \lambda_m h) - F_1(g_m)}{\lambda_m} \geq -\varepsilon_n \|h\|$.

So it can apply 6.14, $\exists\, s_n$ in $\mu(g_n)$ such that

(117) $$\varphi^0(g_n(s_n); v) \geq -\varepsilon_n \|v\|\ \forall\, v \in X.$$

Retain also (the definition!)

(118) $$\varphi(g_n(s_n)) = F_1(g_n).$$

Set $x_n := g_n(s_n)$, $n \in \mathbf{N}$. (115) gives $c = \lim\limits_{n\to\infty} F_1(g_n)$ and combining with (118) we get

(119) $$\lim_{n\to\infty} \varphi(x_n) = c.$$

Let now ζ_n be in $S^* := \{\xi \in X^*\colon \|\xi\| \leq 1\}$. Since $|\zeta(v)| \leq \|v\|\ \forall\, v \in X$, (117) gives

$$\varphi^0(x_n; v) \geq -\varepsilon_n \zeta_n(v)\ \forall\, v \in X,$$

i.e. $\xi_n := -\varepsilon_n \zeta_n \in \partial\varphi(x_n)$. Thus

$$\inf\{\|\xi\|\colon \xi \in \partial\varphi(x_n)\} \leq \|\xi_n\| \leq \varepsilon_n,$$

consequently (see also (119)) it can apply $(PS)_*$ to $(x_n)_{n\geq 1}$, $\exists$ a convergent subsequence $(x_{k_n})_{n\geq 1}$, let be $x_{k_n} \to x_0$. Obviously then $\varphi(x_0) = c$ ((119)). Moreover, we have $\forall\, v \in X$ (φ^0 is u.s.c., 5.16)

$$\varphi^0(x_0; v) \geq \overline{\lim_{n\to\infty}}\, \varphi^0(g_{k_n}(s_{k_n}); v) \overset{(117)}{\geq} -\lim_{n\to\infty} \varepsilon_{k_n} \|v\| = 0,$$

i.e. $0 \in \partial\varphi(x_0)$, x_0 is a critical point for φ. Finally, (112) and (119) give $c \geq \inf_{\mathrm{Fr}\Omega} \varphi$. ■

6.16 Saddle Point Theorem. *Let* X *be a reflexive or separable Banach space and* φ: X $\to$ **R** *locally Lipschitz verifying* $(PS)_*$. *Suppose*

$$X = X_1 \oplus X_2,\ \dim X_1 < +\infty.$$

Let be

$$\sigma_1 := \{x \in X_1 : \|x\| = 1\},\ S_1 := \{x \in X_1 : \|x\| < 1\},$$

$$\sigma_r := \{x \in X_2 : \|x\| = r\}.$$

If for c_0 *and* c_1 *from* **R**, $c_0 > c_1$, *and for* θ *from* $C(\sigma_1; X)$ *we have*

(120) $\varphi(x) \geq c_0$ *on* σ_r,

(121) $\varphi(x) \leq c_1$ *on* $\theta(\sigma_1)$,

(122) *the linking number* $l(\sigma_r, \theta(\sigma_1))$ *is not zero*,

then

$$c := \inf_{g\in\Gamma} \sup_{\overline{S_1}} (\varphi \circ g),\ \textit{where}\ \Gamma = \{g \in C(\overline{S_1}; X):\ g\,|\,\sigma_1 = \theta\},$$

is a critical value for φ *and*

$$c \geq c_0$$

(*Shi Shuzhong* version, [65]).

■ The proof, excepting the beginning, is the same as for 6.15. Take $K = \overline{S_1}$ (which is correct, since $\dim X < +\infty \Rightarrow \overline{S_1}$ compact), $K_0 = \sigma_1$. (122) ensures

(123) $$g(\overline{S_1}) \cap \sigma_r \neq \varnothing\ \forall\, g \in \Gamma.$$

By replacing (111) with (123) we get also

$$\sup_{t\in\overline{S_1}} \varphi(g(t)) \geq \inf_{\sigma_r} \varphi \geq c_0 > c_1 \geq \sup_{\theta(\sigma_1)} \varphi.$$

Consequently

$$\mu(g) \subset S_1 = \overline{S_1} \setminus \sigma_1\ (= K \setminus K_0)\ \forall\, g \in \Gamma.$$

From this point on, the proof remains the same as for 6.15. ■

MINIMAX THEOREM III

Let us begin the preparations needed for the proof of the third minimax theorem.

Let K be a compact space and $E := C(K; \mathbf{R})$ the Banach space with the norm $\|f\| = \sup_{x\in K} |f(x)|$ and $\mathcal{M}(K; \mathbf{R})$ the Banach space, with the norme $\|\mu\| = |\mu|(K)$, of the regular real measures on the σ-algebra of the Borel sets from K ([47], vol III, pages 547 ((6)), 571 (4.7), 576 (4.10)).

The map $\mathcal{M}(K; \mathbf{R}) \to E^*$, $\mu \to \Phi_\mu$, $\Phi_\mu(f) = \int_K f \, d\mu$ is a norm-preserving isomorphism between vector spaces (Riesz representation theorem, [47], vol III, p. 572 and 576, 4.9 and 4.10). Identify $\mathcal{M}(K; \mathbf{R})$ and E^* by this isometrical isomorphism.

Let μ be in $\mathcal{M}(K; \mathbf{R})$. According to the general definitions, μ is *positive*, $\mu \geq 0$, when $f \geq 0 \Rightarrow \mu(f) \geq 0$ and the *support* of μ, supp μ, is the complementary of the biggest open set U in K having the property: the support supp f of f included in $U \Rightarrow \mu(f) = 0$ (the complementary of the biggest open set in K on which μ is *cancelled*).

In the following two statements K is a compact metric space.

6.17 *Let* f *be arbitrary in* E. *If* f (x) = 0 *on* supp μ, *then* μ(f) = 0.

■ Let be $A :=$ supp μ and, for every $\varepsilon > 0$, $A_\varepsilon := \{x \in K : d(x, A) < \varepsilon\}$. A_ε is open. Let n be arbitrary in $\mathbf{N}$. Since $(K \setminus A_{\frac{2}{n}}) \cap \overline{A_{\frac{1}{n}}} = \emptyset$, $\exists\, \varphi_n : K \to [0, 1]$ continuous with $\varphi_n(x) = 1$ on $K \setminus A_{\frac{2}{n}}$, $\varphi_n(x) = 0$ on $\overline{A_{\frac{1}{n}}}$. The sequence $(\varphi_n f)_{n\geq 1}$ is convergent in E to f – calculate and majorize $\sup_{x\in K} |\varphi_n(x) f(x) - f(x)|$ for $x \in$ supp μ, $x \notin$ supp μ and $d(x, \text{supp}\, \mu) < \frac{1}{n}, \frac{2}{n}$ and $d(x, \text{supp}\, \mu) \geq \frac{2}{n}$. Then $\mu(\varphi_n f) \to \mu(f)$ and hence $\mu(f) = 0$, since $\mu(\varphi_n f) = 0\ \forall\, n$: $\text{supp}(\varphi_n f) \subset \text{supp}\, \varphi_n \subset K \setminus A_{\frac{1}{n}} \subset K \setminus \text{supp}\, \mu$. ■

6.18 *Let be* $\Psi : E \to \mathbf{R}$, $\Psi(f) = \sup_K f$. Ψ *is convex, Lipschitz and, for every* f *from* E,

$$\mu \in \partial\Psi(f) \Leftrightarrow \mu \geq 0,\ \mu(1) = 1,\ \text{supp}\, \mu \subset \{x \in K : f(x) = \Psi(f)\}.$$

■ *First statement*. $\alpha, \beta \geq 0 \Rightarrow \alpha f(x) + \beta g(x) \leq \alpha \sup_K f + \beta \sup_K g$ etc.

Second statement. Let f, g be arbitrary in E. $\Psi(f) - \Psi(g) = f(x_0) - \sup_K g$, where $f(x_0) = \sup_K f$, hence $\Psi(f) - \Psi(g) \leq f(x_0) - g(x_0) \leq \sup_{x \in K} |f(x) - g(x)| = \|f - g\|$ etc.

Third statement. Necessity. By definition (§5, (1), see also 5.18), we have

(124) $$\mu \in \partial\Psi(f) \Rightarrow \Psi(g) - \Psi(f) \geq \mu(g - f) \ \forall\, g \in E.$$

Let h be arbitrary in E, $h \geq 0$. For $g = f - h$ in (124) we get $\Psi(f - h) - \Psi(f) \geq -\mu(h)$. But $f - h \leq f$ since $h \geq 0$, hence $\Psi(f - h) - \Psi(f) \leq 0$ and hence $\mu(h) \geq 0$, μ is positive.

Let a be arbitrary in $\mathbf{R}$. For $g = f + a$ in (124) one finds

$$\sup_K (f + a) - \sup_K f \geq \mu(a) = a\mu(1).$$

But $\sup_K (f + a) = \sup_K f + a$, hence $a \geq a\mu(1)$, $\mu(1) = 1$.

Finally, set $M := \{x \in K\colon f(x) = \Psi(f)\}$ and prove $\operatorname{supp}\mu \subset M$, which is equivalent with: μ is cancelled on every open set U included in $K \setminus M$. Let h be arbitrary in E, whose support F is included in U. We must prove (125) $\mu(h) = 0$. We have

$$\rho := \Psi(f) - \sup_F f > 0,$$

since F is compact and $\sup_K f$ is attained only on M. Take $\varepsilon > 0$ such that $\pm\varepsilon h(x) < \rho \ \forall\, x \in$ $\in K$. Then $f(x) \pm \varepsilon h(x) \leq \Psi(f) \ \forall\, x \in F$, *a fortiori* $\forall\, x \in K$ ($h(x) = 0$ on $K \setminus F$), and hence $\Psi(f \pm \varepsilon h) \leq \Psi(f)$. Take in (124) $g = f \pm \varepsilon h$, we get $\pm\varepsilon\mu(h) \leq 0$, and this implies (125).

Sufficiency. We must prove (§5, (1))

(126) $$\Psi(g) \geq \Psi(f) + \mu(g - f) \ \forall\, g \in E.$$

$h := f - \Psi(f) \in E$ and, by hypothesis, $h(x) = 0$ on $\operatorname{supp}\mu$, consequently $\mu(h) \overset{6.17}{=} 0$ since $\mu(1) = 1$ and (126) reverts to

(127) $$\Psi(g) \geq \mu(g) \ \forall\, g \in E.$$

But $u := \Psi(g) - g \in E$, $u \geq 0$ and since $\mu \geq 0$ we have $\mu(\Psi(g) \cdot 1 - g) \geq 0$, and consequently (127) since $\mu(1) = 1$. ∎

It can pass now to

6.19 Minimax theorem. *Let* X *be a Banach space*, K *a compact metric space*, K_0 *a compact nonempty part of* K, θ *from* $C(K_0;\, X)$ *and* $T := \{f \in C(K;\, X)\colon f\,|K_0 = \theta\}$ *the Banach space with Tschebysheff norm.*

Let $\varphi\colon X \to \mathbf{R}$ *be of* C^1 *- Fréchet class and*

$$c := \inf_{f \in T} \sup_{K} (\varphi \circ f), \; c_0 := \sup_{K_0} (\varphi \circ \theta).$$

If

$$\sup_{K} (\varphi \circ f) > c_0 \; \forall f \in T,$$

then for every $\varepsilon > 0$ *there exists* v_ε *in* X *such that*

$$c \le \varphi(v_\varepsilon) \le c + \varepsilon$$

and

$$\|\varphi'(v_\varepsilon)\| \le \varepsilon.$$

(*H. Brezis*).

■ [32]. The statement is correct, we have $T \neq \varnothing$ (Dugundji theorem). Consider the function $\Phi: T \to \mathbf{R}$,

$$\Phi(f) = \sup_{K} (\varphi \circ f).$$

Φ is lower bounded (through c_0, for instance), but also continuous. Indeed, let f be arbitrary fixed in T and $\eta > 0$. There is, for every x from $f(K)$, an open ball $S(x, \delta_x)$ such that

(128) $$y \in S(x, \delta_x) \Rightarrow |\varphi(x) - \varphi(y)| \le \frac{\eta}{2}.$$

The balls $S(x, \frac{1}{2}\delta_x)$, x goes over $f(K)$, cover $f(K)$ compact, let be $S(x_1, \frac{1}{2}\delta_{x_1})$, ... , $S(x_N, \frac{1}{2}\delta_{x_N})$ a finite covering. Then taking $\delta > 0$ with $\delta < \min_{1 \le k \le N} \frac{1}{2}\delta_{x_k}$, we have

(129) $$x \in f(K), y \in X, \|x - y\| \le \delta \Rightarrow |\varphi(x) - \varphi(y)| \le \eta.$$

Indeed, if for instance $x \in S(x_1, \frac{1}{2}\delta_{x_1})$, then

$\|y - x_1\| \le \|y - x\| + \|x - x_1\| \le \delta + \frac{1}{2}\delta_{x_1} < \delta_{x_1}$, hence $|\varphi(y) - \varphi(x_1)| \overset{(128)}{\le} \frac{\eta}{2}$ and since we also have $|\varphi(x) - \varphi(x_1)| \le \frac{\eta}{2}$, it results $|\varphi(x) - \varphi(y)| \le \eta$.

And now let g be arbitrary in T with $\|f - g\| \le \delta$. Then $\Phi(g) - \Phi(f) = \varphi(g(t')) - \sup_{K} (\varphi \circ f) \le \varphi(g(t')) - \varphi(f(t'))$. But $\|g(t') - f(t')\| \le \|g - f\| \le \delta$, hence, according to (129), $\varphi(g(t')) - \varphi(f(t')) \le \eta$, $\Phi(g) - \Phi(f) \le \eta$. Just the same reasoning yields $\Phi(f) - \Phi(g) \le \eta$, i.e. $|\Phi(g) - \Phi(f)| \le \eta$, Φ is continuous at f.

Thus it can apply Ekeland principle 1.1 to Φ and ε with $\lambda = 1$, there is f_ε in T such that

(130) $$c \leq \Phi(f_\varepsilon) \leq c + \varepsilon,$$

(131) $$\Phi(f_\varepsilon) \leq \Phi(f) + \varepsilon \,\|f - f_\varepsilon\| \;\forall\; f \in T.$$

Set $T_0 := \{g \in C(K;X)\colon g\,|K_0 = 0\}$. Let g be arbitrary fixed in $T_0 \setminus \{0\}$ and $\lambda > 0$ arbitrary. We have, applying the local Taylor formula,

(132) $\varphi(f_\varepsilon(t)) + \lambda g(t) = \varphi(f_\varepsilon(t)) + \lambda\varphi'(f_\varepsilon(t))(g(t)) + o(\lambda\,\|g(t)\|)\;\forall\; t \in K$ (if $g(t) = 0$, the relationship reduces to two terms).

Taking into account (132),

(133) $$\sup_K \varphi \circ (f_\varepsilon + \lambda g) \leq \sup_{t\in K}\,[\varphi(f_\varepsilon(t)) + \lambda\varphi'(f_\varepsilon(t))(g(t))] + o(\lambda\,\|g\|)$$

$$\Big(\frac{o(\lambda\|g(t)\|)}{\lambda\|g(t)\|} = \alpha(\lambda),\ \lim_{\lambda\to 0}\alpha(\lambda) = 0,\ o(\lambda\|g(t)\|) = \lambda\|g(t)\|\alpha(\lambda) \leq \lambda\|g\|\alpha(\lambda)\Big).$$

Take $f = f_\varepsilon + \lambda g$ in (131), it results, combining with (133),

(134) $$\sup_K (\varphi \circ f_\varepsilon) \leq \sup_{t\in K}\,[\varphi(f_\varepsilon(t)) + \lambda\varphi'(f_\varepsilon(t))(g(t))] + o(\lambda\|g\|) + \varepsilon\lambda\|g\|.$$

Set (135) $u := \varphi \circ f_\varepsilon$, $v\colon v(t) = \varphi'(f_\varepsilon(t))(g(t))$, $t \in K$, hence $u, v \in C(K; \mathbf{R})$. Thus (134) becomes, with the notation of 6.18,

$$-\,\varepsilon\|g\| \leq \frac{\Psi(u+\lambda v) - \Psi(u)}{\lambda} + \frac{o(\lambda\|g\|)}{\lambda}$$

and passing to the limit for $\lambda \to 0+$ we find (§5. (9)) $-\,\varepsilon\|g\| \leq \Psi'(u; v)$, hence (see 5.5, Ψ is convex continuous, 6.18)

(136) $$-\,\varepsilon\|g\| \leq \sup_{\mu\in\partial\Psi(u)} \mu(v).$$

According to 6.18, $\mu \geq 0$, $\mu(1) = 1$, (137) $\operatorname{supp}\mu \subset K_\varepsilon := \{t \in K\colon \varphi(f_\varepsilon(t)) = \sup_K (\varphi \circ f_\varepsilon)\}$.

Replace g by $-\,g$ in (136), one gets (v becomes $-\,v$)

(138) $$\inf_{\mu\in\partial\Psi(u)} \mu(v) \leq \varepsilon\|g\|.$$

Divide by $\|g\|$ in (138) and take the least upper bound, one finds (abuse of notation, attention to (135))

(139) $$\sup_{\substack{g\in T_0\\ \|g\|\leq 1}}\ \inf_{\mu\in\partial\Psi(u)} \mu(\varphi'(f_\varepsilon(\cdot))(g(\cdot))) \leq \varepsilon,$$

since (138) remains obviously true for $g = 0$ as then $v = 0$ ((135)).

But in (139) it can change sup by inf and conversely (von Neumann min-max theorem, *P. Bates* and *I. Ekeland*, *A Saddle Point Theorem*, Differential Equations, Academic Press, 1980, 123-126),

$$(140)\qquad \inf_{\mu\in\partial\Psi(u)} \sup_{\substack{g\in T_0\\ \|g\|\le 1}} \mu(\varphi'(f_\varepsilon(\cdot))(g(\cdot))) \le \varepsilon .$$

But, as will be shown, we have

$$(141)\qquad \sup_{\substack{g\in T_0\\ \|g\|\le 1}} \mu(\varphi'(f_\varepsilon(\cdot))(g(\cdot))) = \sup_{\substack{g\in C(K;X)\\ \|g\|\le 1}} \mu(\varphi'(f_\varepsilon(\cdot))(g(\cdot))).$$

Indeed, firstly remark that $K_0 \cap K_\varepsilon = \varnothing$: if, *par absurdum* $t' \in K_0 \cap K_\varepsilon$, then $\sup_K (\varphi \circ f_\varepsilon) =$

$= \varphi(f_\varepsilon(t')) \le \sup_{K_0} (\varphi \circ f_\varepsilon)$, a contradiction with the hypothesis. So take $h: K \to [0,1]$

continuous with $h(t) = 1$ on K_ε, $h(t) = 0$ on K_0. Let g be arbitrary in $C(K; X)$, $\|g\| \le 1$. $g_0 := hg$ is in T_0, $\|g_0\| \le 1$ and

$$(142)\qquad \mu(\varphi'(f_\varepsilon(\cdot))(g(\cdot))) = \mu(\varphi'(f_\varepsilon(\cdot))(g_0(\cdot))),$$

since supp $\mu \overset{(137)}{\subset} K_\varepsilon$ and $g = g_0$ on K_ε. Obviously, (142) imposes (141).

Thus (140) becomes, *via* (141),

$$(143)\qquad \inf_{\mu\in\partial\Psi(u)} \sup_{\substack{g\in C(K;X)\\ \|g\|\le 1}} \mu(\varphi'(f_\varepsilon(\cdot))(g(\cdot))) \le \varepsilon .$$

But $\sup_{\substack{g\in C(K;X)\\ \|g\|\le 1}} \mu(\varphi'(f_\varepsilon(\cdot))(g(\cdot))) = \mu(\sup_{\substack{g\in C(K;X)\\ \|g\|\le 1}} \varphi'(f_\varepsilon(\cdot))(g(\cdot)))$ (μ is continuous increasing

with respect to the order relation $f \le h$ and $\{g \in C(K; X)\colon \|g\| \le 1\}$ is right filtrable, Moore-Smith limit) and the second therm is equal to $\mu(\|\varphi'(f_\varepsilon(\cdot))\|)$, consequently, combining with (143), we get

$$(144)\qquad \inf_{\mu\in\partial\Psi(u)} \mu(\|\varphi'(f_\varepsilon(\cdot))\|) \le \varepsilon.$$

Then let μ_0 be in $\partial\,\Psi(u)$ which realizes the infimum. (144) gives $\mu_0(\|\varphi'(f_\varepsilon(\cdot))\|) \le \varepsilon$ and since $\mu_0(1) = 1$ and supp $\mu_0 \subset K_\varepsilon$, it results that there is t_0 in K_ε such that $\|\varphi'(f_\varepsilon(t_0))\| \le \varepsilon$.

Take $v_\varepsilon = f_\varepsilon(t_0)$, we have in addition $c \le \varphi(v_\varepsilon) \le c + \varepsilon$ since $\varphi(v_\varepsilon) \overset{(137)}{=} \sup_K (\varphi \circ f_\varepsilon) = \Phi(f_\varepsilon)$,

take into account (130). ■

Remark. Minimax Theorem III is, as we seen, a variant of Minimax Theorem I. While the requirement for φ is strenghten in III – C^1-Fréchet class, the requirement in I

$$c > c_0$$

was been relaxed in III to

$$\sup_K (\varphi \circ f) > c_0 \ \forall f \in T.$$

Even by this fact the situation has been suddenly complicated.

We obtain now, with the help of the minimax theorem III,

6.20 Generalized Saddle Point theorem. *Let* X *be a Banach space and* $\varphi: X \to \mathbf{R}$ *of* C^1*-Fréchet class. Suppose* $X = V \oplus W$, *direct sum*, V *and* W *closed subspaces*, $\dim V < < +\infty$. *Let be* w_0 *in* W, $0 < \rho < R$, $M := \{v + tw_0 : v \in V, \|v\| \le R, t \in [0, R]\}$ *and* $\sigma_\rho := = \{w \in W: \|w\| = \rho\}$. *Suppose*

$$\sup \varphi(\mathrm{Fr}\, M) < \inf \varphi(\sigma_\rho) \tag{145}$$

and let be $\Gamma := \{g \in C(M; X): g(x) = x$ *on* $\mathrm{Fr}\, M\}$,

$$c := \inf_{g\in\Gamma} \sup_M (\varphi \circ g).$$

If φ *verifies the* $(PS)_c$ *weak condition, then* c *is critical value of* φ (*P. H. Rabinowitz*).

■ Apply 6.19 with $K = M$, $K_0 = \mathrm{Fr}\, M$ (which is correct, since M is compact being closed bounded and $\dim V < +\infty$) and $\varepsilon = \frac{1}{n}$, $n \in \mathbf{N}$. It suffices to prove

$$\sup_M (\varphi \circ g) > \sup_{\mathrm{Fr}M} (\varphi \circ g) \ \forall g \in \Gamma. \tag{146}$$

For every g in Γ, we have

$$\sup_{\mathrm{Fr}M} (\varphi \circ g) \le \sup \varphi(\mathrm{Fr}\, M) \overset{(145)}{<} \inf \varphi(\sigma_\rho),$$

consequently, it is enough to prove that $\forall g \in \Gamma \ \exists x \in M$ such that

$$g(x) \in \sigma_\rho. \tag{147}$$

As for 6.6, use Brouwer topological degree. Fix g and let π be the projection of X in V along of W. Consider the map $h : M \to V \oplus \mathbf{R}w_0$,

$$h(v + tw_0) = \pi g(v + tw_0) + \|(\mathrm{Id}_X - \pi)g(v + tw_0)\| w_0. \tag{148}$$

h is continuous and $v + tw_0 \in \mathrm{Fr}\, M \Rightarrow h(v + tw_0) = \pi g(v + tw_0) + \|(\mathrm{Id}_X - \pi)g(v+tw_0)\| w_0 = = v + tw_0$. Moreover, since $\rho < R$ the point $0 + \rho w_0$ is an inner point of M with respect to $V \oplus \mathbf{R}w_0$. Consequently, there is $v_1 + t_1 w_0$ in M such that $h(v_1 + t_1 w_0) = \rho w_0$. We get then, taking into account (148) (X is a direct sum), $\pi g(v_1 + t_1 w_0) = 0$, hence $g(v_1 + t_1 w_0) \in W$ and in addition

$$\rho = \|(\mathrm{Id}_X - \pi)g(v_1 + t_1 w_0)\| = \|g(v_1 + t_1 w_0)\|$$

and this justifies (147). ■

7

Variational theorems of minimax type on Finsler manifolds

As in § 6, Ekeland principle 1.1 is used to obtain, in the Finsler manifolds, three theorems of minimax type with the afferent Mountain Pass and Saddle Point theorems. Some elements of the theory of this manifolds and, generally, of the differential manifolds theory, are given in the ANNEX.

In this paragraph, excepting the last section, we follow closely chapters **3**, **4**, **5** and **6** of [34] with the necessary corrections of some errors.

The following notion play a central role in this paragraph.

Definition. Let X be a topological space, M a compact subset and $\mathcal{K}$ a nonempty set of compact nonempty subsets of X.

$\mathcal{K}$ is *homotopy-stable* with *boundary* M (*homotopy-stable*, in the case $M = \varnothing$) if

1° $M \subset A \; \forall A \in \mathcal{K}$;

2° For every α in $C([0, 1] \times X; X)$, with the property

$$\alpha(t, x) = x \; \forall \, (t, x) \in (\{0\} \times X) \cup ([0, 1] \times M),$$

and for every A in $\mathcal{K}$ we have

$$\alpha(\{1\} \times A) \in \mathcal{K}.$$

The definition is coherent since $\{1\} \times A$ being compact, $\alpha(\{1\} \times A)$ has the same property.

7.1 Ghoussoub deformation lemma. *Let* X *be a Finsler* C^1*-manifold connected and complete,* $\varphi : X \to \mathbf{R}$ *of* C^1*class, and* A, B *nonempty disjoint subsets,* A *closed and* B *compact.*

If $\|d\varphi(x)\| > 2\varepsilon$, $\varepsilon > 0$, $\forall\, x \in B$, *then* $\forall\, \lambda > 1$ *there is* $g : X \to \mathbf{R}_+$ *continuous,* α *in* $C([0, 1] \times X; X)$ *and* t_0 *in* $(0, 1]$ *such that,* $\forall\, t$ *from* $[0, t_0)$, *we have*

1° $\alpha(t,x) = x\ \forall\, x \in A$, $\alpha(0, x) = x\ \forall\, x \in X$;

2° $\rho(\alpha(t, x), x) \le \lambda t\ \forall\, x \in X$;

3° $\varphi(\alpha(t, x)) - \varphi(x) \le -\varepsilon g(x)t\ \forall\, x \in X$;

4° $g(x) = 1\ \forall\, x \in B$.

Explanation. ρ *is the* Finsler metric d_F on X (*see* ANNEX, X is connected). The elements of $C([0, 1] \times X; X)$ are called *deformations*.

■ For every x_i in B (these is not critical point of φ!) take (U_i, φ_i, E) a local map at x_i, $f_i := \pi^{-1}_{x_i,i} \circ \varphi_i$ ($\pi_{x_i,i}$ the canonical isomorphism, ANNEX, **12**) and v_i in $T_{x_i}X$ (use ANNEX, (32), **21**, (26), take into account the continuity of $(\varphi \circ f_i^{-1})'$) such that

$$(1)\qquad \frac{1}{\lambda}\|\pi^{-1}_{x_i,i}(w)\|_{x_i} \le \|\pi^{-1}_{x,i}(w)\|_x \le \lambda\|\pi^{-1}_{x_i,i}(w)\|_{x_i}\quad \forall\, x \in U_i,\ \forall\, w \in E$$

and

$$(2)\qquad (\varphi \circ f_i^{-1})'(y)\left(\frac{v_i}{\|v_i\|}\right) \ge \frac{1}{2}\|d\varphi(x_i)\|\ \forall\, y \in f_i(U_i).$$

Let V_i be an open neighbourhood of x_i, $V_i \subset U_i$ and, for a δ_i from $(0, 1)$, (3) $B(V_i, \delta_i)$ $(= \{y \in X : \text{dist}(y, V_i) < \delta_i\}) \subset U_i$ and $B(f_i(V_i), \delta_i) \subset f_i(U_i)$ (δ_i - neighbourhoods in the Finsler metric). B being compact, there is $V_{i_1}, \ldots, V_{i_n}$ which cover B. We write $V_1, \ldots, V_n$ and we associate to them $\chi_1, \ldots, \chi_n : X \to [0,1]$ continuous, supp $\chi_k \subset V_k$, $\sum_{k=1}^{n}\chi_k(x) = 1$ on B, $\sum_{k=1}^{n}\chi_k(x) \le 1$ on X. Let $\nu: X \to [0, 1]$ be continuous with

$$(4)\qquad \nu(x) = \begin{cases} 1, & x \in B \\ 0, & x \in (X \setminus \bigcup_{k=1}^{n} V_k) \cup A. \end{cases}$$

Take (5) $t_0 = \dfrac{\delta_0}{1+\lambda^2}$, (6) $\delta_0 := \min\{\delta_{i_1}, \ldots, \delta_{i_n}\}$.

Define by recurrence the continuous functions

(7) $$\alpha_j : [0, 1] \times X \to X, j = \overline{0, n}.$$

j = 0. $\alpha_0(t, x) = x$.

$$j - 1 \Rightarrow j.\ \alpha_j(t, x) = \begin{cases} f_j^{-1}(f_j(\alpha_{j-1}(t, x)) - t\nu(x)\chi_j(x)\dfrac{v_j}{\|v_j\|}),\ \alpha_{j-1}(t, x) \in U_j \\ \alpha_{j-1}(t, x),\ \alpha_{j-1}(t, x) \notin U_j. \end{cases}$$

Prove by recurrence that for $t \in (0, t_0)$ and $1 \le j \le n$ we have

(8) α_j is correctly defined

(i.e. $f_j(\alpha_{j-1}(t, x)) - t\nu(x)\chi_j(x)\dfrac{v_j}{\|v_j\|} \in f_j(U_j)$ if $\chi_j(x) \neq 0$ and $\alpha_{j-1}(t, x)) \in U_j$)

and continuous,

(9) $$\rho(\alpha_{j-1}(t, x), \alpha_j(t, x)) \le \lambda\, \nu(x)\, \chi_j(x)\, t,$$

(10) $$\varphi(\alpha_j(t, x)) - \varphi(\alpha_{j-1}(t, x)) \le -\varepsilon\, \nu(x)\, \chi_j(x)\, t.$$

j = 1. (8). Use (3) ($x \in V_1$, $x \notin V_1$, supp $\chi_1 \subset V_1$, see also (5)).

(9). If $x \notin U_1$, then $x \notin V_1$, $\chi_1(x) = 0$, obvious inequality. Suppose $x \in U_1$. Consider the continuous path in U_1, $\gamma_1 : \gamma_1(s) = \alpha_1(s, x)$, $s \in [0, t]$. This path joins x (s = 0) with $\alpha_1(t, x)$, and its class is C^∞ ($f_1(\alpha_1(s, x)) = f_1(x) - s\nu(x)\chi_1(x)\dfrac{v_1}{\|v_1\|}$) (ANNEX, *Finsler manifold*). Then

$$\rho(x, \alpha_1(t, x)) \le (\text{ANNEX, (37) + (38)}) \int_0^t \|\gamma_1'(s)\|_{\gamma_1(s)}\, ds \le ((1) + \text{ANNEX, (28)})$$

$$\lambda \int_0^t \|(f_1 \circ \gamma_1)'(s)\|_{x_1}\, ds = \lambda \int_0^t \|-\nu(x)\chi_1(x)\frac{v_1}{\|v_1\|}\|_{x_1}\, ds = \lambda\nu(x)\,\chi_1(x)\, t.$$

(10). If $x \notin U_1$, $\varphi(\alpha_1(t, x)) - \varphi(x) = \varphi(x) - \varphi(x) = 0$ and $\chi_1(x) = 0$. Suppose $x \in U_1$ and apply a finite increment formula.

$$\varphi(\alpha_1(t, x)) - \varphi(x) = (\varphi \circ f_1^{-1})(f_1(x) - t\nu(x)\chi_1(x)\frac{v_1}{\|v_1\|}) - (\varphi \circ f_1^{-1})(f_1(x)) = -t\nu(x) \times$$

$$\times \chi_1(x)(\varphi \circ f_1^{-1})'(f_1(x) - \theta t\nu(x)\chi_1(x)\frac{v_1}{\|v_1\|})(\frac{v_1}{\|v_1\|}) \overset{(2)}{\le} -\frac{1}{2}t\nu(x)\chi_1(x)\|d\varphi(x_1)\| \le -\varepsilon\nu(x) \times$$

$\times \chi_1(x)t$ since $\|d\varphi(x)\| \ge 2\varepsilon$ on B.

$j-1 \Rightarrow j$. (8). If $\alpha_{j-1}(t, x) \notin U_j$ – nothing to prove. Suppose $\alpha_{j-1}(t, x) \in U_j$ and in addition $x \in \operatorname{supp} \chi_j$ (hence $x \in V_j$), otherwise we have also nothing to prove. Applying the triangle inequality and (9) with j replaced by $1, 2, ..., j-1$, one finds

$$\rho(x, \alpha_{j-1}(t, x)) = \rho(\alpha_0(t, x), \alpha_{j-1}(t, x)) \le \lambda t \nu(x) \sum_{k=1}^{j-1} \chi_k(x) \le \lambda t \nu(x) \le \lambda t,$$

$$\rho(x, \alpha_{j-1}(t, x)) \le \lambda t \overset{(5)}{<} \frac{\delta_0}{2} \quad \forall x \in X, \forall t \in (0, t_0). \tag{11}$$

Taking into account (11), we get

$$\rho(x, \alpha_{j-1}(t, x)) = \inf \{L(\gamma) : \gamma \text{ in } U_j \text{ joins } x \text{ with } \alpha_{j-1}(t, x)\}, \tag{12}$$

as for a smooth path γ which goes out from U_j we have $L(\gamma) \ge \delta_0$ since $\operatorname{dist}(x, X \setminus U_j) \overset{(3)}{\ge} \delta_0$. Thus when γ is like in (12) (see also ANNEX) $L(\gamma) = \int_0^t \| \gamma'(s) \|_{\gamma(s)} \, ds \overset{(1)}{\ge} \frac{1}{\lambda} \times$

$$\times \int_0^t \| (f_j \circ \gamma)'(s) \|_{X_j} \, ds \ge \frac{1}{\lambda} \int_0^t \| (f_j \circ \gamma)'(s) ds \|_{X_j} = \frac{1}{\lambda} \left\| f_j(\alpha_{j-1}(t, x)) - f_j(x) \right\|_{X_j},$$ consequently,

according to (12),

$$\| f_j(\alpha_{j-1}(t, x)) - f_j(x) \|_{X_j} \le \lambda \, \rho(x, \alpha_{j-1}(t, x)) \overset{(11)}{\le} \lambda^2 t,$$

hence

$$\| f_j(\alpha_{j-1}(t, x)) - t \nu(x) \chi_j(x) \frac{v_j}{\| v_j \|} - f_j(x) \|_{X_j} \le \lambda^2 t + t \overset{(5)}{<} \delta_0$$

and therefore

$$f_j(\alpha_{j-1}(t, x)) - t \nu(x) \chi_j(x) \frac{v_j}{\| v_j \|} \in f_j(U_j),$$

the definition in the case j is correct.

(9). In the case $x \in U_j$ consider the smooth path in U_j joining $\alpha_{j-1}(t, x)$ with $\alpha_j(t, x)$,

$$\gamma : \gamma(s) = f_j^{-1}(f_j(\alpha_{j-1}(t, x)) - s \nu(x) \chi_j(x) \frac{v_j}{\| v_j \|}), \; s \in [0, t]$$

and one continues as for $j = 1$.

(10). Apply, as for $j = 1$, a finite increment formula between $\alpha_{j-1}(t, x)$ and $\alpha_j(t, x)$.

Thus (8), (9) and (10) are true $\forall x \in X$, $\forall t \in (0, t_0)$ and for $j = \overline{1, n}$. It only remains to remark that (8), (9) and (10) are also true for $t = 0$ and then to take

$$\alpha(t, x) = \alpha_n(t, x),\ g(x) = \nu(x)\sum_{k=1}^{n}\chi_k(x)$$

in order to finish the proof ($x \in A \Rightarrow \nu(x) = 0 \Rightarrow \alpha_n(t, x) = x$). ■

Remark. Ghoussoub lemma remains true by removal only of the property 1° $\alpha(t, x) = x$ $\forall\, x \in A$, in the case $A = \varnothing$ (obviously $B \neq \varnothing$) – the only change in the proof being undergone by the function ν (see (4)) : $\nu(x) = 1$ for $x \in B$, $\nu(x) = 0$ for $x \in X \setminus \bigcup_{k=1}^{n} V_k$.

MINIMAX THEOREM I

7.2 Minimax theorem. *Let* X *be a Finsler* C^1*-manifold connected complete,* $\varphi\colon X \to \mathbf{R}$ *of* C^1 *class,* $\mathscr{K}$ *a set of nonempty compact subsets homotopy-stable with boundary* M *and*

$$c = c(\varphi, \mathscr{K}) := \inf_{A\in\mathscr{K}} \sup \varphi(A).$$

If

(F0) $$\sup \varphi(M) < c,$$

then for every sequence $(A_n)_{n\geq 1}$ *from* $\mathscr{K}$ *with* $\lim\limits_{n\to\infty} \sup \varphi(A_n) = c$ *there is a sequence* $(x_n)_{n\geq 1}$ *in* X *with the properties*

1° $\lim\limits_{n\to\infty} \varphi(x_n) = c$,

2° $\lim\limits_{n\to\infty} \|d\varphi(x_n)\| = 0$,

3° $\lim\limits_{n\to\infty} \operatorname{dist}(x_n, A_n) = 0$.

The statement of the theorem remains true in the case $M = \varnothing$ *considering* $\sup \varphi(\varnothing) = -\infty$ ([34] for the C^1 class of Finsler manifold).

■ (*A. Szulkin*). For every A_n $\exists\, \rho > 0$ such that $c \leq \sup \varphi(A_n) \leq c + \rho$, let γ_n be the infimum of this numbers ρ, thus $c \leq \sup \varphi(A_n) \leq c + \gamma_n$. We have $\lim\limits_{n\to\infty} \gamma_n = 0$: supposing *par absurdum* the contrary, one finds $u > 0$ such that $\gamma_{k_n} > u$ $\forall\, n$, (γ_{k_n}) a subsequence of

(γ_n); but for $n \geq N$ sup $\varphi(A_n) \leq c + u$, hence for a $k_n > N$ we have sup $\varphi(A_{k_n}) \leq c + u < c +$

$+ \gamma_{k_n}$, a contradiction . Then denoting $\varepsilon_n := \gamma_n + \frac{1}{n}$, we have

$$\forall n \quad c \leq \sup \varphi(A_n) \leq c + \varepsilon_n,\ \varepsilon_n > 0 \text{ and } \varepsilon_n \to 0.$$

This relation shows that if, by means of (13) $c \leq \sup \varphi(A) < c + \varepsilon^2$, where $A \in \mathscr{K}$ and $\varepsilon > 0$, it can find x_ε in X such that

$$c \leq \varphi(x_\varepsilon) \leq c + \varepsilon^2,\ \|d\varphi(x_\varepsilon)\| \leq 4\varepsilon,\ \text{dist}\,(x_\varepsilon, A) \leq \varepsilon,$$

then the proof is finished.

Consider the subset $\mathscr{D}$ of $C([0, 1] \times X; X)$ of the deformations η with

(14) $$\eta(t, x) = x \text{ for } (t, x) \in (\{0\} \times X) \cup ([0, 1] \times M)$$

and

(15) $$\sup\{\rho(\eta(t, x), x) : (t, x) \in [0, 1] \times X\} < +\infty$$

(ρ the Finsler metric on X).

Endow $\mathscr{D}$ with the distance (use (15) and the triangle inequality)

(16) $$d(\eta_1, \eta_2) = \sup\{\rho(\eta_1(t, x), \eta_2(t, x)) : (t, x) \in [0, 1] \times X\}$$

and one obtains a complete metric space (attention, X is complete!).

Consider the function $\Phi : \mathscr{D} \to \mathbf{R}$,

(17) $$\Phi(\eta) = \sup\{\eta(1, x) : x \in A\}.$$

This definition is correct since $x \to \varphi(\eta(1, x))$ is continuous and A compact. Φ is lower bounded by c ($\eta(\{1\} \times A) \in \mathscr{K}$!) and lower semicontinuous.

Designate by η_1 the deformation from $\mathscr{D}$

$$\eta_1(t, x) = x \text{ on } [0, 1] \times X \text{ (it is correct, } \rho(x, x) = 0).$$

We have

(18) $$\Phi(\eta_1) = \sup \varphi(A) \overset{(13)}{<} c + \varepsilon^2 \leq \inf\{\Phi(\eta) : \eta \in \mathscr{D}\} + \varepsilon^2.$$

Apply Ekeland principle 1.1 with ε^2 and $\lambda = \varepsilon$, $\exists\ \eta_0$ in $\mathscr{D}$ such that

(19) $$\Phi(\eta_0) \leq \Phi(\eta_1),$$

(20) $$d(\eta_0, \eta_1) \leq \varepsilon,$$

(21) $$\Phi(\eta_0) - \varepsilon d(\eta, \eta_0) \leq \Phi(\eta)\ \forall\ \eta \in \mathscr{D}.$$

Let be $B_0 := \{x \in \eta_0(\{1\} \times A) : \varphi(x) = \Phi(\eta_0)\}$, a nonempty compact set ($\eta_0(\{1\} \times A)$ is compact). By the hypothesis (F0) we have

(22) $$B_0 \cap M = \varnothing$$

(*par absurdum*, $\eta_0(\{1\} \times A) \in \mathscr{K}$ (the homotopy-stability), the definition of c). It remains to prove

(23) $$\exists\, x_\varepsilon \text{ in } B_0 \text{ such that } \|d\varphi(x_\varepsilon)\| \le 4\varepsilon,$$

since $x_\varepsilon \in B_0 \Rightarrow c \le \varphi(x_\varepsilon) < c + \varepsilon^2$ ($\varphi(x_\varepsilon) = \Phi(\eta_0) \overset{(19)}{\le} \Phi(\eta_1) \overset{(18)}{<} c + \varepsilon^2$) and dist $(x_\varepsilon, A) \le \varepsilon$ ($d(\eta_0, \eta_1) \overset{(20)}{\le} \varepsilon \Rightarrow \sup\limits_{x \in A} \rho(\eta_0(1, x), x) \le \varepsilon$, but $x_\varepsilon = \eta_0(1, a)$ with $a \in A$, hence $\rho(x_\varepsilon, a) \le \varepsilon$, dist $(x_\varepsilon, A) \le \varepsilon$).

Let us prove (23). Suppose *par absurdum* the contrary, (24) $\|d\varphi(x)\| > 4\varepsilon\ \forall\, x \in B_0$. Let λ be from the interval (1,2). One applies 7.1 to the compact nonempty disjoint sets M and B_0 ((22),(24)), there exists g, α and t_0 with the agreed properties. For every τ from $(0, t_0)$ consider the function $\eta_\tau : [0, 1] \times X \to X$,

(25) $$\eta_\tau(t, x) = \alpha(t\tau, \eta_0(t, x)).$$

(26) $\eta_\tau \in \mathscr{D}$: η_τ is continuous; $(t, x) \in (\{0\} \times X) \cup ([0,1] \times M) \Rightarrow \eta_\tau(t, x) = \alpha(t\tau, \eta_0(t, x)) = \alpha(t\tau, x) = x$ (7.1, 1°); $\sup\{\rho(\eta_\tau(t, x), x) : t \in [0, 1], x \in X\} < +\infty$ (use (25) and 2° of 7.1 with a triangle inequality).

$d(\eta_\tau, \eta_0) = \sup\{\rho(\eta_\tau(t, x), \eta_0(t, x)) : (t, x) \in [0, 1] \times X\} = \sup\{\rho(\alpha(t\tau, \eta_0(t, x)), \eta_0(t, x)) : (t, x) \in [0, 1] \times X\} \overset{7.1, 2^0}{\le} \lambda\tau$ and hence, taking into account (21),

(27) $$\Phi(\eta_\tau) \ge \Phi(\eta_0) - \varepsilon\lambda\tau.$$

Let x_τ be in A such that (28) $\Phi(\eta_\tau) = \varphi(\eta_\tau(1, x_\tau))$ ((17), A is compact). Then taking into account (27), (28) and (17),

(29) $$\varphi(\eta_\tau(1, x_\tau)) - \varphi(\eta_0(1, x)) \ge -\varepsilon\lambda\tau\ \forall\, x \in A.$$

On the other hand, according to 3° of 7.1,

(30) $$\varphi(\eta_\tau(1, x_\tau)) - \varphi(\eta_0(1, x_\tau)) = \varphi(\alpha(\tau, \eta_0(1, x_\tau)) - \varphi(\eta_0(1, x_\tau)) \le -2\varepsilon\tau g(\eta_0(1, x_\tau)).$$

Combining (29) and (30), one finds

(31) $$-\varepsilon\lambda \le -2\varepsilon g(\eta_0(1, x_\tau)).$$

Take $\tau = \dfrac{1}{m}$, $m \in \mathbf{N}$ and let $(x_{\frac{1}{k_m}})_{m \ge 1}$ be a convergent subsequence of $(x_{\frac{1}{m}})_{m \ge 1}$, $x_{\frac{1}{k_m}} \to x_0$. Obviously $x_0 \in A$. Replacing into (25) and passing to the limit for $m \to \infty$, we get

(32) $\lim_{m\to\infty} \eta_{\frac{1}{k_m}}(t, x_{\frac{1}{k_m}}) = \lim_{m\to\infty} \alpha(\frac{t}{k_m}, \eta_0(t, x_{\frac{1}{k_m}})) = \alpha(0, \eta_0(t, x_0)) = \eta_0(t, x_0)$ (take $t = 0$ in 7.1,1°). But $\varphi(\eta_\tau(1, x_\tau)) \overset{(27),(28)}{\geq} \Phi(\eta_0) - \varepsilon\lambda\tau$. Take $\tau = \frac{1}{k_m}$, then the limit for $m \to \infty$ and combining with (32) one finds $\varphi(\eta_0(1, x_0)) \geq \Phi(\eta_0)$, whence $\varphi(\eta_0(1, x_0)) \overset{(17)}{=} \Phi(\eta_0)$, i.e. $\varphi(\eta_0(1, x_0)) \in B_0$. Consequently, $g(\eta_0(1, x_0)) = 1$ (7.1,4°), which, confronted with (31) in which $\tau = \frac{1}{k_m}$ and passing to the limit, gives $\lambda \geq 2$, contradiction, and hence (23).

Finally, in the case $M = \varnothing$, it only remains to point out the remark to 7.1 in order to finish the proof (for instance (26) remains true since $M = \varnothing$). ■

Place us in the situation of the statement 7.2 and give two definitions (the first weaks the Palais-Smale condition, see §4) to formulate then a corollary of this.

Definition 1. Let $(A_n)_{n\geq 1}$ be a sequence in $\mathcal{K}$. φ verifies the $(PS)_c$ *condition along* $(A_n)_{n\geq 1}$ if every sequence $(x_n)_{n\geq 1}$ in X satisfying 1°, 2° and 3° has a convergent subsequence.

Definition 2. The sequence $(A_n)_{n\geq 1}$ from $\mathcal{K}$ is *min-maxing* for φ if

$$\lim_{n\to\infty} \sup \varphi(A_n) = c.$$

And now

7.3 *Corollary. In the conditions of the theorem* 7.2, *if, moreover,* φ *verifies* $(PS)_c$ *along of a min-maxing* $(A_n)_{n\geq 1}$ *from* $\mathcal{K}$, *then there is a critical point* x_0 *of* φ *at the level* c *and*

$$\lim_{n\to\infty} \text{dist}(x_0, A_{k_n}) = 0.$$

■ Let $(x_n)_{n\geq 1}$ be the sequence given by 7.2 and $(x_{k_n})_{n\geq 1}$ a convergent subsequence. $x_0 := \lim_{n\to\infty} x_{k_n}$ verifies the request in the statement. ■

Remark. Using the notations $K_c(\varphi)$ (§4, above of 4.7) and

$$A_\infty := \{x \in X : \underline{\lim}_{n\to\infty} \text{dist}(x, A_n) = 0\},$$

the assertion of 7.3 can be expressed by

$$A_\infty \cap K_c(\varphi) \neq \varnothing.$$

Note

In a recent paper *Minimax theorems on* C^1 - *manifolds via Ekeland variational principle* (Abstract and Applied Analysis 2003: 13(2003) 757-768), *M. Cuesta* presents two minimax theorems, Th. 2.1 and Th. 2.6, to find almost critical points and then, *via* a PS-condition, critical points for functionals $\Phi: X \to \mathbf{R}$ restricted to the „manifold" $M := \{x \in X: F(x) = 1\}$, where $F: X \to \mathbf{R}$ is a C^1 function having 1 as regular value. One supposes that the Banach space X is *uniformly convex* ($\forall \varepsilon$ from $(0,2]$ $\exists\, \delta > 0$ so that $\|x\| \le 1, \|y\| \le 1, \|x-y\| \ge \varepsilon \Rightarrow \left\|\frac{x+y}{2}\right\| \le 1-\delta$) and Φ has also the C^1-Fréchet class.

Some definitions. The *tangent space* $T_{x_0} M$ to M at the point x_0 from M is the set

$$T_{x_0} M = \{x \in X: \mathrm{d}F(x_0)(x)\} = 0 \text{ (Fréchet derivative).}$$

Let u be a point from M. u is *critical point* for the restriction $\varphi := \Phi|M$ if

$$\mathrm{d}\Phi(u)|T_u M = 0.$$

The derivative at u of the restriction φ has the norm

$$\|\varphi'(u)\|_* \overset{\text{def}}{=} \|\mathrm{d}\Phi(u)\|_{(T_u M)^*} \text{ (the dual of } T_u M).$$

In what follows K is a compact metric space and K_0 a closed nonempty subset of K. The first statement is of Mountain pass type and the second of the Ljusternik-Schnirelmann type.

Theorem 2.1 *Let* h_0 *be from* $C(K_0; M)$, $\Gamma := \{h \in C(K; M): h|K_0 = h_0\}$ *nonempty and* $c := \inf_{h\in\Gamma} \sup_{t\in K} \Phi(h(t))$. *If*

$$(*)\ c_0 := \sup_{K_0} (\Phi \circ h_0) < \sup_{K} (\Phi \circ h)\ \forall\, h \in \Gamma,$$

then $\forall\, \varepsilon > 0$ *and* $\forall$ h *from* Γ *with* $\sup_K (\Phi \circ h) < c + \varepsilon^2$ *there exists* u_ε *in* M *such that*

$$c \le \Phi(u_\varepsilon) \le c + \varepsilon^2,\ \mathrm{dist}\,(u_\varepsilon, h(K)) \le \varepsilon,\ \|\varphi'(u_\varepsilon)\|_* \le \varepsilon.$$

The proof follows in general features that of 6.19 from §6 and so this from [32]. One uses Ekeland principle 1.1 with ε^2 and $\lambda = \varepsilon$.

From Th. 2.1 directly results (Φ, Γ, c as above, (*) is verified)

If Φ *verifies* $(PS)_{c,M}$, *then there exists* u *in* M *so that*

$$\Phi(u) = c \text{ and } \varphi'(u) = 0.$$

Explanation. Φ verifies $(PS)_{c,M}$ means: any sequence $(x_n)_{n\geq 1}$ from M with $\lim_{n\to\infty} \Phi(x_n) = c$ and $\lim_{n\to\infty} \|\varphi'(x_n)\|_* = 0$ posseses a convergent subsequence.

Also (Φ, Γ, c as above, (*) is verified),

If there is h *in* Γ *so that* $\sup_K (\Phi \circ h) = c$, *then there is* u *in* h(K) *so that*

$$\Phi(u) = c \text{ and } \varphi'(u) = 0.$$

Th. 2.1 is related to Th. 7.2 from above.

A second minimax theorem will give almost critical points for the restriction φ when one minimizes along continuous odd maps defined on spheres of finite dimension. S^n, $n \geq 1$ will designate the unit sphere of $\mathbf{R}^{n+1}$. The functions Φ and F are supposed to be *even*. $C_0(S^n; M) := \{h \in C(S^n; M): h \text{ odd}\}$.

Theorem 2.6 *Let be* $\rho := \inf_{h\in C_0(S^n;M)} \sup_{t\in S^n} \Phi(h(t))$, $\varepsilon > 0$ *and* h *from* $C_0(S^n; M)$ *so that*

$$\sup_{S^n} (\Phi \circ h) < \rho + \varepsilon^2.$$

Then there exists u_ε *in* M *so that*

$$\rho \leq \Phi(u_\varepsilon) \leq \rho + \varepsilon^2,\ \mathrm{dist}\,(u_\varepsilon, h(S^n)) \leq \varepsilon,\ \|\varphi'(u_\varepsilon)\|_* \leq \varepsilon.$$

The proof goes along the same lines as the proof for Th. 2.1.

MINIMAX THEOREM II

(strong form)

7.4 Minimax theorem. *Let* X *be a complete connected Finsler* C^1*-manifold*, $\varphi: X \to \mathbf{R}$ *of* C^1 *class*, $\mathscr{K}$ *a set of nonempty compact subsets homotopy-stable with boundary* M,

$$c = c(\varphi, \mathscr{K}) := \inf_{A\in\mathscr{K}} \sup \varphi(A) \in \mathbf{R}$$

and F *a closed nonempty subset of* X. *If*

(F1) $$F \cap M = \varnothing \text{ and } F \cap A \neq \varnothing\ \forall A \in \mathscr{K}$$

and

(F2) $$\inf \varphi(F) \geq c,$$

then for every sequence $(A_n)_{n\geq 1}$ *from* $\mathscr{K}$ *which is min-maxing for* φ *there is a sequence* $(x_n)_{n\geq 1}$ *in* X *with the properties*

1° $\lim_{n\to\infty} \varphi(x_n) = c$,

2° $\lim_{n\to\infty} \|d\varphi(x_n)\| = 0$,

3° $\lim_{n\to\infty} \text{dist}\,(x_n, F) = 0$,

4° $\lim_{n\to\infty} \text{dist}\,(x_n, A_n) = 0$.

The assertion of the theorem remains also true in the case $M = \varnothing$ (*N. Ghoussoub*).

Emphasize (F1) $\Rightarrow \inf \varphi(F) \leq c$, since $\forall\, A \in \mathscr{K}$ we have $\inf \varphi(F) \leq \inf \varphi(F \cap A) \leq \leq \sup \varphi(A)$.

Every closed subset H verifying (F1), i.e. $H \cap M = \varnothing$ and $H \cap A \neq \varnothing\ \forall\, A \in \mathscr{K}$, is by definition *dual* of $\mathscr{K}$. *A fortiori* $H \neq \varnothing$.

For 7.4 it is enough to prove (as for 7.2!)

7.5 Theorem. *Let* X, φ, $\mathscr{K}$, M *be as for* 7.4 *and* F *a closed subset verifying* (F1) *and*

$$\text{(F2)}_\varepsilon \qquad \inf \varphi(F) \geq c - \varepsilon,\ \varepsilon > 0.$$

If

$$\varepsilon < \max\{\frac{1}{32}\,\text{dist}^2\,(M, F), \frac{1}{8}[\inf \varphi(F) - \sup \varphi(M)]\},$$

then for every A *from* $\mathscr{K}$ *satisfying*

$$(33) \qquad \sup \varphi(A) \leq c + \varepsilon,$$

there is x_ε *in* X *such that*

1° $c - \varepsilon \leq \varphi(x_\varepsilon) \leq c + 9\varepsilon$,

2° $\|d\varphi(x_\varepsilon)\| \leq 18\sqrt{\varepsilon}$,

3° $\text{dist}\,(x_\varepsilon, F) \leq 5\sqrt{\varepsilon}$,

4° $\text{dist}\,(x_\varepsilon, A) \leq 3\sqrt{\varepsilon}$

(*N. Ghoussoub*).

■ Take $\delta > 0$ defined by (34) $\varepsilon = \dfrac{\delta^2}{8}$. The hypothesis means

$$(35) \qquad \delta < \max\{\frac{1}{2}\,\text{dist}\,(M, F), \sqrt{[\inf\varphi(F) - \sup\varphi(M)]^+}\,\}$$

and

$$(36) \qquad \inf \varphi(F) \geq c - \frac{\delta^2}{8}.$$

Prove the existence of a x_δ in X such that

$$(37) \qquad c - \frac{\delta^2}{4} \leq \varphi(x_\delta) \leq c + \frac{9\delta^2}{8},$$

$$(38) \qquad \operatorname{dist}(x_\delta, F) \leq \frac{3\delta}{2},$$

$$(39) \qquad \|d\varphi(x_\delta)\| \leq 6\delta,$$

$$(40) \qquad \operatorname{dist}(x_\delta, A) \leq \frac{\delta}{2}.$$

7.5 devolves obviously from this relations.

The proof follows step by step that of 7.2. Consider the subset of X

$$(41) \qquad F_\delta := \{x \in X : \operatorname{dist}(x, F) < \delta\}$$

and the subset $\mathcal{D}$ of $C([0, 1] \times X, X)$ built with the deformations η,

$$(42) \qquad \eta(t, x) = x \text{ on } (\{0\} \times X) \cup ([0, 1] \times (M \cup (A \setminus F_\delta)))$$

and

$$(43) \qquad \sup\{\rho(\eta(t, x), x) : (t, x) \in [0, 1] \times X\} < +\infty.$$

Retain (44) $\eta \in \mathcal{D} \Rightarrow \eta(\{1\} \times A) \in \mathcal{K}$. Endow $\mathcal{D}$ with the distance

$$(45) \qquad d(\eta_1, \eta_2) = \sup\{\rho(\eta_1(t, x), \eta_2(t, x)) : (t, x) \in ([0, 1] \times X)\}$$

and one obtains a complete metric space. Associate with φ the Lipschitz function of constant smaller than δ,

$$(46) \qquad \psi : X \to \mathbf{R}_+, \ \psi(x) = \max\{0, \delta^2 - \delta \operatorname{dist}(x, F)\}.$$

Finally, define $\Phi: \mathcal{D} \to \mathbf{R}$,

$$(47) \qquad \Phi(\eta) = \sup\{(\varphi + \psi)(\eta(1, x)) : x \in A\},$$

a correct definition since $\varphi + \psi$ is continuous and A compact. Φ is lower semicontinuous. Let

$$(48) \qquad r := \inf\{\Phi(\eta) : \eta \in \mathcal{D}\}.$$

According to (F1) and (36) we get, according to (44) and because $\psi(x) = \delta^2$ on F,

$\Phi(\eta) \geq \sup\{(\varphi + \psi)(x) : x \in \eta(\{1\} \times A) \cap F\} \geq c - \dfrac{\delta^2}{8} + \delta^2 \ \ \forall \eta \in \mathcal{D}$, consequently ((48))

$$(49) \qquad r \geq c + \frac{7\delta^2}{8}.$$

Let η_1 be in $\mathscr{D}$, $\eta_1(t, x) = x$ on $[0, 1] \times X$. We have

(50) $$r \le \Phi(\eta_1) = \sup\{(\varphi + \psi)(x) : x \in A\} \overset{(33),(46)}{<} c + \frac{9\delta^2}{8}.$$

(49) and (50) give

(51) $$\Phi(\eta_1) < c + \frac{9\delta^2}{8} \le r + \frac{\delta^2}{4} \overset{(48)}{=} \inf\{\Phi(\eta) : \eta \in \mathscr{D}\} + \frac{\delta^2}{4}.$$

Apply Ekeland principle (1.1, $\varepsilon = \frac{\delta^2}{4}$, $\lambda = \frac{\delta}{2}$), $\exists\, \eta_0$ in $\mathscr{D}$ such that

(52) $$\Phi(\eta_0) \le \Phi(\eta_1),$$

(53) $$d(\eta_0, \eta_1) \le \frac{\delta}{2},$$

(54) $$\Phi(\eta_0) - \frac{\delta}{2} d(\eta, \eta_0) \le \Phi(\eta) \quad \forall\, \eta \in \mathscr{D}.$$

Let be (55) $B := \{x \in \eta_0(\{1\} \times A) : (\varphi + \psi)(x) = \Phi(\eta_0)\}$ and (56) $M_1 := M \cup (A \setminus F_\delta)$. Prove (57) $B \cap M_1 = \varnothing$, which is equivalent to $B \cap M = \varnothing$ and $B \cap (A \setminus F_\delta) = \varnothing$, which will follow from

$$(58_1)\quad \sup(\varphi + \psi)(A \setminus F_\delta) \le r - \frac{3\delta^2}{4} \quad \text{and} \quad (58_2)\quad \sup(\varphi + \psi)(M) \le r - \frac{3\delta^2}{4},$$

as on B we have $(\varphi + \psi)(x) = \Phi(\eta_0) \overset{(48)}{\ge} r$.

Since $\psi(x) = 0$ on $X \setminus F_\delta$, $\sup(\varphi + \psi)(A \setminus F_\delta) = \sup \varphi(A \setminus F_\delta) \le \sup \varphi(A) \overset{(33)}{\le} c + \frac{\delta^2}{8} = c + \frac{7\delta^2}{8} - \frac{3\delta^2}{4} \overset{(49)}{\le} r - \frac{3\delta^2}{4}$ and hence (58_1). Pass to (58_2). If $\delta \overset{(35)}{<} \frac{1}{2}$ dist (M, F), then dist $(x, F) > 2\delta\ \forall\, x \in M$, hence $M \overset{(41)}{\subset} A \setminus F_\delta$, which imposes *a fortiori* (58_2). If $\delta \overset{(35)}{<} \sqrt{[\inf\varphi(F) - \sup\varphi(M)]^+}$, then $\sup \varphi(M) < \inf \varphi(F) - \delta^2 \overset{(F1)}{<} c - \delta^2$, consequently $\sup(\varphi + \psi)(M) \overset{(46)}{\le} c - \delta^2 + \delta^2 \overset{(49)}{\le} r - \frac{7\delta^2}{8}$ and hence again (58_2).

It remains to prove

(59) there is x_δ in B such that $\|d\varphi(x_\delta)\| \le 6\delta$.

Indeed,

$x_\delta \in B \Rightarrow c + \frac{7\delta^2}{8} \overset{(49)}{\leq} r \overset{(48)}{\leq} \Phi(\eta_0) = (\varphi + \psi)(x_\delta) \overset{(52)}{\leq} \Phi(\eta_1) \overset{(50)}{<} c + \frac{9\delta^2}{8}$, but $0 \leq \psi(x_\delta) \leq \delta^2$ ((46)), whence $c - \frac{\delta^2}{8} \leq \varphi(x_\delta) \leq c + \frac{9\delta^2}{8}$, i.e. (37).

$x_\delta \in B \Rightarrow x_\delta = \eta_0(1, a)$, $a \in A$. We have $a \in F_\delta$ since $a \notin F_\delta \Rightarrow a \in M_1 \Rightarrow x_\delta \overset{(42)}{=} a \Rightarrow B \cap \cap M_1 \neq \varnothing$, a contradiction ((57)). Thus dist $(a, F) < \delta$. But $\rho(x_\delta, a) = \rho(\eta_0(1, a), a) \overset{(45)}{\leq} d(\eta_0, \eta_1) \overset{(53)}{\leq} \frac{\delta}{2}$, whence, *via* the triangle inequality, dist $(x_\delta, F) \leq \frac{3\delta}{2}$, i.e. (38).

Finally, $\rho(x_\delta, A) \leq \frac{\delta}{2} \Rightarrow$ dist $(x_\delta, A) \leq \frac{\delta}{2}$, i.e. (40).

Pass to the proof of (59). This will repeat the proof for (23). Suppose *par absurdum* $\|d\varphi(x)\| > 6\delta \ \forall x \in B$ and let τ be in the interval (1,2). Apply 7.1 to the sets M_1 and B (see (57)), $\exists\, g$, α and t_0 with the properties agreed. For every τ in $(0, t_0)$ consider the function $\eta_\tau : [0, 1] \times X \to X$,

$$\eta_\tau(t, x) = \alpha(t\tau, \eta_0(t, x)).$$

One has $\eta_\tau \in \mathscr{D}$. Since $d(\eta_\tau, \eta_0) \leq \lambda\tau$, $\Phi(\eta_\tau) \overset{(54)}{\geq} \Phi(\eta_0) - \frac{\delta}{2}\lambda\tau$. Let x_τ be from A such that $\Phi(\eta_\tau) = (\varphi + \psi)(\eta_\tau(1, x_\tau))$. Then

$$(\varphi + \psi)(\eta_\tau(1, x_\tau)) - (\varphi + \psi)(\eta_0(1, x)) \geq -\frac{\delta}{2}\lambda\tau \ \forall x \in A.$$

Since the Lipschitz constant of ψ is smaller than δ, we have

$$\varphi(\eta_\tau(1, x_\tau)) - \varphi(\eta_0(1, x_\tau)) \geq -\frac{3\delta}{2}\lambda\tau.$$

But

$$\varphi(\eta_\tau(1, x_\tau)) - \varphi(\eta_0(1, x_\tau)) \leq -3\delta\tau g(\eta_0(1, x_\tau)),$$

consequently,

$$-\frac{3\delta}{2}\lambda\tau \leq -3\delta\tau g(\eta_0(1, x_\tau)). \tag{60}$$

Just as in the quoted proof, we get a point x_0 in A for which $g(\eta_0(1, x_0)) = 1$ and (60) remains true if we replace x_τ by x_0, and this imposes $\lambda \geq 2$, a contradiction and consequently (59) holds and this ends the proof. ■

Remark. Theorem 7.4 justifies the name of „strong form" with respect to the theorem 7.2, since if (F0) is verified, then $F := \{x \in X : \varphi(x) \geq c\}$ satisfies (F1) and (F2), and consequently there is a sequence $(x_n)_{n\geq 1}$ in X with the properties 1°, 2°, 3° and 4° , and 1°, 2° and 4° are stated in 7.2.

Definition. Let $(A_n)_{n\geq 1}$ be a sequence from $\mathscr{K}$. φ verifies

$$(\mathrm{PS})_{c,\,F} \textit{ condition along of } (A_n)_{n\geq 1}$$

if every sequence $(x_n)_{n\geq 1}$ from X with the properties

(61) $\lim\limits_{n\to\infty} \varphi(x_n) = c,\ \lim\limits_{n\to\infty} \|d\varphi(x_n)\| = 0,\ \lim\limits_{n\to\infty} \mathrm{dist}\,(x_n, F) = 0$ and $\lim\limits_{n\to\infty} \mathrm{dist}\,(x_n, A_n) = 0$

has a convergent subsequence.

And now several corollaries of Ghoussoub theorem.

7.6 *Corollary. In the assumptions of* 7.4, *if in addition* φ *verifies* $(\mathrm{PS})_{c,\,F}$ *along a min-maxing sequence* $(A_n)_{n\geq 1}$ *from* $\mathscr{K}$, *then there is a critical point* x_0 *of* φ *at the level* c *situated in* F *and* $\lim\limits_{n\to\infty} \mathrm{dist}\,(x_0, A_{k_n}) = 0$.

■ Let $(x_n)_{n\geq 1}$ be a sequence in X given by 7.4 and a convergent subsequence $(x_{k_n})_{n\geq 1}$, $x_{k_n} \to x_0$. Then, since $\varphi(x_{k_n}) \to \varphi(x_0)$ and $d\varphi(x_{k_n}) \to d\varphi(x_0)$, we have $\varphi(x_0) = c$ and $d\varphi(x_0) = 0$. Moreover, since $\mathrm{dist}\,(x_0, F) \leq \rho(x_0, x_{k_n}) + \mathrm{dist}\,(x_{k_n}, F)$, we have $x_0 \in F$. It is also obvious that $\mathrm{dist}\,(x_0, A_{k_n}) \to 0$. ■

Remark. Using the notations K_c and A_∞ (see Remark to 7.3), the assertion of 7.6 can be expressed by

$$F \cap A_\infty \cap K_c(\varphi) \neq \varnothing.$$

7.7 *Corollary. In the assumptions of* 7.4 ($M \neq \varnothing$), *but with* (F2) *replaced by*

(F3) $$\sup \varphi(M) \leq \inf \varphi(F),$$

if φ *verifies* $(\mathrm{PS})_c$ *then* c *is a critical value for* φ.

■ It was mentioned above (F1) ⇒ $\sup \varphi(F) \leq c$. So we have either (62) $\sup \varphi(M) < c$, or (63) $\sup \varphi(M) = c$. In the case (62), the set $F' := \{x \in X : \varphi(x) \geq c\}$ verifies (F1) and (F2) and one uses 7.4. In the case (63), F verifies (F1) and (F2) (see (F3)) and hence the conclusion. ■

The forced pendulum equation. Let $h : \mathbf{R} \to \mathbf{R}$ be a 2π-periodic function of C^1 class and $f : \mathbf{R} \to \mathbf{R}$ a T-periodic continuous function. Consider the differential equation

(63_1) $$\ddot{u} + h'(u) = f(t) \text{ a.e.}$$

(the forced pendulum equation for $h(u) = -\cos u$; the point indicates the derivative with respect to t, the time).

We use 7.2 and 7.6 to obtain a proposition about the existence of the solutions of (63_1).

Let H be the Hilbert space of T-periodic absolutely continuous functions u on $\mathbf{R}$ for which $\int_0^T |\dot u(t)|^2\, dt < +\infty$ with the scalar product

$$(63_2)\qquad u\cdot v = \int_0^T (\dot u\dot v + uv)dt\,.$$

H is a connected complete Finsler manifold.

The solutions of (63_1) are the critical points of the action – functional $\varphi : H \to \mathbf{R}$,

$$(63_3)\qquad \varphi(u) = \int_0^T [\frac{1}{2}\dot u^2 - h(u(t)) + f(t)u(t)]\, dt$$

(the coherence of the assertion is provided by the corresponding theorems of Lebesgue, Jordan and Riesz).

It can state

7.7_1 *Suppose* $\int_0^T \mathrm{f(t)\ dt} = 0$. *Then the equation* (63_1) *has at least two* T*-periodic solutions which not differ by a multiple of* 2π ([34], theorem 4.13).

■ Put for every u in H

$$(63_4)\qquad k_u := \frac{1}{T}\int_0^T u(t)\, dt,\; v_u := u - k_u\,.$$

Thus

$$(63_5)\qquad \varphi(u) = \int_0^T (\frac{1}{2}\dot u^2 - h(v_u + k_u) + f\,v_u)\, dt.$$

Obviously $\varphi(u + 2\pi) = \varphi(u)$.

Prove

(63_6) *φ is lower bounded.*

One finds, using Schwarz and Wirtinger inequalities ($a := \sup h(\mathbf{R})$, $\rho > 0$),

$$(63_7)\qquad \varphi(u) \geq \int_0^T \frac{1}{2}\dot v_u^2\, dt - aT - \left(\int_0^T f^2 dt\right)^{\frac{1}{2}} \left(\int_0^T v_u^2 dt\right)^{\frac{1}{2}} \geq \int_0^T \frac{1}{2}\dot v_u^2 dt - aT -$$

$$-\rho\left(\int_0^T f^2 dt\right)^{\frac{1}{2}}\left(\int_0^T \dot{v}_u^2 dt\right)^{\frac{1}{2}} = \left(\int_0^T \dot{v}_u^2 dt\right)^{\frac{1}{2}}\left[\frac{1}{2}\left(\int_0^T \dot{v}_u^2 dt\right)^{\frac{1}{2}} - \rho\left(\int_0^T f^2 dt\right)^{\frac{1}{2}}\right] - aT$$

and consequently (63_6). ■

Consider, for every N in **N**,

$$(63_8) \qquad F_N := \{u \in H : \left|\frac{1}{T}\int_0^T u(t)dt\right| \leq 2N\pi\}.$$

Prove

$$(63_9) \qquad \rho \textit{ verifies } (\mathrm{PS})_{c,F_N} \textit{ for every } c \in \mathbf{R} \textit{ and } N \in \mathbf{N}.$$

Let $(u_n)_{n\geq 1}$ be a sequence in H with (63_{10}) $\lim_{n\to\infty} \varphi(u_n) = c$, $\lim_{n\to\infty} \varphi'(u_n) = 0$ and $\lim_{n\to\infty} \text{dist}\,(u_n, F_N) = 0$. Using again Wirtinger inequality to (63_7) we find that $\left(v_{u_n}\right)_{n\geq 1}$ is bounded in H. Moreover, (63_{10}) allows us to admit that $\left(k_{u_n}\right)_{n\geq 1}$, and consequently also $(u_n)_{n\geq 1}$, is bounded in H. Since H has a compact embedding in $\mathbf{L}^2(\mathbf{R})$, there is a subsequence $(u_n)_{n\geq 1}$, denote it identically, with the properties

$$(63_{11}) \qquad u_n \to u \text{ uniformly on } [0, T],\ \dot{u}_n \to \dot{u} \text{ weakly in } \mathbf{L}^2(\mathbf{R}).$$

Then, for m and n arbitrary,

$$\varphi'(u_n - u_m)(u_n - u_m) = \int_0^T (\dot{u}_n^2 - \dot{u}_m^2)\mathrm{d}t - \int_0^T h'(u_n - u_m)(u_n - u_m)dt \geq \|\dot{u}_n - \dot{u}_m\|_2^2 - \alpha\|u_n - u_m\|_\infty,$$

$\alpha > 0$. Taking into account (63_{10}) and (63_{11}) one gets $\|\dot{u}_n - \dot{u}_m\|_2 \to 0$ for $m, n \to \infty$, this imposes $(u_n)_{n\geq 1}$ Cauchy sequence and consequently convergent in H.

Now combining (63_6) with (63_9) we get that φ has a global minimum u_0 in H and $\varphi(u_0 + 2\pi) = \varphi(u_0)$.

Let $\mathcal{K}$ be the set of continuous paths in H that joint u_0 with $v := u_0 + 2\pi$ ($v \in H$!). $\mathcal{K}$ is homotopy-stable with boundary $M := \{u_0, v\}$. Let $F := S_\rho$ be the sphere centred at u_0 with radius ρ, $\rho < \|u_0 - v\|$. F is dual of $\mathcal{K}$ and (F3) is verified:

$$(63_{12}) \qquad \sup \varphi(M) = \sup \{\varphi(u_0), \varphi(u_0 + 2\pi)\} = \varphi(u_0) \leq \inf \varphi(F),$$

u_0 being a global minimum point.

Let be $c := c(\varphi, \mathcal{K})$. Only the following two situations are possible and in each of these we find two T-periodic solutions of (63_1) which do not differ by a multiple of 2π.

The case $c = \varphi(u_0)$. Let S_ρ be as above, $\rho < \|2\pi\|$. Since for N high enough we have $S_\rho \subset \subset F_N$, (63_9) validates the assertion φ verifies $(PS)_{c, S_\rho}$. As (F1) and (F2) are also verified with respect to S_ρ, 7.6 gives a critical point u_ρ for φ in S_ρ at the level c, i.e. u_ρ is a solution of (63_1). There is ρ_0 such that $u_{\rho_0} \neq u_0 + 2\pi$, in the oposite case, contracting the spheres in u_0, one finds a sequence $(u_n)_{n\geq 1}$, $u_n = u_0 + 2\pi$, with $u_n \to u_0$, contradiction. Thus u_0 and u_{ρ_0} verify the statement.

The case $c > \varphi(u_0)$. Since $\sup \varphi(M) \overset{(63_{12})}{=} \varphi(u_0)$, we have in fact $\sup \varphi(M) < c$, apply 7.2 and one finds $(u_n)_{n\geq 1}$ a sequence in H with $\lim\limits_{n\to\infty} \varphi(u_n) = c$, $\lim\limits_{n\to\infty} \varphi'(u_n) = 0$. But the periodicity of φ allows to admit $u_n \in F_1$ $\forall\, n \geq 1$, then enter in action (63_9) and one gets v_0 a critical point at the level c. Since $c \neq \varphi(u_0)$, *a fortiori* $v_0 \neq u_0 + 2\pi$. ■

The theorem 7.4 may give information, as we see later, on the type of critical point provided by it, even on the structure of the critical points in the Mountain Pass theorem.

For the moment

7.8 *Corollary. Let be* φ *and* $\mathcal{K}$, *but the sets of* $\mathcal{K}$ *are connected, as in* 7.4. *If* φ *verifies* (F0) *and* $(PS)_c$, *then* φ *has a critical point at the level* c *that is not a local minimum point.*

■ $F := \mathrm{Fr}\,\{x \in X : \varphi(x) \geq c\}$ is dual of $\mathcal{K}$ (*par absurdum*), one finds with the help of 7.4 a critical point x_0 at the level c situated in F. x_0 cannot be a local minimum point (*par absurdum*). ■

7.9 *In the hypotheses of* 7.4, *if moreover* $d\varphi$ *is uniformly continuous on the almost critical sequences at the level* c *and around* F, *then*

1° $\forall\, \varepsilon > 0\ \exists\, \delta > 0$ *and* $\exists\, A$ *in* $\mathcal{K}$ *such that*

$$\sup \varphi(A) < c + \delta \text{ and } \|d\varphi(x)\| < \varepsilon\ \forall\, x \in A \cap F^\delta \cap L_{c-\delta}\,;$$

2° *If* $\sup \varphi(M) < c$, *then* $\forall\, \varepsilon > 0\ \exists\, \delta > 0$ *and* $\exists\, A \in \mathcal{K}$ *such that*

$$\sup \varphi(A) < c + \delta \text{ and } \|d\varphi(x)\| < \varepsilon\ \forall\, x \in A \cap L_{c-\delta}\,.$$

Explanations. The assumption on $d\varphi$ means:

$$d(x_n, y_n) \to 0,\ \varphi(x_n) \to c,\ \|d\varphi(x_n)\| \to 0 \text{ and } \mathrm{dist}\,(x_n, F) \to 0 \Rightarrow \|d\varphi(y_n)\| \to 0.$$

$$L_{c-\delta} := \{x \in X : \varphi(x) \geq c - \delta\},\ F^\delta := \{x \in X : \mathrm{dist}\,(x, F) \leq \delta\}.$$

■ Suppose *par absurdum* the contrary of 1°. Then there is $\varepsilon > 0$ such that $\forall\, \delta > 0$ the set

$$F^{\delta} \cap L_{c-\delta} \cap \{x \in X: \|d\varphi(x)\| \geq \varepsilon\}$$

verifies (F1) and $(F2)_\delta$ of 7.5 (ε replaced by δ). So there is x_δ verifying the conclusions of the theorem. Taking $\delta = \frac{1}{n}$, $n \in \mathbf{N}$ one obtains an almost critical sequence $(x_n)_{n\geq 1}$ whose „distance" to F and to the set $\{x \in X : \|d\varphi(x)\| \geq \varepsilon\}$ tends to 0, which contradicts the uniform continuity of dφ. ■

Definition. Let X be a topological space. The closed subset F *separates* the distinct points u, v in X, if these are found in the distinct connected components of $X \setminus F$ (repeated definition, see above of 6.7).

Denote by Γ_u^v the set of the continuous paths in X with the origin in u and the end in v,

$$\Gamma_u^v = \{\gamma \in C([0, 1]; X): \gamma(0) = u, \gamma(1) = v\}.$$

One identifies Γ_u^v with the set $\{\gamma([0, 1]): \gamma \in \Gamma_u^v\}$, i.e. we do not distinguish between the continuous path and its support. Γ_u^v is homotopy-stable with boundary $\{u, v\}$. Obviously F nonempty separates u and v iff it is dual of Γ_u^v.

7.10 Mountain-Pass Theorem. *Let* X *be a Banach space,* $\varphi : X \to \mathbf{R}$ *of* C^1 *- Fréchet class and* u,v *distinct points in* X. *Suppose* $c := \inf_{g\in\Gamma_u^v} \sup_{t\in[0,1]} \varphi(g(t)) \in \mathbf{R}$ *and* F, *a closed nonempty subset of* $\{x \in X : \varphi(x) \geq c\}$, *separates* u *and* v.

If φ *verifies* $(PS)_{c, F}$, *then* $F \cap K_c \neq \varnothing$.

Explanation. $(PS)_{c, F}$ and K_c $(= K_c(\varphi))$ are defined in §4. *A fortiori* K_c is compact.

■ Apply 7.6, F is dual of Γ_u^v and moreover verifies (7.2). ■

Here are now the propositions on the structure of the set of crtical points in the Mountain Pass theorem. They must be connected to the propositions 6.10 and 6.11 (§6). The notations used are φ_X^c, $A_{\varphi, c}$, $M_c(\varphi)$ already known (see 6.10, 6.11) and

$$\mu_c(\varphi) = \mu_c := \{x \in K_c(\varphi) : x \text{ strict local maximum point}\},$$

$$S_c(\varphi) = S_c := \{x \in K_c(\varphi) : x \text{ saddle point}\}$$

(x *saddle point* for φ $\overset{\text{def}}{\Leftrightarrow}$ in every neighbourhood of x there exist x', x'' such that $\varphi(x') < \varphi(x) < \varphi(x'')$).

We have

(64) $$K_c = M_c \cup \mu_c \cup S_c, \text{ partition of the first member.}$$

Indeed, if $x \notin S_c$ $\exists$ V a neighbourhood of x, we can suppose it open connected, such that either $\varphi(x') \geq \varphi(x)$ $\forall$ $x \in V$, or $\varphi(x'') \leq \varphi(x)$ $\forall$ $x \in V$. In the first case $x \in M_c$. Place in the second case and consider the sets $A := \{x'' \in V : \varphi(x'') = \varphi(x)\}$, $B := \{x'' \in V: \varphi(x'') <$ $< \varphi(x)\}$. Obviously $A \cup B = V$, $A \cap B = \varnothing$, A is closed in V, B is open in V, hence open. V being connected, this imposes either $A = \varnothing$, and in this case $x \in \mu_c$, or $B = \varnothing$, and in this case $x \in M_c$.

7.11 Theorem. *In the hypotheses of* 7.10 *with* φ *verifying* $(PS)_{c,\,F}$, *if* $F \cap \mu_c$ *does not include any compact subset which separates* u *and* v, *then*

$$\textit{either } F \cap M_c \neq \varnothing, \textit{ or } F \cap S_c \neq \varnothing$$

(*N. Ghoussoub, D. Preiss*).

■ Suppose *par absurdum* (65) $F \cap M_c = F \cap S_c = \varnothing$. Since F separates u and v, it has a nonempty closed connected subset F_1 which also separates u and v (*Whyburn*). One has, taking into account (64) and (65),

$$F_1 \cap K_c = F_1 \cap \mu_c. \tag{66}$$

But $F_1 \cap K_c$, i.e. $F_1 \cap \mu_c$, is closed in F_1 and also $F_1 \cap \mu_c$ is open in F_1 connected, consequently either $F_1 \cap \mu_c = \varnothing$, i.e. $F_1 \cap K_c \overset{(66)}{=} \varnothing$, or $F_1 \cap \mu_c = F$. The first situation is prohibited through 7.10 since φ, verifying $(PS)_{c,\,F}$, verifies also $(PS)_{c,F_1}$ and consequently $F_1 \cap K_c \neq \varnothing$. It remains $F_1 \cap \mu_c = F_1$ and hence $F_1 \cap \mu_c$, a compact subset of $F \cap \mu_c$ (see (66)), separates u and v, what contradicts the hypothesis. ■

7.12 *Corollary. Let* φ, u, v, c, F *be as in* 7.10. *Suppose* φ *verifies* $(PS)_c$ *and* u, v $\notin \overline{M}_c$. *If* μ_c *does not include any compact nonempty set which separates* u *and* v, *then*

$$S_c \neq \varnothing.$$

(*G. Fang*).

■ Suppose *par absurdum* (67) $S_c = \varnothing$. There is, for every x from M_c, $S(x, r_x)$ an open ball centred at x included in $A_{\varphi,\,c}$, let be $E := \bigcup_{x \in M_c} S(x, r_x)$. Obviously (68) $M_c \subset E \subset A_{\varphi,\,c}$.

Since u, $v \notin \overline{M}_c$, a compact set (K_c is compact, it follows from $(PS)_c$), it can admit (69) u, $v \notin \overline{E}$ (conveniently manipulate the balls and take a finite number from this). Let be

$$F_0 := (F \setminus E) \cup \operatorname{Fr} E.$$

We have inf $\varphi(F_0) \geq c$ (see (68), $\overline{E} \subset A_{\varphi, c}$) and F_0 separates u and v (use (69)). Moreover, $F_0 \cap (M_c \cup S_c) = \varnothing$ since $F_0 \cap S_c \overset{(67)}{=} \varnothing$, $F_0 \cap M_c = \varnothing$ as $F_0 \cap E = \varnothing$. So 7.11 imposes $F_0 \cap \mu_c$, and even more μ_c, has a compact nonempty subset separating u and v, a contradiction and $S_c \neq \varnothing$. ■

Remark to 7.12. If in particular max $(\varphi(u), \varphi(v)) < c$ and X has an infinite dimension, then $S_c \neq \varnothing$.

LJUSTERNIK-SCHNIRELMANN THEOREM

At the end of „Minimax Theorem II (strong form)" we present a very simple proof of Ljusternik - Schnirelmann theorem ([34], p.77). But firstly the definition of $\mathrm{Cat}_X(A)$ and some properties of this.

Let X be a topological space and $A \subset X$. The *Ljusternik - Schnirelmann category* of A *in* X,

$$\mathrm{Cat}_X(A),$$

is defined as follows:

$$A = \varnothing \Rightarrow \mathrm{Cat}_X(A) = 0;$$

When $A \neq \varnothing$, either there is a finite number of closed sets contractible in X covering A, or not and in this case

$$\mathrm{Cat}_X(A) = +\infty.$$

In the first case

$$\mathrm{Cat}_X(A) = k,$$

k the smallest natural number for which there is a covering as above of A with k elements.

Explanations. Let X be a topological space and $A \subset X$ nonempty. The homotopy H: $[0, 1] \times A \to X$, to which the initial map ($t = 0$) is $H(0, x) = x$ $\forall\, x \in A$, is called *continuous deformation of* A *in* X respectively *continuous deformation of* A in the case $A = X$. The continuous deformation H: $[0, 1] \times A \to X$, to which the final map ($t = 1$) is constant, is called *contraction of* A *in* X respectively *contraction of* A if $A = X$. In this case A *is contractible in* X respectively X *is contractible*. If in addition there is x_0 in A such that $H(t, x_0) = x_0$ $\forall\, t \in [0, 1]$, then A is by definition *contractible at* x_0 (for instance $H(t, x) =$ $= (1 - t)x$ is a contraction at 0 of the ball $\{x \in X: \|x\| \leq 1\}$, X a normed space; in fact every nonempty subset of X is contractible). Obviously, every nonempty subset of a contractible set, respectively contractible at x_0 is contractible respectively contractible at x_0 (if x_0 belongs to the subset). Every set with only an element is contractible in itself and so every nonempty set is the union of closed sets contractible at a point.

Let us present some properties of Ljusternik - Schnirelmann category ((C1), (C2), (C3)).

(C1) $\mathrm{Cat}_X(A) = 0 \Rightarrow A = \varnothing$; $A_1 \subset A_2 \Rightarrow \mathrm{Cat}_X(A_1) \le \mathrm{Cat}_X(A_2)$;

$$\mathrm{Cat}_X(A_1 \cup A_2) \le \mathrm{Cat}_X(A_1) + \mathrm{Cat}_X(A_2);\ y \in Y \Rightarrow \mathrm{Cat}_{X\times Y}(A \times \{y\}) = \mathrm{Cat}_X(A).$$

■ *First statement.* Suppose *par absurdum* $A \neq \varnothing$, then the definition imposes $1 \le \le \mathrm{Cat}_X(A) \le +\infty$, contradiction. *Second statement.* It is enough to consider only the case A_1, A_2 nonempty and then $\mathrm{Cat}_X(A_2) < +\infty$. But every covering for A_2 is also a covering for A_1. *Third statement.* It is enough to consider the case A_1, A_2 nonempty and $\mathrm{Cat}_X(A_1)$, $\mathrm{Cat}_X(A_2) < +\infty$ and then take certain coverings of A_1 and A_2. *Fourth statement.* Take into account that the set $\{y\}$ is contractible in Y. ■

(C2) *Let* A_1, A_2 *be nonempty closed sets in* X *and* $H: [0, 1] \times A_1 \to X$ *a continuous deformation of* A_1 *in* X *having the property* $H(\{1\} \times A_1) \subset A_2$. *Then*

$$\mathrm{Cat}_X(A_1) \le \mathrm{Cat}_X(A_2).$$

■ We can admit $\mathrm{Cat}_X(A_2) < +\infty$ and (C1) allows to suppose (69_1) $H(\{1\}\times A_1) = A_2$, since in the opposite case take $A'_2 := H(\{1\}\times A_1)$, a closed set. Let be $k := \mathrm{Cat}_X(A_2)$ and $C_1, \dots, C_k$ closed sets contractible at a point which cover A_2. It can admit, it is the definition of the category that allows, $C_i \cap A_2 \neq \varnothing$, $i = \overline{1,k}$. Consider the sets $D_i := = H_1^{-1}(C_i \cap A_2)$, $i = \overline{1,k}$, where $H_1 : A_1 \to A_2$, $H_1(x) = H(1, x)$. D_i are closed and cover A_1 (see (69_1)). It remains to prove that every D_i is contractible in X, since this imposes, by the definition, the wanted inequality. Let f_i be the contraction in X of C_i, $i = \overline{1,k}$ and $g_i : [0, 1] \times D_i \to X$,

$$(69_2) \qquad g_i(t, x) = \begin{cases} H(2t, x),\ t \in [0, \frac{1}{2}], \\ f_i(2t-1, H(1,x)),\ t \in [\frac{1}{2}, 1]. \end{cases}$$

g_i is a homotopy (the continuity for $t = \frac{1}{2}$ is justified by the remark: $H(1, x) = f_i(0, H(1, x))$ since $x \in D_i \Rightarrow H(1, x) \in C_i$ and f_i is deformation of C_i) and even a contraction in X

of $D_i : x \in D_i \Rightarrow g_i(0, x) = H(0, x) = x$ ($D_i \subset A_1$!); let be $u_i := f_i(1, \cdot)$, then $x \in D_i \Rightarrow$ $\Rightarrow g_i(1, x) = f_i(1, H(1, x)) = u_i$ since $H(1, x) \in C_i$. ■

(C3) *Let* X *be a connected Finsler manifold. For every* A *a closed subset, there is* V *a closed neighbourhood of* A *with the property*

$$\mathrm{Cat}_X(V) = \mathrm{Cat}_X(A).$$

■ We can suppose $k := \mathrm{Cat}_X(A) < +\infty$, in the opposite case take $V = X$. Let $C_1, \dots, C_k$ be closed contractible in X sets covering A and $f_1, \dots, f_k$ the corresponding contractions, $a_i := f_i(1, \cdot)$. Prove, and this will be sufficient, that every C_i has a closed neighbourhood U_i contractible in X (take $V := U_1 \cup \dots \cup U_k$, $\mathrm{Cat}_X(V) \le k$ (the definition) and then (C1)). Consider the closed subset E_i of $[0, 1] \times X$,

$$E_i := ([0, 1] \times C_i) \cup (\{0\} \times X) \cup (\{1\} \times X)$$

and the function $g_i : E_i \to X$,

$$g_i(t, x) = \begin{cases} f_i(t, x), & (t, x) \in [0,1] \times C_i, \\ x, & (t, x) \in \{0\} \times X, \\ a_i, & (t, x) \in \{1\} \times X. \end{cases} \tag{69_3}$$

g_i is continuous (use $[0, 1] = \{0\} \cup (0, 1) \cup \{1\}$). Since X is an absolute neighbourhood retract (Appendix, **20**), g_i has a continuous prolongation h_i on a neighbourhood W_i of E_i which can be supposed closed (every metric space is normal). But [0, 1] being compact, we can find U_i a closed neighbourhood of C_i such that $[0, 1] \times U_i \subset W_i$. Using ($69_3$) one gets that h_i is a contraction of U_i in X. ■

Remark. In (C3) X can be replaced by any metric space ANR (APPENDIX, *Differential manifolds*).

Now we can pass to

Ljusternik - Schnirelmann theorem. *Let* X *be a complete connected* C^1 *- Finsler manifold*, $\mathrm{Cat}_X(X) = N$ *and* $\varphi: X \to \mathbf{R}$ *of* C^1*class lower bounded and verifying* PS. *Consider for every* n *in* **N**, $1 \le n \le N$

$$\mathscr{K}_n := \{A \subset X: A \textit{ compact}, \mathrm{Cat}_X(A) \ge n\}$$

and let be

$$c_n := \inf_{A \in \mathscr{K}_N} \sup \varphi(A).$$

If

$$c_j = c_{j+p}, \textit{ where } 1 \le j \le j + p \le N,$$

then for every min-maxing sequence $(A_m)_{m\geq 1}$ *in* $\mathcal{K}_{j+p}$ *we have*

(*) $$\mathrm{Cat}_X(K_{c_j}\cap A_\infty)\geq p+1.$$

Consequently φ *has at least* N *distinct critical points.*

Explanation. For $K_c (= K_c(\varphi))$ and A_∞ see Remark to 7.3.

■ The statement is correct, $\mathcal{K}_n\neq\varnothing$, $n=\overline{1,N}$. Use 7.6. $(PS)_c$ is verified $\forall\ c$ in **R** (§4, 4.1). We have $c_n>-\infty$, $n=\overline{1,N}$: $-\infty<\inf\varphi(X)\leq\sup\varphi(A)\ \forall\ A\in\mathcal{K}_n$ and thus $-\infty<$ $<c_1\leq c_2\leq\ldots\leq c_N<+\infty$. $\mathcal{K}_n$ is homotopy-stable ($M=\varnothing$): $\alpha\in C([0,1]\times X;X)$, $\alpha(0,x)=$ $=x\ \forall\ x\in X$ (hence in particular $\forall\ x\in A$) $\Rightarrow\alpha(\{1\}\times A)\in\mathcal{K}_n\ \forall\ A$ from $\mathcal{K}_n$, since $\alpha(\{1\}\times A)$ is compact and $\mathrm{Cat}_X\,\alpha(\{1\}\times A)\overset{(C2)}{\geq}\mathrm{Cat}_X(A)\geq n$.

First statement (*the case* p = 0).

Apply 7.6 with Remark to $\mathcal{K}_j$, with $F=\{\varphi\geq c_j\}$, correctly since $F\cap A\neq\varnothing\ \forall\ A\in$ $\in\mathcal{K}_j$ ($c_j\leq\sup\varphi(A)$, supremum attained) and obviously $\inf\varphi(f)\geq c_j$. Thus *a fortiori* $K_{c_j}\cap A_\infty\neq\varnothing$ and hence, according to the definition, we have (*). Retain in particular $K_{c_j}\neq\varnothing$, i.e. φ has at least one critical point at level c_j.

First statement (*the case* p > 0).

Suppose *par absurdum*

(69_4) $$\mathrm{Cat}_X(K_{c_j}\cap A_\infty)\leq p.$$

Using (C3) one finds an open neighbourhood U of $K_{c_j}\cap A_\infty$ such that

(69_5) $$\mathrm{Cat}_X(\overline{U})\leq p.$$

We have

(69_6) $A\in\mathcal{K}_{j+p}\Rightarrow A\setminus U\in\mathcal{K}_j : j+p\leq\mathrm{Cat}_X(A)\overset{(C1)}{\leq}\mathrm{Cat}_X(A\setminus U)+\mathrm{Cat}_X(A\cap U)\overset{(69_5)}{\leq}$ $\leq\mathrm{Cat}_X(A\setminus U)+p$.

According to (69_6) we get (69_7) $A\in\mathcal{K}_{j+p}\Rightarrow(A\setminus U)\cap\{\varphi\geq c_j\}\neq\varnothing$ (*par absurdum*) and (69_7) implies

(69_8) $$F\cap A\neq\varnothing\ \ \forall\ A\in\mathcal{K}_{j+p},\ F:=\{\varphi\geq c_{j+p}\}\setminus U\ \ (=\{\varphi\geq c_j\}\setminus U).$$

Indeed,

$$(\{\varphi\geq c_{j+p}\}\setminus U)\cup(\{\varphi\geq c_{j+p}\})\cap U)=\{\varphi\geq c_{j+p}\},$$

intersect with $A\setminus U$ and take into account (69_7), $c_{j+p}=c_j$.

Apply 7.6 (it is correct, $\varphi(F) \subset [c_{j+p}, +\infty)$),

(69₉) $$K_{c_{j+p}} \cap A_\infty \cap F \neq \emptyset.$$

But the first member of (69_9) is included in ($K_{c_j} \cap A_\infty) \setminus U$, a contradiction since U is a neighbourhood of the first term of the difference. We have been led to this contradiction by (69_4) and this imposes (∗).

Second statement.

Prove by recurrence on i, $2 \leq i \leq N$, that the critical values $c_1, c_2, \ldots, c_N$ (see the case $p = 0$) are given by at least N distinct critical points.

i = 2. We have $c_1 = c_2$ or $c_1 < c_2$. If $c_1 = c_2$ the formula (∗) gives, *via* (C1), $\mathrm{Cat}_X(K_{c_1}) \geq$ ≥ 2, hence K_{c_1} has at least two distinct points. If $c_1 < c_2$, *a fortiori* the critical values c_1 and c_2 are taken in distinct critical points.

$\leq$ i $\Rightarrow$ i + 1. We know $c_1 \leq c_2 \leq \ldots \leq c_i \leq c_{i+1}$. If $c_i < c_{i+1}$, the critical value c_{i+1} is taken in critical points distinct by all the critical points corresponding to the levels $c_1, \ldots, c_i$ and applying the hypothesis of induction we find at least $i + 1$ distinct critical points for φ. Pass to the case $c_i = c_{i+1}$ and let j be the last index for which we have (69_{10}) $c_j = c_{j+1} = \ldots = c_i = c_{i+1}$. The situation is the following:

$$c_1 \leq c_2 \leq \ldots \leq c_{j-1} < c_j = c_{j+1} = \ldots = c_i = c_{i+1}.$$

It only remains to apply to $c_1, c_2, \ldots, c_{j-1}$ the hypothesis of induction and to $c_j, c_{j+1}, \ldots, c_i, c_{i+1}$ the formula (∗) according to (69_{10}) ($c_j = c_{i+1}, p = i + 1 - j$). ■

Corollary. Any functional of C^1 *class bounded from below and verifying* (PS) *on the* N-*dimensional torus* $\mathbf{T}^N$ *has at least* N + 1 *distinct critical points.*

■ $\mathrm{Cat}_{\mathbf{T}^N}(\mathbf{T}^N) = N + 1$ (standard proposition in the algebraic topology), one applies the Ljusternik-Schnirelmann theorem. ■

MINIMAX THEOREM III

(relaxed boundary conditions)

In the following the boundary condition (F0) of 7.2 is relaxed in two ways: in one way one admit that sup $\varphi(M)$ may be equal with the critical value c and in another way M is not required compact nor $M \subset A\ \forall\ A \in \mathcal{K}$.

Precisely

Definition. Let X be topological space, M nonempty closed subset and $\mathcal{K}$ a nonempty set of compact nonempty subsets. $\mathcal{K}$ is *homotopy-stable* with *extended boundary M*, if for any α from $C([0, 1] \times X; X)$, with the property $\alpha(x, t) = x$ on $(\{0\} \times X) \cup ([0, 1] \times M)$, and for any A from $\mathcal{K}$,

(70) $$\alpha(\{1\} \times A) \in \mathcal{K}.$$

We put here the notation

(71) $$\mu(E; f) := \{x \in E: f(x) = \sup f(E)\}.$$

7.13 Theorem. *Let* X *be complete connected Finsler* C^1*-manifold*, $\varphi: X \to \mathbf{R}$ *of* C^1 *class*, $\mathcal{K}$ *homotopy-stable with extended boundary* M, $c = c(\varphi, \mathcal{K}) := \inf\limits_{A \in \mathcal{K}} \sup \varphi(A)$ *and* F *nonempty closed subset. If*

(F′1) $$(A \cap F) \setminus M \neq \emptyset \ \forall\, A \in \mathcal{K}$$

and

(F′2) $$\sup \varphi(M) \leq c \leq \inf \varphi(F),$$

then for any sequence $(A_n)_{n\geq 1}$ *from* $\mathcal{K}$ *min-maxing for* φ, *there exists a sequence* $(x_n)_{n\geq 1}$ *in* $X \setminus M$ *so that*

1° $\lim\limits_{n\to\infty} \varphi(x_n) = c$,

2° $\lim\limits_{n\to\infty} \|d\varphi(x_n)\| = 0$,

3° $\lim\limits_{n\to\infty} \text{dist}\,(x_n, F) = 0$,

4° $\lim\limits_{n\to\infty} \text{dist}\,(x_n, A_n) = 0$.

Particularly, if φ *verifies* $(PS)_{c,F}$ *along of* $(A_n)_{n\geq 1}$, *we have*

$$K_c \cap F \neq \emptyset$$

(*N. Ghoussoub*; *H. Brezis - L. Nirenberg* for $F = \{\varphi \leq c\}$).

■ Obviously (F′2) implies $c \in \mathbf{R}$. The demonstration follows step by step that for 7.4. Let be $\varepsilon > 0$ and A from $\mathcal{K}$ so that

(72) $$c \leq \sup \varphi(A) < c + \frac{\varepsilon^2}{8}.$$

So that the beginning of the demonstration for 7.2 shows, all reverts to prove that from (72) it can obtain a point x_ε in $X \setminus M$ such that

1) $c - \frac{\varepsilon^2}{8} \le \varphi(x_\varepsilon) \le c + \frac{5\varepsilon^2}{4}$,

2) $\|d\varphi(x_\varepsilon)\| \le 10\varepsilon$,

3) dist $(x_\varepsilon, F) \le \frac{3\varepsilon}{2}$,

4) dist $(x_\varepsilon, A) \le \frac{\varepsilon}{2}$.

Let be $F_\varepsilon := \{x \in X : \text{dist}(x, F) < \varepsilon\}$. Consider the subset $\mathscr{D}$ of $C([0, 1] \times X ; X)$ made with the deformations η,

(73) $$\eta(t, x) = x \text{ on } (\{0\} \times X) \cup ([0, 1] \times M_1),\ M_1 := (A \setminus F_\varepsilon) \cup M$$

and

(74) $$\sup\{\rho(\eta(t, x), x) : (t, x) \in [0, 1] \times X\} < +\infty,$$

Retain (75) $\eta \in \mathscr{D} \Rightarrow \eta(K_1) \in \mathscr{K}$, $K_1 := \{1\} \times A$.

Set on $\mathscr{D}$ the distance

(76) $$d(\eta_1, \eta_2) = \sup\{\rho(\eta_1(t, x), \eta_2(t, x)) : t \in [0, 1], x \in X\}$$

and obtain a complete metric space. Associate with φ the functions

(77) $$\varphi_1 : X \to \mathbf{R},\ \varphi_1(x) = \max\{0, \varepsilon^2 - \varepsilon\, \text{dist}(x, F)\},$$

(78) $$\varphi_2 : X \to \mathbf{R},\ \varphi_2(x) = \min\{\frac{\varepsilon^2}{8}, \text{dist}(x, M_1)\}$$

and let be (79) $\psi := \varphi_1 + \varphi_2$. Finally define $\Phi : \mathscr{D} \to \mathbf{R}$,

(80) $$\Phi(\eta) = \sup\{\psi(\eta(1, x)) : x \in A\} (= \sup \psi(\eta(K_1))),$$

a lower semicontinuous function. Let be

(81) $$\delta := \inf\{\Phi(\eta) : \eta \in \mathscr{D}\}.$$

We have $\forall\, \eta \in \mathscr{D}$, because $\eta(K_1) \overset{(75)}{\in} \mathscr{K}$ and according to $\eta(K_1) \cap F \ne \varnothing$ (see (F′1)), $\varphi_1(x) = \varepsilon^2$ on F and $\inf \varphi(F) \ge c$ ((F′2)),

$$\Phi(\eta) \ge \sup\{(\varphi + \varphi_1)(x) : x \in \eta(K_1) \cap F\} \ge c + \varepsilon^2,$$

consequently,

(82) $$\delta \ge c + \varepsilon^2.$$

Let be $\eta_1 : [0, 1] \times X \to X$, (83) $\eta_1(t, x) = x$. Obviously $\eta_1 \in \mathscr{D}$. $\Phi(\eta_1) = \sup\{(\varphi + \varphi_1 +$ $+ \varphi_2)(x) : x \in A\} \overset{(72),(77),(78)}{<} c + \frac{\varepsilon^2}{8} + \varepsilon^2 + \frac{\varepsilon^2}{8}$, $\Phi(\eta_1) < c + \frac{5\varepsilon^2}{4} \overset{(82)}{\le} \delta + \frac{\varepsilon^2}{4}$, that is

$$\Phi(\eta_1) < \inf_{\eta \in \mathcal{D}} \Phi(\eta) + \frac{\varepsilon^2}{4}.$$

Apply Ekeland principle 1.1 with $\frac{\varepsilon^2}{4}$ and $\lambda = \frac{\varepsilon}{2}$, $\exists\ \eta_0$ in $\mathcal{D}$ such that

(84) $$\Phi(\eta_0) \leq \Phi(\eta_1),$$

(85) $$d(\eta_0, \eta_1) \leq \frac{\varepsilon}{2},$$

(86) $$\Phi(\eta) \geq \Phi(\eta_0) - \frac{\varepsilon}{2}\, d(\eta, \eta_0)\ \forall\, \eta \in \mathcal{D}.$$

Let be

(87) $$B := \{x \in \eta_0(K_1) \colon \psi(x) = \sup \psi(\eta_0(K_1))\}.$$

We show that (88) $B \cap M_1 = \varnothing$, this is equivalent to $B \cap (A \setminus F_\varepsilon) = \varnothing$ and $B \cap M = \varnothing$, resulting from

(89) $$\sup \psi(A \setminus F_\varepsilon),\ \sup \psi(M) \leq \sup \psi(\eta_0(K_1)) - r,\ r > 0.$$

There exists x_0 in $\mu(\eta_0(K_1) \cap F; \varphi) \setminus M$ (see (71)) as if *par absurdum* one admits $\mu(\eta_0(K_1) \cap F; \varphi) \subset M$, one obtains a contradiction, according to (F′2), taking a point in $(\eta_0(K_1) \cap F) \setminus M$, nonempty set taking into account (F′1) since $\eta_0(K_1) \overset{(75)}{\in} \mathcal{K}$. $\sup \psi(\eta_0(K_1)) \geq \sup \psi(\eta_0(K_1) \cap F) \geq \varphi(x_0) + \varphi_1(x_0) + \varphi_2(x_0) \geq c + \varepsilon^2 + \varphi_2(x_0)$ because $\varphi_1(x_0) = \varepsilon^2$ on F,

(90) $$\sup \psi(\eta_0(K_1)) - \varphi_2(x_0) \geq c + \varepsilon^2$$

and $\varphi_2(x_0) > 0$ because $x_0 \notin M_1$ as $x_0 \notin M$ and $x_0 \in F$. On the other side, because $\varphi_1(x) = 0$ on $X \setminus F_\varepsilon$ and $\varphi_2(x) = 0$ on $A \setminus F_\varepsilon$, we have

$$\sup \psi(A \setminus F_\varepsilon) \leq \sup \varphi(A) \overset{(72)}{<} c + \frac{\varepsilon^2}{8}$$

and hence (see (90))

$$\sup \psi(\eta_0(K_1)) - \varphi_2(x_0) > \sup \psi(A \setminus F_\varepsilon),$$

that is the first inequality from (89). Moreover, because $\varphi_2(x) = 0$ on M, we have $\sup \psi(M) = \sup(\varphi + \varphi_1)(M) \overset{(F'2)}{\leq} c + \varepsilon^2$ and consequently, *via* (90), it results the second inequality from (89).

Now prove 1),3) and 4) $\forall\, x$ from B and thus it remains to demonstrate

(91) $$\exists\, x_\varepsilon \text{ in } B \text{ so that } \|d\varphi(x_\varepsilon)\| \leq 10\varepsilon.$$

Verify 1). $x_\varepsilon \in B \Rightarrow \delta \le \Phi(\eta_0) = \sup \psi(\eta_0(K_1)) = \psi(x_\varepsilon) = (\varphi + \varphi_1 + \varphi_2)(x_\varepsilon) \overset{(84)}{\le} \Phi(\eta_1) < c + \frac{5\varepsilon^2}{4}$, but $\delta \overset{(82)}{\ge} c + \varepsilon^2$ and $0 \le \varphi_1(x) \le \varepsilon^2$, $0 \le \varphi_2(x) \le \frac{\varepsilon^2}{8}$ on X, consequently

$$c + \varepsilon^2 - \varepsilon^2 - \frac{\varepsilon^2}{8} \le \varphi(x_\varepsilon) < c + \frac{5\varepsilon^2}{4},$$

that is 1).

Verify 3). $x_\varepsilon \in B \Rightarrow x_\varepsilon = \eta_0(1, a)$, $a \in A$. (88) imposes $a \in F_\varepsilon$ (if *par absurdum* $a \notin F_\varepsilon$, we have $a \in M_1$, but $a \overset{(73)}{=} \eta_0(1, a) = x_\varepsilon$ and one contradicts (88)), that is dist $(a, F) < \varepsilon$. On the other side, $\rho(x_\varepsilon, a) = \rho(\eta_0(1, a), a) \le d(\eta_0, \eta_1) \overset{(85)}{\le} \frac{\varepsilon}{2}$, consequently dist $(x_\varepsilon, F) \le \frac{3\varepsilon}{2}$.

Verify 4). Remark only that $a \in A$.

Pass to (91). One repeats the reasoning for (23). Suppose *par absurdum* $\|d\varphi(x)\| > 10\varepsilon$ $\forall\, x \in B$ and let λ be from (1, 2). Apply 7.1 to the sets M_1 and B, $\exists\, g$, α and t_0 with the required properties. Consider for each τ of $(0, t_0)$ the function η_τ: $[0,1] \times X \to X$,

$$\eta_\tau(t, x) = \alpha(t\tau, \eta_0(t, x)).$$

We have $\eta_\tau \in \mathscr{D}$ ((73) and (74) are verified *via* 1° and 2°). Because $d(\eta_\tau, \eta_0) \le \lambda\tau$, $\Phi(\eta_\tau) \overset{(86)}{\ge} \Phi(\eta_0) - \varepsilon\lambda\frac{\tau}{2}$. Let x_τ be in A such that $\Phi(\eta_\tau) = \psi(\eta_\tau(1, x_\tau))$. Then

$$\psi(\eta_\tau(1, x_\tau)) - \psi(\eta_0(1, x)) \ge -\varepsilon\lambda\frac{\tau}{2} \quad \forall\, x \in A.$$

Because the Lipschitz constants of φ_1 and φ_2 are less than ε, we have (see (79))

$$\varphi(\eta_\tau(1, x_\tau)) - \varphi(\eta_0(1, x_\tau)) \ge -5\varepsilon\lambda\frac{\tau}{2}.$$

But

$$\varphi(\eta_\tau(1, x_\tau)) - \varphi(\eta_0(1, x_\tau)) \overset{7.1}{\le} -5\varepsilon\tau g(\eta_0(1, x_\tau)),$$

consequently

(92) $$-5\varepsilon\frac{\lambda}{2} \le -5\varepsilon g(\eta_0(1, x_\tau)).$$

Exactly as in the mentioned proof one obtains a point x_0 in A for which $g(\eta_0(1, x_0)) = 1$ and (92) remains true when one replaces x_τ by x_0, which imposes $\lambda \geq 2$, contradiction and therefore (91). ■

Remark. When $\mathcal{K}$ is homotopy-stable with *boundary* M, the conditions (F′1) and (F′2) are implied by the conditions (F1) and (F2) (see 7.4): $F \cap M = \varnothing$ and $F \cap A \neq \varnothing \Rightarrow (A \cap F) \setminus M \neq \varnothing$ and for $\sup \varphi(M) \leq c$ remark that $M \subset A\ \forall\ A \in \mathcal{K}$.

Remark also that if (F′2) is verified, then (F′1) is equivalent to

$$\mu(A \cap F\,;\varphi) \setminus M \neq \varnothing\ \forall\ A \in \mathcal{K},$$

which has already been used in the proof.

7.14 *Corollary. Let* X, φ, $\mathcal{K}$, M *and* $c \in \mathbf{R}$ *be as in* 7.13 *and* $(A_n)_{n\geq 1}$ *a min-maxing sequence for* φ *along which is verified* $(PS)_c$. *If*

(F′0) $$(A \cap \{x \in X: \varphi(x) \geq c\}) \setminus M \neq \varnothing\ \forall\ A \in \mathcal{K}$$

and

(F″0) $$\sup \varphi(M) \leq c,$$

then

$$K_c \neq \varnothing$$

(*N. Ghoussoub*).

■ Apply 7.13 with $F = \{\varphi \geq c\}$, as (F′0) and (F″0) imply (F′1) and (F′2), $K_c = K_c \cap F \neq$ $\neq \varnothing$. ■

7.15 *Corollary. Let* X, φ, $\mathcal{K}$, M *and* $c \in \mathbf{R}$ *be as in* 7.13 *and* F *nonempty closed set that satisfies* (F′1) *and*

(F3) $$\sup \varphi(M) \leq \inf \varphi(F).$$

If φ *verifies* $(PS)_c$, *then*

$$K_c \neq \varnothing$$

(*N. Ghoussoub*).

■ (F′1) implies $\inf \varphi(F) \leq c$, because $\inf \varphi(F) \leq \inf \varphi(A \cap F) \setminus M) \leq \sup \varphi((A \cap F) \setminus$ $\setminus M) \leq \sup \varphi(A)\ \forall\ A \in \mathcal{K}$. (F3) imposes:

$$\text{either (93) } \sup \varphi(M) < c, \text{ or (94) } \sup \varphi(M) = \inf \varphi(F) = c.$$

In both cases (F′0) and (F″0) are verified. Indeed, the statement relative to (F″0) being evident, pass to (F′0). In the case (94), we have $\varphi(x) \geq c\ \forall\ x \in F$, therefore $F \subset \{\varphi \geq c\}$ and then (F′1) $\Rightarrow$ (F′0). In the case (93), (F′0) equalizes (95) $A \cap \{\varphi \geq c\} \neq \varnothing\ \forall\ A$.

Suppose *par absurdum* $\exists A_1$ in $\mathscr{K}$ so that $A_1 \cap \{\varphi \geq c\} = \varnothing$, that is $\varphi(x) < c \ \forall x \in A_1$. This implies sup $\varphi(A_1) < c$ (A_1 is compact), contradiction, therefore (95). So being, one applies 7.14, $K_c \neq \varnothing$. ∎

Remark. In case (94) we have even $K_c \cap F \neq \varnothing$ (apply 7.13, (F′2) is verified *via* (94) and (F′1) *via* the hypothesis).

Continue with

RELAXED MOUNTAIN PASS AND SADDLE POINT THEOREMS

7.16 *Let* X *be complete connected Finsler* C^1*-manifold,* $\varphi: X \to \mathbf{R}$ *of* C^1 *class,* K *compact space,* K_0 *nonempty closed subset of* K, $\gamma_0 : K_0 \to X$ *continuous and*

$$\Gamma := \{\gamma \in C(K; X): \gamma|K_0 = \gamma_0\}.$$

Let be

$$c := \inf_{\gamma \in \Gamma} \sup \varphi(\gamma(K)) \in \mathbf{R}$$

and F *nonempty closed subset of* X *which satisfies*

$$1^\circ \ F \cap \gamma(K \setminus K_0) \neq \varnothing \ \forall \gamma \in \Gamma$$

and

$$2^\circ \ \sup \varphi(\gamma_0(K_0)) \leq \inf \varphi(F).$$

If φ *verifies* $(PS)_c$, *then*

$$K_c \neq \varnothing.$$

Moreover, if $c = \inf \varphi(F)$, *then*

$$K_c \cap F \neq \varnothing$$

(*N. Ghoussoub*).

∎ Consider the set

$$\mathscr{K} := \{\gamma(K): \gamma \in \Gamma\}.$$

$\mathscr{K}$ is homotopy-stable with boundary $M := \gamma_0(K_0)$ (this is an extended boundary also), one applies 7.15 with the remark : condition 1° is condition (F′1) relative to $\mathscr{K}$ and condition 2° is condition (F3). ∎

In the following, until the end of this section, X is **real Banach space**.

Let u_0, u_1 be distinct points from X and

$$\mathcal{K}_{u_0}^{u_1} := \{\gamma([0, 1]) : \gamma \in C([0, 1]; X), \gamma(0) = u_0, \gamma(1) = u_1\}.$$

$\mathcal{K}_{u_0}^{u_1}$ is homotopy-stable with boundary $M := \{u_0, u_1\}$. Any sphere $\sigma_r(u_0)$ $(= \{x \in X : \|x - u_0\| = r\})$ with $r < \|u_0 - u_1\|$ is dual of $\mathcal{K}_{u_0}^{u_1}$ (above 7.5).

7.17 *Let* X *be Banach space,* $\varphi : X \to \mathbf{R}$ *of* C^1*-Fréchet class,* u_0, u_1 *distinct points from* X *and* $a \in \mathbf{R}$. *Suppose*

1° *There exists* r, $0 < r < \|u_0 - u_1\|$ *so that* $\inf \varphi(\sigma_r(u_0)) \geq a$

and

2° $\max\{\varphi(u_0), \varphi(u_1)\} \leq a$.

Then $c := c(\varphi, \mathcal{K}_{u_0}^{u_1}) \geq a$ *and if* φ *verifies* $(PS)_c$, c *is critical value for* φ. *Moreover, if* $c = a$, *we have*

$$K_c \cap \sigma_r(u_0) \neq \varnothing$$

(*P. H. Rabinowitz, N. Ghoussoub*).

■ We have $\forall\, \gamma([0, 1])$ from $\mathcal{K}_{u_0}^{u_1}$

$$(95_1) \qquad a \leq \inf \varphi(\sigma_r(u_0)) \leq \inf \varphi(\sigma_r(u_0) \cap \gamma([0, 1])) \leq \sup \varphi(\gamma([0, 1])),$$

therefore $a \leq c$. Obviously, $c \in \mathbf{R}$. One applies 7.15 with $\mathcal{K} = \mathcal{K}_{u_0}^{u_1}$, $F = \sigma_r(u_0)$. (F′1) and (F3) are verified (see 1° and 2°), c is critical value for φ.

Suppose $c = a$. Then $c \leq \inf \varphi(\sigma_r(u_0))$, but $\inf \varphi(\sigma_r(u_0)) \leq c$ ((95_1)), therefore $c = \inf \varphi(\sigma_r(u_0))$ and (F′2) is verified, $K_c \cap F \neq \varnothing$ (7.13). ■

Suppose now

$$X = Y \oplus Z, \text{ direct sum, } \dim Y < +\infty$$

and let U be nonempty open bounded subset of Y. $\overline{U}$ is compact set included in Y. Consider the sets

$$\Gamma := \{\gamma \in C(\overline{U}; X) : \gamma(x) = x \text{ on } \mathrm{Fr}\, U\}$$

(correct definition, Fr $U \neq \varnothing$),

$$\mathcal{K}_U := \{\gamma(\overline{U}) : \gamma \in \Gamma\}.$$

$\mathcal{K}_U$ is homotopy-stable with boundary (compact) Fr U. Z is dual of $\mathcal{K}_U$. Indeed, γ be any of Γ and π the projection of X on Y (direct sum!). The application $\pi \circ \gamma : \overline{U} \to Y$ is semicontinuous and equal to the identity on Fr U, one applies the Brouwer theorem (see

Note above 7.4), $\exists\, x'$ in $\overline{U}$ so that $\pi(\gamma(x')) = 0$, therefore $\gamma(x') \in Z$, $Z \cap \gamma(\overline{U}) \neq \varnothing$. Moreover $Z \cap \mathrm{Fr}\, U = \varnothing$. ■

And now

7.18 *Let* X *be Banach space,* $X = Y \oplus Z$, $\dim Y < +\infty$ *and* $\varphi: X \to \mathbf{R}$ *of* C^1*-Fréchet class. Suppose there is* $a \in \mathbf{R}$ *so that*

$$1^\circ \inf \varphi(Z) \geq a$$

and

$$2^\circ \sup \varphi(\mathrm{Fr}\, U) \leq a,\ U \subset Y \textit{ nonempty open bounded.}$$

Then $c := c(\varphi, \mathscr{K}_U) \geq a$ *and if* φ *verifies* $(PS)_c$, c *is critical value for* φ. *Moreover, if* $c = a$, *we have*

$$K_c \cap Z \neq \varnothing$$

(*P. H. Rabinowitz, N. Ghoussoub*).

■ We have $c \geq a$ because, $\forall\, \gamma$ of Γ, $\sup \varphi(\gamma(\overline{U})) \geq a$ as $\gamma(\overline{U}) \cap Z \neq \varnothing$ and $\inf \varphi(Z) \geq$ $\geq a$. Apply 7.15 taking $\mathscr{K} = \mathscr{K}_U$, $M = \mathrm{Fr}\, U$, $F = Z$. (F′1) and (F3) are verified, c is critical value. In the case $c = a$, (F′2) is also verified, $K_c \cap Z \neq \varnothing$. ■

Since in the following the Brouwer degree theory will be used, for the topological boundary of a set E we use ∂E and not Fr E.

7.19 *Let* X *be Banach space,* $X = Y \oplus Z$, $\dim Y < +\infty$ *and* $\varphi: X \to \mathbf{R}$ *of* C^1*-Fréchet class which verifies* PS. *Suppose that for* z_0 *in* Z, $\|z_0\| = 1$ *there is* $\gamma_0: \partial K \to X$ *continuous,* $K \subset X \oplus \mathbf{R}z_0$, $K = \{u = y + tz_0 : y \in Y, t \geq 0, \|u\| \leq R\}$ *with the properties* :

1° $\gamma_0(y) = y$ *for* $y \in Y$, $\|y\| \leq R$ *and* $\|\gamma_0(u)\| \geq r$, $r > 0$ *for* $u \in K$, $\|u\| = R$;

2° $\varphi(\gamma_0(u)) \leq a$, $a \in \mathbf{R}\ \forall\, u \in \partial K$.

Suppose also that

3° $\exists\, \rho$, $0 < \rho < r$ *so that* $\inf \varphi(Z \cap S_\rho(0)) \geq a$.

Then φ *has a critical point at a level* c *and* $c \geq a$. *Besides, if* $c > a$ *then* K_c *contains one point which is not of local minimum and if* $c = a$ *then*

$$K_c \cap Z \cap S_\rho(0) \neq \varnothing.$$

In particular, if $\varphi(0) = a$ *then* φ *has a non zero critical point*

(*P. H. Rabinowitz, H. Brezis - L. Nirenberg*).

■ The statement is coherent, $Z \cap S_\rho(0) \neq \varnothing$. K is compact being closed and bounded and $\dim(Y \oplus \mathbf{R}z_0) < +\infty$. Let be

$$\Gamma := \{\gamma \in C(K;X) : \gamma|\partial K = \gamma_0\},\ \mathcal{K} := \{\gamma(K) : \gamma \in \Gamma\}.$$

$c := \inf\limits_{\gamma \in \Gamma} \sup \varphi(\gamma(K)) \in \mathbf{R}$, because $c > -\infty$ as $\sup \varphi(\gamma_0(\partial K)) \le \sup \varphi(\gamma(K))\ \forall\ \gamma \in \Gamma$.

To apply 7.16 ($\Gamma = \{\gamma \in C(K;X) : \gamma|K_0 = \gamma_0\}$, $\mathcal{K} = \{\gamma(K) : \gamma \in \Gamma\}$) with $K_0 = \partial K$ and $F = Z \cap S_\rho(0)$, we must prove

(96) $$F \text{ is dual of } \mathcal{K}.$$

(96) gives $a \le \inf \varphi(F) \le \inf(\varphi(F \cap A)) \le \sup \varphi(A)\ \forall\ A \in \mathcal{K}$, therefore (97) $a \le c$.

Pass to prove (96). Use the Brouwer degree. We can suppose $R = 1$. Let be $\gamma : K \to X$ continuous verifying

(98) $$\begin{aligned} &y \in Y \text{ and } \| y \| \le 1 \Rightarrow \gamma(y) = y \\ &u \in K \text{ and } \| u \| = 1 \Rightarrow \| \gamma(u) \| \ge r. \end{aligned}$$

Note π the projection of Y along of Z. If we prove

(99) $$\exists\ u_0 \in K \text{ so that } \pi(\gamma(u_0)) = 0 \text{ and } \|\gamma(u_0)\| = \rho,$$

then (96) is obviously verified.

Let be $X_0 := Y \oplus \mathbf{R}z_0$. Consider the application $T : K \to X_0$,

(100) $$T(u) = \pi(\gamma(u)) + \|(\mathrm{Id}_X - \pi) \circ \gamma(u)\| z_0$$

(Id_X the identical application of X).

For (99) it is enough to find u_0 in K with

(101) $$T(u_0) = \rho z_0.$$

Use $\deg(T, \overset{\circ}{K}, \rho z_0)$ (see *Note*, above 7.4), correctly considered as $\rho z_0 \notin T(\partial K)$ (see below, $\rho < r$), showing that

(102) $$\deg(T, \overset{\circ}{K}, \rho z_0) = 1$$

and consequently (101). Obviously it can consider only $T|\partial K$. $\partial K = A \cup B$, $A := \{y + 0 : y \in Y, \|y\| \le 1\}$, $B := \{u \in K : \|u\| = 1\}$. One has $T(u) = u$ on A and $\|T(u)\| \ge r$ on B. Define on ∂K

$$\widetilde{T}(u) = \begin{cases} u, & u \in A \\ \dfrac{T(u)}{\| T(u) \|}, & u \in B. \end{cases}$$

We have $\widetilde{T}(B) \subset B$ and $\widetilde{T} = \mathrm{Id}_X$ on ∂B. But T and $\widetilde{T}$ are homotopic in $X_0 \setminus \{\rho z_0\}$ through

$$T_t u = tT(u) + (1 - t)\widetilde{T}(u),\ t \in [0, 1]$$

(use (98)) and there is, $\forall\, t$ of [0, 1], $\widetilde{T}_t$ which ties $\widetilde{T}$ with the identical application of B having $\widetilde{T}_t = \mathrm{Id}_X$ on ∂B. Consequently $T|\partial K$ is homotopic with Id_X in $X_0 \setminus \{\rho z_0\}$ and thus $\deg(T, \overset{o}{K}, \rho z_0) = \deg(\mathrm{Id}_X, \overset{o}{K}, \rho z_0) = 1$, which imposes (102).

The rest of the proof is similar to what we have already worked many times. ■

CRITICAL ORBITS OF THE REAL PERIODIC FUNCTIONALS

This section finishes the paragraph 7. Here the Ljusternik - Schnirelmann theorem and the considerations on the category with the same name, made at the end of the second section, are used We work on periodic functionals of C^1-Fréchet class.

Definitions. Let X be real normed space, and G subgroup of X. G is *discrete* when

(103) $$r := \inf\{\|x\| : x \in G \setminus \{0\}\} > 0.$$

$f: X \to \mathbf{R}$ is G-*invariant* when

(104) $$f(x + u) = f(x)\ \forall\, x \in X, \forall\, u \in G.$$

In this case, if f is Fréchet differentiable, then f' is also G-invariant. Consequently, if x_0 is critical point of f, that is $f'(x_0) = 0$, then all the points of $\pi^{-1}(\pi(x_0))$ are critical points for f, where

(105) $$\pi: X \to X/G$$

is the canonical surjection ($\pi(x) = x + G$). Indeed, $x \in \pi^{-1}(\pi(x_0)) \Leftrightarrow \pi(x) = \pi(x_0) \Leftrightarrow x - x_0 \in G$.

$\pi^{-1}(\pi(x_0))$ is by definition a *critical orbit* of f.

Retain that (103) implies

7.20 *All the points of the discrete group* G *are isolated and every convergent sequence from* G *is constant from a rank on.*

■ *First affirmation*. Let be $r \overset{(103)}{=} \inf_{x \in G \setminus \{0\}} \|x\|$ and x_0 arbitrary fixed from G. For any x in $G \setminus \{0\}$ we have $\|x - x_0\| \geq r$, that is x does not penetrate in the open ball centred in x_0 with r radius.

Second affirmation. Let $(x_n)_{n\geq 1}$ be Cauchy sequence in G and ε in $(0, r)$, r from (103). There is a rank N from that $\|x_m - x_N\| \leq \varepsilon$, therefore $m \geq N \Rightarrow x_m = x_n$. ■

Let X be Banach space and G discrete subgroup. Taking

(106) $$d(\pi(x), \pi(y)) = \inf_{u\in G} \|x - y - u\|$$

one obtains a distance on X/G. Indeed, firstly the definition is correct: $\pi(x) = \pi(x')$ and $\pi(y) = \pi(y') \Rightarrow \inf_{u\in G} \|x - y - u\| = \inf_{u\in G} \|x' - y' - u\|$ because $x - x' \in G \Rightarrow x = x' + g$, $g \in G$ and $y - y' \in G \Rightarrow y = y' + h$, $h \in G$ and $g - h - u$ goes over G if u goes over G. Now note

(107) $$d(\pi(x), \pi(y)) = d(\pi(x - y), 0) = d(\pi(x) - \pi(y), 0).$$

Verify the axioms of the distance. $d(\pi(x), \pi(x)) = \inf_{u\in G} \|u\| = 0$; $d(\pi(x), \pi(y)) = 0 \Rightarrow$ $\Rightarrow \inf_{u\in G} \|x - y - u\| = 0$, let $(\varepsilon_n)_{n\geq 1}$ be a sequence of strictly positive numbers, $\varepsilon_n \to 0$ and $(u_n)_{n\geq 1}$ the corresponding sequence in G with $\|x - y - u_n\| \leq \varepsilon_n$, consequently $u_n \to x - y$ and according to 7.20 it results $x - y \in G$, that is $\pi(x) = \pi(y)$. The symmetry is obvious and as to the triangle inequality: $\|x - z - (u + v)\| \leq \|x - y - u\| + \|y - z - v\|$ and $\inf_{u,v\in G} \|x - z - (u + v)\| = \inf_{w\in G} \|x - z - w\|$.

We emphasize also the property:

(108) $$\exists\, y \text{ in } \pi(x) \text{ so that } d(\pi(x), 0) \geq \frac{1}{2} \|y\|,$$

because if *par absurdum* we suppose the contrary, putting y in the shape $y = x - v$, $v \in G$, we have $\inf_{u\in G} \|x - u\| < \frac{1}{2} \|x - v\|$ $\forall\, v \in G$, hence $\inf_{u\in G} \|x - u\| \leq \frac{1}{2} \inf_{v\in G} \|x - v\|$, contradiction (it was admitted $\inf_{u\in G} \|x - u\| > 0$, otherwise (108) is verified with $y = 0$ since $\pi(x) = 0$).

7.21 *The canonical surjection π is continuous.*

■ Let be $x_n \to x_0$. $d(\pi(x_n), \pi(x_0)) = \inf_{u\in G} \|x_n - x_0 - u\| \leq \|x_n - x_0\|$, take the limit for $n \to \infty$. ■

Definition. The series $\sum_{n=1}^{\infty} \pi(x_n)$ from X/G is *convergent* if the sequence $\left(\sum_{k=1}^{n} \pi(x_k)\right)_{n\geq 1}$ is convergent, in which case

$$\sum_{n=1}^{\infty} \pi(x_n) \overset{\text{def}}{=} \lim_{n\to\infty} \sum_{k=1}^{n} \pi(x_k).$$

One deduces from 7.21

(109) $$x = \sum_{n=1}^{\infty} x_n \Rightarrow \pi(x) = \sum_{n=1}^{\infty} \pi(x_n),$$

because $x = \lim_{n\to\infty} \sum_{k=1}^{n} x_k \Rightarrow \pi(x) = \lim_{n\to\infty} \pi(\sum_{k=1}^{n} x_k) = \lim_{n\to\infty} \sum_{k=1}^{n} \pi(x_k)$.

And now

7.22 *The metric space* (X/G, d) *is complete.*

■ Let $(\pi(x_n))_{n\geq 1}$ be Cauchy sequence from X/G. Rarefying this sequence, we can make so that (110) $\sum_{n=1}^{\infty} d(\pi(x_{n+1}), \pi(x_n))$ be convergent (similar construction is made in the demonstration for 7.23). Let be, for n any in $\mathbf{N}$, y_n in X so that (111) $\pi(x_{n+1}) - \pi(x_n) = \pi(y_n)$ and d $(\pi(x_{n+1}), \pi(x_n)) \geq \frac{1}{2} \|y_n\|$ (see (107) and (108)). The series $\sum_{n=1}^{\infty} \|y_n\|$ is thus convergent ((110)) and also $\sum_{n=1}^{\infty} y_n$ because X is complete, let be $y = \sum_{n=1}^{\infty} y_n$. Then $\pi(y) \overset{(109)}{=} \sum_{n=1}^{\infty} \pi(y_n) \overset{(111)}{=} \sum_{n=1}^{\infty} [\pi(x_{n+1}) - \pi(x_n)] = \lim_{n\to\infty} \pi(x_n)$, X/G is complete. ■

Let X be Banach space and G discrete subgroup of X. Suppose that the vector subspace V generated by G has finite dimension N. Then X is isomorphic to $\mathbf{R}^N \times W$, where $X = V \oplus W$, G is isomorphic with $\mathbf{Z}^N$ and $\pi(X)$ is isomorphic to $\mathbf{T}^N \times W$, $\mathbf{T}^N$ the N-torus ($= S^1 \times S^1 \times \ldots \times S^1$, N factors, $S^1 := \{x \in \mathbf{R}^2 : \|x\| = 1\}$). So, if $A := [0,1]^N \times \{0\}$, we have $\mathrm{Cat}_{\pi(X)}(\pi(A)) = \mathrm{Cat}_{\mathbf{T}^N \times W}(\mathbf{T}^N \times \{0\})$, but the second member is equal to $\mathrm{Cat}_{\mathbf{T}^N}(\mathbf{T}^N)$ (see (C1), the end of the second section) and $\mathrm{Cat}_{\mathbf{T}^N}(\mathbf{T}^N) = N + 1$ (loc. cit.), hence

(112) $$\mathrm{Cat}_{\pi(X)}(\pi(A)) = N + 1.$$

With (112) one concludes

(113) $$1 \leq i \leq N+1 \Rightarrow \mathscr{A}_i := \{A \subset X : A \text{ is compact, } \mathrm{Cat}_{\pi(X)}(\pi(A)) \geq i\} \neq \varnothing.$$

A fortiori, any set from $\mathscr{A}_i$ is nonempty. In addition

$$(113_1) \qquad \pi(X) \text{ is an ANR,}$$

because $\pi(X)$ is isomorphic to $(S^1)^N \times W$ and one takes into account **3** (3°,4°,5°), ANNEX.

It can assert

7.23 *The set* $\mathscr{A}_i$, $1 \le i \le N+1$, *with the Hausdorff distance*

$$(114) \qquad \delta(A,B) = \max\{\sup_{a\in A} \text{dist}(a, B), \sup_{b\in B} \text{dist}(b, A)\},$$

is complete metric space

(*A. Szulkin*, [68]).

■ Consider the set

$$\mathscr{M} := \{A \subset X : A \text{ nonempty closed bounded}\}.$$

The Hausdorff distance is distance on $\mathscr{M}$: $\delta(A, B) = 0 \Rightarrow \text{dist}(a, B) = 0 \ \forall\, a \in A$ and $\text{dist}(b, A) = 0 \ \forall\, b \in B$, therefore $A \subset B$ and $B \subset A$, $A = B$; $\forall\, a \in A$, $b \in B$, $c \in C$ we have $d(a, c) \le d(a, b) + d(b, c)$, therefore $d(a, c) \le \text{dist}(a, B) + \text{dist}(c, B) \le \sup_{a\in A} \text{dist}(a, B) + \sup_{c\in C} \text{dist}(c, B) \le \delta(A, B) + \delta(B, C)$, $d(a, c) \le \delta(A, B) + \delta(B, C) \ \forall\, a \in A, \forall\, c \in C$, it results $\text{dist}(a, C), \text{dist}(c, A) \le \delta(A, B) + \delta(B, C) \ \forall\, a \in A, \forall\, c \in C$ and so $\delta(A, C) \le$ $\le \delta(A, B) + \delta(B, C)$.

Prove

$$(116) \qquad \text{The metric space } (\mathscr{M}, \delta) \text{ is complete.}$$

Let $(A_n)_{n\ge1}$ be Cauchy sequence in $(\mathscr{M}, \delta)$. Consider the closed set

$$(117) \qquad A_0 := \bigcap_{n=1}^{\infty} \overline{\bigcup_{m=n}^{\infty} A_m}\,.$$

Let $\varepsilon > 0$ be anyone, $\exists\, N_\varepsilon$ in **N** so that

$$(118) \qquad m, n \ge N_\varepsilon \Rightarrow \delta(A_m, A_n) \le \varepsilon.$$

To obtain (116), it remains to prove

$$(119) \qquad \delta(A_0, A_{N_\varepsilon}) \le \varepsilon,$$

because then, according to (118) and (119), $n \geq N_\varepsilon \Rightarrow \delta(A_0, A_n) \overset{(117)}{\leq} 2\varepsilon$. But $A_0 \subset \overline{\bigcup_{m=N_\varepsilon}^{\infty} A_m}$ and this implies

(120) $$\sup_{a\in A_0} \text{dist}\,(a, A_{N_\varepsilon}) \leq \varepsilon.$$

Indeed, let a be anyone from A_0, take $(x_p)_{p\geq 1}$, $x_p \in A_{m_p}$ with $m_p \geq N_\varepsilon$ and $x_p \to a$, we have $\text{dist}\,(a, A_{N_\varepsilon}) \leq \|a - x_p\| + \text{dist}\,(x_p, A_{N_\varepsilon}) \overset{(118)}{\leq} \|a - x_p\| + \varepsilon$, pass to the limit for $p \to \infty$. According to (120) one deduces that A_0 is bounded. On the other side we also have

(121) $$\sup_{b\in A_{N_\varepsilon}} \text{dist}\,(b, A_0) \leq \varepsilon.$$

Indeed, let b be anyone from A_{N_ε}. $(A_n)_{n\geq 1}$ being Cauchy sequence, this permits building a sequence $(b_{k_n})_{n\geq 1}$ with $(k_n)_{n\geq 1}$ strictly increasing, $b_{k_n} \in A_{k_n}$ and $\| b_{k_n} - b_{k_{n-1}} \| \leq \frac{\varepsilon}{2^{n-1}}$ $\forall\, n \geq 1$, where $b_{k_0} = b$. Here it is the construction: for $\frac{\varepsilon}{2}$ $\exists\, N_1$, $N_1 > N_\varepsilon$, from that $\delta(A_m, A_n) \leq \frac{\varepsilon}{2}$, for $\frac{\varepsilon}{2^2}$ $\exists\, N_2$, $N_2 > N_1$, from that $\delta(A_m, A_n) \leq \frac{\varepsilon}{2^2}$, ... , for $\frac{\varepsilon}{2^n}$ $\exists\, N_n$, $N_n > N_{n-1}$, from that $\delta(A_m, A_n) \leq \frac{\varepsilon}{2^n}$ etc.; take $k_1 > N_1$, $\exists\, b_{k_1}$ in A_{k_1} so that $\| b_{k_1} - b\| \leq \varepsilon$ (*par absurdum*, A_{k_1} is compact, one contradicts (118)), take $k_2 > N_2$, $\exists\, b_{k_2}$ in A_{k_2} so that $\| b_{k_1} - b_{k_2} \| \leq \frac{\varepsilon}{2}$ etc.

We continue. $(b_{k_n})_{n\geq 1}$ being Cauchy sequence, $\exists\, a_0$ in X so that $b_{k_n} \to a_0$. Obviously $a_0 \overset{(117)}{\in} A_0$ ((k_n) is strictly increasing !) and so A_0 is nonempty, bounded and closed, i.e. $A_0 \in \mathscr{M}$. Moreover, as $\| b_{k_n} - b_{k_0} \| \leq \varepsilon$ $\forall\, n \geq 1$, it results $\|a_0 - b_{k_0}\| = \|a_0 - b\| \leq \varepsilon$ and consequently (121). (120) and (121) corroborate to obtain (119).

Prove now

(122) $$\mathscr{K} := \{A \in \mathscr{M} : A \text{ is compact}\} \text{ is closed.}$$

Let A be from $\overline{\mathcal{K}}$, $\exists\ (A_n)_{n\geq 1}$ a sequence from $\mathcal{K}$ with (123) $\lim\limits_{n\to\infty} \delta\ (A_n, A) = 0$. The subspace A being complete (X is Banach space), it is enough to show that A is precompact to have $A \in \mathcal{K}$, i.e. for any $r > 0$, A may be covered by a finite number of open balls of radius r. But $\exists\ A_p$ so that $\delta(A_p, A) < \frac{r}{2}$ and A_p being compact, this may be covered by a finite number of open balls of radius $\frac{r}{2}$ and therefore the conclusion.

According to (122) it remains to prove $\mathcal{A}_i = \overline{\mathcal{A}_i}$, $i = \overline{1, N+1}$. But as $\mathcal{A}_i \subset \mathcal{K}$ and therefore $\overline{\mathcal{A}_i} \subset \mathcal{K}$, it remains to show (124) $A \in \mathcal{K}$ and $\lim\limits_{n\to\infty} \delta\ (A, A_n) = 0$, $(A_n)_{n\geq 1}$ a sequence from $\mathcal{A}_i \Rightarrow A \in \mathcal{A}_i$. Because $\pi(X)$ is an ANR ((113$_1$)), $\exists\ U$ a closed neighbourhood of $\pi(A)$ so that (see (C3) with Remark, the end of the second section)

$$\text{(125)} \qquad \mathrm{Cat}_{\pi(X)}(\pi(A)) = \mathrm{Cat}_{\pi(X)}(U).$$

But $\pi^{-1}(U)$ is closed neighbourhood of the compact set A, hence $\exists\ n_0 \in \mathbf{N}$ so that $A_{n_0} \subset \subset \pi^{-1}(U)$ ($\delta(A, A_n) \to 0$!), i.e. $\pi(A_{n_0}) \subset U$ and consequently

$$\text{(126)} \qquad \mathrm{Cat}_{\pi(X)}(U) \geq \mathrm{Cat}_{\pi(X)}(\pi(A_{n_0})) \overset{(117)}{\geq} i.$$

Taking into account (125) and (126) one finds $A \in \mathcal{A}_i$, i.e. (124). ■

We can still state

7.24 *Let be* $\varphi : X \to \mathbf{R}$ *continuous. The function* $F : \mathcal{A}_i \to \mathbf{R}$, $1 \leq i \leq N+1$,

$$\text{(127)} \qquad F(A) = \sup_{x\in A} \varphi(x)$$

is lower semicontinuous ([68]).

■ Let A be fixed anyone from $\mathcal{A}_i$ and $(A_n)_{n\geq 1}$ a sequence in $\mathcal{A}_i$ with $A_n \to A$, i.e. (128) $\lim\limits_{n\to\infty} \delta(A, A_n) = 0$. We have to prove

$$\text{(129)} \qquad F(A) \leq \varliminf_{n\to\infty} F(A_n).$$

Let a be anyone from A. We have

$$\mathrm{dist}\ (a, A_n) \leq \sup_{x\in A} \mathrm{dist}\ (x, A_n) \leq \delta(A, A_n),$$

consequently $\lim_{n\to\infty} \text{dist}(a, A_n) \overset{(128)}{=} 0$. A_n being compact, $\exists\, a_n$ in A_n so that dist $(a, A_n) =$ $= \|a_n - a\|$, therefore we found $(a_n)_{n\geq 1}$ sequence from X with $a_n \in A_n$ $\forall\, n$ and $a_n \to a$. We then have $\varphi(a) = \lim_{n\to\infty} \varphi(a_n)$, but $\varphi(a_n) \overset{(127)}{\leq} F(a_n)$, so $\overline{\lim_{n\to\infty}}\, \varphi(a_n) \leq \overline{\lim_{n\to\infty}}\, F(A_n)$ and hence (130) $\varphi(a) \leq \overline{\lim_{n\to\infty}}\, F(A_n)$. a being arbitrary in A, (130) supplies (129). ■

And now the main statement of this section, the notations are the same as above.

7.25 Theorem. *Let* X *be Banach space,* G *discrete subgroup and* $\varphi : X \to \mathbf{R}$ *of* C^1 *- Fréchet class,* G*-invariant and satisfying the* $(PS)_G$ *condition. If* φ *is bounded from below and the dimension* N *of the vector subspace generated by* G *is finite, then* φ *has at least* N + 1 *distinct critical orbits*

(D. Lupo - S. Solimini, R. Palais, J. Mawhin - M. Willem).

Explanation. φ *verifies* $(PS)_G$ means:

For any sequence $(x_n)_{n\geq 1}$ from X with $(\varphi(x_n))_{n\geq 1}$ bounded and $\varphi'(x_n) \to 0$, the sequence $(\pi(x_n))_{n\geq 1}$ has a convergent subsequence.

■ Consider for any i, $1 \leq i \leq N+1$,

(131) $$c_i := \inf_{A\in\mathcal{A}_i} \sup \varphi(A).$$

Because $\mathcal{A}_{i+1} \subset \mathcal{A}_i$, we have the sequence

$$-\infty < c_1 \leq c_2 \leq \ldots \leq c_N \leq c_{N+1} < +\infty.$$

For the assertion of the statement it is enough to prove (see the demonstration of Ljusternik - Schnirelmann theorem, the end of the second section) that if

$$c_i = c_k, \text{ where } 1 \leq i \leq k \leq N+1$$

then

(132) $$K_{c_i} := \{x \in X : \varphi(x) = c, \varphi'(c_i) = 0\}$$

contains at least k – i + 1 *distinct critical orbits.*

Case i < k.

Set $c := c_i = c_k$ and suppose *par absurdum* that in K_c exist only n, (∗) $n \leq k - i$, critical orbits

$$\pi^{-1}(\pi(a_1)), \pi^{-1}(\pi(a_2)), \ \ldots, \pi^{-1}(\pi(a_n)).$$

Take $\rho > 0$ so that the restriction of π to $S_{2\rho}(a_m) := \{x \in X : \|x - a_m\| \le 2\rho\}$, $1 \le m \le n$, is injective (ρ exists for any x from X, on the contrary $\forall\, n \ge 1$ there are two different points x_n', x_n'' in $S_{\frac{1}{n}}(x)$ with $x_n' - x_n'' \in G$, then $\| x_n' - x_n'' \| \le \frac{2}{n}$, hence $x_n' - x_n'' \to 0$ and so from a rank on the sequence is constant (7.20), contradiction).

Consider the set (open balls !)

$$\text{(133)} \qquad M_\rho := \bigcup_{m=1}^{n} \bigcup_{u \in G} S(a_m + u, \rho).$$

$\pi|S_{2\rho}(a_m + u)$ is also injective: if $a_m + u + x' \ne a_m + u + x''$, where $\|x'\|, \|x''\| \le 2\rho$, then $\varphi(a_m + u + x') = \varphi(a_m + x') \ne \varphi(a_m + x'') = \varphi(a_m + u + x'')$, as $a_m + x', a_m + x'' \in S_{2\rho}(a_m)$.

One asserts

$$\text{(134)} \qquad \exists\, \varepsilon \text{ in } (0, \rho^2) \text{ so that } x \in \varphi^{-1}([c - \varepsilon, c + \varepsilon]) \setminus M_\rho \Rightarrow \|\varphi'(x)\| > \sqrt{\varepsilon}\,.$$

Suppose *par absurdum* the contrary, there is a sequence $(x_j)_{j \ge 1}$,

$$\text{(135)} \qquad x_j \in X \setminus M_\rho,\ c - \frac{1}{j} \le \varphi(x_j) \le c + \frac{1}{j},\ \|\varphi'(x_j)\| \le \sqrt{\frac{1}{j}}\,.$$

$(PS)_G$ permits to affirm, by passage to a subsequence if necessary, that $\pi(x_j) \to \pi(x_0)$, $x_0 \in$ $\in X$, i.e. ((106))

$$\lim_{j \to \infty} \inf_{u \in G} \|x_j - x_0 - u\| = 0.$$

Let $(\delta_j)_{j \ge 1}$ be a strictly decreasing sequence, $\delta_j \to 0$. $\exists\, N_1$ so that $\inf\limits_{u \in G} \| x_{N_1} - x_0 - u \| < \delta_1$, we take u_1 in G so that $\| x_{N_1} - x_0 - u_1\| < \delta_1$, $\exists\, N_2$ so that $\inf\limits_{u \in G} \| x_{N_2} - x_0 - u \| < \delta_2$, we take u_2 in G so that $\| x_{N_2} - x_0 - u_2\| < \delta_2$ etc. , one obtains the sequences $(x_{N_j})_{j \ge 1}$ and $(u_j)_{j \ge 1}$ with $\| x_{N_j} - x_0 - u_j\| \to 0$. Taking $y_j := x_{N_j} - u_j$ one finds a sequence $(y_j)_{j \ge 1}$, (136) $y_j \to x_0$. From (135) and (136) results (φ and φ' are G-invariant) $\varphi(x_0) = c$, $\varphi'(x_0) = 0$, i.e. x_0 is critical point for φ at level c. But $y_j \in X \setminus M_\rho$ $\forall\, j \ge 1$, because as $\| x_{N_j} - (a_m + u)\| \overset{(135)}{\ge} \rho$ $\forall\, u \in G$ and $m = \overline{1, n}$, we have also $\| x_{N_j} - u_j - (a_m + v)\| \ge \rho$ $\forall\, v \in G$ and $m = \overline{1, n}$ (u goes over $G \Rightarrow v := u - u_j$ goes over G). So, as $X \setminus M_\rho$ is closed, it results $x_0 \in X \setminus M_\rho$, contradiction, since M_ρ is neighbourhood of K_c and consequently (134) (all the critical points of φ at level c are $a_m + u$, $m = \overline{1, n}$, u goes over G).

Now comes into the scene the function F ((127)), to which one applies the Ekeland principle. According to the definition of c_k ((131)), there is A in $\mathcal{A}_k$ so that

(137) $$F(A) \leq c + \varepsilon,$$

ε from (134). Take (138) $B := A \setminus M_{2\rho}$. Then (139) $k \leq \mathrm{Cat}_{\pi(X)}(\pi(A)) \overset{(C1)}{\leq} \mathrm{Cat}_{\pi(X)}(\pi(B) \cup \cup \pi(M_{2\rho})) \overset{(C1)}{\leq} \mathrm{Cat}_{\pi(X)}(\pi(B)) + \mathrm{Cat}_{\pi(X)}(\pi(M_{2\rho}))$.

But

(140) $$\mathrm{Cat}_{\pi(X)}(\pi(M_{2\rho})) \leq n.$$

Indeed, as

$$\pi(M_{2\rho}) \overset{(133)}{\subset} \bigcup_{m=1}^{n} \bigcup_{u \in G} \pi(S(a_m + u, 2\rho))$$

and

$$\mathrm{Cat}_{\pi(X)}(\pi(\bigcup_{u \in G} S(a_m + u, 2\rho))) = 1,$$

because $\pi(\bigcup_{u \in G} S(a_m + u, 2\rho)) = \pi(S(a_m, 2\rho))$ ($y = \pi(a_m + u + x)$, $\|x\| < 2\rho \Rightarrow y = \pi(a_m + x)$) and $H : [0, 1] \times \pi(S) \to \pi(X)$, where $S = S_{2\rho}(a_m)$, $H(t, \pi(x)) = \pi((1 - t)x)$ (correct definition, π is injective on S) is contraction of $\pi(S)$ in $\pi(X)$ ($H(1, \pi(x)) = \pi(0)$).

According to (139), (140) and (*), one obtains

$$k \leq \mathrm{Cat}_{\pi(X)}(\pi(B)) + k - i,\ i \leq \mathrm{Cat}_{\pi(X)}(\pi(B)),$$

i.e. $B \in \mathcal{A}_i$. (137) and (138) give

$$F(B) \leq c + \varepsilon.$$

Apply Ekeland principle 1.1 with ε and $\lambda = \sqrt{\varepsilon}$ in the complete metric space $\mathcal{A}_i$ (7.23, 7.24), there is C in $\mathcal{A}_i$ so that

(141) $$\begin{cases} F(C) < F(D) + \sqrt{\varepsilon}\,\delta(C, D)\ \forall D \in \mathcal{A}_i \setminus \{C\}, \\ F(C) \leq F(B) \leq c + \varepsilon, \\ \delta(B, C) \leq \sqrt{\varepsilon}. \end{cases}$$

Because $B \cap M_{2\rho} \overset{(138)}{=} \varnothing$ and $\delta(B, C) \leq \varepsilon \overset{(134)}{\leq} \rho$, we have $C \cap M_\rho = \varnothing$, in particular

(142) $$E := \{x \in C : c - \varepsilon \leq \varphi(x)\} \subset \varphi^{-1}([c - \varepsilon, c + \varepsilon]) \setminus M_\rho,$$

because in addition $\sup \varphi(C) \overset{(141)}{\leq} c + \varepsilon$.

Let x be arbitrary from E. Combining (142) with (134) and taking into account the continuity of φ' it can state:

$\exists\, r_x > 0$ and v_x in X with $\|v_x\| = 1$ so that

(143) $$\|h\| \le r_x \Rightarrow \varphi'(x + u + h)(v_x) < -\sqrt{\varepsilon}\ \ \forall\, u \in G.$$

E is compact, let be $S_1, \dots, S_p$ which cover E, $S_j := \{x \in X : \|x - x_j\| < r_{x_j}\}, j = \overline{1, p}$.

Consider the continuous function $\psi_j : X \to [0, 1]$, $1 \le j \le p$,

(144) $$\psi_j(x) = \begin{cases} \dfrac{\sum\limits_{u\in G} \operatorname{dist}(x+u, X \setminus S_j)}{\sum\limits_{s=1}^{p}\sum\limits_{u\in G} \operatorname{dist}(x+u, X\setminus S_s)}, & x \in \bigcup\limits_{s=1}^{p} S_s \\ 0, & x \in X \setminus \bigcup\limits_{s=1}^{p} S_s. \end{cases}$$

The definition, for instance of the numerator, is correct: if $x + u \in X \setminus S_j$, then dist $(x + u, X \setminus S_j) = 0$ and if $x + u \in S_j$ one takes into account that only a finite number of terms $x + v$, $v \in G$ can be found in S_j as imposed by 7.20.

Besides the functions ψ_j consider also $r := \min\{r_{x_1}, \dots, r_{x_p}\}$ and the G-invariant continuous function $\psi : X \to [0, 1]$ (X is normal)

$$\psi(x) = \begin{cases} 1, & \text{if } c \le \varphi(x), \\ 0, & \text{if } \varphi(x) \le c - \varepsilon. \end{cases}$$

Let now be H: $[0, 1] \times X \to X$ the continuous deformation of X

(145) $$H(t, x) = x + t\, r\, \psi(x) \sum_{j=1}^{p} \psi_j(x) v_{x_j}\,.$$

We remark that

(146) $$H(t, x + u) = H(t, x) + u\ \ \forall\, t \in [0, 1],\ \forall\, x \in X,\ \forall\, u \in G.$$

So, $\eta : [0, 1] \times \pi(X) \to \pi(X)$,

$$\eta(t, \pi(x)) = \pi(H(t, y)),\ y \text{ arbitrary from } \pi(x)$$

is correctly defined because

$$y = x + u \Rightarrow \pi(H(t, x + u)) \overset{(146)}{=} \pi(H(t, x) + u) = \pi(H(t, x)).$$

η is continuous. Indeed, let be $t_n \to t$ and $\pi(x_n) \to \pi(x)$. We must prove $\eta(t_n, \pi(x_n)) \to \to \eta(t, \pi(x))$, i.e. $\pi(H(t_n, y_n)) \to \pi(H(t, y))$, where y_n is an element of $\pi(x_n)$ and y an element of $\pi(x)$. As π is continuous (7.21), it remains to show

$$(146_1) \qquad H(t_n, y_n) \to H(t, y),$$

y_n conveniently chosen from $\pi(x_n)$ and y conveniently chosen from $\pi(x)$. $\pi(x_n) \to \pi(x) \Leftrightarrow$ $\Leftrightarrow \mathrm{d}\,(\pi(x_n - x), 0) \to 0$, but $\exists\, z_n \in \pi(x_n - x)$, i.e. $z_n = x_n - x + u_n$, $u_n \in G$, so that $\mathrm{d}\,(\pi(x_n - x), 0) \geq \frac{1}{2}\|z_n\|$ ((108)), consequently $x_n + u_n \to x$ and (146_1) is verified with $y_n = x_n + u_n$ and $y = x$ because H is continuous.

η is even a continuous deformation of $\pi(X)$ as $\eta(0, \pi(x)) = \pi(H(0, x)) = \pi(x)$. Putting $D := H(\{1\} \times C)$ we have, by applying (C2) as $H(D) = \pi(H(\{1\} \times C)) = \eta(\{1\} \times \pi(C))$,

$$\mathrm{Cat}_{\pi(X)}\,\pi(D) \geq \mathrm{Cat}_{\pi(X)}\,\pi(C) \geq i,$$

therefore $D \in \mathscr{A}_i$, D being also nonempty compact.

Let x be anyone from E. One applies the finite increment formula, $\exists\, \tau$ in (0, 1) so that

$$\varphi(H(1, x)) - \varphi(x) \overset{(145)}{=} \varphi'(\xi)\Big(r\psi(x)\sum_{j=1}^{p}\psi_j(x)v_{x_j}\Big),$$

$$\xi := x + \tau r\psi(x)\sum_{j=1}^{p}\psi_j(x)v_{x_j},$$

$$(147)\ \varphi(H(1, x)) - \varphi(x) = r\psi(x)\sum_{j=1}^{p}\psi_j(x)\,\varphi'\Big(x + \tau\, r\psi(x)\sum_{j=1}^{p}\psi_j(x)v_{x_j}\Big)(v_{x_j}) \overset{(143)}{\leq} -\sqrt{\varepsilon}\, r\,\psi(x),$$

since $\sum_{j=1}^{p}\psi_j(x) = 1$, $\|v_{x_j}\| = 1$. $F(D) \overset{(127)}{=} \sup_{x\in D}\varphi(x) = \sup_{x\in C}\varphi(H(1, x))$, let x_0 be in C so that

(148) $\varphi(H(1, x_0)) = F(D)$, hence $c \leq \varphi(H(1, x_0)) \overset{(147)}{\leq} \varphi(x_0)$, consequently $x_0 \in E$ and $\psi(x_0) = 1$. (147) and (148) give $F(D) - \varphi(x_0) \leq -\sqrt{\varepsilon}\, r$, $F(D) + \sqrt{\varepsilon}\, r \leq \varphi(x_0) \leq F(C)$,

$$(149) \qquad F(D) + \sqrt{\varepsilon}\, r \leq F(C).$$

But

$$(150) \qquad \delta(C, D) \leq r,$$

because for $x \in C$ and $y \in D$ tied by (see (145))

$$y = x + r\psi(x)\sum_{j=1}^{p}\psi_j(x)v_{x_j}$$

we have $\|x - y\| \leq r$.

According to (149) and (150) one finds

$$F(D) + \sqrt{\varepsilon}\,\delta(C, D) \leq F(C),$$

in contradiction with the strict inequality from (141) because $C \neq D$ as attested by (149) – in K_c there are at least $k - i + 1$ distinct critical orbits (see (∗)).

Case i = k

The proof of the precedent case remains valid in its most part. Let be $c := c_i$. Suppose *par absurdum* that φ has no one critical point at level c.

One affirms:

(151) $$\exists\, \varepsilon > 0 \text{ so that } x \in \varphi^{-1}([c - \varepsilon, c + \varepsilon]) \Rightarrow \|\varphi'(x)\| > \sqrt{\varepsilon}\,.$$

Suppose *par absurdum* the contrary, there exists a sequence $(x_j)_{j\geq 1}$ with

$$c - \frac{1}{j} \leq \varphi(x_j) \leq c + \frac{1}{j} \text{ and } \|\varphi'(x_j)\| \leq \sqrt{\frac{1}{j}}\,.$$

$(PS)_G$ gives then, if necessary by passing to a subsequence, $\pi(x_j) \to \pi(x_0)$, that is $\lim\limits_{j\to\infty} \inf\limits_{u\in G} \|x_j - x_0 - u\| = 0$. As in the first case, we find the sequences $(x_{N_j})_{j\geq 1}$ and $(u_j)_{j\geq 1}$ so that $\| x_{N_j} - x_0 - u_j \| \to 0$. Setting $y_j := x_{N_j} - u_j$, we have $y_j \to x_0$, hence $\varphi(x_0) = c$, $\varphi'(x_0) =$ $= 0$, i.e. x_0 is critical point at the level c, in contradiction with the initially admitted hypothesis and therefore (151) is validated.

Let be now A in $\mathcal{A}_i$ so that $F(A) \leq c + \varepsilon$, ε from (151). One applies Ekeland principle 1.1 with ε and $\lambda = \sqrt{\varepsilon}$, there is C in $\mathcal{A}_i$ so that

(152) $$F(C) < F(D) + \sqrt{\varepsilon}\,\delta(C, D)\ \forall\, D \in \mathcal{A}_i \setminus \{C\},$$

$$F(C) \leq F(A),\ \delta(A, C) \leq \sqrt{\varepsilon}\,.$$

Consider the set $E := \{x \in C : c - \varepsilon \leq \varphi(x)\}$. From now on the proof repeats that from previous case before and one gets in collision with (152), consequently φ has at least one critical point x_0 at the level c, $\pi^{-1}(\pi(x_0))$ is critical orbit. ■

Remark. In [45] the authors did not observe that their demonstration has missed the case $i = k$.

8

Ekeland principle for vector-valued functions and minimal points in product spaces

In this paragraph some theorems of the minimal point in product spaces are presented and corresponding versions of the Ekeland principle for functions with vectorial values (*A. Göpfert*, *Chr. Tammer*, *C. Zălinescu*, [35]). The starting point of these propositions, with the afferent variants, is the geometric interpretation, by means of Phelps cone, of the fundamental inequality (1) of the Ekeland principle (§1, before 1.1_0).

The display becomes more abstract and remains so until the end of the paragraph 9.

PRELIMINARIES OF AFFINE GEOMETRY

Let E be real vector space.

The nonempty subset C of E is named *cone* when

(1) $\quad$ for any $\lambda > 0$ we have $\lambda C \subset C$,

which is equivalent to

(1′) $\quad$ for any $\mu > 0$ we have $\mu C = C$.

The cone C is by definition cone with *vertex* when $0 \in C$. This is obviously characterized by

(2) for any $\lambda \geq 0$ we have $\lambda C \subset C$.

A convex cone with vertex is by definition *pointed* when the cone $C \setminus \{0\}$ is convex.

8.1 *Let* C *be cone.* C *is connected iff*

$$C + C \subset C.$$

■ *Necessary*. We have (3) $C + C = 2C$, even more, M nonempty convex, $\alpha > 0$, $\beta > 0$ $\Rightarrow \alpha M + \beta M = (\alpha + \beta)M$: for the inclusion $\subset$ observe that $\forall\, x, y$ from M, $\frac{\alpha}{\alpha+\beta} x + \frac{\beta}{\alpha+\beta} y \in M$; (3) and (1) give $C + C \subset C$. *Sufficiently*. Let be λ , $\mu > 0$, $\lambda + \mu = 1$. Then $\lambda C + \mu C \overset{(1')}{=} C + C \subset C$, that implies C convex. ■

8.2 *Let* C *be a convex cone with vertex.* C *is pointed iff*

$$C \cap (-C) = \{0\}.$$

■ *Necessity*. Let be $C_1 := C \setminus \{0\}$. Suppose *par absurdum* $\exists\, x, y \neq 0$ with $x = -y$, i.e. (4) $x + y = 0$. But $x, y \in C_1$, and $C_1 + C_1 \overset{8.1}{\subset} C_1$, hence $0 \overset{(4)}{\in} C_1$, contradiction. *Sufficiency*. Let x, y be in C_1. Then $x + y \neq 0$ (the hypothesis) and $x + y \overset{8.1}{\in} C$, hence $x + y \in C_1$, that is $C_1 + C_1 \subset C_1$ and therefore C_1 is convex (8.1). ■

8.3 *Let* E *be real vector space.*

1° *If* E *is preordered, then the set*

$$C := \{x \in E : x \geq 0\}$$

is convex cone with vertex.

2° *Let* C *be convex cone with vertex in* E. *The binary relation*

$$x \overset{C}{\leq} y \overset{\text{def}}{\Leftrightarrow} y - x \in C$$

is a preordered relation in E *compatible with the structure of vector space and*

$$C = \{x \in E : x \overset{C}{\geq} 0\}.$$

$\overset{C}{\leq}$ *is an order relation iff* C *is pointed.*

Definition. C is named the *positive cone* of E.

Explanation. Preorder relation means reflexivity + transitivity.

■ 1° Use (2) and 8.1. 2° *First affirmation.* $\overset{C}{\le}$ is preorder relation: $x \overset{C}{\le} x$ as $x - x = 0 \in$ $\in C$, $x \overset{C}{\le} y$ and $y \overset{C}{\le} z \Rightarrow y - x \in C$ and $z - y \in C$, add them, $z - x \in C$ (8.1). $\overset{C}{\le}$ is compatible: $x \overset{C}{\ge} 0 \Rightarrow x \in C$, hence $\lambda \ge 0 \Rightarrow \lambda x \in C$, $\lambda x \overset{C}{\ge} 0$ etc. The formula is obvious.

Second affirmation. Necessary. Let x be from $C \cap (-C)$. Then $x \overset{C}{\ge} 0$ (the first affirmation) and $-x \overset{C}{\ge} 0$, so $0 \overset{C}{\le} x \overset{C}{\le} 0$, $x = 0$, C is pointed (8.2). *Sufficiently.* Let be $x \overset{C}{\le} y$ and $y \overset{C}{\le} x$, i.e. $y - x, x - y \in C$, so $y - x \in C \cap (-C)$, consequently $y - x = 0$, $y = x$. ■

Remark. $x \overset{C}{\le} y \Leftrightarrow y \in x + C$.

Preorder relation $\overset{k_0}{\le}$.

Let Y be real locally convex space. For any C convex cone with vertex in Y (Y^* is the topological dual of Y), we put

(5) $$C^+ := \{\xi \in Y^* : \xi(y) \ge 0 \ \forall \ y \in C\},$$

(6) $$C^{++} := \{\xi \in Y^* : \xi(y) > 0 \ \forall \ y \in C \setminus \{0\}\}.$$

C^+ is convex cone with vertex in Y^* (use (2) and 8.1). Let be

(7) $$k_0 \in C \setminus (-\overline{C}).$$

Retain that $-\overline{C} = \overline{-C}$. $\{k_0\}$ being compact and $-\overline{C}$ closed, there is a closed hyperplane which very strongly separates them, therefore there is ξ_0 in C^+ so that

(8) $$\xi_0(k_0) = 1.$$

Associate with Y the metric space (X, d) and consider the set $X \times Y$. One sets on this the binary relation $\overset{k_0}{\le}$:

(9) $$(x_1, y_1) \overset{k_0}{\le} (x_2, y_2) \overset{\text{def}}{\Leftrightarrow} y_1 + k_0 \, \mathrm{d}(x_1, x_2) \overset{C}{\le} y_2.$$

The definition (9) is clearly inspired (see below) by the order relation established on epi φ by the Phelps cone (§1, above 1.1_0).

$\overset{k_0}{\le}$ is a preorder relation, being reflexive (as $\overset{C}{\le}$ is reflexive) but transitive also: let be $y_2 -$ $- [y_1 + k_0 \, \mathrm{d}(x_1, x_2)] \in C$ and $y_3 - [y_2 + k_0 \, \mathrm{d}(x_2, x_3)] \in C$, then by addition one obtains (take into account 8.1)

$$y_1 + k_0 [\mathrm{d}(x_1, x_2) + \mathrm{d}(x_2, x_3)] \overset{C}{\le} y_3,$$

but the first member is C-minorised by $y_1 + k_0\,\mathrm{d}(x_1, x_3)$ (the triangle inequality, $k_0 \overset{(7)}{\in} C$) and hence the conclusion, $\overset{C}{\leq}$ being transitive. If C is pointed, $\overset{k_0}{\leq}$ is even an order relation in $X \times Y$: $(x_1, y_1) \overset{k_0}{\leq} (x_2, y_2)$ and $(x_2, y_2) \overset{k_0}{\leq} (x_1, y_1) \Rightarrow$ (10) $y_2 - y_1 - k_0\,\mathrm{d}(x_1, x_2) \in C$ and $y_1 - y_2 - k_0\,\mathrm{d}(x_1, x_2) \in C$, add them (see 8.1), $-2k_0\,\mathrm{d}(x_1, x_2) \in C$. If, *par absurdum*, $\mathrm{d}(x_1, x_2) \neq 0$, it results $k_0 \in -C$, this contradicts (7) and consequently $x_1 = x_2$. Then, according to (10), $y_2 - y_1 \in C \cap (-C)$, that imposes $y_2 = y_1$ (8.2).

Retain from the last demonstration (the hypothesis „C pointed" has been used only at the end)

(11) $$(x_1, y_1) \overset{k_0}{\leq} (x_2, y_2) \text{ and } (x_2, y_2) \overset{k_0}{\leq} (x_1, y_1) \Rightarrow x_1 = x_2.$$

If X is real normed space, $\overset{k_0}{\leq}$ coincides with $\overset{C_{k_0}}{\leq}$, $C_{k_0} := \{(x, y) \in X \times Y : \|x\| k_0 \overset{C}{\leq} y\}$, convex cone with vertex $((x_1, y_1), (x_2, y_2) \in C_{k_0} \Rightarrow (x_1, y_1) + (x_2, y_2) \in C_{k_0}$, as $\|x_1 + x_2\| \times k_0 \overset{C}{\leq} y_1 + y_2$ because $k_0 \|x_1 + x_2\| \overset{C}{\leq} \|x_1\|\, k_0 + \|x_2\|\, k_0 \overset{C}{\leq} y_1 + y_2)$. Indeed, $(x_1, y_1) \overset{C_{k_0}}{\leq} (x_2, y_2) \Leftrightarrow (x_2 - x_1, y_2 - y_1) \in C_{k_0} \Leftrightarrow k_0 \|x_2 - x_1\| \overset{C}{\leq} y_2 - y_1 \Leftrightarrow y_2 - y_1 - k_0 \| x_2 - x_1\| \in C \overset{(9)}{\Leftrightarrow} (x_1, y_1) \overset{k_0}{\leq} (x_2, y_2)$.

Retain also

(12) $$C \text{ pointed} \Rightarrow C_{k_0} \text{ pointed}.$$

For (12) use 8.2. Let be $(x, y) \in C_{k_0} \cap (-C_{k_0})$, that is $\|x\| k_0 \overset{C}{\leq} y$ and $\|x\| k_0 \overset{C}{\leq} -y$, hence, by adding, $-2\|x\| k_0 \in C$, i.e. $\|x\| k_0 \in -C$, but $\|x\| k_0 \in C$ ((7)), hence $\|x\| k_0 = 0$ and $x = 0$. Then also $y \in C$ and $-y \in C$, i.e. $y \in C \cap (-C)$ and hence $y = 0$.

Let A be nonempty subset of the set $X \times Y$.

Definition. For every (x, y) from A consider

(13) $$A(x, y) := \{(v, w) \in A : (v, w) \overset{k_0}{\leq} (x, y)\}.$$

$A(x, y)$ is the *lower cut* of A along of (x, y) with respect to $\overset{k_0}{\leq}$. Obviously $(x, y) \in A(x, y)$.

Condition (H1). Let $(x_n, y_n)_{n \geq 1}$ be a k_0-decreasing sequence of points from A. If $x_n \to x$, there is y in Y so that $(x, y) \in A$ and

$$(x, y) \overset{k_0}{\leq} (x_n, y_n)\ \forall\ n \geq 1.$$

Condition (H2). Let $(x_n, y_n)_{n\geq 1}$ be a sequence of points from A with $x_n \to x$ and $(y_n)_{n\geq 1}$ C-decreasing. Then there is y in Y so that $(x, y) \in A$ and

$$y \overset{C}{\leq} y_n\ \forall\ n \geq 1.$$

When Y is separated this two conditions are connected by

8.4 *If* A *verifies* (H2) *and* $C \cap (z - \mathbf{R}_+ k_0)$ *is closed for any* z *from* C, *then* A *verifies also* (H1).

■ Let $(x_n, y_n)_{n\geq 1}$ be a k_0-decreasing sequence of points from A with $x_n \to x$. $(y_n)_{n\geq 1}$ is C-decreasing : $(x_{n+1}, y_{n+1}) \overset{k_0}{\leq} (x_n, y_n) \Rightarrow y_{n+1} + k_0\, d(x_n, x_{n+1}) \overset{C}{\leq} y_n \Rightarrow y_n - y_{n+1} \in k_0\, d(x_n, x_{n+1}) + {}$ $+\, C$ and the last sum is, according to 8.1, contained in C. The hypothesis gives then y in Y so that $(x, y) \in A$ and $y \overset{C}{\leq} y_n\ \forall\ n \geq 1$. It follows

$$y + k_0\, d(x_{n+p}, x_n) \overset{C}{\leq} y_{n+p} + k_0\, d(x_{n+p}, x_n) \overset{C}{\leq} y_n\ \forall\ n, p \geq 1. \tag{14}$$

Fix n. Taking the first and the last term from (14) we find

$$(y_n - y) - k_0\, d(x_{n+p}, x_n) \in C. \tag{15}$$

In (15) pass to the limit for $p \to \infty$ and one obtains $(y_n - y) - k_0\, d(x, x_n) \in C$, because the set $((y_n - y) - \mathbf{R}_+ k_0) \cap C$ is closed, that is $(x, y) \overset{k_0}{\leq} (x_n, y_n)\ \forall\ n \geq 1$. ■

Finaly, designate π_X, π_Y the canonical projections of $X \times Y$, $\pi_X(x, y) = x$, $\pi_Y(x, y) = y$.

From now on

Y is real separated locally convex space

and, except the last section,

X is complete metric space.

MINIMAL POINT THEOREM I

8.5 Theorem I ([35], th. 1). *Let* A *be nonempty subset of the set* X×Y *which verifies* (H1) *and for which there is* y_A *in* Y *having the property*

$$\pi_Y(A) \subset y_A + C. \tag{16}$$

Then for each (x_0, y_0) *from* A *there is* (v, w) *in* A *so that*

$$(v, w) \overset{k_0}{\leq} (x_0, y_0)$$

and if $(x, y) \in A$ *and* $(x, y) \overset{k_0}{\leq} (v, w)$ *then*

$$x = v.$$

■ We have (17) $(x, y) \in A \Rightarrow \pi_X(A(x, y))$ bounded. Indeed, let x' be arbitrary in $\pi_X(A(x, y))$. According to (13) $\exists\, y'$ in Y so that $(x', y') \in A$ and $(x', y') \overset{k_0}{\leq} (x, y)$, i.e. $y - y' - k_0 \mathrm{d}(x, x') = z$, $z \in C$. Apply ξ_0 which verifies (8), one obtains $\mathrm{d}(x, x') \leq \xi_0(y) - \xi_0(y')$. But $y' \overset{(16)}{=} y + t$, $t \in C$, therefore $\mathrm{d}(x, x') \leq \xi_0(y) - \xi_0(y_A)$ and hence (17).

Construct by recurrence a sequence $(x_n, y_n)_{n\geq 0}$ in A as follows. The first term of the sequence is (x_0, y_0) from the statement. Assume (x_n, y_n) defined. The set $B_n := \pi_X(A(x_n, y_n))$ being bounded ((17)), the set $\{\mathrm{d}(x, x_n) : x \in B_n\}$ is also bounded, let ρ_n be its supremum. Then there is x_{n+1} in B_n so that $\mathrm{d}(x_{n+1}, x_n) \geq \frac{1}{2}\rho_n$ $(\rho_n > 0, \rho_n = 0)$, i.e. (18) $\mathrm{d}(x_{n+1}, x_n) \geq \frac{1}{2}\sup\{\mathrm{d}(x, x_n) : x \in B_n\}$. According to the projection's definition $\exists\, y_{n+1}$ in Y so that $(x_{n+1}, y_{n+1}) \in A(x_n, y_n)$. (13) gives $(x_{n+1}, y_{n+1}) \overset{k_0}{\leq} (x_n, y_n)$, that is (19) the sequence $(x_n, y_n)_{n\geq 0}$ is k_0-decreasing.

Return to the sequence $(B_n)_{n\geq 0}$. Obviously (20) $x_n \in B_n$ (the reflexivity !). Because $A(x_{n+1}, y_{n+1}) \subset A(x_n, y_n)$ ($\overset{k_0}{\leq}$ is transitive !), we have (21) $B_{n+1} \subset B_n\ \forall\, n \geq 0$. Moreover, (22) $\lim\limits_{n\to\infty} \operatorname{diam} B_n = 0$. Indeed, in the oposite case $\exists\, \delta > 0$ and $\exists\, (p_n)_{n\geq 1}$ subsequence from $\mathbf{N}$ so that (23) $\operatorname{diam} B_{p_n} \geq \delta\ \forall\, n \geq 1$. Then, $\forall\, (x, y)$ from B_{p_n},

$$4\mathrm{d}(x_{p_n+1}, x_{p_n}) \overset{(18)}{\geq} \rho_{p_n} + \rho_{p_n} \geq \mathrm{d}(x, x_{p_n}) + \mathrm{d}(y, x_{p_n}) \geq \mathrm{d}(x, y),$$

hence (24) $\mathrm{d}(x_{p_n+1}, x_{p_n}) \geq \frac{\delta}{4}\ \forall\, n \geq 1$. Then, using (24) and (19),

$$\frac{1}{4}\delta k_0 \overset{C}{\leq} k_0 \mathrm{d}(x_{p_n+1}, x_{p_n}) \overset{C}{\leq} y_{p_n} - y_{p_n+1} \quad ((19)),$$

by isolating the first and the last last term and appling ξ_0 of (8) one obtains

$$\frac{1}{4}\delta \leq \xi_0(y_{p_n}) - \xi_0(y_{p_n+1}),$$

add these inequalities for $n = 1, 2, \ldots, m$,

$$\frac{1}{4}\delta_m \le \xi_0(y_{p_1}) - \xi_0(y_{p_m+1}) \overset{(16)}{\le} \xi_0(y_{p_1}) - \xi_0(y_A).$$

Pass to the limit for $m \to \infty$ and the contradiction appeared validates (22).

So (20), (21) and (22) permit to apply the Cantor theorem, $\bigcap_{n=0}^{\infty} \overline{B_n}$ has a single element,

$$\bigcap_{n=0}^{\infty} \overline{B_n} = \{v\}. \tag{25}$$

Obviously $x_n \to v$. Apply (H1), there is w in Y so that $(v, w) \overset{k_0}{\le} (x_n, y_n)\ \forall\ n \ge 0$, particularly $(v, w) \overset{k_0}{\le} (x_0, y_0)$. And if for (x, y) from A we have $(x, y) \overset{k_0}{\le} (v, w)$, it results (the transitivity !) $(x, y) \overset{k_0}{\le} (x_n, y_n)$, hence $(x, y) \in A(x_n, y_n)$ and hence $x \in B_n$, $x \in \overline{B_n}\ \forall\ n \ge 0$ and (25) imposes $x = v$. ■

Remark. Despite of its name, (v,w) is not, as one sees, an authentic minimal point. Such a point will be provided by theorem II.

As consequences of theorem 1, we see in the following two vectorial variants of Ekeland principle. But firstly –

Consider *the disjoint reunion* $Y^{\bullet} := Y \cup \{\infty\}$. Take by definition $y \overset{C}{\le} \infty\ \forall\ y \in Y$.

Let be $\varphi : X \to Y^{\bullet}$. The *domain* of φ, dom φ, is $\{x \in X : \varphi(x) \ne \infty\}$, the *epigraph* of φ, epi φ, is $\{(x, y) \in X \times Y : y \overset{C}{\le} \varphi(x)\}$, the *graph* of φ, Gr φ, is $\{(x, \varphi(x)) : x \in X, \varphi(x) \ne \infty\}$. φ is *proper* if dom $\varphi \ne \varnothing$.

8.6 *Vectorial variant of Ekeland principle* ([35]). *Let* $\varphi : X \to Y^{\bullet}$ *be* C - *bounded from below and proper which verifies the condition*

$$\{x' \in X : \varphi(x') + k_0\, d(x', x) \overset{C}{\le} \varphi(x)\} \text{ is closed } \forall\ x \in X. \tag{H3}$$

Then for any x_0 *from* dom φ *there is* v *in* X *so that*

$$\varphi(v) + k_0\, d(v, x_0) \overset{C}{\le} \varphi(x_0) \tag{26}$$

and

$$\varphi(x) + k_0\, d(v, x) \le \varphi(v) \Rightarrow x = v. \tag{27}$$

■ Take $A := \text{Gr}\ \varphi$. We have $A \ne \varnothing$ because φ is proper and show that A verifies (H1). Let $(x_n, y_n)_{n\ge1}$ be a k_0-decreasing sequence from A with $x_n \to x$. Then, according to (9),

(28) $$\forall\, n, p \geq 1\ x_{n+p} \in \{x' \in X : \varphi(x') + k_0\, d(x', x_n) \overset{C}{\leq} \varphi(x_n)\}.$$

Fix n. The set from (28) being closed, it contains also $\lim\limits_{p\to\infty} x_{n+p}$, that is x, otherwise said

$$(x, \varphi(x)) \overset{k_0}{\leq} (x_n, \varphi(x_n))\ \forall\, n \geq 1.$$

Moreover, if y_A C-lower bounds φ, this means, according to the definition (8.3, 2°), $\pi_Y(A) \subset \subset y_A + C$.

So being we can apply 8.5, $\exists\ (v, w)$ in A, $(v, w) \overset{k_0}{\leq} (x_0, \varphi(x_0))$ and $(x, y) \in A$, $(x, y) \overset{k_0}{\leq} (v, w) \Rightarrow x = v$, i.e. (26) and (27). ■

Remark 1. From 8.6 it can indeed obtain a form of Ekeland principle 1.1_0, §1, which justifies the name of 8.6 as vectorial variant.

Let (X, d) *be complete metric space and* $\varphi : X \to (-\infty, +\infty]$ *bounded from below, lower semicontinuous and proper. For any* $\varepsilon > 0$ *and* x_0 *of* X, *there is* v_ε *in* X *so that*

$$\varphi(v_\varepsilon) < \varphi(x) + \varepsilon d(v_\varepsilon, x)\ \forall\ x \in X \setminus \{v_\varepsilon\}$$

and

$$\varphi(v_\varepsilon) < \varphi(x_0) - \varepsilon d(v_\varepsilon, x_0).$$

Indeed, take in 8.6 $Y = \mathbf{R}$, $Y^\bullet = (-\infty, +\infty]$, $C = [0, +\infty)$, $k_0 = \varepsilon$ $(C \setminus (-\overline{C}) = (0, +\infty))$. The preorder relation $\overset{C}{\leq}$ coincides with the order relation on **R** which is total, and so (27) gives the fundamental inequality and (26) the second one. The application of 8.6 is correct because (H3) is ensured by the lower semicontinuity of φ.

Remark 2. If at the hypotheses of 8.6 one adds

(29) $$\varphi(X) \cap [\varphi(x_0) - \lambda k_0 - (C \setminus \{0\})] = \varnothing,$$

where $\lambda \in (0, +\infty)$, then at the conclusions of (29) one adds

(30) $$d(v, x_0) \leq \lambda.$$

Indeed, suppose *par absurdum* (31) $d(v, x_0) > \lambda$. Then for a u from C we have (see (26)), $\varphi(v) = \varphi(x_0) - k_0\, d(v, x_0) - u = \varphi(x_0) - \lambda k_0 - [d(v, x_0) - \lambda]k_0 - u \in \varphi(x_0) - \lambda k_0 - (C \setminus \{0\})$, because $[d(v, x_0) - \lambda]k_0 + u \in C \setminus \{0\}$ (*par absurdum* $k_0\, d(v, x_0) + u = \lambda k_0$, one gets $\varphi(v) = \varphi(x_0) - \lambda k_0$ and, from (26), $k_0\,[d(v, x_0) - \lambda] \overset{C}{\leq} 0$, contradiction if one takes into account (31) and (7)), and we come in collision with (29).

Remark 3. A particular case of 8.6 is obtained in [37], Theorem 4.

Here is a second vectorial variant of the Ekeland principle.

8.7 *Vectorial variant of the Ekeland principle* ([35]). *Let* $\varphi: X \to Y^{\bullet}$ *be lower* C-*bounded and proper verifying*

(H4) *for any sequence* $(x_n)_{n\geq 1}$ *from* dom φ *with* $x_n \to x$ *and* $(\varphi(x_n))_{n\geq 1}$ C-*decreasing, we have* $\varphi(x) \leq \varphi(x_n)\ \forall\ n \geq 1$.

If $C \cap (z - \mathbf{R}_+ k_0)$ *is closed for any* z *from* C, *then are valid the conclusions from* 8.6.

■ Take $A = \text{Gr}\ \varphi$. A verifies (H2). Indeed, $(x_n, y_n)_{n\geq 1}$ being a sequence of points from A with $x_n \to x$ and $(y_n)_{n\geq 1}$ C-decreasing, the hypothesis (H4) gives exactly the condition (H2). But then, according to 8.4, A also verifies (H1).

In the following one repeats the reasoning for 8.6 (now the relation $(x, \varphi(x)) \overset{k_0}{\leq} (x_n, \varphi(x_n))\ \forall\ n \geq 1$ is not obtained with (H3), but with (H1)). ■

Remark. A particular case of 8.7 is found in [51], 1, proposition 1.

Attention. For $Y = \mathbf{R}$, $Y^{\bullet} = (-\infty, +\infty]$ and $C = [0, +\infty)$, 8.7 gives the form of the Ekeland principle from the Remark 1 *without the demand* φ *lower semicontinuous* but with the demand (H4) verified.

For example, this form can be applied to the function $\varphi : \mathbf{R} \to \mathbf{R}$, $\varphi(x) = e^{-|x|}$, $x \neq 0$, $\varphi(0) = 2$: φ is not lower semicontinuous in 0, verifies the demands from 8.7 and $[0, +\infty) \cap \cap (z - [0, +\infty)k_0)$ is closed $\forall$ z from $[0, +\infty)$ ($k_0 \overset{(7)}{\in} (0, +\infty)$).

MINIMAL POINT THEOREM II

Consider on $X \times Y$ the binary relation (ξ_0 is the functional from (8))

(32) $$(x_1, y_1) \overset{k_0,\xi_0}{\leq} (x_2, y_2) \Leftrightarrow \begin{cases} (x_1, y_1) = (x_2, y_2) \\ \text{or} \\ (x_1, y_1) \overset{k_0}{\leq} (x_2, y_2) \text{ and } \xi_0(y_1) < \xi_0(y_2). \end{cases}$$

This is an order relation. The transitivity: let be $(x_1, y_1) \overset{k_0,\xi_0}{\leq} (x_2, y_2) \overset{k_0,\xi_0}{\leq} (x_3, y_3)$; if $(x_1, y_1) = (x_2, y_2)$ – nothing to prove; in the contrary case we have $(x_1, y_1) \overset{k_0}{\leq} (x_2, y_2)$ and $\xi_0(y_1) < \xi_0(y_2)$; if $(x_2, y_2) = (x_3, y_3)$ – nothing to prove again; in the contrary case we have $(x_2, y_2) \overset{k_0}{\le} (x_3, y_3)$ and $\xi_0(y_2) < \xi_0(y_3)$; as „ $\overset{k_0}{\le}$ ” and „<” are transitive, the conclusion is imposed. The reflexivity: obvious. $(x_1, y_1) \overset{k_0,\xi_0}{\le} (x_2, y_2)$ and $(x_2, y_2) \overset{k_0,\xi_0}{\le} (x_1, y_1) \Rightarrow (x_1, y_1) = (x_2, y_2)$: *par absurdum*, one obtains $\xi_0(y_1) < \xi_0(y_2)$ and $\xi_0(y_2) < \xi_0(y_1)$, contradiction.

8.8 Theorem II. *Let* A *be nonempty subset of the set* X × Y *which verifies* (H1) *and* ξ_0 *is lower bounded on* π_Y (A). *Then for each* (x_0, y_0) *from* A *there is* (v, w) *in* A *so that*

$$(v, w) \overset{k_0,\xi_0}{\le} (x_0, y_0)$$

and if (x, y) ∈ A *and* $(x, y) \overset{k_0,\xi_0}{\le} (v, w)$, *then* (x, y) = (v, w)

(*A. Göpfert - Chr. Tammer - C. Zălinescu*).

■ Construct by recurrence a sequence $(x_n, y_n)_{n\ge0}$ of points from A. (x_0, y_0) is the point from the statement. Suppose (x_n, y_n) defined. Take (x_{n+1}, y_{n+1}) from A, $(x_{n+1}, y_{n+1}) \overset{k_0,\xi_0}{\le}$ $\overset{k_0,\xi_0}{\le} (x_n, y_n)$ and C-diminish y_{n+1} to obtain

(33) $$\xi_0(y_{n+1}) < \inf\{\xi_0(y) : (x, y) \in A, (x, y) \overset{k_0,\xi_0}{\le} (x_n, y_n)\} + \frac{1}{n+1},$$

correct demand because the set at which one takes the infimum is nonempty (the reflexivity) and ξ_0 is lower bounded on $\pi_Y(A)$. So

(34) $$(x_{n+1}, y_{n+1}) \overset{k_0,\xi_0}{\le} (x_n, y_n)\ \forall\, n \ge 0,$$

(34) imposes

(35) $(x_n, y_n)_{n\ge0}$ k_0 – decreasing , $(y_n)_{n\ge0}$ C-decreasing , $(\xi_0(y_n))_{n\ge0}$ decreasing.

Suppose ∃ N ≥ 0 *so that* $\lim_{n\to\infty} \xi_0(y_n) = \xi_0(y_N)$. *A fortiori* $\xi_0(y_n) = \xi_0(y_N)\ \forall\, n \ge N$ and, as $(x_n, y_n) \overset{k_0,\xi_0}{\le} (x_N, y_N)\ \forall\, n \ge N$, it results $(x_n, y_n) \overset{(32)}{=} (x_N, y_N)\ \forall\, n \ge N$. Take $v = x_N$, $w = y_N$. We have then $\forall\, n > N$ (see (33))

$$\xi_0(y_n) = \xi_0(w) < \inf\{\xi_0(y) : (x, y) \in A, (x, y) \overset{k_0,\xi_0}{\le} (v, w)\} + \frac{1}{n},$$

pass to the limit for $n \to \infty$, one gets,

(36) $$\xi_0(w) \le \xi_0(y)\ \forall\, (x, y) \in A \text{ with } (x, y) \overset{k_0,\xi_0}{\le} (v, w),$$

i.e.

$$(x, y) = (v, w) \text{ or } (x, y) \overset{k_0}{\leq} (v, w) \text{ and } \xi_0(y) < \xi_0(w).$$

As the second inequality is hindered by (36), it results $(x, y) = (v, w)$, (v, w) is minimal point of A for the relation „ $\overset{k_0,\xi_0}{\leq}$ ". Obviously $(v, w) \overset{k_0,\xi_0}{\leq} (x_0, y_0)$ ((35)).

Suppose $\lim_{n\to\infty} \xi_0(y_n) < \xi_0(y_n)\ \forall\, n \geq 0$. Let be $n, p \geq 1$. As $(x_{n+p}, y_{n+p}) \overset{k_0,\xi_0}{\leq} (x_n, y_n)$ ((34)), by using ξ_0 ((8)) and (33) one obtains

$$d(x_{n+p}, x_n) \leq \xi_0(y_n) - \xi_0(y_{n+p}) \leq \frac{1}{n},$$

i.e. $(x_n)_{n\geq 0}$ is Cauchy sequence, let be $v = \lim_{n\to\infty} x_n$. According to (H1), there is w in Y so that $(v, w) \in A$ and (37) $(v, w) \overset{k_0}{\leq} (x_n, y_n)\ \forall\, n \geq 0$, hence $\xi_0(y_n) - \xi_0(w) \geq d(x_n, w)$, consequently (38) $\xi_0(w) \leq \lim_{n\to\infty} \xi_0(y_n)$ and hence (see the hypothesis) (39) $\xi_0(w) < \xi_0(y_n)\ \forall\, n \geq$ ≥ 0. Consequently (see (32)), (40) $(v, w) \overset{k_0,\xi_0}{\leq} (x_n, y_n)\ \forall\, n \geq 0$. Let (x, y) be any from A with $(x, y) \overset{k_0,\xi_0}{\leq} (v, w)$. Because $(x, y) \overset{k_0,\xi_0}{\leq} (x_n, y_n)\ \forall\, n \geq 0$ (see (40)), we have (apply ξ_0 if neccesary) $0 \leq d(x, v) \leq \xi_0(w) - \xi_0(y) \overset{(38)}{\leq} \xi_0(y_n) - \xi_0(y) \overset{(33)}{\leq} \frac{1}{n}\ \forall\, n \geq 1$, which imposes $\xi_0(w) = \xi_0(y)$ and, *via* the definition (32), $(x, y) = (v, w)$, (v, w) is minimal point of A with $(v, w) \overset{k_0,\xi_0}{\leq} (x_0, y_0)$ ((40)). ■

Remark. Theorem I is immediate consequence of theorem II, since if $\pi_Y(A) \subset y_A + C$, that is for any y of $\pi_Y(A)$ we have $y = y_A + u$ with $u \in C$, then $\xi_0(y - y_A) \overset{(5)}{\geq} 0$, $\xi_0(y) \geq$ $\geq \xi_0(y_A)$, ξ_0 is bounded from below on $\pi_Y(A)$.

MINIMAL POINT THEOREM III

Preliminaries. Let E be vector space on **K, K** = **R** or **K** = **C**, and x, y from E, $x \neq y$. The *closed interval* $[x, y]$ is $[x, y] = \{(1 - \lambda)x + \lambda y : \lambda \in [0, 1]\}$. If $\lambda \in [0, 1)$, $\lambda \in (0, 1]$ or $\lambda \in$ $\in (0, 1)$, one obtains respectively $[x, y)$, $(x, y]$ and (x, y) (*open interval*). The *straight line* d which *passes* through x and y is

$$d = \{(1-\lambda)x + \lambda y : \lambda \in \mathbf{R}\}.$$

If $x_1 \in d$, we have $d = \{x_1 + \lambda(y - x) : \lambda \in \mathbf{R}\}$ ($x_1 \in d \Rightarrow x_1 = (1-\lambda_1)x + \lambda_1 y$, replace).

Let M be nonempty subset of E. x_0 from M is by definition *algebraic interior point*, if on each straight line that passes through x_0 there is an open interval included in M which contains x_0. The set of these points is noted M^{i} or aint M, the *algebraic interior* of M. y_0 from E is by definition *algebraic adherent point* to M if $\exists\, x$ in M so that $[x, y_0) \subset M$. The set of these points, the *algebraic closure* of M, is noted M^{a}. $M^{\mathrm{a}} \setminus M^{\mathrm{i}}$ is the *algebraic boundary* of M.

Obviously $M^{\mathrm{i}} \subset M$, M convex $\Rightarrow M \subset M^{\mathrm{a}}$. If M is convex and $M^{\mathrm{i}} \neq \emptyset$, M is named *algebraic convex body*.

It can affirm

(41) If C is convex cone with vertex and $C^{\mathrm{i}} \neq \emptyset$, then

$$C + C^{\mathrm{i}} = C^{\mathrm{i}}.$$

■ $C^{\mathrm{i}} = C^{\mathrm{i}} + 0 \subset C^{\mathrm{i}} + C$. Reciprocally, let u be from C, v from C^{i} and $\{u + v + \lambda z : \lambda \in \in \mathbf{R}\}$ any straight line that passes through $u + v$. We have $u + v + \lambda z = v + \lambda(z + u) + (1 - - \lambda)u$. But $\{v + \lambda(z + u) : \lambda \in \mathbf{R}\}$ being a straight line that passes through v from C^{i}, $\exists\, \lambda > > 0$ so that $v + \lambda(z + u) \in C$ for $\lambda \in (-\delta, \delta)$, $\delta > 0$. For δ sufficiently small we have also $\lambda \in (-\delta, \delta) \Rightarrow (1-\lambda)u \in C$, consequently $\lambda \in (-\delta, \delta) \Rightarrow u + v + \lambda z \in C$, i.e. $u + v \overset{8.1}{\in} C^{\mathrm{i}}$. ■

Return to the subject of this section.

In this section, the convex cone C with vertex is assumed closed.

So

(42) $k_0 \in C \setminus (-C)$ (see (7)).

For each y from Y, consider the set

(43) $M_y := \{t \in \mathbf{R} : y \in tk_0 - C\}$.

8.9 *If* $M_y \neq \emptyset$, *then there is* t_y *in* $\mathbf{R}$ *so that*

$$M_y = [t_y, +\infty).$$

■ We have

(44) $t \in M_y$ and $t' > t \Rightarrow t' \in M_y$,

because $y \in tk_0 - C = t'k_0 - [(t' - t)k_0 + C]$, but $(t' - t)k_0 + C \overset{8.1}{\subset} C$, hence $y \in t'k_0 - C$, that is $t' \in M_y$.

On the other hand, M_y is lower bounded. Indeed, supposing *par absurdum* the contrary, (44) allows to affirm $-n \in M_y\ \forall\ n \geq 1$, i. e. $y \in -nk_0 - C\ \forall\ n \geq 1$, which is equivalent to $-k_0 - \frac{1}{n}y \in C\ \forall\ n \geq 1$ and by passing to the limit one gets (C is closed) $-k_0 \in C$, which contradicts (42).

So being, take $t_y := \inf M_y$ and it remains to show (45) $t_y \in M_y$. Consider a sequence $(t_n)_{n\geq 1}$, $t_n \in M_y$, $t_n \to t_y$. As $y \in t_n k_0 - C$, it results $-y + t_n k_0 \in C\ \forall\ n \geq 1$ and, by passing to the limit, $-y + t_y x_0 \in C$, that is (45). ■

Consider the **Luc functional** $L : Y \to \overline{\mathbf{R}}$ ([43]),

$$L(y) = \begin{cases} \inf M_y, & M_y \neq \varnothing \\ +\infty, & M_y = \varnothing. \end{cases} \tag{46}$$

8.10 *Lemma.* 1° L *is lower semicontinuous, proper, sublinear and* C-*increasing.*

2° $\{y \in Y: L(y) \leq t\} = tk_0 - C$;

$L(y + \lambda k_0) = L(y) + \lambda\ \forall\ y \in Y, \forall\ \lambda \in \mathbf{R}$;

$L(-y) \geq -L(y)\ \forall\ y \in Y$;

$L(0) = 0,\ L(\lambda k_0) = \lambda\ \forall\ \lambda \in \mathbf{R}$.

3° *Moreover, if* $k_0 \in \operatorname{aint} C$, L *takes its values in* **R** *and*

$$\{y \in Y: L(y) < t\} = t k_0 - \operatorname{aint} C,$$

$$y_2 - y_1 \in \operatorname{aint} C \Rightarrow L(y_1) < L(y_2).$$

■ 1° Let y be arbitrary from Y. Assume $L(y) < +\infty$. Then for any t from **R** we have (47) $L^{-1}((-\infty, t]) = tk_0 - C$, as, for instance, $L(y) \leq t \overset{8.9}{\Rightarrow} y \in tk_0 - C$. Because $tk_0 - C$ is closed, one concludes L l.s.c. in y. Suppose $L(y) = +\infty$, i. e. $y \notin \mathbf{R}k_0 - C$. As $\mathbf{R}k_0 - C$ is closed because $\mathbf{R}k_0$ has a finite dimension, $\exists\ V$ neighbourhood of y so that $V \cap (\mathbf{R}k_0 - C) = \varnothing$, that is $L(z) = +\infty$ on V, L is l.s.c. in y.

From 8.9 and (46) one obtains

$$\operatorname{epi} L \text{ (in } Y \times \mathbf{R}) = T^{-1}(C),\ T: Y \times \mathbf{R} \to Y,\ T(y, t) = tk_0 - y. \tag{48}$$

Because $T^{-1}(C) \neq \varnothing$ ($k_0 = T(0, 1)$!), L is proper. In addition, T being linear and continuous, (48) imposes epi L (in $Y \times \mathbf{R}$) convex cone with vertex and consequently L is sublinear.

Finally, let be $y_1 \overset{C}{\leq} y_2$ and $L(y_2) < +\infty$ (when $L(y_2) = +\infty$ – nothing to prove). According to 8.9, $L(y_2) \in M_{y_2}$, i. e. $y_2 = L(y_2)k_0 - u$, $u \in C$. But $y_2 - y_1 = v$, $v \in C$, hence $y_1 = = L(y_2)k_0 - (u + v)$, that is $y_1 \in L(y_2)k_0 - C$ and hence $L(y_1) \leq L(y_2)$ ((46), 8.9), L is C-increasing.

2° The first formula has been already obtained at (47). Go to the second formula. We have $t \in M_{y+\lambda k_0} \overset{(43)}{\Leftrightarrow} y + \lambda k_0 \in tk_0 - C \Leftrightarrow y \in (t - \lambda)k_0 - C \Leftrightarrow t - \lambda \in M_y \Leftrightarrow t \in \lambda + M_y$, hence $M_{y+\lambda k_0} = \lambda + M_y$ and hence the second formula. Pass to the third relation. Suppose $L(y) < +\infty$, otherwise nothing to prove. $-L(y) = -\inf\{t \in \mathbf{R}: y \in tk_0 - C\} = \sup\{-t \in \mathbf{R}: y \in tk_0 - C\} = \sup\{\tau \in \mathbf{R} : y \in -\tau k_0 - C\} = \sup A$, $A := \{\tau \in \mathbf{R} : -y \in tk_0 + C\}$. $L(-y) = = \inf B$, $B := \{t \in \mathbf{R} : -y \in tk_0 - C\}$. Let τ, t be any from A resp. B. $-y = \tau k_0 + c_1 = tk_0 - - c_2$, c_1 and c_2 from C, hence $(t - \tau)k_0 = c_1 + c_2 \in C$ and hence $t - \tau \overset{8.1}{\geq} 0$ (use (42)), that imposes $\inf B \geq \sup A$. Pass to the last two formulae. $L(0) = \inf\{t \in \mathbf{R} : 0 \in tk_0 - C\} = = \inf\{t \in \mathbf{R} : tk_0 \in C\} = 0$, because if *par absurdum* $\exists\, t' := -t$, $t > 0$ so that $t'k_0 \in C$, then $tk_0 \in -C$, hence $k_0 \in -C$, in contradiction with (42). Finally, $L(\lambda k_0) = L(0 + \lambda k_0) = L(0) + + \lambda = \lambda$.

3° In this case we have (49) $\operatorname{dom} L = Y$, that is (50) $M_y \neq \varnothing\ \forall\, y$ from Y. Indeed (see the preliminaries), the straight line $\{tk_0 + (1 - t)y : t \in \mathbf{R}\}$ passing through k_0 has an open interval contained in C in which is k_0, otherwise expressed, for $\delta > 0$ sufficiently small $t_{k_0} + (1 - t)y = u_t$, $u_t \in C$, $t \in (1 - \delta, 1 + \delta)$, hence $y = \dfrac{t}{t-1}k_0 - \dfrac{u_t}{t-1}$ for $1 < t < 1 + \delta$ and so (50). Moreover, (51) $\inf L(Y) = -\infty$. Indeed, take $t < 0$ and $y \in tk_0 - C$, multiply by n from $\mathbf{N}$, $ny \in ntk_0 - C$, therefore $L(ny) \leq nt$ and pass to the limit for $n \to \infty$.

Pass to the second affirmation of 3°. We have $\{y \in Y : L(y) < t\} \subset \{y \in Y : L(y) \leq t\}^{\mathrm{i}}$: let be $L(y) < t$ and $\{y + \lambda z : \lambda \in \mathbf{R}\}$ any straight line passing through y, we have

$$\text{(52)} \qquad L(y + \lambda z) \overset{1^0}{\leq} L(y) + L(\lambda z), \text{ but } L(\lambda z) = \begin{cases} \lambda L(z), \lambda \geq 0 \\ -\lambda L(-z), \lambda < 0, \end{cases}$$

because $\lim\limits_{\lambda \to 0+} \lambda L(z) = 0$, $\lim\limits_{\lambda \to 0-} -\lambda L(-z) = 0$ ($L(-z) \overset{(49)}{\in} \mathbf{R}$!), from this results $\lim\limits_{\lambda \to 0} L(\lambda z) = = 0$ and hence $\exists\, \delta > 0$ so that $\lambda \in (-\delta, \delta) \Rightarrow L(\lambda z) < t - L(y)$, consequently, taking into account (52), $\lambda \in (-\delta, \delta) \Rightarrow L(y + \lambda z) < t$. Reciprocally, $\{y \in Y : L(y) \leq t\}^{\mathrm{i}} \subset \{y \in Y :$

$L(y) < t\}$: let y be an element from the first member, the straight line $\{y + \lambda k_0 : \lambda \in \mathbf{R}\}$ and $\lambda \in (-\delta, \delta) \Rightarrow L(y + \lambda k_0) \le t$, then $y + \lambda k_0 \in tk_0 - C$, $y \in (t - \lambda)k_0 - C \ \forall \lambda \in (-\delta, \delta)$, take λ from $(0,\delta)$ so that $t - \lambda < t$, hence $L(y) \le t - \lambda < t$, y is an element from the second member. So

$$\{y \in Y : L(y) < t\} = \{y \in Y : L(y) \le t\}^{\mathrm{i}} \overset{2^0}{=} (tk_0 - C)^{\mathrm{i}} = tk_0 - C^{\mathrm{i}}.$$

Finally, pass at the third affirmation of 3°. $y_2 = L(y_2)k_0 - u$, $u \in C$, $y_1 = y_2 - v$, $v \in C^{\mathrm{i}}$, hence $y = L(y_2)k_0 - (u + v)$ with $u + v \overset{(41)}{\in} C^{\mathrm{i}}$ and the conclusion is imposed by considering the straight line through $u + v$ and k_0, both from C^{i} (already $L(y_1) \le L(y_2)$). ■

Remark 1. In [35] the demonstration of lemma 7, here 8.10, is full of errors.

Remark 2. When $k_0 \in C^{\mathrm{i}}$ and int $C \neq \varnothing$, L is continuous, because on the $k_0 - C$ neighbourhood of 0 we have $L(y) \le 1$.

Remark 3. The attentive analysis of the proofs of 8.9 and 8.10 shows that, if Y is only a vector topological space and if (42) is replaced by $k_0 \in C \setminus (-\overline{C})$, C convex cone with vertex, and in the definition (43) of M_y C is replaced by $\overline{C}$ (convex cone with vertex!), then remain true 8.9 and also the properties 1° and 2° of the Luc functional from 8.10, where C is replaced by $\overline{C}$.

This remark is used in §9, I.

For any nonempty subset A of $X \times Y$ and for each x from X and y from Y consider the sets

(53) $$A_x := \{y \in Y : (x, y) \in A\}, A_y := \{x \in X : (x, y) \in A\}.$$

Obviously

(54) $$\pi_X(A) = \{x \in X : A_x \neq \varnothing\}, \pi_Y(A) = \{y \in Y : A_y \neq \varnothing\}.$$

And now

8.11 Theorem III ([35], th. 8). *Let* (x_0, y_0) *be from* A. *Suppose*:

a) *there is* t_0 *in* $\mathbf{R}$ *so that* $\pi_Y(A) \cap (y_0 - t_0 k_0 - C) = \varnothing$;

b) *for any* r *from* $\mathbf{R}$

$$A^r := \{x \in X : \exists\, y \overset{C}{\le} y_0 + r k_0 \textit{ and } (x, y) \in A\} \textit{ is closed};$$

c) *any function* $Y \to \overline{\mathbf{R}}$ *lower semicontinuous, proper, sublinear and* C*-increasing attains its infimum on* $A_x - y_0 \ \forall x \in \pi_X(A)$.

Then there is (v, w) *in* A *so that*

(55) $$w + k_0\, d(v, x_0) \overset{C}{\le} y_0$$

and

(56) $$(x, y) \in A,\ y + k_0\, d(v, x) \overset{C}{<} w \Rightarrow x = v \textit{ and } y \in (w - C) \setminus (w - (0, +\infty)k_0 - C).$$

If $k_0 \in C^i$ (resp. $k_0 \in \text{int } C$),

c) *may be replaced by*

d) *any lower semicontinuous functional* $Y \to \mathbf{R}$ (resp. *continuous*), *sublinear and C-increasing attains its infimum on* $A_x - y_0\ \forall\ x \in \pi_X(A)$

and (56) *by*

(57) $$(x, y) \in A,\ y + k_0\, d(v, x) \overset{C}{\le} w \Rightarrow x = v \textit{ and } y \notin w - C^i \text{ (resp. } y \notin w - \text{int } C).$$

■ Use the Luc functional ((46)). Consider the function $T: X \to \overline{\mathbf{R}}$,

(57_1) $$T(x) = \begin{cases} \inf\{L(y - y_0) : y \in A_x\}, & \text{for } x \in \pi_X(A) \\ +\infty, & \text{for } x \notin \pi_X(A). \end{cases}$$

(58) T *is bounded from below by* $-\, t_0$. Indeed, suppose *par absurdum* $\exists\ x$ in X so that $T(x) < -\, t_0$. Then, according to (57_1) ($T(x) < +\infty$), $\exists\ y$ in Y with $(x, y) \in A$ and $L(y - y_0) < -\, t_0$, therefore $y - y_0 \overset{8.9}{\in} -\, tk_0 - C$ and one gets in contradiction with a).

(59) T *is lower semicontinuous*. Let r be anyone from $\mathbf{R}$. Take r' from $\mathbf{R}$ with $r < r'$. We have

(60) $$A^r \subset \{x \in X : T(x) \le r\} \subset A^{r'}.$$

Indeed, $x \in A^r \overset{b)}{\Rightarrow} \exists\, y$ with $(x, y) \in A$ and $y \overset{C}{\le} y_0 + rk_0$, then $T(x) \le L(y - y_0) \overset{8.10,1^0}{\le} L(r, k_0) \overset{8.10,2^0}{=} r$, and therefore the first inclusion from (60). Pass to the second inclusion. Let x be from X so that $T(x) \le r$. As $r' > r$, $\exists\ y$ in A_x so that $L(y - y_0) \overset{(57)}{<} r'$, hence $y \overset{(46)}{\in} y_0 + r'k_0 - C$, i.e. $y \overset{C}{\le} y_0 + r'k_0$ and $x \in A^{r'}$.

So, taking (60) into account,

$$\{x \in X : T(x) \le r\} \subset \bigcap_{r'>r} A^{r'} \subset \bigcap_{r'>r} \{x \in X : T(x) \le r'\} = \{x \in X : T(x) \le r\},$$

hence $\{x \in X : T(x) \le r\} = \bigcap_{r'>r} A^{r'}$, consequently, $A^{r'}$ being closed (b)), $\{x \in X : T(x) \le r\}$ is closed, T is l.s.c.

So being, (58), (59) together with the remark T proper (see (57_1), $(x_0, y_0) \in A, L(0)=0$) allows to apply to T a variant of Ekeland principle (Remark 1 to 8.6) with $\varepsilon = 1$, $\exists\, v$ in X so that

$$(61) \qquad T(v) < T(x) + \mathrm{d}(v, x) \;\forall\, x \in X \setminus \{v\}$$

and

$$(62) \qquad T(v) + \mathrm{d}(v,x_0) \le T(x_0) \;(\overset{(57)}{\le} L(y_0 - y_0) = 0).$$

Taking into account 8.10, 1° and the hypothesis c), we find w in A_v so that (63) $T(v) = L(w - y_0)$. Then combining 8.10, 2° with (62) it results $L(w - y_0) + \mathrm{d}(v, x_0) = L(w - y_0 + k_0\, \mathrm{d}(v, x_0)) \le 0$, which imposes (55) *via* 8.9 and (46).

Pass to (56). Let (x, y) be from A with $y + k_0\, \mathrm{d}(v, x) \overset{C}{\le} w$. Then

$$T(x) + \mathrm{d}(v, x) \overset{(57)}{\le} L(y - y_0) + \mathrm{d}(v, x) \overset{8.10}{=} L(y - y_0 + k_0\, \mathrm{d}(v, x)) \overset{8.10}{\le} L(w - y_0) \overset{(63)}{=} T(v)$$

and applying (61) one obtains $x = v$. And if, *par absurdum*, $y \in w - (0, +\infty)k_0 - C$, i. e. $y = w - tk_0 - u$ with $t > 0$ and $u \in C$, then

$$T(v) = T(x) \le L(y - y_0) = L(w - y_0 - tk_0 - u) \overset{8.10,1^0}{\le} L(w - y_0 - tk_0) \overset{8.10,2^0}{=} L(w - y_0) - t \overset{(63)}{<} T(v)$$

and consequently a contradiction.

For the last affirmation of the statement take into account that $k_0 \in C^{\mathrm{i}}$ (resp. $k_0 \in \mathrm{int}C) \Rightarrow L$ finite, 8.10, 3° (resp. continuous, Remark 2 to 8.10) and $C^{\mathrm{i}} \subset (0, +\infty)k_0 + C$. ■

Remark 1. Under the conditions from 8.11 we have also

$$\mathrm{d}(v, x_0) \le t_0,$$

because $T(v) \overset{(58)}{\ge} -t_0$ and one applies (62).

Remark 2. The point (v, w) supplied by the theorem III is not, like in the theorem II, an authentic minimal point.

And now

8.12 *Vectorial variant of Ekeland principle* ([35]). *Let be* $\varphi : \mathrm{X} \to \mathrm{Y}^\bullet$ *proper*, $\mathrm{x}_0 \in \mathrm{dom}\, \varphi$ *and* $\varepsilon > 0$. *If*

$$(64) \qquad \textit{for any } \mathrm{r} \textit{ from } \mathbf{R},\ \{\mathrm{x} \in \mathrm{X} : \varphi(\mathrm{x}) \overset{C}{\le} \varphi(\mathrm{x}_0) + \mathrm{r}\, \mathrm{k}_0\} \textit{ is closed}$$

and

$$(65) \qquad \varphi(\mathrm{X}) \cap (\varphi(\mathrm{x}_0) - \varepsilon \mathrm{k}_0 - \mathrm{C}) = \varnothing,$$

then there is v *in* dom φ *so that*

$$(66)\qquad \varphi(v) + \sqrt{\varepsilon}\, k_0\, d(v, x_0) \overset{C}{\leq} \varphi(x_0),\ d(v, x_0) \leq \sqrt{\varepsilon}$$

and

$$(67)\qquad \varphi(x) + \sqrt{\varepsilon}\, k_0\, d(v, x) \overset{C}{\leq} \varphi(v) \Rightarrow x = v.$$

■ Apply 8.11 with A = epi φ, $y_0 = \varphi(x_0)$, $t_0 = \sqrt{\varepsilon}$ and k_0 replaced by $\sqrt{\varepsilon}\, k_0$. a) is verified *via* (65), b) is verified *via* (64) and concerning c), we observe that $A_x - y_0 = \{\varphi(x) - \varphi(x_0)\}$, that is a single element, ∀ x from X. There is $(v, \varphi(v))$ which verifies (55) and (56). So (66) are obvious (see (55) and Remark 1), and (67) as well, *via* (56). ■

Remark. The variant 8.12 does not require „φ lower C-bounded" as 8.6 and 8.7 do this.

Taking in 8.12 $Y = \mathbf{R}$, $Y^{\bullet} = (-\infty, +\infty]$, $C = [0, +\infty)$ and $k_0 = 1$ one finds (φ l. s. c. ⇒ (64))

Let $\varphi : X \to (-\infty, +\infty]$ *be lower semicontinuous and proper. For any* x_0 *from* dom φ *and* ε > 0, *if*

$$\varphi(X) \cap (\varphi(x_0) - \varepsilon - [0, +\infty)) = \varnothing,$$

then there is v *in* X *so that*

$$\varphi(v) < \varphi(x) + \sqrt{\varepsilon}\, d(v, x)\ \forall\, x \in X \setminus \{v\}$$

and

$$\varphi(v) \leq \varphi(x_0) - \sqrt{\varepsilon}\, d(v, x_0),\ d(v, x_0) \leq \sqrt{\varepsilon}\,.$$

MINIMAL POINT THEOREM IV

In this section is presented a minimal point theorem in a product space $X \times Y$ that uses „distances" on the set X with values into a cone C, C-*valued metrics*. It is inspired by a vectorial variant of Ekeland principle, obtained by A. B. Nemeth, where such metrics appear ([51], 2).

Definitions. Let X be nonempty set, Y real separated locally convex space and C convex cone with vertex in Y. C is named *normal* if 0 has a basis of full neighbourhoods with respect to C (the subset M of Y is *full* with respect to C if (68) $M = (M + C) \cap (M - C)$).

The map $r : X \times X \rightarrow C$ is, by definition, C-*valued metric on* X or C-*metric on* X if it verifies the usual distance axioms:

$$r(x_1, x_2) = 0 \Leftrightarrow x_1 = x_2,\ r(x_1, x_2) = r(x_2, x_1),$$

$$r(x_1, x_2) \overset{C}{\leq} r(x_1, x_3) + r(x_2, x_3).$$

Recapitulation. Let A be nonempty set and „≤" a binary relation in A. This is called *preorder relation* if it is reflexive and transitive. In this case A becomes *preordered.* „ ≤ " is *filtrable to right* when $\forall\ x, y$ from $A\ \exists\ z$ in A so that $x \leq z, y \leq z$.

Let E be a set and A filtrable to right preordered set. The family $(x_\alpha)_{\alpha\in A}$, $x_\alpha \in E$ is named *generalized sequence*. Suppose E topological space. The generalized sequence $(x_\alpha)_{\alpha\in A}$ from E *converges* to x from E, $\lim\limits_{\alpha\in A} x_\alpha = x$ or $x_\alpha \xrightarrow[\alpha\in A]{} x$, if for any V neighbourhood of $x\ \exists\ \alpha_V$ in A so that

$$\alpha \geq \alpha_V \Rightarrow x_\alpha \in V \text{ (Moore - Smith).}$$

Return to the subject of this section and let be (X, r), r C-metric on X, and $(x_i)_{i\in I}$ generalized sequence in X. $(x_i)_{i\in I}$ r-*converges* to x,

$$x_i \xrightarrow[i\in I]{r} x,$$

if $r(x_i, x) \xrightarrow[i\in I]{} 0$ ($(r(x_i, x))_{i\in I}$ is generalized sequence in Y). $(x_i)_{i\in I}$ is r-*fundamental* or r-*Cauchy* if for any V neighbourhood of 0 in Y there is i_V in I so that

$$i, j \geq i_V \Rightarrow r(x_i, x_j) \in V.$$

(X, r) is named *complete* when every generalized sequence r-fundamental from X is r-convergent.

Suppose C normal. Then

(69) the r-limit of any generalized sequence is unique

and

(70) any r-convergent generalized sequence is r-fundamental.

Indeed, note $\mathscr{V}(0)$ the set of neighbourhoods of 0 from Y and let be $r(x_i, x) \xrightarrow[i\in I]{} 0$ and $r(x_i, y) \xrightarrow[i\in I]{} 0$, V from $\mathscr{V}(0)$ which verifies (68), W symmetrical from $\mathscr{V}(0)$ with $W \subset V$, W_1, W_2 from $\mathscr{V}(0)$ with $W_1 + W_2 \subset W$. $\exists\ i_1, i_2$ in I so that $i \geq i_1 \Rightarrow r(x_i, x) \in W_1$, $i \geq i_2 \Rightarrow r(x_i, y) \in W_2$, so if $i_3 \geq i_1, i_2$ (I is filtrable to right) then $i \geq i_3 \Rightarrow r(x_i, x) + r(x_i, y) \in W$. Suppose $i \geq i_3$. $r(x, y) \overset{C}{\leq} r(x_i, x) + r(x_i, y)$, i. e. $r(x, y) - (r(x_i, x) + r(x_i,$

$y)) \in -C$, hence $r(x, y) \in W - C \subset V - C$. Likewise, as $r(x, y) + r(x_i, x) - r(x_i, y) \in C$, $r(x, y) \in C + r(x_i, y) - r(x_i, x) \overset{8.1}{\subset} C - r(x_i, x) \subset C - W = C + W$, hence $r(x, y) \in C + W \subset$ $\subset C + V$, which implies $r(x, y) \overset{(68)}{\subset} V$. Consequently $r(x, y) \in \bigcap_{U \in \mathscr{V}(0)} U$, therefore $r(x, y) - 0$ because Y is separated, $x = y$, i.e. (69).

The proof for (70) is identical: let be $r(x_i, x) \to 0$, use as above the relations $r(x_i, x) +$ $+ r(x_j, x) - r(x_i, x_j) \in C$ and $r(x_i, x_j) + r(x_j, x) - r(x_i, x) \in C$.

Add to these

$$(71) \qquad x_i \xrightarrow[i \in I]{r} x \Rightarrow r(x_i, y) \xrightarrow[i \in I]{} r(x, y).$$

Indeed, let V be from $\mathscr{V}(0)$ full with respect to C and W symmetrical from $\mathscr{V}(0)$ with $W \subset$ $\subset V$. $\exists\, i_W$ in I so that $i \geq i_W \Rightarrow r(x_i, x) \in W$. But $r(x_i, y) - r(x, y) \in C - r(x_i, x)$, hence $i \geq i_W \Rightarrow r(x_i, y) - r(x, y) \in C - W = W + C \subset V + C$. On the other hand, $i \geq i_W \Rightarrow r(x_i,$ $y) - r(x, y) = r(x_i, y) + r(x_i, x) - r(x_i, x) - r(x, y) \in W - C \subset V - C$, hence $i \geq i_W \Rightarrow r(x_i,$ $y) - r(x, y) \in V$ and consequently (71).

Example. Let d be a distance on the set X set and y_0 from $Y \setminus \{0\}$. $C := [0, +\infty)y_0$ is convex cone with vertex. The map $r : X \times X \to C$, $r(x_1, x_2) = \mathrm{d}(x_1, x_2)y_0$ is C-metric on X because, for instance, $r(x_1, x_2) \overset{C}{\leq} r(x_1, x_3) + r(x_2, x_3)$ as $\mathrm{d}(x_1, x_3) + \mathrm{d}(x_2, x_3) - \mathrm{d}(x_1, x_2) \geq 0$. Observe that a generalized sequence from X is r-convergent resp. r - fundamental iff it is convergent resp. fundamental.

Definitions. Let C_0 and C be, $C_0 \subset C$ convex cones with vertex from Y. C_0 is *sequentially C-bound*, *C-seq*-b, if any sequence from C_0 which is C_0-increasing and C-bounded is fundamental. In this case (72) $C_0 \cap (-C) = \{0\}$. In particular C_0 is pointed (8.2).

Let now be C_0 and C, $C_0 \subset C$, convex cones with vertex from Y and r C_0-metric on X. Consider the binary relation $\overset{r}{\leq}$ in $X \times Y$,

$$(73) \qquad (x_1, y_1) \overset{r}{\leq} (x_2, y_2) \overset{\text{def}}{\Leftrightarrow} y_1 + r(x_1, x_2) \overset{r}{\leq} y_2.$$

„$\overset{r}{\leq}$" is a preorder relation and, in the case C pointed, even an order relation.

Condition (H5). Let A be a nonempty subset of $X \times Y$. For any generalized sequence $(x_i, y_i)_{i\in I}$ from A $\overset{r}{\leq}$-decreasing and with $x_i \xrightarrow[i\in I]{r} x$, there is y in Y so that

$$(74) \qquad (x, y) \in A \text{ and } (x, y) \overset{r}{\leq} (x_i, y_i) \ \forall\, i \in I.$$

Condition (H6). Let A be nonempty subset of $X \times Y$. For any generalized sequence $(x_i, y_i)_{i\in I}$ from A with $x_i \xrightarrow[i\in I]{r} x$ and $(y_i)_{i\in I}$ C-decreasing, there is y in Y so that

$$(75) \qquad (x, y) \in A \text{ and } y \overset{C}{\leq} y_i \ \forall\, i \in I.$$

8.13 *If* C_0 *is normal and* $C \cap (y - C_0)$ *closed for any* y *from* C, *then*

$$(H6) \Rightarrow (H5).$$

■ Let $(x_i, y_i)_{i\in I}$ be r-decreasing generalized sequence from A with $x_i \xrightarrow[i\in I]{r} x$. Then $(y_i)_{i\in I}$ is C-decreasing because $i_1 \leq i_2 \Rightarrow y_{i_1} + r(x_{i_1}, x_{i_2}) \overset{C}{\leq} y_{i_2} \Rightarrow y_{i_2} - y_{i_1} \in C + r(x_{i_1}, x_{i_2}) \subset \subset C + C_0 \overset{8.1}{\subset} C + C \subset C$. So, acording to (H6), $\exists\, y$ in Y so that $(x, y) \in A$ and (76) $y \overset{C}{\leq} y_i \ \forall\, i \in \in I$. We have then for i and j arbitrary from I, $i \leq j$ (see (76)), $y + r(x_i, x_j) \overset{C}{\leq} y_j + r(x_i, x_j) \overset{C}{\leq} y_i$ (r-decreasing sequence). Take the first and the last term,

$$y_i - y - r(x_i, x_j) \in C \cap (y_i - y - C_0),$$

fix i and by passing to the limit for $j \in I$ one obtains, according to (71), the second member being closed set,

$$y_i - y - r(x_i, x) \in C, \text{ that is } (x, y) \overset{r}{\leq} (x_i, y_i). \ ■$$

8.14 Theorem IV ([35], th. 10). *Let* C *be pointed convex cone with vertex in* Y, C_0 *convex cone with vertex normal* C-*seq*-b, r C_0-*metric on* X *with* (X, r) *complete and* A *nonempty subset of* X × Y *with* $\pi_Y(A)$ *lower* C-*bounded and which verifies* (H5). *Then, for any* (x_0, y_0) *from* A, *there is* (v, w) *in* A $\overset{r}{\leq}$-*minimal point so that*

$$(v, w) \overset{r}{\leq} (x_0, y_0).$$

■ C being pointed, $\overset{r}{\leq}$ is an order relation in $X \times Y$. Let be $B := \{(x, y) \in A : (x, y) \overset{r}{\leq} \overset{r}{\leq} (x_0, y_0)\}$, a nonempty r-ordered set. Apply Hausdorff theorem to the r-totally ordered set $\{(x_0, y_0)\}$, there is M r-totally ordered subset of B with $(x_0, y_0) \in M$ and maximal with respect to the inclusion. Put on the set $I := M$ the totally order relation $\leq$:

(77) $$(x, y) \leq (x', y') \overset{\text{def}}{\Leftrightarrow} (x', y') \overset{r}{\geq} (x, y)$$

and consider the family $(x_i, y_i)_{i\in I}$, where $i = (x, y) \Rightarrow (x_i, y_i) = (x, y)$. This family is even generalized sequence, because I is filtrable to the right being totally ordered, and, according to (77), r-decreasing. Prove

(78) the generalized sequence $(x_i)_{i\in I}$ is r-fundamental.

Suppose *par absurdum* the contrary. Then there is V neighbourhood of 0 in Y so that

$$\forall\, i \text{ of } I\ \exists\, j, k \text{ in } I, j, k \geq i \text{ and } r(x_j, x_k) \notin V.$$

So for $i_0 := (x_0, y_0)$ $\exists\, i_1, i_2$ in I, $i_2 \geq i_1$, $i_1 \geq i_0$ and $r(x_{i_1}, x_{i_2}) \notin V$. For i_2 $\exists\, i_3, i_4$ in I, $i_4 \geq i_3$, $i_3 \geq i_2$ and $r(x_{i_4}, x_{i_3}) \notin V$ etc. One obtains a sequence $\leq$-increasing $(i_n)_{n\geq 0}$ with

(79) $$r(x_{i_{2n+2}}, x_{i_{2n+1}}) \notin V\ \forall\, n \geq 0.$$

Let be $i_n := (x_n, y_n)$. Then the sequence $(x_n, y_n)_{n\geq 0}$ from M is r-decreasing ((77)), that is $y_{n+1} + r(x_{n+1}, x_n) \overset{C}{\leq} y_n\ \forall\, n \geq 0$, from which we obviously obtain

(80) $\forall\, n \geq 0$ $s_n := \sum_{k=0}^{n} r(x_{k+1}, x_k) \overset{C}{\leq} \sum_{k=0}^{n} (y_k - y_{k+1}) = y_0 - y_{n+1} \overset{C}{\leq} y_0 - y_\mu$, y_μ C-minorant of $\pi_Y(A)$. Observe that the sequence $(s_n)_{n\geq 0}$ from C_0 is C_0-increasing and upper C-bounded ((80)), consequently it is fundamental because C_0 is C-seq-b. So for the neighbourhood V ((79)) $\exists\, N \geq 0$ so that $m, n \geq N \Rightarrow s_m - s_n \in V$, in particular $r(x_{2N+2}, x_{2N+1}) = s_{2N+1} - s_{2N} \in V$, which contradicts (79) and imposes (78).

Then let be $x_i \xrightarrow[i\in I]{r} v$ ((X, r) is complete). According to (H5) $\exists\, w$ in Y so that $(v, w) \in A$ and $(v, w) \overset{r}{\leq} (x_i, y_i)\ \forall\, i \in I$, i. e.

(81) $$(v, w) \overset{r}{\leq} (x, y)\ \forall\, (x, y) \in M.$$

In particular, $(v, w) \overset{r}{\leq} (x_0, y_0)$, hence $(v, w) \in B$ and even (82) $(v, w) \in M$, otherwise using $M \cup \{(v, w)\}$, which *via* (81) appears r-totally ordered, one gets in collision with M maximal with respect to the inclusion. (81) and (82) give (v, w) the smallest element of

M. (v, w) is even r-minimal element of A. Indeed, suppose *par absurdum* $(v_1, w_1) \in A$ and

$$(83) \qquad (v_1, w_1) \overset{r}{<} (v, w).$$

Obviously $(v_1, w_1) \overset{r}{\leq} (x_0, y_0)$ ((81)), hence $(v_1, w_1) \in B$ and therefore $(v_1, w_1) \in M$, in the contrary case considering the set $M \cup \{(v_1, w_1)\}$ we are leading, *via* (81) and (83), to a contradiction. So $(v, w) \overset{r}{\leq} (v_1, w_1)$ and the confrontation with (83) gives the final contradiction. ■

Theorem IV easily supplies the following

8.15 *Vectorial variant of Ekeland principle* ([35]). *Let be* $\varphi : X \to Y^{\bullet}$, C *pointed convex cone with vertex in* Y, C *normal convex cone with vertex* C-*seq*-b *with* $C \cap (y - - C_0)$ *closed* $\forall\, y \in C$ *and* r C_0-*metric on* X, (X, r) *complete.*

Suppose φ C-*bounded from below and* $\forall\ (x_i)_{i \in I}$ *generalized sequence from* X *with* $x_i \xrightarrow[i \in I]{r} x$ *and* $(\varphi(x_i))_{i \in I}$ C-*decreasing we have*

$$\varphi(x) \leq \varphi(x_i)\ \forall\, i \in I.$$

Then for any x_0 *from* dom φ *there is* v *in* dom φ *so that*

$$(84) \qquad \varphi(v) + r(v, x_0) \overset{C}{\leq} \varphi(x_0)$$

and

$$(85) \qquad \varphi(x) + r(v, x) \overset{C}{\leq} \varphi(v) \Rightarrow x = v.$$

■ Let be $A := \operatorname{Gr} \varphi$, the graph of φ. A obviously verifies (H6) and even (H5) (according to 8.13). Apply 8.14 with $(x_0, y_0) = (x_0, \varphi(x_0))$. There is (v, w) $\overset{r}{\leq}$-minimal point in A with $(v, w) \overset{r}{\leq} (x_0, \varphi(x_0))$ and from this (84) and (85) (see (73)). ■

9

Other variants of Ekeland principle

A. VECTORIAL VARIANT OF EKELAND PRINCIPLE ([70])

Some definitions and notations. Let X be nonempty set.

The subset M of X is called *proper* if $\varnothing \neq M \neq X$.

Let be $\zeta : X \to \mathbf{R}$ and A nonempty subset of X.

(1) $$\zeta(A) \geq 0,\ \zeta(A) > 0,\ \zeta(A) \leq 0,\ \zeta(A) < 0$$

means:

$\forall x$ from A we have respectively

$$\zeta(x) \geq 0,\ \zeta(x) > 0,\ \zeta(x) \leq 0,\ \zeta(x) < 0.$$

Let Y be real vector space and $M \subset Y$, $M \neq \varnothing$. The functional $\zeta : Y \to \mathbf{R}$ is by definition M - *monotonous* (resp. M - *strictly monotonous*) if $\forall y_1, y_2$ from Y we have

(2) $$y_1 \in y_2 + (M \setminus \{0\}) \Rightarrow \zeta(y_1) \geq \zeta(y_2) \text{ (resp. } \zeta(y_1) > \zeta(y_2)).$$

We will use a separation theorem of two sets from which only one is necessarily convex ([33]).

9.1 Theorem. *Let* Y *be a real topological vector space and* D *convex open proper set from* Y *with the property that there is* k_0 *in* $Y \setminus \{0\}$ *so that*

(3) $$\overline{D} - \alpha k_0 \subset \overline{D} \ \ \forall \alpha \geq 0,$$

(4) $$Y = \bigcup_{\alpha \in \mathbf{R}} (\overline{D} + \alpha k_0) .$$

Then there is $\zeta : Y \to \mathbf{R}$ *convex continuous*, $\zeta(Y) = \mathbf{R}$, $\zeta(y) < 0 \Leftrightarrow y \in D$, $\zeta(y) = 0 \Leftrightarrow y \in \text{Fr } D$.

Moreover:

(5) *For any nonempty subset* A *of* Y,

$$A \cap D = \varnothing \Leftrightarrow \zeta(A) \geq 0 \textit{ and } \zeta(\text{int } A) > 0;$$

(6) *If* $\text{FrD} - M \subset \overline{D}$ resp $\text{FrD} - (M \setminus \{0\}) \subset D$, ζ *is* M - *monotonous* resp. M - *strictly monotonous*;

(7) *If* $\text{FrD} + \text{FrD} \subset \overline{D}$, *then* ζ *is subadditive*

(*Chr.Gerth* (*Tammer*) - *P. Weidner*).

We call ζ the **Tammer functional** (see (7_{19})).

■ Set, for each α from **R**,

(7_1) $$S_\alpha : = \overline{D} + \alpha k_0 .$$

Then (4) becomes

(7_2) $$Y = \bigcup_{\alpha \in \mathbf{R}} S_\alpha .$$

But since $\overline{D} - \alpha k_0 \subset \overline{D}$ $\forall\, \alpha \geq 0$, in fact we have, *via* (7_2),

(7_3) $$Y = \bigcup_{\alpha \in \mathbf{R}_+} S_\alpha .$$

Prove

I. $y \in \overline{D} + \lambda k_0 \Rightarrow y \in D + \mu k_0 \ \forall\, \mu > \lambda$.

Obviously (7_4) $a_1 : = y - \lambda k_0 \in \overline{D}$. We have also (7_5) $a_2 : = y - \mu k_0 \in \overline{D}$, because (7_6) $a_2 = a_1 - t_1 k_0$, $t_1 : = \mu - \lambda > 0$ (see (3)).

Let now z_1 be *anyone* from Y and (7_7) $z_2 : = z_1 + a_2$. According to (7_3), $\exists\, a_3$ in $\overline{D}$ and $r_1 \geq 0$ so that (7_8) $z_2 = a_3 + r_1 k_0$. If $r_1 = 0$, then (7_9) $z_2 = a_2 + z_1 \in \overline{D}$. In the case $r_1 > 0$, putting (7_{10}) $t_2 : = \dfrac{r_1}{r_1 + t_1}$, hence $t_2 \in (0, 1)$, we have $t_2 a_2 + (1 - t_2) z_2 \overset{(7_8),(7_6)}{=} t_2 (a_1 - t_1 k_0) +$

$+ (1 - t_2)(a_3 + r_1 k_0) = t_2 a_1 + (1 - t_2) a_3 \in \overline{D}$ ($\overline{D}$ is convex !), hence $a_2 + (1 - t_2)(z_2 - a_2) \in$ $\in \overline{D}$ and so (see (7_9))

(7_{11}) $$a_2 + t_3 z_1 \in \overline{D} ,\ t_3 : = 1 - t_2 \in (0, 1).$$

It follows, according to (7_{11}), as $\overline{D}$ is convex (graphic representation in $\mathbf{R}^2$),

(7_{12}) $$a_2 + tz_1 \in \overline{D} \ \forall t \in [0, t_3],$$

that is a_2 has the property $\forall\ z_1$ from $Y \ \exists\ t_3$ in (0, 1) so that (7_{12}) is verified, $a_2 \in \overline{D}^i$ (the algebraic interior, §8, „Minimal point theorem III", preliminaries), but $\overline{D}^i = \text{int}\,\overline{D}$ (*R. B. Holmes*, *Geometric functional analysis and its applications*, Springer Verlag, New York, 1975) and int $\overline{D} = D$, consequently $a_2 \in D$, i. e. (see (7_5)) $y - \mu k_0 \in D$, I is proved.

From (7_3) and I, we also find

(7_{13}) $$Y = \bigcup_{\alpha \in \mathbf{R}_+} (D + \alpha k_0)\,.$$

In particular one retains

(7_{14}) $$\lambda < \mu \Rightarrow S_\lambda \subset S_\mu\,.$$

II. *For any* y *from* Y $\exists\ \lambda$ *in* **R** *so that*

$$y \notin S_\lambda\,.$$

Suppose *par absurdum* $\exists\ y_0$ in Y so that $y_0 \in S_\lambda\ \forall\ \lambda \in \mathbf{R}$. But then I gives ($7_{15}$) $y_0 \in D + \lambda k_0\ \forall\ \lambda \in \mathbf{R}$. Let y be any from Y. If we prove

(7_{16}) $$y \in D,$$

this means that $Y = D$, a contradiction because D is proper. Take (7_{17}) $z := 2y - y_0$. But (7_{18}) $z = a + rk_0$, $a \in D$, $r \in \mathbf{R}$ (see (7_{13})) and then $y = \frac{1}{2}(2y - y_0 - rk_0) + \frac{1}{2}(y_0 + rk_0) \overset{(7_{17})}{=} \frac{1}{2}(z - rk_0) + \frac{1}{2}(y_0 + rk_0) \overset{(7_{18})}{=} \frac{1}{2}a + \frac{1}{2}(y_0 + rk_0) \in D$, because a, $y_0 + rk_0 \in D$, D convex, consequently (7_{16}).

III. *Definition and properties of the functional* ζ.

For any y from Y take

(7_{19}) $$\zeta(y) = \inf\,\{\lambda \in \mathbf{R} : y \in S_\lambda\}.$$

The set which appears in 7.19 is not empty, (7_2) this ensures. (7_{14}) imposes $\zeta(y) < +\infty$ and II $\zeta(y) > -\infty$, i. e. ζ takes values in **R**.

Definition (7_{19}) is clearly inspired by the Luc functional (§8, Minimal point theorem III).

These are the properties of ζ, r anyone from **R**.

III 1. $\zeta(y) < r \Leftrightarrow y \in D + rk_0$;

III 2. $\zeta(y) \le r \Leftrightarrow y \in \overline{D} + rk_0$;

III 3. $\zeta(y) = r \Leftrightarrow y \in \mathrm{Fr}D + rk_0$;

III 4. $\zeta(y) \geq r \Leftrightarrow y \notin D + rk_0$;

III 5. $\zeta(y) > r \Leftrightarrow y \notin \overline{D} + rk_0$.

Pass to the proof of these properties.

III 1. *Necessary.* $\zeta(y) < r \overset{(7_{19})}{\Rightarrow} \exists\, \lambda < r$ so that $y \in S_\lambda$, hence $y \in D + rk_0$ (see I). *Sufficient.* $y \in D + rk_0 \Rightarrow y = a + rk_0$, $a \in D$, $\exists\, t > 0$ so that $a + tk_0 \in D$ (open set !), taking $\lambda := r - t$ we have $y = a + rk_0 = a + (r - \lambda)k_0 + \lambda k_0 = (a + tk_0) + \lambda k_0 \in D + \lambda k_0$, hence $y \in S_\lambda$ with $\lambda < r$ and so $\zeta(y) \overset{(7_{19})}{<} r$.

III 2. *Necessary.* Let be $\zeta(y) \leq r$ and U some neighbourhood of rk_0. $\exists\, \lambda > r$ so that $\lambda k_0 \in U$. Because $\zeta(y) \leq r$, we have $y \in S_\lambda$, that is $\lambda k_0 \in y - \overline{D}$, so $\lambda k_0 \in U \cap (y - \overline{D})$, rk_0 is adherent point of $y - \overline{D}$. But $\overline{y - \overline{D}} = y - \overline{D}$, hence $rk_0 \in y - \overline{D}$, $y \in rk_0 + \overline{D}$. *Sufficient.* $y \in \overline{D} + rk_0 \overset{(7_{19})}{\Rightarrow} \zeta(y) \leq r$.

III 4 results from III 1, III 5 results from III 2, III 3 results from III 2 and III 4, the properties III 1 – III 5 are demonstrated.

III 3 obviously supplies

$$\zeta(Y) = \mathbf{R}. \tag{7_{20}}$$

(7_{19}) cam be completed:

$$y \in \overline{D} + \zeta(y)k_0. \tag{7_{21}}$$

Indeed, let be $(\lambda_n)_{n\geq 1}$, $y \in S_{\lambda_n}$, decreasing sequence with $\lim\limits_{n\to\infty} \lambda_n = \zeta(y)$. But $y = a_n + \lambda_n k_0$, $a_n \in \overline{D}$ $\forall\, n \geq 1$, pass to the limit, $y = a + \zeta(y)k_0$, $a := \lim\limits_{n\to\infty} a_n$ and it remains to remark $a \in \overline{D}$ (closed set !).

ζ *is continuous* as $\forall\, u, v$ from $\mathbf{R}$, $u < v$ we have

$$\zeta^{-1}((u, v)) = (D + vk_0) \cap (Y \setminus (\overline{D} + uk_0))$$

(III 1, III 5), an open set.

ζ *is convex*: let y_1, y_2 be from Y and $\lambda_1, \lambda_2 \geq 0$, $\lambda_1 + \lambda_2 = 1$, we have

$$\lambda_1 y_1 + \lambda_2 y_2 \in \overline{D} + [\lambda_1 \zeta(y_1) + \lambda_2 \zeta(y_2)]k_0, \text{ since } y_i \overset{(7_{21})}{\in} \overline{D} + \zeta(y_i)k_0,\ i = 1, 2 \tag{7_{22}}$$

and hence

$\lambda_1 y_1 + \lambda_2 y_2 \in \lambda_1 \overline{D} + \lambda_2 \overline{D} + [\lambda_1 \zeta(y_1) + \lambda_2 \zeta(y_2)]k_0 \subset \overline{D} + [\lambda_1 \zeta(y_1) + \lambda_2 \zeta(y_2)]k_0$ ($\overline{D}$ is convex). Combine (7_{22}) with III 2.

(7_{23}) $$y \in D \Leftrightarrow \zeta(y) < 0,\ y \in \mathrm{Fr}\, D \Leftrightarrow \zeta(y) = 0.$$

Apply III 1 resp. III 3 with $r = 0$.

IV. *Prove* (5). $A \cap D = \varnothing \overset{(7_{23})}{\Rightarrow} \zeta(A) \geq 0$. Moreover , as int $A \cap D = \varnothing$ it results int $A \cap \cap \overline{D} = \varnothing$, apply III 5 with $r = 0$ and obtain $\zeta(\text{int } A) > 0$. Reciprocally , if only $\zeta(A) \geq 0$, this attracts, v*ia* (7_{23}), $A \cap D = \varnothing$.

V. *Prove* (6). Suppose (7_{24}) Fr $D - M \subset \overline{D}$. Let y_1 , y_2 be with $y_1 \in y_2 + (M \setminus \{0\})$ and $r := \zeta(y_1)$, then $y_1 \overset{\text{III 3}}{\in} \mathrm{Fr}\, D + rk_0$, that is $y_1 = a + rk_0$, $a \in \mathrm{Fr}\, D$. So

$$y_2 \in y_1 - M = (a + rk_0) - M \overset{(7_{24})}{\subset} \overline{D} + rk_0$$

and hence $\zeta(y_2) \leq r$ (III 2).

Suppose (7_{25}) Fr $D - (M\setminus \{0\}) \subset D$ and let y_1 , y_2 be as above. $y_2 \in y_1 - (M\setminus\{0\}) = (a + rk_0) - (M \setminus \{0\}) \overset{(7_{25})}{\subset} D + rk_0$ and, according to III 1, one obtains $\zeta(y_2) < \zeta(y_1)$.

VI. *Prove* (7). Let y_1 , y_2 be from Y and $r_i := \zeta(y_i)$, i=1, 2. Then $y_i \overset{\text{III 3}}{\in} \mathrm{Fr}\, D + r_i k_0$, so $y_1 + y_2 \in \mathrm{Fr}\, D + \mathrm{Fr}\, D + (r_1 + r_2)k_0 \overset{\text{hypothesis}}{\subset} \overline{D} + (r_1 + r_2)k_0$ and consequently (III 2)

$$\zeta(y_1 + y_2) \leq r_1 + r_2 = \zeta(y_1) + \zeta(y_2).\ \blacksquare$$

Remark. Theorem 9.1 is Theorem 2.2 from [33]. Its wording was modified because the assertion from a), p.306:

„ $A \cap C = \varnothing \Rightarrow$ there is a continuous convex functional $z : Y \to \mathbf{R}$..."

is false. The existence of z results from (ii)+(iii)+(iv).

The formula (7_{21}) $y \in \overline{D} + \zeta(y)k_0$ has been observed by the author. Only with this it can prove the convexity of ζ (z in Gerth - Weidner notation).

Retain from the demonstration of 9.1 the expression of the Tammer functional

$$\zeta(y) = \inf \{\alpha \in \mathbf{R} : y \in S_\alpha\},\ S_\alpha := \overline{D} + \alpha k_0$$

and its properties:

(8) $\zeta(y) < r \Leftrightarrow y \in D + rk_0$, (9) $\zeta(y) \leq r \Leftrightarrow y \in \overline{D} + rk_0$,

(10) $\zeta(y) \geq r \Leftrightarrow y \notin D + rk_0$, (11) $\zeta(y) > r \Leftrightarrow y \notin \overline{D} + rk_0$,

(12) $$\zeta(y) = r \Leftrightarrow y \in \mathrm{Fr}\, D + rk_0 .$$

From now on Y *is a real topological vector space.*

Definitions and notations. Let E , B be proper subsets of Y.

The element y_1 from E is an *efficient point* of E *with respect* to B if

(13) $$y_1 \notin y + (B \setminus \{0\}) \ \forall\, y \in E.$$

Let k_0 be from $Y \setminus \{0\}$ with (14) $B + \alpha k_0 \subset B \ \forall\, \alpha > 0$, $\varepsilon > 0$ and $B_{\varepsilon k_0} : -\varepsilon k_0 + (B \setminus \{0\})$. The element y_1 from E is an εk_0-*efficient point* of E *with respect* to B if

(15) $$y_1 \notin y + B_{\varepsilon k_0} \ \ \forall\, y \in E.$$

The set of points from E which verify (13) resp (15) are denoted by

$$\text{Eff}\,(E, B) \text{ resp Eff}\,(E, B_{\varepsilon k_0}).$$

We obviously have

(16) $$\text{Eff}\,(E, B) \subset \text{Eff}\,(E, B_{\varepsilon k_0}).$$

Let E, B, M be proper subsets of Y and k_0 from $Y \setminus \{0\}$. Further, at a moment or another, the following conditions appear:

(C1) B is convex open with $0 \in \overline{B} \setminus B$ and

$$Y = \bigcup_{\alpha \in \mathbf{R}} (\overline{B} + \alpha k_0)\,;$$

(C2) $\alpha k_0 \in M \setminus \{0\} \ \forall\, \alpha > 0$ and $0 \in \overline{M} \setminus M$;

(C3) $\overline{B} + (M \setminus \{0\}) \subset B$;

(C4) $\overline{B} + \overline{B} = \overline{B}$.

Moreover, let X be real normed space, F nonempty closed subset of X and $f : X \to Y$.

f is M-*bounded from below on* F if there is y_M in Y so that

$$f(F) \subset y_M + M.$$

f is *lower semicontinuous on* F *with respect to* k_0 *and* B, (k_0, B)-l. s. c. , if the set

(17) $$T^{F,B}_{r,k_0}(f) := \{x \in F : f(x) \in rk_0 - \overline{B}\,\}$$

is closed for any r from $\mathbf{R}$.

In the particular case $Y = \mathbf{R}$, $B = (0, +\infty)$ and $k_0 = 1$ we have $T^{F,B}_{r,k_0}(f) = \{x \in F : f(x) \le r\}$, we come to usual definition of the lower semicontinuity for $f\,|\,F$.

For instance, when $Y = \mathbf{R}^2$, $F = [-2, 2]$, $B = \{(x, y) \in \mathbf{R}^2 : x \ge 0, y \ge 0\}$ and $k_0 \in B \setminus \{(0, 0)\}$, the function $f = (f_1, f_2): F \to \mathbf{R}^2$, $x \in F \Rightarrow f_1(x) = x$, $x \in [-2, -1] \Rightarrow f_2(x) = x^2$, $x \in (-1, 1) \Rightarrow f_2(x) = x + 3$, $f_2(1) = \dfrac{1}{2}$, $x \in (1, 2] \Rightarrow f_2(x) = x^2$ is (k_0, B) - l. s. c.

9.2 *Suppose the conditions* (C1) - (C4) *verified. If* $f : X \to Y$ *is* (k_0, B) - l. s. c. *on* F *and* ζ *is the Tammer functional from* 9.1 *with* $D = y - (B \setminus \{0\})$, y *from* Y, *then* $\zeta \circ f|F$ *is lower semicontinuous.*

■ The statement is correct, D verifies (3) and (4) from 9.1: $\overline{D} = y - \overline{B}$, we have $\forall\ \alpha \geq 0$ $\overline{D} - \alpha k_0 \subset \overline{D}$, i. e. $y - \overline{B} - \alpha k_0 \subset y - \overline{B}$, equivalently $\overline{B} + \alpha k_0 \subset \overline{B}$, as it results from (C2) and (C3); $\bigcup_{\alpha\in\mathbf{R}}(\overline{D} + \alpha k_0) = \bigcup_{\alpha\in\mathbf{R}}(y - \overline{B} + \alpha k_0) = y + \bigcup_{\alpha\in\mathbf{R}}(-\overline{B} + \alpha k_0) = y + \bigcup_{\alpha\in\mathbf{R}}(-\overline{B} - \alpha k_0) \overset{(C1)}{=} y - Y = Y.$

Let r be anyone from **R**.

$$\{x \in F : \zeta(f(x)) \leq r\} = \{x \in F : f(x) \in \{y \in Y : \zeta(y) \leq r\}\} \overset{(9)}{=} \{x \in F : f(x) \in \overline{D} + rk_0\} =$$

$$= \{x \in F : f(x) \in y - \overline{B} + rk_0\} \overset{(C1)}{=} \{x \in F : f(x) \in -r'k_0 - \overline{B} - \overline{B} + rk_0\} \overset{(C4)}{=} \{x \in F : f(x) \in$$

$$\in (r - r')k_0 - \overline{B}\} \overset{(17)}{=} T^{F,B}_{r-r',k_0}(f),$$

a closed set by hypothesis, and therefore the conclusion. ■

The following two propositions prepare the announced Ekeland variant.

9.3 *Let* B, M *be which verify the conditions* (C1), (C2), (C3) *and* y_ε *from* $\mathrm{Eff}(E, B_{\varepsilon k_0})$. *Then there exists* $\zeta : Y \to \mathbf{R}$ *convex continuous strictly* M - *monotonous and*

(18) $$\zeta(y_\varepsilon) \leq \zeta(y + \varepsilon k_0)\ \forall\ y \in E.$$

If in addition $\mathrm{Fr}\,B + \mathrm{Fr}\,B \subset y_\varepsilon + \overline{B}$, *then* ζ *is subadditive and so*

(19) $$\zeta(y_\varepsilon) \leq \zeta(y) + \zeta(\varepsilon k_0)\ \forall\ y \in E.$$

■ *First assertion.* From (C2) and (C3),

(20) $$\overline{B} + \alpha k_0 \subset B\ \forall\ \alpha > 0,\ \overline{B} + \alpha k_0 \subset \overline{B}\ \ \forall\ \alpha \geq 0$$

(the first inclusion shows that $\mathrm{Eff}(E, B_{\varepsilon k_0})$ has been correctly considered).

According to (15) we have $E \cap [y_\varepsilon - \varepsilon k_0 - (B \setminus \{0\})] = \varnothing$, that is

(21) $$(E + \varepsilon k_0) \cap [y_\varepsilon - (B \setminus \{0\})] = \varnothing.$$

Take (22) $A = E + \varepsilon k_0$, $D = y_\varepsilon - (B \setminus \{0\})$ and apply 9.1, correctly because $\overline{D} - \alpha k_0 \subset \overline{D}$ $\forall\ \alpha \geq 0$ ((3)), which is equivalent to $-\overline{B} - \alpha k_0 \subset -\overline{B}$ $\forall\ \alpha \geq 0$ validated by (20), and moreover $\bigcup_{\alpha\in\mathbf{R}}(\overline{D} + \alpha k_0) = \bigcup_{\alpha\in\mathbf{R}}(y_\varepsilon - \overline{B} + \alpha k_0) = y_\varepsilon + \bigcup_{\alpha\in\mathbf{R}}(-\overline{B} - \alpha k_0) = y_\varepsilon - \bigcup_{\alpha\in\mathbf{R}}(\overline{B} + \alpha k_0) \overset{(C1)}{=} y_\varepsilon -$

$-Y = Y$ and consequently (4). We have $\zeta(\overline{D}) \le 0$ (9.1) and, as $A \cap D \overset{(21)}{=} \varnothing$, $\zeta(A) \ge 0$ (see (5)). So, in particular (see (22)) $\zeta(y + \varepsilon k_0) \ge 0 \ \forall\, y \in E$ and $\zeta(y_\varepsilon) \le 0$ since $y_\varepsilon \in \overline{D}$ because $\overline{D} = y_\varepsilon - \overline{B}$ and $0 \overset{(1)}{\subset} \overline{B}$, (18) is proved. ζ is strictly M - monotonous (see (6)) as Fr $D -$ $-(M \setminus \{0\}) \subset D$ (use (22) and (C3)).

Second assertion. Fr B + Fr $B \subset y_\varepsilon + \overline{B} \Rightarrow$ Fr D + Fr $D \overset{(22)}{=} y_\varepsilon -$ Fr $B + y_\varepsilon -$ Fr $B = 2y_\varepsilon -$ $-$ Fr $B -$ Fr $B \subset 2y_\varepsilon - y_\varepsilon - \overline{B} \overset{(22)}{=} \overline{D}$, therefore, according to (7), ζ is subadditive. ■

Remark. The function ζ from 9.3 verifies, for any $\varepsilon > 0$, the relations

$$\zeta(0) = 0,\ \zeta(k_0) = 1,\ \zeta(\varepsilon k_0) = \varepsilon,\ \zeta(-\varepsilon k_0) = -\varepsilon.$$

Indeed, we have $D = y_\varepsilon - (B \setminus \{0\})$, $D \cup$ Fr $D = \overline{D} = y_\varepsilon - \overline{B} = y_\varepsilon - (B \cup \text{Fr } B)$ and because $D \cap$ Fr $D = B \cap$ Fr $B = \varnothing$ it results Fr $D = -$ Fr B. But $0 \overset{(C1)}{\in}$ Fr B, hence $0 \in -$ Fr B, $0 \in$ $\in$ Fr D. It only remains to apply formula (12) to obtain the mentioned relations.

9.4 *Let* M *be which verifies* (C2), $\varepsilon > 0$ *and* y_ε *from* E *so that*

$$\zeta(y_\varepsilon) \le \zeta(y) - \zeta(-\varepsilon k_0)\ \forall\, y \in E, \tag{23}$$

where $\zeta : Y \to \mathbf{R}$ *is continuous, subadditive and strictly* M - *monotonous.*

Then there is B *proper open set, with* $M \setminus \{0\} \subset B$, $0 \in \overline{B} \setminus B$ *and* $\overline{B} + (M \setminus \{0\}) \subset B$, *for which*

$$y_\varepsilon \in \mathrm{Eff}\,(E, B_{\varepsilon k_0}). \tag{24}$$

■ Take

$$B := \{y \in Y : \zeta(-y + y_\varepsilon - \varepsilon k_0) < \zeta(y_\varepsilon - \varepsilon k_0)\}. \tag{25}$$

Consider the functional $\xi : Y \to \mathbf{R}$,

$$\xi(y) = \zeta(y + y_\varepsilon - \varepsilon k_0). \tag{26}$$

ξ is also strictly M - monotonous (the definition (2)).

Let y be from E. $\xi(y + \varepsilon k_0 - y_0) \overset{(26)}{=} \zeta(y) \overset{(23)}{\ge} \zeta(y_\varepsilon) + \zeta(-\varepsilon k_0)$, hence, ζ being subadditive, (27) $y \in E \Rightarrow \zeta(y + \varepsilon k_0 - y_\varepsilon) \ge \xi(0)$. Let y be from B. $\xi(-y) = \zeta(-y + y_\varepsilon -$ $- \varepsilon k_0) \overset{(25)}{<} \zeta(y_\varepsilon - \varepsilon k_0) = \xi(0)$ and comparing with (27) one finds $(E + \varepsilon k_0 - y_\varepsilon) \cap (-B) = \varnothing$, i.e. (28) $E \cap (y_\varepsilon - \varepsilon k_0 - B) = \varnothing$. (28) expresses exactly (15).

Let y be from $M \setminus \{0\}$. ζ being strictly M - monotonous, $\zeta(-y + y_\varepsilon - \varepsilon k_0) < \zeta(y_\varepsilon - \varepsilon k_0)$, hence $y \overset{(25)}{\in} B$, $M \setminus \{0\} \subset B$. This gives also $B \neq \varnothing$, as $M \setminus \{0\} \overset{(C2)}{\neq} \varnothing$. B is even proper as, for instance, $0 \overset{(25)}{\notin} B$. Moreover, ζ being continuous, B is open. Pass to verify (29) $\overline{B} + (M \setminus \{0\}) \subset M$. Transcribe the definition (25) of B taking (26) into account:

(30) $$B = \{z \in Y : \xi(-z) < \xi(0)\}.$$

From (30),

(31) $$\overline{B} = \{z \in Y : \xi(-z) \leq \xi(0)\}.$$

Let now x be any from $\overline{B} + (M \setminus \{0\})$, that is $x \in y + (M \setminus \{0\})$ with y from $\overline{B}$. ξ being strictly M - monotonous, $\xi(-x) < \xi(-y)$ and as $\xi(-y) \overset{(31)}{\leq} \xi(0)$, we have $\xi(-x) < \xi(0)$, hence $x \overset{(30)}{\in} B$ and therefore (29).

It remains to verify (32) $0 \in \overline{B} \setminus B$, as $B + \alpha k_0 \subset B\ \forall\ \alpha > 0$, i. e. (14), is implied by (29) according to (C2), and we already have (15). Obviously $0 \in \overline{M \setminus \{0\}}$, but $M \setminus \{0\} \subset B$, hence $0 \in \overline{B}$ and consequently (32) because $0 \notin B$ (see (25)).

We can go now to

9.5 Theorem (*Vectorial variant of Ekeland principle*). *Let* B, M *be which verify* (C1)–(C4) *and* φ: X $\to$ Y, X *real Banach space*, (k_0, B) - *lower semicontinuous and* M-*lower bounded on the nonempty closed set* F.

Then for any $\varepsilon > 0$ *and* u *from* F *with*

$$\varphi(u) \in \mathrm{Eff}\,(\varphi(F), B_{\varepsilon k_0}),$$

there is v_ε *in* F *so that*

(33) $$\varphi(v_\varepsilon) \in \mathrm{Eff}\,(\varphi(F), M_{\varepsilon k_0}),$$

(34) $$\|v_\varepsilon - u\| \leq \sqrt{\varepsilon}\,,$$

(35) $$\varphi_{\varepsilon k_0}(v_\varepsilon) \in \mathrm{Eff}\,(\varphi_{\varepsilon k_0}(F), M),$$

where $\varphi_{\varepsilon k_0}(x) = \varphi(x) + \sqrt{\varepsilon}\ \|v_\varepsilon - x\| k_0$

(*Chr. Tammer*).

■ The condition to which u is submitted by hypothesis means:

(36) $$\varphi(F) \cap [\varphi(u) - \varepsilon k_0 - (B \setminus \{0\})] = \varnothing.$$

Consider the function $\psi := \varphi - \varphi(u)$. (36) gives $\psi(F) \cap [\psi(u) - \varepsilon k_0 - (B \setminus \{0\})] = \varnothing$, i.e.

(37) $$\psi(u) \in \mathrm{Eff}\,(\psi(E), B_{\varepsilon k_0}).$$

As $\psi(u) = 0$, (38) $\text{Fr } B + \text{Fr } B \overset{(C4)}{\subset} \psi(u) + \overline{B}$. Apply 9.3 with Remark (see (37)), there exists $\zeta : Y \to \mathbf{R}$ continuous strictly M - monotonous subadditive (see (38)) and $\zeta(\psi(u)) \leq$

$\overset{(19)}{\leq} \zeta(\psi(x)) + \zeta(\varepsilon k_0) = \zeta(\psi(x)) + \varepsilon \ \forall\, x \in F$, therefore

$$\zeta(\psi(u)) \leq \inf_{F} (\zeta \circ \psi) + \varepsilon. \tag{39}$$

As φ is (k_0, B) - l. s. c. , ψ has the same property. Indeed, let r be any from $\mathbf{R}$. $x \in$

$\in \psi^{-1}(rk_0 - \overline{B}) \Rightarrow \varphi(x) \in \varphi(u) + rk_0 - \overline{B}$. But $-\varphi(u) \overset{(C1)}{\in} r'k_0 + \overline{B}$, hence $\varphi(u) \in -r'k_0 -$ $- \overline{B}$, $\varphi(x) \in (r - r')k_0 - \overline{B} - \overline{B} = (r - r')k_0 - \overline{B}$, hence $\psi^{-1}(rk_0 - \overline{B}) \subset \varphi^{-1}((r - r')k_0 -$ $- \overline{B}$). The inequality of contrary sens being also true, it results $\psi^{-1}(rk_0 - \overline{B})$ closed , ψ is (k_0, B) - l. s. c. ψ is also M - bounded from below on F, because $\exists\, y_M$ in Y so that $\varphi(F) \subset$ $\subset y_M + M$ and then (40) $\psi(F) \subset \varphi(u) + y_M + M$. Consequently, $\zeta \circ \psi | F$ is l. s. c. (9.2, the function ζ from 9.3 is that from 9.1), but also bounded from below, because ζ being strictly M - monotonous, from (40), $\zeta(\psi(x)) \geq \zeta(\varphi(u) + y_M) \ \forall\, x \in F$.

F is a complete metric subspace, apply to $\zeta \circ \psi | F$ Ekeland principle 1.1 with $\lambda = \sqrt{\varepsilon}$, there is v_ε in F so that

$$\zeta(\psi(v_\varepsilon)) \leq \zeta(\psi(u)) \overset{(39)}{\leq} \inf_{F} (\zeta \circ \psi) + \varepsilon, \tag{41}$$

$$\|v_\varepsilon - u\| \leq \sqrt{\varepsilon}, \tag{42}$$

$$\zeta(\psi(v_\varepsilon)) \leq \zeta(\psi(x)) + \sqrt{\varepsilon}\ \|v_\varepsilon - x\| \ \forall\, x \in F. \tag{43}$$

From (41) and Remark to 9.3,

$$\zeta(\psi(v_\varepsilon)) \leq \inf_{F} (\zeta \circ \psi) + \varepsilon \leq \zeta(\psi(x)) + \zeta(\varepsilon k_0) = \zeta(\psi(x)) - \zeta(-\varepsilon k_0) \ \forall\, x \in F.$$

Apply 9.4 with $E = \psi(F)$ and $y_\varepsilon = \psi(v_\varepsilon)$ (see (23)), let B_1 be the set supplied by it. We have

$$\psi(F) \cap [\psi(v_\varepsilon) - \varepsilon k_0 - (B_1 \setminus \{0\})] \overset{(24)}{=} \varnothing$$

and in particular, because $M \setminus \{0\} \subset B_1$,

$$\psi(F) \cap [\psi(v_\varepsilon) - \varepsilon k_0 - (M \setminus \{0\})] = \varnothing. \tag{44}$$

From (44), by reduction of terms, one obtains

$$\varphi(F) \cap [\varphi(v_\varepsilon) - \varepsilon k_0 - (M \setminus \{0\})] = \varnothing,$$

i.e. (33). (34) being already obtained (see (42)), pass to (35). Suppose *par absurdum* $\exists\, x$

in F so that $\varphi_{\varepsilon k_0}(x) \in \varphi_{\varepsilon k_0}(v_\varepsilon) - (M \setminus \{0\})$ (so $x \neq v_\varepsilon$), i.e.

$$\varphi(x) \in \varphi(v_\varepsilon) - \sqrt{\varepsilon}\ \|v_\varepsilon - x\| k_0 - (M \setminus \{0\}). \tag{45}$$

(45) gives

$$\psi(x) \in \psi(v_\varepsilon) - \sqrt{\varepsilon}\ \|v_\varepsilon - x\| k_0 - (M \setminus \{0\}). \tag{46}$$

ζ being strictly M - monotonous and subadditive, from (46),

$$\zeta(\psi(x)) < \zeta(\psi(v_\varepsilon) - \sqrt{\varepsilon}\ \|v_\varepsilon - x\| k_0) \leq \zeta(\psi(v_\varepsilon)) + \zeta(-\sqrt{\varepsilon}\ \|v_\varepsilon - x\| k_0) \overset{\substack{9.3 \\ \text{Rem.}}}{=} \zeta(\psi(v_\varepsilon)) - \sqrt{\varepsilon}\ \|v_\varepsilon - x\|$$

and we come in collision with (43), which proves (35). ■

Remark 1. In [70] the condition (C4) is

$$\text{Fr } B + \text{Fr } B \subset \overline{B}.$$

In order to remove all doubts which hover some demonstrations from [70] (see for instance lemma 4.1 and „$\tilde{f}$ is (k_0, B) - l.s.c.", p.138), the author has been obliged to replace the original (C4) by the stronger demand

$$\overline{B} + \overline{B} = \overline{B}.$$

This solves all misunderstandings and it is easy to accept it.

Remark 2. A rapid analysis of the demonstration shows that X from 9.5 can be only complete metric space (the definitions for „φ M - bounded from below on F" and „φ (k_0, B) - l. s. c." remain unchanged in the case X topological space).

Remark 3. We have to mention that all the considerations from section A, hence in particular the Gerth - Weidner separation theorem and the Tammer version of Ekeland principle, are made with a Y real topological vector space to which is not required to be separated. This is allowed by the sentence:

E topological vector space on the *topological body* K, $\lambda \in K \setminus \{0\}$, $a \in E$, $M \subset E \Rightarrow \overline{\lambda M + a} = \lambda \overline{M} + a$, M open (closed) $\Rightarrow \lambda M$ open (closed).

(see for instance, *C. Meghea*, *Foundations of mathematical analysis*, p. 368, Bucharest, 1977).

B. VECTORIAL VARIANT OF EKELAND PRINCIPLE ([64])

We present another vectorial variant of Ekeland principle. This improves a G. Isac proposition ([37]). The proof will need ([24])

9.6 Theorem. *Let* (X, d) *be complete metric space and* $\Gamma: X \to 2^X$. *Suppose*:

a) $\Gamma(x)$ *is closed for any* x *from* X;

b) $x \in \Gamma(x)\ \forall$ x *from* X;

c) $y \in \Gamma(x) \Rightarrow \Gamma(y) \subset \Gamma(x)\ \forall$ x, y *from* X;

d) *for any sequence* $(x_n)_{n\geq 1}$ *from* X, *when* $x_{n+1} \in \Gamma(x_n)\ \forall\, n \geq 1$, *then we have* $\lim\limits_{n\to\infty} d(x_{n+1}, x_n) = 0$.

In this case, Γ *has a critical point in* Γ(x) *for any* x *from* X

(*S. Dancs - M. Hegedus - P. Medvegyev*).

Explanation. x^* is by definition *critical point* for the multivalued function Γ when

$$\Gamma(x^*) = \{x^*\}.$$

∎ We may assume $d \leq 1$, because in the contrary case one takes the distance on X $d' := \frac{d}{d+1}$ and for this it is obviously verified the condition $d'(x_{n+1}, x_n) \to 0$.

Let x be any fixed from X. Define by recurrence a sequence $(x_n)_{n\geq 1}$ so that, $\forall\, n \geq 1$,

$$(46_1)\quad x_{n+1} \in \Gamma(x_n), \quad (46_2)\quad d(x_{n+1}, x_n) \geq \frac{1}{2}\operatorname{diam}\Gamma(x_n) - \frac{1}{2^n}:$$

$x_1 = x$; suppose x_n defined, $\exists\, u_n, v_n$ in $\Gamma(x_n)$ so that $d(u_n, v_n) \geq \operatorname{diam}\Gamma(x_n) - \frac{1}{2^n}$, then we have $d(x_n, u_n) + d(x_n, v_n) \geq \operatorname{diam}\Gamma(x_n) - \frac{1}{2^n}$, hence

$$[d(x_n, u_n) - \frac{1}{2}\operatorname{diam}\Gamma(x_n)] + [d(x_n, v_n) - \frac{1}{2}\operatorname{diam}\Gamma(x_n) + \frac{1}{2^n}] \geq 0$$

and so at least a bracket is ≥ 0 (*par absurdum*); when the first bracket is ≥ 0 one takes $x_{n+1} = u_n$, when the second bracket is ≥ 0 one takes $x_{n+1} = v_n$, in both cases (46_2) is verified.

(46_1) and (46_2) give, taking into account c) resp. d),

$$(46_3)\qquad \Gamma(x_{n+1}) \subset \Gamma(x_n)\ \forall\, n \geq 1,$$

$$(46_4)\qquad \lim_{n\to\infty} \operatorname{diam}\Gamma(x_n) = 0.$$

So it can apply Cantor theorem,

$$(46_5) \qquad \bigcap_{n=1}^{\infty} \Gamma(x_n) = \{x_0\}.$$

Obviously $x_0 \in \Gamma(x)$. x_0 is even critical point of Γ because, as $x_0 \in \Gamma(x_n)\ \forall\ n \geq 1$, we have $\Gamma(x_0) \overset{c)}{\subset} \Gamma(x_n)\ \forall\ n \geq 1$, which imposes, *via* (46_5), $\Gamma(x_0) \subset \{x_0\}$ and hence $\Gamma(x_0) = \{x_0\}$. ■

Remark. If the metric space X is not complete, there is Γ with the properties from the statement, but which however has no critical point. Indeed, X being not complete, the reciprocal Cantor theorem ([47], vol. I, p.218, 1.7) supplies a decreasing sequence $(S_n)_{n\geq 1}$ of closed balls from X with $\lim\limits_{n\to\infty} \operatorname{diam} S_n = 0$ and with (46_6) $\bigcap\limits_{n=1}^{\infty} S_n = \varnothing$. Complete this sequence with $S_0 := X$. Consider the multivalued function $\Gamma: X \to 2^X$,

$$(46_7) \qquad \Gamma(x) = S_{n+1} \cup X, \text{ if } x \in S_n \setminus S_{n+1}.$$

The definition is, according to (46_6), correct ($S_0 = X$!). Obviously $\Gamma(x)$ is closed and $x \in \Gamma(x)$. We have $y \in \Gamma(x) \Rightarrow \Gamma(y) \subset \Gamma(x)$: for $y = x$ – nothing to prove; suppose $y \neq x$, hence $y \overset{(46_7)}{\in} S_{n+1}$, let p be with $y \in S_{n+p} \setminus S_{n+p+1}$, then $\Gamma(y) = S_{n+p+1} \cup \{y\}$ and so $\Gamma(y) \subset \Gamma(x)$ because $S_{n+p+1} \subset S_{n+1}$. Finally, if $x_{n+1} \in \Gamma(x_n)$, let be $\Gamma(x_n) = S_{p_n+1} \cup \{x_n\}$, it results (see (46_7)) $x_n, x_{n+1} \in S_{p_n}$ and hence $\lim\limits_{n\to\infty} d(x_n, x_{n+1}) = 0$. Nevertheless Γ has no fixed point, because if *par absurdum* $\Gamma(x_0) = \{x_0\}$, then $\exists\ q \geq 0$ so that $\{x_0\} = S_{q+1} \cup \{x_0\}$, that implies $S_{q+1} = \varnothing$, contradiction.

Pass to the subject of this section.

In the following Y *is separated real locally convex space and* C *closed convex cone with vertex in* Y. $\leq$ *is the preorder relation defined in* Y *by* C (§8, „*preliminaries of affine geometry*”). *Besides this it also appears the relation* $\leq$ *in* **R**. *Confusion is not possible.*

Definition. The function $\delta: X \times X \to Y$ is *half-distance* on the set X if

1) $\delta(x, x) = 0\ \forall\ x \in X$,

2) $\delta(x, y) \leq \delta(x, z) + \delta(z, y)\ \forall\ x, y, z \in X$.

For instance, any distance is half-distance too. Also, for $f: X \to Y$ any, $\delta: \delta(x, y) = f(y) - f(x)$ is half-distance on X. Also, V being real vector space and $T: V \to Y$ subadditive with $T(0) = 0$, for any $f: X \to V$, $\delta: \delta(x, y) = T(f(y) - f(x))$ is half-distance on X.

9.7 Theorem (*Vectorial variant of Ekeland principle*). *Let* (X, d) *be metric space,* $\delta: X \times X \to Y$ *half-distance on* X *and* k_0 *from* $C \setminus (-C)$. *Suppose*

(∗) *for any* x *from* X, *the set*

$$\{y \in X : \delta(x, y) + d(x, y)k_0 \in -C\}$$

is closed;

(∗∗) *there exists* u_0 *in* X *and* v_0 *in* Y *so that*

$$\delta(u_0, x) \geq v_0 \ \forall\, x \in X.$$

Then there is x^* *in* X *so that*

$$\delta(x^*, x) + d(x^*, x)k_0 \notin -C \ \forall\, x \textit{ from } X \setminus \{x^*\}$$

(*Shen Jie Li - Hiao Qi Yang - Gung-ya Chen*).

■ Consider the multivalued function $\Gamma: X \to 2^X$,

(47) $$\Gamma(x) = \{y \in X : \delta(x, y) + d(x, y)k_0 \in -C\}.$$

The assertion from the statement is equivalent to „ $\exists\, x^*$ so that $\Gamma(x^*) = \{x^*\}$". Certainly, we use 9.6 and pass to verify its requests.

a) Obvious (see (∗)).

b) Obvious as $0 \in -C$.

c) Let y be from $\Gamma(x)$, i.e. (see (47)) (48) $\delta(x, y) + d(x, y)k_0 \in -C$, and let z be from $\Gamma(y)$, that is (49) $\delta(y, z) + d(y, z)k_0 \in -C$. We must prove that $z \in \Gamma(x)$, i.e. (50) $\delta(x, z) + d(x, z)k_0 \in -C$.

(48) and (49) give respectively

(51) $$\delta(x, y) + d(x, y)k_0 = -c_1,\ c_1 \in C,$$

(52) $$\delta(y, z) + d(y, z)k_0 = -c_2,\ c_2 \in C.$$

Moreover (see above 2)),

(53) $$\delta(x, z) = \delta(x, y) + \delta(x, z) - c_3,\ c_3 \in C$$

and also (triangle inequality for d, C is cone with vertex)

(54) $$d(x, z)k_0 = d(x, y)k_0 + d(y, z)k_0 - c_4,\ c_4 \in C.$$

Then

$$\delta(x, z) + d(x, z)k_0 \overset{(53,)(54)}{=} \delta(x, y) + \delta(y, z) - c_3 + d(x, y)k_0 + d(y, z)k_0 - c_4 \overset{(51),(52)}{=} -c_1 - c_2 - $$
$$- c_3 - c_4 \in -C \ (\S 8,\ 8.1)$$

and consequently (50).

d) Let $(x_n)_{n\geq 1}$ be a sequence from X with the property $x_{n+1} \in \Gamma(x_n)$ $\forall$ $n \geq 1$. We must prove

$$\lim_{n\to\infty} d(x_n, x_{n+1}) = 0.$$

We have $\forall$ $n \geq 1$ (see (47))

$$\delta(x_n, x_{n+1}) + d(x_n, x_{n+1})k_0 \in -C,$$

that is

$$\delta(x_n, x_{n+1}) + d(x_n, x_{n+1})k_0 \leq 0. \tag{55}$$

Apply the Luc functional to (55) (see 8.10, 1°,2°), one finds

$$L(\delta(x_n, x_{n+1}) + d(x_n, x_{n+1})k_0) \leq L(0) = 0 \ \forall\ n \geq 1. \tag{56}$$

By succesive application of 2) one obtains

$$\delta(x_1, x_{n+1}) \leq \sum_{i=1}^{n} \delta(x_i, x_{i+1}),$$

consequently

$$L(\delta(x_1, x_{n+1}) + [\sum_{i=1}^{n} d(x_i, x_{i+1})]\ k_0) \leq L(\sum_{i=1}^{n} [\delta(x_i, x_{i+1}) + d(x_i, x_{i+1})k_0]) \overset{8.10,1^0}{\leq} \sum_{i=1}^{n} L(\delta(x_i, x_{i+1}) + d(x_i, x_{i+1})k_0),$$

hence, taking into account (56),

$$L(\delta(x_1, x_{n+1}) + [\sum_{i=1}^{n} d(x_i, x_{i+1})]\ k_0) \leq 0. \tag{57}$$

So

$$L(-\delta(x_1, x_{n+1}) - [\sum_{i=1}^{n} d(x_i, x_{i+1})]\ k_0) \overset{8.10,2^0}{\geq} -L(\delta(x_1, x_{n+1}) + [\sum_{i=1}^{n} d(x_i, x_{i+1})]k_0) \overset{(57)}{\geq} 0,$$

take the first and the last term, apply the formula from 8.10, 2° and obtain

$$\sum_{i=1}^{n} d(x_i, x_{i+1}) \leq L(-\delta(x_1, x_{n+1}))\ \forall\ n \geq 1. \tag{58}$$

But

$$-\delta(x_1, x_{n+1}) \overset{2)}{\leq} \delta(u_0, x_1) - \delta(u_0, x_{n+1}) \tag{59}$$

and then, from (58) and (59),

(60) $$\sum_{i=1}^{n} d(x_i, x_{i+1}) \leq L(\delta(u_0, x_1)) + L(-\delta(u_0, x_{n+1})) \leq L(\delta(u_0, x_1)) + L(-v_0)$$

((**) from the hypothesis).

(60) shows that the series $\sum_{n=1}^{\infty} d(x_n, x_{n+1})$ is convergent, which implies $\lim_{n\to\infty} d(x_n, x_{n+1}) =$

$= 0.$

The proof finishes by application of 9.6. ■

Remark. The author has eliminated the condition „int $C \neq \varnothing$" from [64]. Concerning the condition „$k_0 \in \text{int}\, C$", [64], Theorem 3, this has been replaced by „$k_0 \in C \setminus (-C)$".

C. VECTORIAL VARIANT OF EKELAND PRINCIPLE ([13])

In this section we present, principally, a variant of Ekeland principle.

In preparation – two sentences without proofs and one definition.

Let Y be real locally convex space and C pointed convex cone with vertex from Y (§8, „preliminaries of affine geometry"). C is normal (§8, „Minimal Point Theorem IV") iff there exists a family $(p_i)_{i\in I}$ of semi-norms generating the topology of Y with the property ([60], p. 215)

(61) $$0 \overset{C}{\leq} y_1 \overset{C}{\leq} y_2 \Rightarrow p_i(y_1) \leq p_i(y_2)\ \forall\, i \in I.$$

Moreover (the definition of C^+ is to (5) of §8)

(62) $$C \text{ normal cone} \Rightarrow Y^* = C^+ - C^+$$

([60], p. 218).

Definition. Let X be topological space, Y real topological vector space, $f: X \to Y$, C convex cone with vertex and x_0 from X. f is C-*continuous in* x_0, if for any neighbourhood V of $f(x_0)$ there is U neighbourhood of x_0 so that

(63) $$f(U) \subset V + C.$$

And now

9.8 Theorem (*Vectorial variant of Ekeland principle*). *Let* (X, d) *be complete metric space,* Y *separated real locally convex space,* C *normal cone in* Y *and* φ: X → Y C-*con-*

tinuous and C-*bounded from below on* X.

Then, for any k_0 *from* C \ {0}, *the set*

$$E_{\varphi,k_0} := \{v \in X : \varphi(x) + d(x, v)k_0 \notin \varphi(v) - C \ \forall\, x \in X \setminus \{v\}\}$$

is nonempty

(*F. Cammaroto - A. Chinni - G. Sturiale*).

Explanation. φ *C*-bounded from below on *X* means

$$\exists\, y_0 \text{ in } Y \text{ so that } f(x) \overset{C}{\geq} y_0 \ \forall\, x \in X.$$

■ As $k_0 \neq 0$ we have $k_0 \notin \overline{\{0\}}$ (*Y* is separated), consequently there exists ξ in Y^* with $\xi(k_0) > 0$. But $\xi \overset{(62)}{=} \xi_1 - \xi_2$, $\xi_1, \xi_2 \in C^+$, hence $\xi_1(k_0) = \xi(k_0) + \xi_2(k_0) > 0$. So, set $T := \xi_1$, we have

(64) $$T(k_0) > 0.$$

Since φ is *C*-continuous on *X*, the functional $T \circ \varphi$ is lower semicontinuous on *X*. Indeed, let x_0 be any from *X* and $T(\varphi(x_0)) > r$, $r \in \mathbf{R}$. $\exists\, V$ neighbourhood of $\varphi(x_0)$ so that $y \in V \Rightarrow T(y) > r$ and for $V\ \exists\, U$ neighbourhood of x_0 so that $x \in U \Rightarrow \varphi(x) \overset{(63)}{\in} V + C$, it results $x \in U \Rightarrow T(\varphi(x)) > r$, *T* being linear also in C^+. Moreover, as there is y_0 so that $\varphi(X) \subset y_0 + C$, we have

(65) $$T(\varphi(x)) \geq T(y_0) \ \forall\, x \in X.$$

So $T \circ \varphi : X \to \mathbf{R}$ is l. s. c. and bounded from below on *X* ((65)). Apply Ekeland principle 1.1 with $\varepsilon = T(k_0)$ (correctly, see (64)) and $\lambda = 1$, there exists *v* in *X* so that

$$T(\varphi(v)) < T(\varphi(x)) + T(k_0)d(v, x) \ \forall\, x \in X \setminus \{v\},$$

i.e.

(66) $$T(\varphi(x) - \varphi(v) + d(v, x)k_0) > 0 \ \forall\, x \in X \setminus \{v\},$$

consequently

(67) $$\varphi(x) + d(v, x)k_0 \notin \varphi(v) - C \ \forall\, x \in X \setminus \{v\},$$

because if *par absurdum* $\exists\, x' \neq v$ such that $\varphi(x') + d(v, x')k_0 = \varphi(v) - y$, $y \in C$, then

$$T(\varphi(v) - \varphi(x') - d(v, x')k_0) = T(y) \geq 0,$$

in contradiction with (66). (67) expresses that $v \in T_{\varphi,k_0}$. ■

Remark. Take the particular case $Y = \mathbf{R}$, $C = [0, +\infty)$. We have φ *C*-bounded on $X \Leftrightarrow \varphi$ bounded from below on *X*, as, for „$\Rightarrow$" for instance, $\varphi(X) \subset y_0 + [0, +\infty) \Rightarrow \varphi(x) \geq y_0 \ \forall\, x \in X$. We also have φ *C*-continuous in $x_0 \Leftrightarrow \varphi$ lower semicontinuous in x_0: for

„⇒" for instance, let be $r < \varphi(x_0)$, for $(r', +\infty)$ with $r < r' < \varphi(x_0)$ there is U neighbourhood of x_0 so that $\varphi(U) \subset r' + [0, +\infty)$ and then $x \in U \Rightarrow \varphi(x) \geq r' > r$. Because for $\varepsilon = k_0$ $E_{\varphi,\varepsilon} \neq \varnothing$ means $\exists\, v$ in X so that

$$\varphi(v) < \varphi(x) + \varepsilon d(v, x) \ \forall\, x \in X \setminus \{v\},$$

in the particular case considered $Y = \mathbf{R}$ and $C = [0, +\infty)$, 9.8 is Ekeland principle 1.1 (the fundamental inequality only).

Further extend the formula

$$\overline{\text{conv}}\ E_{\varphi,k_0} = X$$

(§1, (11_5)).

9.9 *Suppose verified the demands from* 9.8 *with* X *Banach space. Then, for any* k_0 *from* $C \setminus \{0\}$ *with the properties*

a) *there exists* h *in* Y *with* $0 \neq k_0 - h \in C$

and

b) *there exists* $\rho > 0$ *so that* $\|x\| > \rho \Rightarrow \|x\|\, h - \varphi(x) \in C$,

we have

$$\overline{\text{conv}}\ E_{\varphi,k_0} = X.$$

■ C being normal, there exists T_1, T_2 in C^+ so that $T_1(k_0 - h) > 0$ resp. $T_2(k_0) > 0$ (as for (64)). For $T := T_1 + T_2$ we have so

(68) $$T(k_0 - h) > 0,\ T(k_0) > 0.$$

$T \circ \varphi$ is l. s. c. and bounded from below (see the proof for 9.8). Moreover , from b), $\|x\| \geq$ $\geq \rho \Rightarrow T(\varphi(x)) \leq \|x\| T(h)$, which implies

$$\overline{\lim_{\|x\|\to+\infty}} \frac{T(\varphi(x))}{\|x\|} \leq T(h) \overset{(68)}{<} T(k_0).$$

Apply (11_5), §1 with $\varepsilon = T(k_0)$ (which is correct, see (68)), we have

$$\overline{\text{conv}}\ E_{T\circ\varphi, T(k_0)} = X$$

and it remains only to remark (*par absurdum*, T is linear)

$$E_{T\circ\varphi, T(k_0)} \subset E_{\varphi,k_0}. \ ■$$

Remark. 9.9 is Theorem 3.2 from [13]. The erroneous demonstration of this has been retrieved by the author.

In the following it is given a complement at the variant 9.5 of Ekeland principle.

9.9₁ *Let* X *be real Banach space,* Y *real topological vector space,* B, M *proper sub-*

sets of Y *and* k_0 *from* Y \ {0}.

Suppose verified the properties (C1) – (C4) (section A).

Let be $\varphi : X \to Y$ (k_0, B) - l. s. c. *and* M-*bounded from below on* X *with the property*

(68₁) $\exists\, \delta > 0$ *and* L *in* **R** *so that* $\|x\| \geq \delta \Rightarrow \varphi(x) \in L\|x\|k_0 - \overline{B}$.

Then there exists r *in* **R** *so that*

(68₂) $\varepsilon > \max(0, r) \Rightarrow \overline{\text{conv}}\,\{v \in X : \varphi(v) \in \text{Eff}\,(\varphi_{v,\varepsilon k_0}(X), M)\} = X$,

where $\varphi_{y,\varepsilon k_0}(x) = \varphi(x) + \varepsilon\|x - y\|k_0$, $x \in X$ ([13]).

■ Apply 9.1 with $D = -B$, ζ the Tammer functional ((7_{19})). So ζ is *M*-strictly monotonous (and subadditive) and consequently $\zeta \circ \varphi : X \to \mathbf{R}$ is bounded from below as φ is *M*- bounded from below on *X*. Moreover, $\zeta \circ \varphi$ is, following 9.2, l. s. c. on *X*. But (7_{19}) and (68_1) give $\|x\| \geq \delta \Rightarrow \zeta(\varphi(x)) \leq L\|x\|$, consequently

$$(68_3)\qquad \overline{\lim_{\|x\|\to+\infty}} \frac{\zeta(\varphi(x))}{\|x\|} < +\infty .$$

So, according to (11_5), §1, taking $r := \overline{\lim\limits_{\|x\|\to+\infty}} \dfrac{\zeta(\varphi(x))}{\|x\|}$ we have $\forall\, \varepsilon > \max(0, r)$

$$(68_4)\qquad \overline{\text{conv}}\ E_{\zeta o\varphi,\varepsilon} = X,$$

where

$$(68_5)\qquad E_{\zeta o\varphi,\varepsilon} = \{w \in X : \zeta(\varphi(w)) < \zeta(\varphi(x)) + \varepsilon\|w - x\|\ \forall\, x \in X \setminus \{w\}\}.$$

Taking into account (68_5), it remains to prove

$$(68_6)\qquad E_{\zeta o\varphi,\varepsilon} \subset \{v \in X : \varphi(v) \in \text{Eff}\,(\varphi_{v,\varepsilon k_0}(X), M)\}.$$

Suppose *par absurdum* $\exists\, w_0$ in $E_{\zeta o\varphi,\varepsilon}$ so that

$$\varphi(w_0) \notin \text{Eff}\,(\varphi_{w_0,\varepsilon k_0}(X), M),$$

i.e. $\exists\, x_0$ in *X* so that

$$\varphi(w_0) \in \varphi_{w_0,\varepsilon k_0}(x_0) + (M \setminus \{0\}),$$

$$(68_7)\qquad \varphi(w_0) \in \varphi(x_0) + \varepsilon\|x_0 - w_0\|k_0 + (M \setminus \{0\}).$$

From (68_7), ζ being *M* - strictly monotonous and subadditive,

$$(68_8)\qquad \zeta(\varphi(w_0)) + \zeta(-\varepsilon\|x_0 - w_0\|k_0) > \zeta(\varphi(x_0)).$$

But $\zeta(-\varepsilon\|x_0 - w_0\|k_0) = -\varepsilon\|x_0 - w_0\|$ (9.1 has been applied with $D = -B$, because Fr $D =$ $= -$ Fr B we have $0 \overset{(C1)}{\in}$ Fr D, then see formula (12)), hence (68_8) becomes

$$\zeta(\varphi(w_0)) > \zeta(\varphi(x_0)) + \varepsilon\|x_0 - w_0\|,$$

a contradiction in the case $w_0 = x_0$ and also in the case $w_0 \neq x_0$ ($w_0 \in E_{\zeta o\varphi,\varepsilon}$!), (68_6) is proved. ■

In the following – a variant of 9.9_1, we obtain this using 9.9.

9.9_2 *Let* X *be complete metric space*, Y *separated real locally convex space*, C *normal cone in* Y *and* φ : X → Y C-*continuous and* C-*bounded from below on* X. *Then for any* $\varepsilon > 0$ *and* k_0 *from* C \ {0} *with the properties* a) *and* b) *from* 9.9, *we have*

$$\overline{\text{conv}}\,\{v \in X : \varphi(v) \in \text{Eff}\,(\varphi_{v,\varepsilon k_0}(X), C)\} = X,$$

$$\varphi_{y,\varepsilon k_0}(x) := \varphi(x) + \varepsilon\|x - y\|k_0,\ x \in X\ ([13]).$$

■ Let $E_{\varphi,\varepsilon k_0}$ be the set from 9.9. According to 9.9, it remains to show

(68_9) $$E_{\varphi,\varepsilon k_0} \subset \{v \in X : \varphi(v) \in \text{Eff}\,(\varphi_{v,\varepsilon k_0}(X), C)\}.$$

Suppose *par absurdum* the contrary. Then there exists (68_{10}) $v_0 \in E_{\varphi,\varepsilon k_0}$ and x_0 in X so that

(68_{11}) $$\varphi(v_0) \in \varphi(x_0) + \varepsilon\|x_0 - v_0\|k_0 + (C \setminus \{0\}).$$

If $v_0 = x_0$, (68_{11}) gives the contradiction and if $v_0 \neq x_0$, (68_{11}) comes in collision with (68_{10}), consequently (68_9) is proved. ■

D. VECTORIAL GENERALIZATION OF THE EKELAND AND CARISTI - KIRK THEOREMS ([39])

This section contains generalizations of the theorems from the title and also their equivalence.

We recall the equivalence Ekeland principle ⇔ Caristi - Kirk theorem, §2, section B.

Preliminaries. Let X be nonempty set and Ω nonempty subset of the cartesian product $X^{\mathbf{N}} \times X$. The couple (X, Ω) is named L-*space* (*Fréchet*) if

(69) $$x_n = x\ \forall\, n \in \mathbf{N} \Rightarrow ((x_n)_{n\geq 1}, x) \in \Omega;$$

(70) $$((x_n)_{n\geq 1}, x) \in \Omega \Rightarrow ((x_{k_n})_{n\geq 1}, x) \in \Omega,$$

for any subsequence $(x_{k_n})_{n\geq 1}$ of (x_n).

To facilitate the writing, the *L*-space (X, Ω) will be noted sometimes simply *X*. The relation $((x_n)_{n\geq 1}, x) \in \Omega$ is designated by

(71) $$x_n \to x$$

and it will be expressed by

$(x_n)_{n\geq 1}$ *converges to* x.

Let *X* be *L*-space, *Y* ordered real vector space, *C* the positive cone of *Y* (§8, 8.3, *C* pointed convex cone with vertex) and d: $X \times X \to C$.

d is by definition *half-distance* on the set *X* if $\forall\, x, y, z$ from *X*

(72) $$d(x, y) = 0 \Leftrightarrow x = y;$$

(73) $$d(x, y) \leq d(x, z) + d(z, y).$$

In the section B there is an analogous concept, with the same name „half-distance".

X is named d-*complete* when, for any sequence $(x_n)_{n\geq 1}$ from *X*,

(74) $$\sum_{i=1}^{n} d(x_{i+1}, x_i) \leq k_0,\ k_0 \in C,\ \forall\, n \geq 1 \Rightarrow \exists\, x \text{ in } X \text{ with } x_n \to x.$$

Finally, let *X* be *L*-space, *Y* ordered real vector space and $J : X \to Y$. *J* is by definition *lower semicontinuous* (l. s. c.) when

(76) $$x_n \to a \text{ in } X \text{ and } J(x_n) \leq y\ \forall\, n \geq 1 \Rightarrow J(a) \leq y.$$

Let *Y* be ordered real vector space.

Y is *complete* if for any nonempty subset *A*, which is lower and upper bounded, there exists inf *A* and sup *A*. *Y* is *vector lattice* if for any subset $\{x, y\}$ there is

$$\inf\{x, y\} \text{ and } \sup\{x, y\}.$$

And now –

9.10 Theorem (*Vectorial generalization of Caristi - Kirk theorem*). *Let* X, X_1 *be* L-*spaces,* Y *separable complete vector lattice with* C *the positive cone,* $B \subset X$ *nonempty,* $\mathscr{F}$ *a set of applications* $f : B \to 2^X_*$, $\varphi : B \to X_1$, d *and* d_1 *half-distances on* B *and* $\varphi(B)$ *for which* B *is* d-*complete.*

Suppose that there exists $J : \varphi(B) \to Y$ *so that*

(77) $$J(\varphi(B)) \text{ is C-lower bounded};$$

(78) *for any* x *from* B, *the function* $B \to Y$,

$$z \to \sup\{d(z, x), d_1(\varphi(z), \varphi(x))\} + J(\varphi(z)) \text{ is } l.s.c.;$$

(79) $\forall$ f *from* $\mathscr{F}$ *and* $\forall$ x *from* B *there is* z *in* f (x) *which verifies*

$$\sup\{d(z,x),\ d_1(\varphi(z),\ \varphi(x))\} \leq J(\varphi(x)) - J(\varphi(z)).$$

Then

(80) *for any* u *from* B *there is* a *in* B *a fixed point for all functions from* $\mathscr{F}$ *so that*

$$\sup\{d(a,\ u),\ d_1(\varphi(a),\ \varphi(u))\} \leq J(\varphi(u)) - J(\varphi(a))$$

(*P. Q. Khanh*).

Explanations. 2^X_* is the set of the nonempty subsets of X.

x^* is fixed point for f if $x^* \in f(x^*)$ (see §2, B).

Specification. The demand „Y vector lattice" can be replaced, much more weaked, by

(81) $$\exists \sup\{d(u,\ x),\ d_1(\varphi(u),\ \varphi(x))\}\ \forall\ u, x \in B.$$

But the statement is already overloaded.

■ Consider in B the order relation

(82) $$z \leq x \overset{\text{def}}{\Leftrightarrow} \sup\{d(z,\ x),\ d_1(\varphi(z),\ \varphi(x))\} \leq J(\varphi(x)) - J(\varphi(z)).$$

This is correct: 1° $z = x \Rightarrow z \leq x$ and $x \geq z$ because $d(z, x) = d_1(\varphi(z), \varphi(x)) \overset{(72)}{=} 0$,`$J(\varphi(x)) =$ $= J(\varphi(z))$, reciprocally $z \leq x$ and $x \leq z \Rightarrow x = z$ as, because d and d_1 take positive values (in C !), we have $0 \leq J(\varphi(x)) - J(\varphi(z))$, $0 \leq J(\varphi(z)) - J(\varphi(x))$, hence $J(\varphi(z)) = J(\varphi(x))$, which implies, *via* the supremum, $d(z, x) = 0$ and hence $z \overset{(72)}{=} x$; 2° $z \leq x$ and $x \leq y \Rightarrow z \leq y$ – write (82) for $z \leq x$, then for $x \leq y$, add taking into account (73) ($a \leq b + c$ and $a_1 \leq b_1 +$ $+ c_1$, $b + c$, $b_1 + c_1 \leq \max(b, b_1) + \max(c, c_1)$, this majorizes a and a_1).

Retain

(83) $$z \leq x \Rightarrow J(\varphi(z)) \leq J(\varphi(x))$$

(d and d_1 take positive values).

According to (77) there exists h_0 in Y so that $J(\varphi(B)) - h_0 \subset C$. This allows to suppose

(84) $$J(\varphi(B)) \subset C,$$

because if neccesary we replace J by $J - h_0$ and this also verifies (77), (78) and (79) and as for the affirmation (80), this being true for $J - h_0$, is true for J as well.

Prove (80).

Let u be any fixed from B. Consider the set

(85) $$A := \{x \in B : x \leq u\}.$$

A is d-complete. Indeed, let $(x_n)_{n\geq 1}$ be a sequence from A with $\sum_{i=1}^{n} d(x_{i+1}, x_i) \leq k_0$, $k_0 \in C$, $\forall\ n \geq 1$ (see (74)). Then, B being d-complete, $\exists\ x'$ in B so that $x_n \to x'$ (see (71)). Because $x_n \leq u\ \forall\ n \geq 1$, i. e.

$$\sup\{d(x_n, u), d_1(\varphi(x_n), \varphi(u))\} + J(\varphi(x_n)) \overset{(82)}{\leq} J(\varphi(u)),$$

(78) gives (see (76))

$$\sup\{d(x', u), d_1(\varphi(x'), \varphi(u))\} + J(\varphi(x')) \leq J(\varphi(u)),$$

that is $x' \leq u$, hence $x' \in A$, A is d-complete.

Retain also

(86) $$x \in A \Rightarrow z \text{ given by (79) is in } A \cap f(x),$$

since „z verifies (79)" means $z \overset{(82)}{\leq} x$ and $x \leq u$, hence $z \overset{(85)}{\in} A$.

The particular case „any f *from* $\mathscr{F}$ *is singlevalued,* f : A → A"

(so (79) is verified with $z = f(x)$, i. e. $f(x) \leq x$).

Let M be any totally ordered subset of A. Prove

(87) $$M \text{ is lower bounded.}$$

Indeed, $S := \{J(\varphi(x)) : x \in M\}$ is totally ordered: let be $J(\varphi(x))$, $J(\varphi(x'))$, x and x' from M; M being totally ordered, we have $x \leq x'$ or $x' \leq x$, which implies, *via* (83), $J(\varphi(x)) \leq J(\varphi(x'))$ or $J(\varphi(x')) \leq J(\varphi(x))$. Moreover, S is lower bounded by 0 (see (84)). So, as Y is complete, there exists $b := \inf S$. Moreover, S, monotonous net, converges to b in the order topology ([60], p.230) and as Y is separable, in S there are the terms of a decreasing sequence $(J(\varphi(x_n)))_{n\geq 1}$ which converges to b. We have

(88) $$\sup\{d(x_{i+1}, x_i), d_1(\varphi(x_{i+1}), \varphi(x_i))\} \leq J(\varphi(x_i)) - J(\varphi(x_{i+1})).$$

Indeed, when $x_{i+1} = x_i$, (88) is verified ((72)). Suppose $x_{i+1} \neq x_i$. We have either $x_{i+1} \leq x_i$, or $x_i \leq x_{i+1}$. But the first case means (88) and the second case leads, *via* (82), to $J(\varphi(x_{i+1})) - J(\varphi(x_i)) > 0$ because $d(x_{i+1}, x_i) \overset{(72)}{>} 0$, contradiction since $(J(\varphi(x_n)))_{n\geq 1}$ is decreasing.

So, taking into account (88), $\forall\ n \geq 1$

$$\sum_{i=1}^{n} d(x_{i+1}, x_i) \leq \sum_{i=1}^{n} \sup\{d(x_{i+1}, x_i), d_1(\varphi(x_{i+1}), \varphi(x_i))\} \leq J(\varphi(x_1)) - J(\varphi(x_2)) + J(\varphi(x_2)) - J(\varphi(x_3)) + \ldots + J(\varphi(x_n)) - J(\varphi(x_{n+1})) \leq J(\varphi(x_1)),$$

hence, A being d-complete, there exists a_0 in A so that (89) $x_n \to a_0$.

Let now x be an element from M. Two cases are possible (S is totally ordered !):

(90) $$J(\varphi(x)) \leq J(\varphi(x_n)) \ \forall\, n \geq 1,$$

(91) $$\exists\, N \text{ in } \mathbf{N} \text{ with } J(\varphi(x_N)) < J(\varphi(x)).$$

In the case of (90) it results $J(\varphi(x)) \leq b$, hence $J(\varphi(x)) = b$ and so x is minorant of M (*par absurdum* $\exists\, x' \in M$ with $x' < x$, then $J(\varphi(x')) \overset{(83)}{\leq} J(\varphi(x)) = b$ and, *via* (82) as $d(x', x) > 0$, $J(\varphi(x')) < b$, contradiction). Pass to the case (91). It results $n \geq N \Rightarrow J(\varphi(x_n)) \leq J(\varphi(x))$. x_n and x being in M which is totally ordered, we have one of the situations

$$x_n = x,\ x_n < x,\ x_n > x.$$

The last situation leads, *via* (82) and (72), to $J(\varphi(x_n)) > J(\varphi(x))$, contradiction as $(J(\varphi(x_n)))_{n\geq 1}$ is decreasing, consequently (92) $n \geq N \Rightarrow x_n \leq x$. (92) leads, using the lower semicontinuity assumed to (78), to $a_0 \leq x$: setting $J_1 := \sup\{d(\cdot, x), d_1(\varphi(\cdot), \varphi(x))\} + J(\varphi(\cdot))$, we have $n \geq N \Rightarrow J_1(x_n) \overset{(92)}{\leq} J(\varphi(x))$ and because $x_n \xrightarrow{(89)} a_0$, it results $J_1(a_0) \leq J(\varphi(x))$, i. e. $a_0 \leq x$. So (87) is proved.

Let now M_1 be a totally ordered subset of A and maximal with respect to inclusion (Hausdorff theorem). According to (87), there is a in A minorant of M_1. a verifies the demand from (80), as obviously $a \leq u$, but also $f(a) = a\ \forall\, f \in \mathscr{F}$. Indeed, suppose *par absurdum* $\exists\ f_1$ in $\mathscr{F}$ so that $f_1(a) \neq a$. Then the hypothesis assummed in the particular case we are given $f_1(a) \leq a$. On the other hand, $f_1(a)$ being so comparable with all elements of M_1, we have $f_1(a) \in M_1$ (*par absurdum*, M_1 is maximal with respect to inclusion), hence $a \leq f_1(a)$, $a = f_1(a)$, contradiction and (80) is proved.

The general case. Suppose *par absurdum* the contrary. It can thus affirm in particular $\forall\, x \in A\ \exists\ f_x$ in $\mathscr{F}$ so that $x \in A \setminus f_x(x)$. Then let $\zeta(x)$ be from $A \cap f_x(x)$ given by (79) (see (86)). With $\mathscr{F} = \{\zeta\}$ we are in the demonstrated particular case, so $\exists\ a'$ in A such that $a' = \zeta(a')$, hence $a' \in A \cap f_{a'}(a')$ and we come in collision with (93). Consequently $\exists\ a$ in A so that $a \overset{(85)}{\in} f(a)\ \forall\ f$ from $\mathscr{F}$. As in addition $a \in A$ means $a \leq u$, (80) is proved. ■

Particular case

Caristi - Kirk theorem (§2, B). *Let* X *be complete metric space and the multivalued* *function* $f : X \to 2^X_*$. *If there is* $J : X \to \mathbf{R}$ *bounded from below and l. s. c. with the property* $\forall$ x *from* X *there exists* y *in* f(x) *so that*

$$d(x, y) \le J(x) - J(y),$$

then f *has a fixed point*

can be obtained from 9.10, where one takes X complete metric space, $X = X_1 = B$, $Y = \mathbf{R}$, $C = [0, +\infty)$, φ the identical map of X, $d = d_1$ = the distance on X. Indeed, the couple (X, Ω), $\Omega := \{((x_n)_{n\ge1}, x) : x_n \to x\}$ subset of $X^{\mathbf{N}} \times X$, is L-space as (69) and (70) are obviously verified. Moreover, X being complete, it is d-complete also according to the definition (74): $\sum_{n=1}^{\infty} d(x_n, x_{n+1})$ results convergent, consequently $\forall\, \varepsilon > 0\ \exists\, n$ in $\mathbf{N}$ so that $m, n \ge N \Rightarrow d(x_n, x_m) \le d(x_n, x_{n+1}) + d(x_{n+1}, x_{n+2}) + \ldots + d(x_{m-1}, x_m) \le \varepsilon$, i.e. $(x_n)_{n\ge1}$ is fundamental, and hence convergent. Moreover, J is l. s. c. even according to the definition (76): if *par absurdum* $J(a) > y$, it results $\overline{\lim_{n\to\infty}}\ J(x_n) > y$, in contradiction with $J(x_n) \le y$ $\forall\, n \ge 1$.

So (77) and (78) from 9.10 are verified, also (79) with $\mathcal{F} = \{f\}$ *via* the inequality from the statement C - K , because $\overset{C}{\le}$ in $\mathbf{R}$ is the usual order relation. According to (80), f has a fixed point.

Remark. In the inequality from C - K theorem

$$d(x, y) \le J(x) - J(y)$$

we recognize the origin of the order relation (82), which is the basis of the statements from this section.

Note. Theorem 9.10 is Theorem 2.1 from [39], hardly recovered by the author : the statement has been modified and the demonstration, in many parts erroneous, has been corrected. Only then, with much difficulty, it has been obtained a real generalization of Caristi - Kirk theorem, which is not true for Th. 2.1, [39]. Precisely –

1. φ closed and $\varphi(B)$ d_1 -complete are not required.
2. (iii) from the statement has been replaced with

„for each $f \in \mathcal{F}$ and for each $x \in B$ there is a $z = z(x, f) \in f(x)$ satisfying (∗)."

3. The conclusion b) of the statement has been removed as false and the same thing has to be done for b), Theorem 3 also.

4. Zorn theorem does not apply, but the Hausdorff theorem does. The fact that any totally ordered part of A has a minorant must not be used to validate the application of Zorn theorem, but to find an infimum relative to C.

5. The „justification" for b) shows the incorrect knowledge of Zorn theorem.

Pass now at the other generalization announced in the start of the section.

9.11 Theorem (*Vectorial generalization of Ekeland principle*). *Let be* X, X_1, Y, B, φ, d *and* d_1 *as in Theorem* 9.10. *Suppose that the application* $J : \varphi(B) \to Y$ *verifies* (77) *and* (78).

Then for any u *of* B *there exists* v *minimal point in* B *for the order relation* (82) *with the properties*:

(94) $$J(\varphi(v)) \leq J(\varphi(u)) - \sup\{d(v, u), d_1(\varphi(v), \varphi(u))\}$$

and

(95) $$J(\varphi(v)) \neq J(\varphi(x)) + \sup\{d(x, v), d_1(\varphi(x), \varphi(v))\} \ \forall\, x \in B \setminus \{v\}$$

(*P. Q. Khanh*).

■ Suppose *par absurdum* $\exists\, u$ in B so that for each v from the set

$$A := \{x \in B : J(\varphi(x)) \leq J(\varphi(u)) - \sup\{d(x, u), d_1(\varphi(x), \varphi(u))\}\}$$

there is v_1 in $A \setminus \{v\}$ with $v_1 \leq v$, i. e.

$$J(\varphi(v_1)) + \sup\{d(v_1, v), d_1(\varphi(v_1), \varphi(v))\} \leq J(\varphi(v)).$$

Observe $A = \{x \in B : x \leq u\}$. Consider the application $f : A \to A$, (96) $f(v) = v_1$. Apply 9.10 with B replaced by A and with $\mathcal{F}$ a set of singlevalued applications $A \to A$ which verifies (79) $\forall\, x \in A$, correctly because A is d-complete (see (85)) and (77), (78) are verified according to the hypotheses ($A \subset B$!). Because $f \in \mathcal{F}$ (see (96)), $\exists\, a$ in A so that $f(a) = a$, hence (see (96)) $a = a_1$ with $a_1 \in E \setminus \{a\}$, contradiction and (94) is proved with v minimal point for B. Pass to (95). Let *par absurdum* x' be in $B \setminus \{v\}$ so that

(97) $$J(\varphi(v)) = J(\varphi(x')) + \sup\{d(x', v), d_1(\varphi(x'), \varphi(v))\}.$$

(97) expresses that $x' \leq v$ and hence $x' < v$, contradiction since v is minimal point of B. ■

Particular case. Ekeland principle 1.1_0, §1 *with* $\varepsilon = \lambda = 1$ *and without the inequality* (3) can be obtained from 9.11 where one takes X complete metric space, $X = X_1 = B$, $Y = \mathbf{R}$, $C = [0, +\infty)$, φ the identical map of X, $d = d_1 =$ the distance on X. Indeed, (94) gives $J(v) \leq J(u) - d(v, u)$, that is (2_0). Moreover, it results from (95) ($\mathbf{R}$ is totally ordered)

$$x \in X \setminus \{v\} \Rightarrow J(v) < J(x) + \ldots \text{ or } J(v) > J(x) + \ldots .$$

But the second situation gives $x \leq v$ and since $x \neq v$, results $x < v$, contradiction as v is minimal point.

In fact,

9.12 *Theorem* 9.10 *and Theorem* 9.11 *are equivalent* (*P. Q. Khanh*).

■ 9.10 $\Rightarrow$ 9.11. See above.

9.11 $\Rightarrow$ 9.10. Let u be anyone fixed from B. As (77) and (78) are verified, 9.11 provides v as minimal point in B for the order relation (82). Moreover, according to (94), we have $v \leq u$. So, for any z from $B \setminus \{v\}$,

(98) $$J(\varphi(z)) + \sup\{d(z, v), d_1(\varphi(z), \varphi(v))\} \not\leq J(\varphi(v))$$

(*par absurdum*, v is minimal point). But (98) forces, *via* (79) with $x = v$, $v \in f(v)$ $\forall f$ from $\mathscr{F}$ (*par absurdum*). It remains only to take $a = v$. ■

E. MULTIVALUED VARIANT OF EKELAND PRINCIPLE ([19])

Definitions. Let X, Y be nonempty sets and $f: X \to 2^Y$.

$$\text{dom } f := \{x \in X : f(x) \neq \varnothing\}.$$

f is named *proper* if dom $f \neq \varnothing$.

Suppose Y topological space. f has *compact values* when $f(x)$ is compact $\forall x \in X$.

Suppose X and Y topological spaces and let x_0 be from X. f is *upper semicontinuous* (u.s.c.) *in* x_0 if for any neighbourhood V of $f(x_0)$ there exists U neighbourhood of x_0 so that

(99) $$f(x) \subset V \ \forall x \in U.$$

Let be X set, Y real vector space and C cone from Y (see §8, „preliminaries of affine geometry"). f is C-*lower bounded* if there is y_f in X so that

(100) $$f(x) \subset y_f + C \ \forall x \in X.$$

When dom $f = \varnothing$, f is by definition C- *lower bounded.*

Finally let X be set, Y real locally convex space, C closed pointed convex cone with vertex in Y and k_0 from int C. x^* from X is ε-*solution* of f, where $\varepsilon > 0$, if there is y^* in $f(x^*)$ so that

(101) $$[f(x^*) - y^*] \cap [-C \setminus \{0\}] = \varnothing$$

and

(102) $$[f(x) - y^* + \varepsilon k_0] \cap [-C \setminus \{0\}] = \varnothing \ \forall\, x \in X \setminus \{x^*\}.$$

Denote the set of ε-solutions of f by

$$\varepsilon \text{ - } S(f).$$

Remark. For ε = 0 (101) defines x^* as *efficient solution* of f (*H. W. Corely*). The adjective „efficient" is in accordance to the definition (13), section A, because this can be put in the form

$$y - y_1 \notin -B \setminus \{0\} \ \forall\, y \in E.$$

Next, some considerations for which we appeal to [43].

Definition. Let Y be real topological vector space and C convex cone in Y. The nonempty subset M of Y has the *dominant property*, (DP), if for each y from M ∃ z in Eff (M, C) (section A, (13), here is not required M and C proper) so that

$$z \overset{C}{\leq} y$$

(§8, „preliminaries of affine geometry"). Obviously (103) M has the property (DP) ⇔ $M \subset \text{Eff}\,(M, C) + C$.

It can affirm

9.13 *If* Y *is locally convex and* C *closed pointed convex cone with vertex, any nonempty compact subset of* Y *has the* (DP) *property.*

■ Let M be the same as in the statement and x anyone from M. $M_x := M \cap (x - C)$ being compact is C-complete (ch. 2, Lemma 3.5, 2)), apply Theorem 4.3, i), ch. 2, correctly as $\overline{C} + C \setminus \{0\} \subset C$ (page 2), because C is closed and convex (§8, 8.1). ■

And now let X be set, Y real locally convex space, C closed pointed convex cone with vertex and k_0 from int C.

9.14 *If* $f : X \to 2^Y$ *is proper, with compact values and* C*-bounded from below, then for any* ε > 0 *there exists* x^* *in* dom f *and* y^* *in* $f(x^*)$ *so that*

(104) $$[f(x^*) - y^*] \cap [-C \setminus \{0\}] = \varnothing$$

and

(105) $$[f(x) - y^* + \varepsilon k_0] \cap (-C) = \varnothing \ \forall\, x \in \text{dom } f$$

(*Guang Ya Chen - X. X. Huang*).

■ We begin by observing

„$\forall\, x^* \in \text{dom } f \ \exists\, y^* \in f(x^*)$ so that (104) is verified".

Indeed, $f(x^*)$ being nonempty compact , it follows *via* 9.13 in particular $\exists\, y^* \in$

$\in \text{Eff}\,(f(x^*), C)$, i. e. (see (13))

$$y - y^* \notin -C \setminus \{0\} \ \forall\, y \in f(x^*),$$

which is equivalent to (104).

Pass to the proof of the statement and suppose *par absurdum* the contrary, that is, taking into account the remark from the begining, $\exists\ \varepsilon_0 > 0$ so that $\forall\, x \in \text{dom}\ f$ and $\forall\, y \in$ $\in f(x)\ \exists\, z \in \text{dom}\ f$ having the property

(106) $$[f(z) - y + \varepsilon_0 k_0] \cap (-C) \neq \varnothing.$$

Let x_1 be from dom f and y_1' from $\text{Eff}\,(f(x_1), C)$ (9.13). According to (106) $\exists\, z_1 \in \text{dom}\ f$ and $\exists\, y_2 \in f(z_1)$ so that

(107) $$y_2 - y_1' + \varepsilon_0 k_0 \in -C.$$

$f(z_1)$ being nonempty compact, for $y_2\ \exists\ y_2'$ in $\text{Eff}\,(f(z_1), C)$ so that (9.13)

(108) $$y_2' - y_2 \in -C.$$

Add (107) and (108), one finds (see §8, 8.1)

$$y_2' - y_1' + \varepsilon_0 k_0 \in -C.$$

Repeat the reasoning with y_2' and $f(z_1)$, $\exists\, z_2 \in \text{dom}\ f$ and $y_3 \in f(z_0)$ so that $y_3 - y_2' +$ $+ \varepsilon_0 k_0 \in -C$. Take in addition y_3' from $\text{Eff}\,(f(z_2), C)$ so that $y_3' - y_3 \in -C$ and add etc. So we obtain a sequence $(y_i')_{i\geq 1}$ having the property $y_i' - y_{i-1}' + \varepsilon_0 k_0 \in -C\ \forall\, i \geq 2$ and hence $\sum_{i=2}^{n} (y_i' - y_{i-1}' + \varepsilon_0 k_0) = y_n' - y_1' + (n-1)\varepsilon_0 k_0 \in -C\ \forall\, n \geq 2$,

(109) $$\frac{1}{n-1}(y_n' - y_1') + \varepsilon_0 k_0 \in -C\ \forall\, n \geq 2.$$

But f being C-lower bounded, $\exists\, y_f$ in Y so that $y_n' - y_f \in C\ \forall\, n \geq 1$, consequently, taking into account (109),

$$\frac{1}{n-1}(y_f - y_1') + \varepsilon_0 k_0 \in -C\ \forall\, n \geq 2$$

(take 8.1 into account).

Pass to the limit for $n \to \infty$, one obtains, as C is closed, $\varepsilon_0 k_0 \in -C$. But $\varepsilon_0 k_0 \in C$, consequently $\varepsilon_0 k_0 = 0$, C being pointed (§8, 8.2). As $\varepsilon_0 > 0$, *a fortiori* $k_0 = 0$, contradiction since $k_0 \in \text{int}\ C$. ■

Remark 1 (relative to Theorem 2.1, [19]). In b) the assertion is true $\forall\, x \in \text{dom}\ F$ *not*

$\forall\, x \in X \setminus \{x^*\}$. Besides and moreover, the demonstration was completed and corrected by the author.

Remark 2. The reference to [43] for Lemma 2.1, [19] (here 9.13) is fancy („see (5)"), one must combine several statements from [43] in order to obtain it.

9.14 may be reformulated, *X*, *Y* and *C* are as in 9.14,

9.15 *If* $f : X \to 2^Y$ *is proper, with compact values and* C-*lower bounded, then for any* $\varepsilon > 0$ *we have*

$$\varepsilon\text{ - }S(f) \neq \varnothing.$$

Pass to a multivalued variant of Ekeland principle. In the following (*X*, d) is complete metric space, *Y* metrizable real locally convex space, *C* closed pointed convex cone with vertex and $k_0 \in \operatorname{int} C$. The order relation $\overset{C}{\leq}$ is that from §8, „preliminaries of affine geometry".

9.16 Theorem. *Let* $f : X \to 2^Y$ *be proper, with compact values*, C-*bounded from below and u. s. c. on* X.

For any $\varepsilon > 0$ *and* x_1 *from* dom f, y_1 *from* $f(x_1)$ *which verify*

$$[f(x_1) - y_1] \cap [-C \setminus \{0\}] = \varnothing \tag{110}$$

and

$$[f(x) - y_1 + \varepsilon k_0] \cap [-C \setminus \{0\}] = \varnothing \;\; \forall\, x \in \operatorname{dom} f \tag{111}$$

and for any $\lambda > 0$, *there exist* x_2 *in* dom f *and* y_2 *in* $f(x_2)$ *so that*

$$y_2 \overset{C}{\leq} y_1; \tag{112}$$

$$d(x_1, x_2) \leq \lambda; \tag{113}$$

$$[f(x_2) - y_2] \cap [-C \setminus \{0\}] = \varnothing; \tag{114}$$

$$[f(x) - y_2 + \varepsilon k_0] \cap [-C \setminus \{0\}] = \varnothing \;\; \forall\, x \in \operatorname{dom} f; \tag{115}$$

$$[f(x) - y_2 + \frac{\varepsilon}{\lambda} d(x_2, x) k_0] \cap (-C) = \varnothing \;\; \forall\, x \in \operatorname{dom} f \setminus \{x_2\} \tag{116}$$

(*Guang Ya Chen - X. X. Huang*).

■ The statement has consistency: for any $\varepsilon > 0$ $\exists\, x_1 \in \operatorname{dom} f$, $\exists\, y_1 \in f(x_1)$ so that (110) and (111) are verified (see 9.15).

Consider the auxiliary multivalued function $f_1 : X \to 2^Y$,

$$f_1(x) = \{y \in f(x) : y \overset{C}{\leq} y_1\} = f(x) \cap (y_1 \setminus C). \tag{117}$$

f_1 is proper ($y_1 \in f(x_1)$, $f_1(x_1) \neq \varnothing$), with compact values ($y_1 - C$ is closed since C is closed) and C-bounded from below ($f_1(x) \subset f(x)$!)).

Now consider the function $\varphi : X \to (-\infty, +\infty]$, where L is the Luc functional (§8, minimal point theorem III),

(118)
$$\varphi(x) = \begin{cases} \inf_{y \in f_1(x)} L(y - y_1), \text{ when } x \in \operatorname{dom} f_1 \\ +\infty, \text{ when } x \notin \operatorname{dom} f_1. \end{cases}$$

Underline that L is continuous because $k_0 \in \operatorname{int} C$ (Remark 2 to 8.10). Prove that φ is lower bounded, l.s.c. and proper in order to apply Ekeland principle 1.1.

f_1 is proper, let be $x \in \operatorname{dom} f_1$. Then $\varphi(x) \in \mathbf{R}$ as L is continuous on the nonempty compact set $f_1(x) - y_1$. Here it is one more argument, we have (119) $\varphi(x_1) = 0$. Indeed, (110) gives

$y - y_1 \notin -C \setminus \{0\}$ $\forall y \in f(x_1)$ and on the other side $y - y_1 \in -C$ $\forall y \in f_1(x_1)$,

consequently $y - y_1 = 0$ on $f_1(x_1)$ and hence $L(y - y_1) = 0$ on $f_1(x_1)$ (§8, 8.10).

Let y_f be from Y so that $f_1(x) \subset y_f + C$ on X, i.e. $y \in y_f + C$ $\forall y \in f_1(x)$, $\forall x \in$ $\in \operatorname{dom} f_1$, consequently, L being C-increasing (§8, 8.10), $L(y) \geq L(y_f)$ $\forall y \in f_1(x)$, $\forall x \in$ $\in \operatorname{dom} f_1$, φ is bounded from below.

Pass to the lower semicontinuity of φ. Let t be any from $\mathbf{R}$ and $M := \{x \in X : \varphi(x) \leq$ $\leq t\}$. It must prove that M is closed. Let $(a_n)_{n\geq 1}$ be sequence from M with $a_n \to x^*$. $\varphi(a_n) \leq$ $\leq t$ implies (see (118), L atteins its infimum on $f_1(a_n)$) $\exists\, b_n \in f_1(a_n)$ so that (120) $b_n \overset{C}{\leq}$ $\overset{C}{\leq} y_1$, $L(b_n - y_1) \leq t$. Let be $\delta > 0$. Take a covering of $f(x^*)$ with open balls with diameter $\frac{\delta}{2}$ so that each of these to cut $f(x^*)$ (Y is metrizable). For the reunion V of these there exists U neighbourhood of x^* having the property relative to the upper semicontinuity of f in x^*, let a_{k_1} be from U ($a_n \to x^*$!). Then $f(a_{k_1}) \subset V$ and hence b_{k_1} is in a ball S of the covering ($f_1(a_{k_1}) \subset f(a_{k_1})$!). Take an element c_{k_1} from $S \cap f(x^*)$ and then dist$(b_{k_1}$, $c_{k_1}) < \frac{\delta}{2}$. By repetition we find $(b_{k_n})_{n\geq 1}$ subsequence of $(b_n)_{n\geq 1}$ and $(c_{k_n})_{n\geq 1}$ sequence from $f(x^*)$ so that dist $(b_{k_n}, c_{k_n}) < \frac{\delta}{2^n}$. So (121) $b_{k_n} - c_{k_n} \to 0$. But $(c_{k_n})_{n\geq 1}$ has a convergent subsequence $(c_{l_{k_n}})_{n\geq 1}$ ($f(x^*)$ is compact), $c_{l_{k_n}} \to y^* \in f(x^*)$ and consequently $b_{l_{k_n}} \to y^*$ ((121)). Then, from (120), $y^* \overset{C}{\leq} y_1$ (C is closed), i.e. $y^* \in f_1(x^*)$. Moreover, once again *via* (120), $L(y^{*} - y_1) \leq t$ (L is continuous), consequently $\varphi(x^{*}) \leq t$, $x^{*} \in M$, M is closed.

In the following show that (122) $\varphi(x) + \varepsilon \geq 0 \ \forall\, x \in X$. Suppose *par absurdum* $\exists\, x'$ so that $\varphi(x') + \varepsilon < 0$, then $\exists\, y' \in f_1(x')$ so that $L(y' - y_1) + \varepsilon \overset{8.10}{=} L(y' - y_1 + \varepsilon k_0) < 0 \overset{8.10}{=} L(0)$, which implies $y' - y_1 + \varepsilon k_0 \in -C \setminus \{0\}$ (see §8, 8.9), in contradiction with (111).

So, taking into account (122) and (119),

$$\varphi(x_1) \leq \inf \varphi(X) + \varepsilon,$$

apply Ekeland principle, $\exists\, x_2$ in X so that

$$\varphi(x_2) \leq 0, \tag{123}$$

$$d(x_1, x_2) \leq \lambda, \tag{124}$$

$$\varphi(x_2) < \varphi(x) + \frac{\varepsilon}{\lambda} d(x_2, x) \ \forall\, x \in X \setminus \{x_2\}. \tag{125}$$

(123) gives $f_1(x_2) \neq \varnothing$ (see (118)). $\exists\, y_2'$ in $f_1(x_2)$ so that (125_1) $\varphi(x_2) = L(y_2' - y_1)$, hence $L(y_2' - y_1) \overset{(123)}{\leq} 0$, and $\exists\, y_2$ in Eff $(f_1(x_2), C)$ so that (125_2) $y_2 \overset{C}{\leq} y_2'$ (9.13), consequently $y_2 - y_1 \overset{C}{\leq} y_2' - y_1$, $L(y_2 - y_1) \overset{8.10}{\leq} L(y_2' - y_1) \leq 0$. Then, according to 8.9, §8, $y_2 - y_1 = L(y_2 - y_1)k_0 - u$, $u \in C$ and as $L(y_2 - y_1)k_0 \in -C$, it results $y_2 - y_1 \in -C - C \overset{8.1}{\subset}$ $\subset -C$, $y_2 \overset{C}{\leq} y_1$.

So it remains to verify (114), (115), (116). Begin with (114), which means

$$y_2 \notin y + (C \setminus \{0\}) \ \forall\, y \in f(x_2). \tag{126}$$

(126) is true $\forall\, y \in f_1(x_2)$, because $y_2 \in$ Eff $(f_1(x_2), C)$. Let now be $y \in f(x_2) \setminus f_1(x_2)$, i.e. $y \in f(x_2)$ and $y \overset{(117)}{\notin} y_1 - C$, and suppose *par absurdum* $y_2 \in y + (C \setminus \{0\})$. But then $y_1 \underset{(112)}{\overset{C}{\geq}} y_2$, $y_2 \overset{C}{\geq} y$, hence $y_1 \overset{C}{\geq} y$, that is $y \in y_1 - C$, contradiction and (126) is proved.

Pass to (115). Suppose $y_2 \neq y_1$, for the contrary case we send to (111), and suppose *par absurdum* $\exists\, x' \in \operatorname{dom} f$, $y' \in f(x')$ so that

$$y' - y_2 + \varepsilon k_0 \in -C \setminus \{0\}. \tag{127}$$

(112) gives $y_1 = y_2 + u$, $u \in C \setminus \{0\}$ and combining with (127), $y' - y_1 + \varepsilon k_0 \in -C \setminus \{0\}$, which contradicts (111).

Finally, pass to (116) and let x be from $[(\text{dom } f \setminus \text{dom } f_1) \cup \text{dom } f_1] \setminus \{x_2\}$, for instance $x \in \text{dom } f \setminus \text{dom } f_1$. Suppose *par absurdum* $\exists y' \in f(x) \setminus f_1(x)$, that is (128) $y' \notin$ $\notin y_1 - C$ (see (117)), so that

(129) $$y' - y_2 + \frac{\varepsilon}{\lambda} \mathrm{d}(x_2, x) k_0 \in -C.$$

But $-y_2 \overset{(112)}{=} -y_1 + u$, $u \in C$ and combining with (129) one finds, taking into account §8, 8.1, $y' \in y_1 - C$, in contradiction with (128). The remaining case is $x \in \text{dom } f_1 \setminus \{x_2\}$. Suppose *par absurdum*

(130) $$\exists y' \in f_1(x) \text{ so that } y' - y_2 + \frac{\varepsilon}{\lambda} \mathrm{d}(x_2, x) k_0 \in -C.$$

On the other side, (125) and (125_1) give (see also (118))

$$\inf_{y \in f_1(x)} L(y - y_1) - L(y_2' - y_1) + \frac{\varepsilon}{\lambda} \mathrm{d}(x_2, x) > 0,$$

(131) $$L(y - y_1) - L(y_2' - y_1) + \frac{\varepsilon}{\lambda} \mathrm{d}(x_2, x) > 0, \ \forall\, y \in f_1(x).$$

As $y_2' - y_1 \overset{C}{\geq} y_2 - y_1$ (see (125_2)), we have $L(y_2' - y_1) \overset{8.10}{\geq} L(y_2 - y_1)$ and then, *via* (131),

$$L(y - y_1) - L(y_2 - y_1) + \frac{\varepsilon}{\lambda} \mathrm{d}(x_2, x) > 0 \ \forall\, y \in f_1(x),$$

$$L(y - y_2) + \frac{\varepsilon}{\lambda} \mathrm{d}(x_2, x) > 0 \ \forall\, y \in f_1(x),$$

as $L(y - y_1) \overset{10.8}{\leq} L(y - y_2) + L(y_2 - y_1)$, consequently (see §8, (46))

$$y - y_2 + \frac{\varepsilon}{\lambda} \mathrm{d}(x_2, x) k_0 \notin -C \ \forall\, y \in f_1(x),$$

in contradiction to (130). ■

Remarks relative to Theorem 3.1, [19], alias 9.16 here.

1) The topological vector space Y must be *metrizable* (it has been used the affirmation „any sequence which is in a compact part of Y has a convergent subsequence").

2) Demand VI must be replaced by „$\forall\, x \in \text{dom } F$" and the assertions X and XI are true $\forall\, x \in \text{dom } F$ respectively $\forall\, x \in \text{dom } F \setminus \{x_2\}$.

3) The assertion (4) is false, it is facilitated by the absence of the justification. In fact it is true for a subsequence $(\overline{y}_{l_{k_n}})_{n \geq 1}$ of a subsequence $(\overline{y}_{k_n})_{n \geq 1}$ of the sequence $(\overline{y}_n)_{n \geq 1}$.

4) Elsewhere, the author gives in [19] „demonstrations" for affirmations which are, obviously, unwarranted.

F. MULTIVALUED VARIANTS OF EKELAND PRINCIPLE ([70$_1$])

These variants involve the Clarke normal conc to thc graph of a multivalued function. Some preparations are necessary.

In this section

X and Y are real Banach spaces.

The Clarke tangent cone. Let M be nonempty subset of X, $d_M : X \to \mathbf{R}$ the function $x \to d(x, M)$ (d the distance afferent to the norm) and x_0 from M. The vector v from M is *tangent to* M *in* x_0 if

$$d_M^0(x_0 ; v) = 0$$

(the Clarke derivative of d_M at x_0 in the direction v, §5, section F; the function d_M is Lipschitz of constant 1 and convex). The set of tangent vectors to M at x_0 is denoted

$$T(x_0 ; M),$$

the *Clarke tangent cone* to M at x_0.

$T(x_0 ; M)$ is closed convex cone with vertex (see §8, preliminaries of affine geometry).

9.17 *Let* M *be convex. Then*

$$v \in T(x_0 ; M) \Leftrightarrow d_M^0(x_0 ; v) = d_M'(x_0 ; v) = 0.$$

■ d_M being convex it is regular (5.31, 2°) and hence the conclusion. ■

9.18 *Intrinsic characterization.* v *is tangent to* M *at* $x_0 \Leftrightarrow \forall$ *sequence* $(x_n)_{n\geq 1}$ *from* M *with* $x_n \to x_0$ *and* $\forall$ *sequence* $(t_n)_{n\geq 1}$ *from* $(0, +\infty)$ with $t_n \downarrow 0$ $\exists$ *a sequence* $(v_n)_{n\geq 1}$ *from* X *with* $v_n \to v$ *such that* $x_n + t_n v_n \in M$ $\forall$ $n \geq 1$ ([20], 1, 2.4.5).

■ *Necessary.* Let (x_n), (t_n) be sequences as in the statement. Because $d_M^0(x_0 ; v) = 0$, we have $\lim\limits_{n\to\infty} \dfrac{d_M(x_n + t_n v)}{t_n} = \lim\limits_{n\to\infty} \dfrac{d_M(x_n + t_n v) - d_M(x_n)}{t_n} \leq 0$ ([47], vol.II, V, §5, 11.9$_1$), hence $\lim\limits_{n\to\infty} \dfrac{d_M(x_n + t_n v)}{t_n} = 0$. Let a_n be from M so that $\|x_n + t_n v - a_n\| \leq d_M(x_n + t_n v) + \dfrac{t_n}{n}$. Taking $v_n := \dfrac{a_n - x_n}{t_n}$, we have $\|v_n - v\| \to 0$. Moreover, $x_n + t_n v_n = a_n \in Y$.

Sufficient. Let v be with the stated properties concerning sequences. Choose a sequence (y_n), $y_n \to x_0$ and a sequence (t_n), $t_n \downarrow 0$ so that

(132) $$\lim_{n\to\infty} \frac{d_M(y_n + t_n v) - d_M(y_n)}{t_n} = d_M^0(x_0; v).$$

It is enough to prove that $d_M^0(x_0; v) \le 0$. Let a_n be from M so that (133) $\|a_n - y_n\| \le \le d_M(y_n) + \frac{t_n}{n}$. (133) involves $a_n \to x_0$ because $d_M(y_n) \to 0$. The hypothesis gives a sequence convergent to v so that $x_n + t_n v_n \in M\ \forall\ n \ge 1$. But, d_M being Lipschitz,

$$d_M(y_n + t_n v) \le d_M(a_n + t_n v_n) + \|y_n - a_n\| + t_n\|v - v_n\| \overset{(133)}{\le} d_M(y_n) + t_n(\|v - v_n\| + \frac{1}{n}).$$

Combining with (132) we find $d_M^0(x_0; v) \le 0$. ■

9.19 *Corollary. Let be* $X = X_1 \times X_2$, X_1 *and* X_2 *Banach spaces, and* $x = (x_1, x_2)$ *from* $M_1 \times M_2$, M_i *subset of* X_i, $i = 1, 2$. *Then*

$$T(x; M_1 \times M_2) = T(x_1; M) \times T(x_2; M),$$
$$N(x; M_1 \times M_2) = N(x_1; M_1) \times N(x_2; M_2)$$

(for the second relation see (134)).

Clarke normal cone. Let M be nonempty subset of X and x_0 from M. The *Clarke normal cone to* M *at* x_0, $N(x_0; M)$, is

(134) $$N(x_0; M) = \{\xi \in X^* : \xi(v) \le 0\ \forall\ v \in T(x_0; M)\}.$$

We have (Clarke subderivative, 5.18, d_M is convex and Lipschitz)

(135) $$N(x_0; M) = \mathrm{cl}\{\bigcup_{\lambda>0} \lambda\partial d_M(x_0)\},$$

where cl denotes the *-weak closure ([20], 1, 2.4.2).

Clarke contingent cone and regular sets. Let M be nonempty subset of X and x_0 from M. The *Clarke contingent cone* $K(x_0; M)$ is ($S_1(0) = \{x \in X : \|x\| \le 1\}$)

(136) $$K(x_0; M) = \{v \in X : \forall\ \varepsilon > 0\ \exists\ t \in (0, \varepsilon) \text{ and } \exists\ w \in v + \varepsilon S_1(0) \text{ such that } x_0 + tw \in \in M\}.$$

Obviously $x_0 \in \overline{M}$. We have (137) $T(x_0; M) \subset K(x_0; M)$ (use 9.18).

Definition. M is *regular at* x_0 when

(138) $$T(x_0; M) = K(x_0; M).$$

Any convex nonempty set is regular at each of its points (use 9.17).

9.20 *Let* $f : X \to \mathbf{R}$ *be Lipschitz around* x_0 *and* $0 \notin \partial f(x_0)$. *Then*

(139) $$\{v \in X : f^0(x_0; v) \le 0\} \subset T(x_0; M),$$

where $M = \{x \in X : f(x) \le f(x_0)\}$.

If f *is regular at* x_0 (§5, (64)), *then equality holds and* M *is regular at* x_0 ([20], 1, 2.4.7).

■ *First affirmation.* Because $0 \notin \partial f(x_0)$, there exists, according to 5.17, 2° and to (41), v_1 in X so that $f^0(x_0; v_1) < 0$. Let v be with $f^0(x_0; v) \le 0$. We have $f^0(x_0; v+\varepsilon v_1) <$ $< 0 \; \forall \varepsilon > 0$ (use 5.15, 1°). Consequently it is enough to prove

(140) $$f^0(x_0; v) < 0 \Rightarrow v \in T(x_0; M).$$

Let v be so that $f^0(x_0; v) < 0$. According to the definition $\exists\, \varepsilon > 0, \delta > 0$ so that

(141) $$y \in S(x_0; \varepsilon),\, t \in (0, \varepsilon) \Rightarrow f(y + tv) - f(y) \le -\delta t.$$

Take a sequence (x_n) from M, $x_n \to x_0$ and a sequence (t_n), $t_n \downarrow 0$. For $n \ge N$ (141) gives $f(x_n + t_n v) \le f(x_n) - \delta t_n \le f(x_0) - \delta t_n$ ($x_n \in M$!), hence $n \ge N \Rightarrow x_n + t_n v \in M$ and hence $\overset{9.18}{\Rightarrow}$ $v \in T(x_0; M)$, i. e. (140).

Second affirmation. It is enough to prove

(142) $$v \in K(x_0; M) \Rightarrow f^0(x_0; v) \le 0,$$

as combining with (137) and (139) one obtains that these three sets coincide. We have, *via* the definition (136),

$$\lim_{t \to 0+} \frac{d_M(x_0 + tv)}{t} = 0.$$

So, $\varepsilon > 0$ being any, we can find a sequence (t_n), $t_n \downarrow 0$ so that (143) $n \ge N \Rightarrow d_M(x_0 + t_n v) \le \varepsilon t_n$. (143) gives a point x_n in M which verifies

(144) $$\|x_0 + t_n v - x_n\| \le 2\varepsilon t_n.$$

Then

$$\frac{f(x_0 + t_n v) - f(x_0)}{t_n} \le \frac{f(x_n) + 2\varepsilon t_n L - f(x_0)}{t_n} \le 2\varepsilon L,$$

L the Lipschitz constant around x_0 (use (144), $f(x_n) \le f(x_0)$). Take the limit and as ε is any results $f'(x_0; v) \le 0$, hence $f^0(x_0; v) = f'(x_0; v) \le 0$ and so (142). ■

9.21 *Corollary. Let* $f : X \to \mathbf{R}$ *be Lipschitz around* x_0, $0 \notin \partial f(x_0)$ *and* M *as in* 9.20. *Then*

$$N(x_0; M) \subset \bigcup_{\lambda > 0} \lambda \partial f(x_0).$$

If f *is regular at* x_0, *then equality holds.*

■ The polar cone to the left-hand side of (139) is the *-weak closed convex cone generated by $\partial f(x_0)$. Take into account 5.17, 1° from §5 and 9.20 (taking polars, this reverses inclusions). ■

The epigraph intervention. The epigraph of the function $f: X \to \mathbf{R}$, epi f, is the set

$$\{(x, r) \in X \times \mathbf{R} : r \geq f(x)\}.$$

9.22 *Let* $f : X \to \mathbf{R}$ *be Lipschitz around* x_0. *Then*

a) $\operatorname{epi} f^0(x_0; \cdot) = T((x_0, f(x_0)); \operatorname{epi} f)$;

b) f *is regular at* $x_0 \Leftrightarrow$ epi f *is regular at* $(x_0, f(x_0))$ ([20], 1, 2.4.9).

■ a) „⊃". Let (v, r) be from $T((x_0, f(x_0)); \operatorname{epi} f)$. Take sequences $(y_n)_{n\geq 1}$, $y_n \to x_0$ and $(t_n)_{n\geq 1}$, $t_n \downarrow 0$ so that

(145)
$$\lim_{n\to\infty} \frac{f(y_n + t_n v) - f(y_n)}{t_n} = f^0(x_0; v).$$

As $(y_n, f(y_n))_{n\geq 1}$ is a sequence from epi f converging to $(x_0, f(x_0))$, there exists, according to 9.18, a sequence $(v_n, r_n)_{n\geq 1}$ in $X \times \mathbf{R}$ converging to (v, r) so that $f(y_n) + t_n r_n \geq f(y_n + t_n v_n)$, i.e. $\dfrac{f(y_n + t_n v_n) - f(y_n)}{t_n} \leq r_n$, $\forall\, n \geq 1$. Pass to the limit and taking into account (145) one obtains $f^0(x_0; v) \leq r$, that is $v \in \operatorname{epi} f^0(x_0; \cdot)$.

„⊂". It is enough to prove

$$(v, f^0(x_0; v) + \delta) \in T((x_0, f(x_0)); \operatorname{epi} f)\ \forall\, v \in X,\ \forall\, \delta \geq 0.$$

Use 9.18. Let $(x_n, r_n)_{n\geq 1}$ be a sequence from epi f converging to $(x_0, f(x_0))$ and $(t_n)_{n\geq 1}$ a sequence with $t_n \downarrow 0$. It follows to find a sequence $(v_n, s_n)_{n\geq 1}$ converging to $(v, f^0(x_0; v) + \delta)$ with the property $(x_n, r_n) + t_n(v_n, s_n) \in \operatorname{epi} f$, i.e.

(146)
$$r_n + t_n s_n \geq f(x_n + t_n v_n)\ \forall\, n \geq 1.$$

Take $v_n = v$ and $s_n = \max\{f^0(x_0; v) + \delta, \dfrac{f(x_n + t_n v) - f(x_n)}{t_n}\}$. We have $s_n \to f^0(x_0; v) + \delta$ since $\displaystyle\overline{\lim_{n\to\infty}} \frac{f(x_n + t_n v) - f(x_n)}{t_n} \leq f^0(x_0; v)$. It remains to verify (146). But $r_n + t_n s_n \geq r_n + [f(x_n + t_n v) - f(x_n)]$ and consequently (146), because $r_n \geq f(x_n)$ as $(x_n, r_n) \in \operatorname{epi} f$.

b) *Suppose* f *regular at* x_0. Then the function $g: X \times \mathbf{R} \to \mathbf{R}$, $g(x, r) = f(x) - r$ is regular at $(x_0, f(x_0))$ and $\operatorname{epi} f = M := \{(x, r) \in X \times \mathbf{R} : g(x, r) \leq 0\}$. But $0 \notin \partial g(x_0, f(x_0))$ because the second member equals to $\partial f(x_0) \times \{-1\}$. Apply 9.20, M, i.e. epi f, is regular at $(x_0, f(x_0))$.

Suppose epi f *regular at* $(x_0, f(x_0))$. Consider the Dini derivative $f_b'(x_0; \cdot)$,

$$f_b'(x_0; v) = \underline{\lim}_{t\to 0+} \frac{f(x_0+tv)-f(x_0)}{t}.$$

We have $K((x_0, f(x_0)); \text{epi} f) = \text{epi}\, f_b'(x_0; \cdot)$ (demonstration as for a)), hence epi $f_b'(x_0;$ $\cdot) \overset{(138)}{=} T((x_0, f(x_0)); \text{epi} f)$ and so epi $f_b'(x_0; \cdot) \overset{a)}{=} \text{epi} f^0(x_0; \cdot)$, consequently $f_b'(x_0; v) =$ $= f^0(x_0; v)\ \forall\, v \in X$. But, $\forall\, t \in (0, r)$ and $\forall\, V \in \mathscr{V}(x_0)$,

$$\frac{f(x_0+tv)-f(x_0)}{t} \le \sup_{\substack{x\in V\\ t\in(0,r)}} \frac{f(x+tv)-f(x)}{t},$$

hence (see (31) from §5, section F)

$$\overline{\lim_{t\to 0+}} \frac{f(x_0+tv)-f(x_0)}{t} \le f^0(x_0; v),$$

which combined with the last equality gives $\exists\, f'(x_0; v)$ and $f'(x_0; v) = f^0(x_0; v)\ \forall\, v \in X$.■

9.23 *Corollary. Let* $f : X \to \mathbf{R}$ *be Lipschitz around* x_0. *Then*

$$\xi \in \partial f(x_0) \Leftrightarrow (\xi, -1) \in N((x_0, f(x_0)); \text{epi } f).$$

■ We have $\xi \in \partial f(x_0) \overset{\text{def}}{\Leftrightarrow} f^0(x_0; v) \ge \xi(v)\ \forall\, v \Leftrightarrow (\xi, -1)((v, r)) \le 0,\ \forall\, v, \forall\, r \ge f^0(x_0;$ $v)$. But

$\text{epi} f^0(x_0; \cdot) \overset{a)}{=} T((x_0, f(x_0)); \text{epi} f)$, consequently $\xi \in \partial f(x_0) \Leftrightarrow (\xi, -1)((v, r)) \le 0,\ \forall\, (v,$ $r) \in T((x_0, f(x_0)); \text{epi} f)$, and we take into account the definition (134). ■

Generalized Clarke subderivative. Will be accepted fot subderivation also functions with infinite values. The generalization is made *via* the corollary 9.23.

Definition. Let be $f : X \to \mathbf{R} \cup \{\pm\infty\}$ and x_0 from X with $f(x_0)$ finite. The *generalized Clarke subderivative* of f at x_0, $\partial f(x_0)$, is

(147) $$\partial f(x_0) = \{\xi \in X^* : (\xi, -1) \in N((x_0, f(x_0)); \text{epi} f)\}.$$

The keeping of the notation $\partial f(x_0)$ from §5, G is justified by 9.23.

For instance, for the function $x \to \|x\|$ we have $(147_1)\ \partial \|\cdot\|(0) = S^*$, where $S^* = \{\xi \in$ $\in X^* : \|\xi\| \le 1\}$.

f is by definition *regular at* x_0, $f(x_0) \in \mathbf{R}$, provided epi f is regular at $(x_0, f(x_0))$ (the epigraph definition remains the same for the functions with values in $\overline{\mathbf{R}}$).

Some properties of the generalized Clarke subderivative.

9.24 *Let* $f : X \to \mathbf{R} \cup \{\pm\infty\}$ *be with* $f(x_0)$ *finite. If* $f(x) \geq f(x_0)$ *on a neighbourhood of* x_0, *then*

$$0 \in \partial f(x_0).$$

■ It follows to show, according to the definition, $(0, -1) \in N((x_0, f(x_0));\ \text{epi}\, f)\}$, equivalently (see (134)) $(0, -1)((v, r)) \leq 0$, that is $r \geq 0$, $\forall\ (v, r) \in T((x_0, f(x_0));\ \text{epi}\, f)\}$. Take such an element (v, r). According to 9.18, for a sequence $(t_n)_{n\geq 1}$ with $t_n \downarrow 0$ $\exists$ $(v_n, r_n)_{n\geq 1}$ a sequence with $(v_n, r_n) \to (v, r)$ and $(x_0, f(x_0)) + t_n(v_n, r_n) = (x_0 + t_n v_n, f(x_0) + t_n r_n) \in$ $\in \text{epi}\, f$ $\forall\ n \geq 1$, hence $f(x_0) + t_n r_n \geq f(x_0 + t_n v_n)$ $\forall\ n \geq 1$. The hypothesis gives a rank N from which $f(x_0 + t_n v_n) \geq f(x_0)$ $(t_n v_n \to 0\ !)$, so $n \geq N \Rightarrow t_n r_n \geq 0 \Rightarrow r_n \geq 0$, which imposes $r \geq 0$. ■

9.25 *Corollary to the Rockafellar theorem* ([20], 1, 2.9.8). *Let be* $f_1, f_2 : X \to \mathbf{R} \cup$ $\cup \{\pm\infty\}$. *If* f_1 *is finite at* x_0 *and* f_2 *Lipschitz around* x_0, *then*

$$\partial(f_1 + f_2)(x_0) \subset \partial f_1(x_0) + \partial f_2(x_0).$$

When f_1, f_2 *are also regular at* x_0, *we have equality.*

Definition. The *indicator* of a subset M of X is the function $\psi_M : X \to \mathbf{R} \cup \{+\infty\}$,

$$\psi_M(x) = \begin{cases} 0, & x \in M \\ +\infty, & x \in X \setminus M \end{cases}.$$

9.26 *Let* x_0 *be a point from* M. *Then*

a) $\partial \psi_M(x_0) = N(x_0; M)$,

b) ψ_M *is regular at* $x_0 \Leftrightarrow$ M *is regular at* x_0.

■ a) $\xi \in \partial\psi_M(x_0) \overset{(147)}{\Leftrightarrow} (\xi, -1) \in N((x_0, 0);\ \text{epi}\,\psi_M)$. But $\text{epi}\,\psi_M = M \times [0, +\infty)$, hence, applying 9.19, $\xi \in \partial\psi_M(x_0) \Leftrightarrow \xi \in N(x_0; M)$ and $\eta \in N(0, [0, +\infty))$, where $\eta(r) = -r$ $\forall\, r \in$ $\in \mathbf{R}$. As, indeed, $\eta \in N(0, [0, +\infty))$, a) results demonstrated.

b) Use the definitions. ■

The coderivative. Let be $f: X \to 2^Y_*$. The graph of f, $\text{Gr}\, f$, is, one knows, $\text{Gr}\, f = \{(x, y) \in X \times Y : y \in f(x)\}$. Let (x_0, y_0) be from $\text{Gr}\, f$. The *coderivative of* f *at* (x_0, y_0) is the set-valued map $D^* f(x_0, y_0) : Y^* \to 2^{X^*}$,

(147_2) $$D^* f(x_0, y_0)(\eta) = \{\xi \in X^* : (\xi, -\eta) \in N((x_0, y_0);\ \text{Gr}\, f)\}.$$

In the following

C is closed pointed convex cone with vertex from Y.

Recall the notations C^+, C^{++} (§8, (5), (6)). Also the order relation $y_1 \overset{C}{\leq} y_2$ from Y (8.3).

Definitions. Let A be a nonempty subset of Y and a_0 from A. Begin with the remark

(148) $$a_0 \text{ is a } C\text{-minimal element of } A \Leftrightarrow a \notin a_0 - (C \setminus \{0\}) \ \forall\, a \in A$$

($\Rightarrow$: $a \in a_0 - (C \setminus \{0\}) \Rightarrow a = a_0 - u$, $u \in C \setminus \{0\}$, hence $a_0 = a + u$, i.e. $a \overset{C}{<} a_0$, a contradiction; $\Leftarrow$: similarly).

The set of the C-minimal elements of A is denoted by $\mu_C(A)$.

(149) a_0 is a C-*properly positive minimal element* of A if there is ξ in C^{++} such that $\xi(a) \geq$ $\geq \xi(a_0)$ $\forall\, a \in A$. The set of these is denoted by $\pi\mu_C(A)$.

We have

(149_1) $$\pi\mu_C(A) \subset \mu_C(A) \text{ and, in the case } A \text{ compact,}$$

(149_2) $$\mu_C(A) \neq \varnothing,\ A \subset \mu_C(A) + \mathrm{C},\ \pi\mu_C(A) \neq \varnothing \text{ ([43]).}$$

(150) A is C-*bounded from below* if $\exists\, v \in Y$ such that $A \supset v + C$, i. e. $v \overset{C}{\leq} a$ $\forall\, a \in A$. A is C-*weak bounded from below* if $\exists\, M$ bounded subset of Y such that $A \subset M + C$. Every set C-bounded from below is C-weak bounded from below, but there are sets which are C-weak bounded from below but not C-bounded from below. These concepts coincide when int $C \neq \varnothing$.

(151) Let be $f : X \to 2^Y_*$. f is C-*bounded from below* resp. C-*weak bounded from below* when $f(X) := \bigcup_{x \in X} f(x)$ is C-bounded from below (see also section E, (100)) resp C-weak bounded from below.

(152) Let be $f : X \to 2^X_*$ and (x_0, y_0) from Gr f.

a) (x_0, y_0) is a C-*minimizer* of f if $y_0 \in \mu_C(f(x_0))$;

b) (x_0, y_0) is a C-*properly positive minimizer* of f if $y_0 \in \pi\mu_C(f(x_0))$.

(153) Let be $f : X \to 2^Y_*$, (x_0, y_0) from Gr f and $\varepsilon > 0$, k_0 from $C \setminus \{0\}$.

a) (x_0, y_0) is an εk_0-*minimizer* of f if $y_0 \in \mu_C(f(x_0))$ and $y + \varepsilon k_0 \overset{C}{\nleq} y_0$ $\forall\, y \in f(X)$;

b) (x_0, y_0) is a C-*properly positive* εk_0-*minimizer* of f if there is ξ in C^{++} with $\xi(k_0) = 1$ and $\xi(y_0) \leq \xi(y)$ $\forall\, y \in f(x_0)$ and $\xi(y) + \varepsilon > \xi(y_0)$ $\forall\, y \in f(X)$.

The Hiriart - Urruty functional ([35_2]). Let A be a *proper closed* subset of Y. The *Hiriart - Urruty functional* $\Delta_A : \mathrm{Y} \to \mathbf{R}$ is

$$\Delta_A(y) = \mathrm{d}(y, A) - \mathrm{d}(y, Y \setminus A),$$

d the distance on Y afferent to the norm.

It can affirm ([35_2]):

(154) Δ_A *is Lipschitz of constant* 1;

(155) Δ_A *is convex when* A *is convex and* Δ_A *is positive homogeneous when* A *is a cone with vertex*;

(156) $\Delta_A(y) > 0 \Leftrightarrow y \notin A$ (from definition, *A* is closed);

(157) $\partial \Delta_A(a) = N(a; A) \cap S_Y^* \setminus \{0\}$ *when* $a \in \operatorname{Fr} A$ *and* $\operatorname{int} A \neq \varnothing$ ($S_Y^* = \{\xi \in Y^* : \|\xi\| \leq$ $\leq 1\}$).

We have

$$(158)\quad \begin{cases} \Delta_{-C}(k) = d(k,-C) > 0\ \forall k \in C \setminus \{0\} (k \notin -C \text{ because } C \text{ is pointed}); \\ \partial\Delta_{-C}(0) = C^+ \cap S_Y^* \setminus \{0\} \text{ when } \operatorname{int} C \neq \varnothing \text{ (take } a = 0 \text{ in (157))}; \\ \Delta_{-C}(y_1 + y_2) \leq \Delta_{-C}(y_1) + \Delta_{-C}(y_2)\ \forall y_1, y_2 \in Y \end{cases}$$

(consider the cases $y_1, y_2 \in -C$ (use then §8, 8.1), $y_1, y_2 \notin -C$, $y_1 \in -C$ and $y_2 \notin -C$).

The minimal element may be characterized with the Hiriart - Urruty functional:

$$(159)\qquad a_0 \in \mu_C(A) \Leftrightarrow \Delta_{-C}(a - a_0) > 0\ \forall\, a \in A \setminus \{a_0\}$$

(*A. Zaffaroni*, *Degrees of efficiency and degrees of minimality*, SIAM J.Control optimal 42 (2003), 1071-1086).

Another two propositions are necessary ([70_1], lemmas 3.1 and 3.2).

9.27 *Let* k_0 *be from* $C \setminus \{0\}$ *with* $d(k_0, -C) = 1$ *and* $\varepsilon > 0$. *Then*

$$y + \varepsilon k_0 \overset{C}{\not\leq} y' \Rightarrow 0 < \Delta_{-C}(y - y') + \varepsilon.$$

■ As $y - y' + \varepsilon k_0 \notin -C$ we have (160) $\Delta_{-C}(y - y' + \varepsilon k_0) > 0$ ((156)). Moreover, (161) $\Delta_{-C}(\varepsilon k_0) \overset{(155)}{=} \varepsilon\Delta_{-C}(k_0) \overset{(158)}{=} \varepsilon d(k_0, -C) = \varepsilon$, apply the triangle inequality to $\Delta_{-C}((y - y') + \varepsilon k_0)$ and take into account (160) and (161). ■

9.28 *Let* $f : X \to 2_*^Y$ *be with compact values, upper semicontinuous* (*section* E, (99)) *and* C-*bounded from below. For any* $\varepsilon > 0$ *and* k_0 *from* $C \setminus \{0\}$ *and for any* $\overline{y}$ *from* f(X), *there exists an* εk_0-*minimizer* (x_0, y_0) *of* f *such that*

$$y_0 \leq \overline{y}.$$

■ Firstly prove

(162) A *nonemtpy subset of* Y C-*bounded from below* $\Rightarrow \exists\, \bar{a} \in A$ *such that* $a + \varepsilon k_0 \overset{C}{\not\leq}$ $\overset{C}{\not\leq} \bar{a}\ \ \forall\, a \in A$.

▲ Assuming *par absurdum* the contrary, we find a sequence $(a_n)_{n\geq 1}$ in A with $a_{n+1} +$ $+ \varepsilon k_0 \overset{C}{\leq} a_n\ \forall\, n \geq 1$. By adding one obtains $a_{n+1} + n\varepsilon k_0 \overset{C}{\leq} a_1\ \forall\, n \geq 1$. Let v be from Y such that $v \overset{C}{\leq} a\ \forall\ a \in A$ ((150)). Then (163) $v + n\,\varepsilon k_0 \overset{C}{\leq} a_1\ \forall\ n \geq 1$. $\{k_0\}$ is compact, $-C$ is closed, $\{k_0\} \cap -C \neq \varnothing$ (C is pointed !), apply a separation theorem, there exists a closed hyperplane which strictly separates these, hence there is ξ in C^+ so that $\xi(k_0) > 0$. Using (163), we find $n\,\varepsilon\,\xi(k_0) \leq \xi(a_1 - v)\ \forall\ n \geq 1$ and passing to the limit one obtains the contradiction. ▲

Continue the proof of the proposition. Apply (162) to $A := f(X) \cap (\bar{y} - C)$, correctly as $A \neq \varnothing$ ($y \in \bar{y} - C$!) and A is C-bounded from below since $f(X)$ has this property, $\exists\, y' \in A$ such that (164) $y + \varepsilon k_0 \overset{C}{\not\leq} y'\ \forall\ y \in A$. But (164) is also verified $\forall\ y$ from $f(X) \setminus$ $\setminus (\bar{y} - C)$: if for an y from this set we have $y + \varepsilon k_0 \overset{C}{\leq} y'$, then, as $y' \overset{C}{\leq} \bar{y}$ ($y' \in \bar{y} - C$!), it results $y + \varepsilon k_0 \overset{C}{\leq} \bar{y}$, i. e. $y \in \bar{y} - (\varepsilon k_0 + C) \subset \bar{y} - C$ (§8, 8.1), a contradiction. So

$$(165) \qquad\qquad y + \varepsilon k_0 \overset{C}{\not\leq} y'\ \ \forall\ y \in f(X).$$

Let now be x_0 from X with $y' \in f(x_0)$. $f(x_0)$ being compact, we have (166) $f(x_0) \subset$ $\subset \mu_C(f(x_0)) + C$ (see (149_2)), hence $\exists\ y_0 \in \mu_C(f(x_0))$ so that $y_0 \overset{C}{\leq} y'$. But then (165) imposes *a fortiori* $y + \varepsilon k_0 \overset{C}{\not\leq} y_0\ \forall\ y \in f(X)$. Moreover, $y_0 \overset{C}{\leq} \bar{y}$ since $y' \overset{C}{\leq} \bar{y}$. ■

We pass to multivalued variants of Ekeland principle.

9.29 *Variant of Ekeland principle.* Let $f : X \to 2^Y_*$ *be with compact values, upper semicontinuous,* C-*bounded from below and* $\operatorname{int} C \neq \varnothing$. *For any* $\varepsilon > 0$ *and* k_0 *from* $C \setminus \{0\}$ *with* $d(k_0, -C) = 1$ *and for any* $\bar{y}$ *from* $f(X)$, *there exists an* εk_0*-minimizer* $(x_\varepsilon, y_\varepsilon)$ *of* f *such that*

1) $y_\varepsilon \overset{C}{\leq} \bar{y}$;

2) $y + \varepsilon\|x - x_0\|k_0 \overset{C}{\not\leq} y_\varepsilon \ \forall\ (x, y) \in \mathrm{Gr}\, f \setminus \{(x_\varepsilon, y_\varepsilon)\}$;

3) *there exists* $(\xi_\varepsilon, -\eta_\varepsilon)$ *in* $N((x_\varepsilon, y_\varepsilon); \mathrm{Gr}\, f)$ *with* $\eta_\varepsilon \in (C^+ \cap S_Y^*) \setminus \{0\}$ *and* $\|\xi_\varepsilon\| \leq \varepsilon$

(*Truong Xuan Duc Ha*).

The assertion 3) may be expressed using the coderivative ((147₂), $S_X^* = \{\xi \in X^* : \|\xi\| \leq \leq 1\}$):

$$\exists\ \eta_\varepsilon \text{ in } (C^+ \cap S_Y^*) \setminus \{0\} \text{ such that } D^* f(x_\varepsilon, y_\varepsilon)(\eta_\varepsilon) \cap \varepsilon S_X^* \neq \emptyset.$$

■ Note from the begining that Gr f is closed.

Let (x_0, y_0) be an εk_0-minimizer of f with (167) $y_0 \overset{C}{\leq} \bar{y}$ (9.28). Applying (116) from 9.16 (section E) one finds an εk_0-minimizer $(x_\varepsilon, y_\varepsilon)$ of f which verifies assertion 2) and also (168) $y_\varepsilon \overset{C}{\leq} y_0$. From (167) and (168) results also assertion 1).

Now combine assertion 2) with 9.27, one obtains

(169) $$\Delta_{-C}(y - y_\varepsilon) + \varepsilon\|x - x_\varepsilon\| > 0 \ \forall\ (x, y) \in \mathrm{Gr}\, f \setminus \{(x_\varepsilon, y_\varepsilon)\}.$$

(169) expresses that $(x_\varepsilon, y_\varepsilon)$ is unique minimizer of the function $g : X \times Y \to \mathbf{R} \cup \{+\infty\}$,

$$g(x, y) = \Delta_{-C}(y - y_\varepsilon) + \varepsilon\|x - x_\varepsilon\| + \psi_{\mathrm{Gr}\, f}(x, y)$$

(the indicator of Gr f) and hence, applying successively 9.24, 9.25, 9.26 a), (147_1) and (158), one obtains

$$0 \in \partial g(x_\varepsilon, y_\varepsilon)$$

and

$$\partial g(x_\varepsilon, y_\varepsilon) \subset \{0\} \times \partial\Delta_{-C}(0) + \varepsilon S_X^* \times \{0\} + N((x_\varepsilon, y_\varepsilon); \mathrm{Gr}\, f) = \{0\} \times (C^+ \cap S_Y^* \setminus \{0\}) +$$
$$+ \varepsilon S_X^* \times \{0\} + N((x_\varepsilon, y_\varepsilon); \mathrm{Gr}\, f),$$

consequently $\exists\ \xi_\varepsilon \in \varepsilon S_X^*$ and $\exists\ \eta_\varepsilon \in (C^+ \cap S_Y^*) \setminus \{0\}$ such that $(\xi_\varepsilon, -\eta_\varepsilon) \in N((x_\varepsilon, y_\varepsilon);$ Gr $f)$, that is assertion 3). ■

9.30 *Variant of Ekeland principle. Let* $f : X \to 2_*^Y$ *be with compact values, upper semicontinuous and* C-*weak bounded from below. For any* $\varepsilon > 0$, k_0 *from* $C \setminus \{0\}$ *and* η *from* C^{++} *with* $\eta(k_0) = 1$, *there exists a properly positive* εk_0-*minimizer* $(x_\varepsilon, y_\varepsilon)$ *of* f *such that*

1) $y + \varepsilon\|x - x_\varepsilon\|k_0 \overset{C}{\not\leq} y_\varepsilon \ \forall\ (x, y) \in \mathrm{Gr}\, f \setminus \{(x_\varepsilon, y_\varepsilon)\}$;

2) $\exists\ \xi_\varepsilon \in X^*$ *such that* $(\xi_\varepsilon, -\eta) \in N((x_\varepsilon, y_\varepsilon); \mathrm{Gr}\, f)$ *and* $\|\xi_\varepsilon\| \leq \varepsilon$

(*Troung Xuan Duc Ha*).

The assertion 2) may be expressed using the coderivative:

$$D^* f(x_\varepsilon, y_\varepsilon)(\eta) \cap \varepsilon S_X^* \neq \varnothing.$$

■ We have

(170) $$\inf\{\eta(y) : y \in f(X)\} > -\infty.$$

Indeed, let y be any from $f(X)$. But $f(X) \subset M + C$, M bounded, hence $y = u + k$, $u \in M$ and $k \in C$. η being linear and continuous, it is bounded on M, so $\exists\, m_0$ such that $\eta(v) \geq m_0$, $\forall\, v \in M$ and then, as $\eta \in C^+$, $\eta(y) = \eta(u) + \eta(k) \geq \eta(u) \geq m_0$, i.e. (170).

Because η is bounded from below and attains its infimum on any $f(x)$, $x \in X$ as this is compact, it can consider the function $\varphi : X \to \mathbf{R}$, $\varphi(x) = \min_{y \in f(x)} \eta(y)$. φ is lower semicontinuous (use f is compact-valued and u.s.c.) and in addition (170) gives φ is lower bounded. Apply Ekeland principle, §1, 1.1 with $\lambda = 1$, $\exists\, x_\varepsilon$ in X such that

(171) $$\varphi(x_\varepsilon) < \varphi(x) + \varepsilon \;\forall\, x \in X$$

and

(172) $$\varphi(x) + \varepsilon\|x - x_\varepsilon\| > \varphi(x_\varepsilon) \;\forall\, x \in X \setminus \{x_\varepsilon\}.$$

Let y_ε be from $f(x_\varepsilon)$ so that $\varphi(x_\varepsilon) = \eta(y_\varepsilon)$ ($\leq \eta(y)\; \forall\, y \in f(x_\varepsilon)$). (171) and (172) give respectively

(173) $$\eta(y_\varepsilon) < \eta(y) + \varepsilon \;\forall\, y \in f(X)$$

and

(174) $$\eta(y) - \eta(y_\varepsilon) + \varepsilon\|x - x_\varepsilon\| > 0 \;\forall\, (x, y) \in \mathrm{Gr}\, f \setminus \{(x_\varepsilon, y_\varepsilon)\}.$$

The last bracket and (173) tell us that $(x_\varepsilon, y_\varepsilon)$ is a properly positive εk_0-minimizer of f ((153), b)).

Prove 1). Suppose *par absurdum* that $\exists\, (x', y')$ in $\mathrm{Gr}\, f \setminus \{(x_\varepsilon, y_\varepsilon)\}$ such that

$$y' + \varepsilon\|x' - x_\varepsilon\| k_0 - y_\varepsilon \overset{C}{\leq} 0.$$

η being from C^+ with $\eta(k_0) = 1$, we have then

$$\eta(y') - \eta(y_\varepsilon) + \varepsilon\|x' - x_\varepsilon\| \leq 0,$$

in contradiction with (174).

Prove 2). (174) expresses that $(x_\varepsilon, y_\varepsilon)$ is minimizer unique of the function $g : X \times Y \to$ $\to \mathbf{R} \cup \{+\infty\}$,

$$g(x, y) = \eta(y - y_\varepsilon) + \varepsilon\|x - x_\varepsilon\| + \psi_{\mathrm{Gr}\, f}(x, y).$$

In the following the proof repeats this for assertion 3) from 9.29. ■

G. EKELAND PRINCIPLE IN SEQUENTIALLY COMPLETE LOCALLY CONVEX SPACES ([35_1], 1)

Two locally convex variants of Ekeland principle are presented (the aim function φ preserves its values in $(-\infty, +\infty]$, but the place of the norm is taken by the Minkowski functional respectively by a seminorm family.

The idea on which A. Hamel based his paper ([35_1], 1) is to reduce the locally convex case to the Banach spaces case. The locally convex space which appears is not that given by a positive semi-norm family, but by a family of „semi-norms" with values in $(-\infty, +\infty)$, which obliges to submit this family under some supplementary conditions. The author was separated of this position and was passing all in positive semi-norms.

As usual – the preparations.

Balanced set. Let E be real vector space. The nonempty subset M of E is named *balanced* or *circled* when

(175) $$\lambda M \subset M \ \forall \lambda \in [-1, 1].$$

For instance, $\{0\}$ is balanced. Any balanced set contains 0. In fact, (176) M balanced $\Rightarrow$ $M = -M$ ($x \in M \Rightarrow -x \in M, x \in -M$ etc.). Any reunion resp. nonempty intersection of balanced sets is balanced. SM, $S := \{\lambda \in \mathbf{R} : |\lambda| \leq 1\}$, is the smallest balanced set which contains M (*balanced hull of* M). The convex hull of the balanced hull of M is named *convex balanced hull*, it is the smallest convex symmetric set which includes M and its elements are of the form

$$\sum_{i=1}^{n} \lambda_i x_i, \ n \in \mathbf{N}, x_i \in M, \lambda_i \in \mathbf{R}, \ \sum_{i=1}^{n} |\lambda_i| \leq 1.$$

Absorbing set. Let E be real vector space and A nonempty subset. By definition A *absorbs* the nonempty subset B of E if there exists $r > 0$ so that

(177) $$|\lambda| \geq r \Rightarrow \lambda A \supset B.$$

A is named *absorbing* when it absorbs all the points of E. A absorbs $B \Leftrightarrow \exists\ \rho > 0$ so that (178) $\mu \neq 0$, $|\mu| \leq \rho \Rightarrow \mu B \subset A$. A balanced absorbs B if $\exists\ \lambda_0$ so that $\lambda_0 A \supset B$. In particular, A balanced is absorbing $\Leftrightarrow \forall\, x \in E\ \exists\ \mu \neq 0$ so that $\mu x \in A$. Any finite nonempty intersection of absorbing sets is absorbing.

Minkowski functional. Let E be real vector space and M convex subset with $0 \in M$. Define, for each x from E, $\mu_M(x)$ in the following manner:

$$\text{when } \nexists\, \rho > 0 \text{ so that } x \in \rho M,\ \mu_M(x) = +\infty;$$

$$\text{in the contrary case } \mu_M(x) = \inf\{\rho > 0 : x \in \rho M\}.$$

The application $E \to [0, +\infty], x \to \mu_M(x)$ is the *Minkowski functional* of M. Obviously $A \subset \subset B \Rightarrow \mu_B \leq \mu_A$. We have also $\mu_{\alpha M} = \dfrac{1}{\alpha}\mu_M$, $\alpha > 0$. Moreover,

(179) $\mu_M(\lambda x) = \lambda \mu_M(x), \lambda > 0$, (180) $\mu_M(0) = 0$, (181) $\mu_M(x + y) \leq \mu_M(x) + \mu_M(y)$.

When M is absorbing, μ_M takes finite values, hence, according to (179), (180) and (181), μ_M is sublinear and hence convex ($f : E \to \mathbf{R}_+$ sublinear $\overset{\text{def}}{\Leftrightarrow}$ f positive homogeneous, convex and $f(0) = 0$; the condition „f convex" can be replaced by „$f(x + y) \leq f(x) + f(y)$").

Let M be as above. Set

(182) $$U(\mu_M, 1) := \{x \in E : \mu_M(x) < 1\},\ V(\mu_M, 1) := \{x \in E : \mu_M(x) \leq 1\}.$$

We have

(183) $$U(\mu_M, 1) \subset M \subset V(\mu_M, 1).$$

Real topological vector space (*real t.v.s.*). Let X be real topological vector space. There exists a fundamental system $\mathcal{V}$ of closed neighbourhoods of 0 with the properties:

(184) Any set of $\mathcal{V}$ is balanced and absorbing;

(185) $\lambda U \in \mathcal{V}\ \forall\, \lambda \neq 0, \forall\, U$ from $\mathcal{V}$;

(186) For each U of $\mathcal{V}$ there exists V in $\mathcal{V}$ so that

$$V + V \subset U.$$

Let X be real t.v.s.

The sequence $(a_n)_{n\geq 1}$ from X is by definition *Cauchy sequence* if $\forall\ V$ balanced neighbourhood of $0\ \exists\ N$ in $\mathbf{N}$ so that

(187) $$m, n \geq N \Rightarrow a_m - a_n \in V.$$

The definition is correct since $V \overset{(176)}{=} -V$.

Let M be nonempty subset of X. M is by definition *sequentially complete* (s-*complete*) if any Cauchy sequence from M is convergent in M.

The nonempty subset M of X is by definition *bounded*, if M is absorbed by any neighbourhood of 0, equivalently $\forall$ V neighbourhood of 0 $\exists\ \lambda > 0$ so that

(188) $$\lambda M \subset V.$$

It can suppose $0 < \lambda < 1$ ((185)).

Let M be bounded. Then any nonempty subset is bounded, λM is also bounded $\forall\ \lambda$ of **R**. Any finite reunion of bounded sets is bounded. M is bounded $\Leftrightarrow$ $\forall$ sequence (x_n) from M and $\forall$ sequence (λ_n), $\lambda_n \geq 0$, $\lambda_n \to 0$ we have $\lambda_n x_n \to 0$. The image of a bounded set by a linear continuous map is bounded.

Suppose the topology of E defined by the family of semi-norms $(p_i)_{i\in I}$ (see below). In this case M is bounded iff $p_i(M)$ is bounded for any i from I.

9.31 *The set of the points of a Cauchy sequence from a t.v.s. is bounded.*

■ Let $(a_n)_{n\geq 1}$ be Cauchy sequence, $M := \{a_n : n \geq 1\}$ and U neighbourhood of 0. $\exists\ V$ balanced neighbourhood of 0 so that $V + V \subset U$ ((186)). Because we have, begining from a rank N on, $a_m \overset{(187)}{\in} a_n + V$, $\exists\ a_1, \ldots, a_p$ so that $M \subset \bigcup_{i=1}^{p}(a_i + V)$. But the set $A := \{a_i : i = \overline{1, p}\ \}$ being finite, it is bounded, hence $\exists\ \lambda$ in (0, 1) so that $\lambda A \overset{(188)}{\subset} V$. So we have ($V$ is balanced) $\lambda M \subset \lambda A + \lambda V \subset V + V \subset U$, $\lambda M \subset U$. ■

Still two definitions.

The nonempty subset M of X is *sequentially closed* (s-*closed*) if the limit of each sequence of points from M is in M. $\varnothing$ is *sequentially closed.*

$f: X \to \overline{\mathbf{R}}$ is *sequentially lower semicontinuous* (s-*lower semicontinuous*, s-*l.s.c.*) resp. *sequentially upper semicontinuous* (s-*upper semicontinuous*, s-*u.s.c.*) at the point a, if for any sequence $(x_n)_{n\geq 1}$ with $x_n \to a$ we have

$$\underline{\lim}_{n\to\infty} f(x_n) \geq f(a) \text{ resp. } \overline{\lim}_{n\to\infty} f(x_n) \leq f(a).$$

Semi-norm. Let E be real vector space. The application $p : E \to \mathbf{R}_+$ is named **semi-norm**, if $\forall\ x, y$ from E and $\forall\ \lambda$ from **R**

(189) $$p(x + y) \leq p(x) + p(y),$$

(190) $$p(\lambda x) = |\lambda| p(x).$$

If $p_1, \ldots, p_n$ are semi-norms on E, then $\max_{1\leq i\leq n} p_i$ has the same property.

Let p be semi-norm on E. Consider the set $\mathscr{V}$ of the sets $V(p, \rho)$, $\rho > 0$,

(191) $$V(p, \rho) = \{x \in E : p(x) \le \rho\}.$$

$V(p, \rho)$ is balanced and absorbing. There exists a single topology $\mathcal{T}$ on E compatible with the structure of vector space and for which $\mathcal{V}$ is a fundamental system of neighbourhoods of 0. p is continuous relative to $\mathcal{T}$, hence $V(p, \rho)$ is closed and $U(p, \rho) := \{x \in E: p(x) < \rho\}$ is open.

Let now be $(p_i)_{i \in I}$ a family of semi-norms on E. The supremum $\mathcal{T}$ of the topologies on E defined by p_i, i goes over I, is a topology compatible with the structure of vector space of E and the sets of the form

(192) $$\bigcap_{k=1}^{n} V(p_{i_k}, \rho),\ n \in \mathbf{N},\ \rho > 0$$

form a fundamental system of neighbourhoods of 0 for $\mathcal{T}$. The closure of $\{0\}$ in $\mathcal{T}$ is equal to the set of the points x of E for which $p_i(x) = 0\ \forall\ i$ from I. Consequently, $\mathcal{T}$ is separated $\Leftrightarrow \forall\ x \neq 0\ \ \exists\ i$ in I so that $p_i(x) \neq 0$. All the semi-norms p_i are continuous relative to $\mathcal{T}$.

Real locally convex space. A real topological vector space, in which 0 has a fundamental system of convex neighbourhoods, is named *real locally convex space* and its topology *locally convex*. In this case, 0 has a fundamental system of symmetrical convex closed neighbourhoods.

The topology $\mathcal{T}$ of a real topological vector space is locally convex, iff $\mathcal{T}$ is defined by a family of semi-norms.

Pass to the subject of this section.

All the locally convex spaces which appear are real.

Firstly, some preparatory propositions.

Let X be separated locally convex space, $(p_i)_{i \in I}$ the family of semi-norms which defines its topology and M nonempty convex balanced bounded subset. Consider the Minkowski functional μ_M of M.

9.32 μ_M *is norm on the vector subspace* W_M *generated by* M ([35_1, 1]).

■ Indeed, because μ_M is positive homogeneous and M balanced we have $\mu_M(\lambda x) = |\lambda| \times \mu_M(x)\ \forall\ x \in W_M,\ \forall\ \lambda \in \mathbf{R}$ ($\lambda > 0 \Rightarrow \mu_M(-\lambda x) = \inf\{\rho > 0: -\lambda x \in \rho M\} \overset{(176)}{=} \inf\{\rho > 0: x \in \frac{1}{\lambda}\rho M\} = \mu_{\frac{1}{\lambda}M}(x) = \lambda\mu_M(x)$). Taking into account (180) and (181), it remains to prove

$\mu_M(x) = 0 \Rightarrow x = 0$. Let U be any neighbourhood of 0. M being bounded, $\exists\ \lambda > 0$ so that $\lambda M \subset U$ ((188)). But $x \in \lambda M$ since $\mu_M(x) = 0$ ($\exists\ \rho$, $0 < \rho < \lambda$ so that $x \in \rho M$, $\frac{\rho}{\lambda} M \subset M$, M balanced), hence $x \in U$ and because U is anyone and X separated, results $x = 0$. ■

Note the norm μ_M on W_M by $\|\cdot\|_M$. Suppose moreover X and M *sequentially complete*. Then

9.33 *The normed space* $(W_M, \|\cdot\|_M)$ *is complete*

(*A. Hamel*).

■ Firstly prove

(193) $$M \text{ is complete according to } \|\cdot\|_M.$$

Let $(x_n)_{n\geq 1}$ be Cauchy sequence from M according to $\|\cdot\|_M$. Then $(x_n)_{n\geq 1}$ is also Cauchy sequence in X ((187)). Indeed, let be $V := \bigcap_{k=1}^{N} V_{i_k}$, $V_{i_k} := V(p_{i_k}, r)$ as in (192). M being bounded, $\exists\ \alpha > 0$ so that $M \subset \alpha V$, hence $M \subset \alpha V_{i_k}$ and hence $\frac{1}{\alpha}\mu_{V_{i_k}} \leq \mu_M$, (194) $\mu_{V_{i_k}} \leq \alpha\mu_M$, $k = \overline{1, N}$. Set $U_{i_k} := \{x \in X: p_{i_k}(x) \leq 1\}$. Then $U_{i_k} = \frac{1}{r} V_{i_k}$ and on the other hand $p_{i_k} = \mu_{U_{i_k}}$ and then

$$p_{i_k}(x_m - x_n) = \mu_{U_{i_k}}(x_m - x_n) = \mu_{\frac{1}{r}V_{i_k}}(x_m - x_n) = r\,\mu_{V_{i_k}}(x_m - x_n) \overset{(194)}{\leq} r\alpha\mu_M(x_m - x_n).$$

But (x_n) being Cauchy sequence in M $\exists\ N_1$ so that $m, n \geq N_1 \Rightarrow \mu_M(x_m - x_n) \leq \frac{1}{\alpha}$, hence $m, n \geq N_1 \Rightarrow p_{i_k}(x_m - x_n) \leq r$, $x_m - x_n \in V_{i_k}$, $k = \overline{1, N}$, $x_m - x_n \in V$.

So, as X is s-complete $\exists\ x_0$ in X such that (195) $x_n \xrightarrow{\text{in X}} x_0$. M being s-closed, it results $x_0 \in M$. It remains to prove, for having (193),

(196) $$x_n \xrightarrow{\|\cdot\|_M} x_0.$$

Let be $\varepsilon > 0$. $\exists\ N$ in $\mathbf{N}$ so that $m, n \geq N \Rightarrow \mu_M(x_m - x_n) < \varepsilon$ (Cauchy sequence in W_M !). $\exists\ \alpha$ so that $\mu_M(x_m - x_n) < \alpha < \varepsilon$ and $x_m - x_n \in \alpha M$, but $\alpha M \subset \varepsilon M$ because $\frac{\alpha}{\varepsilon} M \subset M$ (balanced set), hence $m, n \geq N \Rightarrow x_m - x_n \in \varepsilon M$. Taking $m \to \infty$, one obtains $n \geq N \overset{(195)}{\Rightarrow} x_n - x_0 \in \varepsilon M$ (M is s-closed in X), hence $\mu_M(x_n - x_0) \leq \varepsilon$ and hence (196).

Let be $(x_n)_{n\geq 1}$ now any Cauchy sequence in W_M. Show that this sequence is convergent in W_M. Because

$$|\,\|x_m\|_M - \|x_n\|_M\,| \leq \|x_m - x_n\|_M,$$

the sequence $(\|x_n\|_M)_{n\geq 1}$ is fundamental hence convergent, let be (197) $\alpha = \lim_{n\to\infty} \|x_n\|_M$.

When $\alpha = 0$, we have $x_n \xrightarrow{\|\cdot\|_M} 0$, the proof is finished. Suppose $\alpha > 0$, then from a rank on we have $\|x_n\|_M \geq \rho > 0$. When $(\frac{x_n}{\rho})_{n\geq 1}$ is convergent in W_M, the sequence $(x_n)_{n\geq 1}$ has the same property, consequently we can accept

$$(198)\ \|x_n\|_M \geq 1\ \forall\, n \geq 1.$$

$(\frac{x_n}{\|x_n\|_M})_{n\geq 1}$ is a sequence from M ($\frac{1}{\|x_n\|_M} \leq 1$, M is balanced), even a Cauchy sequence.

Indeed, $\left\| \frac{x_m}{\|x_m\|_M} - \frac{x_n}{\|x_n\|_M} \right\|_M = \left\| \frac{x_m}{\|x_m\|_M} - \frac{x_n}{\|x_m\|_M} + \frac{x_n}{\|x_m\|_M} - \frac{x_n}{\|x_n\|_M} \right\|_M \leq \frac{1}{\|x_m\|_M} \|x_m - x_n\|_M +$

$+ \|x_n\|_M \left| \frac{1}{\|x_m\|_M} - \frac{1}{\|x_n\|_M} \right| \overset{(198)}{\leq} \|x_m - x_n\|_M + \frac{1}{\|x_m\|_M} \left|\,\|x_n\|_M - \|x_m\|_M\,\right| \leq 2\|x_m - x_n\|_M$. So

$(\frac{x_n}{\|x_n\|_M})_{n\geq 1}$ is a convergent sequence in M, let be (199) $y_0 = \lim_{n\to\infty} \frac{x_n}{\|x_n\|_M}$ ((193)).

Obviously $\alpha y_0 \in W_M$ and we prove $x_n \xrightarrow{\|\cdot\|_M} \alpha y_0$.

$\|x_n - \alpha y_0\|_M = \| (x_n - \|x_n\|_M y_0) + (\|x_n\|_M y_0 - \alpha y_0)\|_M \leq \|x_n - \|x_n\|_M y_0\|_M + \|\, \|x_n\|_M y_0 -$

$- \alpha y_0\|_M = \|x_n\|_M \left\| \frac{x_n}{\|x_n\|_M} - y_0 \right\|_M + \|y_0\|_M\, |\, \|x_n\|_M - \alpha|$

and one takes into account (197) and (199). ■

Retain from the last proof:

$$(200)\ x_n \xrightarrow{\|\cdot\|_M} x_0 \Rightarrow x_n \xrightarrow{\text{in } X} x_0.$$

Indeed, (x_n) being Cauchy sequence according to $\|\cdot\|_M$, it is also Cauchy sequence in X, then $x_n \xrightarrow{\text{in } X} x'$ and because we have also $x_n \xrightarrow{\|\cdot\|_M} x'$ ((196)), *a fortiori* $x' = x_0$.

9.33₁ *Let* A *be* s-*closed subset of* X. *Then* $A \cap W_M$ *is closed with respect to* $\|\cdot\|_M$.

■ Let (x_n) be a sequence in $A \cap W_M$ convergent in $\|\cdot\|_M$. (x_n) is so Cauchy sequence in $\|\cdot\|_M$ from W_M which is complete (9.33), let be $\lim_{n\to\infty} x_n = a \in W_M$. But (x_n) is also a

Cauchy sequence in the locally convex topology (the proof for 9.33), consequently (x_n) is convergent to a in X, hence $a \in A$ because A is s-closed. ∎

9.33$_2$ *Let* B *be a nonempty subset of* M *with* $0 \notin \overline{B}$ *and* $C := \mathbf{R}_+ B$. *If the subset* A *of* $x_0 + C$, $x_0 \in W_M$, *is bounded in the locally convex topology, then* A *is bounded in* $\|\cdot\|_M$.

∎ Let $(p_i)_{i\in I}$ be the family of semi-norms which defines the topology of X. A being bounded in X, $A - x_0$ has the same property ($p_i(A)$ bounded $\forall\, i \Rightarrow p_i(A - x_0)$ bounded $\forall\, i$). So we can assume $x_0 = 0$.

Let x be any from $A \setminus \{0\}$. As $x \in C$ $\exists\, t > 0$ and b in B so that $x = tb$. Because $0 \notin \overline{B}$ and X is separated, there exists $i \in I$ and $\delta > 0$ so that $p_i(y) \geq \delta$ on B. So (200$_1$) $p_i(b) \geq \delta$. Moreover, A being bounded in X, $\exists\, \rho > 0$ so that (200$_2$) $p_i(y) \leq \rho$ $\forall\, y \in A$. Or $p_i(x) = p_i(tb) = tp_i(b)$, hence $t = \dfrac{p_i(x)}{p_i(b)} \leq \dfrac{\rho}{\delta}$ ((200$_1$), (200$_2$)). As $B \subset M$ we have, according to (183), $\|b\|_M \leq 1$ and then $\|x\|_M = t\|b\|_M \leq \dfrac{\rho}{\delta}$. ∎

Remark. 9.33$_2$ is Proposition 4 from [35$_1$] in which x_0 was introduced. This is the form which will be used later on.

And now

s-*complete version of the Ekeland principle*

9.34 Theorem Let X *be Hausdorff* s-*complete locally convex space*, $\varphi : X \to (-\infty, +\infty]$ *proper bounded from below* s-*lower semicontinuous and* S *bounded convex* s-*closed with* $0 \in S$. *For any* $\varepsilon > 0$ *and* u *from* dom φ, *there exists* v_ε *in* X *so that*

(201) $$\varphi(v_\varepsilon) < \varphi(x) + \varepsilon\mu_S(x - v_\varepsilon)\ \forall\, x \in X \setminus \{v_\varepsilon\},$$

(202) $$\varphi(v_\varepsilon) + \varepsilon\mu_S(v_\varepsilon - u) \leq \varphi(u)$$

(*A. Hamel*).

∎ Firstly prove

(203) $$\mu_S \text{ is } s\text{-l.s.c. on } X.$$

Let x_0 be a point of X and $(x_n)_{n\geq 1}$ sequence from X, $x_n \to x_0$. If $x_0 = 0$, obviously $\mu_S(x_0) = 0 \leq \underline{\lim}_{n\to\infty} \mu_S(x_n)$. Then let be $x_0 \neq 0$. Suppose *par absurdum* $\underline{\lim}_{n\to\infty} \mu_S(x_n) < \mu_S(x_0)$ and let r be,

(204) $$\underline{\lim}_{n\to\infty} \mu_S(x_n) < r < \mu_S(x_0).$$

Then there exists a subsequence $(x_{k_n})_{n\geq 1}$ so that $\mu_S(x_{k_n}) < r < \mu_S(x_0)$ $\forall$ $n \geq 1$ and let be $\rho_{k_n} > 0$ with $x_{k_n} \in \rho_{k_n} S$ and $\rho_{k_n} < r$ (the definition !). So (205) $x_{k_n} = \rho_{k_n} y_n$, $y_n \in S$ $\forall$ $n \geq 1$. $(\rho_{k_n})_{n\geq 1}$ being bounded, it has a convergent subsequence, we likewise denote, $\rho_{k_n} \to \rho_0$. Obviously $\rho_0 \leq r$. We also have $\rho_0 > 0$. Indeed, suppose *par absurdum* $\rho_0 = 0$. Let U be a balanced neighbourhood of 0 and V a neighbourhood of x_0 so that (206) $U \cap \cap V = \varnothing$ ($x_0 \neq 0$, X is separated, (184)). We have, from a rank on, $\rho_{k_n} y_n \in V$. S being bounded, $\exists$ $\lambda > 0$ so that $\lambda S \subset U$ ((188)) and suppose that from a rank N on we have $\frac{\rho_{k_n}}{\lambda} < 1$. Then, for $n \geq N$, $\rho_{k_n} y_n = \frac{\rho_{k_n}}{\lambda} \lambda y_n \in U$ since U is balanced and we come in collision with (206), $\rho_0 > 0$. So in (205) it can take the limit for $n \to \infty$, $\lim\limits_{n\to\infty} y_n = \frac{x_0}{\rho_0} =: y_0 \in$ $\in S$ (s-closed set). Then $x_0 = \rho_0 y_0 \in \rho_0 S$, hence, from the definition, $\mu_S(x_0) \leq \rho_0$ and hence $\mu_S(x_0) \leq r$, in contradiction with (204), (203) is proved.

Consider the set M the sequential closure of the convex hull of the set

$$S \cup -S \cup \{\pm u\}$$

and let W_M be the vector subspace generated by M. W_M with the norm $\|\cdot\|_M$ is Banach space (9.33).

Consider the nonempty set ($u \in A$)

(207) $$A := \{x \in W_M : \varphi(x) + \varepsilon\mu_S(x-u) \leq \varphi(u)\}.$$

A is closed in W_M: let $(x_n)_{n\geq 1}$ be a sequence from A and $x_n \xrightarrow{\|\cdot\|_M} x'$, then $x_n \xrightarrow{\text{in X}} x'$ ((200)) and since φ and μ_S are s-l.s.c. we have $\varphi(x') + \varepsilon\mu_S(x'-u) \leq \underline{\lim}_{n\to\infty} \varphi(x_n) + \underline{\lim}_{n\to\infty} \varepsilon\mu_S(x_n) \leq \underline{\lim}_{n\to\infty} (\varphi(x_n) + \varepsilon\mu_S(x_n - u)) \overset{(164)}{\leq} \varphi(u)$, hence $x' \in A$.

Take the function $f: W_M \to (-\infty, +\infty]$,

(208) $$f(x) = \begin{cases} \varphi(x), & x \in A \\ +\infty, & x \in W_M \setminus A. \end{cases}$$

f is l.s.c. in the norm topology (take into account that φ is finite on A (see (207)) and l.s.c. on A in the norm topology (use (200)), bounded from below and proper. Apply the Ekeland principle §1, 1.1, $\exists$ v_ε in W_M so that

(209) $$f(v_\varepsilon) < f(x) + \varepsilon\,\|x - v_\varepsilon\|_M \;\forall\; x \in W_M \setminus \{v_\varepsilon\}.$$

A fortiori (210) $f(v_\varepsilon) < +\infty$, (211) $v_\varepsilon \in A$. (211) and (207) give (202).

Pass to verify (201). Let x be anyone from $X \setminus \{v_\varepsilon\}$. If $x \in X \setminus W_M$, (201) is verified because $\mu_S(y) = +\infty$ on $X \setminus W_M$ ($W_M \supset \rho M\ \forall\ \rho > 0$). Suppose $x \in W_M$. For $x \in A$ we have $f(x) \overset{(208)}{=} \varphi(x)$ and also $f(v_\varepsilon) = \varphi(v_\varepsilon)$ ((211)), replacing in (209), it results (201), because

$$\|x - v_\varepsilon\|_M = \mu_M(x - v_\varepsilon) \le \mu_S(x - v_\varepsilon).$$

Finally, suppose $x \in W_M \setminus A$. Then ((207))

$$\varphi(x) + \varepsilon\mu_S(x - u) \le \varphi(u),$$

but also ((202))

$$\varphi(v_\varepsilon) + \varepsilon\mu_S(v_\varepsilon - u) \le \varphi(u),$$

hence

$$\varphi(v_\varepsilon) + \varepsilon\mu_S(v_\varepsilon - u) < \varphi(x) + \varepsilon\mu_S(x - u) \overset{(181)}{\le} \varphi(x) + \varepsilon\mu_S(v_\varepsilon - u) + \varepsilon\mu_S(x - v_\varepsilon).$$

Take the first and the last member and reduce the term $\varepsilon\mu_S(v_\varepsilon - u)$, this is finite by (202). ■

REMARK. *Theorem* 9.34 *and Theorem* **E** *in norm* (§1, REMARK 2 to 1.1_0) *are equivalent.*

■ *Th.* 9.34 ⇒ *Th.* **E** (*in norm*). Indeed, because X being a real normed space and $S :=$ $= \{y \in X : \|y\| \le 1\}$, bounded convex closed set, we have $\|x\| = \mu_S(x)\ \forall\ x \in X$ (see the definition, if $x = \alpha y$ with $\alpha > 0$ and $y \in S$ then $\|x\| = \alpha\|y\| \le \alpha$).

Th. **E** (*in norm*) ⇒ *Th.* 9.34. In the demonstration for 9.34 the Ekeland principle 1.1 is used. ■

The s-complete version 9.34 and also the s-complete variant of the Ekeland principle which follows are connected to the Phelps lemma and now we pass to their study.

Let X be real normed space, f from X^* with $\|f\| = 1$ and $\lambda \in (0, 1)$.

(212) $$C_{\lambda, f} := \{x \in X : \lambda\|x\| \le f(x)\} \text{ ([56], 1)}$$

is closed convex cone with vertex (§8, „preliminaries of affine geometry", $\lambda\|x + y\| \le$ $\le \lambda(\|x\| + \|y\|) \le \lambda(f(x) + f(y)) = \lambda f(x + y)$ etc.).

Let be $B := \{x \in X : \|x\| \le 1,\ f(x) \ge \lambda\}$. B is nonempty ($0 < \lambda < 1 = \|f\| \Rightarrow \exists\ x'$, $\|x'\| \le 1$ and $\lambda < |f(x')|$, take x' so that $f(x') > 0$) bounded closed convex. We have

$$C_{\lambda, f} = \mathbf{R}_+ B.$$

Indeed, $x \in C_{\lambda,f} \setminus \{0\} \Rightarrow \lambda\|x\| \le f(x)$, $\lambda \le f(\frac{x}{\|x\|})$, $y := \frac{x}{\|x\|} \in B$, $x = \|x\|y \in \mathbf{R}_+B$; $x \in$ $\in \mathbf{R}_+B \setminus \{0\} \Rightarrow x = \alpha y$, $\alpha > 0$ and $y \in B$, hence $\|y\| \le 1$, $f(y) \ge \lambda$ and so $\lambda\|x\| = \lambda\alpha\|y\| \le$ $\le \alpha f(y) = f(x)$.

9.35 *Let* X *be real Banach space,* A *closed nonempty subset,* $\lambda \in (0, 1)$ *and* f *from* X^* *with* $\|f\| = 1$ *and* $\sup f(A) < +\infty$. *For any* x_0 *from* A *there exists* x^* *in* A *such that*

$$x^* \in x_0 + C_{\lambda,f} \text{ and } A \cap (x^* + C_{\lambda,f}) = \{x^*\}$$

([56], 1, lemma 1.1).

■ One defines by recurrence the sequences $(x_n)_{n\ge1}$ and $(A_n)_{n\ge1}$. $x_1 = x_0$, $A_1 = A \cap (x_1 + C_{\lambda,f})$; suppose defined x_n and $A_n := A \cap (x_n + C_{\lambda,f})$, take x_{n+1} from A_n so that $\sup f(A_n) < f(x_{n+1}) + \frac{1}{n}$. Let be $n \ge 1$. Because $x_{n+1} \in x_n + C_{\lambda,f}$ and $C_{\lambda,f}$ is convex, it results (see 8.1) $x_{n+1} + C_{\lambda,f} \subset x_n + C_{\lambda,f} + C_{\lambda,f} \subset x_n + C_{\lambda,f}$, hence $A_{n+1} \subset A_n$. So if $y \in A_{n+1}$ then $f(y) \le$ $\le \sup f(A_n)$ and in addition $y - x_{n+1} \in C_{\lambda,f}$, hence

$$\lambda\|y - x_{n+1}\| \overset{(212)}{\le} f(y - x_{n+1}) = f(y) - f(x_{n+1}) \le \sup f(A_n) - f(x_{n+1}) < \frac{1}{n}$$

and hence $\operatorname{diam} A_{n+1} \le \frac{2}{\lambda n}$. Apply the Cantor theorem, $\bigcap_{n=1}^{\infty} A_n = \{x^*\}$. $x^* \in A_1$ means $x^* \in$ $\in A$ and $x^* \in x_0 + C_{\lambda,f}$. Moreover, $\forall\, n \ge 1$ $x^* \in x_n + C_{\lambda,f}$, hence $x^* + C_{\lambda,f} \subset x_n + C_{\lambda,f} +$ $+ C_{\lambda,f} \overset{8.1}{\subset} x_n + C_{\lambda,f}$ and so $A \cap (x^* + C_{\lambda,f}) \subset \bigcap_{n=1}^{\infty} A_n = \{x^*\}$, $A \cap (x^* + C_{\lambda,f}) = \{x^*\}$. ■

It can now affirm

9.36 *Phelps lemma. Let* X *be real Banach space,* A *nonempty closed subset,* B *nonempty closed convex bounded subset with* $0 \notin B$ *and the cone* $C := \mathbf{R}_+B$. *For every* x_0 *from* A *with* $A \cap (x_0 + C)$ *bounded there exists* x^* *such that*

$$x^* \in A \cap (x_0 + C) \text{ and } \{x^*\} = A \cap (x^* + C).$$

■ Use 9.35 with A replaced by $A \cap (x_0 + C)$. Look for $\lambda > 0$ and f from X^* with $\|f\| =$ $= 1$ so that

$$C \subset C_{\lambda,f}. \tag{213}$$

As $0 \notin B$ $\exists$ $S_\delta(0) := \{x \in X : \|x\| \le \delta\}$ so that $S_\delta(0) \cap B = \varnothing$. Apply Hahn - Banach separation theorem, the second geometrical form (§1, *Memento*, before (11_3)), $\exists f$ in X^*,

$\|f\| = 1$ (it can suppose) so that $r := \sup f(S_\delta(0)) \le \inf f(B)$. So $r \le f(y)\ \forall\ y \in B$. B being bounded, $\exists\ \rho > 0$ so that $\|y\| \le \rho\ \forall\ y \in B$, consequently $y \in B \Rightarrow \frac{r}{\rho}\|y\| \le f(y)$, hence $B \subset C_{\lambda,\mathrm{f}}$ where $\lambda = \frac{r}{\rho}$ and hence $\mathbf{R}_+ B \subset C_{\lambda,\mathrm{f}}$.

Apply 9.35 with A replaced by $A \cap (x_0 + C)$, correctly, as $A \cap (x_0 + C)$ being bounded, f is bounded on this. There exists x^* in $A \cap (x_0 + C)$ such that

(214) $$\{x^*\} = A \cap (x_0 + C) \cap (x^* + C_{\lambda,\mathrm{f}}).$$

But $x^* + C_{\lambda,\mathrm{f}} \overset{(213)}{\supset} x^* + C$ and $x^* \in x_0 + C \overset{8.1}{\Rightarrow} x^* + C \subset x_0 + C$, hence (214) gives $\{x^*\} \supset A \cap \cap (x^* + C) \supset \{x^*\}$ and so equality. ■

Here it is a locally convex version of 9.36.

9.37 *Phelps lemma in* s-*complete spaces* ([35$_1$], 1). *Let* X *be Hausdorff* s-*complete locally convex space,* A *nonempty* s-*closed subset,* B *nonempty* s-*closed bounded convex subset with* $0 \notin \overline{\mathrm{B}}$ *and* $\mathrm{C} := \mathbf{R}_+\mathrm{B}$. *For any* x_0 *from* A *such that* $\mathrm{A} \cap (\mathrm{x}_0 + \mathrm{C})$ *is bounded, there is* x^* *such that*

$$\mathrm{x}^* \in \mathrm{A} \cap (\mathrm{x}_0 + \mathrm{C}) \textit{ and } \{\mathrm{x}^*\} = \mathrm{A} \cap (\mathrm{x}^* + \mathrm{C})$$

(*A. Hamel*).

■ Consider the set M the sequential closure of the convex hull of the set

(215) $$B \cup -B \cup \{\pm x_0\}.$$

W_M with $\|\cdot\|_\mathrm{M}$ is Banach space (9.33). Let be $A_1 := A \cap W_\mathrm{M}$. A_1 is nonempty ($x_0 \in A_1$, see (215)) closed in $\|\cdot\|_\mathrm{M}$ (see 9.33$_1$). The same property has also B ($\subset W_\mathrm{M}$!). Besides, $C \subset W_\mathrm{M}$ and $A_1 \cap (x_0 + C)$ is bounded in W_M as subset of $A \cap (x_0 + C)$, which, being bounded in X, is bounded also in W_M (9.33$_2$).

Apply Phelps lemma (9.36) in W_M, there is x^* in W_M such that

$$x^* \in A_1 \cap (x_0 + C) \text{ and } \{x^*\} = A_1 \cap (x^* + C).$$

We have even $\{x^*\} = A \cap (x^* + C)$, because $x^* + C \subset W_\mathrm{M}$. ■

In fact

9.38 *Phelps lemma in Banach spaces and its* s-*complete version are equivalent.*

■ Indeed, 9.37 is obviously a generalization of 9.36 and in the proof of 9.37 is used 9.36. ■

And now

s-complete version of Ekeland principle

9.39 Theorem. *Let* X *be Hausdorff* s-*complete locally convex space*, $\varphi : X \to (-\infty, +\infty]$ *lower bounded proper* s-*lower semicontinuous and* $(p_i)_{i\in I}$ *the family of semi-norms afferent to* X. *For any* $(\varepsilon_i)_{i\in I}$ *a family of strictly positive numbers and for any* u *from* dom φ, *there exists* v^* *in* X *so that*

(216) $\forall$ x *from* $X \setminus \{v^*\}$ *there exists* j *in* I *for which*

$$\varphi(v^*) < \varphi(x) + \varepsilon_j p_j (x - v^*)$$

and

(217) $$\varphi(v^*) + \varepsilon_i p_i (v^* - u) \le \varphi(u) \ \forall\, i \in I$$

(*J. - X. Fang* [30_1], *A. Hamel*).

■ Let be $Y := X \times \mathbf{R}$. Consider the family $(q_i)_{i\in I}$ of semi-norms on Y, $q_i (x, r) = \max(p_i (x), |r|)$. This makes Y a s-complete separated locally convex space. $A := \text{epi}\ \varphi$ is nonempty (φ is proper) and *s*-closed (φ is *s*-l.s.c.). $B := \{(x, -1) \in Y : \varepsilon_i p_i (x) \le 1 \ \forall\, i \in I\}$ is nonempty ($(0, -1) \in B$), *s*-closed (use the definition, p_i are continuous), convex, bounded (use the definitions of q_i and B) and $(0,0) \notin \overline{B}$. Obviously

(218) $$C := \mathbf{R}_+ B = \{(x, s) \in Y : s + \varepsilon_i p_i (x) \le 0 \ \forall\, i \in I\}.$$

Let $(x_0, \varphi(x_0))$ be from A. $A_1 := A \cap ((x_0, \varphi(x_0)) + C)$ is bounded. Indeed, let (x, r) be any from A_1 and i abitrary fixed in I. Then we have, as $r \ge \varphi(x) \ge m := \inf \varphi(X)$,

$$m - \varphi(x_0) + \varepsilon_i p_i (x - x_0) \le r - \varphi(x_0) + \varepsilon_i p_i (x - x_0) \overset{(218)}{\le} 0,$$

and hence $p_i (x - x_0) \le \dfrac{\varphi(x_0) - m}{\varepsilon_i}$, $r \le m$.

Apply 9.37,

(219) $$\exists\, (x^*, r^*) \in A \cap ((x_0, \varphi(x_0)) + C) \text{ so that}$$

(220) $$\{(x^*, r^*)\} = A \cap ((x^*, r^*) + C).$$

Take $u = x_0$, $v^* = x^*$ and verify (216), (217). (219) gives $r^* \ge \varphi(v^*)$ and $(v^* - u, r^* - \varphi(u)) \in C$, i.e. (see (218)) $r^* - \varphi(u) + \varepsilon_i p_i (v^* - u) \le 0 \ \forall\, i \in I$ and so (217).

Let be $x' \in X \setminus \{v^*\}$. It can assume $x' \in \text{dom}\ \varphi$, in the contrary case (216) being verified. So being, $(x', \varphi(x')) \in A$, which implies, *via* (220), $(x', \varphi(x')) \notin (v^*, r^*) + C$, i.e. (see (218))

$$\exists\, j \in I \text{ so that } \varphi(x') - r^* + \varepsilon_j p_j (x' - v^*) > 0$$

and hence (216), as $r^* \geq \varphi(v^*)$. ■

REMARK. In [35_1], 1 one asserts that, reciprocally, 9.39 ⇒ 9.37 (Theorem 3 ⇒ Theorem 1, p.3031). The proof of this statement is *wrong*: the relations of the page 3032

$$l(x - x^*) + \gamma_\mu p_\mu(x - x^*) > 0$$

and

$$\gamma_\mu p_\mu(x - x^*) \leq l(x - x^*) \text{ (i.e. } x - x^* \in K_\Lambda)$$

can coexist, hence the relation $x \notin x^* + K_\Lambda$, and so in particular $x \notin x^* + K$, is not proved.

Also, the assertion Theorem 2 ⇒ Theorem 1 (p. 3032) has in the proof exactly the same vice.

So being, the assertion of Theorem 6:

Theorem 3 (9.39) *and Theorem* 2 (9.34) *are mutually equivalent*

is without covering. However this is true. Indeed, in the following section H the propositions 9.44 and 9.45 will be demonstrated. These are *generalizations* to locally complete locally convex spaces of 9.39 respectively 9.34. But there also is demonstrated (see 9.46) that the two propositions are equivalent, consequently, taking into account also REMARK to 9.34, it can be anticipated

9.39$_1$ s-*complete versions* 9.34 *and* 9.39 *of Ekeland principle are equivalent with Ekeland principle* 1.1 *in norm.*

Now prove

9.39$_2$ 9.34 ⇒ 9.37.

■ Let S be the convex hull of $\{0\} \cup B$, so $S = \{ty : y \in B, t \in [0, 1]\}$. $0 \in S$, S is s-closed as B has this property and in addition S is bounded since B is bounded: let $(p_i)_{i\in I}$ be the semi-norm family afferent to X and i any fixed, we have, on S, $p_i(ty) = tp_i(y) \leq p_i(y)$ and hence p_i is bounded on S.

As $0 \notin \overline{B}$ $\exists$ ξ in X^* and $\exists$ $\rho > 0$ so that (220_1) $\xi(y) \geq \rho$ $\forall$ $y \in B$ (Dunford - Schwartz theorem, §1, *Memento*, before (11_3)). Let be $C := \mathbf{R}_+B$. We have, $\forall$ $x \in C$ (hence $x = \lambda y$, $\lambda \geq 0$, $y \in B$),

$$\rho\mu_S(x) = \rho\mu_S(\lambda y) = \lambda\rho\mu_S(y) \overset{(183)}{\leq} \lambda\rho \overset{(220_1)}{\leq} \lambda\xi(y) = \xi(\lambda y) = \xi(x),$$

hence

(220_2) $$C \subset C_\rho := \{x \in X : \rho\mu_S(x) \leq \xi(x)\}.$$

C is s-closed and hence also $x_0 + C$ and $A_0 := A \cap (x_0 + C)$, nonempty set. Consider the function $\varphi : X \to (-\infty, +\infty]$,

$$\varphi(x) = \begin{cases} -\xi(x), & x \in A_0 \\ +\infty, & x \notin A_0. \end{cases}$$

φ is s-lower semicontinuous, bounded from below ($-\xi$, linear continuous functional, is bounded on the bounded set A_0 (under (188))) and proper. Apply to them 9.34 with $\varepsilon = \rho$ and $u = x_0$ (dom $\varphi = A_0$), $\exists\, x^*$ so that

(220₃) $$\varphi(x^*) < \varphi(x) + \rho\mu_S(x - x^*) \quad \forall\, x \in X \setminus \{x^*\},$$

(220₄) $$\varphi(x^*) + \varepsilon\mu_S(x^* - x_0) \le \varphi(x_0).$$

(220₄) imposes $\varphi(x^*) \le \varphi(x_0) < +\infty$, i.e. $x^* \in$ dom $\varphi = A \cap (x_0 + C)$. It remains to show

(220₅) $$\{x^*\} = A \cap (x^* + C).$$

Let be *par absurdum* $x \in A \cap (x^* + C)$ and $x \neq x^*$.

Assume $x \notin A_0$. As $x \in A$ it results $x \notin x_0 + C$. On the other hand, $x \in x^* + C$ and as $x^* \in x_0 + C$ it results $x \in x_0 + C + C \overset{8.1}{\subset} x_0 + C$, $x \in x_0 + C$, contradiction.

Assume $x \in A_0$. (220₃) gives $-\xi(x^*) < -\xi(x) + \rho\mu_S(x - x^*)$, i.e. $\xi(x - x^*) < \rho\mu_S(x - x^*)$, i.e. $x - x^* \overset{(220_2)}{\notin} C_\rho$, *a fortiori* $x - x^* \notin C$, $x \notin x^* + C$, contradiction because $x \in A \cap (x^* + + C) \subset x^* + C$. ■

So being, combining 9.39_1 with 9.39_2, we find, as 9.37⇒ 9.39 (see the proof of 9.39),

(220₆) 9.34 ⇔ 9.39 ⇔ 9.37 ⇔ *Ekeland principle* 1.1 *in norm.*

On the other hand, theorem 9.46 from the following section H permits to insert also in the equivalences sequence from (220₆) and

9.39₃ *Daneš drop theorem in* s-*complete spaces. Let* X *be Hausdorff s-complete locally convex space,* A *nonempty* s-*closed subset and* B *nonempty* s-*closed bounded convex subset with* $0 \notin \overline{A - B}$. *For any* x_0 *from* A $\exists\, x^*$ *in* $A \cap D(x_0, B)$ *so that*

$$\{x^*\} = A \cap D(x^*, B).$$

Remark that in particular it was obtained an indirect proof of Daneš drop theorem 2.2_1 from §2.

Express once again the things established above:

The s-*complete versions* 9.34 *and* 9.39 *of the Ekeland principle* (*A. Hamel*), *the* s-*complete version* 9.37 *of the Phelps lemma* (*A. Hamel*) *and the* s-*complete version of Daneš drop theorem* (particularized from 9.40, *J. - H. Kiu*) *are equivalent with Ekeland principle* 1.1 *in norm.*

At the end of this section here is another proof of version 9.34 using Phelps lemma 9.37 ([35_1], 1).

■ Let $(p_i)_{i\in I}$ be the family of semi-norms which defines the locally convex topology of X and $Y := X \times \mathbf{R}$. Consider the family $(q_i)_{i\in I}$ of semi-norms on Y, $q_i(x, r) = \max(p_i(x), |r|)$. This makes from Y a s-complete separated locally convex space. $A := \text{epi}\,\varphi$ is nonempty (φ is proper) and s-closed (φ is s-l.s.c.). $B := \{(x,-1) \in Y: \varepsilon\mu_S(x) \leq 1\}$ is non-empty ($(0,-1) \in B$), s-closed, convex, bounded ($x \in p_1(B)$, $\frac{1}{\varepsilon} \geq \mu_S(x) = \inf\{\alpha > 0 : x \in \alpha S\}$, $x \in \alpha S \Rightarrow p_i(x) = \alpha p_i(y) \leq \alpha \sup p_i(S)$, $p_i(x) \leq \mu_S(x) \sup p_i(S) \leq \frac{1}{\varepsilon} \sup p_i(S)$) and $(0,0) \notin B$. Obviously

$$C := \mathbf{R}_+ B = \{(x, s) \in Y: s + \varepsilon\mu_S(x) \leq 0\}. \tag{220_7}$$

Let $(x_0, \varphi(x_0))$ be from A. $A_1 := A \cap ((x_0, \varphi(x_0)) + C)$ is bounded because S is bounded, apply 9.37, (220_8) $\exists\, (x^*, r^*) \in A \cap ((x_0, \varphi(x_0)) + C)$ so that

$$\{(x^*, r^*)\} = A \cap ((x^*, r^*) + C). \tag{220_9}$$

Take $u = x_0$, $v_\varepsilon = x^*$ and verify (201) and (202), (220_8) gives $r^* \geq \varphi(v_\varepsilon)$ and $(v_\varepsilon - u, r^* - \varphi(u)) \in C$, i.e. (see ($220_7$)) $r^* - \varphi(u) + \varepsilon\mu_S(v_\varepsilon - u) \leq 0$ and consequently (202). Let be $x' \in X \setminus \{v_\varepsilon\}$. It can assume $x' \in \text{dom}\varphi$, in the contrary case (201) being verified. So being, $(x', \varphi(x')) \in A_1$, this imposes, *via* ($220_9$), $(x', \varphi(x')) \notin (v_\varepsilon, r^*) + C$, that is (see ($220_7$)) $\varphi(x') - r^* + \varepsilon\mu_S(x' - v_\varepsilon) > 0$ and consequently (201) as $r^* \geq \varphi(v_\varepsilon)$. ■

H. EKELAND PRINCIPLE IN LOCALLY COMPLET LOCALLY CONVEX SPACES ([57_1], 3)

In this section two equivalent versions of the Ekeland principle in locally complete locally convex spaces are presented. These *generalize* the s-complete versions 9.34 and 9.39 of the same principle (A.Hamel). In particular these permit to prove that 9.34 and 9.39 are equivalent with the Ekeland principle in the normed form **E** (§1, REMARK 2 to 1.1_0), while these are authentic generalizations of this.

Because some propositions from [57_1], 3 are based on key-assertions with suspicious insufficient justifications (see for instance in [57_1], 3, p.33, the 7th line, $\{x \in X: f(x) \leq$

$\leq f(x_0)\} = p(x_0)\ G \cap D(x_0, B) \cap A)$, the author replaced everywhere in the statements the condition „either A, or B is locally complete" (A and B subsets of X locally convex space) with „ X locally complete".

In this section the main role belongs to ([57_1], 1, see §2, (2))

9.40 *Daneš drop theorem in locally complete spaces. Let* X *be Hausdorff locally complete locally convex space,* A *nonempty locally closed subset,* B *nonempty locally closed bounded convex subset and* $0 \notin \overline{A-B}$. *Then* $\forall\ x_0$ *from* A $\exists$ a *in* $A \cap D(x_0, B)$ *such that*

$$\{a\} = A \cap D(a, B)$$

(*J. - H. Qiu*).

Retain

$$(*)\ 0 \notin \overline{A-B} \Leftrightarrow \exists\, j \in I \text{ and } \exists\, \delta > 0 \text{ so that } p_j(a-b) \geq \delta\ \forall\, a \in A,\ \forall\, b \in B$$

($(p_i)_{i \in I}$ the family of semi-norms afferent to X).

Remark. The same statement, but in the particular case *X sequentially complete*, was obtained by *L. Cheng - Y. Zhou - F. Zhang* (*Daneš drop theorem in locally convex spaces*, Proc. Amer. Math. Soc., 124 (12): 3699-3702, 1996) and by *A. Hamel* ([35_1], 1, Theorem 7). The equivalence of these two was proved in Lemma 1, [35_1], 1, p.3033.

All the locally convex spaces which apear in this section are real and separated.

LOCAL COMPLETNESS [*)]

Let X be a locally convex space.

The nonempty subset M of X is by definition *absolutely convex*, if $\forall\ x, y$ from M we have

(221) $$\lambda x + \mu y \in M, \text{ where } |\lambda| + |\mu| \leq 1.$$

In this case, M is obviously also convex and also balanced ((175)).

Definitions. A nonempty subset of X bounded and absolutely convex is called *disk.*

[*)] See *P. Pérez Carreras*, *J. Bonnet*, *Barrelled locally convex spaces*, North-Holland, Amsterdam, 1987.

Let B be a disk from X. Then we can consider the Minkovski functional μ_B of B (section G, $0 \in B$). This is, according to 9.32, a norm on the vector subspace W_B of X generated by B. If W_B is Banach space, then B is called *Banach disk.*

The sequence $(x_n)_{n\geq 1}$ from X is *locally convergent* (*l-convergent*) if $\exists\ x_0$ and $\exists\ B$ disk in X so that $x_n \to x_0$ in W_B (*convergent sequence in Mackey sense*).

The sequence $(x_n)_{n\geq 1}$ from X is *locally Cauchy* (*l-Cauchy*) if $\exists\ B$ disk in X so that (x_n) is Cauchy sequence in W_B.

X is called *locally complete* (*l-complete*) if any *l*-Cauchy sequence is *l*-convergent, equivalently any bounded subset of X is included in a Banach disk.

Let A be nonempty subset of X. A is *locally complete* (*l-complete*) when any *l*-Cauchy sequence from A is *l*-convergent in A. A is *locally closed* (*l-closed*) when the limit of each *l*-convergent sequence from A is in A. X *l*-complete and A *l*-closed $\Rightarrow$ A *l*-complete (as any *l*-Cauchy sequence from A is *l*-Cauchy sequence in X), A closed $\Rightarrow$ A *l*-closed (let (x_n) be sequence from A, $x_n \to x_0$ in a W_B, $A \cap W_B$ is closed in W_B (9.33_1), so $x_0 \in A$).

9.41 *Every* s-*complete disk is a Banach disk.*

9.42 *Any locally convex space which is* s-*complete is also l-complete.*

The converse is not true (Example 5.1.12 from loc.cit.*)).

9.43 Any s-*closed set is also l-closed.*

The converse is not true ([57_1], 1, Example 3.1).

Let $f: X \to (-\infty, +\infty]$ be proper, f is by definition *locally lower semicontinuous* (*l-l.s.c.*) if the set

(222) $$\{x \in X : f(x) \leq a\}$$

is locally closed for any a from **R**. Obviously f *s*-l.s.c. $\Rightarrow$ f *l*-l.s.c. The converse is not true.

TWO *l*-COMPLETE VARIANTS OF EKELAND PRINCIPLE

9.44 Theorem. *Let* X *be l-complete locally convex space,* $\varphi : X \to (-\infty, +\infty]$ *lower bounded proper l-l.s.c. and* $(p_i)_{i\in I}$ *the family of semi-norms afferent to* X. *For any* $(\varepsilon_i)_{i\in I}$ *a family of strictly positive numbers and for any* u *from* dom φ, $\exists\ v^*$ *in* X *so that*

(223) $\forall$ x *from* $X \setminus \{v^*\}$ *there is* j *in* I *for which*

$$\varphi(v^*) < \varphi(x) + \varepsilon_j\, p_j(x - v^*)$$

and

(224) $$\varphi(v^*) + \varepsilon_i\, p_i(v^* - u) \le \varphi(u)\ \forall\, i \in I$$

(*J. - H. Qiu*).

■ Let be $f: X \to (-\infty, +\infty]$, $f(x) = \varphi(x + u) - \varphi(u)$. We have $f(0) = 0$ and f has the properties of φ. Besides, $\{x \in X: f(x) \le 0\} = \{x \in X: \varphi(x + u) \le \varphi(u)\}$ and $\{x \in X: \varphi(x + u) - \varphi(u) \le 0\} + u = \{y \in X: \varphi(y) \le \varphi(u)\}$.

Consider the locally convex space $Y: = X \times \mathbf{R}$ with the family of semi-norms $(q_i)_{i\in I}$, $q_i(x, t) = p_i(x) + |t|$ and in this the subsets

(225) $$A: = \{(x, t) \in Y: f(x) \le t \le 0\},$$

let be $m: = \inf\{t \in \mathbf{R}: (x, t) \in A\}$ (hence $-\infty < m \le 0$),

(226) $$B: = \{(x, r) \in Y: p_i(x) \le \frac{-r}{\varepsilon_i}\ \forall\, i \in I\},$$

r arbitrary *fixed*, $r < m$ and $C: = \mathbf{R}_+ B$, hence

(227) $$C = \{(x, t) \in Y: \varepsilon_i p_i(x) \le -t\ \forall\, i \in I\}.$$

We have A l-closed, B l-closed bounded convex and $0 \notin \overline{A - B}$ [see (*), x, $t \in A$ and $(y, r) \in B \Rightarrow q_i((x, t) - (y, r)) = q_i(x - y, t - r) = p_i(x - y) + |t - r| \ge t - r \ge m - r > 0$].

Apply 9.40 with (0,0) from A,

(228) $$\exists\, (v_1, t_1) \in A \cap D((0,0), B)\ (\subset A \cap C) \text{ so that}$$

(229) $$(v_1, t_1) = A \cap D((v_1, t_1), B).$$

From (228), (225) and (227) results

(230) $$f(v_1) \le t_1 \le 0,$$

(231) $$\varepsilon_i p_i(v_1) \le -t_1\ \forall\, i \in I$$

and from (230) and (231),

(232) $$-f(v_1) \ge -t_1 \ge \varepsilon_i p_i(v_1)\ \forall\, i \in I.$$

Taking $v^* = v_1 + u$ and passing to the function φ one obtains, *via* (232), $\varphi(u) - \varphi(v^*) \ge \varepsilon_i p_i(v^* - v)\ \forall\, i \in I$, i.e. (224) from the statement.

Pass to the proof of (223) and return to f. Firstly show

(233) $$t_1 = f(v_1).$$

Extend (230), $r < m \le f(v_1) \le t_1$ and take $\delta: = \frac{f(v_1) - r}{t_1 - r}$, hence $\delta \in (0,1)$. We have $\delta t_1 + (1 - \delta)r = \frac{f(v_1) - r}{t_1 - r} t_1 + \frac{t_1 - f(v_1)}{t_1 - r} r = f(v_1)$, so

(234) $(v_1, f(v_1)) = (v_1, \delta t_1 + (1-\delta)r) = \delta(v_1, t_1) + (1-\delta)(v_1, r)$.

From (232), $p_i(v_1) \le \frac{-f(v_1)}{\varepsilon_i} < \frac{-r}{\varepsilon_i}$ $\forall\, i \in I$, that is $(v_1, r) \in B$, hence (234) expresses that $(v_1, f(v_1)) \in D((v_1, t_1), B)$ and so $(v_1, f(v_1)) \in A \cap D((v_1, t_1), B)$. Remains only to look at (229) to accept (233).

Let now be x from $X \setminus \{v^*\}$. As $v^* = v_1 + u$, we have (235) $x_1 := x - u \neq v_1$. If $(x_1, f(x_1)) \notin A$, *a fortiori* $f(x_1) > 0$, hence $f(x_1) + \varepsilon_i p_i(x_1 - v_1) \ge f(x_1) > 0 \ge f(v_1)$ $\forall\, i \in I$,

$$\varphi(x) - \varphi(u) + \varepsilon_i p_i(x - v^*) > \varphi(v_1 + u) - \varphi(u) \ \forall\, i \in I$$

and hence (223) *con brio*. So being,

Pass to the case $(x_1, f(x_1)) \in A$. We prove

(236) $(x_1, f(x_1)) \notin (v_1, t_1) + C$, i.e. $(x_1 - v_1, f(x_1) - t_1) \notin C$,

that is $\exists\, j \in I$ so that $\varepsilon_j p_j(x_1 - v_1) > t_1 - f(x_1)$, or still, taking into account (233),

$$\varepsilon_j p_j(x - v^*) > f(v_1) - f(x_1) = \varphi(v_1 + u) - \varphi(u) - \varphi(x_1 + u) + \varphi(u),$$

which, according to (235), is even (223).

Suppose *par absurdum* (237) $(x_1, f(x_1)) \in (v_1, t_1) + C$, i.e. (238) $(x_1 - v_1, f(x_1) - t_1) \in C$, $t_1 - f(x_1) \ge \varepsilon_i p_i(x_1 - v_1)$ $\forall\, i \in I$. Since X is separated and $x_1 - v_1 \neq 0$ ((235))$\exists\, i'$ so that $p_{i'}(x_1 - v_1) \neq 0$ (section G, under (192)), which implies, *via* (238), (239) $t_1 - f(x_1) > 0$. Let be $\rho := \frac{t_1 - f(x_1)}{t_1 - r}$. We have $\rho \in (0,1)$ ((239), (230) extended, $r < m$). C being a cone, (238) gives $\left(\frac{x_1 - v_1}{\rho}, \frac{f(x_1) - t_1}{\rho}\right) = \left(\frac{x_1 - v_1}{\rho}, r - t_1\right) \in C$. Moreover, $(v_1, t_1) \overset{(228)}{\in} C$, consequently, C being convex, $(v_1, t_1) + \left(\frac{x_1 - v_1}{\rho}, r - t_1\right) \overset{8.1}{\in} C$, that is $\left(v_1 + \frac{x_1 - v_1}{\rho}, r\right) \overset{(227)}{\in} B$. But

$$(1-\rho)t_1 + \rho r = \frac{f(x_1) - r}{t_1 - r} t_1 + \frac{t_1 - f(x_1)}{t_1 - r} r = f(x_1),$$

hence

$$(x_1, f(x_1)) = (1-\rho)(v_1, t_1) + \rho\left(v_1 + \frac{x_1 - v_1}{\rho}, r\right) \in D((v_1, t_1), B).$$

So $(x_1, f(x_1)) \in A \cap D((v_1, t_1), B)$ and confronting with (229) one finds $(x_1, f(x_1)) = (v_1, t_1)$, hence in particular $x_1 = v_1$, in contradiction with (235) and so (236) is demonstrated.■

Remark. As *s*-complete $\Rightarrow$ *l*-complete but *l*-complete $\not\Rightarrow$ *s*-complete, theorem 9.44 is a *generalization* of the theorem 9.39 (J. X. Fang, A. Hamel).

With a demonstration similar to that for 9.44, using 9.40, one obtains

9.45 Theorem. *Let* X *be l-complete locally convex space,* $\varphi : X \to (-\infty, +\infty]$ *lower bounded proper l-lower semicontinuous and* S *bounded convex l-closed with* $0 \in S$. *For any* $\varepsilon > 0$ *and* u *from* dom φ, *there exists* v_ε *in* X *so that*

(240) $$\varphi(v_\varepsilon) < \varphi(x) + \varepsilon\mu_S(x - v_\varepsilon)\ \forall\, x \in X \setminus \{v_\varepsilon\}$$

and

(241) $$\varphi(v_\varepsilon) + \varepsilon\mu_S(v_\varepsilon - u) \le \varphi(u)$$

(*J. - H. Qiu*).

Remark 1. As 9.44 for 9.39, theorem 9.45 is a *generalization* of theorem 9.34 (A. Hamel).

REMARK 2. As it results from what was mentioned in REMARK to 9.34 and was anticipated in the statement 9.39_1, the *l*-complete versions 9.44 and 9.45 are indeed *generalizations* of the Ekeland principle 1.1. The confrontation with the F. Sullivan statement (§1, after REMARK 2 to 1.1_0) emphasizes their importance.

THE *l*-COMPLETE VERSIONS OF EKELAND PRINCIPLE FACE TO THE *l*-COMPLETE VERSION OF DANEŠ DROP THEOREM

9.46 Theorem. *The propositions*

9.40, 9.44 and 9.45

are equivalent

(*J. - H. Qiu*).

■ **9.45 $\Rightarrow$ 9.44**

Let be $S := \bigcap_{i\in I}\{x \in X : \varepsilon_i p_i(x) \le 1\}$. S is bounded, absolutely convex (hence $0 \in S$) and *l*-closed (being *s*-closed, 9.43). We have

(242) $$\mu_S(x) = \sup_{i\in I} \varepsilon_i p_i(x)\ \forall\, x \in X.$$

The hypothesis from 9.44 used in 9.45 gives $v^* := (v_\varepsilon, \varepsilon = 1)$ which verifies (240) and (241), which implies, *via* (242), (223) (*par absurdum*) and (224).

9.44 ⇒ 9.45

Let M be the absolute convex hull of $S \cup \{u\}$. μ_M is a norm on W_M (9.32). The set S, being l-closed and X l-complete, is l-complete, which imposes M Banach disk, i.e. W_M is Banach space. Consider the subset of W_M

(243) $$A := \{x \in W_M : \varphi(x) + \varepsilon\mu_S(x-u) \leq \varphi(u)\}.$$

A is nonempty ($u \in A$) and closed in W_M because φ and μ_S are l-l.s.c.

Consider the function $f: W_M \to (-\infty, +\infty]$,

(244) $$f(x) = \begin{cases} \varphi(x), & x \in A \\ +\infty, & x \in W_M \setminus A. \end{cases}$$

f is bounded from below, l.s.c., proper and $u \in \operatorname{dom} f$ ($u \in A$!). Apply Ekeland principle 1.1 in norm, the accepted hypothesis 9.44 with X Banach space allows us this, $\exists\, v_\varepsilon$ in W_M so that

(245) $$x \in W_M \setminus \{v_\varepsilon\} \Rightarrow f(v_\varepsilon) < f(x) + \varepsilon\,\|x - v_\varepsilon\|_M,$$

(246) $$f(v_\varepsilon) + \varepsilon\,\|v_\varepsilon - u\|_M \leq f(u)\ (= \varphi(u)).$$

(246) imposes $f(v_\varepsilon) < +\infty$, consequently $v_\varepsilon \in A$, i.e. $\varphi(v_\varepsilon) + \varepsilon\mu_S(v_\varepsilon - u) \leq \varphi(u)$, which is (241).

Pass to the proof of (240).

Suppose $x \neq v_\varepsilon$ *and* $x \in A$. Then (245) and (244) give $\varphi(v_\varepsilon) < \varphi(x) + \varepsilon\,\|x - v_\varepsilon\|_M$ and as $\mu_M(x - v_\varepsilon) \leq \mu_S(x - v_\varepsilon)$ because $S \subset M$, it results (240).

Suppose $x \neq v_\varepsilon$ *and* $x \in W_M \setminus A$. Then

$$\varphi(x) + \varepsilon\mu_S(x - v_\varepsilon) + \varepsilon\mu_S(v_\varepsilon - u) \overset{(181)}{\geq} \varphi(x) + \varepsilon\mu_S(x-u) \overset{(243)}{>} \varphi(u) \overset{(241)}{\geq} \varphi(v_\varepsilon) + \varepsilon\mu_S(v_\varepsilon - u)$$

and it remains only to observe that $\varepsilon\mu_S(v_\varepsilon - u) \overset{(241)}{<} +\infty$ so that by reducing this term to obtain (240).

Suppose $x \neq v_\varepsilon$ *and* $x \notin W_M$. Then, since $v_\varepsilon \in W_M$, *a fortiori* $\rho(x - v_\varepsilon) \notin W_M$, in particular $\rho(x - v_\varepsilon) \notin S\ \forall\, \rho > 0$, hence $\mu_S(x - v_\varepsilon) = +\infty$ and (240) is *con brio* verified. ■

As 9.40 ⇒ 9.45 (in the demonstration of 9.45 is used 9.40), it remains only to show

9.45 ⇒ 9.40

It can assume $0 \in B$.

As $0 \notin \overline{A-B}$, $\exists\, p$ semi-norm and $\delta > 0$ so that $p(a-b) \geq \delta\ \forall\, a \in A, \forall\, b \in B$. $V :=$ $= \{x \in X : p(x) \leq \frac{\delta}{4}\}$ is closed (and hence l-closed, 9.43) absorbent absolutely convex 0-neighbourhood and $(A-B) \cap 3V = \varnothing$. This implies

(247) $$(B + 3V) \cap A = \varnothing$$

($x \in A \cap (B + 3V) \Rightarrow a = b + y$ with $a \in A$, $b \in B$, $y \in 3V$, hence $a - b \in 3V$, contradiction). As X is l-complete, V absorbs any bounded set. Let $\Delta := \Delta(x_0, B)$ be the absolutely convex hull of $\{x_0\} \cup B$. This being bounded, $\exists\, \rho$ in $(0,1)$ so that (248) $\rho\Delta \subset V$ (see (178)). Let F be the local closure of $B + \rho\Delta$. F is bounded convex set and $0 \in F$. Δ being in particular balanced we have $\frac{\rho^2}{2}\Delta \subset \frac{\rho}{2}\Delta \overset{(248)}{\subset} \frac{1}{2}V$, as also $\frac{\rho}{2}B \subset \frac{1}{2}V$ it results $\frac{\rho}{2}(B +$ $+ \rho\Delta) \subset \frac{1}{2}V + \frac{1}{2}V = V$, $B + \rho\Delta \subset \frac{2}{\rho}V$, then, V being l-closed, $F \subset \frac{2}{\rho}V$, $\frac{\rho}{2}F \subset V$. As in addition $F \subset B + \rho\Delta + V$, it results $F + \frac{\rho}{2}F \subset B + \rho\Delta + V + V \overset{(248)}{\subset} B + V + V + V = B +$ $+ 3V$, which, confronted with (247), gives

(249) $$\left(F + \frac{\rho}{2}F\right) \cap A = \varnothing.$$

Thus $x \in A \overset{(249)}{\Rightarrow} x \notin F + \frac{\rho}{2}F \overset{(183)}{\Rightarrow} \mu_{(1+\frac{\rho}{2})F}(x) \geq 1$, therefore $\mu_{(1+\frac{\rho}{2})F}(x) = (1 + \frac{\rho}{2})^{-1}\mu_F(x) \geq 1$ and hence

(250) $$x \in A \Rightarrow \mu_F(x) \geq 1 + \frac{\rho}{2}.$$

Let x_0 be from A. Consider the function $\varphi : X \to (-\infty, +\infty]$,

$$\varphi(x) = \begin{cases} \mu_F(x), & x \in A \cap D(x_0, B) \\ +\infty, & x \notin A \cap D(x_0, B). \end{cases}$$

Remark that φ is finite on $A \cap D(x_0, B)$, since in this situation $x \in \Delta$, $\rho x \in \rho\Delta \subset F$, $\rho\mu_F(x) =$ $= \mu_F(\rho x) \overset{(183)}{\leq} 1$, (251) $\mu_F(x) \leq \frac{1}{\rho}$. φ is bounded from below (by 0) and l-l.s.c. Apply to this 9.45 with $S = F$ and $\varepsilon = \frac{\rho^2}{4}$, $\exists\, v^*$ (in dom $\varphi \overset{(251)}{=} A \cap D(x_0, \rho)$, (241) imposes this since u $\in$ dom φ) so that

(252) $$\varphi(v^*) < \varphi(x) + \frac{\rho^2}{4}\mu_F(x - v^*) \;\; \forall\, x \in X \setminus \{v^*\}.$$

It remains to prove

(253) $$\{v^*\} = A \cap D(v^*, B).$$

Firstly prove

(254) $$a \in D(x_0, M) \Rightarrow D(a, M) \subset D(x_0, M)$$

(§2, definition (2)).

Let x be from $D(a, M)$, $x = ta + (1-t)y$, $y \in M$, $t \in [0,1]$. $a = \tau x_0 + (1-\tau)y'$, $y' \in M$, $\tau \in [0,1]$, hence $x = t\tau x_0 + t(1-\tau)y' + (1-t)y$. It can suppose $t\tau \neq 1$, in the contrary case we have $t = \tau = 1$ and hence $x = x_0 \in D(x_0, M)$. Take y'' defined by $(1 - t\tau)y'' = t(1-\tau)y' + (1-t)y$. Dividing by $1 - t\tau$, we find $y'' \in M$ and hence $x \in D(x_0, M)$.

Let now x be arbitrary from $A \cap D(v^*, B)$. We have $x = tv^* + (1-t)b$, $b \in B$ (so $b \in F$), $t \in [0,1]$. Also, $\rho(\frac{1}{2}b - \frac{1}{2}v^*) \in \rho\Delta$ (absolutely convex hull !) $\subset F$, consequently

(255) $\mu_F(\rho(\frac{1}{2}b - \frac{1}{2}v^*)) \leq 1$, $\mu_F(b) \leq 1$ ((183)). Then, as $D(v^*, B) \overset{(254)}{\subset} D(x_0, B)$ and hence $\varphi(x) = \mu_F(x)$, we have

$$\varphi(x) + \frac{\rho^2}{4}\mu_F(x - v^*) = \mu_F(tv^* + (1-t)b) + \frac{\rho^2}{4}\mu_F((1-t)(b - v^*)) \overset{(179,(181))}{\leq} t\mu_F(v^*) + (1-t)\times$$
$$\times\, \mu_F(b) + \frac{\rho}{2}(1-t)\mu_F(\frac{\rho}{2}(b - v^*)) \overset{(255)}{\leq} t\mu_F(v^*) + (1-t) + \frac{\rho}{2}(1-t) = t\mu_F(v^*) + (1-t)(1 +$$
$$+ \frac{\rho}{2}) \overset{(250)}{\leq} t\mu_F(v^*) + (1-t)\mu_F(v^*) = \mu_F(v^*),$$

consequently $\varphi(x) + \frac{\rho^2}{4}\mu_F(x - v^*) \leq \varphi(v^*)$, which confronted with (252) imposes $x = v^*$ and so (253). ■

I. EKELAND PRINCIPLE IN UNIFORM SPACES ([35₁], 2; [3₁])

In this section two theorems of the mininimal element are exposed, 9.55 and 9.64, and then with the aid of these two variants of the Ekeland principle are obtained, 9.58 resp.

9.65, in the uniform spaces defined by a family resp. generalized family of cuasi-distances. In the first case, the aim function φ takes values in $(-\infty, +\infty]$ and in the second case in a topological vector space.

INTRODUCTION

NOTATIONS

Let X be nonempty set.

For M subset of $X \times X$ set

(256) $$\overset{-1}{M} := \{(x, y) \in X \times X : (y, x) \in M\}.$$

M is called *symmetrical* if $M = \overset{-1}{M}$ (it is enough $M \subset \overset{-1}{M}$; for instance $P \cup \overset{-1}{P}$ and $P \cap \overset{-1}{P}$, with $P \subset X \times X$, are symmetrical).

Let M, N be subsets of $X \times X$. Note

(257) $$M \circ N := \{(x, y) \in X \times X : \exists\, z \in X \text{ so that } (x, z) \in M, (z, y) \in N\}.$$

In particular, $\overset{2}{M} := M \circ M$, $\overset{3}{M} := M \circ M \circ M$ etc. (we have associativity). Put $\overset{1}{M} := M$. We have

(258) $$\overset{-1}{\overbrace{M \circ N}} = \overset{-1}{N} \circ \overset{-1}{M}.$$

Also, (259) $M \subset N$ and $P \subset Q \Rightarrow M \circ P \subset N \circ Q$. In particular (260) $M \subset N \Rightarrow \overset{n}{M} \subset \overset{n}{N}\ \forall\, n$.

Let be $A \subset X$, $M \subset X \times X$. Note

$$M(A) := \{y \in X : \exists\, x \in A \text{ so that } (x, y) \in M\}.$$

In particular, for $A = \{a\}$

$$M(a) = \{y \in X : (a, y) \in M\} \text{ – the } \textit{cut} \text{ along } a.$$

Obviously $M \subset N \Rightarrow M(a) \subset N(a)$.

The *diagonal* Δ of the set $X \times X$ is the subset of this

$$\Delta = \{(x, x) : x \in X\}.$$

We have, $\forall\, M \subset X \times X$, (261) $M \circ \Delta = \Delta \circ M = M$ ($(x, y) \in M \circ \Delta \Rightarrow \exists\, z$ so that $(x, z) \in$ $\in M$ and $(z, y) \in \Delta$, hence $z = y$, $(x, y) \in M$ etc.).

UNIFORM SPACE

Let X be nonempty set. The set $\mathscr{U}$ of nonempty subsets of $X \times X$ is by definition a *uniform structure* on X if

(F_I) $M \in \mathscr{U}$ and $N \supset M \Rightarrow N \in \mathscr{U}$;
(F_{II}) $M, N \in \mathscr{U} \Rightarrow M \cap N \in \mathscr{U}$; (*H. Cartan axioms*)

(U_I) The diagonal Δ of $X \times X$ is included in each set from $\mathscr{U}$;

(U_{II}) $U \in \mathscr{U} \Rightarrow \overset{-1}{U} \in \mathscr{U}$;

(U_{III}) $\forall\ U \in \mathscr{U}\ \exists\ V \in \mathscr{U}$ so that $\overset{2}{V} \subset U$ (*A. Weil axiom*).

The set X with the uniform structure $\mathscr{U}$ define a *uniform space* $(X, \mathscr{U})$ (*A. Weil*). The sets from $\mathscr{U}$ are the *entourages* of $(X, \mathscr{U})$. Retain:

(262) $$U \text{ entourage} \Rightarrow \overset{n}{U} \subset \overset{n+1}{U} \quad \forall\, n.$$

Indeed, $\overset{n}{U} \circ \Delta \subset \overset{n}{U} \circ U$ ((U_I) and (259)) and $\overset{n}{U} \circ \Delta \overset{(261)}{=} \overset{n}{U}$.

Let $(X, \mathscr{U})$ be uniform space. A subset $\mathscr{B}$ of $\mathscr{U}$ is by definition *basis* of $\mathscr{U}$ if any entourage includes a set from $\mathscr{B}$. In this case

(B_I) the intersection of two sets from $\mathscr{B}$ includes a set from $\mathscr{B}$;

(U'_I) Δ is included in any set from $\mathscr{B}$;

(U'_{II}) $\forall\ U$ from $\mathscr{B}$ $\exists\ V$ in $\mathscr{B}$ so that $V \subset \overset{-1}{U}$;

(U'_{III}) $\forall\ U$ from $\mathscr{B}$ $\exists\ V$ in $\mathscr{B}$ so that $\overset{2}{V} \subset U$.

Reciprocally, let X be nonempty set and $\mathscr{B}$ a set of nonempty subsets of $X \times X$ which verifies (B_I), (U'_I), (U'_{II}) and (U'_{III}). In this case there exists $\mathscr{U}$ uniform structure unique on X with the property that $\mathscr{B}$ is basis of $\mathscr{U}$. Indeed, $\mathscr{U}$ is the set of the parts of $X \times X$ which include a set from $\mathscr{B}$.

Examples. 1) The set of the symmetrical entourages of a uniform structure form a basis for this. Indeed, let U be any entourage, then $\overset{-1}{U}$ is an entourage ((U_{II})) and also $U \cap \overset{-1}{U}$ ((F_{II})) and $U \cap \overset{-1}{U} \subset U$.

2) For each real number $r > 0$ let be $U_r := \{(x, y) \in \mathbf{R}^2 : |x - y| < r\}$. The sets U_r, $r > 0$ verify $(B_I) - (U'_{III})$ ($U_{r_1} \cap U_{r_2} = U_{\min(r_1, r_2)}$, $U_r = \overset{-1}{U_r}$, $\overset{2}{U_r} \subset U_{2r}$), hence they form a basis for a uniform structure on $\mathbf{R}$.

3) Let (X, d) be a metric space and, for each real number $r > 0$, $U_r := \{(x, y) \in X \times X : d(x, y) < r\}$. The sets U_r, $r > 0$ form a basis for a uniform structure on X.

Definition. The subset $\mathscr{S}$ of a uniform structure $\mathscr{U}$ is *prebasis* of this when the set of finite intersections of the sets from $\mathscr{S}$ forms a basis for $\mathscr{U}$.

Let be $X \neq \varnothing$. The set $\mathscr{S}$ of nonempty subsets of $X \times X$ forms a prebasis for a uniform structure on X iff

(U''_I) Δ is included in each set from $\mathscr{S}$;

(U''_{II}) For any U from $\mathscr{S}$, $\overset{-1}{U}$ inludes a set from $\mathscr{S}$;

(U''_{III}) $\forall\, U$ from $\mathscr{S}$ $\exists\, V$ in $\mathscr{S}$ so that $\overset{2}{V} \subset U$.

■ Take into account that $V_i \subset \overset{-1}{U_i}$ resp. $\overset{2}{V_i} \subset U_i$, $i = \overline{1,n}$ implies $V \subset \overset{-1}{U}$ resp. $\overset{2}{V} \subset U$, where $U = \bigcap_{i=1}^{n} U_i$, $V = \bigcap_{i=1}^{n} V_i$. ■

So the reunion of a set of uniform structure S on X is prebasis for a uniform structures on X.

Remark. (U''_I), (U''_{II}) and (U'''_{III}) are even the basis axioms, but without (B_I).

THE TOPOLOGY OF A UNIFORM SPACE

Let $(X, \mathscr{U})$ be a uniform space. Consider, for each x from X,

$$\mathscr{V}(x) := \{U(x) : U \in \mathscr{U}\}.$$

$\mathscr{V}(x)$ has the properties:

(V_I) Each part of X which includes a set from $\mathscr{V}(x)$ is in $\mathscr{V}(x)$;

(V_{II}) Each finite intersection of sets from $\mathscr{V}(x)$ is in $\mathscr{V}(x)$;

(V_{III}) x is in all the sets from $\mathscr{V}(x)$;

(V_{IV}) For any $U(x)$ $\exists\, V(x)$ so that $y \in V(x) \Rightarrow U(x) \in \mathscr{V}(y)$.

■ (V_I): let be $A \supset U(x)$, take the entourage $V := (A \times A) \cup U$, then $A = V(x)$ ($y \in A \Rightarrow (x, y) \in A \times A$ (as $x \in U(x)$) $\Rightarrow (x, y) \in V \Rightarrow y \in V(x)$; $y \in V(x) \Rightarrow (x, y) \in V \Rightarrow$ either $(x, y) \in A \times A$ and so $y \in A$, or $(x, y) \in U$ and hence $y \in U(x)$, $y \in A$); (V_{II}): $U(x) \cap V(x) = (U \cap V)(x)$; ($V_{III}$): $x \in U(x)$ (the diagonal !); (V_{IV}): let V be an entourage with $\overset{2}{V} \subset U$, then $U(x) \in \mathscr{V}(y)$ $\forall$ y from $V(x)$ because $V(y) \subset U(x)$. ■

So being, there exists a unique topology on X with the property that $\mathscr{V}(x)$ is, for any x from X, the set of neighbourhoods of x (Hausdorff neighbourhoods theorem). This is noted $\tau(\mathscr{U})$. Retain : $\mathscr{B}$ basis for $\mathscr{U} \Rightarrow \mathscr{B}(x)$ basis for $\mathscr{V}(x)$.

Each uniform space $(X, \mathscr{U})$ is considered to be endowed with the topology $\tau(\mathscr{U})$ and any topological property of this will be expressed by an added adjective to „uniform space", for instance separated uniform space, compact uniform space etc.

9.47 (X, $\mathscr{U}$) *is separated* $\Leftrightarrow \bigcap_{U \in \mathscr{U}} U = \Delta$.

■ *Necessary.* Let be $(x, y) \in (X \times X) \setminus \Delta$. As $x \neq y$ $\exists$ U an entourage so that $y \notin U(x)$, that is $(x, y) \notin U$. *Sufficient.* Let be $x \neq y$, $\exists$ U an entourage so that $(x, y) \notin U$. Let V be a symmetrical entourage with $\overset{2}{V} \subset U$, then $V(x) \cap V(y) = \varnothing$: if *par absurdum* $z \in V(x) \cap V(y)$, then $(x, z) \in V$ and $(z, y) \in V$ (the symmetry !), so $(x, y) \in U$, a contradiction. ■

9.48 *Main proposition. Let* (X; $\mathscr{U}$) *be uniform space and* M *nonempty subset of the topological space* X × X. *For every entourage* U, U o M o U *is a neighbourhood of* M *and*

(263)
$$\overline{M} = \bigcap_{U \in \mathscr{U}} U \circ M \circ U .$$

■ *First assertion.* It can assume U symmetrical, this is permitted by ex. 1 and (259). Let (x, y) be any point from M. $U(x) \times U(y)$ is a neighbourhood of (x, y). This is included in $U \circ M \circ U$. Indeed, $(u, v) \in U(x) \times U(y) \Rightarrow (u, x) \in U$ and $(y, v) \in U$, hence $(u, v) \in U \circ M \circ U$. So $U \circ M \circ U$ is a neighbourhood of each point from M and consequently is a neighbourhood of M.

Second assertion. $\subset$. Let (x, y) be any point from $\overline{M}$, $\mathscr{S}$ the set of the symmetrical entourages of X and U any from $\mathscr{S}$. $U(x) \times U(y)$ is a neighbourhood of (x, y), let (u, v) be a point from $(U(x) \times U(y)) \cap M$. Then we have $(x, u) \in U$ and $(v, y) \in U$, so $(x, y) \in$

$\in U \circ M \circ U$. $\supset$. Let (x, y) be a point arbitrary from $\bigcap_{U \in \mathcal{S}} U \circ M \circ U$. Then, $\forall$ U from $\mathcal{S}$,

(264) $(U(x) \times U(y)) \cap M \neq \varnothing$, because $\exists$ u, v so that $(x, u) \in U$ (hence $u \in U(x)$), $(u, v) \in$ $\in M$ and $(v, y) \in U$ (so $(y, v) \in U$, $v \in U(y)$). But when U goes over $\mathcal{S}$, $U(x) \times U(y)$ form a basis of neighbourhoods of (x, y). Indeed, let be $V(x) \times W(y)$, V and W entourages, $\exists$ S symmetrical entourage so that $S \subset V \cap W$ ((F_{II}) and ex. 1), then $S(x) \times S(y) \subset (V \cap W)(x) \times$ $\times (V \cap W)(y) \subset V(x) \times W(y)$. So being, (264) gives $(x, y) \in \overline{M}$, $\bigcap_{U \in \mathcal{S}} U \circ M \circ U \subset \overline{M}$. In

conclusion $\overline{M} = \bigcap_{U \in \mathcal{S}} U \circ M \circ U = \bigcap_{U \in \mathcal{U}} U \circ M \circ U$ (ex. 1). ■

It follows a sequence of consequences.

(265) U *entourage* $\Rightarrow$ $\overset{3}{U}$ *entourage and neighbourhood of* U.

■ Take $M = U$ in 9.48, see also (F_I). ■

(266) U *entourage* $\Rightarrow$ $\overset{n}{U}$ *entourage* $\forall$ n.

■ $\overset{3}{U}$ is an entourage ((265)), hence $\overset{n}{U}$ is an entourage $\forall$ $n \geq 3$ ((262) and (F_I)). $\exists$ V entourage so that $\overset{2}{V} \subset U$ ((U_{III})), hence $\overset{4}{V} \overset{(259)}{\subset} \overset{2}{U}$ and $\overset{3}{V} \overset{(262)}{\subset} \overset{4}{V}$, so $\overset{3}{V} \subset \overset{2}{U}$ and as $\overset{3}{V}$ is an entourage, $\overset{2}{U}$ has the same property. ■

(267) *The open entourages form a basis.*

■ Let U be entourage, $\exists$ V entourage with $\overset{2}{V} \subset U$ and $\exists$ W entourage with $\overset{2}{W} \subset V$, hence $\overset{3}{W} \subset \overset{4}{W} \subset \overset{2}{V} \subset U$. But $\overset{3}{W}$ is a neighbourhood of W (9.48), $\exists$ G open so that $W \subset$ $\subset G \subset \overset{3}{W}$. So G is an entourage ((F_I)) and $G \subset U$. ■

(268) *Let* $\mathcal{B}$ *be a basis of entourages and* n *from* **N**. *Then*

$$\{\overset{n}{U} : U \in \mathcal{B}\}$$

is a basis of entourages.

■ The statement is correct since U entourage $\overset{(266)}{\Rightarrow}$ $\overset{n}{U}$ entourage. Recurrence over n. n = 2. Let V be entourage, $\exists$ U in $\mathcal{B}$ so that $U \subset V$ and $\exists$ W in $\mathcal{B}$ so that $\overset{2}{W} \subset U$

((U'_{III})), hence $\overset{2}{W} \subset V$. n $\Rightarrow$ n + 1. Let V be entourage, $\exists$ U from $\mathcal{B}$ so that $\overset{n}{U} \subset V$ and $\exists$ W in $\mathcal{B}$ so that $\overset{2}{W} \subset U$. So $\overset{n+1}{W} \subset \overset{2n}{W} \subset \overset{n}{U} \subset V$. ■

(269) U *entourage* $\Rightarrow \overset{o}{U}$ *entourage.*

■ There exists G open entourage so that $G \subset U$ ((267)), *a fortiori* $G \subset \overset{o}{U}$ and hence the conclusion ((F_I)). ■

(270) *Let* $\mathcal{B}$ *be basis of entourages. Then* $\{ \overset{o}{U} : U \in \mathcal{B}\}$ *and* $\{ \overline{U} : U \in \mathcal{B} \}$ *are basis of entourages.*

■ *First affirmation.* Use (269). *Second affirmation.* Let V be entourage, $\exists$ U in $\mathcal{B}$ so that $\overset{3}{U} \subset V$ ((268)), but $\overline{U} \overset{9.48}{\subset} \overset{3}{U}$ and so $\overline{U} \subset V$. ■

(271) *The set of the closed symmetrical entourages forms a basis of entourages.*

■ The set from the statement is equal to $\{\overline{U} : U \in \mathcal{S}\}$, $\mathcal{S}$ the set of symmetrical entourages (M symmetrical $\Rightarrow \overline{M}$ symmetrical, $(x, y) \to (y, x)$ being a homeomorphism of $X \times X$ on $X \times X$). $\mathcal{S}$ is a basis, apply (270). ■

9.49 *Let* (X, $\mathcal{U}$) *be uniform space and* A *a nonempty subset of* X. *For any entourage* U, U(A) *is a neighbourhood of* A *and*

(272)
$$\overline{A} = \bigcap_{U \in \mathcal{U}} U(A).$$

■ *First affirmation.* Ex. 1 permits to assume U symmetrical. Then we have (273) $U \circ (A \times A) \circ U = U(A) \times U(A)$ and so, according to 9.48, $U(A) \times U(A)$ is a neighbourhood of $A \times A$, $\exists$ G open so that $A \times A \subset G \subset U(A) \times U(A)$, which gives $A \subset p_1(G) \subset U(A)$, p_1 the first projection of $X \times X$. Remain only to remark that $p_1(G)$ is open in X.

Second affirmation. Take $M = A \times A$ in the formula (263) and take into account (273) and $\overline{A \times A} = \overline{A} \times \overline{A}$. ■

9.50 *Any uniform space verifies the axiom*:

„ *The set of closed neighbourhoods of an arbitrary point forms a basis of neighbourhoods of this*".

Every separated uniform space is regular.

■ Let X be uniform space and x a point from X. An arbitrary neighbourhood of x is of the form $U(x)$, U entourage. According to (271) there exists V closed entourage with $V \subset$ $\subset U$. We have $V(x) \subset U(x)$ and $V(x)$ is closed as V is closed. ■

PSEUDO-DISTANCE

Let X be nonempty set. The application $q : X \times X \to [0, +\infty)$ is by definition *pseudo-distance* on X if

(P1) $q(x, x) = 0 \ \forall\, x \in X$;

(P2) $q(x, y) = q(y, x) \ \forall\, x, y \in X$;

(P3) $q(x, y) \leq q(x, z) + q(z, y) \ \forall\, x, y, z \in X$.

A set endowed with a pseudo-distance is called *pseudo-metric space*. $q\,(x, y)$ is the *pseudo-distance* between x and y.

Let (X, q) be a pseudo-metric space. For a a point from X and $r > 0$ the *open ball* $S(a, r)$ resp. the *closed ball* $S_r(a)$ with *a centre* and *r radius* is the set $\{x \in X : q\,(a, x) < r\}$ resp. $\{x \in X : q\,(a, x) \leq r\}$. Every point from the intersection of two open balls is the centre of an open ball included in this: $x \in S(a, r_1) \cap S(b, r_2) \Rightarrow S(x, r) \subset S(a, r_1) \cap S(b, r_2)$, where $r = \min\,(r_1 - q\,(a, x), r_2 - q\,(b, x))$. So being, the set of open balls forms a basis of a topology on X – *the topology defined by* q. In this topology the closed balls are closed.

Let (X, q) be a pseudo-metric space. The *pseudo-distance* $q\,(x, A)$ beween the point x and the nonempty subset A is $q\,(x, A) = \inf\,\{q\,(x, y) : y \in A\}$. The function $x \to q\,(x, A)$ is continuous on X. We have $\overline{A} = \{x \in X : q(x, A) = 0\}$.

Every pseudo-metric space X is normal and verifies the first axiom of countability. The second axiom of countability is verified $\Leftrightarrow$ X is separable.

Let (X, q) be pseudo-metric space. Consider, for each $r > 0$, the subset of $X \times X$ $U_r :=$ $= \{(x, y) \in X \times X : q\,(x, y) < r\}$. The sets U_r verify the axioms (B_I), (U'_I), (U'_{II}) and (U'_{III}): $U_{r_1} \cap U_{r_2} = U_r$, $r := \min\,(r_1, r_2)$; $\Delta \subset U_r \ \forall\, r > 0$; $\overset{-1}{U}_r = U_r \ \forall\, r > 0$; $\overset{2}{U}_{\frac{r}{2}} \subset U_r$ $\forall\, r > 0$. So being, there is a unique uniform structure $\mathscr{U}$ on X having $\mathscr{B}$ as basis. $\mathscr{U}$ is by definition the uniform structure *defined* by q.

A uniform space $(X, \mathscr{U})$ is by definition *pseudo-metrizable* if $\mathscr{U}$ is defined by a pseudo-distance on X. A uniform space X is pseudo-metrizable $\Leftrightarrow$ the uniform structure of X has a countable basis ([38_2], Ch.6, **13**).

Let X be nonempty set and π a set of *pseudo-distances* on X. Consider, for each q from π and for each $r > 0$, the subset of $X \times X$

$$U_{q,r} := \{(x, y) \in X \times X : q(x, y) < r\}.$$

The sets $U_{q,r}$, $q \in \pi$ and $r > 0$ form a prebasis of a uniform structure $\mathscr{U}$ on X, because verify (U''_I), (U''_{II}) and (U''_{III}): $\Delta \subset U_{q,r}$ $\forall q \in \pi$, $\forall r > 0$; $\overset{-1}{U}_{q,r} = U_{q,r}$; $\overset{2}{U}_{q,\frac{r}{2}} \subset U_{q,r}$. $\mathscr{U}$ is by definition the uniform structure *defined* by π.

9.51 (X, $\mathscr{U}$) *is separated* $\Leftrightarrow$ ($q(x, y) = 0$ $\forall q \in \pi \Rightarrow x = y$).

■ *Necessary*. $(X, \mathscr{U})$ separated $\overset{9.47}{\Rightarrow}$ $\bigcap_{q\in\pi,r>0} U_{q,r} = \Delta$. Suppose *par absurdum* $x \neq y$. So $(x, y) \notin$ $\notin \Delta$, hence $\exists q_1 \in \pi$, $r_1 > 0$ so that $q_1(x, y) \geq r_1$ and hence $q_1(x, y) > 0$, a contradiction. *Sufficient*. Let (x, y) be from $\bigcap_{q\in\pi,r>0} U_{q,r}$, hence for q arbitrary we have $(x, y) \in U_{q,r}$ $\forall r >$ > 0, that is $q(x, y) < r$ $\forall r > 0$ and so $q(x, y) = 0$. The hypothesis implies $x = y$, that is $(x, y) \in \Delta$, $\bigcap_{q\in\pi,r>0} U_{q,r} = \Delta$ and X is separated (9.47). ■

9.51 justifies the denomination π *separating* used when π has the property:

$$q(x, y) = 0 \ \forall q \in \pi \Rightarrow x = y.$$

In fact

9.52 Theorem. *Every uniform structure on a set* X *is defined by a set of pseudo-distances on* X ([38_2], Ch.6, **15**).

FAMILY OF QUASI-DISTANCES

Let X be a nonempty set and I an *ordered* set. The family of functions $(q_i)_{i\in I}$, $q_i : X \times X \to$ $\to [0, +\infty)$ is called *family of quasi-distances* on X if

(Q1) $q_i(x, x) = 0$ $\forall i \in I$, $\forall x \in X$;

(Q2) $q_i(x, y) = q_i(y, x)$ $\forall i \in I$, $\forall x, y \in X$;

(Q3) $\forall i \in I$, $\exists j \in I$, $i \leq j$ so that $q_i(x, y) \leq q_j(x, z) + q_j(z, y)$ $\forall x, y, z \in X$.

If in addition

(Q4) $q_i(x, y) = 0 \ \forall \ i \in I \Rightarrow x = y$,

then the family $(q_i)_{i\in I}$ is by definition *separating* (*J. - X. Fang*, [30_1].

REMARK 1. The author remove from the axioms system from [30_1] (p. 399) the axiom

$$i \le j \Rightarrow q_i(x, y) \le q_j(x, y) \ \forall \ x, y \in X$$

being without connection with the subject of this monograph.

Remark 2. Apart from the „pseudo-distance", the concept „quasi-distance" was not defined, but that of „family of quasi-distances" was.

Let X be nonempty set and $(q_i)_{i\in I}$ a family of quasi-distances on X. Consider, for each i from I and for each r from the interval $(0, +\infty)$, the subset of $X \times X$

(274) $$U_{i,r} := \{(x, y) \in X \times X : q_i(x, y) < r\}.$$

The sets $U_{i,r}$ form, when i goes over I and r over $(0, +\infty)$, a prebasis of a uniform structure $\mathscr{U}$ on X, because they verify (U''_I), (U''_{II}), (U''_{III}): $\Delta \subset U_{i,r} \ \forall \ i \in I, \forall \ r > 0$ (take into account (Q1)); $\overset{-1}{U}_{i,r} = U_{i,r} \ \forall \ i \in I, \forall \ r > 0$ (take into account (Q2)); let be $U_{i,r}$, $\exists \ j \in I$ so that $q_i(x, y) \le q_j(x, z) + q_j(z, x) \ \forall \ x, y, z \in X$ ((Q3)), which implies $U_{j,\frac{r}{2}} \circ U_{j,\frac{r}{2}} \subset U_{i,r}$. $\mathscr{U}$ is, by definition, the uniform structure *defined* by the family $(q_i)_{i\in I}$.

In the topology $\tau(\mathscr{U})$, for each point x from X the sets $\bigcap_{k=1}^{n} U_{i_k, r_k}(x)$, $n \in \mathbf{N}$, $i_1, \dots, i_n \in I$, $r_1, \dots, r_n > 0$ form a basis of neighbourhoods of x.

With the same proof as for 9.51 one obtains

9.53 (X, $\mathscr{U}$) *is separated* $\Leftrightarrow$ $(q_i)_{i\in I}$ *is separating.*

In fact

9.54 *Any uniform structure on a set is defined by a quasi-distance family on this set.*

■ Let $(X, \mathscr{U})$ be a uniform space. There exists a set π of pseudo-distances on X which defines $\mathscr{U}$ (9.52) and let $i \to q_i$ be a bijection of a set I on π. A family of applications $(q_i)_{i\in I}$, $q_i : X \times X \to [0, +\infty)$ is obtained. Setting on I the order relation $i \preceq j \overset{\text{def}}{\Leftrightarrow} i = j$, one obtains a family of quasi-distances on X: (P1) $\Rightarrow$ (Q1), (P2) $\Rightarrow$ (Q2), (P3) $\Rightarrow$ (Q3) ($j \succeq i \Leftrightarrow j = i$). Obviously, $\mathscr{U}$ is defined by this family. ■

EKELAND PRINCIPLE IN UNIFORM SPACES (FAMILIES OF QUASI-DISTANCES)

Definitions. Let X be uniform space.

The sequence $(x_n)_{n\geq 1}$ from X is a *Cauchy sequence* if for any symmetrical entourage V of X there is N in **N** so that

$$m, n \geq N \Rightarrow (x_m, x_n) \in V.$$

X is *sequentially complete* if any Cauchy sequence from X is convergent.

Let $\preceq$ be a preorder relation in X (i.e. the relation is reflexive and transitive). The sequence $(x_n)_{n\geq 1}$ from X is *decreasing* (with respect to $\preceq$) if $x_{n+1} \preceq x_n$ $\forall\ n \geq 1$. The interval $I(x) := \{y \in X : y \preceq x\}$ is *sequentially lower closed* when for any decreasing sequence $(x_n)_{n\geq 1}$ from $I(x)$, if $x_n \to x_0$ then $x_0 \in I(x)$. The preorder relation $\preceq$ is *sequentially lower closed* if, for any x from X, the interval $I(x)$ is sequentially lower closed.

The following statement has the principal role in this subsection.

9.55 Theorem (*Minimal element*). *Let* X *be sequentially complete separated uniform space,* $\preceq$ *sequentially lower closed preorder in* X *and* φ: X $\to (-\infty, +\infty]$ *proper, bounded from below and increasing with respect to* $\preceq$. *In addition,* $\forall$ *entourage* U $\exists\ \delta >$ > 0 *so that*

(275) $$x \preceq y \text{ and } \varphi(y) - \varphi(x) < \delta \Rightarrow (x, y) \in U.$$

Then for any $I(x_0)$, x_0 *from* dom φ, *there is* x^* *in* $I(x_0)$ *so that*

(276) $$\{x^*\} = I(x^*)$$

(*A. Brønstedt, N. Mizoguchi, A. Hamel*).

■ Define by recurrence a decreasing (with respect to $\preceq$) sequence $(x_n)_{n\geq 1}$ as follows:

(277) $$x_1 = x_0\,;\ x_{n+1} \in I(x_n) \text{ and } \varphi(x_{n+1}) \leq \inf_{x\in I(x_n)} \varphi(x) + \frac{1}{n},$$

correct definition because φ is lower bounded. $(\varphi(x_n))_{n\geq 1}$ is decreasing, as φ is increasing, and bounded from below, let be $\alpha = \lim\limits_{n\to\infty} \varphi(x_n)$. Obviously (278) $\alpha \leq \varphi(x_n)$ $\forall\ n \geq 1$.

$(x_n)_{n\geq 1}$ is Cauchy sequence. Indeed, let V be any symmetrical entourage of X. Take N from **N** so that $p \geq N \Rightarrow \varphi(x_p) < \alpha + \delta$, δ that from (275). Suppose $m, n \geq N$. Then, for

$m \geq n$ for instance, we have $x_m \preceq x_n$ and $\varphi(x_n) - \varphi(x_m) \overset{(278)}{<} \alpha + \delta - \alpha = \delta$, so $(x_m, x_n) \in V$ ((275)).

Let be $x^* = \lim_{n\to\infty} x_n$. Because $\preceq$ is sequentially lower closed we have $x^* \in I(x_n)\ \forall\ n \geq 1$, in particular $x^* \in I(x_0)$ (of course $x^* \in \text{dom}\ \varphi$). Let x be from $I(x^*)$, it follows to prove $x = x^*$. We have (279) $\varphi(x) \leq \varphi(x^*)$. In addition, as $x \preceq x^* \preceq x_n\ \forall\ n \geq 1$, it results $\varphi(x^*) \leq$ $\leq \varphi(x_{n+1}) \overset{(277)}{<} \varphi(x) + \dfrac{1}{n}$, hence, by passing to the limit, $\varphi(x^*) \leq \alpha \leq \varphi(x)$ and, combining with (279), $\varphi(x^*) = \alpha = \varphi(x)$.

Let U be any symmetrical entourage. As $\varphi(x_n) \to \varphi(x)$, $\exists\ N$ in **N** so that $n \geq N \Rightarrow$ $\Rightarrow \varphi(x_n) - \varphi(x) < \delta$, hence, as $x \preceq x_n$, $n \geq N \overset{(275)}{\Rightarrow} (x, x_n) \in U$ and so $x_n \to x$. X being separated, *a fortiori* $x = x^*$ ■

9.55 may be completed and namely

9.56 *Consider, in the assumptions of* 9.55, *the multivalued function* $\text{T}: \text{X} \to 2^{\text{X}}_*$.

If $\forall$ x *from* dom φ $\exists$ y *in* T(x) *so that* $\text{y} \preceq \text{x}$, *then* T *has a fixed point.*

If $\forall$ x *from* dom φ *and* $\forall$ y *from* T(x) *we have* $\text{y} \preceq \text{x}$, *then* T *has an invariant point* ([35_1], 2, Th. 2).

■ *First assertion.* The hypothesis means $T(x) \cap I(x) \neq \varnothing\ \forall\ x \in \text{dom}\ \varphi$. Then, in particular for x^*, the point supplied by 9.55, we have, according to (276), $x^* \in T(x^*)$. *Second assertion.* The hypothesis means $T(x) \subset I(x)\ \forall\ x \in \text{dom}\ \varphi$, hence for the point x^* supplied by 9.55 we have $T(x^*) = \{x^*\}$. ■

Look now *another formulation* of 9.55.

9.57 *In the assumptions of* 9.55, *for any* x_0 *from* dom φ, *if the subset* M *satisfies*

$$\forall\ \text{x} \in \text{I}(\text{x}_0) \setminus \text{M}\ \exists\ \text{y} \in \text{I}(\text{x}) \setminus \{\text{x}\},$$

then $\text{I}(\text{x}_0) \cap \text{M} \neq \varnothing$ ([35_1], 2, th. 3).

■ 9.55 gives x^* from $I(x_0)$ so that $\{x^*\} = I(x^*)$. This imposes $x^* \in M$, the hypostasis $x^* \notin M$ comes to a contradiction. ■

9.57 is indeed another formulation of 9.55: accept 9.57 and let $I(x_0)$ be with $x_0 \in$ $\in \text{dom}\ \varphi$, the set $M := \{x \in X : \{x\} = I(x)\}$ verifies the condition from the hypothesis, hence $I(x_0) \cap M \neq \varnothing$, every element from this intersection verifies the demand from 9.55.

Remark. 9.57 is a variant for uniform spaces of the Oettli-Théra theorem (§2, C).

Definition. $f: X \to (-\infty, +\infty]$, X topological space, is *sequentially lower monotone*, if for any sequence $(x_n)_{n\geq 1}$ from X convergent to x_0 and $f(x_{n+1}) \leq f(x_n)$ $\forall\ n \geq 1$, we have $f(x_0) \leq f(x_n)$ $\forall\ n \geq 1$.

Sequential lower monotonicity is weaker than sequential lower semicontinuity.

And now

9.58 *Uniform generalization of Ekeland principle. Let* X *be sequentially complete separated uniform space,* $(q_i)_{i\in I}$ *an afferent family of cuasi-distances* (9.54) *and* φ: X $\to$ $\to (-\infty, +\infty]$ *lower bounded, proper, sequentially lower monotone. For any* $\varepsilon: I \to (0, +\infty)$ *increasing and* x_0 *from* dom φ, *there exists* v^* *in* dom φ *so that*

(280) $$\forall\ x \text{ from } X \setminus \{v^*\}\ \exists\ j \text{ in } I \text{ for which}$$

$$\varphi(v^*) < \varphi(x) + \varepsilon(j) q_j(x, v^*)$$

and

(281) $$\varphi(v^*) + \varepsilon(i)\, q_i(v^*, x_0) \leq \varphi(x_0)\ \forall\ i \in I$$

(*J. - X. Fang, A. Hamel*).

■ Use 9.55. Define on X the binary relation $\preceq$,

(282) $$x \preceq y \overset{\text{def}}{\Leftrightarrow} \varphi(x) + \varepsilon(i)\, q_i(x, y) \leq \varphi(y)\ \forall\ i \in I.$$

$\preceq$ is obviously reflexive, show that it is also transitive, i.e. a preorder relation. Let be $x \preceq y$, $y \preceq z$. If $\varphi(z) = +\infty$, (282) tells that $x \preceq z$. Suppose $\varphi(z) < +\infty$. *A fortiori* $\varphi(x)$, $\varphi(y) <$ $< +\infty$. According to (282),

(283) $$\varphi(x) + \varepsilon(i)\, q_i(x, y) \leq \varphi(y)\ \forall\ i \in I,$$

(284) $$\varphi(y) + \varepsilon(i)\, q_i(y, z) \leq \varphi(z)\ \forall\ I \in I.$$

Applying (Q3), one finds $j \geq i$ (the order relation from I) with a certain property and then $\forall\ i$ we have, combining with (283) and (284),

$$\varphi(x) + \varepsilon(i)\, q_i(x, z) \leq \varphi(x) + \varepsilon(j)[q_j(x, y) + q_j(y, z)] \leq \varphi(x) + \varphi(y) - \varphi(x) + \varphi(z) - \varphi(y) =$$
$$= \varphi(z), \text{ i.e. } x \preceq z.$$

Prove in the following that $\preceq$ is sequentially lower closed. Let $(x_n)_{n\geq 1}$ be decreasing sequence with $x_n \to x'$. It is enough to prove

$$\exists\ N \text{ in } \mathbf{N} \text{ so that } x' \preceq x_N. \tag{285}$$

According to (282) we have

$$\varphi(x_{n+1}) + \varepsilon(i)\, q_i(x_{n+1}, x_n) \leq \varphi(x_n)\ \forall\ n \geq 1. \tag{286}$$

(286) gives $\varphi(x_{n+1}) \leq \varphi(x_n)\ \forall\ n \geq 1$, hence, according to the hypothesis, (287) $\varphi(x') \leq \varphi(x_n)$ $\forall\ n \geq 1$.

There are possible cases: $\varphi(x_n) = +\infty\ \forall\ n \geq 1$ and $\exists$ rank from which $\varphi(x_n) < +\infty$. In the first case it results $x' \preceq x_n\ \forall\ n \geq 1$. Pass to the second case. It can suppose $\varphi(x_n) < +\infty$ $\forall\ n \geq 1$. Fix N in $\mathbf{N}$. The transitivity gives $m > N \Rightarrow x_m \preceq x_N$, i.e.

$$m > N \Rightarrow q_i(x_m, x_N) \leq \frac{1}{\varepsilon(i)}[\varphi(x_N) - \varphi(x_m)]\ \forall\ i \in I. \tag{288}$$

Let i be arbitrary fixed. $\exists\ j \geq i$ so that $q_i(x', x_N) \leq q_j(x', x_m) + q_j(x_m, x_N)$, hence

$$m > N \Rightarrow q_i(x', x_N) \overset{(288)}{\leq} q_j(x', x_m) + \frac{1}{\varepsilon(j)}[\varphi(x_N) - \varphi(x_m)] \overset{(287)}{\leq} q_j(x', x_m) + \frac{1}{\varepsilon(i)}[\varphi(x_N) - \varphi(x')]$$

(ε is increasing).

Taking the first and the third term and then the limit for $m \to +\infty$ it results

$$q_i(x', x_N) \leq \frac{1}{\varepsilon(i)}[\varphi(x_N) - \varphi(x')].$$

As i is arbitrary, (285) is obtained.

Finally, verify (275). Let U be any entourage of X. $\exists\, U_{i_k, r_k}$, $k = \overline{1, n}$ (see (274)) so that

$\bigcap_{k=1}^{n} U_{i_k, r_k} \subset U$. Taking $\delta = \min_{1 \le k \le n} r_k\, \varepsilon(i_k)$, if $x \preceq y$ and $\varphi(y) - \varphi(x) < \delta$, then in particular

$\varepsilon(i_k)\, q_{i_k}(x, y) \le \varphi(y) - \varphi(x) < r_k\, \varepsilon(i_k)$, $k = \overline{1, n}$ and hence $(x, y) \in \bigcap_{k=1}^{n} U_{i_k, r_k} \subset U$.

Apply 9.55, $\exists\, v^*$ in dom φ so that

$$v^* \in I(x_0) \text{ and } \{v^*\} = I(v^*).$$

The first relation gives $v^* \preceq x_0$, i.e. (281). The second relation gives (280), because $x \ne v^* \Rightarrow x \npreceq v^*$. ■

Remark 1. For $(X, \mathscr{U}) = (X, \mathrm{d})$, complete metric space, and $\varepsilon_i = \varepsilon\ \forall\, i \in I$ in 9.58, one recovers the form **E** of Ekeland principle (after REMARK 2 to 1.1_0).

REMARK 2. In $[35_1]$, 2, pag. 918 there are two mistakes:

1) In the statement of Th. 4 (here 9.58) it must „for every $x_0 \in \operatorname{dom} f$" instead of „for $x_0 \in X$".

2) The proof for Th. 4 it was made for f with values only in **R**.

Another smaller mistakes are tacitly corrected. The author was able to recover everything.

Two statements obtained with 9.58 are following.

9.59 *Uniform generalization of Caristi-Kirk theorem. Let* X *be sequentially complete separated uniform space,* $(\mathrm{q}_i)_{i \in I}$ *an afferent family of cuasi-distances* (9.54) *and* $\mathrm{T}\colon \mathrm{X} \to 2^{\mathrm{X}}_*$. *If* $\exists\, \varphi\colon \mathrm{X} \to (-\infty, +\infty]$ *lower bounded proper sequentially lower monotone and* $\exists\, \alpha\colon \mathrm{I} \to (0. +\infty)$ *decreasing so that*

(289) $\quad \forall\, \mathrm{i} \in \mathrm{I}, \forall\, \mathrm{x} \in \mathrm{X}\ \exists\, \mathrm{y} \in \mathrm{T(x)}$ *so that* $\mathrm{q_i(x, y)} \le \alpha(\mathrm{i})[\varphi(\mathrm{x}) - \varphi(\mathrm{y})]$,

respectively

(290) $\quad \mathrm{q_i(x, y)} \le \alpha(\mathrm{i})[\varphi(\mathrm{x}) - \varphi(\mathrm{y})]\ \forall\, \mathrm{i} \in \mathrm{I}, \forall\, \mathrm{x} \in \mathrm{X}, \forall\, \mathrm{y} \in \mathrm{T(x)}$,

then T *has a fixed point* respectively T *has an invariant point*

(*A. Hamel*).

■ Take the function $\varepsilon: I \to (0, +\infty)$, obviously increasing , $\varepsilon = \frac{1}{\alpha}$ and apply 9.58, $\exists\, v^*$ in dom φ so that $y \neq v^* \Rightarrow \exists\, j$ so that $\varphi(v^*) < \varphi(y) + \varepsilon(j)q_j(y, v^*)$, what is equivalent (even in the case $\varphi(y) = +\infty$!) to

(291) $$y \neq v^* \Rightarrow \exists\, j \text{ so that } q_j(y, v^*) > \alpha(j)[\varphi(v^*) - \varphi(y)].$$

Transcribe (289) with $x = v^*$ and $i = j$,

(292) $$\exists\, y \in T(v^*) \text{ so that } q_j(v^*, y) \leq \alpha(j)[\varphi(v^*) - \varphi(y)].$$

If *par absurdum* $y \neq v^*$, (292) comes in collision with (291), consequently $y = v^*$ and hence $v^* \in T(v^*)$.

Transcribe now (290) with $x = v^*$ and $i = j$,

(293) $$q_j(v^*, y) \leq \alpha(j)[\varphi(v^*) - \varphi(y)]\ \forall\, y \in T(v^*).$$

If *par absurdum* $\exists\, y \in T(v^*)$ and $y \neq v^*$, (293) comes in collision with (291), consequently $y \in T(v^*) \Rightarrow y = v^*$ and hence $\{v^*\} = T(v^*)$. ■

Remark. For $(X, \mathcal{U}) = (X, \mathrm{d})$, complete metric space, and $\alpha(i) = 1\ \forall\, i \in I$, (289) gives indeed the C – K theorem from §2, B, but with φ even with values in $(-\infty, +\infty]$, obviously proper.

9.60 *Let* X *be sequentially complete separated uniform space,* $(\mathrm{q_i})_{\mathrm{i}\in \mathrm{I}}$ *an afferent family of cuasi-distances* (9.54) *and* $\varphi : \mathrm{X} \to \mathbf{R}$ *lower bounded, sequentially lower monotone. If* $\exists\, \varepsilon : \mathrm{I} \to (0, +\infty)$ *increasing with the property* $\forall\, \mathrm{x}$ *from* X *with* $\varphi(\mathrm{x}) > \inf \varphi(\mathrm{X})\ \exists\, \mathrm{y} \in \mathrm{X} \setminus \setminus \{\mathrm{x}\}$ *so that*

$$\varphi(\mathrm{y}) + \varepsilon(\mathrm{i})\mathrm{q_i}(\mathrm{y}, \mathrm{x}) \leq \varphi(\mathrm{x})\ \forall\, \mathrm{i} \in \mathrm{I},$$

then $\exists\, \mathrm{x}^*$ *in* X *so that*

$$\varphi(\mathrm{x}^*) = \inf \varphi(\mathrm{X})$$

(*W. Takahashi, A. Hamel*).

■ It can obviously suppose φ not constant. Consider the multivalued function $T : X \to \to 2^X_*$,

$$T(x) = \{y \in X : \varphi(y) + \varepsilon(i)q_i(y, x) \leq \varphi(x)\ \forall\, i \in I\}.$$

$T(x)$ is, according to the hypothesis, indeed nonempty. T verifies (290) with $\alpha = \frac{1}{\varepsilon}$ (here occurs the hypothesis "φ with values in $\mathbf{R}$"), apply 9.59, $\exists\, x^*$ in X so that $\{x^*\} = T(x^*)$.

We have $\varphi(x^*) = \inf \varphi(X)$. Indeed, if *par absurdum* $\varphi(x^*) > \inf \varphi(X)$, then according to the hypothesis $\exists\, y \in T(x^*) \setminus \{x^*\}$, a contradiction. ■

REMARK. In th. 6 from $[35_1]$, 2 the hypothesis $f : X \to \mathbf{R} \cup \{+\infty\}$ is wrong, the presented proof does not permit to f infinite values.

EKELAND PRINCIPLE IN UNIFORM SPACES

(GENERALIZED FAMILIES OF QUASI-DISTANCES)

GENERALIZED FAMILY OF QUASI-DISTANCES ($[3_1]$)

Let X be nonempty set. The function family $(q_i)_{i\in I}$, $q_i : X \times X \to [0, +\infty)$ is called *generalized family of quasi-distances* if

(Q′1) $q_i(x, x) = 0 \;\forall\, i \in I, \forall\, x \in X$;

(Q′2) $q_i(x, y) = q_i(y, x) \;\forall\, i \in I, \forall\, x, y \in X$;

(Q′3) There exists $(j_1(i))_{i\in I}$, $(j_2(i))_{i\in I}$ subfamilies of I so that

$$z \neq y \text{ and } z \neq x \Rightarrow q_i(x, y) \leq q_{j_1(i)}(x, z) + q_{j_2(i)}(z, y) \;\forall\, i \in I.$$

If in addition

(Q′4) $q_i(x, y) = 0 \;\forall\, i \in I \Rightarrow x = y$,

the family $(q_i)_{i\in I}$ is by definition *separating*.

Let X be a nonempty set and $(q_i)_{i\in I}$ generalized family of quasi-distances on X. Consider, for each $i_1, \ldots, i_n$ from I and r from the interval $(0, +\infty)$, the subset of $X \times X$

$$(294)\; U(i_1, \ldots, i_n, r) := \{(x, y) \in X \times X : q_{i_k}(x, y) < r, k = \overline{1, n}\}.$$

The sets $U(i_1, \ldots, i_n, r)$, where $n \in \mathbf{N}$, $i_1, \ldots, i_n \in I$ and $r \in (0, +\infty)$, form a basis of a uniform structure $\mathscr{U}$ on X. Indeed, because (B_I), (U'_I), (U'_{II}) and (U'_{III}) are verified: let be $U(i_1, \ldots, i_m, r)$, $U(j_1, \ldots, j_n, s)$, then if $t = \min(r, s)$ we have $U(i_1, \ldots, i_m, j_1, \ldots, j_n, t) \subset U(i_1, \ldots, i_m, r) \cap U(j_1, \ldots, j_n, s)$; for (U_1') take into account (Q′1); $\overline{U(i_1,\ldots,i_n,r)}^{-1} =$ $= U(i_1, \ldots, i_n, r)$ – use (Q′2); let be $U := U(i_1, \ldots, i_n, r)$, for

$$V := U(i_1, .., i_n, j_1(i_1), j_2(i_1), \ldots, j_1(i_n), j_2(i_n), \frac{r}{2})$$

(see (Q′3)) we have (295) $V \circ V \subset U$. Indeed, let be $(x, y) \in V^2$, $\exists z \in X$ so that $(x, z), (z, y) \in V$. If either $z = y$, or $z = x$, we have $(x, y) \in V$, hence in particular $q_{i_k}(x, y) < \frac{r}{2} < r$, $k = \overline{1, n}$ and hence $(x, y) \in U$. In the contrary case, $q_{i_k}(x, y) \le q_{j_1(i_k)}(x, z) + q_{j_2(i_k)}(z, y) < \frac{r}{2} + \frac{r}{2} = r$, $k = \overline{1, n}$, so again $(x, y) \in U$ and hence (295).

$\mathscr{U}$ is by definition the uniform structure *defined* by $(q_i)_{i \in I}$. In the topology $\tau(\mathscr{U})$, for each point x from X the sets $U(i_1, \ldots, i_n, r)(x)$, $n \in \mathbf{N}$, $i_1, \ldots, i_n \in I$ and $r \in (0, +\infty)$, form a basis of neighbourhoods of x.

With the same proof as for 9.51 one obtains

9.61 $(X, \mathscr{U})$ *is separated* $\Leftrightarrow$ $(q_i)_{i \in I}$ *is separating.*

Reciprocally,

9.62 *Every uniform structure on a set is defined by a generalized family of quasi-distances on this set.*

■ Let $(X, \mathscr{U})$ be uniform space, π a set of pseudo-distances on X which defines $\mathscr{U}$ (9.54) and I a set which goes bijective on π by $i \to q_i$. $(q_i)_{i \in I}$ is a generalized family of quasi-distances on X (for (Q′3) take $j_1(i) = j_2(i) = i$) which defines $\mathscr{U}$ (take into account (294). ■

MINIMAL POINT THEOREM

Let Y be a real vector space and C a convex cone with vertex in Y. The binary relation $y_1 \overset{C}{\le} y_2 \Leftrightarrow y_2 - y_1 \in C$ is preorder relation in Y. $\overset{C}{\le}$ is order relation $\Leftrightarrow$ C is pointed (§8, 8.3).

Let X be separated uniform space, $(q_i)_{i\in I}$ an afferent generalized family of quasi-distances on X (9.62) and (296) $k_0 \in C \setminus (-\overline{C})$. Put on the set $X \times Y$ the binary relation (inspired by §8, (9))

(297) $$(x_1, y_1) \overset{k_0}{\leq} (x_2, y_2) \overset{\text{def}}{\Leftrightarrow} y_1 + q_i(x_1, x_2)k_0 \overset{C}{\leq} y_2 \ \forall\, i \in I.$$

9.63 $\overset{k_0}{\leq}$ *is a preorder relation. If in addition* C *is pointed, then this is an order relation.*

■ *First assertion.* Obviously $(x_1, y_1) \overset{k_0}{\leq} (x_1, y_1)$ (see (Q′1)). Let be $(x_1, y_1) \overset{k_0}{\leq} (x_2, y_2)$ and $(x_2, y_2) \overset{k_0}{\leq} (x_3, y_3)$. These give, $\forall\, i \in I$, $y_1 + q_i(x_1, x_2)k_0 \overset{C}{\leq} y_3 - q_i(x_2, x_3)k_0$, $y_1 + [q_i(x_1, x_2) + q_i(x_2, x_3)]k_0 \overset{C}{\leq} y_3$. If either $x_2 = x_3$, or $x_2 = x_1$, it results $(x_1, y_1) \overset{k_0}{\leq} (x_3, y_3)$. In the contrary case apply (Q′3), we have, $\forall\, i \in I$,

$$y_1 + q_i(x_1, x_3)k_0 \leq y_1 + [\,q_{j_1(i)}(x_1, x_2) + q_{j_2(i)}(x_2, x_3)]k_0 \overset{C}{\leq} y_1 + (y_2 - y_1) + (y_3 - y_2) = y_3$$

(it was used $\alpha \leq \beta \Rightarrow \alpha k_0 \overset{C}{\leq} \beta k_0$, because $(\beta - \alpha)k_0 \in C$), i.e. $(x_1, y_1) \overset{k_0}{\leq} (x_3, y_3)$.

Second assertion. $(x_1, y_1) \overset{k_0}{\leq} (x_2, y_2)$ and $(x_2, y_2) \overset{k_0}{\leq} (x_1, y_1) \Rightarrow$

(298) $$y_2 - y_1 - q_i(x_1, x_2)k_0 \in C \text{ and } y_1 - y_2 - q_i(x_1, x_2)k_0 \in C \ \ \forall\, i \in I.$$

By adding one obtains $-2q_i(x_1, x_2)k_0 \in C \ \forall\, i\in I$, which imposes $q_i(x_1, x_2) = 0 \ \forall\, i \in I$, in the contrary case it results $k_0 \in -C \subset -\overline{C}$, in contradition with (296). X being separated, this imposes $x_1 = x_2$ (9.61). Then (298) gives $y_2 - y_1 \in C$ and $y_1 - y_2 \in C$, that is $y_2 - y_1 \in C \cap (-C)$ and hence $y_2 = y_1$ (8.2). ■

9.64 Theorem (*Minimal point*). *Let* X *be separated uniform space,* $(q_i)_{i\in I}$ *an afferent generalized family of quasi-distances on* X (9.62), Y *real topological vector space,* C *convex cone with vertex in* Y *and* k_0 *from* $C \setminus (-\overline{C})$.

Let A *be nonempty subset of* $X \times Y$ *and* (x_0, y_0) *from* A *for which* $A_0 := \{(x, y) \in A: (x, y) \overset{k_0}{\leq} (x_0, y_0)\}$ *has the properties*:

a) $\pi_Y(A_0)$ *is* $\overset{C}{\leq}$*-lower bounded*;

b) *Each* $\overset{k_0}{\leq}$ *- decreasing sequence from* A_0 *has a* k_0 *- minorant.*

Then there exists $(\bar{x}, \bar{y})$ *in* A *so that*

$$1^\circ\ (\bar{x}, \bar{y}) \overset{k_0}{\leq} (x_0, y_0),\ 2^\circ\ (x, y) \in A \text{ and } (x, y) \overset{k_0}{\leq} (\bar{x}, \bar{y}) \Rightarrow x = \bar{x}$$

(*A. Benbrik - A. Mbarki - S. Lahrech - A. Ouahab*).

Explanation. See the notations $\pi_X(A)$ and $\pi_Y(A)$ from (54), §8.

■ Use the Brezis-Browder statement (§1, 1.1_2) and the Luc functional (§8, 8.10 with Remark 3).

A_0 is obviously preordered by $\overset{k_0}{\leq}$. Consider the map $\varphi : A_0 \to (-\infty, +\infty]$, $\varphi(x, y) = L(y - y_0)$. φ is increasing: $(x_1, y_1) \overset{k_0}{\leq} (x_2, y_2) \overset{(297)}{\Rightarrow} y_1 - y_0 + q_i(x_1, x_2)k_0 \overset{C}{\leq} y_2 - y_0\ \forall\, i \in I$, apply L (this is C-increasing), $L(y_1 - y_0 + q_i(x_1, x_2)k_0) \overset{8.10,2^\circ}{=} L(y_1 - y_0) + q_i(x_1, x_2) \leq L(y_2 - y_0)$, hence $\varphi(x_1, y_1) \leq \varphi(x_2, y_2)$. Also, φ is lower bounded, because $\pi_Y(A_0)$ is C-lower bounded and L is C-increasing. Note in addition that φ takes values in $\mathbf{R}$: $(x, y) \in A_0 \Rightarrow (x, y) \overset{k_0}{\leq} (x_0, y_0) \Rightarrow y \overset{C}{\leq} y_0$, then $y - y_0 \in -C \subset -\bar{C}$ and hence $\varphi(x, y) = L(y - y_0) \leq 0$ (see §8, (43) and (46)). Apply 1.1_2, $\exists\, (\bar{x}, \bar{y}) \in A_0$ so that

$$(\bar{x}, \bar{y}) \overset{k_0}{\leq} (x_0, y_0) \text{ and } (299)\ (x, y) \overset{k_0}{\leq} (\bar{x}, \bar{y}) \Rightarrow L(y - y_0) = L(\bar{y} - y_0).$$

Verify 2°. Let (x, y) be from A with $(x, y) \overset{k_0}{\leq} (\bar{x}, \bar{y})$. Then $y - y_0 + q_i(x, \bar{x})k_0 \overset{C}{\leq} \bar{y} - y_0$ $\forall\, i$, apply L, $L(y - y_0) + q_i(x, \bar{x}) \leq L(\bar{y} - y_0)$ $\forall\, i$, then (finite values!) $q_i(x, \bar{x}) \leq L(\bar{y} - y_0) - L(y - y_0) \overset{(299)}{=} 0$ $\forall\, i$ and so $x = \bar{x}$ as X is separated (9.61). ■

9.65 *Uniform variant of Ekeland principle* ([3_1], th. 5). *Let* X *be sequentially complete separated uniform space,* $(q_i)_{i \in I}$ *an afferent generalized family of quasi-distances on* X (9.62), Y *real topological vector space,* C *convex cone with vertex in* Y, $k_0 \in C \setminus (-\bar{C})$ *and* $\varphi : X \to Y^\bullet$ $\overset{C}{\leq}$*-lower bounded proper.*

If $\forall$ x *from* dom φ *the set*

$$S(x) := \{u \in X: \varphi(u) + q_i(u, x)k_0 \overset{C}{\leq} \varphi(x)\ \forall\ i \in I\}$$

is sequentially closed, then $\forall\ x_0$ *from* dom φ $\exists\ x^*$ *in* dom φ *so that*

(300) $$x \in X \setminus \{x^*\} \Rightarrow \exists\ j \in I \text{ so that } \varphi(x) + q_j(x, x^*)k_0 \overset{C}{\not\leq} \varphi(x^*)$$

and

(301) $$\varphi(x^*) + q_i(x^*, x_0)k_0 \overset{C}{\leq} \varphi(x_0)\ \forall\ i \in I.$$

Explanation. For $Y^\bullet$ see §8, before 8.6.

■ Let be $A := \{(x, y) \in X \times Y : y = \varphi(x)\}$, a set obviously nonempty because φ is proper, (x_0, y_0) from A and $A_0 := \{(x, y) \in A : (x, y) \overset{k_0}{\leq} (x_0, y_0)\}$. Verify a) and b) from 9.64. a) is ensured as φ is $\overset{C}{\leq}$-lower bounded. Pass to verify b) and let $(x_n, y_n)_{n\geq 1}$ be a $\overset{C}{\leq}$-decreasing sequence from A_0. Prove (302) $(x_n)_{n\geq 1}$ is Cauchy sequence in X. Taking into account the definition and (294), it follows to prove: $\forall\ U_{i_1,\ldots,i_n,r}$ $\exists\ N$ in $\mathbf{N}$ so that $m, n \geq N$ $\Rightarrow q_{i_k}(x_m, x_n) < r$, $k = \overline{1, n}$ and for this it is enough to prove (*) $\forall\ i \in I$ and $\forall\ \varepsilon > 0\ \exists\ N$ in $\mathbf{N}$ so that $m, n \geq N \Rightarrow q_i(x_m, x_n) \leq \varepsilon$. Fix i and let be $m \geq n$. We have $(x_m, y_m) \overset{k_0}{\leq} (x_n, y_n)$, so

(303) $$y_m - y_0 + q_i(x_m, x_n)k_0 \overset{C}{\leq} y_n - y_0.$$

Apply the Luc functional to (303) (see the proof for 9.64) and denoting $a_p := L(y_p - y_0)$, $p \geq 1$ one obtains

(304) $$a_m + q_i(x_m, x_n) \leq a_n.$$

But, for p any, we have $y_p + q_i(x_p, x_0)k_0 \overset{C}{\leq} y_0$, hence $y_p - y_0 \overset{C}{\leq} 0$ and hence $a_p \leq 0$ (8.10, 1° and 2°), so $(a_p)_{p\geq 1}$ is a sequence of real numbers. Moreover, this is decreasing ((304)) and lower bounded because φ is C-lower bounded and L C-increasing. So being, (a_p) is convergent, hence Cauchy sequence, this imposes, *via* (304), (*) and hence (302). Let be $x' = \lim_{n\to\infty} x_n$ and n arbitrary. As $m \geq n \Rightarrow x_m \in S(x_n)$ (remove y_0 from (303)!), passing to the limit one obtains $x' \in S(x_n)$, i.e. $(x', \varphi(x')) \overset{k_0}{\leq} (x_n, \varphi(x_n))\ \forall\ n \geq 1$, b) is verified.

Apply 9.64, $\exists\ (x^*, y^*)$ in A so that

$$(305)\quad (x^*, y^*) \overset{k_0}{\leq} (x_0, y_0), \qquad (306)\quad (x, y) \in A \text{ and } (x, y) \overset{k_0}{\leq} (x^*, y^*) \Rightarrow x = x^*.$$

(305) is even (301), Let be $x \in X \setminus \{x^*\}$. Then, from (306), $(x, \varphi(x)) \overset{k_0}{\not\leq} (x^*, \varphi(x^*))$, hence $\exists\, j \in I$ so that $\varphi(x) + q_j(x, x^*)k_0 \overset{C}{\not\leq} \varphi(x^*)$, i.e. (300). ■

Remark 1. 9.65 is indeed a variant of the Ekeland principle, as show (300) and (301) in the particular case $(X, \mathscr{U}) = (X, \mathrm{d})$, complete metric space, $Y = \mathbf{R}$ and $C = [0, +\infty)$, when therefore $k_0 \overset{(296)}{\in} (0, +\infty)$ and $\overset{C}{\leq}$ is even the order relation $\leq$ between the real numbers: (300) becomes $x \neq x^* \Rightarrow \varphi(x) + k_0\, \mathrm{d}(x, x^*) \not\leq \varphi(x^*)$, hence $\varphi(x^*) < \varphi(x) + k_0\, \mathrm{d}(x, x^*)$ and (301) becomes $\varphi(x^*) + k_0\, \mathrm{d}(x^*, x_0) \leq \varphi(x_0)$, that is ($e_2$) respectively ($e_1$) from the form **E** of the Ekeland principle (after REMARK 2 to 1.1_0).

Remark 2. We point out the deficiencies in the article [3_1]:

1) In the statement of th.5 it was forgotten the condition „X sequentially complete";

2) In the proof of th. 5 the consideration of the multivalued function F is useless.

J. EKELAND PRINCIPLE FACE TO THE w-DISTANCE ([54], 2)

This section is the continuation of the section C from §2 of which statements it generalizes.

Definitions. Let (X, d) be a metric space. The map $\delta : X \times X \to [0, +\infty)$ is a w-*distance on* X if

(w1) $\delta(x, y) \leq \delta(x, z) + \delta(z, y)\ \forall\, x, y, z \in X$;

(w2) $\delta(x, \cdot)$ is lower semicontinuous $\forall\, x \in X$;

(w3) $\forall\, \varepsilon > 0\ \exists\, \eta > 0$ so that $\delta(x, y) \leq \eta$ and $\delta(x, z) \leq \eta \Rightarrow \mathrm{d}(y, z) \leq \varepsilon$

([38_1]).

Let (X, d) be a metric space. Consider the function $f : X \times X \to (-\infty, +\infty]$ with the properties

(f1) $f(x, y) \leq f(x, z) + f(z, y)\ \forall\, x, y, z \in X$;

(f2) $f(x, \cdot)$ is lower semicontinuous $\forall\, x \in X$;

In every nonmempty set X, preordered by $\preceq$, $J(x)$, x element from X, designates the interval $\{y \in X : x \preceq y\}$.

And now –

9.66 Theorem. *Let* (X, d) *be a metric space,* δ w-*distance on* X, $f: X \times X \to (-\infty, +\infty]$ *with the properties* (f1), (f2) *and the preorder relation* $\preceq$ *on* X

$$x \preceq y \overset{\text{def}}{\Leftrightarrow} \textit{either } x = y, \textit{ or } f(x,y) + \delta(x,y) \le 0.$$

Let v_0 *be from* X *so that* $\inf_{y\in X} f(v_0, y) > -\infty$ *and* $J(v_0)$ *is* $\preceq$*- complete. Then the following statements hold and are equivalent*:

1° $J(v_0)$ *has a maximal element*;

2° *If* $T: J(v_0) \to 2^X$ *has the property*

$$v \in J(v_0) \setminus T(v) \Rightarrow \exists\, w \in X \setminus \{v\} \textit{ so that } v \preceq w,$$

then T *has a fixed point*;

3° *Every function* $g : J(v_0) \to X$ *with the property*

$$v \preceq g(v)\ \forall\, v \in J(v_0)$$

has a fixed point;

4° *If* $T: J(v_0) \to 2^X_*$ *has the property*

$$v \in J(v_0) \textit{ and } w \in T(v) \Rightarrow v \preceq w,$$

then T *has an invariant point*;

5° *For every set* $\mathscr{G}$ *of functions* $g: J(v_0) \to X$, *with the property*

$$v \in J(v_0) \Rightarrow v \le g(v),$$

there is a common fixed point;

6° *If the subset* M *of* X *has the property* $v \in J(v_0) \setminus M \Rightarrow J(v) \setminus \{v\} \ne \varnothing$, *then*

$$J(v_0) \cap M \ne \varnothing;$$

7° *If* $v \in J(v_0)$ *and* $\inf_{y\in X} f(v,y) < 0 \Rightarrow J(v) \setminus \{v\} \ne \varnothing$, *then* $\exists\, v^*$ *in* $J(v_0)$ *so that*

$$\inf_{y\in X} f(v^*, y) \ge 0$$

(*S. Park*).

Explanation. The interval $J(x_0)$ is by definition $\preceq$*-complete* when every increasing Cauchy sequence from the interval $J(x_0)$ is convergent in this.

■ ([54], 2). $\preceq$ is indeed a preorder relation on X, as one confirms using (w1) and (f1).

Proof of 1° (inspired from the proof for th. 4, §2, section C). Construct by recurrence a sequence $(v_n)_{n\geq 0}$ from $J(v_0)$ with $J(v_n) \neq \emptyset$ and $\gamma_n := \inf_{v\in J(v_n)} f(v_n, v) > -\infty \quad \forall\, n \geq 0$. v_0 is the element from the statement. Suppose defined v_{n-1}, $J(v_{n-1}) \neq \emptyset$ and $\gamma_{n-1} > -\infty$. Take v_n from $J(v_{n-1})$ so that (307) $f(v_{n-1}, v_n) \leq \gamma_{n-1} + \dfrac{1}{n}$. Obviously $J(v_n) \neq \emptyset$. Show (308) $J(v_n) \subset J(v_{n-1})$: let be $v \in J(v_n)$, if $v = v_n$ – nothing to prove, if $v \neq v_n$ then $f(v_{n-1}, v) + \delta(v_{n-1}, v) \overset{(w1),(f1)}{\leq} f(v_{n-1}, v_n) + \delta(v_{n-1}, v_n) + f(v_n, v) + \delta(v_n, v) \overset{v_n \in J(v_{n-1})}{\leq} f(v_n, v) + \delta(v_n, v) \leq \leq 0$, so $v \in J(v_{n-1})$. In addition, (309) $\gamma_n \geq -\dfrac{1}{n}$, hence in particular $\gamma_n > -\infty$. Indeed, $\gamma_n \overset{(f1)}{\geq} \inf_{v\in J(v_n)} [f(v_{n-1}, v) - f(v_{n-1}, v_n)] \overset{(308)}{\geq} \inf_{v\in J(v_{n-1})} [f(v_{n-1}, v) - f(v_{n-1}, v_n)] = \gamma_{n-1} - f(v_{n-1}, v_n) \overset{(307)}{\geq} -\dfrac{1}{n}$, hence (309).

The sequence $(v_n)_{n\geq 0}$ is thus built. (308) shows that all terms of this are in $J(v_0)$ and a look back leads us to the conclusion (310) $\gamma_n \geq -\dfrac{1}{n} \quad \forall\, n \geq 1$.

Prove now (311) $\lim_{n\to\infty} \operatorname{diam} J(v_n) = 0$. Observe that for v any from $J(v_n) \setminus \{v_n\}$ we have $\delta(v_n, v) \leq -f(v_n, v) \leq -\gamma_n \overset{(310)}{\leq} \dfrac{1}{n}$, this conclusion is kept also when $v = v_n$, so (312) $v \in \in J(v_n) \Rightarrow \delta(v_n, v) \leq \dfrac{1}{n}$. Let be $\varepsilon > 0$, $\exists\, \eta > 0$ with the property from (w3) and $\exists$ a rank N from which $\dfrac{1}{n} \leq \eta$, hence $n \geq N \Rightarrow \forall\, v', v''$ from $J(v_n)$ we have $\delta(v_n, v') \overset{(312)}{\leq} \dfrac{1}{n} \leq \eta$, $\delta(v_n, v'') \leq \dfrac{1}{n} \leq \eta$, consequently $d(v', v'') \leq \varepsilon$, $\operatorname{diam} J(v_n) \leq \varepsilon$, i.e. (311). (311) imposes $(v_n)_{n\geq 0}$ Cauchy sequence: let be $\varepsilon > 0$, $\exists\, N$ in $\mathbf{N}$ so that $n \geq N \Rightarrow \operatorname{diam} J(v_n) \leq \varepsilon$, then $m \geq n \geq N \Rightarrow J(v_m) \subset J(v_n) \Rightarrow d(v_m, v_n) \leq \varepsilon$. Moreover, the sequence (v_n) is, according to its definition, $\preceq$-increasing.

Now let v^* be from $J(v_0)$, $v^* = \lim_{n\to\infty} v_n$. Fix $n \geq 0$. According to the definition, we have $J(v_n) = \{v_n\} \cup \{x \in X : f(v_n, x) + \delta(v_n, x) \leq 0\}$, hence $J(v_n)$ is closed as $f(v_n, \cdot)$ and

$\delta(v_n, \cdot)$ are l.s.c. Consequently, as for $p \geq n$ we have $v_p \in J(v_n)$, it results $v^* \in J(v_n)$ and so (*) $v^* \in \bigcap_{n=0}^{\infty} J(v_n)$. (311) imposes even (313) $\{v^*\} = \bigcap_{n=0}^{\infty} J(v_n)$.

It remains to show that v^* is a maximal element in $J(v_0)$. Let be *par absurdum* $\bar{v}$ from $J(v_0) \setminus \{v^*\}$ with $v^* \preceq \bar{v}$, that is $f(v^*, \bar{v}) + \delta(v^*, \bar{v}) \leq 0$. Suppose $v^* \neq v_n\ \forall\ n \geq 0$. Then $f(v_n, v^*) + \delta(v_n, v^*) \leq 0\ \forall\ n \geq 0$ ((*)), hence by applying (f1) and (w1) one obtains $f(v_n, \bar{v}) + \delta(v_n, \bar{v}) \leq 0\ \forall\ n \geq 0$, i.e. $\bar{v} \in \bigcap_{n=0}^{\infty} J(v_n)$, in contradiction with (313).

Pass to the case $\exists\ N \geq 0$ so that $v^* = v_N$ (not observed in [54], 2, th. 1, pag. 884). (313) imposes $v_N \in J(v_n)\ \forall\ n \geq N$, so $J(v_N) \subset J(v_n)\ \forall\ n \geq N$ and combining with (308) one obtains $J(v_N) = J(v_n)\ \forall\ n \geq N$, which leads, *via* (313), to $\{v_N\} = \bigcap_{n=0}^{N} J(v_n) \overset{(308)}{=} J(v_N)$, $\{v_N\} = J(v_N)$. But $v_N \preceq \bar{v}$, i.e. $\bar{v} \in J(v_N)$, hence $\bar{v} = v_N = v^*$, again a contradiction.

1°$\Rightarrow$ 2°. Suppose *par absurdum* $v \notin T(v)\ \forall\ v \in J(v_0)$. Then in particular $v^* \notin T(v^*)$ and hence $\exists\ v \in X \setminus \{v^*\}$ so that $v^* \preceq v$, a contradiction, v^* being maximal element (the last inequality gives also $v \in J(v_0)$).

Attention! The hypothesis does not request $T(v) \neq \varnothing\ \forall\ v \in J(v_0)$.

2° $\Rightarrow$ 3°.Take $T : J(v_0) \to 2^X$, $T(v) = \{g(v)\}$.

3° $\Rightarrow$ 4°. Suppose *par absurdum* $T(v) \neq \{v\}\ \forall\ v$. We have then, because $T(v) \neq \varnothing\ \forall\ v$, $T(v) \setminus \{v\} \neq \varnothing\ \forall\ v$ and so it can consider a function g defined on $J(v_0)$ with $g(v) \in T(v) \setminus \{v\}$. Obviously (314) $g(v) \neq v\ \forall\ v$. On the other hand, the hypothesis gives $v \preceq g(v)\ \forall\ v$ and hence, according to 3°, g has a fixed point, this contradicts (314).

4° $\Rightarrow$ 5°. Consider the function $T : J(v_0) \to 2^X_*$, $T(v) = \{g(v): g \in \mathscr{G}\}$. Because, $\forall\ v$, $w \in T(v) \Rightarrow v \preceq w$, T has, according to 4°, an invariant point. This is obviously a common fixed point.

5° $\Rightarrow$ 1°. Suppose *par absurdum* the contrary, i.e. $\forall\ v \in J(v_0)\ \exists\ w \in X \setminus \{v\}$ so that (315) $v \preceq w$. Consider the function $g: J(v_0) \to X$, $g(v) = w$, w verifying (315). So g has not a fixed point. On the other hand, it can apply 5° to the set $\mathscr{G} := \{g\}$, this imposes g has a fixed point, contradiction.

1° ⇒ 6°. Let v^* be maximal element of $J(v_0)$. Then $J(v^*) \setminus \{v^*\} = \varnothing$ (*par absurdum*, use the transitivity) and the hypothesis imposes $v^* \in M$, that is $v^* \in J(v_0) \cap M$.

6° ⇒ 1°. Consider, for any v from X, $A(v) := \{w \in X\setminus\{v\} : v \preceq w\}$ and let be $M := \{v \in \subset X : A(v) = \varnothing\}$. Observe that $\bar{v} \in J(v_0) \cap M \Rightarrow \bar{v}$ maximal element of $J(v_0)$. So in the case $J(v_0) \subset M$ the proof is finished. Let then be $v \in J(v_0) \setminus M$. This gives $A(v) \neq \varnothing$, i.e. $J(v) \setminus \{v\} \neq \varnothing$ and hence 6° imposes $J(v_0) \cap M \neq \varnothing$.

1° ⇒ 7°. Suppose *par absurdum* the contrary, i.e. $\inf_{y\in X} f(v, y) < 0 \ \forall\ v \in J(v_0)$. The hypothesis gives (316) $J(v) \setminus \{v\} \neq \varnothing \ \forall\ v \in J(v_0)$. But (316) prevents $J(v_0)$ to have a maximal element, contradiction.

7° ⇒ 1°. Suppose *par absurdum* the contrary, hence (317) $J(v) \setminus \{v\} \neq \varnothing \ \forall\ v \in J(v_0)$. 7° gives v^* from $J(v_0)$ so that $\inf_{y\in X} f(v^*, y) \geq 0$. From (317), $J(v^*) \setminus \{v^*\} \neq \varnothing$, let v_1 be from the first member, then (318) $v_1 \neq v^*$ and

(319) $$\inf_{y\in X} f(v^*, y) + \delta(v^*, v_1) \leq f(v^*, v_1) + \delta(v^*, v_1) \leq 0.$$

(319) imposes $\inf_{y\in X} f(v^*, y) = 0$ (*par absurdum*, δ takes positive values), so $\delta(v^*, v_1) \leq 0$ and so (320) $\delta(v^*, v_1) = 0$. Apply once again (317) and let be (321) $v_2 \in J(v_1) \setminus \{v_1\}$, v_1 from (318) and (319). $f(v^*, v_2) \overset{(f1)}{\leq} f(v^*, v_1) + f(v_1, v_2)$, $\delta(v^*, v_2) \overset{(w1)}{\leq} \delta(v^*, v_1) + \delta(v_1, v_2)$, by adding and taking into account (319) and (321) one obtains $f(v^*, v_2) + \delta(v^*, v_2) \leq 0$. But since $\inf_{y\in X} f(v^*, y) \geq 0$ it results $f(v^*, v_2) = \delta(v^*, v_2) = 0$. As also $\delta(v^*, v_1) \overset{(320)}{=} 0$, one concludes $d(v_1, v_2) = 0$ (use (w3), take $\varepsilon = \dfrac{1}{n}$, $n \in \mathbf{N}$ and $\eta > 0$ any) and hence $v_1 = v_2$ in cotradiction with (321). ■

REMARK. The implication 7° ⇒ 1° *without any condition* (see the condition from [54], 2, th. 1′) was obtained by the author.

Look another simplified form of 9.66.

9.67 *Let* X *be complete metric space,* δ w-*distance on* X, $f : X \times X \to (-\infty, +\infty]$ *with the properties* (f1), (f2) *and the preorder relation* ⪯ *on* X,

$$x \preceq y \overset{\text{def}}{\Leftrightarrow} \textit{either } x = y, \textit{ or } f(x, y) + \delta(x, y) \leq 0.$$

Let v_0 *be from* X *so that* $\inf_{y\in X} f(v_0, y) > -\infty$. *Then the following statements hold and are equivalent*:

1° $J(v_0)$ *has a maximal element*;

2° *If* $T: X \to 2^X$ *has the property*

$$x \in X \setminus T(x) \Rightarrow \exists\, y \in X \setminus \{x\} \text{ so that } x \preceq y,$$

then T *has a fixed point*;

3° *Any function* $g: X \to X$, *with the property* $x \preceq g(x)\ \forall\, x \in X$, *has a fixed point*;

4° *If* $T: X \to 2^X_*$ *has the property*

$$x \in X \text{ and } y \in T(x) \Rightarrow x \preceq y,$$

then T *has an invariant point*;

5° *For any set* $\mathscr{G}$ *of functions* $g: X \to X$ *with the property*

$$x \preceq g(x)\ \forall\, x \in X$$

there exists a common fixed point;

6° *If the subset* M *of* X *has the property*

$$x \in X \setminus M \Rightarrow \exists\, y \in X \setminus \{x\} \text{ with } y \succeq x,$$

then

$$J(v_0) \cap M \neq \varnothing;$$

7° *If* $\forall$ v *from* X *with* $\inf_{y\in X} f(v, y) < 0$ *we have* $J(v) \setminus \{v\} \neq \varnothing$, *then* $\exists\, v^*$ *in* X *so that* $\inf_{y\in X} f(v^*, y) \geq 0$

([54], 2).

■ $J(v_0) = \{v_0\} \cup \{x \in X : f(v_0, x) + \delta(v_0, x) \leq 0\}$, hence $J(v_0)$ is closed, so complete (X is complete) and hence in particular $\preceq$ – complete. Apply 9.66. ■

Note. Theorem 9.66 (S. Park) is the generalization of the theorem 2.6 (W. Oettli - M. Théra) from §2, C. 9.66, 6° corresponds to the theorem 4 (W. Oettli - M. Théra), § 2, C and 9.66, 7° corresponds to the theorem 2 (W. Takahashi) from §2, C *via* 9.67, 7° (*E. Blum - W. Oettli*), From optimization and variational inequalities to equilibrium problems, Math. Students 63 (1994), 123-145).

In 9.67, 2° are also involved

H. Tuy, A fixed point theorem involving a hybrid inwardness-contraction condition, Math. Nachr. 102 (1981), 271-275,

N. Mizoguchi - W. Takahashi, Fixed point theorems for multivalued mappings on complete metric spaces, J. Math. Anal. Appl. 141 (1989), 177-188.

In 9.67, 4° are also involved

M. Maschler - B. Peleg, Stable sets and stable points of set-valued dynamic systems, SIAM J. Control, 14 (1976), 985-995.

In 9.67, 5° is also involved

S. Kasahara, On fixed points in partially ordered sets and Caristi-Kirk theorem, Math. Sem. Notes Kobe Univ. 3(1975), 229-232.

And now –

9.68 9.66, 1° *generalizes Ekeland principle* 1.1 *with the aim function finite* ([54], 2).

■ Keep the notations and the numbering from §1, 1.1. Take δ: $\delta(x, y) = \frac{\varepsilon}{\lambda} d(x, y)$, f: $f(x, y) = \varphi(y) - \varphi(x)$. For $v_0 = u$ we have $\inf_{y \in X} f(u, y) = \inf \varphi(X) - \varphi(u) \geq -\varepsilon > -\infty$ and $J(u)$ is $\preceq$-complete being closed subset, and hence complete, of X. Apply 9.66, according with 1° $J(u)$ has a maximal element v_ε, hence $f(v_\varepsilon, w) + \delta(v_\varepsilon, w) > 0\ \forall\ w \in X \setminus \{v_\varepsilon\}$ (*par absurdum*) and consequently (1). Remain to verify (2) and (3). If $v_\varepsilon = u$, the proof is finished. Suppose $v_\varepsilon \neq u$. Then $v_\varepsilon \in J(u)$ imposes (322) $f(u, v_\varepsilon) + \delta(u, v_\varepsilon) \leq 0$. Because $\delta(u, v_\varepsilon) \geq 0$, it results $0 \geq f(u, v_\varepsilon) = \varphi(v_\varepsilon) - \varphi(u)$ and hence (2). Finally, $\frac{\varepsilon}{\lambda} d(u, v_\varepsilon) = \delta(u, v_\varepsilon) \overset{(322)}{\leq} -f(u, v_\varepsilon) = \varphi(u) - \varphi(v_\varepsilon) \leq \varphi(u) - \inf \varphi(X) \leq \varepsilon$ and consequently (3). ■

REMARK. In the statement of th. 3 from [54], 2 it is mistakenly permitted to the aim function φ (F in the article) to take the value $+\infty$, because, in accordance with 9.66, it is forbidden the value $-\infty$ to the auxiliary function f from the proof (Φ in the article).

With 9.67, 3° it can quickly obtain the following generalization ([54], 2, th. 4) of the Downing-Kirk theorem (§2, 2.13).

9.69 *Let* X *and* Y *be complete metric spaces*, δ_X *and* δ_Y w-*distances on* X resp. Y *and* h: X → X. *Suppose there exists* g: X → Y *closed and* f: g(X) × g(X) → (− ∞, + ∞], *which verifies* (f1), (f2) *relative to* g(X) × g(X) + ∃ v_0 *with* $\inf_{y \in g(X)} f(v_0, y) > -\infty$, *and so that* $f(g(x), gh(x)) + \max\{\delta_X(x, h(x)), \delta_Y(g(x), gh(x))\} \leq 0\ \forall\ x \in X$.

Then h *has a fixed point.*

■ $\delta: X \times X \to (-\infty, +\infty]$,

$$\delta(u, v) = \max \{\delta_X(u, v), \delta_Y(g(u), g(v))\}$$

is w-distance on X and the function $F: X \times X \to (-\infty, +\infty]$,

$$F(x, y) = f(g(x), g(y))$$

verifies (f1) and (f2) on $X \times X$. We have, according to the hypothesis,

$$F(x, h(x)) + \delta(x, h(x)) \leq 0 \ \forall \ x \in X,$$

hence $x \preceq h(x) \ \forall \ x \in X$. Apply 9.67, 3°. ■

Remark. In [54], 2 S. Park considered everywhere *quasi-metric spaces*. Here the author considered „en toute tranquillité” *metric spaces*.

10

Ekeland principle and Caristi - Kirk theorem in probabilistic metric spaces

In this paragraph, the two propositions from the title are extended to probabilistic metric spaces and one establishes here an equivalence between these variants ([17]).

A. MENGER PM - SPACE

DISTRIBUTION FUNCTION

A function $\delta : \mathbf{R} \to [0,1]$ increasing continuous to the left with $\inf \delta(\mathbf{R}) = 0$, $\sup \delta(\mathbf{R}) = 1$ is called *distribution function*. $\mathcal{D}$ will denote the set of those distribution functions having the property

$$\delta(0) = 0,$$

which imposes *a fortiori*

(1) $$t \leq 0 \Rightarrow \delta(t) = 0.$$

The function $H : \mathbf{R} \to [0, 1]$,

(2) $$H(t) = \begin{cases} 0, & t \leq 0 \\ 1, & t > 0 \end{cases}$$

is, obviously, in $\mathscr{D}$, the *specific distribution function*.

t-NORM

An application $T: [0,1] \times [0,1] \to [0,1]$ commutative, associative ($T(T(x, y), z)) = T(x, T(y, z))$), increasing in every variable and

(3) $$T(a,1) = a \ \forall \ a \in [0,1],$$

(4) $$T(0,0) = 0$$

is by definition t-*norm* (*K. Menger*, *B. Schweizer - A. Sklar*).

MENGER PM - SPACE

A triplet (X, F, T) is by definition *Menger probabilistic metric space* (*Menger* PM-*space*) if X is nonempty set, T t-norm and

$$F: X \times X \to \mathscr{D}$$

has the properties ($F(x, y)$ is noted F_{xy})

(PM_1) $$F_{xy} = H \Leftrightarrow x = y,$$

(PM_2) $$F_{xy} = F_{yx} \ \ \forall \ x, y \in X,$$

(PM_3) $$F_{xz}(t_1 + t_2) \geq T(F_{xy}(t_1), F_{yz}(t_2)) \ \forall \ x, y, z \in X, \forall \ t_1, t_2 \in \mathbf{R}$$

(*B. Schweizer - A. Sklar*).

Suppose $F_{xy}(t_1) = 1$, $F_{yz}(t_2) = 1$. Then $F_{xz}(t_1 + t_2) \overset{(PM_3)}{\geq} T(1,1) \overset{(3)}{=} 1$ and since F_{xz} takes values in [0,1], one obtains

(4) $$F_{xy}(t_1) = 1 \text{ and } F_{yz}(t_2) = 1 \Rightarrow F_{xz}(t_1 + t_2) = 1.$$

Let now be $\mathrm{d}: X \times X \to [0, +\infty)$ having the property

(5) $$F_{xy}(t) = H(t - \mathrm{d}(x, y)) \ \forall \ x, y \in X, \forall \ t \in \mathbf{R}.$$

Then

(6) d *is distance on* X.

Indeed, let be $x = y$ and *par absurdum* $\mathrm{d}(x, y) > 0$. Take t, $0 < t < \mathrm{d}(x, y)$, we have on one side $F_{xx}(t) \overset{(PM_1),(2)}{=} 1$ and on the other side $F_{xx}(t) \overset{(5)}{=} 0$, contradiction. Reciprocally, $\mathrm{d}(x, y) \overset{(5)}{=} 0 \Rightarrow F_{xy}(t) = H(t) \ \forall \ t \in \mathbf{R} \overset{(PM_1)}{\Rightarrow} x = y$. Obviously $\mathrm{d}(x, y) \overset{(PM_2)}{=} \mathrm{d}(y, x)$ (take $t = \mathrm{d}(x, y)$, $t = \mathrm{d}(y, x)$). Finally, let be x, y, z any in X and t_1, t_2 any with $t_1 > \mathrm{d}(x, y)$, $t_2 > \mathrm{d}(y, z)$. Then

$$F_{xy}(t_1) \overset{(5)}{=} H(t_1 - d(x,y)) = 1,\ F_{yz}(t_2) = H(t_2 - d(y,z)) = 1,$$

consequently $F_{xz}(t_1 + t_2) \overset{(4)}{=} 1 = H(t_1 + t_2 - d(x, z))$, hence $t_1 + t_2 \overset{(2)}{>} d(x, z)$ and hence $d(x, y) + d(y, z) \geq d(x, z)$, (6) is proved.

On the other hand, when d is distance on the set X, the map F defined on $X \times X$ by

(7) $$F_{xy}(t) = H(t - d(x, y))\ \forall\ t \in \mathbf{R}$$

takes values in $\mathscr{D}$ and verifies (PM$_1$) ($x = y \overset{(7)}{\Rightarrow} F_{xy} = H$, $F_{xy} = H \overset{(7)}{\Rightarrow} x = y$ (*par absurdum*, take t with $0 < t < d(x, y)$)), (PM$_2$) and $F_{xy}(t_1) = 1$, $F_{yz}(t_2) = 1 \Rightarrow F_{xz}(t_1 + t_2) \overset{(7)}{=} 1$ [$t_1 >$ $> d(x, y)$, $t_2 > d(y, z) \Rightarrow t_1 + t_2 > d(x, y) + d(y, z) \geq d(x, z)$].

So, *via* (6), we have the guarantee that the metric spaces theory may be included in the Menger PM - spaces theory.

Let (X, F, T) be Menger PM - space. Suppose that T has the property

(8) $$\sup_{t\in[0,1]} T(t, t) = 1.$$

Consider for every x from X and for every $\varepsilon > 0$, $\lambda \in (0,1)$ the set

(9) $$U_x(\varepsilon, \lambda) := \{y \in X : F_{xy}(\varepsilon) > 1 - \lambda\}$$

(*B. Schweizer - A. Sklar*).

There is a Hausdorff topology $\mathscr{T}_0$ on X so that, for any x from X, the sets $U_x(\varepsilon, \lambda)$, $\varepsilon > 0$ and $\lambda \in (0, 1)$, form a basis of neighbourhoods of x. $\mathscr{T}_0$ is metrizable ([62]).

From here on

Any Menger PM-space (X, F, T) will satisfy the condition

(10) $$\lim_{t\to 1-} T(t, a) = a\ \forall\ a \in [0, 1].$$

The condition (10) *implies the condition* (8) ([18]).

10.1 *Let* (X, F, T) *be Menger* PM-*space and* $(x_n)_{n\geq 1}$, $(y_n)_{n\geq 1}$ *sequences from* X. *If* $x_n \xrightarrow{\mathscr{T}_0} x$ and $y_n \xrightarrow{\mathscr{T}_0} y$, *then*

a) $\varliminf_{n\to\infty} F_{x_n y_n}(t) \geq F_{xy}(t)\ \forall\ t \in \mathbf{R}$;

b) *When* F_{xy} *is continuous in* t, *we have*

$$\lim_{n\to\infty} F_{x_n y_n}(t) = F_{xy}(t)$$

([18], [61]).

10.2 *Let* (X, F, T) *be* $\mathcal{T}_0$ *- complete Menger* PM-*space*, $\varphi : X \to \mathbf{R}$ *bounded from below lower semicontinuous and* $\mu > 0$. *Then the binary relation in* X

$$(*) \qquad x \le y \overset{\text{def}}{\Leftrightarrow} F_{xy}(t) \ge H(t - \mu(\varphi(x) - \varphi(y)))\ \forall\ t > 0$$

is an order relation. This has at least one maximal element

(*S. S. Chang - Y. Q. Chen - J. L. Guo*).

■ *First assertion.*

Prove firstly

$$(11) \qquad x \le y \Rightarrow \varphi(y) \le \varphi(x).$$

Suppose *par absurdum* $\varphi(y) > \varphi(x)$. This implies (12) $x \ne y$. (*) gives $F_{xy}(t) \overset{(2)}{\ge} 1\ \forall\ t > 0$, hence $F_{xy}(t) = 1\ \forall\ t > 0$. As $F_{xy}(t) = 0\ \forall\ t \le 0$ ((1)), it results $F_{xy} = H$ and hence $x \overset{(PM_1)}{=} y$, in contradiction with (12).

$\le$ *is reflexive* – use (PM_1).

$\le$ *is antisymmetric*. Let be $x \le y$ and $y \le x$. This implies, *via* (11), $\varphi(x) = \varphi(y)$ and then, taking into account (*), $F_{xy}(t) = 1\ \forall\ t > 0$, consequently, as above, $F_{xy} = H$ and $x = y$.

$\le$ *is transitive*. Let be $x \le y$ and $y \le z$. Then, according to (11),

$$(13) \qquad \varphi(z) \le \varphi(y) \le \varphi(x).$$

Let be $t > 0$ anyone. Show that (*) is verified.

Suppose $t \le \mu(\varphi(x) - \varphi(z))$. Then $H(t - \mu(\varphi(x) - \varphi(z))) \overset{(2)}{=} 0$ and hence $F_{xz}(t) \ge H(t - \mu(\varphi(x) - \varphi(z)))$.

Suppose $t > \mu(\varphi(x) - \varphi(z))$. Take $t_1, t_2 > 0$ so that $t_1 + t_2 = t$ and $t_1 \ge \mu(\varphi(x) - \varphi(y))$, $t_2 \ge \mu(\varphi(y) - \varphi(z))$. We have

$$F_{xz}(t) \overset{(PM_3)}{\ge} T(F_{xy}(t_1), F_{yz}(t_2)) \ge T(H(t_1 - \mu(\varphi(x) - \varphi(y))), H(t_2 - \mu(\varphi(y) - \varphi(z)))) = T(1,1) \overset{(3)}{=} 1 \ge H(t - \mu(\varphi(x) - \varphi(z))).$$

Second assertion.

Use Zorn theorem. Let $(x_\alpha)_{\alpha \in I}$ be a totally ordered part of *X*. We must prove that this is upper bounded. Define in *I* an order relation, we note also this $\le$,

$$(14) \qquad \alpha \le \beta \overset{\text{def}}{\Leftrightarrow} x_\alpha \le x_\beta .$$

$(\varphi(x_\alpha))_{\alpha \in I}$ is a decreasing net in **R** (see (11), (14)), let then be (15) $r := \lim\limits_{\alpha \in I} \varphi(x_\alpha)$. φ being bounded from below, we have $r \in \mathbf{R}$. Prove $(x_\alpha)_{\alpha\in I}$ is a Cauchy net in X. We take into account (9). Let $\varepsilon > 0$ be anyone and λ anyone from $(0,1)$ with $\varepsilon > \mu\lambda$. From (15), $\exists\ \alpha_0$ in I so that

(16) $$\alpha \geq \alpha_0 \Rightarrow r \leq \varphi(x_\alpha) < r + \lambda,$$

hence, taking into account (16) and (11),

$$\alpha_0 \leq \alpha \leq \beta \Rightarrow 0 \leq \varphi(x_\alpha) - \varphi(x_\beta) \leq \lambda$$

and hence, H being increasing,

$$F_{x_\alpha x_\beta}(\varepsilon) \geq H(\varepsilon - \mu(\varphi(x_\alpha)) - \varphi(x_\beta))) \geq H(\varepsilon - \mu\lambda) = 1 > 1 - \lambda,$$

$\alpha_0 \leq \alpha \leq \beta \Rightarrow F_{x_\alpha x_\beta}(\varepsilon) > 1 - \lambda$, $(x_\alpha)_{\alpha\in I}$ is Cauchy net (here we see that if λ does not verify $\varepsilon > \mu\lambda$, we can take $\lambda' < \lambda$ with λ' verifying $1 - \lambda' > 1 - \lambda$). X being $\mathscr{T}_0$-complete, $\exists\ x^*$ in X so that

(17) $$x^* = \lim_{\alpha\in I} x_\alpha .$$

Show

(18) $$x_\alpha \leq x^* \ \forall\ \alpha \in I.$$

As φ is l.s.c. we have

$$\varphi(x^*) \leq \underline{\lim}_{\alpha\in I}\ \varphi(x_\alpha) = \lim_{\alpha\in I} \varphi(x_\alpha) \overset{(15)}{=} r \leq \varphi(x_\alpha),$$

(19) $$\varphi(x^*) \leq \varphi(x_\alpha)\ \forall\ \alpha \in I.$$

Let now α be anyone from I and t anyone in $(0, +\infty)$. According to (*), to obtain (18) we must show

(20) $$F_{x_\alpha x^*}(t) \geq H(t - \mu(\varphi(x_\alpha) - \varphi(x^*))).$$

Suppose $F_{x_\alpha x^*}$ *continuous in* t. Then we have, taking into account the total order (moreover, H is continuous to left),

$$F_{x_\alpha x^*}(t) \overset{10.1}{=} \lim_{\beta\in I} F_{x_\alpha x_\beta}(t) \geq \lim_{\beta\in I} H(t - \mu(\varphi(x_\alpha) - \varphi(x_\beta))) \overset{(19)}{\geq} H(t - \mu(\varphi(x_\alpha) - \varphi(x^*)))$$

and consequently (20).

Suppose $F_{x_\alpha x^*}$ *discontinuous in* t. $F_{x_\alpha x^*}$ being increasing, $\exists\ (t_n)_{n\geq 1}$ strictly increasing sequence of points of continuity of this with $\lim_{n\to\infty} t_n = t$. Then

$$F_{x_\alpha x^*}(t_n) \geq H(t_n - \mu(\varphi(x_\alpha) - \varphi(x^*)))\ \forall\ n \geq 1$$

(the particular case, we can suppose $t_n > 0\ \forall\ n$). Pass to the limit for $n \to \infty$ and as $F_{x_\alpha x^*}$ and H are continuous to left one obtains (20). ■

B. CARISTI - KIRK THEOREM IN PROBABILISTIC METRIC SPACES

Let (X, F, T) be a $\mathcal{T}_0$ - complete Menger PM-space having the property (10).

10.3 Theorem. *Let be* $f: E \to X$ *surjective*, $E \subset X$, $\varphi: X \to \mathbf{R}$ *bounded from below lower semicontinuous*, $(\Phi_\alpha)_{\alpha\in I}$, $\Phi_\alpha : E \to 2_*^X$ *and* $\mu > 0$.

Suppose $\forall\ x \in E$ *with* $f(x) \notin \bigcap_{\alpha\in I}\Phi_\alpha(x)$ *there exists* α_0 *in* I *and* y *in* $\Phi_{\alpha_0}(x) \setminus \{f(x)\}$ *with the property*

(21) $$F_{f(x),y}(t) \geq H(t - \mu(\varphi(f(x)) - \varphi(y)))\ \forall\ t > 0.$$

Then $\exists$ u *in* E *so that*

(22) $$f(u) \in \bigcap_{\alpha\in I}\Phi_\alpha(u)$$

(*S. S. Chang - I. J. Cho - J. K. Kim*).

Explanation. 2_*^X is the set of nonempty subsets of X.

■ Consider the order relation in X (see (*))

$$x \leq y \overset{\text{def}}{\Leftrightarrow} F_{xy}(t) \geq H(t- \mu(\varphi(x) - \varphi(y)))\ \forall\ t > 0.$$

X has a maximal element v (10.2). f being surjective, $\exists\ u$ in X so that (23) $v = f(u)$. Suppose *par absurdum* (24) $f(u) \notin \bigcap_{\alpha\in I}\Phi_\alpha(u)$. Then, according to the hypothesis, there exists α_0 in I and y in $\Phi_{\alpha_0}(u) \setminus \{f(u)\}$ so that

$$F_{f(u),y}(t) \geq H(t - \mu(\varphi(f(u)) - \varphi(y)))\ \forall\ t > 0,$$

i.e. $f(u) \leq y$. (23) imposes $f(u) = y$ and hence $f(u) \in \Phi_{\alpha_0}(u) \setminus \{f(u)\}$, contradiction and (24) is false. ■

10.4 *Corollary* ([17]). *Let be* f, φ, μ *as in* 10.3 *and* $\Phi: E \to 2_*^X$. *Suppose that* $\forall$ x *from*

E *with* $f(x) \notin \Phi(x)$ *there exists* y *in* $\Phi(x) \setminus \{f(x)\}$ *with the property*

$$F_{f(x),\,y}(t) \geq H(t - \mu(\varphi(f(x)) - \varphi(y)))\ \forall\ t > 0.$$

Then there is u *in* E *so that*

$$f(u) \in \Phi(u).$$

■ Take in 10.3 $\Phi_\alpha = \Phi\ \forall \alpha \in I$. ■

REMARK. In Corollary 3.2, [17], there is a mistake, we must have $y \in S(x) \setminus \{f(x)\}$ (notation from the original).

10.5 *Corollary. Let* $\varphi : X \to \mathbf{R}$ *be bounded from below l. s. c. If* $\Phi : X \to X$ *satisfies*

(25) $$F_{x,\,\Phi(x)}(t) \geq H(t - \mu(\varphi(x) - \varphi(\Phi(x))))\ \forall\ x \in X,\ \forall\ t > 0$$

then Φ *has a fixed point.*

■ Demonstration of Cantorian type. Suppose *par absurdum* (26) $x \neq \Phi(x)\ \forall\ x \in X$. Take $f = \mathrm{Id}_X$ and apply 10.4, this is correct according to (26) and (25). One obtains Φ has a fixed point, contradiction. ■

10.6 *Corollary* ([17]). *Let* (X, d) *be complete metric space,* $\varphi : X \to \mathbf{R}$ *bounded from below l. s. c.,* $\Phi : X \to 2^X_*$, $f : X \to X$ *surjective and* $\mu > 0$. *Suppose* $\forall$ x *from* X $\exists$ y *in* $\Phi(x) \setminus \{f(x)\}$ *having the property*

(27) $$d(f(x), y) \leq \mu(\varphi(f(x)) - \varphi(y)).$$

Then there exists u *in* X *so that*

$$f(u) \in \Phi(u).$$

■ The application $T_1 : [0, 1] \times [0, 1] \to [0, 1]$, $T_1(x, y) = \min\{x, y\}$ is *t*-norm and the application $F : X \times X \to \mathcal{D}$,

(28) $$F_{xy}(t) = H(t - d(x, y))\ \forall\ t \in \mathbf{R}$$

verifies (PM_1), (PM_2) (see (7)), but also (PM_3) relative to T_1, that is $\forall\ x, y, z \in X$ and $\forall\ t_1, t_2 \in \mathbf{R}$

(29) $$H(t_1 + t_2 - d(x, z)) \geq \min\{H(t_1 - d(x, y)), H(t_2 - d(y, z))\}.$$

When the first member takes the value 1, (29) is verified. Suppose that it takes the value 0, i.e.

(30) $$t_1 + t_2 \leq d(x, z).$$

If the second member is equal to 0, (29) is verified. Suppose *par absurdum* that it is equal to 1, that is $t_1 > d(x, y)$ and $t_2 > d(y, z)$. We add, $t_1 + t_2 > d(x, y) + d(y, z) \geq d(x, z)$, in contradiction with (30).

So (X, F, T_1) is a $\mathscr{T}_0$- complete Menger PM-space (attention to d !). φ remains l.s.c. But (27) and (28) give $\forall\, x$ of X $\exists\, y$ in $\Phi(x) \setminus \{f(x)\}$ so that

$$F_{f(x),\,y}(t) \ge H(t - \mu(\varphi(f(x)) - \varphi(y)))\ \forall\, t > 0.$$

Apply 10.4 with $E = X$ and one obtains the conclusion. ■

REMARK. *Mistakes* in [17]: φ must be *bounded from below*; $y \in S(x) \setminus \{f(x)\}$, *not only* $y \in S(x)$.

10.7 *Caristi - Kirk theorem* (§2, B) *is consequence of* 10.6.

■ Preserve the notations from §2, B. Supose *par absurdum* $x \notin T(x)$, i.e. $T(x) = T(x) \setminus \setminus \{x\}$, $\forall\, x \in X$, consequently the hypothesis affirms in fact that $\forall\, x$ of X $\exists\, y$ in $T(x) \setminus \{x\}$ so that

$$d(x, y) \le \varphi(x) - \varphi(y).$$

Apply 10.6 with $f = \mathrm{Id}_X$ and $\mu = 1$, $\exists\, u$ in X so that $u \in T(u)$, contradiction and hence the conclusion. ■

Remark 1. The proof for 10.7, and also for 10.5, is of Cantorian type.

Remark 2. 10.7 permits to assert that Theorem 10.3 is a probabilistic variant of Caristi - Kirk theorem.

C. EKELAND PRINCIPLE IN PROBABILISTIC METRIC SPACES

Let (X, F, T) be a $\mathscr{T}_0$ - complete Menger PM-space having the property (10).

10.8 Theorem (*Probabilistic variant of Ekeland principle*, [17]). *Let* $\varphi : \mathrm{X} \to \mathbf{R}$ *be bounded from below lower semicontinuous and* $\mu > 0$.

For any $\varepsilon > 0$ *and* u *from* X *with*

$$\varphi(\mathrm{u}) \le \inf \varphi(\mathrm{X}) + \varepsilon, \tag{31}$$

there exists v_ε **-maximal between the *-majorants of* u *so that*

$$\mathrm{F}_{\mathrm{uv}_\varepsilon}(\mathrm{t}) \ge \mathrm{H}(\mathrm{t} - \varepsilon\mu)\ \forall\, \mathrm{t} > 0 \tag{32}$$

and $\forall$ x *from* $\mathrm{X} \setminus \{\mathrm{v}_\varepsilon\}$ *there exists* $\mathrm{t}_\mathrm{x} > 0$ *so that*

$$\mathrm{F}_{\mathrm{v}_\varepsilon \mathrm{x}}(\mathrm{t}_\mathrm{x}) < \mathrm{H}(\mathrm{t}_\mathrm{x} - \mu(\varphi(\mathrm{v}_\varepsilon) - \varphi(\mathrm{x}))). \tag{33}$$

■ *First assertion.*

Consider the set

$$(34) \qquad X_u := \{x \in X: F_{ux}(t) \geq H(t - \mu(\varphi(u) - \varphi(x)))\ \forall\, t > 0\}.$$

$X_u \neq \varnothing$ because $u \in X_u$ ((PM$_1$)). Prove

$$(35) \qquad X_u \text{ is closed.}$$

Let $(x_n)_{n\geq 1}$ be sequence of points from X_u with (36) $x_n \xrightarrow{\mathcal{T}_0} x^*$. For obtaining (35), we must show $x^* \in X_u$, that is

$$(37) \qquad F_{ux^*}(t) \geq H(t - \mu(\varphi(u) - \varphi(x^*)))\ \forall\, t > 0.$$

Fix $t > 0$.

Suppose F_{ux^*} *continuous in* t. We have, $\forall\, n \geq 1$,

$$F_{ux_n}(t) \geq H(t - \mu(\varphi(u) - \varphi(x_n))).$$

Pass to the limit for $n \to \infty$, taking into account (36), 10.1 and the lower semicontinuity of φ (H is increasing),

$$F_{ux^*}(t) \geq \underline{\lim}_{n\to\infty} H(t - \mu(\varphi(u) - \varphi(x_n))) \geq H(\underline{\lim}_{n\to\infty} (t - \mu(\varphi(u) - \varphi(x_n)))) \geq H(t - \mu(\varphi(u) - \varphi(x^*)))$$

and hence (37) [justify the relation $\underline{\lim}_{n\to\infty} H(y_n) \geq H(\underline{\lim}_{n\to\infty} y_n)$: $l := \underline{\lim}_{n\to\infty} y_n$, $l > 0$ ($H(l) = 1$, $n \geq N \Rightarrow y_n > 0 \Rightarrow H(y_n) = 1 \to 1$), $l \leq 0$ ($H(l) = 0$, $H(y_n) \geq 0\ \forall\, n \geq 1$, $\underline{\lim}_{n\to\infty} H(y_n) \geq 0$)].

Suppose F_{ux^*} *discontinuous in* t. Take a strictly increasing sequence $(t_n)_{n\geq 1}$ of continuity points of F_{ux^*} with $t_n \to t$ and work as for 10.2, (35) is proved.

So (X_u, F, T) is a $\mathcal{T}_0$ - complete Menger PM-space. Consequently, according to 10.2, the binary relation in X_u

$$x \leq y \overset{\text{def}}{\Leftrightarrow} F_{xy}(t) \geq H(t - \mu(\varphi(x) - \varphi(y)))\ \forall\, t > 0,$$

obvious restriction to X_u of the order relation (*) in X, is order relation and it has a maximal element, let it be v_ε, and the elements from X_u are *-majorants of u. Moreover, according to (31),

$$(38) \qquad 0 \leq \varphi(u) - \varphi(v_\varepsilon) \leq \varepsilon$$

and combining (34) with (38) one obtains (32), H being increasing.

Second assertion.

Suppose *par absurdum*

(40) $\exists\, x_0 \in X \setminus \{v_\varepsilon\}$ so that $F_{v_\varepsilon x_0}(t) \geq H(t - \mu(\varphi(v_\varepsilon) - \varphi(x_0)))\ \forall\, t > 0$.

But (40) means $x_0 \geq v_\varepsilon$ and because $v_\varepsilon \geq u$, it results $x_0 \overset{(34)}{\in} X_u$, which imposes $x_0 = v_\varepsilon$, since v_ε is maximal in X_u, in contradiction with (40). ■

REMARK. Theorem 10.8 is Theorem 4.1 from [17] with *the statement changed* and with *the demonstration corrected* by the author: (1) and (2) are without value, for $u = x_0$ both relations are verified because H is increasing; the justification for (3) is false, since the opposite of the assertion (3) is not what one affirms to be.

Retain an abbreviated statement for Theorem 10.8, (X, F, T) is $\mathscr{T}_0$ - complete Menger PM-space.

10.8_1 Theorem. *If* $\varphi : X \to \mathbf{R}$ *is bounded from below l. s. c. and* $\mu > 0$, *there exists* v *in* X *with the property*:

$$\forall\, x \in X \setminus \{v\}\ \exists\, t_x > 0 \text{ so that } F_{vx}(t_x) < H(t_x - \mu(\varphi(v) - \varphi(x))).$$

■ Suppose *par absurdum* the contrary, i. e. $\forall\, x$ of $X\ \exists\, y$ in $X \setminus \{x\}$ so that $\forall\, t > 0$ we have $F_{xy}(t) \geq H(t - \mu(\varphi(x) - \varphi(y)))$. Consider the application $\Phi : X \to X$, $\Phi(x) = y \neq x$, where y verifies the last relation. Apply 10.5, Φ has a fixed point x_0, hence $\Phi(x_0) = x_0 \neq$ $\neq x_0$, contradiction. ■

10.9 *Theorem* 10.8_1 *and Theorem* 10.3 *are equivalent.*

■ *Th.* $10.8_1 \Rightarrow$ *Th.*10.3. The hypothesis supplies v in X having the property $\forall\, x \in X \setminus$ $\setminus \{v\}\ \exists\, t_x > 0$ so that

(40_1) $$F_{vx}(t_x) < H(t_x - \mu(\varphi(v) - \varphi(x))).$$

f being surjective, $\exists\, x_0$ in E with $f(x_0) = v$. If $f(x_0) \in \bigcap_{\alpha\in I} \Phi_\alpha(x_0)$, the proof is finished.

Suppose $f(x_0) \notin \bigcap_{\alpha\in I} \Phi_\alpha(x_0)$. The hypothesis of 10.3 supplies α_0 and y_0, (41) $y_0 \in \Phi_\alpha(x_0) \setminus$ $\setminus \{f(x_0)\} = \Phi_\alpha(x_0) \setminus \{v\}$, so that (see (21), $f(x_0) = v$)

(42) $$F_{vy_0}(t) \geq H(t - \mu(\varphi(v) - \varphi(y_0)))\ \forall\, t > 0.$$

But $y_0 \overset{(41)}{\neq} v$ and comparing (40_1), where $x = y_0$, with (42), where $t = t_x$, the contradiction is obtained and $f(x_0) \in \bigcap_{\alpha\in I} \Phi_\alpha(x_0)$. ■

Th. $10.3 \Rightarrow$ *Th.* 10.8_1. Indeed, because $10.3 \Rightarrow 10.4 \Rightarrow 10.5$ and 10.5 has been used to prove 10.8_1. ■

REMARK. The equivalence Th $10.8_1 \Leftrightarrow$ Th. 10.3 was observed by the author. The equivalence theorem 5.1, [17], i.e. $3.1 \Leftrightarrow 4.1$, is false (see REMARK to 10.8).

Some more things about [17]. The obsessive presence of the function $k : (0,1) \to \to (0, +\infty)$, in Lemma 2.2 and in the sections **3** and **4**, is incomprehensible, *it appears nowhere*, M is simply a real strictly positive number.

It follows a sketch which visualizes the established interdependences between the statements.

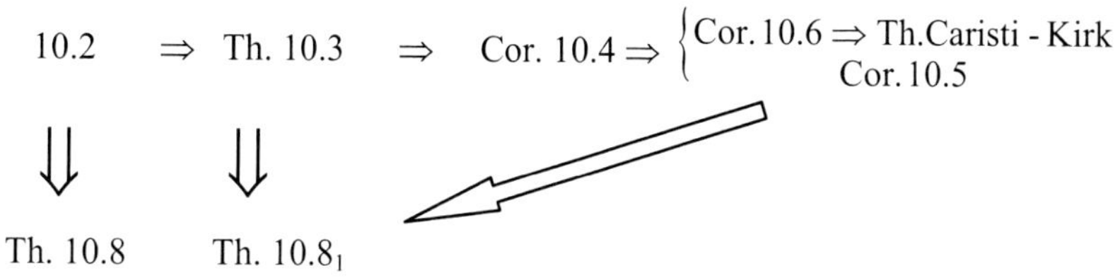

One still deduces from the sketch:

Th. 10.3, *Cor.*10.4, *Cor.*10.5 *and Th.* 10.8_1 *are equivalents.*

We finish the section with

10.10 *Theorem* 10.8 *is indeed a variant of the abbreviated Ekeland principle for Menger* PM-*spaces.*

■ The abbreviated Ekeland principle is equivalent to the Caristi - Kirk theorem (§2, 2.4), for the rest – consult the sketch. ■

Chapter II

Smooth perturbed variational principles

1

Borwein - Preiss principle

This is a generalization of Ekeland principle Ch.I, 1.1, perturbing function is smooth in the case of a Banach space. We follow [6].

All the Banach spaces are real.

Begin the preliminaries. Let X be a real normed space. For any p from $[1, +\infty)$ consider the set Σ_p of the functions $\Delta_p: X \to \mathbf{R}$,

$$(1) \qquad \Delta_p(x) = \sum_{n=0}^{\infty} \mu_n \|x - v_n\|^p, \mu_n \geq 0, \sum_{n=0}^{\infty} \mu_n = 1, \lim_{n\to\infty} v_n = v,$$

a convergent series since we have, $\forall n \geq 0$, $\left|\mu_n \|x - v_n\|^p\right| \leq M^p \mu_n$, $M > 0$. Moreover,

1.1 *Let* β *be bornology on* X (I, §3). *If* $p > 1$ *and* ν: $\nu(x) = \|x\|$ *is* β *- differentiable on* $X \setminus \{0\}$, *then* Δ_p *is* β *- differentiable on* X.

■ Let be ν_a : $\nu_a(x) = \|x - a\|^p$. We have

$$(2) \qquad \begin{cases} x \neq a \Rightarrow \nabla_\beta \nu_a(x) = p\|x - a\|^{p-1} \nabla_\beta \nu(x - a), \\ x = a \Rightarrow \nabla_\beta \nu_a(a) = 0, \end{cases}$$

Indeed, for the first relation apply the formula from 3.11, Ch.I and for the second – the definition (I, §3, "β - derivative"): $\lim_{\substack{t\to 0+ \\ v\in S}} \frac{\nu_a(a + tv) - \nu_a(a)}{t} = \lim_{\substack{t\to 0+ \\ v\in S}} \frac{t^p}{t} \|v\|^p = 0$, as S is bounded.

From (2), the series of β - derivatives of the terms from $\Delta_p(x)$ is

(3) $$p\sum_{n=0}^{\infty}\mu_n\|x-v_n\|^{p-1}\nabla_\beta v(x-v_{n)}\,,$$

where for $x = v_n$ corresponding term is equal to 0 ((2)). As for $x \neq v_n$ we have $\|\nabla_\beta\, v(x - - v_n)\| = 1$ (I, 5.7$_1$), the series from (3) is uniformly convergent on any open ball σ of X (Weierstrass rule), consequently Δ_p is β - differentiable on σ (I, §3, 3.12) and hence on X. ■

And now

1.2 Borwein - Preiss principle (smooth variational principle). *Let* X *be Banach space,* $\varphi : X \to (-\infty, +\infty]$ *bounded from below, lower semicontinuous and proper. For any* $\varepsilon > 0$, $\lambda > 0$, $p \geq 1$ *and* x_0 *of* X *with*

(I) $$\varphi(x_0) < \inf\varphi(X) + \varepsilon,$$

there exists Δ_p *in* Σ_p *and* v_ε *in* X *so that*

(II) $$\varphi(v_\varepsilon)+\frac{\varepsilon}{\lambda^p}\Delta_p(v_\varepsilon)\leq\varphi(x)+\frac{\varepsilon}{\lambda^p}\Delta_p(x)\ \ \forall x\in X\,,$$

(III) $\varphi(v_\varepsilon) < \inf\varphi(X) + \varepsilon$, (IV) $\|v_\varepsilon - x_0\| < \lambda$.

Moreover, if the norm of X *is* β *- differentiable on* $X \setminus \{0\}$ *and* $p > 1$, *then*

(V) $$0\in\partial_\beta\varphi(v_\varepsilon)+p(\frac{\varepsilon}{\lambda}S^*)\,,\ S^*:=\{\xi\in X^*:\|\xi\|\leq 1\}.$$

Δ_p is called *perturbing function.*

■ (I) gives $0 \leq \varphi(x_0) - \inf\varphi(X) < \varepsilon$. Take ε_1, ε_2 so that

(4) $$\varphi(x_0) - \inf\varphi(X) < \varepsilon_2 < \varepsilon_1 < \varepsilon$$

and then μ with

(5) $$0 < \mu < 1 - \frac{\varepsilon_1}{\varepsilon},$$

θ with

(6) $$0<\frac{\theta}{\mu}<\left[1-\left(\frac{\varepsilon_2}{\varepsilon_1}\right)^{\frac{1}{p}}\right]^p$$

and finally

(7) $$\delta:=\frac{(1-\mu)\varepsilon}{\lambda^p}.$$

We have $\theta \in (0,1)$ as $\mu\left[1-\left(\frac{\varepsilon_2}{\varepsilon_1}\right)^{\frac{1}{p}}\right]^p < 1$. The series of Δ_p is defined by recurrence:

(8) $$v_0 := x_0,\ \varphi_0 := \varphi.$$

v_n and φ_n defined $\Rightarrow$ $\begin{cases}(9)\ \varphi_{n+1}(x) = \varphi_n(x) + \delta\mu^n\|x - v_n\|^p,\\ (10)\ v_{n+1} : \varphi_{n+1}(v_{n+1}) \le \theta\varphi_n(v_n) + (1-\theta)\inf \varphi_{n+1}(X).\end{cases}$

The definition is correct: on one hand φ_{n+1} is bounded from below and on the other hand v_{n+1} from (10) may be chosen so that the inequality is verified. Indeed, because $x_0 \in \mathrm{dom} f$ (see (I)), we have $\varphi_{n+1}(v_n) \overset{(9)}{=} \varphi_n(v_n) \in \mathbf{R}$. (10) becomes

(11) $$\varphi_{n+1}(v_{n+1}) \le \theta\varphi_{n+1}(v_n) + (1-\theta)\inf \varphi_{n+1}(X),$$

hence

(12) $$\inf \varphi_{n+1}(X) \le \theta\varphi_{n+1}(v_n) + (1-\theta)\inf \varphi_{n+1}(X).$$

When we have equality in (12), i. e. $\inf \varphi_{n+1}(X) = \varphi_{n+1}(v_n)$ ($\theta > 0$!), take $v_{n+1} = v_n$ and (10) is verified. In the contrary case $\exists\ v_{n+1}$ in X so that

$$\inf \varphi_{n+1}(X) \le \varphi_{n+1}(v_{n+1}) < \theta\, \varphi_{n+1}(v_n) + (1-\theta)\inf \varphi_{n+1}(X),$$

(10) is also verified. Retain that $v_n \in \mathrm{dom}\ \varphi_n\ \forall\ n \ge 0$. Moreover, $(\varphi_n)_{n\ge 0}$ is increasing ((9)) and, $\forall\ n \ge 0$, φ_n is l.s.c. and $\mathrm{dom}\ \varphi_n \overset{(9)}{=} \mathrm{dom}\ \varphi$.

In the hypothesis $x \in \mathrm{dom}\ \varphi$ replace in (9) n by k, $k = 0, 1, \ldots, n$ and add,

(13) $$\varphi_{n+1}(x) = \varphi(x) + \delta\sum_{k=0}^{n}\mu^k\|x - v_k\|^p,\ \forall n \ge 0.$$

Set $a_n := \varphi_n(v_n)$, $b_n := \inf \varphi_n(X)$, $n \ge 0$. We have

(14) $$b_n \le b_{n+1} \le a_{n+1} \overset{(10)}{\le} \theta a_n + (1-\theta)b_{n+1} \le a_n\ \forall n \ge 0$$

(for the last inequality: $b_{n+1} \le \varphi_{n+1}(v_n) = \varphi_n(v_n)$).

As $-b_{n+1} \le -b_n$, (14) gives $a_{n+1} - b_{n+1} \le \theta(a_n - b_n)$ and by repetition

(15) $$0 \le a_{n+1} - b_{n+1} \le \theta^{n+1}(a_0 - b_0),\ n \ge 0.$$

Take $x = v_{n+1}$ in (9), one obtains ($\varphi_n \le \varphi_{n+1}$!) $a_n \ge a_{n+1} = \varphi_n(v_{n+1}) + \delta\mu^n\|v_{n+1} - v_n\|^p \ge b_n +$ $+ \delta\mu^n\|v_{n+1} - v_n\|^p$, hence $a_n - b_n \ge \delta\mu^n\|v_{n+1} - v_n\|^p$ and combining with (15), (16) $\delta\mu^n \times$ $\times \|v_{n+1} - v_n\|^p \le \theta^n(a_0 - b_0) \overset{(4)}{\le} \theta^n\varepsilon_2$.

(16) gives $\|v_{n+1} - v_n\| \le \left(\frac{\varepsilon_2}{\delta}\right)^{\frac{1}{p}} \left(\frac{\theta}{\mu}\right)^{\frac{n}{p}}$, hence $m \ge n \Rightarrow \|v_m - v_n\| \le \left(\frac{\varepsilon_2}{\delta}\right)^{\frac{1}{p}} \times$

$\times \left[\left(\frac{\theta}{\mu}\right)^{\frac{n}{p}} + \ldots + \left(\frac{\theta}{\mu}\right)^{\frac{m-1}{p}}\right]$, hence $(v_n)_{n\ge 0}$ is Cauchy sequence as $\sum_{n=0}^{\infty}\left(\frac{\theta}{\mu}\right)^{\frac{n}{p}} = \frac{1}{1-\left(\frac{\theta}{\mu}\right)^{\frac{1}{p}}}$.

Moreover,

$$\|v_m - v_n\| \le \left(\frac{\varepsilon_2}{\delta}\right)^{\frac{1}{p}} \frac{1}{1-\left(\frac{\theta}{\mu}\right)^{\frac{1}{p}}} \overset{(6)}{\le} \left(\frac{\varepsilon_1}{\delta}\right)^{\frac{1}{p}} \tag{17}$$

and hence also

$$\|v_m - v_n\| \le \left(\frac{\varepsilon_1}{\delta}\right)^{\frac{1}{p}} \overset{(5),(7)}{<} \left(\frac{\varepsilon}{\delta}\right)^{\frac{1}{p}} (1-\mu)^{\frac{1}{p}} = \lambda. \tag{18}$$

Let be $v_\varepsilon := \lim_{n\to\infty} v_n$ and, $\forall\, x \in X$, $\Delta_p(x) = \sum_{n=0}^{\infty} \mu_n \|x - v_n\|^p$, $\mu_n := \mu^n(1-\mu)$. Obviously $\Delta_p \in \Sigma_p$.

Prove (II). (II) is obviously verified if $\varphi(x) = +\infty$. Suppose $x \in \operatorname{dom}\varphi$. (13) and (7) give (the series is convergent)

$$\varphi(x) + \frac{\varepsilon}{\lambda^p}\Delta_p(x) = \lim_{n\to\infty} \varphi_n(x). \tag{19}$$

But $\varphi_n(x) \ge b_n$, hence ($\exists$ limit, see(14)) $\lim_{n\to\infty} \varphi_n(x) \ge \lim_{n\to\infty} b_n$ and $\lim_{n\to\infty} b_n = \lim_{n\to\infty} \varphi_n(v_n)$ (use (15), $\theta \in (0, 1)$), consequently

$$\varphi(x) + \frac{\varepsilon}{\lambda^p}\Delta_p(x) \ge \lim_{n\to\infty} \varphi_n(v_n). \tag{20}$$

But $\varphi_n(v_n) \ge \varphi_m(v_n)$ $\forall\, n \ge m$, hence $\forall\, m$ $\lim_{n\to\infty} \varphi_n(v_n) \ge \underline{\lim}_{n\to\infty} \varphi_m(v_n) \ge \varphi_m(v_\varepsilon)$ (the lower semicontinuity) and hence

$$\lim_{n\to\infty} \varphi_n(v_n) \ge \sup_{m\in\mathbf{N}} \varphi_m(v_\varepsilon) \overset{(13),(7)}{=} \varphi(v_\varepsilon) + \frac{\varepsilon}{\lambda^p}\Delta_p(v_\varepsilon)$$

and by confrontation with (20) one obtains (II).

Prove (III). $\Delta_p(x_0) \overset{(8)}{=} \sum_{n=1}^{\infty} \mu_n \|v_0 - v_n\|^p \overset{(18)}{\leq} \frac{\varepsilon_1}{\delta} \sum_{n=1}^{\infty} \mu_n = \frac{\varepsilon_1}{\delta}\mu \overset{(18)}{<} \lambda^p\mu$, hence

$$\varphi(x_0) + \varepsilon\mu > \varphi(x_0) + \frac{\varepsilon}{\lambda^p}\Delta_p(x_0) \overset{(II)}{\geq} \varphi(v_\varepsilon),$$

consequently

$$\varphi(v_\varepsilon) \overset{(4)}{<} \inf\varphi(X) + \varepsilon_1 + \varepsilon\mu \overset{(5)}{<} \inf\varphi(X) + \varepsilon.$$

Prove (IV). (18) gives $\lim_{m\to\infty} \|v_m - x_0\| \leq \left(\frac{\varepsilon_1}{\delta}\right)^{\frac{1}{p}} < \lambda$, $\|v_\varepsilon - x_0\| < \lambda$.

Prove (V). The demand is correct, $v_\varepsilon \in$ dom φ (see (III)). Let be $\eta > 0$ and S from β anyone. Δ_p being β - differentiable in v_ε (1.1), it is also β - superdifferentiable in v_ε (I, 5.10_1, Remark), hence for $\eta_1 := \frac{\lambda^p}{\varepsilon}\eta$ $\exists\, \rho > 0$ so that

$$\text{(21)} \qquad t \in (0, \rho) \Rightarrow \frac{\Delta_p(v_\varepsilon + tv) - \Delta_p(v_\varepsilon)}{t} - \xi_1(v) \leq \eta_1 \ \forall\, v \in S,\ \xi_1 := \nabla_\beta\Delta_p(v_\varepsilon).$$

Using (21) and by means of (II), one obtains

$$t \in (0, \rho) \Rightarrow \frac{\varphi(v_\varepsilon + tv) - \varphi(v_\varepsilon)}{t} - \xi(v) \geq -\eta \ \forall\, v \in S,$$

where $\xi = -\frac{\varepsilon}{\lambda^p}\xi_1 = -\frac{\varepsilon}{\lambda^p}\nabla_\beta\Delta_p(v_\varepsilon)$, hence $-\frac{\varepsilon}{\lambda^p}\nabla_\beta\Delta_p(v_\varepsilon) \in \partial_\beta\varphi(v_\varepsilon)$.

But

$$\nabla_\beta\Delta_p(v_\varepsilon) \overset{(3)}{=} p\sum_{n=0}^{\infty}\mu_n \|v_\varepsilon - v_n\|^{p-1}\nabla_\beta \nu(v_\varepsilon - v_n),$$

hence $\|\nabla_\beta\Delta_p(v_\varepsilon)\| \leq p\lambda^{p-1}$ ((18), I, 5.7_1) and hence $\left\|\frac{\varepsilon}{\lambda^p}\nabla_\beta\Delta_p(v_\varepsilon)\right\| \leq \frac{p\varepsilon}{\lambda}$ and consequently (V). ■

R*emark.* Condition (I) from 1.2 makes the demand "f bounded from below" tautological. The same (I) implies "f proper ".

1.2_1 Borwein-Preiss principle (complete metric space). *Let* (X, d) *be complete metric space,* $\varphi : X \to (-\infty, +\infty]$ *bounded from below l. s. c. proper. For any* $\varepsilon > 0$, $\lambda > 0$, $p \geq 1$ *and* x_0 *of* X *with*

$$\text{(I)} \qquad \varphi(x_0) < \inf\varphi(X) + \varepsilon,$$

there exists Δ_p *in* Σ_p *and* v_ε *in* X *so that*

(II) $$\varphi(v_\varepsilon) + \frac{\varepsilon}{\lambda^p}\Delta_p(v_\varepsilon) \le \varphi(x) + \frac{\varepsilon}{\lambda^p}\Delta_p(x)\ \forall x \in X,$$

(III) $\varphi(v_\varepsilon) < \inf \varphi(x) + \varepsilon,$ (IV) $d(v_\varepsilon, x_0) < \lambda.$

Explanation.

$$\Sigma_p := \{\Delta_p : X \to \mathbf{R} : \Delta_p(x) = \sum_{n=0}^{\infty} \mu_n\, d^p(x, v_n),\ \mu_n \ge 0,\ \sum_{n=0}^{\infty} \mu_n = 1,\ (v_n)_{n\ge 0} \text{ convergent sequence}\}.$$

■ Exactly the same demonstration as for 1.2, but replacing $\|x - y\|$ by $d(x, y)$ everywhere. ■

REMARK 1. There is *almost* coincidence between Borwein - Preiss principle for $p = 1$ and Ekeland principle I, 1.1. Indeed, the demand (I) is strict, while at 1.1 is not so. Then (II) gives

$$\varphi(v_\varepsilon) \le \varphi(x) + \frac{\varepsilon}{\lambda}[\Delta_1(x) - \Delta_1(v_\varepsilon)]\ \forall x \in X,$$

and

$$\Delta_1(x) - \Delta_1(v_\varepsilon) = \sum_{n=0}^{\infty} \mu_n\ [d(x, v_n) - d(v_\varepsilon, v_n)] \le d(x, v_\varepsilon)\sum_{n=0}^{\infty}\mu_n = d(x, v_\varepsilon),\ \varphi(v_\varepsilon) \le \varphi(x) + \frac{\varepsilon}{\lambda} d(v_\varepsilon, x)\ \forall x \in X,$$

which is (1), but weaker. (III) cannot give (2), while this can. Finally, IV gives (3), this is weaker.

Remark 2. A simpler demonstration of the B - P principle, and which improves its statement, results from the proof of 4.1.

Remark 3. In the case of a Hilbert space, perturbing function Δ_2 can be taken in the form

$$\Delta_2(x) = \|x - w\|^2.$$

Here is an immediate consequence of B - P principle (J. Borwein - D. Preiss).

1.3. *Let* X *be Banach space with equivalent norm* β *- differentiable,* β *bornology on* X, *and* $f : X \to (-\infty, +\infty]$ *l.s.c. proper and with the property* $f(x) \ge \gamma(\|x\|)$, $\gamma : [0, +\infty) \to \mathbf{R}$ *continuous and* $\lim_{\|x\|\to+\infty} \frac{\gamma(\|x\|)}{\|x\|} = +\infty.$

Then the value set of the multivalued function $v \to \partial_\beta f(v)$ *is dense in* X^*.

■ Let φ be from X^* and $\varepsilon > 0$ anyone. We must find w in X and ξ in $\partial_\beta f(w)$ so that

$$\|\varphi - \xi\| \le \varepsilon. \tag{22}$$

Note by $\|\cdot\|_1$ the β - differentiable equivalent norm on X and in the same manner the corresponding norm on X^*. As $\|\cdot\|$ on X^* and $\|\cdot\|_1$ on X^* are equivalent, (22) in fact reverts to the proof of (except a constant factor)

$$\|\varphi - \xi\|_1 \le \varepsilon. \tag{23}$$

Let be $g := f - \varphi$, obvious l. s. c. and proper function. g is also bounded from below. Indeed, $g(x) \ge \gamma(\|x\|) - \|\varphi\|\,\|x\| := h(\|x\|)$. $h := [0, +\infty) \to \mathbf{R}$ is continuous and (24) $\lim\limits_{\|x\|\to+\infty} \dfrac{h(\|x\|)}{\|x\|} = +\infty$. Suppose *par absurdum* h not bounded from below. Then $\forall\, n$ of $\mathbf{N}\ \exists\, x_n$ in X so that (25) $h(\|x_n\|) \le -n$. If the sequence $(\|x_n\|)_{n\ge1}$ is not upper bounded, it has a subsequence, denote this identically, with the limit $+\infty$ and then on one hand (25) gives $\dfrac{h(\|x_n\|)}{\|x_n\|} \le -\dfrac{n}{\|x_n\|} \le \dfrac{1}{\|x_n\|}$, hence $\overline{\lim\limits_{n\to\infty}} \dfrac{h(\|x_n\|)}{\|x_n\|} \le 0$ and on the other hand $\lim\limits_{n\to\infty} \dfrac{h(\|x_n\|)}{\|x_n\|} \overset{(24)}{=} +\infty$, contradiction. In the contrary case, i.e. $(\|x_n\|)_{n\ge1}$ upper bounded, this has a convergent subsequence, denote this identically, $\|x_n\| \to \rho \ge 0$. h being continuous, (25) gives $h(\rho) = -\infty$, once again a contradiction, hence h is bounded from below and consequently also g.

Consider X with $\|\cdot\|_1$. g being proper, we have $\inf g(X) < +\infty$, hence $\exists\, x_0$ in X so that $g(x_0) < \inf g(X) + \varepsilon$. Apply 1.2, for $\lambda = p = 2\ \exists\, v_\varepsilon$ in X so that $0 \in \partial_\beta g(v_\varepsilon) + \varepsilon S^*$, S^* closed unit ball with respect to $\|\cdot\|_1$. Let μ be in $\partial_\beta g(v_\varepsilon)$ and ν in S^* so that $\mu + \varepsilon\nu = 0$. So (26) $\|\mu\|_1 \le \varepsilon$. But

$$\mu \in \partial_\beta f(v_\varepsilon) - \varphi \tag{27}$$

and (23) results from (26) and (27). ■

Application to the β-subderivative

1.4 *Let* X *be Banach space with* β *- differentiable equivalent norm,* β *bornology on* X*, and* $f : X \to (-\infty, +\infty]$ *l. s. c. with* $\operatorname{dom} f \ne \varnothing$*. Then* f *is* β *- subdifferentiable on a subset* A *of* $\operatorname{dom} f$ *with the property* $\{(x, f(x)) : x \in A\}$ *is dense in the graph of* f (*J. Borwein -*

D. Preiss).

■ Let x_0 be from dom f and $\varepsilon > 0$ abitrary. Take λ from $(0, \varepsilon)$ so that

$$\inf\{f(x) : \|x - x_0\| < \lambda\} > f(x_0) - \varepsilon, \tag{28}$$

which is possibly: take ε_1 in $(0, \varepsilon)$, then $f(x) > f(x_0) - \varepsilon_1$ on an open ball $\|x - x_0\| < \lambda$ with $\lambda < \varepsilon$ (lower semicontinuity in x_0), consequently

$$\inf_{x \in S(x_0, \lambda)} f(x) \geq f(x_0) - \varepsilon_1 > f(x_0) - \varepsilon.$$

Consider the function $f_\lambda := f + \chi_\lambda$, $\chi_\lambda(x) = 0$ for $\|x - x_0\| < \lambda$, $\chi_\lambda(x) = 1$ for $\|x - x_0\| > \lambda$. χ_λ being l.s.c., f_λ has the same property. Moreover,

$$f_\lambda(x_0) < \inf f_\lambda(X) + \varepsilon,$$

as $\inf f_\lambda(X) = \inf f(S(x_0, \lambda))$, $f_\lambda(x_0) = f(x_0)$ and one takes into account (28). So apply 1.2 with Remark to f_λ, the equivalent norm $\|\cdot\|_1$ being on X, for ε, λ and $p = 2$ for instance, $\exists\, v_\varepsilon$ in X so that

(29) $f_\lambda(v_\varepsilon) < \inf f_\lambda(X) + \varepsilon$, (30) $\|x_0 - v_\varepsilon\|_1 < \lambda$ and (31) $\partial_\beta f_\lambda(v_\varepsilon) \neq \varnothing$.

But $f_\lambda(x) = f(x)$ on the open ball $S(v_\varepsilon, \lambda - \|x_0 - v_\varepsilon\|_1)$, hence $\partial_\beta f_\lambda(v_\varepsilon) = \partial_\beta f(v_\varepsilon)$, $\partial_\beta f(v_\varepsilon) \overset{(31)}{\neq} \varnothing$. In particular $v_\varepsilon \in \operatorname{dom} f$. Moreover, $f(x_0) - \varepsilon \overset{(28)}{\leq} f(v_\varepsilon) = f_\lambda(v_\varepsilon) \overset{(29)}{<} \inf f_\lambda(X) + \varepsilon \leq \leq f(x_0) + \varepsilon$, hence $|f(x_0) - f(v_\varepsilon)| < \varepsilon$, which, adjoining to (30) in wich λ is replaced by ε, finishes the proof. ■

Remark. The condition from1.4 "equivalent norm β - differentiable" is replaced in 2.4 by another strict weaker condition.

The case X *reflexive space*

In the following we present a connection between the reflexivity of a Banach space and the perturbing function Δ_p of the B - P principle.

Definition. The norm of a normed space X has the *Kadec property* when the strong and weak topologies coincide on $\{x \in X : \|x\| = 1\}$.

1.5 *Let* X *be reflexive space and* $\|\cdot\|$ *equivalent Kadec norm on* X. *Then Borwein - Preiss principle is true with*

$$\Delta_p : \Delta_p(x) = \|x - v_1\|^p, \ v_1 \in X$$

(*J. Borwein - D. Preiss*).

■ Consider the functional $h : X \to \mathbf{R}$,

(32) $$h(v) = \inf \{\varphi(x) + \frac{\varepsilon}{\lambda^p} \|x - v\|^p : \|x - x_0\| \le 2\lambda \},$$

correct definition as φ is bounded from below. Theorem 11 from [5] shows that, because x from (32) is restrained to a bounded set, there is A a dense set in X having the property

(33) $$v \in A \Rightarrow \exists\, v_w \in S_{2\lambda}(x_0) \text{ so that } h(v) = \varphi(v_w) + \frac{\varepsilon}{\lambda^p} \|v_w - v\|^p$$

(closed ball, infimum attained).

Let $\varepsilon_1, \varepsilon_2$ be as to (4) (proof of 1.2). Take v_1 in A so that

(34) $$\|x_0 - v_1\| \le \lambda \left[1 - \left(\frac{\varepsilon_1}{\varepsilon}\right)^{\frac{1}{p}}\right],$$

(35) $$\|x_0 - v_1\| \le \lambda \left(\frac{\varepsilon_1 - \varepsilon_2}{\varepsilon}\right)^{\frac{1}{p}}$$

and $\Delta_p : \Delta_p(x) = \|x - v_1\|^p$. We have

(36) $\|x_0 - v_1\|^p + \frac{\lambda^p}{\varepsilon}[\varphi(x_0) - \inf \varphi(X)] < \lambda^p$. Indeed, take into account (4), 1.2 and

$$\|x_0 - v_1\|^p \overset{(34)}{\le} \lambda^p \left[1 - \left(\frac{\varepsilon_1}{\varepsilon}\right)^{\frac{1}{p}}\right]^p, \left[1 - \left(\frac{\varepsilon_1}{\varepsilon}\right)^{\frac{1}{p}}\right]^p + \frac{\varepsilon_1}{\varepsilon} \le 1:$$

$$1 - \left(\frac{\varepsilon_1}{\varepsilon}\right)^{\frac{1}{p}} \le 1 - \frac{\varepsilon_1}{\varepsilon} \le \left(1 - \frac{\varepsilon_1}{\varepsilon}\right)^{\frac{1}{p}}, \left[1 - \left(\frac{\varepsilon_1}{\varepsilon}\right)^{\frac{1}{p}}\right]^p \le 1 - \frac{\varepsilon_1}{\varepsilon}.$$

(36) gives in particular

(37) $$\|x_0 - v_1\| < \lambda.$$

As $v_1 \in A$ $\exists\, v_\varepsilon$ in X so that

(38) $$h(v_1) = \varphi(v_\varepsilon) + \frac{\varepsilon}{\lambda^p} \|v_\varepsilon - v_1\|^p, \|v_\varepsilon - v_1\| \le 2\lambda.$$

On the other hand, we also have

(39) $$h(v_1) = \inf \{\varphi(x) + \frac{\varepsilon}{\lambda^p} \|x - v_1\|^p : x \in X \}.$$

Indeed, let x_1 be any with $\|x_1 - x_0\| > 2\lambda$, then $2\lambda - \lambda \overset{(37)}{<} \|x_1 - x_0\| - \|x_0 - v_1\| \le \|x_1 - v_1\|$, $\|x_1 - v_1\| > \lambda$ and hence

$$(40) \qquad \varphi(x_1) + \frac{\varepsilon}{\lambda^p}\|x_1 - v_1\|^p > \inf \varphi(X) + \varepsilon.$$

Multiply by $\frac{\varepsilon}{\lambda^p}$ to (36), one finds

$$(41) \qquad h(v_1) \overset{(32)}{\leq} \varphi(x_0) + \frac{\varepsilon}{\lambda^p}\|x_0 - v_1\|^p < \inf \varphi(X) + \varepsilon \overset{(40)}{<} \varphi(x_1) + \frac{\varepsilon}{\lambda^p}\|x_1 - v_1\|^p,$$

which imposes (39) *via* (32). Comparing (38) and (39) one obtains II, 1.2. Pass to verify (III):

$$\varphi(v_\varepsilon) \leq \varphi(v_\varepsilon) + \frac{\varepsilon}{\lambda^p}\|v_\varepsilon - v_1\|^p \overset{(38)}{=} h(v_1) \overset{(41)}{<} \inf\varphi(X) + \varepsilon.$$

Finally verify (IV). We have

$$(42) \qquad \|v_\varepsilon - v_1\| < \lambda\left(\frac{\varepsilon_1}{\varepsilon}\right)^{\frac{1}{p}}.$$

Indeed, if *par absurdum* one supposes the contrary, then

$$h(v_1) = \varphi(v_\varepsilon) + \frac{\varepsilon}{\lambda^p}\|v_\varepsilon - v_1\|^p \geq \inf \varphi(X) + \varepsilon_1 \overset{(4)}{>} \varphi(x_0) + \varepsilon_1 - \varepsilon_2 \overset{(35)}{\geq} \varphi(x_0) + \frac{\varepsilon}{\lambda^p}\|x_0 - v_1\|^p$$

and we come in collision with (32).

So

$$\|x_0 - v_\varepsilon\| \leq \|x_0 - v_1\| + \|v_\varepsilon - v_1\| \overset{(34),(42)}{<} \lambda\left[1 - \left(\frac{\varepsilon_1}{\varepsilon}\right)^{\frac{1}{p}}\right] + \lambda\left(\frac{\varepsilon_1}{\varepsilon}\right)^{\frac{1}{p}} = \lambda$$

and hence (IV). ■

Remark to 1.5. Any reflexive space admits an equivalent Kadec norm Fréchet differentiable ([28], Ch.V, §2, Corollary 2, p.148).

Somewhat reciprocally

1.6 *Let* X *be Banach space with the norm Gâteaux differentiable. If for* X *is valid the Borwein - Preiss principle with* $\Delta_p : \Delta_p(x) = \|x - v\|^p$, $p > 1$, *then* X *is reflexive* (*J. Borwein - D. Preiss*).

■ Use James theorem ([47], vol. III, ch. VIII, §6, 4.33_1).

Set $S := \{x \in X : \|x\| \leq 1\}$ and let f be from X^*. We must prove

$$(43) \qquad \exists\, x_1 \in S \text{ so that } f(x_1) = \|f\|.$$

It can obviously suppose $f \neq 0$ and (44) $\|f\| = 1$. Consider the function $g : X \to (-\infty, +\infty]$,

(45) $$g(x) = \begin{cases} f(x), x \in S \\ +\infty, x \in X \setminus S. \end{cases}$$

g is l.s.c., bounded from below and proper. Obviously (46) $\inf g(X) = \inf f(S)$. Apply to g B - P principle with $x_0 = 0$, $\varepsilon = 2$, $\lambda = 1$, correctly as $g(0) \overset{(45)}{=} 0 \overset{(46)}{<} 2 + \inf f(S)$, because $\inf f(S) \geq -1$ since $|f(x)| \overset{(44)}{\leq} \|x\| \leq 1$. So there exist v_1 and v_ε in X having the properties

(47) $$g(v_\varepsilon) + 2\|v_\varepsilon - v_1\|^p \leq g(x) + 2\|x - v_1\|^p \ \forall\, x \in X,$$

(48) $$\|v_\varepsilon\| < 1.$$

(47), (48) and (45) give

(49) $$f(v_\varepsilon) + 2\|v_\varepsilon - v_1\|^p \leq f(x) + 2\|x - v_1\|^p \ \forall\, x \in S.$$

The function $F: x \to f(x) + 2\|x - v_1\|^p$ is Gâteaux differentiable on $\overset{o}{S}$ ([47], vol. III, §5, 7.5), $v_\varepsilon \overset{(48)}{\in} \overset{o}{S}$, consequently, according to (49), $F'_w(v_\varepsilon) = 0$ ([47],vol. I, ch. IV, §6, 4.1), i.e.

(50) $$f + 2\,(\|x - v_1\|^p)'_w\,(v_\varepsilon) = 0.$$

$$(\|x - v_1\|^p)'_w\,(v_\varepsilon) \overset{(2)}{=} \begin{cases} p\|v_\varepsilon - v_1\|^{p-1} v'_w(v_\varepsilon - v_1), v_\varepsilon \neq v_1 \\ 0, v_\varepsilon = v_1\,. \end{cases}$$

The situation $v_\varepsilon = v_1$ must be eliminated because it implies, *via* (50), $f = 0$, contradiction. So (50) becomes

(51) $$f + 2p\|v_\varepsilon - v_1\|^{p-1}\, v'_w\,(v_\varepsilon - v_1) = 0.$$

(51) gives, *via* I, 5.7_1, $2p\|v_\varepsilon - v_1\|^{p-1} = 1$ and hence, in particular,

(52) $$f(v_\varepsilon - v_1) = -v'_w\,(v_\varepsilon - v_1)(v_\varepsilon - v_1).$$

But using I, 5.7 and the example 1 from §5, A, one finds $v'_w\,(v_\varepsilon - v_1)(v_\varepsilon - v_1) = \|v_\varepsilon - v_1\|$ and hence, *via* (52), $f(v_1 - v_\varepsilon) = \|v_1 - v_\varepsilon\|$, which confirms (43) since $\left\| \frac{v_1 - v_\varepsilon}{\|v_1 - v_\varepsilon\|} \right\| = 1$. ■

Remark. The demand "norm Gâteaux differentiable" from 1.6 has been added by the author. The proof in [6] for Proposition 5.3 is insufficient and suspect.

2

Deville - Godefroy - Zizler principle

This is a smooth perturbed variational principle which extends Borwein - Preiss principle (§1). We follow, modulo necessary corrections, [27], 1.

All the Banach spaces are real.

Let X be real normed space and β bornology on X (I, §3, "β - derivative").

Definition. The topology τ_β on X^* is the topology of the uniform convergence on the sets of β.

So in particular when $\beta = \beta_F$ resp. $\beta = \beta_G$, τ_β coincides with the strong topology resp. the $*$ - weak topology on X^*.

In I, §5, section E are defined the property (H_β) of X and the Banach space $D_\beta(X)$, the norm of this being noted by $\|\cdot\|_\beta$. We adjoin to these "the property (H_β^*)" and "the Banach space $D_\beta^*(X)$".

Definitions. The real vector space of the functions f from $D_\beta(X)$, with the property $\nabla_\beta f : X \to X^*$ is continuous relative to the topologies strong on X and τ_β on X^*, is provided with the norm $\|\cdot\|_\beta$ (loc. cit., (22_6)) and one obtains the *Banach space* $D_\beta^*(X)$.

X has the *property* (H_β^*) if there is a bump function in $D_\beta^*(X)$.

And now

2.1 Theorem. *Let* X *be Banach space having the property* (H_β) resp. (H_β^*) *and* φ : X $\to$ ($-\infty$, $+\infty$] *lower bounded, lower semicontinuous and proper. Then the set* $\mathcal{M}$ *of functions* f *from* D_β (X) resp. D_β^* (X), *with the property* φ + f *has a strong minimum point, is of* G_δ *type and everywhere dense* ([27], 1, Theorem II.3).

■ Apply theorem 1.2, Ch.I, §1 with $\mathcal{F} = D_\beta(X)$ resp. $\mathcal{F} = D_\beta^*(X)$. ■

Remark. In the variational context in which we are, $\mathcal{M}$ becomes a *set of perturbations.*

2.2 *Corollary* ([27], 1)**.** *Let* X *be Banach space with the property* (H_β) resp. (H_β^*) *and, for every* n *of* **N**, φ_n : X $\to$ ($-\infty$, $+\infty$] *bounded from below l. s. c. proper. For any* $\varepsilon > 0$ *there exists* f *in* D_β (X) resp. D_β^* (X) *so that*

$$\| f \|_\infty \le \varepsilon, \ \|\nabla_\beta f\|_\infty \le \varepsilon$$

and each function φ_n + f *attains a strong minimum on* X.

■ Use 2.1, the same justification as for 5.13, ch. I, §5. ■

Remark. 2.2 differs from 5.13, Ch. I only by the appearance of $D_\beta^*(X)$ in the statement.

The following statement, a "bump" variant of Borwein - Preiss principle, localizes the strong minimum point of the perturbed function.

2.3 Deville - Godefroy - Zizler principle. *Let* X *be Banach space with the property* (H_β) resp. (H_β^*). *Then there exists* $\rho > 0$ *so that, for any* φ : X $\to$ ($-\infty$, $+\infty$] *bounded from below, lower semicontinous and proper and for any* ε *from* (0, 1) *and* x_0 *from* X *which verify*

(I) $$\varphi(x_0) < \inf \varphi(x) + \rho\varepsilon^2,$$

there is f *in* D_β (X) resp. D_β^* (X) *and* v_ε *in* X *with the properties*:

(II) v_ε *is a strong minimum point for* φ + f;

(III) $$\| f \|_\infty < \varepsilon, \ \|\nabla_\beta f\|_\infty < \varepsilon;$$

(IV) $$\|v_\varepsilon - x_0\| < \varepsilon.$$

■ Let b be from $D_\beta(X)$ (resp. $D_\beta^*(X)$) a bump function with (open ball)

(1) $$\text{supp}\, b \subset S(0, 1),\ b(0) \ge 1.$$

It can suppose

(2) $$0 \le b(x) \le 1 \text{ on } X,\ b(0) = 1.$$

Indeed, let be $F: \mathbf{R} \to [0, 1]$ of C^1 class increasing with (3) $F((-\infty, 0]) = \{0\}$, $F(1) = 1$. Then $F \circ b$ has a bounded β-derivative (use 3.11, ch. I), is equal to 0 on $X \setminus \operatorname{supp} b$ ((1), (3)) and $F(b(0)) = 1$ since $1 \overset{(3)}{=} F(1) \overset{(1)}{\le} F(b(0)) \le 1$, hence $F \circ b$ is a bump function from $D_\beta(X)$ (resp. $D_\beta^*(X)$) which verifies (2).

Let be

(4) $$M := \max\{\|\nabla_\beta b\|_\infty, 1\},\ \rho := \frac{1}{4M}$$

and

(5) $$g: X \to (-\infty, +\infty],\ g(x) = \varphi(x) - 2\rho\varepsilon^2 b\left(\frac{x - x_0}{\varepsilon}\right).$$

g is bounded from below l.s.c. proper. Apply to this 2.2, with $\rho\frac{\varepsilon^2}{2}$ instead of ε, there exists h in $D_\beta(X)$ (resp. $D_\beta^*(X)$) with (6) $\|h\|_\infty < \rho\frac{\varepsilon}{2}$, $\|\nabla_\beta h\|_\infty < \frac{\varepsilon}{2}$ ($\rho\frac{\varepsilon^2}{2} < \frac{\varepsilon}{2}$!) and $g + h$ has a strong minumum point v_ε in X.

Verify (IV). Suppose *par absurdum* $\|v_0 - x_0\| \ge \varepsilon$, then $\frac{v_\varepsilon - x_0}{\varepsilon} \notin S(0,1)$, hence $b\left(\frac{v_\varepsilon - x_0}{\varepsilon}\right) \overset{(1)}{=} 0$ and hence (7) $g(v_\varepsilon) \overset{(5)}{=} \varphi(v_\varepsilon) \ge \inf \varphi(X)$. We have also

(8) $$g(x_0) \overset{(5),(2)}{=} \varphi(x_0) - 2\rho\varepsilon^2 \overset{(I)}{<} \inf \varphi(X) - \rho\varepsilon^2.$$

But $(g + h)(v_\varepsilon) \le (g + h)(x_0)$, hence $g(v_\varepsilon) \le g(x_0) + h(x_0) - h(v_\varepsilon) \le g(x_0) + 2\|h\|_\infty \overset{(6)}{<} g(x_0) + \rho\varepsilon^2$, consequently, $\inf \varphi(X) \overset{(7)}{<} g(x_0) + \rho\varepsilon^2$, in contradiction with (8), (IV) is verified.

Take now $f: X \to \mathbf{R}$,

(9) $$f(x) = -2\rho\varepsilon^2 b\left(\frac{x - x_0}{\varepsilon}\right) + h(x).$$

We have $f \in D_\beta(X)$ (resp. $f \in D_\beta^*(X)$) and $\varphi + f \overset{(5)}{=} g + h$, consequently v_ε is a strong minimum point for $\varphi + f$, (II) is verified.

Verify (III). $\| f \|_\infty \overset{(9)}{\leq} 2\rho\varepsilon^2 \|b\|_\infty + \|h\|_\infty \overset{(2),(6)}{<} 2\rho\varepsilon^2 + \rho\frac{\varepsilon^2}{2}$. But $2\rho \overset{(4)}{=} \frac{1}{2M} \overset{(4)}{\leq} \frac{1}{2}$, hence

$2\rho\varepsilon^2 + \rho\frac{\varepsilon^2}{2} \leq \frac{1}{2}\varepsilon^2 + \frac{1}{8}\varepsilon^2 < \varepsilon^2 < \varepsilon$ and hence $\| f \|_\infty < \varepsilon$.

Finally,

$$\|\nabla_\beta f\|_\infty \overset{(9)}{\leq} 2\rho\varepsilon^2 \|\nabla_\beta b\|_\infty + \|\nabla_\beta h\|_\infty \overset{(4),(6)}{<} 2\rho\varepsilon^2 M + \frac{\varepsilon}{2} \overset{(4)}{=} \frac{1}{2}\varepsilon^2 + \frac{\varepsilon}{2} < \varepsilon. \blacksquare$$

Remark. D - G - Z Principle extends Borwein - Preiss principle for Banach spaces with β- differentiable norm, because the demand „β-differentiable norm on $X \setminus \{0\}$" is replaced by „X has the property (H_β)", which is strictly weaker, as is shown by the R. Haydon example: a Banach space X which does not admit equivalent β-differentiable on $X \setminus \{0\}$ norm and which has however a Lipschitz bump function Fréchet differentiable on X (*A counterexample in several questions about scattered compact spaces*, Bull. London Math. Soc., **22**, 1990, p.261-268).

Application to the β-subderivative

2.4 *Let* X *be Banach space with the property* (H_β) *and* $f : X \to (-\infty, +\infty]$ *bounded from below lower semicontinuous proper. Then* f *is β-subdifferentiable on a dense subset of* dom f (*R. Deville - G. Godefroy - V. Zizler*).

■ Let x' be from dom f and $r > 0$. We must show that there exists a point in the open ball $S(x', r)$ in which f is β-subdifferentiable. Let b be a bump function on X β-differentiable and Lipschitz with $b(0) \neq 0$ (see the demonstration for 2.3). Consider the function $\varphi : X \to (-\infty, +\infty]$,

(10) $$\varphi(x) = \begin{cases} \dfrac{1}{b^2(\rho(x - x'))}, & b(\rho(x - x')) \neq 0 \\ +\infty, & b(\rho(x - x')) = 0, \end{cases}$$

where $\rho > 0$. ρ can be taken sufficiently big to have

(23) $$\operatorname{dom} f \subset S(x', r).$$

Indeed, $U := \{y \in X : b(y) \neq 0\}$ is bounded, suppose $y \in U \Rightarrow \|y\| \leq \alpha$. Take ρ so that

$\frac{\alpha}{\rho} < r$. Then $x \in \operatorname{dom} \varphi \overset{(10)}{\Rightarrow} b(\rho(x - x')) \neq 0 \Rightarrow \rho(x - x') \in U \Rightarrow \rho\|x - x'\| \leq \alpha \Rightarrow \|x - x'\| \leq$

$\frac{\alpha}{\rho} < r$, that is (11). The function $\varphi + f$ is bounded from below (attention to *b*), l. s. C. (U is open !) and proper ($x' \in \text{dom}(\varphi + f)$). Apply 2.1, $\exists\, g$ in $D_\beta(X)$ so that $(\varphi + f) + g$ has a minimum point, let it be x_0. We have (12) $x_0 \in S(x', r)$, since $x_0 \in \text{dom}\,(\varphi + f + g)$ (*par absurdum*) and $\text{dom}\,(\varphi + f + g) \subset \text{dom}\,\varphi \overset{(11)}{\subset} S(x', r)$. (12) finishes the proof because f is β-subdifferentiable in x_0. Indeed, $x_0 \in \text{dom}\,\varphi$, $\text{dom}\,\varphi = x' + \frac{1}{\rho} U$, an open set, φ is β-differentiable on $\text{dom}\,\varphi$ (I, §3, 3.11) just as g. Taking $\xi := -\nabla_\beta\varphi(x_0) - \nabla_\beta g(x_0)$, we have (13) $\xi \in \partial_\beta f(x_0)$ (I, §5, variant of the definition). Indeed, for $t > 0$

$$\frac{f(x_0 + tv) - f(x_0)}{t} - \xi(v) \geq -\left[\frac{\varphi(x_0 + tv) - \varphi(x_0)}{t} - \nabla_\beta\,\varphi(x_0)(v)\right] - \left[\frac{g(x_0 + tv) - g(x_0)}{t} - \nabla_\beta\, g(x_0)(v)\right]$$

(x_0 is minimum point for $\varphi + f + g$) and φ, g being finite and β-differentiable on U open set, they are also β-superdifferentiable (I, §5, 5.10_1). ∎

Remark. Compare 2.4 with 1.4 taking into account the R. Haydon example.

2.5 *Corollary. Let* X *be Banach space with the property* (H_β) *and* f : U $\to$ **R** *continuous convex,* U *convex open part of* X. *Then* f *is β-differentiable on a dense subset of* U ([27], 1).

∎ f is β - subdifferentiable in every point of U (I, §5, 5.9). From I, 5.10_1 it is sufficient to show that f is β - superdifferentiable on a dense subset of U. Let x_0 be a point of U and $S := S(x_0\,;\, r)$ any open ball centred in x_0 and included in U. This can be supposed sufficiently small for f to be bounded on it. Consider the function $g : X \to (-\infty, +\infty]$,

$$g(x) = \begin{cases} -f(x), & x \in S \\ +\infty, & x \in X \setminus S. \end{cases} \tag{14}$$

g is bounded from below l. s. c. proper, (15) $\text{dom}\, g = S$. Apply to this 2.4, g is β-subdifferentiable on a dense subset of S (see (15)), that is f is β-superdifferentiable on a dense subset of S, consequently in a small ball centred in x_0 included in $S\ \exists\, x_1$ in which f is β-superdifferentiable and from here the conclusion, as $x_1 \in S$. ∎

Remark. The Corollary 2.5 is Corollary 2.8 from [27], 1. The demonstration given there is not sufficient.

Viscosity β-solutions of the differential equation

$$u + F(u') = f$$

(Hamilton – Jacobi type)

Let X be real normed space, β bornology on X, $f: X \to (-\infty, +\infty]$ resp. $f: X \to [-\infty, +\infty)$ proper and x_0 from $D(f) := \{x \in X : f(x) \in \mathbf{R}\}$.

The *strong* β-*subderivative* resp. the *strong* β-*superderivative of* f *at* x_0, $d_\beta^- f(x_0)$ resp. $d_\beta^+ f(x_0)$, is the set of the functionals ξ from X^* so that there exists $\varphi : X \to [-\infty, +\infty)$ resp. $\varphi : X \to (-\infty, +\infty]$ with the following properties ($D(\varphi) \neq \varnothing$):

1° φ is β-differentiable on $D(\varphi)$ and upper semicontinuous resp. lower semicontinuous;

2° $x_0 \in D(\varphi)$ and $f - \varphi$ attains its minimum resp. its maximum at x_0;

3° $\nabla_\beta \varphi(x_0) = \xi$ and $\nabla_\beta \varphi : X \to X^*$ is continuous at x_0 relative to the strong topology on X and the τ_β topology on X^*.

When $d_\beta^- f(x_0) \neq \varnothing$ resp. $d_\beta^+ f(x_0) \neq \varnothing$, then f is by definition *strongly* β-*subdifferentiable* resp. *strongly* β-*superdifferentiable at* x_0.

If $\xi \in d_\beta^- f(x_0)$ resp. $\xi \in d_\beta^+ f(x_0)$, then

$$\varliminf_{t \to 0+} \left[\frac{f(x_0 + th) - f(x_0)}{t} - \xi(h)\right] \geq 0$$

resp.

$$\varlimsup_{t \to 0+} \left[\frac{f(x_0 + th) - f(x_0)}{t} - \xi(h)\right] \leq 0$$

uniformly in h on the sets from β, consequently

(16) f strongly β-subdifferentiable resp. strongly β-superdifferentiable at $x_0 \Rightarrow f$ β-subdifferentiable resp. β-superdifferentiable at x_0 (I, §5, section D, (16) resp. (22_1)).

Let X be real normed space, β bornology on X, Ω nonempty open subset of X and $H: \Omega \times \mathbf{R} \times X^* \to \mathbf{R}$ continuous. Consider the differential equation

(∗) $$H(x, u, u') = 0.$$

A function $u : \Omega \to \mathbf{R}$ is a *classical* β-*solution* of (∗) if u is β-differentiable on Ω, $\nabla_\beta u$ is continuous on Ω relative to the topologies strong on X and τ_β on X^* and

(17) $$H(x, u(x), \nabla_\beta u(x)) = 0 \ \forall\, x \in \Omega.$$

A function $u : \Omega \to \mathbf{R}$ is a *viscosity* β-*subsolution* of (∗), v. β-subsolution, resp. a *viscosity* β-*supersolution* of (∗), v. β-supersolution, when u is u. s. c. + strongly β-superdifferentiable resp l. s. c. + strongly β-subdifferentiable and $\forall\, x \in \Omega$ and $\forall\, \xi \in d^+_\beta\, u(x)$ resp. $d^-_\beta\, u(x)$ we have

$$(18) \qquad H(x, u(x), \xi) \le 0, \text{ resp. } H(x, u(x), \xi) \ge 0.$$

When u is in both situations, it is by definition a *viscosity* β-*solution* of (∗).

2.6 1° u *classical* β-*solution of* (∗) *on* $\Omega \Rightarrow$ u *viscosity* β-*solution of* (∗) *on* Ω;

2° *If* $\Omega = \Omega_1 \cup \Omega_2$, Ω_1 *and* Ω_2 *nonempty open, and* u *is a viscosity* β-*subsolution* resp. β-*supersolution of* (∗) *on* Ω_i $i = \overline{1,2}$, *then* u *is a viscosity* β-*subsolution* resp. β-*supersolution of* (∗) *on* Ω.

■ 1° Take into account I, §5, 5.10_1.

2° Obvious. ■

Notations. Let X be real Banach space and β bornology on X.

$$\mathrm{BUC}(X) \overset{\text{def}}{=} \{f : X \to \mathbf{R} : f \text{ bounded and uniformly continuous}\},$$

$$\mathrm{UC}(X^*) \overset{\text{def}}{=} \{f : X^* \to \mathbf{R} : f \text{ uniformly continuous}\},$$

$$\mathrm{UC}_\beta(X^*) \overset{\text{def}}{=} \{f \in \mathrm{UC}(X^*) : f\ \tau_\beta \text{ continuous}\}.$$

The first two sets are obviously endowed with a structure of vector space on **R**. The third set becomes a subspace of $\mathrm{UC}(X^*)$.

Our goal is to prove

2.7 Theorem. *Let* X *be Banach space which verifies* (H^*_β) *and* f $\in$ BUC(X), F $\in$ $\in \mathrm{UC}_\beta(\mathrm{X})$. *Then the equation*

$$(**) \qquad \mathrm{u} + \mathrm{F}(\mathrm{u}') = \mathrm{f}$$

has a unique bounded viscosity β-*solution on* X

(*R. Deville - G. Godefroy - V. Zizler*).

A sequence of propositions leads us to the proof of this theorem.

The concept and the denomination *viscosity solution* for (∗∗) are owed to M. Crandall and P. Lions. Their study has been made in the case $\beta = \beta_F$ ([23], 1,2).

2.8 *Let* X *be Banach space which verifies* (H^*_β) *and* u, v : X $\to$ **R** *bounded,* u *u.s.c.*

and v *l. s. c. Then* $\forall\ \varepsilon > 0\ \exists\ x_0, y_0$ *in* X *and* $\xi \in d_\beta^+ u(x_0)$, $\eta \in d_\beta^- v(y_0)$ *so that*

1° $\|x_0 - y_0\| < \varepsilon$ and $\|\xi - \eta\| < \varepsilon$;

2° $v(z) - u(z) \geq v(y_0) - u(x_0) - \varepsilon\ \forall\ z \in X$

([27], 1).

■ Let b be a bump function in $D_\beta^*(X)$ with $b(0) = 1$ and supp $b \subset \{x \in X: \|x\| < 1\}$. It can suppose $0 \leq b(x) \leq 1$ on X, in the contrary case one takes $\varphi \circ b$, $\varphi: \mathbf{R} \to [0,1]$ of C^1 class, $\varphi(0) = 0$, $\varphi(1) = 1$ (see the begining of the proof for 2.3).

Take $\lambda > 0$ and $w: X \times X \to \mathbf{R}$,

$$w(x, y) = v(y) - u(x) - \lambda b\left(\frac{x-y}{\varepsilon}\right), \tag{19}$$

$$\lambda > 4\max(\|u\|_\infty, \|v\|_\infty) + \varepsilon. \tag{20}$$

w is bounded and l. s. c. on $X \times X$ and $X \times X$ verifies (H_β^*) ($B: X \times X \to [0,1]$, $B(x, y) = b(x)b(y)$ is a bump function which is in $D_\beta^*(X \times X)$), apply 2.2, $\exists\ g \in D_\beta^*(X \times X)$ and $\exists\ (x_0, y_0)$ in $X \times X$ so that

$$\|g\|_\infty < \frac{\varepsilon}{2},\ \|\nabla_\beta g\|_\infty < \frac{\varepsilon}{2} \tag{21}$$

and

$$v(y) - u(x) - \lambda b\left(\frac{x-y}{\varepsilon}\right) + g(x, y) \geq v(y_0) - u(x_0) - \lambda b\left(\frac{x_0-y_0}{\varepsilon}\right) + g(x_0, y_0) \tag{22}$$

$$\forall\ (x, y) \in X \times X.$$

Taking $x = x_0$ in (22), this shows that $v - \varphi$, $\varphi(y) := \lambda b\left(\frac{x_0-y}{\varepsilon}\right) - g(x_0, y)$, has a minimum point at y_0, consequently $\eta := \nabla_\beta \varphi(y_0) = -\frac{\lambda}{\varepsilon} b'\left(\frac{x_0-y_0}{\varepsilon}\right) - \nabla_{\beta, y}\, g(x_0, y_0) \in d_\beta^- v(y_0)$.

In the same manner one finds (take $y = y_0$ and change the sign)

$$\xi := -\frac{\lambda}{\varepsilon} b'\left(\frac{x_0-y_0}{\varepsilon}\right) + \nabla_{\beta, x}\, g(x_0, y_0) \in d_\beta^+ u(x_0).$$

Verify 1°. $\|\xi - \eta\| = \|\nabla_{\beta, x}\, g(x_0, y_0) + \nabla_{\beta, y}\, g(x_0, y_0)\| \overset{(21)}{<} \varepsilon$. Suppose *par absurdum* $\|x_0 -$

$-\, y_0\| \geq \varepsilon$. Then $b\left(\dfrac{x_0 - y_0}{\varepsilon}\right) = 0$ and (22) gives, since $b(0) = 1$, $v(z) - u(z) - \lambda + g(z, z) \geq$

$\geq v(y_0) - u(x_0) + g(x_0, y_0)\ \forall\, z \in X$, consequently, taking into account also (21),

$$\lambda < 2\|u\|_\infty + 2\|v\|_\infty + \varepsilon \leq 4\max(\|u\|_\infty, \|v\|_\infty) + \varepsilon,$$

in contradiction with (20).

Verify 2°. Take $x = y = z$ in (22) and taking into account also (21) one finds

$$v(z) - u(z) \geq v(y_0) - u(x_0) + \lambda[b(0) - b\left(\frac{x_0 - y_0}{\varepsilon}\right)] - \varepsilon$$

and consequently 2°, as $b(0) = 1 \geq b\left(\dfrac{x_0 - y_0}{\varepsilon}\right)$. ■

2.9 *Let* X *be Banach space which verifies* (H_β^*), $\mathrm{F} \in \mathrm{UC}(\mathrm{X}^*)$, f, g $\in$ BUC(X) *and* u_0 resp. v_0 *a bounded viscosity* β*-subsolution* resp. *bounded viscosity* β*-supersolution on* X *of* $\mathrm{u} + \mathrm{F}(\mathrm{u}') = \mathrm{f}$ resp. $\mathrm{v} + \mathrm{F}(\mathrm{v}') = \mathrm{g}$. *Then*

$$\mathrm{v}_0 - \mathrm{u}_0 \geq \inf_{\mathrm{x} \in \mathrm{X}} (\mathrm{g}(\mathrm{x}) - \mathrm{f}(\mathrm{x})).$$

■ Let n be any from **N** and x_n, y_n, ξ_n, η_n given by 2.8 relative to u_0, $v_0, \dfrac{1}{n}$. According to the hypothesis, we have

(23) $u(x_n) + F(\xi_n) \leq f(x_n)$, $v(y_n) + F(\eta_n) \geq g(y_n)$,

hence $\forall\, z \in X$ (2.8)

$$v_0(z) - u_0(z) \geq v(y_n) - u(x_n) - \frac{1}{n} \overset{(23)}{\geq} g(y_n) - f(x_n) + F(\xi_n) - F(\eta_n) - \frac{1}{n} \geq \inf_{x \in X} (g(x) -$$

$$f(x)) + [f(y_n) - f(x_n)] + [F(\xi_n) - F(\eta_n)] - \frac{1}{n}.$$

Pass to the limit for $n \to \infty$ taking into account that $\|x_n - y_n\|$, $\|\xi_n - \eta_n\| \leq \dfrac{1}{n}$ and f, F are uniformly continuous. ■

Now prove the *uniqueness* in 2.7. Let u_1, u_2 be bounded viscosity β-solutions on X for (**). Then, according to 2.9 where $f = g$, $u_1 - u_2 \geq 0$ and $u_2 - u_1 \leq 0$, hence $u_1 = u_2$.

Continue with the proof of the *existence* in 2.7. But firstly

Definition. Let X be real normed space, Ω nonempty open set of X and $u : \Omega \to \mathbf{R}$. By definition

$$
(24) \qquad \begin{aligned} u^* &= \inf\{v:\Omega \to \mathbf{R} \text{ continuous} \mid v(x) \geq u(x) \text{ on } \Omega\}, \\ u_* &= \sup\{v:\Omega \to \mathbf{R} \text{ continuous} \mid v(x) \leq u(x) \text{ on } \Omega\}. \end{aligned}
$$

In both cases we supposed that the set which appears in the definition is not empty. u^* is u.s.c. and u_* is l.s.c.

Pass to prove the existence in 2.7. Let be $\rho := \| f \|_\infty$, $u_0 := -\rho - F(0)$, $v_0 := \rho - F(0)$. Since $u_0' = v_0' = 0$ (Fréchet derivative), we have $\nabla_\beta u_0'(x) = \nabla_\beta v_0'(x) = 0$ on X (I, §3, 3.9 and 3.10), consequently u_0 resp. v_0 is a classical β-subsolution resp. a classical β-supersolution of (∗∗) on X and hence u_0 resp. v_0 is a viscosity β-subsolution resp. a viscosity β-supersolution of (∗∗) on X. Obviously $u_0 \leq v_0$. Proposition III.7, [27], 1 affirms that there exists $w: X \to \mathbf{R}$, (25) $u_0(x) \leq w(x) \leq v_0(x)$ on X so that w^* resp. w_* is a viscosity β-subsolution resp. a viscosity β-supersolution of (∗∗) on X. (24) and (25) give $u_0 \leq w_* \leq w^* \leq v_0$, hence w_* and w^* are bounded. Applying 2.9 with $f = g$ one obtains $w^* \leq w_*$, hence $w^* = w_*$, $w_0 := w^* = w_*$ is a viscosity β-solution of (∗∗) on X.

3

Ghossoub - Maurey theorems

Among the theorems of the title, one is a generalization of the D - G - Z theorem (I, §1, 1.2) and the others – variational principles of which one, for the reflexive spaces, presents a linear perturbation. Generally, we follow [34] with the necessary corrections of the errors (there is painfull assertions) and with some improvements of the statements.

All the Banach spaces are real.

Let (X, d) be metric space and $(\mathcal{M}, \delta)$ a metric space of real functions defined on X. For each nonempty subset A of X consider

(1) $$\mathcal{M}_A := \{f \in \mathcal{M} : f \text{ upper bounded on } A\}$$

and for each f from $\mathcal{M}_A$ and $t > 0$ the *slice* of A,

(2) $$S(A, f, t) \overset{\text{def}}{=} \{x \in A : f(x) > \sup f(A) - t\},$$

a set obviously nonempty when $\mathcal{M}_A \neq \varnothing$.

Definition. The metric space (X, d) is *uniformly* $\mathcal{M}$*-dentable* if for each nonempty subset A of X, for each f from $\mathcal{M}_A$ and for each $\varepsilon > 0$ there exists g in $\mathcal{M}_A$ and $t > 0$ so that

$$(3) \quad \delta(f, g) \leq \varepsilon \quad \text{and} \quad (4) \quad \operatorname{diam} S(A, g, t) \leq \varepsilon.$$

The definition is extended to the case X pseudo-metric space.

Associate with the metric space (X, d) the set $\widetilde{X} := X \times \mathbf{R}$ equipped with a pseudo-metric $\widetilde{\mathrm{d}}$,

(5) $$\widetilde{\mathrm{d}}\,((x, \lambda), (y, \mu)) = \mathrm{d}\,(x, y).$$

Moreover, associate with the metric space $(\mathcal{M}, \delta)$ the set $\widetilde{\mathcal{M}}$ of the functions $\widetilde{f}$, couples of the form $\widetilde{f} = (f, \lambda)$, $f \in \mathcal{M}$, defined on $\widetilde{X}$ by

(6) $$\widetilde{f}\,(x, \lambda) = f(x) - \lambda.$$

Retain $\widetilde{f} = \widetilde{g} \Leftrightarrow f = g$. Provide $\widetilde{\mathcal{M}}$ with the distance $\widetilde{\delta}$,

(7) $$\widetilde{\delta}\,(\widetilde{f}, \widetilde{g}) = \delta(f, g).$$

Definition. Let (X, d) be metric space and $(\mathcal{M}, \delta)$ metric space of real bounded continuous functions on X. $\mathcal{M}$ is an *admissible family* on X if

(I) $$\mathcal{M} \text{ is complete metric space,}$$

(II) $$\exists\, K > 0 \text{ with the property}$$

$$f, g \in \mathcal{M} \Rightarrow \delta(f, g) \geq K \sup\{|f(x) - g(x)| : x \in X\},$$

(III) The metric space $(\widetilde{X}, \widetilde{d}\,)$ is uniformly $\widetilde{\mathcal{M}}$ - dentable.

And now

3.1 Theorem. *Let* X *be a Banach space,* $(\mathcal{M}, \delta)$ *an admissible family of functions on* X *and* $\varphi : \mathrm{X} \to (-\infty, +\infty]$ *bounded from below, lower semicontinuous and proper. Then the set*

$$\mathcal{M}_1 := \{\mathrm{f} \in \mathcal{M} : \varphi - \mathrm{f} \textit{ attains a strong minimum on } \mathrm{X}\}$$

includes a subset of G_δ *type everywhere dense*

(*N. Ghoussoub - B. Maurey*).

■ Consider for each n from $\mathbf{N}$ the set

(8) $$\mathcal{U}_n := \{f \in \mathcal{M} : \exists\, x_n \in X \text{ with } (\varphi - f)(x_n) < \inf\{(\varphi - f)(x) : \|x - x_n\| \geq \frac{1}{n}\}\}.$$

$\mathcal{U}_n$ *is open*. Indeed, let f_0 be any from $\mathcal{U}_n$. Suppose *par absurdum* $\forall\, \varepsilon > 0\ \exists\, f$ in $\mathcal{U}_n$ so that

(9) $$\delta(f_0, f) < \varepsilon$$

and at the same time

(10) $$(\varphi - f)(x_n) \geq \inf\{(\varphi - f)(x) : \|x - x_n\| \geq \frac{1}{n}\}.$$

$(\varphi - f)(x_n) = (\varphi - f_0)(x_n) + (f_0 - f)(x_n)$, combine (10), (9) and (II) (definition of the admissible family), one obtains

(11) $$(\varphi - f_0)(x_n) + \frac{\varepsilon}{K} \geq \inf\{(\varphi - f)(x) : \|x - x_n\| \geq \frac{1}{n}\}.$$

But $\forall\, x$ with $\|x - x_n\| \geq \frac{1}{n}$ we have

$$(\varphi - f)(x) = (\varphi - f_0)(x) + (f_0 - f)(x) \overset{\mathrm{II}}{\geq} (\varphi - f_0)(x) - \frac{\varepsilon}{K} \geq \inf\{(\varphi - f_0)(x) : \|x - x_n\| \geq \frac{1}{n}\} - \frac{\varepsilon}{K},$$

consequently

$$\inf\{(\varphi - f)(x) : ...\} \geq \inf\{(\varphi - f_0)(x) : ...\} - \frac{\varepsilon}{K},$$

combine with (11), take the limit for $\varepsilon \to 0+$ and one obtains a contradiction since $f_0 \in \mathscr{U}_n$, consequently f_0 is an inner point of $\mathscr{U}_n$.

Prove now

$\mathscr{U}_n$ is everywhere dense (in particular $\mathscr{U}_n \neq \varnothing$). Let f be from $\mathscr{M}$ and $\varepsilon > 0$. We must find g in $\mathscr{M}$ with (12) $\delta(f, g) < \varepsilon$ and x_n in X with

(13) $$(\varphi - g)(x_n) < \inf\{(\varphi - g)(x) : \|x - x_n\| \geq \frac{1}{n}\}.$$

The functional $\tilde{f}$ is upper bounded on the epigraph E of φ in $\operatorname{dom}\varphi \times \mathbf{R}$, a nonempty set because φ is proper: $\lambda \geq \varphi(x) \Rightarrow \tilde{f}(x, \lambda) = f(x) - \lambda \leq f(x) - \varphi(x)$ $\forall\, x \in \operatorname{dom}\varphi$ and f is bounded, $-\varphi$ upper bounded. $(\tilde{X}, \tilde{d})$ being uniformly $\tilde{\mathscr{M}}$-dentable, for $\tilde{f}$ and $\forall\, \varepsilon' > 0$ $\exists\, \tilde{g}$ in $\tilde{\mathscr{M}}_E$ and $t > 0$ so that $\delta(f, g) < \varepsilon'$ and

(14) $$\operatorname{diam} S(E, \tilde{g}, t) < \varepsilon'.$$

Suppose $\varepsilon' < \min\{\varepsilon, \frac{1}{n}\}$. Take (15) (x_n, λ_n) in $S := S(E, \tilde{g}, t)$. Let x be any in X with $\|x - x_n\| \geq \frac{1}{n}$. Suppose $x \in \operatorname{dom}\varphi$. Then $(x, \varphi(x)) \in E \setminus S$ (see (5), (14), (15)) and consequently

(16) $$g(x_n) - \lambda_n > \sup \tilde{g}(E) - t \geq g(x) - \varphi(x).$$

(16) obviously remains true when $x \notin \operatorname{dom}\varphi$. It remains only to change the signs in (16)

and then to take the greatest lower bound to obtain (13), since $\lambda_n \geq \varphi(x_n)$.

Let be $\mathcal{U} := \bigcap_{n=1}^{\infty} \mathcal{U}_n$. $\mathcal{U}$ is dense in $\mathcal{M}$ (Baire theorem). It remains to show that $\mathcal{U} \subset \mathcal{M}_1$, i.e.

(17) $$f \in \mathcal{U} \Rightarrow \varphi - f \text{ attains a strong minimum on } X.$$

For each n from $\mathbf{N}$ $\exists$ x_n in X so that (definition of $\mathcal{U}_n$!)

(18) $$(\varphi - f)(x_n) < \inf\{(\varphi - f)(x) : \|x - x_n\| \geq \frac{1}{n}\}$$

(obviously $x_n \in \operatorname{dom} \varphi$ $\forall$ $n \geq 1$).

A fortiori (19) $p > n \Rightarrow \|x_p - x_n\| < \frac{1}{n}$, otherwise $(\varphi - f)(x_p) \overset{(18)}{>} (\varphi - f)(x_n)$ and on the other hand, as $\|x_p - x_n\| \geq \frac{1}{n} > \frac{1}{p}$, it results ($f \in \mathcal{U}_p$!) $(\varphi - f)(x_n) > (\varphi - f)(x_p)$, contradiction. So (19) imposes $(x_n)_{n\geq 1}$ Cauchy sequence, let be $x_0 = \lim_{n\to\infty} x_n$ and we liquidate (17) by proving

x_0 *is strong minimum point for* $\varphi - f$. φ is l. s. c. in x_0 and taking also into account the property of x_n we have

(20) $$(\varphi - f)(x_0) \leq \varliminf_{n\to\infty} (\varphi - f)(x_n) \leq \varliminf_{n\to\infty} \inf\{(\varphi - f)(x) : \|x - x_n\| \geq \frac{1}{n}\} =: L.$$

But

(21) $$L \leq \inf\{(\varphi - f)(x) : x \in X \setminus \{x_0\}\} =: m.$$

Indeed, suppose *par absurdum* $L > m$. Then there exists x' in $X \setminus \{x_0\}$, a in $\mathbf{R}$ and $(x_{k_n})_{n\geq 1}$ subsequence so that

(22) $$\inf\{(\varphi - f)(x) : \|x - x_{k_n}\| \geq \frac{1}{k_n}\} \geq a > (\varphi - f)(x').$$

But we have not $\|x' - x_{k_n}\| < \frac{1}{k_n}$ $\forall$ $n \geq 1$, since in this case $x' = \lim_{n\to\infty} x_{k_n} = x_0$, contradiction. Let then be k_N so that $\|x' - x_{k_N}\| \geq \frac{1}{k_N}$. This implies, *via* (22), $(\varphi - f)(x') \geq a > (\varphi - f)(x')$, contradiction and hence (21). Combining (20) and (21), it results x_0 minimum point of $\varphi - f$ on X.

It remains to show that, for any sequence $(y_n)_{n\geq 1}$,

(23) $$\lim_{n\to\infty} (\varphi - f)(y_n) = (\varphi - f)(x_0) \Rightarrow (24)\ \lim_{n\to\infty} y_n = x_0 .$$

Suppose *par absurdum* the contrary. Then there exist $\varepsilon > 0$ and $(y_{k_n})_{n\geq 1}$ subsequence of $(y_n)_{n\geq 1}$ so that

(25) $$\| y_{k_n} - x_0 \| \geq \varepsilon\ \forall\, n \geq 1.$$

(25) gives

$$\| y_{k_n} - x_p \| + \|x_p - x_0 \| \geq \varepsilon\ \forall\, n, p \geq 1.$$

As $x_p \to x_0$, one finds N from **N** so that

(26) $$\| y_{k_n} - x_p \| + \frac{\varepsilon}{2} \geq \varepsilon\ \forall\, n \geq 1,\ \forall\, p \geq N.$$

Take p_1 sufficiently big to have $\frac{1}{p_1} \leq \frac{\varepsilon}{2}$ and $p_1 \geq N$. Then, from (26),

(27) $$\| y_{k_n} - x_{p_1} \| \geq \frac{1}{p_1}\quad \forall\, n \geq 1.$$

So, as x_0 is minimum point, we have

(28) $$(\varphi - f)(x_0) \leq (\varphi - f)(x_{p_1}) < \inf\{(\varphi - f)(x) : \|x - x_{p_1}\| \geq \frac{1}{p_1}\} \overset{(27)}{\leq} (\varphi - f)(y_{k_n})$$

$\forall\, n \geq 1.$

(28) imposes $(\varphi - f)(y_{k_n}) \not\to (\varphi - f)(x_0)$ and we come in collision with (23), which attests (24). ■

REMARK. Theorem 3.1 is Theorem 1.27 (*General variational principle*) from [34], Chapter 1, 1.7. This theorem was recovered by making some changes. In Th. 1.27 X is complete metric space. But then, in this case, the definition $\mathscr{U}_n := \{g \in \mathscr{A} : \exists\, x_n \in X$ with $(\varphi - g)(x_n) < \inf\{(\varphi - g)(x) : \mathrm{d}(x, x_n) \geq \frac{1}{n}\}\}$ is not correct, since the inequation

$$\mathrm{d}(x, x_n) \geq \frac{1}{n}$$

has not compulsory solutions in X. The author has passed to a Banach space, in this case the inequation $\|x - x_0\| \geq R$ has always solutions, and correcting also two mistakes was retrieved th. 1.27 as theorem 3.1.

So the Ekeland principle remains without rival as variational principle in metric spaces.

Continue with an example of admissible family.

3.2 *Let* X *be Banach space and* $\mathscr{F}$ *a Banach space of real functions defined, bounded and continuous on* X *having the properties*:

1° $f \in \mathscr{F} \Rightarrow \|f\|_\infty \le \|f\|$;

2° $f \in \mathscr{F}$ *and* $x \in X \Rightarrow$ *the function* $y \to f_x(y) := f(x+y)$ *is in* $\mathscr{F}$ *and* $\|f_x\| = \|f\|$;

3° $f \in \mathscr{F}$ *and* $\lambda \in \mathbf{R} \Rightarrow$ *the function* $y \to f(\lambda y)$ *is in* $\mathscr{F}$;

4° *there exists a bump function in* $\mathscr{F}$ (I, §1, 1.2).

Then $\mathscr{F}$ *is admissible family on* X ([34]).

∎ Verify the properties (I), (II), (III) from the definition. (I) and (II) are given by the hypothesis ($K = 1$). Pass to (III). Let $\tilde{f}$ be any from $\tilde{\mathscr{M}}_A$ and $\varepsilon > 0$. We must find $\tilde{g}$ in $\tilde{\mathscr{M}}_A$ and $t > 0$ so that

(29) $\quad \tilde{\delta}(\tilde{f}, \tilde{g}) \le 2\varepsilon \quad$ and $\quad$ (30) $\quad \operatorname{diam} S(A, \tilde{g}, t) \le 2\varepsilon.$

With 4° and 2° one finds b bump function with $b(0) \ne 0$. Then with 3° we replace $b(x)$ by $\lambda_1 b(\lambda_2 x)$, λ_1 and λ_2 conveniently chosen in $\mathbf{R}$ so that

(31) $\quad b(0) > 0, \qquad$ (32) $\quad \|b\|_{\mathscr{F}} \le \varepsilon, \qquad$ (33) $\quad \|x\| \ge \varepsilon \Rightarrow b(x) = 0.$

Let (x_0, λ_0) be from A so that (see (31))

(34) $$\tilde{f}(x_0, \lambda_0) = f(x_0) - \lambda_0 > \sup \tilde{f}(A) - b(0).$$

Consider the function (35) $h : X \to \mathbf{R}$, $h(x) = b(x - x_0)$ and take $\tilde{g}$ from $\tilde{\mathscr{M}}$,

$$\sim$$

(36) $$\tilde{g} := f + g$$

$\tilde{g} \in \tilde{\mathscr{M}}_A$ because on one hand $f + h \in \mathscr{M}$ and on the other hand $(x, \lambda) \in A \Rightarrow \tilde{g}(x, \lambda) =$

$= [f(x) - \lambda] + h(x)$ and $\tilde{f}$, h are upper bounded on A resp. X. We have $\|h\|_{\mathscr{F}} \overset{(2)}{=} \|b\|_{\mathscr{F}}$, hence

$$\tilde{\delta}(\tilde{f}, \tilde{g}) \overset{(7)}{=} \delta(f, g) = \|f - g\|_{\mathscr{F}} = \|h\|_{\mathscr{F}} \overset{(32)}{\le} \varepsilon$$

and consequently (29).

Transcribe (34):

(37) $$\tilde{g}(x_0, \lambda_0) > \sup \tilde{f}(A).$$

Obviously $\tilde{g}(x_0, \lambda_0) \le \sup \tilde{g}(A)$.

Take now $t > 0$ so that

(38) $$\tilde{g}(x_0, \lambda_0) - t > \sup \tilde{f}(A)$$

(see (37)). It remains to prove (30). Let (x, λ) be any from $S := S(A, \tilde{g}, t)$. Then (39) $\|x - x_0\| < \varepsilon$. Indeed, suppose *par absurdum* $\|x - x_0\| \ge \varepsilon$, this gives $h(x) \overset{(33),(35)}{=} 0$, but $\tilde{g}(x, \lambda) > \sup \tilde{g}(A) - t$, consequently $f(x) - \lambda > \tilde{g}(x_0, \lambda_0) - t \overset{(38)}{>} \sup \tilde{f}(A)$, contradiction since $(x, \lambda) \in A$.

So, if $(x_1, \lambda_1), (x_2, \lambda_2) \in S$ then

$$\tilde{d}((x_1, \lambda_1), (x_2, \lambda_2)) \overset{(5)}{=} d(x_1, x_2) = \|x_1 - x_2\| \overset{(39)}{\le} 2\varepsilon$$

and consequently (30). ■

Remark. Proposition 3.2 shows that Ghoussoub - Maurey theorem (3.1) generalizes Deville - Godefroy - Zizler theorem (I, §1, 1.2; $\mathscr{F}$ being Banach space, $\mathscr{F} = \{-f : f \in \mathscr{F}\}$).

And now

3.3 Ghoussoub - Maurey principle. *Let* X *be Banach space and* $\mathscr{M}$ *a Banach space of real bounded continuous functions on* X *with the properties* 1°, 2°, 3° *and* 4° (3.2). *Then there is* $\rho > 0$ *so that, for any* $\varphi : X \to (-\infty, +\infty]$ *bounded from below l. s. c. proper and for any* ε *from* (0, 1) *and* x_0 *from* X *which verifies*

(I) $$\varphi(x_0) < \inf \varphi(X) + \rho\varepsilon^2,$$

there exists f *in* $\mathscr{M}$ *and* v_ε *in* X *with the properties*:

(II) v_ε *is strongly minimum point for* $\varphi + f$,

(III) $$\|f\|_{\mathscr{M}} < \varepsilon,$$

(IV) $$\|v_\varepsilon - x_0\| < \varepsilon.$$

■ The same demonstration as for 2.3.

It can suppose that there exists b bump function in $\mathscr{M}$ with

(40) $b(0) = 1$, (41) $b(X) \subset [0, 1]$, (42) $\operatorname{supp} b \subset S(0, 1)$

(indications in the proof of 3.2).

Set $M := \|b\|_{\mathscr{M}}$. We have (43) $M \ge 1$ as $M \overset{1^0}{\ge} \|b\|_\infty \overset{(40),(41)}{=} 1$. Take (44) $\rho := \dfrac{1}{4M}$ and consider the function

(45) $$g: X \to (-\infty, +\infty],\ g(x) = \varphi(x) - 2\rho\varepsilon^2 b\left(\frac{x - x_0}{\varepsilon}\right).$$

g is bounded from below, l. s. c. and proper. Apply 3.2 and 3.1 to it, there exists h in $\mathcal{M}$ with (46) $\|h\|_{\mathcal{M}} < \frac{\rho\varepsilon^2}{2}$ and $g + h$ has in X a strong minimum point v_ε ($g + h = g - (-h)$, $\mathcal{M}$ is Banach space).

Verify IV. Suppose *par absurdum* $\|v_0 - x_0\| \geq \varepsilon$, then $\frac{v_\varepsilon - x_0}{\varepsilon} \notin S(0,1)$ (open ball), hence $b\left(\frac{v_\varepsilon - x_0}{\varepsilon}\right) \overset{(42)}{=} 0$ and hence (47) $g(v_\varepsilon) \overset{(45)}{=} \varphi(v_\varepsilon) \geq \inf \varphi(X)$. Moreover,

(48) $$g(x_0) \overset{(45),(40)}{=} \varphi(x_0) - 2\rho\varepsilon^2 \overset{(\mathrm{I})}{<} \inf \varphi(X) - \rho\varepsilon^2.$$

But

$$(g + h)(v_\varepsilon) \leq (g + h)(x_0),$$

hence

$$g(v_\varepsilon) \leq g(x_0) + h(x_0) - h(v_\varepsilon) \leq g(x_0) + 2\|h\|_\infty \overset{(46),1^0}{<} g(x_0) + \rho\varepsilon^2,$$

consequently $\inf \varphi(X) \overset{(47)}{<} g(x_0) + \rho\varepsilon^2$, which contradicts (48), (IV) is verified.

Take now $f: X \to \mathbf{R}$,

(49) $$f(x) = -2\rho\varepsilon^2 b\left(\frac{x - x_0}{\varepsilon}\right) + h(x).$$

We have $f \in \mathcal{M}$ and $\varphi + f = g + h$, consequently v_ε is a strongly minimum point for $\varphi + f$, (II) is verified.

Verify (III). $\|f\|_{\mathcal{M}} \overset{(49)}{\leq} 2\rho\varepsilon^2 \|b\|_{\mathcal{M}} + \|h\|_{\mathcal{M}} \overset{(46)}{<} 2\rho\varepsilon^2 M + \rho\frac{\varepsilon^2}{2}$, but $2\rho \overset{(44)}{=} \frac{1}{2M} \overset{(43)}{\leq} \frac{1}{2}$ and then $2\rho\varepsilon^2 M + \rho\frac{\varepsilon^2}{2} \leq \frac{1}{2}\varepsilon^2 + \frac{1}{8}\varepsilon^2 < \varepsilon^2 < \varepsilon$ and consequently $\|f\|_{\mathcal{M}} < \varepsilon$. ■

Remark. Compare the hypotheses of G-M principle (3.3) with those from D-G-Z principle (2.3).

Pass to the second variational principle. This will be *linear* perturbed.

Definition. Let X be a real normed space, $f: X \to (-\infty, +\infty]$, C nonempty subset of X and $x_0 \in C$. f *strongly exposes* C *from below in* x_0, when

1° $f(x_0) = \inf f(C) < +\infty$

and

2° $x_n \in C\ \forall\, n \geq 1,\ f(x_n) \to f(x_0) \Rightarrow x_n \to x_0$.

When $C = X$, we fall on the definition of strongly minimum point (I, above 1.2).

In the same manner define

f *strongly exposes* C *from above in* x_0.

3.4 Ghoussoub - Maurey linear principle. *Let* X *be reflexive separable space and* φ: X → (− ∞, + ∞] *lower semicontinuous and proper.*

(I) *If* φ *is bounded from below on the closed bounded nonempty subset* C, *the set*

$$\{\xi \in X^* : \varphi + \xi \text{ strongly exposes } C \text{ from below}\}$$

is of G_δ *type and everywhere dense.*

(II) *If, for any* ξ *from* X^*, φ + ξ *is bounded from below, the set*

$$\{\xi \in X^* : \varphi + \xi \text{ strongly exposes } X \text{ from below}\}$$

is of G_δ *type and everywhere dense.*

Theorem 3.4 devolves (see the continuation to 3.10) from a more general theorem 3.10 and we pass to the preparation of this with definitions and some auxiliary propositions.

Definition. Let X be real normed space and C, D with $C \subset D$ nonempty subsets of X^*. C is *strict* w-H_δ *set in* D resp. *strict* w^*-H_δ *set in* D if

$$D \setminus C = \bigcup_{n=1}^{\infty} K_n, \tag{50}$$

$\operatorname{dist}(K_n, C) > 0$, K_n convex and weakly compact resp. ∗-weakly compact.

For instance,

3.5 *Any nonempty closed set* C *of a separable reflexive space* X, C ≠ X, *is strict* w-H_δ *set in* X.

In particular, if φ : X → (− ∞, + ∞] *is l. s. c. and proper, then the epigraph of* φ *in* X × × **R** *is strict* w-H_δ *set in* X × **R**.

■ Let $(x_n)_{n\geq 1}$ be a sequence with the set of the terms dense in $X \setminus C$ (open set). Take, for each n from **N**, K_n the closed ball centred in x_n with radius $r_n := \frac{1}{4}\operatorname{dist}(x_n, C)$. K_n is convex, weakly compact (Kakutani - Šmulian theorem, [47], vol III, p.750) and $\operatorname{dist}(K_n, C) > r_n$: let x be from K_n, $\operatorname{dist}(x, C) \geq \operatorname{dist}(x_n, C) - \operatorname{dist}(x, x_n) \geq 4r_n - r_n = 3r_n$. Take the greatest lower bound. Moreover, $X \setminus C = \bigcup_{n=1}^{\infty} K_n$: let x be any from $X \setminus C$, $\exists$ $(x_{p_n})_{n\geq 1}$ subsequence of $(x_n)_{n\geq 1}$ so that $x_{p_n} \to x$, which also implies dist$(x_{p_n}, C) \to$ dist(x, C), and if u, v are taken so that $0 < u < v <$ dist(x, C), from a rank on we have $4\,\text{dist}(x_{p_n}, x) < u$, but on the other side $4r_n = \text{dist}(x_{p_n}, C) > v$, hence $x \in K_{p_n}$. ■

Let X be reflexive space and C, D subsets of X^*, $C \subset D$. Set

$$M(C, D) := \{x \in X : \exists\, \xi \in C \text{ so that } Jx(\xi) \geq Jx(\eta)\ \forall\, \eta \in D\},$$

otherwise expressed, $M(C, D)$ is the set of x from X for which Jx is upper bounded on D and the least upper bound is attained at a point of C, J the Hahn imbedding of X in X^{**}. So, X being reflexive, J is an isomorphism of vector spaces which preserves the norms.

In the following, to abridge writing, sometimes x *designates* Jx.

Retain that if C is *-weakly compact, $M(C, D)$ is closed.

Notations. $S_X(x_0, r) \equiv$ the closed ball centred in x_0 of radius r in the normed space X.

$S_X \equiv S_X(0,1)$.

$\overline{E}^* \equiv$ the closure of the subset E from X^* for the *-weak topology.

Pass to the auxiliary propositions.

3.6 *Let* X *be reflexive space,* $D \subset X^*$ *and* $K \subset D$, K *convex *-weakly compact. If*

$$S_X(x, \alpha) \subset M(K, D),$$

then, for any $\varepsilon > 0$,

$$S(D, Jx, \varepsilon) \subset K + \frac{\varepsilon}{\alpha} S_{X^*}.$$

In particular, when $C \subset D \subset \overline{\text{conv}}^{\,*} C$, *we have*

$$\text{dist}(K, C) = 0.$$

■ *First assertion*. This reverts to

(51) $$\xi \notin K + \frac{\varepsilon}{\alpha} S_{X^*} \Rightarrow \xi \notin S(D, x, \varepsilon).$$

Suppose *par absurdum* $\xi \in S(D, x, \varepsilon)$, i. e. (see (2))

(52) $x(\xi) > \sup x(D) - \varepsilon$ ($x \in M(K, D) \Rightarrow Jx(D)$ upper bounded).

The first member of (51) gives

(53) $$\|\xi - \eta\|_{X^*} > \frac{\varepsilon}{\alpha}\ \ \forall\, \eta \in K.$$

Let z be from X so that

(54) $$\|Jz\|_{X^{**}} (= \|z\|) = 1 \text{ and } Jz(\xi - \eta) = \|\xi - \eta\|_{X^*}$$

(Hahn lemma, [47], vol. III, p.697).

Combining with (53) and taking the least upper bound, one obtains

(55) $$\sup z(K) \le z(\xi) - \frac{\varepsilon}{2}.$$

But $x + \alpha z \overset{(54)}{\in} S_X(x, \alpha) \subset M(K, D)$, hence $\exists\ \eta_0$ in K so that

(56) $$(x + \alpha z)(\eta_0) \ge (x + \alpha z)(\xi).$$

But

$$(x + \alpha z)(\xi) = x(\xi) + \alpha z(\xi) \overset{(52),(55)}{>} [\sup x(D) - \varepsilon] + \alpha[\sup z(K) + \frac{\varepsilon}{\alpha}] \ge x(\eta_0) + z(\eta_0)$$

and we come in collision with (56), (51) is validated.

Second assertion. This results from

(57) $$\bigcap_{\varepsilon>0} S(D, x, \varepsilon) \subset K,$$

(58) $$\bigcap_{\varepsilon>0} S(D, x, \varepsilon) \cap C \ne \varnothing.$$

For (57): $\xi \in \bigcap_{\varepsilon>0} S(D, x, \varepsilon) \overset{(51)}{\Rightarrow} \xi = \eta_\varepsilon + \frac{\varepsilon}{\alpha} u_\varepsilon$, $\eta_\varepsilon \in K$, $\|u_\varepsilon\| \le 1$, hence $\|\xi - \eta_\varepsilon\| \le \frac{\varepsilon}{\alpha}$, ξ is strong adherent point of K and the strong closure of K is included in the *-weak closure of this, that is in K. ■

3.7 *Let* X *be reflexive space,* $C \subset X^*$ *nonempty and* $U \subset X$ *nonempty open having the property*

$$\sup Jx\,(C) < +\infty\ \forall\ x \in U.$$

Then Jx, *for any* x *from* U, *is upper bounded on* $\overline{\text{conv}}^{\,*} C$ *and attains its least upper bound.*

■ Set $D := \overline{\text{conv}}^{\,*} C$. We have

(59) $$\sup Jx\,(C) = \sup Jx\,(D)$$

[$\xi \in \text{conv}\, C \Rightarrow \xi = \lambda_1 \xi_1 + \lambda_2 \xi_2$, $\xi_1, \xi_2 \in C$, $\lambda_1 + \lambda_2 = 1$, $\lambda_1, \lambda_2 \ge 0 \Rightarrow Jx(\xi) = \lambda_1 Jx(\xi_1) + \lambda_2 Jx(\xi_2) \le \sup Jx\,(C)$; $\xi \in D \Rightarrow \exists\ \xi_n \in \text{conv}\, C$, $\xi_n \xrightarrow{*\text{-weak}} \xi \Rightarrow \xi_n(x) \to \xi(x)$, $\xi_n(x) \le \sup Jx\,(C)\ \forall\ n \ge 1$, hence $\xi(x) \le \sup Jx\,(C)$]

and so (60) $\sup Jx(D) < +\infty \ \forall\, x \in U$, the first assertion is proved.

Pass to the second assertion. Fix x from U, $\exists\, \varepsilon > 0$ with $x + \varepsilon z \in U \ \forall\, z$ in S_X. Then

$$\sup J(x + \varepsilon z)(D) = \sup (Jx + \varepsilon Jz)(D) < +\infty \ \forall\, z \in S_X \ ((60)),$$

consequently, using once again (60),

$$\sup Jz(D) < +\infty \ \forall\, z \in S_X, \tag{61}$$

which implies (62) $\sup Jz(D) < +\infty \ \forall\, z \in X$ [for z any fixed, $z \neq 0$ replace in (61) z by $\frac{z}{\|z\|}$, $Jy(\xi) = \xi(y)$]. Replacing z by $-z$ in (62) one finds (63) $\inf Jz(D) > -\infty \ \forall\, z \in X$. But X is reflexive, hence (62) and (63) express that D is weakly bounded, consequently D is even bounded. D being also *-weak closed, it is *-weak compact ([47], vol.III, p. 744), hence the conclusion by applying Weierstrass theorem. ■

Remark. The proposition 3.7 is Lemma 2.7 from [34], Ch.2. The proof of this is painful.

3.8 *Let* X *be reflexive space,* $C \subset X^*$ *nonempty and* U *nonempty open from* X *so that*

$$\sup Jx(C) < +\infty \ \forall\, x \in U.$$

If C *is strict* w*-H_δ *set in* $D := \overline{\text{conv}}^*\, C$, *then the set*

$$V := \{x \in U : Jx \textit{ attains } \sup Jx(D) \textit{ in } C\}$$

includes a set of G_δ *type dense in* U.

■ According to the definition

$$D \setminus C = \bigcup_{n=1}^{\infty} K_n\,,\ K_n \text{ convex *-weakly compact and } \operatorname{dist}(K_n, C) > 0. \tag{64}$$

Every $Jx, x \in U$ is upper bounded on D and attains its supremum on this (3.7). As $\operatorname{dist}(K_n, C) > 0$, 3.6 prevents $M(K_n, D)$ to include any nonempty ball, in other words $\forall\, n \geq 1$ $\operatorname{int} M(K_n, D) = \varnothing$, and so $M(K_n, D)$, being also closed, it is thin and $U_n := X \setminus M(K_n, D)$ is open and dense in X. Then $\bigcap_{n=1}^{\infty} U_n$ is dense in X (Baire theorem), hence $\bigcap_{n=1}^{\infty} (U \cap U_n)$, a set of G_δ type, is dense in U. We see that if $x \in \bigcap_{n=1}^{\infty} (U \cap U_n)$, then Jx, which is upper bounded on D, attains *a fortiori* its least upper bound on C, as $x \notin M(K_n, D) \ \forall\, n \geq 1$ and one takes into account (64).

3.9 *Let* X *be reflexive space,* C *subset of* X^* *separable,* $D := \overline{\text{conv}}^*\, C$ *and* U *nonempty* *open subset of* X *so that* sup Jx (C) < + ∞ ∀ x ∈ U. *Suppose that* M(C, D) *includes a dense and of* G_δ *type subset of* U.

Then, for any K ⊂ D **-weakly compact with* K ∩ C = ∅ *and for any* ε > 0, *the set*

$$G(K, \varepsilon) := \{x \in U : \exists\, r > 0 \textit{ so that } \overline{S}^*(D, Jx, r) \cap K = \varnothing \textit{ and } \operatorname{diam} \overline{S}^*(D, Jx, r) < \varepsilon\}$$

is open and dense in U.

■ G(K, ε) *is open.* Use the fact that, *D* being bounded (proof for 3.7), the subset *S*(*D*, *Jx*, *r*) is also bounded.

G(K, ε) *is dense in* U. Let be $V \subset U$, *V* nonempty open arbitrary. *C* being separable, we can find a sequence $(C_n)_{n\geq 1}$, $C_n \subset D$, C_n convex *-weakly compact, with the properties

$$(65)\quad C \subset \bigcup_{n=1}^{\infty} C_n, \qquad (66)\quad \operatorname{dist}(C_n, K) > 0\ \forall\, n, \qquad (67)\quad \operatorname{diam} C_n \leq \frac{\varepsilon}{2}\ \forall\, n.$$

We have obviously $V \supset \bigcup_{n=1}^{\infty} M(C_n, D) \cap V \overset{(65)}{\supset} M(C, D) \cap V$. But the last member includes, according to the hypothesis, a set of G_δ type dense in *V*, which forces, *via* the Baire theorem, the second member to have at least one term, let this be $M(C_{n_0}, D) \cap V$, with nonempty interior. From this interior take a point *x*. It remains to show

$$(68)\qquad x \in G(K, \varepsilon).$$

Apply 3.6, for every ρ > 0 there exists *r* > 0 so that

$$(69)\qquad S(D, Jx, r) \subset C_{n_0} + \rho S_{X^*}$$

($\exists\, \alpha > 0$ so that $S_X(x, \alpha) \subset M(C_{n_0}, D)$, take $r = \alpha\rho$). Take (70) $\rho < \min\{\frac{\varepsilon}{4}, \operatorname{dist}(C_{n_0}, K)\}$. (69) gives

$$(71)\qquad \overline{S}^*(D, Jx, r) \subset \overline{C_{n_0} + \rho S_{X^*}}^{\,*} = C_{n_0} + \rho S_{X^*},$$

because the last term being *-weakly compact, it is *-weakly closed also. But

$$(72)\qquad (C_{n_0} + \rho S_{X^*}) \cap K = \varnothing:$$

let be *par absurdum* $\xi + \rho u = \zeta$, $\xi \in C_{n_0}$, $u \in S_{X^*}$, $\zeta \in K$, then $\rho = \|\xi - \zeta\| \geq \operatorname{dist}(C_{n_0}, K)$ and we come in collision with (70). So (71) and (72) give $\overline{S}^*(D, Jx, r) \cap K = \varnothing$. Moreover,

$$\operatorname{diam} \overline{S}^*(D, Jx, r) \overset{(71)}{\leq} \operatorname{diam} C_{n_0} + 2\rho \overset{(67),(70)}{<} \frac{\varepsilon}{2} + \frac{\varepsilon}{2} = \varepsilon,$$

which finishes the proof. ■

And now

3.10 Theorem. *Let* X *be reflexive space*, C *separable subset of* X^* *which is strict* w*-H_δ *set in* $D := \overline{\text{conv}}^* C$ *and* U *open subset of* X *so that* $\sup Jx(C) < +\infty\ \forall\, x \in U$. *Then*

(I) *The set*

$$\{x \in U : Jx \textit{ strongly exposes } D \textit{ from above at a point of } C\}$$

is of G_δ *type and dense in* U;

(II) *If* $\varphi : C \to (-\infty, +\infty]$ *is proper lower semicontinuous and* $\varphi + Jx$ *is*, $\forall$ x *from* X, *bounded from below on* C, *then the set*

$$\{x \in X : \varphi + Jx \textit{ strongly exposes } C \textit{ from below}\}$$

is of G_δ *type and dense in* X

(*N. Ghoussoub, B. Maurey*).

■ (I). According to the hypothesis, $D \setminus C = \bigcup_{n=1}^{\infty} K_n$, K_n convex *-weakly compact, $\operatorname{dist}(K_n, D) > 0$. *M(C, D)* includes a subset of G_δ type dense in *U* (3.8), but then, for each *n* from **N**, the set $V_n := G(K_1 \cup K_2 \cup \ldots \cup K_n, \frac{1}{n})$ is open and dense in *U* (3.9), consequently $\bigcap_{n=1}^{\infty} V_n$ is dense in *U* (relativized Baire theorem) and it remains only to observe that $\bigcap_{n=1}^{\infty} V_n = \{x \in X : Jx$ strongly exposes D from above at a point of $C\}$.

(II). $C \times \mathbf{R}$, separable subset of $X^* \times \mathbf{R}$, is a strict w*-H_δ set in $D \times \mathbf{R}$ and then, φ being l. s. c., the epigraph epi φ in $C \times \mathbf{R}$ (nonempty set, φ is proper) is a strict w*-H_δ set in $D \times \mathbf{R}$ and hence also in $\overline{\text{conv}}^*$ epi φ. $W := \{(x, \alpha) : x \in X, \alpha < 0\}$ is open in $X \times \mathbf{R}$ and $\sup (Jx, \alpha)(\text{epi}\ \varphi) < +\infty\ \forall\ (x, \alpha) \in W$ [(Jx, α), continuous linear functional, acts on $C \times \mathbf{R}$ by the rule $(Jx, \alpha)(\xi, \lambda) = Jx(\xi) + \alpha\lambda$]. Indeed, $Jx(\xi) + \alpha\lambda \leq Jx(\xi) + \alpha\varphi(\xi)$, $\exists\ a$ in **R** so that $J\frac{x}{\alpha}(\xi) + \varphi(\xi) \geq a\ \forall\ \xi \in C$ (the hypothesis), hence $Jx(\xi) + \alpha\varphi(\xi) \leq \alpha a\ \forall\ \xi \in C$. Show that for each $\varepsilon > 0\ \exists\ y_0$ with $\|y_0\| \leq 2\varepsilon$ and $\varphi + Jy_0$ strongly exposes C from below, which is enough to validate (II). Apply (I), $\exists\ (x_\varepsilon, \alpha_\varepsilon)$ in W so that

(73) $$\|(x_\varepsilon, \alpha_\varepsilon) - (0, -1)\| \le \varepsilon$$

($(0, -1) \in W$!) and $(Jx_\varepsilon, \alpha_\varepsilon)$ strongly exposes epi φ from above in a point (ξ_0, λ_0). Then $\forall\ (\xi, \lambda)$ from epi φ with $\xi \neq \xi_0$ we have

$$Jx_\varepsilon(\xi_0) + \alpha_\varepsilon \lambda_0 > Jx_\varepsilon(\xi) + \alpha_\varepsilon \lambda,$$

consequently taking $y_0 := \dfrac{x_\varepsilon}{\alpha_\varepsilon}$ we have in particular

$$\varphi(\xi_0) + Jy_0(\xi_0) < \varphi(\xi) + Jy_0(\xi)\ \forall\ \xi \in C \setminus \{\xi_0\},$$

ξ_0 is a strict global minimum point for $\varphi + Jy_0$. Moreover, as it can suppose $\varepsilon < \dfrac{1}{2}$, we have $\|y_0\| = \dfrac{\|x_\varepsilon\|}{|\alpha_\varepsilon|} \le 2\varepsilon$, because, *via* (73), $\|x_\varepsilon\| \le \varepsilon$ and $|\alpha_\varepsilon + 1| \le \varepsilon$, hence $\alpha_\varepsilon \in \left(-\dfrac{3}{2}, -\dfrac{1}{2}\right)$.

Finally, let $(\xi_n)_{n\ge1}$ be a minimizing sequence for $\varphi + Jy_0$ on C, then $(\xi_n, \varphi(\xi_n))_{n\ge1}$ is maximizing sequence for $(Jx_\varepsilon, \alpha_\varepsilon)$ which strongly exposes epi φ in (ξ_0, λ_0), which impo-ses $\xi_n \to \xi_0$. ■

Proof of Ghoussoub - Maurey linear principle

■ (I). Set $Y := X^*$, a separable reflexive space ([47], vol.III, p.761). Then $Y^* = X$ (identification *via* the Hahn imbedding, X is reflexive). C is separable and strict w*-H_δ set in Y^* (3.5, the weak and *-weak topologies coincide) and hence also in $D := \overline{\text{conv}}^* C$ ($X \setminus C = \bigcup_{n=1}^{\infty} K_n$ with the properties from (50), take the intersection with D). Apply (II), 3.10 transcribed with Y replaced by X, this is correct as $\varphi + \xi$, $\xi \in X^* = Y^{**}$, is bounded from below ($|\xi(x)| \le \|\xi\|\,\|x\|$ and C is bounded).

(II). The epigraph epi φ of φ in $X \times \mathbf{R}$ is strict w-H_δ set in $X \times \mathbf{R}$ (3.5). In the following use the proof for (II), 3.10 begining from (73), epi φ is that considered above. ■

3.11 *Corollary. Let* X *be reflexive space,* C *a subset of* X^* *separable bounded strict* w*-H_δ *set in* $D := \overline{\text{conv}}^* C$ *and* $\varphi : X \to (-\infty, +\infty]$ *bounded from below l. s. c. proper. For any* $\varepsilon > 0$ *there exists* x_0 *in* X *with* $\|x_0\| \le \varepsilon$ *and* ξ_0 *in* C *so that*

1° $(\varphi + Jx_0)(\xi_0) < (\varphi + Jx_0)(\xi)\ \forall\ \xi \in C \setminus \{\xi_0\}$,

2° *Any minimizing sequence from* C *for* $\varphi + Jx_0$ *converges to* ξ_0 ([34]).

■ C bounded implies $\varphi + Jx$ bounded from below $\forall\ x \in X$, consequently II, 3.10 can intercede to obtain 1° and 2°. ■

Application

We imply theorem 3.4 in a minimization problem of the form (*H. Brezis - L. Nirenberg*, *A minimization problem with critical exponent and non-zero data*, Symmetry in Nature, Scuola Norm. Sup. Pisa (1989), p. 129-140)

(74) $$C_f := \min\{ \int_\Omega (\frac{1}{2}|\nabla u|^2 - f(u))\, dx : u \in H_0^1(\Omega), \|u\|_{2^*} = 1\},$$

where Ω is open set of C^1 class in $\mathbf{R}^N$, $N \geq 3$, $f \in H^{-1}(\Omega)$ ($= (H_0^1(\Omega))^*$), $2^* = \frac{2N}{N-2}$ – the critical exponent for the Sobolev imbedding (for the necessary explanations, here and in the following, see I, §4, last section).

Let Ω be an open bounded set of C^1 class in $\mathbf{R}^N$, $N \geq 3$. Consider the problem

(75) $$\begin{cases} -\Delta u = f(x,u) \text{ in } \Omega \\ u = 0 \text{ on } \partial\Omega, \end{cases}$$

where $f : \Omega \times \mathbf{R} \to \mathbf{R}$ is a Carathéodory function with the growth condition

(76) $$|f(x, s)| \leq c|s|^{p-1} + b(x),$$

$$c > 0,\ 2 \leq p \leq \frac{2N}{N-2},\ b \in \mathbf{L}^{p'}(\Omega), \frac{1}{p} + \frac{1}{p'} = 1.$$

The functional $\varphi : H_0^1(\Omega) \to \mathbf{R}$,

(77) $$\varphi(u) = \int_\Omega (\frac{1}{2} | \nabla u |^2 - F(x,u(x)))\, dx,$$

$$F(x, s) := \int_0^s f(x,t)\, dt,$$

is of C^1 - Fréchet class and its critical points are the weak solutions of (75).

Let λ_1 be the first eigenvalue of $-\Delta$ in $H_0^1(\Omega)$ with homogeneous boundary conditions. We have

(78) $$\lambda_1 = \inf\{ \frac{\|u\|^2_{H_0^1}}{\|u\|_2^2} : u \in H_0^1(\Omega) \setminus \{0\}\}$$

(the Rayleigh - Ritz quotient).

And now – an answer for (74).

3.12 *Suppose moreover verified the growth condition*

(79) $$F(x, s) \le c_1 \frac{s^2}{2} + \alpha(x)s,$$

$0 < c_1 < \lambda_1$ *and* $\alpha \in \mathbf{L}^{q'}(\Omega)$ *for some* $2 \le q \le \frac{2N}{N-2}$.

Then

1° *The set of functions* h *from* $H^{-1}(\Omega)$, *having the property the functional* $\varphi_h : H_0^1(\Omega) \to \mathbf{R}$,

$$\varphi_h(u) = \int_\Omega (\frac{1}{2} |\nabla u|^2 - F(x, u(x)) + h(u(x)))\, dx$$

has in only a point attained minimum, includes a G_δ *set everywhere dense*;

2° *The set of functions* h *from* $H^{-1}(\Omega)$, *having the property*

the problem $\begin{cases} -\Delta u = f(x, u) + h(u) \text{ in } \Omega \\ u = 0 \text{ on } \partial\Omega \end{cases}$ *has solutions,*

includes a G_δ *set everywhere dense*;

3° *Moreover, if* $s \to f(x, s)$ *is increasing, then the set of functions* h *from* $H^{-1}(\Omega)$, *having the property*

the problem $\begin{cases} -\Delta u = f(x, u) + h(u) \textit{ in } \Omega \\ u = 0 \textit{ on } \partial\Omega \end{cases}$ *has a unique solution,*

includes a G_δ *set everywhere dense* ([34], theorem 2.13).

■ It is sufficient to justify 1°. Consider, for each h from $H^{-1}(\Omega)$, the functional ξ_h from $H^{-1}(\Omega)$,

$$\xi_h(u) = \int_\Omega h(u(x))\, dx .$$

One sees that $\varphi_h = \varphi + \xi_h$ (see (77)). Consequently, according to (II), 3.4, if we show that φ_h is bounded from below for any h from $H^{-1}(\Omega)$ (this is enough, representation Riesz theorem), then 1° is proved. But, taking into account the Sobolev imbedding and (79), we have $\forall\, u \in H_0^1(\Omega)$

$$(\varphi + \xi_h)(u) \ge \int_\Omega \frac{1}{2}(|\nabla u|^2 - c_1|u(x)|^2)\, dx - \|\alpha\|_{q'}\ \|u\|_q - \|h\|_{H^{-1}}\ \|u\|_{H_0^1} \ge \frac{1}{2}\left(1 - \frac{c_1}{\lambda_1}\right) \times$$

$$\times \int_{\Omega} |\nabla u|^2 \, dx - r\|u\|_{H_0^1}, \; r \in \mathbf{R},$$

and hence the conclusion because $1 - \frac{c_1}{\lambda_1} > 0$. ■

4

Generalization and unification of the Ekeland and Borwein-Preiss principles

The remark which follows to 1.2_1 from §1 shows that the Ekeland principle cannot be recovered in all its force from the Borwein-Preiss principle. The following theorem, due to Li Yongxin and Shi Shuzhong [41]), generalizes and unifies this two principles.

Definition. Let (X, d) be a metric space. $\rho: X \times X \to [0, +\infty]$ is called *gauge function* on X if

(1) for any x from X, $\rho(x, x) = 0$;

(2) for any sequence $(x_n, y_n)_{n\geq 1}$ from $X \times X$, $\rho(x_n, y_n) \to 0 \Rightarrow \mathrm{d}(x_n, y_n) \to 0$;

(3) for any y from X, $x \to \rho(x, y)$ is lower semicontinuous.

For instance, $f \circ \mathrm{d}$ with $f: \mathbf{R}_+ \to \mathbf{R}_+$ continuous strictly increasing and $f(0) = 0$ is a gauge function on X (attention to the inverse function of f).

4.1 Li Yongxin - Shi Shuzhong theorem. *Let* (X, d) *be a complete metric space and* $\varphi: \mathrm{X} \to (-\infty, +\infty]$ *bounded from below, lower semicontinuous and proper. For any* $\varepsilon > 0$ *and* x_0 *from* X *with*

(4) $$\varphi(\mathrm{x}_0) \leq \inf \varphi(\mathrm{X}) + \varepsilon,$$

for any gauge function $\rho: \mathrm{X} \times \mathrm{X} \to [0, +\infty]$ *and for any sequence* $(\delta_n)_{n\geq 0}$, $\delta_0 > 0$ *and* $\delta_n \geq 0 \;\forall\, n \geq 1$, *there exists a convergent sequence* $(\mathrm{x}_n)_{n\geq 1}$ *from* X, $\lim\limits_{n\to\infty} \mathrm{x}_n = \mathrm{v}_\varepsilon$, *so that*

(5) $$\rho(v_\varepsilon, x_n) \le \frac{\varepsilon}{2^n \delta_0} \quad \forall\, n \ge 0;$$

If $\delta_n > 0$ *for an infinity of indices, then*

(6) $$\varphi(v_\varepsilon) \le \varphi(x_0) - \sum_{n=0}^{\infty} \delta_n\, \rho(v_\varepsilon, x_n)$$

and

(7) $$\varphi(v_\varepsilon) + \sum_{n=0}^{\infty} \delta_n\, \rho(v_\varepsilon, x_n) < \varphi(x) + \sum_{n=0}^{\infty} \delta_n\, \rho(x, x_n) \ \forall\, x \in X \setminus \{v_\varepsilon\};$$

If $\delta_N > 0$, *where* $N \ge 0$, *and* $\delta_k = 0$ *for* $k > N$, *there exists* $m > N$ *so that* (6) *is preserved and* (7) *is replaced by*

(8) $$\varphi(v_\varepsilon) + \sum_{k=0}^{N-1} \delta_k\, \rho(v_\varepsilon, v_k) + \delta_N\, \rho(v_\varepsilon, x_m) < \varphi(x) + \sum_{k=0}^{N-1} \rho_k\, \rho(x, x_k) + \delta_N\, \rho(x, x_m),$$

$\forall\, x \in X \setminus \{v_\varepsilon\}$.

Specification. If $N = 0$ the terms with $\sum_{k=0}^{N-1}$ disappear.

■ The idea of the demonstration is found in [2].

The sequence $(\delta_n)_{n\ge 0}$ has two possibilities: $\delta_n > 0$ for an infinity of indices or $\delta_n > 0$ for a finite number of indices only. In the first case, the attentive reading of the statement shows that the supposition $\delta_n > 0\ \forall\, n \ge 0$ is not restrictive, and therefore we make it.

Case I ($\delta_n > 0\ \forall\, n \ge 0$). Consider the set

(9) $$T(x_0) := \{x \in X : \varphi(x) + \delta_0 \rho(x, x_0) \le \varphi(x_0)\}.$$

We have (10) $x_0 \in T(x_0)$ and $T(x_0)$ is closed ($\varphi + \delta_0\, \rho(\,\cdot\,, x_0)$ is l.s.c., see (3)). So

$$y \in T(x_0) \Rightarrow \delta_0 \rho(y, x_0) \le \varphi(x_0) - \varphi(y) \le \varphi(x_0) - \inf \varphi(X) \overset{(4)}{\le} \varepsilon,$$

(11) $$y \in T(x_0) \Rightarrow \delta_0\, \rho(y, x_0) \le \varepsilon.$$

Take x_1 in $T(x_0)$ so that (at the left to (12) is an element of the set to which one takes the infimum, this is bounded from below)

(12) $$\varphi(x_1) + \delta_0 \rho(x_1, x_0) < \inf_{x \in T(x_0)} [\varphi(x) + \delta_0 \rho(x, x_0)] + \delta_1 \frac{\varepsilon}{2\delta_0}$$

and then

$$(13)\qquad T(x_1) := \left\{ x \in T(x_0) : \varphi(x) + \sum_{i=0}^{1} \delta_i \rho(x, x_i) \le \varphi(x_1) + \delta_0 \rho(x_1, x_0) \right\} .$$

We have (14) $x_1 \in T(x_1)$ and $T(x_1)$ is closed.

Suppose defined x_{n-1} in $T(x_{n-2})$ and $T(x_{n-1})$,

$$(15)\qquad T(x_{n-1}) = \left\{ x \in T(x_{n-2}) : \varphi(x) + \sum_{i=0}^{n-1} \delta_i \rho(x, x_i) \le \varphi(x_{n-1}) + \sum_{i=0}^{n-2} \delta_i \rho(x_{n-1}, x_i) \right\} .$$

We have (16) $x_{n-1} \in T(x_{n-1})$ and $T(x_{n-1})$ is closed. Then one takes x_n from $T(x_{n-1})$ so that

$$(17)\qquad \varphi(x_n) + \sum_{i=0}^{n-1} \delta_i \rho(x_n, x_i) < \inf_{x \in T(x_{n-1})} \left[\varphi(x) + \sum_{i=0}^{n-1} \delta_i \rho(x, x_i) \right] + \delta_n \frac{\varepsilon}{2^n \delta_0}$$

and then

$$(18)\qquad T(x_n) := \left\{ x \in T(x_{n-1}) : \varphi(x) + \sum_{i=0}^{n} \delta_i \rho(x, x_i) \le \varphi(x_n) + \sum_{i=0}^{n-1} \delta_i \rho(x_n, x_i) \right\} .$$

We have (19) $x_n \in T(x_n)$ and $T(x_n)$ is closed.

So the sequences $(x_n)_{n\ge 0}$ and $(T(x_n))_{n\ge 0}$ are defined by recurrence. The second sequence is decreasing with the terms nonempty and closed ((19)).

We have

$$(20)\qquad q > n \ge 0 \Rightarrow \varphi(x_q) + \sum_{i=0}^{q-1} \delta_i \rho(x_q, x_i) \le \varphi(x_n) + \sum_{i=0}^{n-1} \delta_i \rho(x_n, x_i)$$

(when $n = 0$ the second term of the second member disappears).

Indeed, as $T(x_{n+1}) \subset T(x_n)$ and $x_{n+1} \in T(x_{n+1})$, in (18) it can take $x = x_{n+1}$, it follows

$$\varphi(x_{n+1}) + \sum_{i=0}^{n} \delta_i \rho(x_{n+1}, x_i) \le \varphi(x_n) + \sum_{i=0}^{n-1} \delta_i \rho(x_n, x_i) .$$

Repeat taking $x = x_{n+2}$ in $T(x_{n+1})$ etc. and one obtains (20).

Let now y be any in $T(x_n)$, $n \ge 1$. We have

$$\delta_n \rho(y, x_n) \overset{(18)}{\le} \varphi(x_n) + \sum_{i=0}^{n-1} \delta_i \rho(x_n, x_i) - \left[\varphi(y) + \sum_{i=0}^{n-1} \delta_i \rho(y, x_i) \right] \overset{T(x_n)\subset T(x_{n-1})}{\le} \varphi(x_n) + \sum_{i=0}^{n-1} \delta_i \rho(x_n, x_i) -$$

$$- \inf_{x \in T(x_{n-1})} \left[\varphi(x) + \sum_{i=0}^{n-1} \delta_i \rho(x, x_i) \right] \overset{(17)}{<} \delta_n \frac{\varepsilon}{2^n \delta_0} ,$$

hence, taking also into account (11),

(21) $$y \in T(x_n) \Rightarrow \rho(y, y_n) \le \frac{\varepsilon}{2^n \delta_0} \quad \forall\, n \ge 0 .$$

For each set $T(x_n)$, $n \ge 0$, consider whatever element from it, let be this y_n. (21) corroborated with (2) gives (22) $\lim\limits_{n\to\infty} d(y_n, x_n) = 0$. Then (23) $\lim\limits_{n\to\infty} \operatorname{diam} T(x_n) = 0$. Indeed, suppose *par absurdum* the contrary. Then there exists a subsequence $(T(x_{k_n}))_{n\ge 0}$ and $r > 0$ so that (24) $\operatorname{diam} T(x_{k_n}) > r \quad \forall\, n \ge 0$. According to (24), there are, in each $T(x_{k_n})$, two points y'_{k_n}, y''_{k_n} so that (25) $d(y'_{k_n}, y''_{k_n}) > r$. But $d(y'_{k_n}, y''_{k_n}) \le$ $\le d(y'_{k_n}, x_{k_n}) + d(y''_{k_n}, x_{k_n})$, this imposes, *via* (21), $d(y'_{k_n}, y''_{k_n}) \to 0$, in contradiction with (25) and so (23) is proved.

Apply Cantor theorem to the sequence $(T(x_n))_{n\ge 0}$, $\bigcap\limits_{n=0}^{\infty} T(x_n)$ has a single element,

(26) $$\bigcap_{n=0}^{\infty} T(x_n) = \{v_\varepsilon\} .$$

As $v_\varepsilon \in T(x_n)$ $\forall\, n \ge 0$, (21) and (2) give $\lim\limits_{n\to\infty} x_n = v_\varepsilon$. For the same reason, (21) supplies also (5) from the statement.

Now let x be any from X, $x \ne v_\varepsilon$. In accordance with (26), $\exists\, m \ge 0$ so that

(27) $$\varphi(x_m) + \sum_{i=0}^{m-1} \delta_i \rho(x_m, x_i) < \varphi(x) + \sum_{i=0}^{m} \delta_i \rho(x, x_i)$$

(when $m = 0$ the second term of the first member disappears).

For any $q \ge 0$ we have

$$\varphi(x_0) \overset{(20)}{\ge} \varphi(x_q) + \sum_{i=0}^{q-1} \delta_i \rho(x_q, x_i) \overset{v_\varepsilon \in T(x_{q+1}),(18)}{\ge} \varphi(v_\varepsilon) + \sum_{i=0}^{q} \delta_i \rho(v_\varepsilon, x_i) .$$

Isolate the first and the third term and taking the limit one obtains (6).

Finally, let be $x \ne v_\varepsilon$. As $v_\varepsilon \in T(x_{q+1})$, (18), (20) and (27) give for $q > n$

$$\varphi(v_\varepsilon) + \sum_{i=0}^{q+1} \delta_i \rho(v_\varepsilon, x_i) \le \varphi(x_{q+1}) + \sum_{i=0}^{q} \delta_i \rho(x_{q+1}, x_i) \le \varphi(x_m) + \sum_{i=0}^{m-1} \delta_i\, \rho(x_m, x_i) <$$

$$< \varphi(x) + \sum_{n=0}^{\infty} \delta_n\, \rho(x, x_n)$$

(correctly considered series, all the terms are positive).

Take the limit for $q \to \infty$ and find (7) (the series of the first member is perforce convergent by the intermediate term; otherwise (6) imposes the same conclusion).

Case II ($\delta_N > 0$, $k > N \Rightarrow \delta_k = 0$). Obviously it can suppose $\delta_i > 0$ even for $i < N$. The sequences $(x_n)_{n \geq 0}$ and $(T(x_n))_{n \geq 0}$ are defined as follows : for $n \leq N$ as in the first case (see (9), (12), (17), (18)); for $n > N$ $(N = 0, N > 0)$ take x_n in $T(x_{n-1})$ so that

$$\varphi(x_n) + \sum_{i=0}^{N-1} \delta_i \rho(x_n, x_i) < \inf_{x \in T(x_{n-1})} \left[\varphi(x) + \sum_{i=0}^{N-1} \delta_i \rho(x, x_i) \right] + \delta_N \frac{\varepsilon}{2^n \delta_0}$$

and then

$$T(x_n) := \left\{ x \in T(x_{n-1}) : \varphi(x) + \sum_{i=0}^{N-1} \delta_i \rho(x, x_i) + \delta_N \rho(x, x_n) \leq \varphi(x_n) + \sum_{i=0}^{N-1} \delta_i \rho(x_n, x_i) \right\} .$$

The reasoning remains the same as in the case I. One proves $y \in T(x_n) \Rightarrow \rho(y, x_n) \leq$ $\leq \dfrac{\varepsilon}{2^n \delta_0}$ $\forall\, n \geq 0$ and then, using this, $\lim_{n \to \infty} \operatorname{diam} T(x_n) = 0$, (28) $\bigcap_{n=0}^{\infty} T(x_n) = \{v_\varepsilon\}$, $x_n \to v_\varepsilon$ and then (5). (20) remains true for $0 \leq n < q \leq N$, hence in particular ($n = 0$, $q = N \geq 1$)

$$\varphi(x_0) \geq \varphi(x_N) + \sum_{i=0}^{N-1} \delta_i \rho(x_N, x_i) \overset{v_\varepsilon \in T(x_N)}{\geq} \varphi(v_\varepsilon) + \sum_{i=0}^{N} \delta_i \rho(v_\varepsilon, x_i),$$ from which we find (6) as $i > N \Rightarrow \delta_I = 0$. In the case $N = 0$, (6) is given by (9).

Pass to verify (8). Let x be from $X \setminus \{v_\varepsilon\}$. (28) supplies $m \geq 0$ so that $x \notin T(x_m)$. It can suppose $m > N$, because if $m \leq N$ *a fortiori* $x \notin T(x_q)$ $\forall\, q > N$ (decreasing sequence !) and we replace m by q in the statement.

So let be $m > N$. Then we have

$$(29) \qquad \varphi(x_m) + \sum_{i=0}^{N-1} \delta_i \rho(x_m, x_i) < \varphi(x) + \sum_{i=0}^{N-1} \delta_i \rho(x, x_i) + \delta_N \rho(x, x_m).$$

But $v_\varepsilon \overset{(28)}{\in} T(x_m)$ and taking $x = v_\varepsilon$ in the definition of $T(x_m)$ one finds that the first member from (29) is minorized by

$$\varphi(v_\varepsilon) + \sum_{i=0}^{N-1} \delta_i \rho(v_\varepsilon, x_i) + \delta_N \rho(v_\varepsilon, x_m) ,$$

(8) is proved. ■

Remark 1. (8) from 4.1 is verified for $m = N$ (use (18)).

Remark 2. Retain that the sequence of positive numbers $(\delta_n)_{n\geq 0}$ is not constrained by *any condition*, for instance it is not at least bounded.

Remark 3. The proof of 4.1 is simple, ingenious and elementary.

Remark 4. Some embarrassments from [41], which are in the demonstration of the case I, have been eliminated by the author.

Ekeland principle I, §1, 1.1_0 *is a particular case of the Li-Shi theorem.*

Indeed, place us in the conditions and notations from I, §1, 1.1 and apply 4.1 with $x_0 = u$, $\rho(x, y) = \frac{\varepsilon}{\lambda} d(x, y)$ and $(\delta_n)_{n\geq 0}$ with $\delta_0 = 1$, $\delta_n = 0$ for $n \geq 1$, hence $N = 0$. (5) from 4.1 gives, for $n = 0$, $\frac{\varepsilon}{\lambda} d(v_\varepsilon, u) \leq \varepsilon$, that is (3) from 1.1. (6) from 4.1 gives (2_0) from 1.1_0 and (8) from 4.1 becomes

$$x \in X \setminus \{v_\varepsilon\} \Rightarrow \varphi(v_\varepsilon) + \frac{\varepsilon}{\lambda} d(v_\varepsilon, x_m) < \varphi(x) + \frac{\varepsilon}{\lambda} d(x, x_m) ,$$

hence

$$\varphi(v_\varepsilon) < \varphi(x) + \frac{\varepsilon}{\lambda}[d(x, x_m) - d(v_\varepsilon, x_m)]$$

and consequently (1) from 1.1.

Borwein-Preiss principle (§1, 1.2_1) *is a particular case of the Li-Shi theorem.*

Place us in the conditions and the notations from 1.2_1 and apply 4.1 with $\varphi(x_0) = \inf \varphi(X) + \varepsilon'$, $\rho(x, y) = \frac{\varepsilon}{\lambda^p} d^p(x, y)$ and $(\delta_n)_{n\geq 0} = (\mu_n)_{n\geq 0}$, where $\mu_n > 0\ \forall\ n$, $\sum_{n=0}^{\infty} \mu_n = 1$.

There is $(v_n)_{n\geq 0}$ a convergent sequence, $\lim_{n\to\infty} v_n = v_{\varepsilon'}$ $(=: v_\varepsilon)$, so that $\frac{\varepsilon}{\lambda^p} d^p(v_\varepsilon, x_0) \overset{(5)}{\leq} \varepsilon' < \varepsilon$, hence $d(v_\varepsilon, x_0) < \lambda$, i.e. (IV) of 1.2_1,

$$\varphi(v_\varepsilon) + \frac{\varepsilon}{\lambda^p} \sum_{n=0}^{\infty} \mu_n d^p(v_\varepsilon, v_n) \overset{(6)}{\leq} \varphi(x_0) < \inf \varphi(X) + \varepsilon$$

and hence (III) from 1.2_1 (improved !).

$$x \in X \setminus \{v_\varepsilon\} \Rightarrow \varphi(v_\varepsilon) + \frac{\varepsilon}{\lambda^p} \sum_{n=0}^{\infty} \mu_n d^p(v_\varepsilon, v_n) \overset{(17)}{<} \varphi(x) + \frac{\varepsilon}{\lambda^p} \sum_{n=0}^{\infty} \mu_n d(x, v_n) ,$$

that is (II) from 1.2_1 (improved !).

Remark. As it was observed above, Li-Shi theorem gives an improved statement of Borwein-Preiss principle.

Pass to a variant of the theorem 4.1 for Banach spaces.

4.2. *Let* X *be Banach space and* $\varphi: X \to (-\infty, +\infty]$ *lower bounded l. s. c. proper. For any* $\varepsilon > 0$ *and* x_0 *from* X *so that*

$$\varphi(x_0) \le \inf \varphi(X) + \varepsilon, \tag{30}$$

for any sequence $(\delta_n)_{n\ge0}$ *with* $\delta_0 > 0$, $\delta_n \ge 0\ \forall\, n \ge 1$ *and for any* $r : X \to [0, +\infty]$ *l. s. c. with the properties* $r(0) = 0$, $r(x_n) \to 0 \Rightarrow \|x_n\| \to 0$, *there exists* $(x_n)_{n\ge0}$ *convergent sequence in* X, $\lim\limits_{n\to\infty} x_n = v_\varepsilon$, *so that*

$$r(v_\varepsilon - x_n) \le \frac{\varepsilon}{2^n \delta_0} \quad \forall\, n \ge 0; \tag{31}$$

When $\delta_n > 0$ *for an infinity of indices, we have*

$$\varphi(v_\varepsilon) \le \varphi(x_0) - \sum_{n=0}^{\infty} \delta_n r(v_\varepsilon - x_n) \tag{32}$$

and

$$\varphi(v_\varepsilon) + \sum_{n=0}^{\infty} \delta_n r(v_\varepsilon - x_n) < \varphi(x) + \sum_{n=0}^{\infty} \delta_n r(x - x_n) \quad \forall\, x \in X \setminus \{v_\varepsilon\}; \tag{33}$$

When $\delta_N > 0$, *where* $N \ge 0$, *and* $\delta_k = 0$ *for* $k > N$, *there exists* $m > N$ *so that* (32) *is preserved and* (33) *is replaced by*

$$\varphi(v_\varepsilon) + \sum_{k=0}^{N-1} \delta_k r(v_\varepsilon - x_k) + \delta_N r(v_\varepsilon - x_m) < \varphi(x) + \sum_{k=0}^{N-1} \delta_k r(x - x_k) + \delta_N r(x - x_m) \tag{34}$$

$\forall\, x \in X \setminus \{v_\varepsilon\}$

([41]).

Specification. If $N = 0$, the terms $\sum\limits_{k=0}^{N-1}$ disappear.

■ Apply 4.1 with the hypotheses from 4.2, where r is replaced by the gauge function $\rho: X \times X \to [0, +\infty]$,

$$\rho(x, y) = r(x - y). \ ■$$

And now pass to a generalization, *via* 4.2, of the Borwein-Preiss principle for Banach spaces (§1, 1.2) in the following variant. But first, as in §1, 1.1 (β is bornology on X, I, §3, "β - derivative"),

4.3. *Let* X *be real normed space,* r *a real Lipschitz function* β *- differentiable on* $S_\alpha := \{x \in X : \|x\| < \alpha\}$, v_ε *from* X, $(x_n)_{n\geq 0}$ *a sequence of points from* X *with*

$$\| v_\varepsilon - x_n \| < \frac{\alpha}{2} \quad \forall\, n \geq 0, \tag{35}$$

$(\delta_n)_{n\geq 0}$ *with* $\delta_n > 0\ \forall$ n *and* $\sum_{n=0}^{\infty} \delta_n$ *convergent and*

$$\Phi : \Phi(x) = \sum_{n=0}^{\infty} \delta_n r(x - x_n). \tag{36}$$

Then Φ *is* β *- differentiable in* v_ε *and*

$$\nabla_\beta \Phi(v_\varepsilon) = \sum_{n=0}^{\infty} \delta_n \nabla_\beta r(v_\varepsilon - x_n). \tag{37}$$

■ Use the rule 3.12 from I, §3 with $U = v_\varepsilon + S_{\frac{\alpha}{2}}$. $x_n \overset{(35)}{\in} U\ \forall\ n \geq 0$, hence $x \in U \Rightarrow \|x - x_n\| < \alpha\ \forall\ n \geq 0$, which imposes, r being Lipschitz on S_α, that the functions $x \to r(x - x_n)$, $\forall\ n \geq 0$ are uniformly bounded on U and consequently the series from (36) is convergent on U.

Show that the series $\sum_{n=0}^{\infty} \delta_n \xi_n(x)$, $\xi_n(x) := \nabla_\beta r(x - x_n)$ converges uniformly on U. Let S be any from β.

$$\xi_n(x)(h) = \underset{\substack{t\to 0\\ h\in S}}{\text{limu}} \frac{r(x - x_n + th) - r(x - x_n)}{t}.$$

But for t sufficiently small and $x \in U$ we have $x - x_n + th \in S_\alpha\ \forall\ n \geq 0$ (S is bounded), hence

$$\left| \frac{r(x - x_n + th) - r(x - x_n)}{t} \right| \leq L \| h \| \quad \forall\, n \geq 0,$$

L the Lipschitz constant, consequently

$$| \xi_n(x)(h) | \leq L \| h \| \quad \forall x \in U,\ \forall h \in S,\ \forall n \geq 0.$$

As $X = \bigcup_{S\in\beta} S$, it results

$$|\xi_n(x)(h)| \leq L\,\|h\| \quad \forall h \in X,\ \forall x \in U,\ \forall n \geq 0\ ,$$

that imposes

(38) $$\|\nabla_\beta r(x - x_n)\| \leq L \quad \forall x \in U,\ \forall n \geq 0\,.$$

Applying the Weierstrass rule we find that the demand (2) from I, 3.12 is also verified for the series of (36), it can β - differentiate on U term by term. ■

REMARK. In [41] the demonstration for the proposition 1, 4.3 here, must be repelled as totally false.

And now

4.4 *Let* X *be Banach space,* $\varphi : X \to (-\infty, +\infty]$ *lower bounded l. s. c. proper,* $\alpha, \lambda, k >$ > 0 *and* $r : X \to [0, +\infty]$ *l. s. c. with the properties*

(39) $$r(0) = 0,$$

(40) $$r(x_n) \to 0 \Rightarrow \|x_n\| \to 0,$$

(41) r *is Lipschitz with the constant* L *and* β - *differentiable on*

$$S_{r,\lambda} := \{x \in X : r(x) < \lambda\},$$

(42) $$\|x\| < \alpha \Rightarrow x \in S_{r,\lambda},\quad r(x) \leq \frac{\lambda}{k} \Rightarrow \|x\| < \frac{\alpha}{2}.$$

Then, for any $\varepsilon > 0$ *and* x_0 *from* X *with*

(43) $$\varphi(x_0) \leq \inf \varphi(X) + \varepsilon,$$

there exists v_ε *in* X *so that*

(44) $$\|v_\varepsilon - x_0\| \leq \frac{\alpha}{2},$$

(45) $$\varphi(v_\varepsilon) \leq \varphi(x_0),$$

(46) $$0 \in \partial_\beta \varphi(v_\varepsilon) + \frac{2\varepsilon k L}{\lambda} S^*,$$

$S^* := \{\xi \in X^* : \|\xi\| \leq 1\}$

([41]).

■ Take a sequence $(\delta_n)_{n\geq 0}$, $\delta_n > 0$ $\forall\, n$ and apply 4.2, $\exists$ $(x_n)_{n\geq 0}$ a convergent sequence from X, $\lim_{n\to\infty} x_n = v_\varepsilon$, so that

(47) $$r(v_\varepsilon - x_n) \leq \frac{\varepsilon}{2^n \delta_0} \quad \forall\, n \geq 0,$$

(48) $$\varphi(v_\varepsilon) + \Phi(v_\varepsilon) \le \varphi(x_0),$$

and

(49) $$\varphi(v_\varepsilon) + \Phi(v_\varepsilon) < \varphi(x) + \Phi(x) \ \forall\, x \in X \setminus \{v_\varepsilon\},$$

where $\Phi(x) = \sum_{n=0}^{\infty} \delta_n r(x - x_n)$.

(48) obviously supplies (45). Particularize the sequence $(\delta_n)_{n\ge 0}$ taking (50) $\delta_0 = = \sum_{n=1}^{\infty} \delta_n = \frac{k\,\varepsilon}{\lambda}$. Then (47) with $n = 0$ and (50) give (44) *via* (42).

Pass to (46). The definition (16) from I, §5, D tells, *via* (49), that

$$0 \in \partial_\beta (\varphi + \Phi)(v_\varepsilon) \subset \partial_\beta\, \varphi(v_\varepsilon) + \partial_\beta\, \Phi(v_\varepsilon).$$

But, as $S_\alpha \overset{(42)}{\subset} S_{r,\lambda}$, Φ is β - differentiable in v_ε and taking also into account I, §3, 5.10_1, one obtains

(51) $$0 \in \partial_\beta \varphi(v_\varepsilon) + \nabla_\beta \Phi(v_\varepsilon).$$

But

$$\| \nabla_\beta \Phi(v_\varepsilon) \| \overset{(37)}{\le} \sum_{n=0}^{\infty} \delta_n \| \nabla_\beta r(v_\varepsilon - x_n) \| \overset{(38),(50)}{\le} \frac{2\varepsilon k}{\lambda} L\ .$$

and so (51) offers (46). ∎

We discuss now about the connexion between the conditions

$$(\mathrm{P}) := (39) + (40) + (41) + (\|x\| < \alpha \Rightarrow x \in S_{r,\lambda})$$

(see 4.4) and

(H) *There exists* $\Phi : \mathrm{X} \to [0, 1]$ *Lipschitz* β *- differentiable having the properties*

(52) $\Phi(0) = 1,$ (53) $\|\mathrm{x}\| > 1 \Rightarrow \Phi(\mathrm{x}) = 0.$

∎ (H) $\Rightarrow$ (39) + (40) + (41). Take $r : X \to [0, +\infty)$,

$$r(x) = \sum_{n=1}^{\infty} \frac{1}{2^{2n}} \left[1 - \Phi(2^n x)\right].$$

The definition is correct, each term is majorized in absolute value by $\frac{1}{2^{2n}}$. r is Lipschitz on X with the constant L the same as for Φ:

$$|r(x)-r(y)|\le\sum_{n=1}^{\infty}\frac{1}{2^{2n}}|\Phi(2^n x)-\Phi(2^n y)|\le\sum_{n=1}^{\infty}\frac{1}{2^n}L\,\|x-y\|\text{ etc.}$$

r is β - differentiable on X : apply I, 3.12, the series of β - derivatives is $-\sum_{n=1}^{\infty}\frac{1}{2^n}\times$ $\times\nabla_\beta\Phi(2^n x)$ (see in the following the calculus of the β - derivatives) and the calculus continues as for (37), Φ being Lipschitz and β - differentiable on X. So (41) is *con brio* ensured. Obviously $r(x)\ge 0$ and $r(0)\overset{(52)}{=}0$. Moreover,

$$\|x\|>\frac{1}{2^n}\Rightarrow r(x)\ge\frac{1}{2^{2n}}[1-\Phi(2^n x)]\overset{(53)}{=}\frac{1}{2^{2n}}$$

and this implication justifies (*par absurdum*!) (40).

Complete the demonstration. For the function Ψ: $\Psi(x)=\Phi(2^n x)$ we have

$$\nabla_\beta\Psi(x)=2^n\nabla_\beta\Phi(2^n x)\ .$$

Indeed, putting $\xi:=\nabla_\beta\Phi(2^n x)$ we have

$$\frac{\Psi(x+th)-\Psi(x)}{t}-2^n\xi(h)=2^n\left[\frac{\Phi(2^n x+2^n th)-\Phi(2^n x)}{2^n t}-\xi(h)\right]$$

and the uniform limit on S of the bracket for $\tau:=2^n t\to 0$ is equal to 0.

(P) $\Rightarrow$ (H). Let α' be from $(0,\alpha)$. $\exists\,\delta>0$ so that

$$\|x\|>\alpha'\Rightarrow r(x)>\delta$$

(*par absurdum*, use (40)). Take $f:\mathbf{R}_+\to[0,1]$ of the C^1 class, $t>\delta\Rightarrow f(t)=0, f(0)=1$. The function $\Phi: X\to[0,1]$, $\Phi(x)=f(r(\alpha' x))$ verifies (H). Indeed, the conditions (52) and (53) are obviously verified. Φ is Lipschitz on X : consider the cases $\|x\|\le 1$ and $\|y\|\le$ ≤ 1, $\|x\|>1$ and $\|y\|>1$, $\|x\|\le 1$ and $\|y\|>1$ which respectively give the cases $\|\alpha' x\|\le\alpha'$ and $\|\alpha' y\|\le\alpha'$, $\|\alpha' x\|>\alpha'$ and $\|\alpha' y\|>\alpha'$, $\|\alpha' x\|\le\alpha'$ and $\|\alpha' y\|>\alpha'$; apply the elementary finite increment formula, f' is bounded. Φ is β - differentiable on X : use the rule for the β - derivative of the composed function (I, §3, 3.11). ■

APPLICATION TO THE β - SUBDERIVATIVE

4.5 *Let* X *be Banach space,* β *bornology on* X *and* $\varphi: X\to(-\infty,+\infty]$ *lower bounded l. s. c. proper. If* X *has the property* (H), *then* φ *is* β*-subdifferentiable on a set* A *so that*

$$\{(x, f(x)) : x \in A\}$$

is dense in the graph of φ *in* $X \times \mathbf{R}$ ([41]).

■ Let x_0 be from dom φ and $\varepsilon > 0$, $\varepsilon < f(x_0)$. The hypothesis (H) supplies $r : X \to$ $\to [0, +\infty)$ Lipschitz and β - differentiable on X (see the demonstration) which verifies (39) and (40). Choose $\lambda > 0$, $\alpha \in (0, 2\varepsilon)$ and $k > 1$ so that (42) to be verified, which is possible. Indeed, fix $\lambda > 0$. $S_{r,\lambda}$ being open and $r(0) = 0$, we have for α positive sufficiently small $\|x\| < \alpha \Rightarrow x \in S_{r,\lambda}$. Pass to the second implication and *par absurdum* suppose it false for every $k > 1$. We find then sequences $(k_n)_{n\geq 1}$, $k_n \uparrow +\infty$ and $(x_n)_{n\geq 1}$ so that, $\forall\ n \geq 1$, $r(x_n) \leq \dfrac{\lambda}{k_n}$ and $\| x_n \| \geq \dfrac{\alpha}{2}$, but then $r(x_n) \to 0$ which implies $\| x_n \| \xrightarrow{(40)} 0$ and the contradiction is obtained. Moreover, there is λ_1 in $(0, \lambda)$ small enough to have

(54) $$\inf \{\varphi(x) : r(x - x_0) \leq \lambda_1\} \geq \varphi(x_0) - \varepsilon.$$

Indeed, supposing *par absurdum* the contrary, one finds sequences $(\eta_n)_{n\geq 1}$, $\eta_n \downarrow 0$ and $(x'_n)_{n\geq 1}$ so that $r(x'_n - x_0) \leq \eta_n$ (which gives $x'_n \xrightarrow{(40)} x_0$) and $\varphi(x'_n) \leq \varphi(x_0) - \varepsilon$ $\forall\ n \geq 1$ and we obtain a contradiction because on a neighbourhood of x_0 we have $\varphi(x) >$ $\varphi(x_0) - \varepsilon$ (φ is l.s.c. at x_0).

Consider the function g_1,

(55) $$g_1(x) = \begin{cases} 0, & \text{if } \ r(x - x_0) \leq \lambda_1 \\ +\infty, & \text{if } \ r(x - x_0) > \lambda_1. \end{cases}$$

g_1 is l. s. c. So $\varphi_1 : X \to (-\infty, +\infty]$,

(56) $$\varphi_1(x) = \varphi(x) + g_1(x)$$

is lower bounded l. s. c. proper. (54), (55) and (56) give

(57) $$\varphi_1(x_0) - \varepsilon < \inf \varphi_1(X)$$

Apply 4.4 to φ_1 with α, λ_1, k as above, there exists v_ε in X so that (see (47) and (50))

(58) $$r(v_\varepsilon - x_0) \leq \frac{\lambda_1}{k} < \lambda_1,$$

(59) $\varphi_1(v_\varepsilon) \leq \varphi_1(x_0)$, (60) $\partial_\beta \varphi_1(v_\varepsilon) \neq \varnothing$.

(58) gives, *via* (42), $\| v_\varepsilon - x_0 \| < \dfrac{\alpha}{2} < \varepsilon$. Moreover, as $\varphi_1(v_\varepsilon) = \varphi(v_\varepsilon)$ and $\varphi_1(x_0) = \varphi(x_0)$ (use (58) and (55)), (57) and (59) give $\varphi(x_0) - \varepsilon < \varphi(v_\varepsilon) \leq \varphi(x_0)$. It remains only to

observe that, as φ_1 and φ coincide on a neighbourhood of v_ε ($\{x \in X: r(x - x_0) < \lambda_1\}$ is open neighbourhood of v_ε, (58)), (60) imposes $\partial_\beta \varphi(v_\varepsilon) \neq \varnothing$. ■

Remark. Compare 4.5 with §1, 1.4.

4.6 *Corollary. Let* X *be Banach space,* β *bornology on* X *and* φ : X → **R** *continuous convex upper bounded.*

If X *has the property* (H), *then* φ *is* β *- differentiable on a subset everywhere dense.*

■ As dom $(-\varphi) = X$, $-\varphi$ is β - subdifferentiable on a set A dense in X (4.5), i.e. φ is β-superdifferentiable on A (I, §5, D). But φ is β - subdifferentiable on X (I, 5.9) and applying I, 5.10_1 one finds φ β - differentiable on A. ■

FINIS

ANNEX

Differential manifolds, Finsler manifolds, Riemann manifolds

DIFFERENTIAL MANIFOLDS

Isometric imbedding in a normed space. 1. *Let* (X, d) *be a metric space. Then* X *can be isometrically imbedded in a real normed space as a closed subset of this* (*R. Arens - J. Eells*).

■ (*E. Michael*). Let be $Y := X \cup \{y_0\}$, where $y_0 \notin X$. Extend d to Y in the following manner: fix a point x' of X and take $\mathrm{d}(x, y_0) = \mathrm{d}(y_0, x) = 1 + \mathrm{d}(x, x')$ $\forall\, x \in X$ and $\mathrm{d}(y_0, y_0) = 0$. Let E be the real vector space of Lipschitz functions $\varphi : Y \to \mathbf{R}$ with $\varphi(y_0) = 0$. Taking on E the norm

$$(1) \qquad ||\varphi|| = \left\{\frac{|\,\varphi(x) - \varphi(y)\,|}{\mathrm{d}(x, y)} : x \neq y\right\},$$

one obtains a Banach space. Consider the map $F: X \to E^*$ with $F(x) : E \to \mathbf{R}$,

$$F(x)(\varphi) = \varphi(x).$$

The definition is correct, $F(x) \in E^*$; $F(x)$ is obviously linear; $F(x)$ is also continuous as for $y = y_0$ in (1) we find $|F(x)(\varphi)| \leq \mathrm{d}(x, y_0)||\varphi||$ $\forall\, y \in E$.

F *is injective.* $F(x_1) = F(x_2) \Rightarrow \varphi(x_1) = \varphi(x_2)$ $\forall \varphi \in E$, take ψ from E, $\psi(x) = \mathrm{d}(x, x_2) - \mathrm{d}(y_0, x_2)$, then $\psi(x_1) = \psi(x_2)$, hence $\mathrm{d}(x_1, x_2) = 0$, $x_1 = x_2$.

F *is isometrical*. Indeed, let x_1, x_2 be any from X, $x_1 \neq x_2$. (1) gives $d(x_1, x_2) \geq \geq \frac{|\varphi(x_1)-\varphi(x_2)|}{\|\varphi\|}$ $\forall\ \varphi \neq 0$, hence $d(x_1, x_2) \geq \|F(x_1) - F(x_2)\|$. Reciprocally, let ψ be from E as above, $\psi(x) = d(x, x_2) - d(y_0, x_2)$. We have (2) $\|\psi\| = 1$: $\|\psi\| \overset{(1)}{=}$

$$= \sup\left\{\frac{|d(x,x_2)-d(y,x_2)|}{d(x,y)} : x \neq y\right\},\ \frac{d(x,x_2)-d(y,x_2)}{d(x,y)} \leq \frac{d(x,y)}{d(x,y)} = 1,$$ hence $\|\psi\| \leq 1$; for $y = x_2$ and $x \neq x_2$ one obtains $\|\psi\| \geq 1$. By using (2),

$$\|F(x_1) - F(x_2)\| \geq |F(x_1)(\psi) - F(x_2)(\psi)| = |\psi(x_1) - \psi(x_2)| = d(x_1, x_2),$$

consequently $\|F(x_1) - F(x_2)\| = d(x_1, x_2)$.

The set F(X) *is free*. Indeed, let $x_1, \ldots, x_n, x_{n+1}$ be distinct points from X. Take $\varphi: Y \to \to \mathbf{R}$, $\varphi(x) = \text{dist}(x, \{y_0, x_1, \ldots, x_n\})$. We have $\varphi \in E$ and $F(x_1)(\varphi) = \ldots = F(x_n)(\varphi) = 0$ and $F(x_{n+1})(\varphi) \neq 0$, consequently $F(x_{n+1})$ is not a linear combination of $F(x_i)$, $i = \overline{1,n}$. It remains only to consider the vector subspace V of E^* generated by $F(X)$. $F(X)$ is closed in V. ■

Locally Lipschitz partition of the unit. Let X be topological space and $\mathscr{U} := (U_\alpha)_{\alpha\in A}$ a covering of X. The family of functions $(f_\alpha)_{\alpha\in A}$, $f_\alpha: X \to [0,1]$ is by definition *locally Lipschitz partition of the unit on* X *relative to* $\mathscr{U}$, if

1° f_α is locally Lipschitz $\forall\ \alpha \in A$;

2° the family of supports $(\text{supp } f_\alpha)_{\alpha\in A}$ is locally finite and $f_\alpha(x) = 0$ on $X \setminus U_\alpha$ $\forall\ \alpha \in \in A$;

3° $\sum_{\alpha\in A} f_\alpha(x) = 1\ \forall\ x \in X$.

2 *If* X *is metric space and* $\mathscr{U} := (U_\alpha)_{\alpha\in A}$ *an open covering of* X, *there exists a locally Lipschitz partition of the unit on* X *relative to* $\mathscr{U}$.

■ Let be $\mathscr{V} := (V_i)_{i\in I}$, $V_i \neq \varnothing\ \forall\ i$, open locally finite covering of X thinner than $\mathscr{U}$ (Stone theorem). Consider for each i from I the real functions ψ_i, φ_i defined on X

(3) $$\psi_i(x) = \begin{cases} 0, & x \notin V_i \\ \text{dist}(x, \text{Fr } V_i), & x \in V_i \end{cases}$$

(4) $$\varphi_i(x) = \frac{\psi_i(x)}{\sum_{j\in I} \psi_j(x)},\ x \in X.$$

The definition of φ_i is correct as, $\mathscr{V}$ being locally finite, the sum makes sense and moreover it is > 0 because $V_i \cap \operatorname{Fr} V_i = \varnothing$. ψ_i is locally Lipschitz and also φ_i (quotient of two functions locally Lipschitz). We obviously have $\sum_{i \in I} \varphi_i(x) = 1$ on X ((4)) and in addition

$x \in X \setminus V_i \Rightarrow \varphi_i(x) = 0$ ((3),(4)), i.e. $(\varphi_i)_{i \in I}$ is a locally Lipschitz partition of the unity on X relative to $\mathscr{V}$. It remains only to choose $\omega : I \to A$ with the property $V_i \subset U_{\omega(i)}$ ($\mathscr{V}$ is thinner than $\mathscr{U}$!) and then to take

$$\alpha \in A \Rightarrow f_\alpha = \begin{cases} \sum_{i \in \omega^{-1}(\alpha)} \varphi_i, & \omega^{-1}(\alpha) \neq \varnothing \\ 0, & \omega^{-1}(\alpha) = \varnothing \end{cases}$$

to obtain the partition of the unit required in the statement. ■

Absolute neighbourhood retract. The metric space X is by definition *absolute retract* (AR) respectively *absolute neighbourhood retract* (ANR) if for any metric space M, any closed nonempty subset F of M and any $f : F \to X$ continuous, f admits a continuous extension to M respectively to a neighbourhood of F.

For exemple, any real normed space is, according to the Dugundji theorem, an AR.

3 *Let* X *be metric space.*

1° *If* X *is an* ANR, *any open nonempty subset of* X *is an* ANR *and any metric space homeomorphic to* X *is an* ANR.

2° X *is an* ANR $\Leftrightarrow$ *there exists* U *open nonempty subset of a normed real space and* $\mathrm{f} : \mathrm{X} \to \mathrm{U}$, $\mathrm{g} : \mathrm{U} \to \mathrm{X}$ *continuous so that*

$$\mathrm{g}(\mathrm{f}(\mathrm{x})) = \mathrm{x} \ \forall \ \mathrm{x} \in \mathrm{X}.$$

3° A *finite product of* ANR *is an* ANR.

4° *Any convex nonempty subset of a real normed space is an* AR.

5° $\mathrm{S}^{n-1} := \{\mathrm{x} \in \mathbf{R}^n : \|\mathrm{x}\| = 1\}$, $n \geq 1$, *is an* ANR (*the norm is the euclidean one*)

■ 1° *First affirmation.* Let G be open nonempty set in X, M metric space, $F \subset M$ closed nonempty and $f : F \to G$ continuous. There exists $W \supset F$ open and $g : W \to X$ continuous with $g | F = f$. Then $g^{-1}(G)$ is open, $F \subset g^{-1}(G) \subset W$ ($x \in F \Rightarrow g(x) = f(x)$, hence $g(x) \in G$), $g | g^{-1}(G)$ is the wanted extension. *Second affirmation.* Let Y be metric space, $\varphi : Y \to X$ homeomorphism, M arbitrary metric space, $F \subset M$ closed nonempty and $f : F \to Y$ continuous. $\varphi \circ f$ being continuous, $\exists\ V$ neighbourhood of F and $g : V \to X$ continuous with $g | F = \varphi \circ f$. Then $\varphi^{-1} \circ g : V \to Y$ is continuous and $\varphi^{-1} \circ g | F = f$.

2° $\Rightarrow$. According to **1** $\exists$ E real normed space, F closed nonempty subset and $h: F \to X$ homeomorphism. X being an ANR, $\exists$ $g: U \to X$ continuous, U open neighbourhood of F, with $g|F = h$. Seting $f: = h^{-1}$, we have $g(f(x)) = h(f(x)) = x$ on X. $\Leftarrow$. Let M be metric space, F closed nonempty subset and $\varphi: F \to X$ continuous. U being an ANR (1°), $f \circ \varphi$ extended in a function h continuous on V neighbourhood in M of F. For $g \circ h: V \to X$ we have $x \in F \Rightarrow g(h(x)) = g(f(\varphi(x))) = \varphi(x)$.

3° It is obvious.

4° Let X be real normed space, C convex nonempty subset, M metric space, F closed nonempty subset and $f: F \to C$ continuous. Dugundji theorem gives $g: M \to X$ continuous with $g|F = f$ and $g(x)$, $\forall$ x from X, convex combination of elements from $f(F)$, consequently $g(x) \in C$ $\forall$ $x \in X$, that is in fact g takes its values in C, C is an AR.

5° Let M be metric space, F closed nonempty subset and $f: F \to S^{n-1}$ continuous. Dugundji theorem gives $g: M \to \mathbf{R}^n$ continuous with $g|F = f$. Let be $U: = \{x \in \mathbf{R}^n: 0 <$ $< \|x\| < 2\}$ and $r: U \to S^{n-1}$, $x = (x_1, \dots, x_n) \to r(x) = \left(\frac{x_1}{\|x\|}, \dots, \frac{x_n}{\|x\|}\right)$. r is continuous and $r(x) = x$ on S^{n-1}. It remains only to consider the application $r \circ g$ defined on $g^{-1}(U)$, as $F \subset g^{-1}(U)$ ($x \in F \Rightarrow f(x) \in S^{n-1} \Rightarrow g(x) \in S^{n-1} \subset U$) and $r \circ g|F = f$ ($x \in F \Rightarrow g(x) =$ $= f(x) \Rightarrow r(g(x)) = r(f(x)) = f(x)$, as $f(x) \in S^{n-1}$). ■

4 *Let* X *be metric space. If* $X = \bigcup_{n=1}^{\infty} U_n$, U_n *open and* ANR $\forall$ $n \geq 1$, *then* X *is an* ANR (*O. Hanner*).

■ *The particular case* U_n, $n \geq 1$ *disjoints two by two.*

Let M be metric space, $F \subset M$ closed nonempty and $f: F \to X$ continuous. $F_n: =$ $= f^{-1}(U_n)$, $n \geq 1$ (the definition imposes $U_n \neq \varnothing$!) are closed because U_n are also closed. Obviously $F = \bigcup_{n=1}^{\infty} F_n$ with the terms two by two disjoint. Let be $f_n: = f|F_n$, so $f_n: F_n \to$ $\to U_n$. U_n being an ANR, $\exists$ A_n open neighbourhood of F_n and $g_n: A_n \to U_n$ continuous which extends f_n. The sets $B_n: = \{x \in M: \operatorname{dist}(x, F_n) < \operatorname{dist}(x, F \setminus F_n)\}$, $n \geq 1$ are disjoint (*par absurdum*, for $m \neq n$ one finds $\operatorname{dist}(x, F_n) < \operatorname{dist}(x, F_m)$ and $\operatorname{dist}(x, F_m) < \operatorname{dist}(x, F_n)$)

and B_n is open neighbourhood of F_n. It remains only to consider the open neighbourhood of F $W := \bigcup_{n=1}^{\infty}(A_n \cap B_n)$ and the function $g: W \to X$, $g \mid A_n \cap B_n = g_n$, $n \geq 1$.

We worked in the hypothesis $F_n \neq \varnothing$ $\forall$ $n \geq 1$. In the contrary case the empty sets are eliminated, the reasoning remains the same.

Emphasize that, in the particular case considered, the assertion of the statement remains true if one replaces the set **N** of the indices by an arbitrary set.

The particular case $X = U_1 \cup \ldots \cup U_p$.

The reccurence permits to suppose $p=2$. Moreover, it can obviously suppose $U_1 \neq U_2$, and also $U_1 \cap U_2 \neq \varnothing$ (the previous case). Let M be metric space, $F \subset M$ closed nonempty and $f: F \to X$ continuous. $F \setminus f^{-1}(U_1)$, $F \setminus f^{-1}(U_2)$ are closed disjoint, suppose them nonempty (in the contrary case – nothing to prove). So, these two sets have open disjoint neighbourhoods M_2, M_1 respectively (M is normal). We have (4_0) $F = (F \setminus \setminus f^{-1}(U_1)) \cup (F \setminus f^{-1}(U_2)) \cup f^{-1}(U_1 \cap U_2)$ and the images through f of the first and the second term are included in U_2 respectively U_1, as $F = f^{-1}(U_1) \cup f^{-1}(U_2)$ because $U_1 \cup \cup U_2 = X$.

Suppose $F \subset M_1 \cup M_2$. $F_i := F \cap M_i$, $i = 1, 2$ are nonempty (see (4_0)). As F_i is closed in M_i and $f(F_i) \subset U_i$, an ANR set, $f | F_i$ is extended in a continuous function $f_i: W_i \to U_i$, W_i open neighbourhood of F_i in M_i, hence open because M_i is open. $g: W_1 \cup W_2 \to X$, $g | W_i = f_i$, $i = 1,2$ is continuous, as $W_1 \cap W_2 = \varnothing$, and extends f. Moreover, $W_1 \cup W_2$ is neighbourhood of F and so the particular case considered is liquidated. *Suppose* $F \not\subset \not\subset M_1 \cup M_2$. Consider the closed set (4_1) $M_0 := M \setminus (M_1 \cup M_2)$ and let be (4_2) $F_0 := F \cap \cap M_0$, closed nonempty set. Obviously (4_3) $F = F_0 \cup F_1 \cup F_2$, (4_4) $f(F_i) \subset U$, $i = 1, 2$ (see above), (4_5) $f(F_0) \subset U_1 \cap U_2$ (see (4_0)). $U_1 \cap U_2$ being an ANR (**3**), $\exists$ V_0 closed neighbourhood of F_0 in M_0, hence closed, and $f_0 : V_0 \to U_1 \cap U_2$ continuous, which extends $f | F_0$. Let be (4_6) $g: V_0 \cup F \to X$, $g | V_0 = f_0$, $g | F = f$. The definition is correct as $V_0 \cap F = F_0$ ((4_1), (4_3)) and $f_0(x) = f(x)$ on F_0. g is continuous (use the definition). We have (4_7) $V_0 \cup F = (V_0 \cup F_1) \cup (V_0 \cup F_2)$ (see (4_3), $F_0 \subset V_0$). $V_0 \cup F_i$ is closed in $V_0 \cup M_i$, $i = 1,2$ (take the intersection with $V_0 \cup F$), (4_8) $g(V_0 \cup F_i) \overset{(4_6)}{\subset} f_0(V_0) \cup f(F_i) \overset{(4_4)}{\subset} (U_1 \cap \cap U_2) \cup U_i = U_i$, $i = 1, 2$, hence, according to (4_8), $\exists$ V_i open neighbourhood in $V_0 \cup M_i$ of $V_0 \cup F_i$ and $g_i : V_i \to U_i$ which extends $g | (V_0 \cup F_i)$, $i = 1, 2$. Finally, let be $V := V_1 \cup V_2$,

$h: V \to X$, $h|V_i = g_i$, $i = 1, 2$. The definition is correct as $V_1 \cap V_2 = V_0$ and $g_1|V_0 = g_2|V_0 = f_0$. h is continuous (use the definition). As $h|F = f$ and $V \supset F$ since $V \supset V_0 \cup F_1 \cup F_2$, it remains only to show that V is open. We have

$$(4_9) \qquad M \setminus V = [(M_0 \cup M_1) \setminus V_1] \cup [(M_0 \cup M_2) \setminus V_2].$$

Indeed, the inclusion $\subset$ being obvious, pass to the contrary inclusion. Let be, for instance, $x \in (M_0 \cup M_1) \setminus V_1$. If $x \in M_0$ then $x \notin M_2$ and as $x \notin V_0$ because $x \notin V_1$, *a fortiori* $x \notin V_2$, $x \in M \setminus V$. If $x \in M_1$, then $x \notin M_2$ and one obtains the same conclusion. For the second term – the same reasoning. But V_1, being open in $V_0 \cup M_1$, it is open also in $M_0 \cup M_1$, hence $(M_0 \cup M_1) \setminus V_1$ is closed in $M_0 \cup M_1$ and hence closed, since $M \setminus M_2$ is closed. The second term from (4_9) has the same property, this finishes the proof.

The general case

It can obviously suppose that we are in the presence of an infinity of distinct terms of X (in the contrary case – nothing to prove). Consider the sets

$$(4_{10}) \quad A_n := \bigcup_{i \le n} U_i, \qquad (4_{11}) \quad B_n := \{x \in A_n : \text{dist}(x, X \setminus A_n) > \frac{1}{n}\}.$$

The definition (4_{11}) is correct, $X \setminus A_n \neq \varnothing$ $\forall\, n \ge 1$. Each A_n is an ANR (the previous case) and each nonempty B_n is also an ANR (**3**). We have (4_{12}) $B_n \subset B_{n+1}$ $[x \in A_n \Rightarrow x \in A_{n+1}$, (4_{13}) $\text{dist}(x, X \setminus A_{n+1}) \overset{(4_{10})}{\ge} \text{dist}(x, X \setminus A_n)]$ and (4_{14}) $X = \bigcup_{n=1}^{\infty} B_n$. Indeed, let x be from X, $\exists\, p$ so that $x \in U_p$, hence $x \in A_n$ $\forall\, n \ge p$. Suppose *par absurdum* $\text{dist}(x, X \setminus A_n) \le \frac{1}{n}$ $\forall\, n \ge p$. Combining with (4_{13}), one finds $\text{dist}(x, X \setminus A_p) \le \frac{1}{q}$ $\forall\, q \ge p$, hence $\text{dist}(x, X \setminus A_p) = 0$, $x \in X \setminus A_p$, contradiction and consequently $\exists\, m \ge p$ so that $\text{dist}(x, X \setminus A_m) > \frac{1}{m}$, hence $x \in B_m$ and so (4_{14}).

It remains only to consider the sequence of open sets $(C_n)_{n \ge 1}$,

$$C_1 = B_1,\ C_2 = B_2,\ C_n = B_n \setminus \overline{B}_{n-2} \ \ \forall\, n \ge 3,$$

as $X = C_1 \cup C_2 \cup (C_3 \cup C_5 \cup \ldots) \cup (C_2 \cup C_4 \cup \ldots)$, the empty terms are eliminated and one applies the two liquidated particular cases corroborated with **3**. ■

The Hanner proposition is generalized to arbitrary sets of indices.

5 *Let* X *be metric space. If* $X = \bigcup_{\alpha \in A} U_\alpha$, U_α *open* ANR $\forall \alpha$ *of* A, *then* X *is an* ANR (*R. S. Palais*).

■ *The particular case* $\alpha \neq \beta \Rightarrow U_\alpha \cap U_\beta = \varnothing$. It was already considered, in a remark, in the demonstration for **4**.

The general case

It can suppose the considered covering $\mathscr{U} := (U_\alpha)_{\alpha \in A}$ as locally finite, in the contrary case take an open locally finite covering of X (with all the terms nonempty) thinner than $\mathscr{U}$ (X is paracompact), any set of this is an ANR (**3**).

Let $(f_\alpha)_{\alpha \in A}$ be locally Lipschitz partition of the unit on X relative to $\mathscr{U}$ (**2**) and, for each n from **N**, the set

$$\mathscr{A}_n := \{\{\alpha_1, \ldots, \alpha_n\} : \alpha_i \in A, i = \overline{1, n}\}.$$

Consider, for each n from **N** and λ from $\mathscr{A}_n$, the set

(5) $$V_{n\lambda} := \{x \in X : \alpha \in \lambda \Rightarrow f(x) > 0 \text{ and } f_\alpha(x) > f_\beta(x)\ \forall\, \beta \notin \lambda\}.$$

$V_{n\lambda}$ is open ((supp $f_\alpha)_{\alpha \in A}$ is locally finite, all f_α take positive values) and (6) $\lambda \neq \mu \Rightarrow$ $\Rightarrow V_{n\lambda} \cap V_{n\mu} = \varnothing$. Indeed, let be *par absurdum* $x \in V_{n\lambda} \cap V_{n\mu}$. The hypothesis implies either $\exists\, \alpha \in \lambda \setminus \mu$ or $\exists\, \beta \in \mu - \lambda$, because in the contrary case we have $\lambda \subset \mu$ and $\mu \subset \lambda$, hence $\lambda = \mu$, contradiction. Suppose $\exists\, \alpha \in \lambda \setminus \mu$, then if $\nexists\, \beta \in \mu \setminus \lambda$ it results $\mu \subset \lambda$ and as $\alpha \in \lambda \setminus \mu$, λ has more elements than μ, contradiction because $\lambda, \mu \in \mathscr{A}_n$ and in conclusion $\exists\, \alpha \in \lambda \setminus \mu$, and $\exists\, \beta \in \mu \setminus \lambda$, which implies (see (5)) $f_\alpha(x) < f_\beta(x)$ and $f_\beta(x) < f_\alpha(x)$, the final contradiction and consequently (6). Moreover, $V_{n\lambda}$, if it is not empty, is an ANR since if $\lambda = \{\alpha_1, \ldots, \alpha_n\}$ then $V_{n\lambda} \subset U_{\alpha_1} \cap \ldots \cap U_{\alpha_n}$ (we have $f_\alpha(x) = 0$ on $X \setminus U_\alpha$) and one applies **3** twice.

So, if we prove

(7) $$X = \bigcup_{n=1}^{\infty} \Big(\bigcup_{\lambda \in \mathscr{A}_n} V_{n\lambda} \Big),$$

then eliminating in (7) the empty terms and taking into account the particular case considered (see (6)) combined with **4** one obtains the conclusion X is an ANR.

Let x_0 be any from X. As $\sum_{\alpha\in A} f_\alpha(x_0)=1$, $\exists\ \alpha$ in A so that $f_\alpha(x_0)>0$. Let then be $\alpha_1, \ldots,$ α_m *all* the indices of A for which $f_{\alpha_i}(x_0)>0$, $i=\overline{1,m}$. Consider the set of strictly positive numbers

$$E:=\{f_{\alpha_1}(x_0), \ldots, f_{\alpha_m}(x_0)\}.$$

Suppose all the elements of E are equal. Then for $\lambda:=\{\alpha_1, \ldots, \alpha_m\}$ we have $x_0\in V_{m\lambda}$ since $\beta\notin\lambda \Rightarrow f_\beta(x_0)=0<f_{\alpha_i}(x_0)$, $i=\overline{1,m}$, (7) is proved. In the contrary case, let $f_{\alpha_{k_1}}(x_0)$ be the greatest among the elements of E. The situation is presented as follows:

$$(8)\qquad f_{\alpha_{k_1}}(x_0)=f_{\alpha_{k_2}}(x_0)=\ldots=f_{\alpha_{k_j}}(x_0)>f_{\alpha_{k_{j+1}}}(x_0)\geq f_{\alpha_{k_{j+2}}}(x_0)\geq\ldots\geq f_{\alpha_{k_m}}(x_0).$$

For $\lambda:=\{\alpha_{k_1}, \ldots, \alpha_{k_j}\}$ we have $x_0\in V_{j\lambda}$ – take into account (8) and the fact that $f_\alpha(x_0)=$ $=0\ \forall\ \alpha\neq\alpha_i$, $i=\overline{1,m}$, (7) is proved. ■

Differential C ^{r}-manifold. Let X be nonempty set and $\mathscr{E}:=(E_\alpha)_{\alpha\in A}$ a family of real Banach spaces.

Definitions. The family $\mathscr{A}:=(U_\alpha, \varphi_\varphi, E_\alpha)_{\alpha\in A}$, U_α nonempty subset of X and $\varphi_\alpha: U_\alpha\to$ $\to E_\alpha$ injective function, is C ^{r} - *atlas on the set* X, $0\leq r\leq\infty$ if

(A$_1$) $X=\bigcup_{\alpha\in A} U_\alpha$,

(A$_2$) $\varphi_\alpha(U_\alpha\cap U_\beta)$ is open $\forall\ \alpha, \beta\in A$,

(A$_3$) $\forall\ \alpha, \beta$ from A $U_\alpha\cap U_\beta\neq\varnothing \Rightarrow \varphi_\alpha\circ\varphi_\beta^{-1}: \varphi_\beta(U_\alpha\cap U_\beta)\to\varphi_\alpha(U_\alpha\cap U_\beta)$ is of C^r class.

When $r\geq 1$, because $(\varphi_\alpha\circ\varphi_\beta^{-1})^{-1}=\varphi_\beta\circ\varphi_\alpha^{-1}$, in (A$_3$) we have in fact a diffeomorphism of C^r class. $(U_\alpha, \varphi_\varphi, E_\alpha)$, α goes over A, are the *local charts*. If $x\in U_\alpha$, the local chart $(U_\alpha, \varphi_\varphi, E_\alpha)$ is *at* x.

The set X together with the atlas $\mathscr{A}$ form a *differential* C ^{r}*-manifold.* If $\mathscr{A}_1$ is another C ^{r} - atlas on X, the couple $(X, \mathscr{A}_1)$ defines *the same* differential C ^{r}-manifold, if $\mathscr{A}\cup$ $\cup\ \mathscr{A}_1$ is also C ^{r}-atlas on X. In this case $\mathscr{A}$ and $\mathscr{A}_1$ are, by definition, *equivalent*. In the case $E_\alpha=E\ \forall\ \alpha$ of A, X has the name *differential manifold modelled on* E (E-*modelled*).

Let X be differential C ^{r}-manifold and $\mathscr{A}:=(U_\alpha, \varphi_\varphi, E_\alpha)_{\alpha\in A}$ an atlas on this. We designate by τ_α the topology of the Banach space E_α (the set of the open sets).

6 *There exists on the set* X *a unique topology* $\tau(\mathscr{A})$ *having the properties*:

1° U_α *is open relative to* $\tau(\mathscr{A})$ $\forall\ \alpha \in A$;

2° $\varphi_\alpha : U_\alpha \to \varphi_\alpha(U_\alpha)$ *is a homeomorphism from* $\tau(\mathscr{A})$ *to* τ_α;

$\tau(\mathscr{A})$ is the topology on *X defined* by the atlas $\mathscr{A}$.

■ *Existence*. Take

(9) $$\tau(\mathscr{A}) = \{V \in X : \varphi_\alpha(V \cap U_\alpha) \text{ open in } E_\alpha\ \ \forall\ \alpha \in A\}.$$

$X, \varnothing \in \tau(\mathscr{A})$. $\varphi_\alpha(X \cap U_\alpha) = \varphi_\alpha(U_\alpha)$, open set in E_α ((A$_2$)), $\varphi_\alpha(\varnothing \cap U_\alpha) = \varnothing$.

Any reunion of sets from $\tau(\mathscr{A})$ *is in* $\tau(\mathscr{A})$. Let be $(V_i)_{i\in I}$, $V_i \in \tau(\mathscr{A})$. $\varphi_\alpha((\bigcup\limits_{i\in I} V_i) \cap$ $\cap\, U_\alpha) = \varphi_\alpha(\bigcup\limits_{i\in I}(V_i \cap U_\alpha)) = \bigcup\limits_{i\in I}\varphi_\alpha(V_i \cap U_\alpha) \in \tau_\alpha$.

Any finite intersection of sets from $\tau(\mathscr{A})$ *is in* $\tau(\mathscr{A})$. Let V_1, V_2 be from $\tau(\mathscr{A})$. $\varphi_\alpha((V_1 \cap V_2) \cap U_\alpha) = \varphi_\alpha(V_1 \cap U_\alpha) \cap \varphi_\alpha(V_2 \cap U_\alpha) \in \tau_\alpha$.

So $\tau(\mathscr{A})$ defines a topology on X. U_α is open in $\tau(\mathscr{A})$ as, $\forall\ \beta$ from A, $\varphi_\beta(U_\alpha \cap U_\beta) \in$ $\in \tau_\beta$ ((A$_2$)), 1° is proved. Pass to 2°.

$\varphi_\alpha : U_\alpha \to \varphi_\alpha(U_\alpha)$ *is continuous*. Let G be open arbitrary from E_α. $\varphi_\alpha^{-1}(\varphi_\alpha(U_\alpha) \cap G) =$ $= U_\alpha \cap \varphi_\alpha^{-1}(G)$, an $\tau(\mathscr{A})$ open set as, $\forall\ \beta \in A$ (see (9)),

$$\varphi_\beta(U_\alpha \cap \varphi_\alpha^{-1}(G) \cap U_\beta) = \varphi_\beta(U_\alpha \cap U_\beta) \cap \varphi_\beta(\varphi_\alpha^{-1}(G))$$

and the two factors are open in E_β because $\varphi_\beta \circ \varphi_\alpha^{-1}$ is of C^r class ((A$_3$)).

$\varphi_\alpha^{-1} : \varphi_\alpha(U_\alpha) \to U_\alpha$ *is continuous*. Let U be arbitrary from $\tau(\mathscr{A})$, $U \subset U_\alpha$. Then $(\varphi_\alpha^{-1})^{-1}(U) = \varphi_\alpha(U \cap U_\alpha) \overset{(9)}{\in} \tau_\alpha$, 2° is proved.

Uniqueness. Let τ_1 be a topology on X with the properties 1° and 2°.

$\tau_1 \subset \tau(\mathscr{A})$. Let U be arbitrary from τ_1. Then $U \cap U_\alpha \overset{1^0}{\in} \tau_1\ \forall\ \alpha$, hence $\varphi_\alpha(U \cap U_\alpha) \overset{2^0}{\in} \tau_\alpha$ $\forall\ \alpha$ and so $U \overset{(9)}{\in} \tau(\mathscr{A})$.

$\tau(\mathscr{A}) \subset \tau_1$. Let V be any from $\tau(\mathscr{A})$, i.e. $\varphi_\alpha(V \cap U_\alpha) \in \tau_\alpha\ \forall\ \alpha$, hence, according to 2°, $\varphi_\alpha^{-1}(\varphi_\alpha(V \cap U_\alpha)) = V \cap U_\alpha \in \tau_1$ and then $V \overset{(A_1)}{=} \bigcup\limits_{\alpha\in A}(V \cap U_\alpha) \in \tau_1$. ■

Remark. Every local chart is locally connected, this is imposed by 2°.

7 *Corollary. Let $\mathcal{A}$ and $\mathcal{B}$ be atlases on the same differential manifold. Then*

$$\tau(\mathcal{A}) = \tau(\mathcal{B}).$$

■ Let be $\mathcal{A} := (U_\alpha, \varphi_\alpha, E_\alpha)_{\alpha\in A}$, $\mathcal{B} := (V_\beta, \psi_\beta, F_\beta)_{\beta\in B}$ atlases on X. According to the definition, $\mathcal{A}\cup\mathcal{B}$ is atlas on X. The local charts of this arc $(U_\alpha, \varphi_\alpha)$, (V_β, ψ_β), α goes over A and β goes over B. $\tau(\mathcal{A}\cup\mathcal{B})$ verifies 1° and 2° of **6** $\forall\ \alpha \in A$, $\forall\ \beta \in B$, which implies $\tau(\mathcal{A}\cup\mathcal{B}) = \tau(\mathcal{A})$ and $\tau(\mathcal{A}\cup\mathcal{B}) = \tau(\mathcal{B})$, hence $\tau(\mathcal{A}) = \tau(\mathcal{B})$. ■

7 enables the definition:

The *topology* of a differential manifold is that defined by an atlas on it. This is locally connected (see the last remark).

8 *Let* X *be differential* C^r*-manifold,* $r \geq 1$ *and* $\mathcal{A} := (U_\alpha, \varphi_\alpha, E_\alpha)_{\alpha\in A}$ *an atlas on this. If the topology of* X *is connected,* E_α *and* E_β *are homeomorphic for any* α *and* β.

■ We design *ad hoc* the homeomorphism by the notation $\simeq$.

The case $U_\alpha \cap U_\beta \neq \varnothing$. Let x be from $U_\alpha \cap U_\beta$. As $(\varphi_\beta \circ \varphi_\alpha^{-1})^{-1} \circ (\varphi_\beta \circ \varphi_\alpha^{-1})$ = the identic map, we have $(\varphi_\beta \circ \varphi_\alpha^{-1})'(\varphi_\beta(x)) = [(\varphi_\beta \circ \varphi_\alpha^{-1})'(\varphi_\alpha(x))]^{-1}$, the first member is found in $L(E_\beta; E_\alpha)$, the second in $L(E_\alpha; E_\beta)$ and so $E_\alpha \simeq E_\beta$.

The general case. Fix α_0 from A, set $A_0 := \{\alpha \in A : E_\alpha \simeq E_{\alpha_0}\}$ and consider the set

$$U := \bigcup_{\alpha\in A_0} U_\alpha. \tag{10}$$

Show (11) $\overline{U} = U$. Let x be anyone from $\overline{U}$. $\exists\ \beta$ in A with $x \in U_\beta$ ((A₁)) and so $U \cap U_\beta \neq \varnothing$, that is $\varnothing \overset{(10)}{\neq} \bigcup_{\alpha\in A_0}(U_\alpha \cap U_\beta)$, let α' be from A_0 so that $U_{\alpha'} \cap U_\beta \neq \varnothing$, but then (the particular case) $E_{\alpha'} \simeq E_\beta$ and hence $E_{\alpha_0} \simeq E_\beta$, consequently $\beta \in A_0$, $x \overset{(10)}{\in} U$ and hence (11). X being connected, (11) imposes (12) $U = X$. Let now be α arbitrary from A. It remains to show (13) $E_\alpha \simeq E_{\alpha_0}$. Let x be a point from U_α. It follows, taking into account (12) and (10), $\exists\ \alpha'$ in A_0 so that $x \in E_{\alpha'}$, consequently (the particular case) $E_\alpha \simeq E_{\alpha'}$ and hence (13). ■

Attention! As it results by the proof, the implication $U_\alpha \cap U_\beta \neq \varnothing \Rightarrow E_\alpha \simeq E_\beta$ is true for X connected or not.

Exemple. Any Banach space E can be *canonically* equipped with a structure of differential C^∞-manifold: $(U_\alpha)_{\alpha\in A}$ being the family of the open nonempty sets of E and $\varphi_\alpha : U_\alpha \to E$, $\varphi_\alpha(x) = x$, $(U_\alpha, \varphi_\alpha, E)_{\alpha\in A}$ is an C^∞-atlas on E.

9 *Every differential manifold, whose topology is metric, is an* ANR.

■ Let X be differential manifold as in the statement and $\mathscr{A} := (U_\alpha, \varphi_\alpha, E_\alpha)_{\alpha\in A}$ an atlas of X. $\varphi_\alpha : U_\alpha \to \varphi_\alpha(U_\alpha)$ is homeomorphism from $\tau(\mathscr{A})$ to the topology τ_α of E_α (**6**). But $\varphi_\alpha(U_\alpha)$ being open ((A_2)) and E_α an AR (Dugundji theorem), $\varphi_\alpha(U_\alpha)$ is an ANR (**3**) and hence U_α has the same property (**3**). Combining (A_1) with **5** one concludes that X is an ANR. ■

Functions of C^l class. Let X be differential C^r-manifold, $r \geq 1$, $f : X \to \mathbf{R}$ and x a point in X. By definition f is of C^l *class*, $1 \leq l \leq r$, *in the neighbourhood of* x if there exists $(U_\alpha, \varphi_\alpha, E_\alpha)$ a local chart at x so that $f \circ \varphi_\alpha^{-1}$ is of C^l class on a neighbourhood of the point $\varphi_\alpha(x)$. In this case, for any local chart $(U_\beta, \varphi_\beta, E_\beta)$ at x, $f \circ \varphi_\beta^{-1}$ is of C^l class on a neighbourhood of the point $\varphi_\beta(x)$, since on $\varphi_\beta(U_\alpha \cap U_\beta)$ we have $f \circ \varphi_\beta^{-1} = (f \circ \varphi_\alpha^{-1}) \circ$ $\circ (\varphi_\alpha \circ \varphi_\beta^{-1})$ and the second factor is of C^r class ((A_3)).

f is of C^l *class on a subset* G of X, if it is of C^l class in the neighbourhood of each point of G.

Let be $\mathscr{C}(X) := \{f : X \to \mathbf{R} \mid f \text{ of } C^l \text{ class}\}$. $\mathscr{C}(X)$ is an algebra with respect to the operations $f + g$, fg, λf with $\lambda \in \mathbf{R}$. Fix a point x in X. Consider in $\mathscr{C}(X)$ the equivalence relation $f \sim g \overset{\text{def}}{\Leftrightarrow} f(y) = g(y)$ on a neighbourhood of x. The equivalence class $\hat{f}$ of f is by definition C^l-*germ* at x. The set of these germs is designated by $\mathscr{C}(x)$ and it is equipped in an obvious manner to become an algebra on $\mathbf{R}$.

Let now be X, Y differential C^r-manifolds, $r \geq 1$, $f : X \to Y$ and $\mathscr{A} := (U_\alpha, \varphi_\alpha, E_\alpha)_{\alpha\in A}$, $\mathscr{B} := (V_\beta, \psi_\beta, F_\beta)_{\beta\in B}$ atlases on X resp. Y.

The local charts $(U_\alpha, \varphi_\alpha)$, (V_β, ψ_β) on X resp. Y are *compatible relative to* f if $f(U_\alpha) \subset V_\beta$. The atlases $\mathscr{A}$ and $\mathscr{B}$ are *compatible relative to* f if for each local chart of $\mathscr{A}$ there exists a local chart of $\mathscr{B}$ with which it is compatible relative to f. Being given $\mathscr{B}$ and f continuous, there is $\mathscr{A}_1$ atlas of X compatible with $\mathscr{B}$ relative to f. Indeed, take $\mathscr{A}_1 = (U_{\alpha,\beta}, \varphi_{\alpha,\beta}, E_{\alpha,\beta})_{(\alpha,\beta)\in(A\times B)_1}$, where $U_{\alpha,\beta} = U_\alpha \cap f^{-1}(V_\beta)$ ($(A \times B)_1$ is the subset of $A \times B$ which supplies nonempty $U_{\alpha,\beta}$), $\varphi_{\alpha,\beta} = \varphi_\alpha | U_\alpha \cap f^{-1}(V_\beta)$, $E_{\alpha,\beta} = E_\alpha$. $\mathcal{A}_1$ is atlas on the set X: $\bigcup_{(\alpha,\beta)\in(A\times B)_1} U_{\alpha\beta} = X$ ($\bigcup_{\beta\in B} f^{-1}(V_\beta) = X$ etc.); $\varphi_{\alpha,\beta}(U_{\alpha,\beta} \cap U_{\alpha',\beta'})$ is open in $E_{\alpha,\beta}$ $(=E_\alpha)$, being equal to $\varphi_\alpha(U_\alpha \cap U_{\alpha'} \cap f^{-1}(V_\beta \cap V_{\beta'}))$ and $\varphi_\alpha : U_\alpha \to \varphi_\alpha(U_\alpha)$ is homeomorphism (**1**, 2°) and $f^{-1}(V_\beta \cap V_{\beta'})$ open in X; $\varphi_{\alpha,\beta} \circ \varphi^{-1}_{\alpha',\beta'} : \varphi_{\alpha',\beta'}(U_{\alpha,\beta} \cap U_{\alpha',\beta'}) \to \varphi_{\alpha,\beta}(U_{\alpha,\beta} \cap U_{\alpha',\beta'})$ is of C^r class, since $U_{\alpha,\beta} \cap U_{\alpha',\beta'} = U_\alpha \cap U_{\alpha'} \cap f^{-1}(V_\beta \cap V_{\beta'})$ and one takes into account the diffeomorphism (if $r \geq 1$) from (A_3). Moreover, $\mathcal{A}_1$ and $\mathcal{A}$ are equivalent, $\mathcal{A}_1 \cup \mathcal{A}$ is atlas on the set X [$\varphi_{\alpha'}(U_\alpha \cap f^{-1}(V_\beta) \cap U_{\alpha'})$ is open in $\varphi_{\alpha'}(U_\alpha \cap U_{\alpha'})$ etc.]. Obviously $\mathcal{A}_1$ and $\mathcal{B}$ are compatible relative to f.

And now

$f: X \to Y$ is by definition of C^l *class*, $1 \leq l \leq r$, if for any atlases $\mathcal{A}$, $\mathcal{B}$ of X resp. Y compatible relative to f and for any charts (U, φ), (V, ψ) of $\mathcal{A}$ resp. $\mathcal{B}$ compatible relative to f,

(14) $$\psi \circ f \circ \varphi^{-1} : \varphi(U) \to \psi(V) \text{ is of } C^l \text{ class.}$$

In this case f is continuous as $f|U = \psi^{-1} \circ (\psi \circ f \circ \varphi^{-1}) \circ \varphi$. If for two atlases $\mathcal{A}$ and $\mathcal{B}$ of X resp. Y, compatible relative to f, the property (14) is verified, then f is of C^l class, since for any atlases $\mathcal{A}_1$ and $\mathcal{B}_1$ of X resp. Y compatible relative to f, $\mathcal{A}_1$ equivalent to $\mathcal{A}$ and $\mathcal{B}_1$ equivalent to $\mathcal{B}$, we have the relation (without other explanations)

$$(\psi_{\beta_1} \circ \psi_\beta^{-1}) \circ (\psi_\beta \circ f \circ \varphi_\alpha^{-1}) \circ (\varphi_\alpha \circ \varphi_{\alpha_1}^{-1}) = \psi_{\beta_1} \circ f \circ \varphi_{\alpha_1}^{-1}.$$

Exemple. The function $f: X \to \mathbf{R}$ is of C^l class according to the above given definition $\Leftrightarrow$ f is of C^l class accordind to the definition from (14): take as atlas on $\mathbf{R}$ the unique local chart $(\mathbf{R}, \mathrm{id}_{\mathbf{R}})$, $\mathrm{id}_{\mathbf{R}}$ the identic map of $\mathbf{R}$.

The definition (14) is still simplified by the proposition below.

10 *If for each* (U, φ) *from* $\mathcal{A}$ *there exists* (V, ψ) *in* $\mathcal{B}$ *compatible with it relative to* f *and*

$$\psi \circ f \circ \varphi^{-1} : \varphi(U) \to \psi(V)$$

is of C^l *class, then* f *is of* C^l *class.*

■ It must prove

(V', ψ') any from $\mathscr{B}$ compatible relative to f with $(U, \varphi) \Rightarrow \psi' \circ f \circ \varphi^{-1} : \varphi(U) \to$ $\to \psi'(V')$ is of C^l class, and this reverts to

(15) $\varphi(x)$ any from $\varphi(U) \Rightarrow \varphi' \circ f \circ \varphi^{-1}$ is of C^l class on a neighbourhood of $\varphi(x)$.

But we see that $f(x) \in V \cap V'$ and that on $\varphi(U \cap f^{-1}(V \cap V'))$ we have

$$\psi' \circ f \circ \varphi^{-1} = (\psi' \circ \psi^{-1}) \circ \psi \circ f \circ \varphi^{-1},$$

which imposes (15) *via* the hypothesis and (A_3). ■

11 *Let* X, Y, Z *be differential* C^r*-manifolds*, $r \geq 1$ *and* $f : X \to Y$, $g : Y \to Z$ *of* C^l *class. Then* $g \circ f$ *is of* C^l *class.*

■ Let be $\mathscr{A} := (U_\alpha, \varphi_\alpha, E_\alpha)_{\alpha \in A}$, $\mathscr{C} := (W_\gamma, \chi_\gamma, G_\gamma)_{\gamma \in C}$ atlases of X resp. Z compatible relative to $h := g \circ f$ and $(U_{\alpha'}, \varphi_{\alpha'})$ any chart from $\mathscr{A}$ and $(W_{\gamma'}, \chi_{\gamma'})$ a chart from $\mathscr{C}$ compatible with the first relative to h. It must prove that

$$\chi_{\gamma'} \circ h \circ \varphi_{\alpha'}^{-1} : \varphi_{\alpha'}(U_{\alpha'}) \to \chi_{\gamma'}(W_{\gamma'})$$

is of C^l class, i.e. of C^l class in a neighbourhood of each point $\varphi_{\alpha'}(x)$ from $\varphi_{\alpha'}(U_{\alpha'})$. Fix x in $U_{\alpha'}$. So $h(x) = g(f(x)) \in W_{\gamma'}$. Let be $\mathscr{B} := (V_\beta, \psi_\beta, F_\beta)_{\beta \in B}$ an atlas of Y and $\mathscr{B}_1$ the atlas given by $\mathscr{B}$, $\mathscr{C}$ and g (see above), $\mathscr{B}_1 = (V_\beta \cap g^{-1}(W_\gamma), \psi_\beta \mid V_\beta \cap g^{-1}(W_\gamma))_{(\beta,\gamma) \in (B \times C)_1}$ and $\mathscr{A}_1$ the atlas of X given by $\mathscr{A}$, $\mathscr{B}_1$ and f, $\mathscr{A}_1 = (U_\alpha \cap f^{-1}(V_\beta) \cap h^{-1}(W_\gamma)$, $\varphi_\alpha \mid \ldots)_{(\alpha,\beta,\gamma) \in (A \times B \times C)_1}$. Take β' in B so that $f(x) \in V_{\beta'}$, then $x \in U_{\alpha'} \cap f^{-1}(V_{\beta'}) \cap h^{-1}(W_{\gamma'})$, open set included in $U_{\alpha'}$ and on $\varphi_{\alpha'}(U_{\alpha'} \cap f^{-1}(V_{\beta'}) \cap h^{-1}(W_{\gamma'}))$ we have

$$\chi_{\gamma'} \circ h \circ \varphi_{\alpha'}^{-1} = (\chi_{\gamma'} \circ g \circ \varphi_{\beta'}^{-1}) \circ (\psi_{\beta'} \circ f \circ \varphi_{\alpha'}^{-1}),$$

hence the conclusion since the two factors are of C^l class on some open sets. ■

Tangent space. Let X be differential C^1-manifold, $\mathscr{A} := (U_\alpha, \varphi_\alpha, E_\alpha)_{\alpha \in A}$ atlas of X and x any point fixed in X. Set $A_x := \{\alpha \in A : x \in U_\alpha\}$, a nonempty set according to (A_1). The set

(16) $$T_x X := \{\xi = (\xi_\alpha)_{\alpha \in A_x} \in \prod_{\alpha \in A_x} E_\alpha : \xi_\gamma = (\varphi_\gamma \circ \varphi_\beta^{-1})'(\varphi_\beta(x))(\xi_\beta)\ \forall\ \beta, \gamma \in A_x\}$$

is by definition the **tangent space** *to* X *in the point* x. Its elements are the **tangent vectors** *to* X *in the point* x.

$T_x X$ *is vector space.*

■ Indeed, in the product vector space $\prod_{\alpha \in A_x} E_\alpha$ the set $T_x X$ is a vector subspace: $\xi = (\xi_\alpha)_{\alpha \in A_x}$, $\eta = (\eta_\alpha)_{\alpha \in A_x}$, $\lambda \in \mathbf{R} \Rightarrow \xi + \eta = (\xi_\alpha + \eta_\alpha)_{\alpha \in A_x} \in T_x X$, $\lambda \xi = (\lambda \xi_\alpha)_{\alpha \in A_x} \in T_x X$ – use the remark: for any β, γ from A_x the map $(\varphi_\gamma \circ \varphi_\beta^{-1})'(\varphi_\beta(x))$ is linear. ■

Consider, for each α from A_x, the map

(17) $$\pi_{x,\alpha} : T_x X \to E_\alpha,\ \xi = (\xi_\beta)_{\beta \in A_x} \Rightarrow \pi_{x,\alpha}(\xi) = \xi_\alpha .$$

12 $\pi_{x,\alpha}$ *is linear and bijective.*

■ $\pi_{x,\alpha}$ *is linear*. This is obvious. $\pi_{x,\alpha}$ *is injective*. Let be $\xi := (\xi_\beta)_{\beta \in A_x}$ in Ker $\pi_{x,\alpha}$, i.e. $\xi_\alpha = 0$. Then $\forall\ \beta$ from A_x we have $\xi_\beta \overset{(16)}{=} (\varphi_\beta \circ \varphi_\alpha^{-1})'(\varphi_\alpha(x))(\xi_\alpha) = 0$. $\pi_{x,\alpha}$ *is surjective.* Let ξ_α be from E_α. Take $\xi := (\xi_\beta)_{\beta \in A_x}$ defined by $\xi_\beta = (\varphi_\beta \circ \varphi_\alpha^{-1})'(\varphi_\alpha(x))(\xi_\alpha)$. We have $\xi \in T_x X$ and obviously $\pi_{x,\alpha}(\xi) = \xi_\alpha$ ($(\mathrm{Id}_{E_\alpha})'(y) = \mathrm{Id}_{E_\alpha}$). For the first assertion : let γ be any from A_x, we have $\xi_\gamma = (\varphi_\gamma \circ \varphi_\alpha^{-1})'(\varphi_\alpha(x))(\xi_\alpha) = (\varphi_\gamma \circ \varphi_\beta^{-1} \circ \varphi_\beta \circ \varphi_\alpha^{-1})'(\varphi_\alpha(x))(\xi_\alpha) = (\varphi_\gamma \circ \varphi_\beta^{-1})'(\varphi_\beta(x))((\varphi_\beta \circ \varphi_\alpha^{-1})'(\varphi_\alpha(x))(\xi_\alpha)) = (\varphi_\gamma \circ \varphi_\beta^{-1})'(\varphi_\beta(x))(\xi_\alpha)$ and hence (16). ■

For each α from A_x define

$$\|\cdot\|_\alpha : T_x X \to \mathbf{R}_+,\ \xi = (\xi_\beta)_{\beta \in A_x} \Rightarrow \|\xi\|_\alpha = \|\xi_\alpha\|.$$

$\|\cdot\|_\alpha$ is a norm on the vector space $T_x X$. ($\|\xi\|_\alpha = 0 \Rightarrow \xi_\alpha = 0 \overset{(16)}{\Rightarrow} \xi_\beta = 0\ \forall\ \beta \in A_x$, hence $\xi = 0$ etc.).

Any two norms $\|\cdot\|_\alpha$, $\|\cdot\|_\beta$ *on* $T_x X$ *are equivalent,*

as

$$\|\xi\|_\beta = \|\xi_\beta\| = \|(\varphi_\beta \circ \varphi_\alpha^{-1})'(\varphi_\alpha(x))(\xi_\alpha)\| \le \|(\varphi_\beta \circ \varphi_\alpha^{-1})'(\varphi_\alpha(x))\|\ \|\xi_\alpha\| = \|(\varphi_\beta \circ \varphi_\alpha^{-1})'(\varphi_\alpha(x))\| \times \|\xi\|_\alpha .$$

Finally,

$\pi_{x,\alpha} : (T_x X, \|\cdot\|_\alpha) \to E_\alpha$ *is an isometry* (see **12**, $\|\pi_{x,\alpha}(\xi)\| = \|\xi_\alpha\| = \|\xi\|_\alpha$).

This isometry is called the *canonical isometry.*

Consequently

$(T_x X, \|\cdot\|_\alpha)$ *is Banach space.*

So, the canonical isometries

$$\pi_{x,\alpha} : (T_x X, \|\cdot\|_\alpha) \to E_\alpha,\ \alpha \in A_x$$

elucidate the manner in which the definition of $T_x X$ depends on the considered atlas of the manifold. Moreover, because $(T_x X, \|\cdot\|_\alpha)$ and E_α are in particular homeomorphic and, for $\beta \in A_x$, E_β and E_α are also (since $U_\alpha \cap U_\beta \neq \emptyset$, **8**), it follows that the topology of $T_x X$ is unique until a homeomorphism.

Let now be X, Y differential C^r-manifolds, $r \geq 1$, $\mathcal{A} := (U_\alpha, \varphi_\alpha, E_\alpha)_{\alpha \in A}$ and $\mathcal{B} := (V_\beta, \psi_\beta, F_\beta)_{\beta \in B}$ atlases of X resp. Y, $f: X \to Y$ of C^1 class and x a point from X. Define the map

$$T_x(f): T_x X \to T_{f(x)} Y:$$

$$\xi = (\xi_\alpha)_{\alpha \in A_x} \overset{\text{the choice of } \alpha \text{ is indifferent}}{\Rightarrow} ((\psi_\beta \circ f \circ \varphi_\alpha^{-1})'(\varphi_\alpha(x))(\xi_\alpha))_{\beta \in B_{f(x)}}.$$

$T_x(f)$ is linear, as $f \to (\psi_\beta \circ f \circ \varphi_\alpha^{-1})'(\varphi_\alpha(x))$ is linear, and continuous because

$$\|T_x(f)(\xi)\|_\beta = \|(\psi_\beta \circ f \circ \varphi_\alpha^{-1})'(\varphi_\alpha(x))(\xi_\alpha)\| \leq \|(\psi_\beta \circ f \circ \varphi_\alpha^{-1})'(\varphi_\alpha(x))\| \, \|\xi_\alpha\| =$$

$$= \|(\psi_\beta \circ f \circ \varphi_\alpha^{-1})'(\varphi_\alpha(x))\| \, \|\xi\|_\alpha .$$

13 *Let* X, Y, Z *be differential* C^r*-manifolds*, $r \geq 1$, $f: X \to Y$, $g: Y \to Z$ *of* C^1 *class and* x *from* X. *Then*

$$T_x(g \circ f) = T_{f(x)}(g) \circ T_x(f).$$

■ Take the atlases $\mathcal{A} = (U_\alpha, \varphi_\alpha, E_\alpha)_{\alpha \in A}$, $\mathcal{B} = (V_\beta, \psi_\beta, F_\beta)_{\beta \in B}$, $\mathcal{C} = (W_\gamma, \chi_\gamma, G_\gamma)_{\gamma \in C}$ of X, Y and Z respectively. Use the notation $F \cdot y := F(y)$, F linear.

$$\xi = (\xi_\alpha)_{\alpha \in A_x} \in T_x X \Rightarrow (T_x(f)(\xi))_\beta = (\psi_\beta \circ f \circ \varphi_\alpha^{-1})'(\varphi_\alpha(x))(\xi_\alpha)$$

and then

$$(T_{f(x)}(g)(T_x(f)(\xi)))_\gamma = (\chi_\gamma \circ g \circ \psi_\beta^{-1})'(\psi_\beta(f(x))) \cdot (T_x(f)(\xi))_\beta = (\chi_\gamma \circ g \circ \psi_\beta^{-1})'(\psi_\beta(f(x))) \cdot$$

$$\cdot (\psi_\beta \circ f \circ \varphi_\alpha^{-1})'(\varphi_\alpha(x))(\xi_\alpha) = (\chi_\gamma \circ (g \circ f) \circ \varphi_\alpha^{-1})'(\varphi_\alpha(x))(\xi_\alpha) = (T_x(g \circ f)(\xi))_\gamma ,$$

and from here the conclusion. ■

Another definition of the tangent space. Let X be differential C^1-manifold, $\mathcal{A} := (U_\alpha, \varphi_\alpha, E_\alpha)_{\alpha \in A}$ atlas of X and x arbitrary fixed point from X. Consider the set Γ of *continuous paths* $\gamma : J \to X$, J open interval of $\mathbf{R}$, with $\gamma(0) = x$, of C^1 class in the sense of the definition (14) (J with local charts (G, Id_G), $G \subset J$ nonempty open). Define in Γ the equivalence relation

$$\gamma_1 \sim \gamma_2 \overset{\text{def}}{\Leftrightarrow} (\varphi_\alpha \circ (\gamma_1 | \gamma_1^{-1}(U_\alpha)))'(0) = (\varphi_\alpha \circ (\gamma_2 | \gamma_2^{-1}(U_\alpha)))'(0),$$

$(U_\alpha, \varphi_\alpha, E_\alpha)$ local chart at x. The definition does not depend on the considered local chart, as $(U_\beta, \varphi_\beta, E_\beta)$ being another local chart at x, we have (it notes the restrictions of γ_i by γ_i)

$$(\varphi_\beta \circ \gamma_i)'(0) = (\varphi_\beta \circ \varphi_\alpha^{-1} \circ \varphi_\alpha \circ \gamma_i)'(0) = (\varphi_\beta \circ \varphi_\alpha^{-1})'(\varphi_\alpha(\gamma_i(0)))(\varphi_\alpha \circ \gamma_i)'(0)$$

and $\varphi_\alpha(\gamma_i(0)) = \varphi_\alpha(x)$. The equivalence class $\hat{\gamma}$ $(= [\gamma]_x)$ of γ is the **tangent vector** *to* X *at the point* x and the set $T_x X$: $\{\hat{\gamma} : \gamma \in \Gamma\}$ is the **tangent space** *to* X *at the point* x. Retain that for every local chart $(U_\alpha, \varphi_\alpha, E_\alpha)$ at x and for any vector v from E_α there exists γ in Γ so that

$$(17_1) \qquad (\varphi_\alpha \circ \gamma)'(0) = v.$$

Indeed, $\gamma : (-\delta, \delta) \to X$, $\gamma(t) = \varphi_\alpha^{-1}(vt + \varphi_\alpha(x))$ (δ sufficiently small) answers to the request.

$T_x X$ becomes vector space on **R** with the operations:

1) $\hat{\gamma}_1 + \hat{\gamma}_2 = \hat{\gamma}_3$, where γ_3 is chosen so that $(\varphi_\alpha \circ \gamma_3)'(0) = (\varphi_\alpha \circ \gamma_1)'(0) + (\varphi_\alpha \circ \gamma_2)'(0)$ (see (17_1)),

2) $\lambda \in \mathbf{R} \Rightarrow \lambda\hat{\gamma} = \hat{\sigma}$, where σ is so that $(\varphi_\alpha \circ \sigma)'(0) = \lambda(\varphi_\alpha \circ \gamma)'(0)$.

The map (18) $\Phi_\alpha(x) : [\gamma]_x \to (\varphi_\alpha \circ \gamma)'(0)$ is a bijection, even a homeomorphism of $T_x X$ on E_α (see (17_1)). This transfers also the norm of E_α to $T_x X$ and consequently the topology of $T_x X$ is unique until a homeomorphism (see the considerations after **12**).

Cotangent space. Let X be differential C^1-manifold, $\mathcal{A} := (U_\alpha, \varphi_\alpha, E_\alpha)_{\alpha \in A}$ atlas of X and x arbitrary fixed point from X. Consider the set Φ of the functions $\varphi : U(x) \to \mathbf{R}$ of C^1 class, $U(x)$ a neighbourhood of x in X, with $\varphi(x) = 0$. Define in Φ the equivalence relation

$$\varphi_1 \sim \varphi_2 \Leftrightarrow (\varphi_1 \circ \gamma)'(0) = (\varphi_2 \circ \gamma)'(0) \ \forall \gamma \in \Gamma$$

(for Γ see **Tangent space**).

The definition is, taking into account **11**, coherent. The equivalence class $[\varphi]_x$ of φ is by definition **cotangent vector** *of* X *at the point* x and

$$T_x^* X := \{[\varphi]_x : \varphi \in \Phi\}$$

is the **cotangent space** *of* X *at the point* x.

Let $(U_\alpha, \varphi_\alpha, E_\alpha)$ be a local chart at x. We have, $\forall \varphi$ from Φ and $\forall \gamma$ from Γ,

$$(19) \qquad (\varphi \circ \gamma)'(0) = (\varphi \circ \varphi_\alpha^{-1})'(\varphi_\alpha(x))(\varphi_\alpha \circ \gamma)'(0),$$

since $\varphi \circ \gamma = (\varphi \circ \varphi_\alpha^{-1}) \circ (\varphi_\alpha \circ \gamma)$.

Consider the map of $T_x^* X$ in E_α^*,

(20) $$\Phi_\alpha^*(x) : [\varphi]_x \to (\varphi \circ \varphi_\alpha^{-1})'(\varphi_\alpha(x)).$$

$\Phi_\alpha^*(x)$ is injective: $\Phi_\alpha^*(x)([\varphi]_x) = \Phi_\alpha^*(x)([\psi]_x) \overset{(20)}{\Rightarrow} (\varphi \circ \varphi_\alpha^{-1})'(\varphi_\alpha(x)) = (\psi \circ \varphi_\alpha^{-1})'(\varphi_\alpha(x)) \Rightarrow$

$\overset{\forall\gamma\in\Gamma}{\Rightarrow} (\varphi \circ \varphi_\alpha^{-1})'(\varphi_\alpha(x))(\varphi_\alpha \circ \gamma)'(0) = (\psi \circ \varphi_\alpha^{-1})'(\varphi_\alpha(x))(\varphi_\alpha \circ \gamma)'(0) \overset{(19)}{\Rightarrow} (\varphi \circ \gamma)'(0) = (\psi \circ \gamma)'(0) \Rightarrow [\varphi]_x = [\psi]_x$.

$\Phi_\alpha^*(x)$ is surjective: let ξ be from E_α^*, take $\varphi = \xi \circ (\varphi_\alpha - \varphi_\alpha(x))$, $\varphi(x) = 0$, $\varphi \circ \varphi_\alpha^{-1} = \xi \circ (\mathrm{id}_{E_\alpha} - \varphi_\alpha(x))$, $(\varphi \circ \varphi_\alpha^{-1})'(\varphi_\alpha(x)) = \xi \circ \mathrm{id}_{E_\alpha} = \xi$.

So $\Phi_\alpha^*(x)$ *is bijective* and consequently $\Phi_\alpha^*(x)$ transports the linear structure and the topology of E_α^* to $T_x^* X$. Moreover, $T_x^* X$ may be considered the dual $(T_x X)^*$ of $T_x X$ by

$$<\cdot, \cdot> : T_x X \times T_x^* X \to \mathbf{R},\ <[\gamma]_x, [\varphi]_x> = (\varphi \circ \gamma)'(0).$$

Tangent sheaf. Let X be differential C^r-manifold, $r \geq 1$ and $\mathcal{A} := (U_\alpha, \varphi_\alpha, E_\alpha)_{\alpha\in A}$ atlas of X.

The disjoint reunion $\bigcup_{x\in X} T_x X$ is by definition the **tangent sheaf** *to* X, one notes TX,

(21) $$TX \overset{\text{def}}{=} \bigcup_{x\in X} T_x X.$$

This will be equipped with a C^{r-1} - atlas. Consider the surjective map

$$\pi : TX \to X,\ \pi(\xi) = x \text{ where } \xi \in T_x X,$$

correct definition since the terms of the reunion (21) are disjoint. Obviously

$$\pi^{-1}(x) = T_x X \ \ \forall x \in X$$

and from this reason $T_x X$ is by definition the *fibre over* x.

For any local chart $(U_\alpha, \varphi_\alpha, E_\alpha)$ of X consider the triplet $(\pi^{-1}(U_\alpha), \chi_\alpha, \varphi_\alpha(U_\alpha) \times E_\alpha)$, where

(22) $$\chi_\alpha : \pi^{-1}(U_\alpha) = \bigcup_{x\in U_\alpha} T_x X \to \varphi_\alpha(U_\alpha) \times E_\alpha$$

is defined as follows:

$$\xi \in \pi^{-1}(U_\alpha),\ \xi \in T_x X \text{ with } x \text{ (unique)} \in U_\alpha,\ \xi \in \prod_{\beta\in A_x} E_\beta \ \ ((16)) \Rightarrow \chi_\alpha(\xi) \in (\varphi_\alpha(x), \xi_\alpha).$$

χ_α is bijective. Indeed, χ_α is injective: let be $\xi \neq \xi'$; if $\xi \in T_xX$, $\xi' \in T_{x'}X$ with $x \neq x'$, then $\varphi_\alpha(x) \neq \varphi_\alpha(x')$, $\chi_\alpha(\xi) \neq \chi_\alpha(\xi')$; if $\xi, \xi' \in T_xX$, then *a fortiori* $\xi_\alpha \neq \xi'_\alpha$ (*par absurdum*, use (16)), $\chi_\alpha(\xi) \neq \chi_\alpha(\xi')$. ξ_α is surjective: let (y, ξ_α) be from $\varphi_\alpha(U_\alpha) \times E_\alpha$, $y = \varphi_\alpha(x)$ with $x \in \subset U_\alpha$, $\exists\, \xi$ in T_xX so that $\pi_{x,\alpha}(\xi) = \xi_\alpha$ **(12)**, hence $\chi_\alpha(\xi) = (y, \xi_\alpha)$. ∎

So

14 *When* α *goes over* A, $(\pi^{-1}(U_\alpha), \chi_\alpha, \varphi_\alpha(U_\alpha) \times E_\alpha)$ *are local charts of an atlas of* C^{r-1} *class on* TX.

∎ Verify the axioms (A_1), (A_2), (A_3).

(A_1). $\xi \in TX \Rightarrow \exists\, x \in X$ so that $\xi \in T_xX$, take α so that $x \in U_\alpha$, then $\xi \in \pi^{-1}(U_\alpha)$ etc.

(A_2) is verified because $\chi_\alpha(\pi^{-1}(U_\alpha) \cap \pi^{-1}(U_\beta)) = \chi_\alpha(\pi^{-1}(U_\alpha \cap U_\beta)) = \varphi_\alpha(U_\alpha \cap U_\beta) \times \times E_\alpha$, an open set: let $(\varphi_\alpha(x), \xi_\alpha)$ be from $\varphi_\alpha(U_\alpha \cap U_\beta) \times E_\alpha$ and ξ from T_xX so that $\pi_{x,\alpha}(\xi) = = \xi_\alpha$ **(12)**, then $\xi \in \pi^{-1}(U_\alpha \cap U_\beta)$ and $\chi_\alpha(\xi) \overset{(22)}{=} (\varphi_\alpha(x), \xi_\alpha)$.

(A_3). $\chi_\beta \circ \chi_\alpha^{-1}$, application of $\chi_\alpha(\pi^{-1}(U_\alpha) \cap \pi^{-1}(U_\beta))$ $[= \chi_\alpha(\pi^{-1}(U_\alpha \cap U_\beta))]$ in $\chi_\beta(\pi^{-1}(U_\alpha) \cap \pi^{-1}(U_\beta))$ $[= \chi_\beta(\pi^{-1}(U_\alpha \cap U_\beta))]$, is of C^{r-1} class. Indeed, let ξ be from $\pi^{-1}(U_\alpha \cap U_\beta)$, $\xi \in T_xX$. $\chi_\alpha(\xi) = (y, \xi_\alpha)$ with $y = \varphi_\alpha(x)$, i.e. $x = \varphi_\alpha^{-1}(y)$. $\chi_\beta(\chi_\alpha^{-1}(y, \xi_\alpha)) =$

$= \chi_\beta(\xi) = (\varphi_\beta(x), \xi_\beta) \overset{(16)}{=} ((\varphi_\beta \circ \varphi_\alpha^{-1})(y), (\varphi_\beta \circ \varphi_\alpha^{-1})'(y)(\xi_\alpha))$,

consequently the two components of $\chi_\beta \circ \chi_\alpha^{-1}$ are $(y, \xi_\alpha) \to (\varphi_\beta \circ \varphi_\alpha^{-1})(y)$, of C^r class, and $(y, \xi_\alpha) \to (\varphi_\beta \circ \varphi_\alpha^{-1})'(y)(\xi_\alpha)$, of C^{r-1} class. ∎

Cotangent sheaf. Let X be differential C^r-manifold, $r \geq 1$ and $\mathscr{A} := (U_\alpha, \varphi_\alpha, E_\alpha)_{\alpha \in A}$ an atlas of X. The disjoint reunion $\bigcup_{x \in X} T_x^* X$, we note T^*X,

(23)
$$T^*X = \bigcup_{x \in X} T_x^* X$$

is by definition the **cotangent sheaf** *to* X. Consider, for any α from A,

(24)
$$W_\alpha := \bigcup_{x \in U_\alpha} T_x^* X,$$

(25)
$$\psi_\alpha^* : W_\alpha \to \varphi_\alpha(U_\alpha) \times E_\alpha^*,$$

$$\psi_\alpha^*([\varphi]_x) = (\varphi_\alpha(x), (\varphi \circ \varphi_\alpha^{-1})'(\varphi_\alpha(x))).$$

ψ_α^* is bijective (an analogous justification as for χ_α, **Tangent sheaf**).

15 *When* α *goes over* A, $(W_\alpha, \psi^*_\alpha, \varphi_\alpha(U_\alpha)\times E^*_\alpha)$ *are the local charts of an atlas on* T^*X *of* C^{r-1} *class.*

■ Analogous demonstration as for **14.** ■

Differential of a real function. Let X be differential C^1-manifold, $f: X\to \mathbf{R}$ of C^1 class and x_0 a point from X. The map

$$F: T_{x_0}X\to \mathbf{R},\ F(v) = (f \circ \varphi_\alpha^{-1})'(\varphi_\alpha(x_0))(v), \tag{26}$$

where $(U_\alpha, \varphi_\alpha, E_\alpha)$ is a local chart at x_0, is called the *differential of* f *at* x_0, it is designated by

$$\mathrm{d}f(x_0).$$

The canonical isometry $\pi_{x_0,\alpha}$ (**12**) makes coherent the definition. This is also correct as, by the change of local chart at x_0, $F(v)$ not change: $v = \pi^{-1}_{x_0,\beta}(\xi_\beta) \Rightarrow (f\circ\varphi_\beta^{-1})'(\varphi_\beta(x_0))(v) =$

$= (f\circ\varphi_\alpha^{-1}\circ\varphi_\alpha\circ\varphi_\beta^{-1})'(\varphi_\beta(x_0))(\xi_\beta) = (f\circ\varphi_\alpha^{-1})'(\varphi_\alpha(x_0))((\varphi_\alpha\circ\varphi_\beta^{-1})'(\varphi_\beta(x_0))(\xi_\beta)) \overset{(16)}{=} (f \circ\varphi_\alpha^{-1})'(\varphi_\alpha(x_0))(\xi_\alpha) \overset{(26)}{=} F(v).$

So

$$v\in T_{x_0}X \Rightarrow \mathrm{d}f(x_0)(v) = (f\circ\varphi_\alpha^{-1})'(\varphi_\alpha(x_0))(v). \tag{27}$$

We see that $\mathrm{d}f(x_0)\in (T_{x_0}X)^*$. Moreover, $x\to \mathrm{d}f(x_0)$, a map of X in $(TX)^*$, is continuous because f having C^1 class, $(f\circ\varphi_\alpha^{-1})'$ is continuous on a neighbourhood of $\varphi_\alpha(x)$ and φ_α is homeomorphism.

FINSLER MANIFOLDS

Derivative of a smoth path. Let X be differential C^1-manifold modelled on E, $\mathcal{A} := (U_\alpha, \varphi_\alpha, E)_{\alpha\in A}$ atlas of X, $\gamma: [a, b]\to X$ a path from X and t_0 from $[a, b]$. Consider $(U_\alpha, \varphi_\alpha, E)$ local chart at $x_0 := \gamma(t_0)$ and suppose $\varphi_\alpha\circ\gamma: [a, b]\to E$ with derivative at t_0 (lateral derivatives at a and b). Then by definition γ is *differentiable at* t_0 and

$$\gamma'(t_0) \overset{\text{def}}{=} (\pi^{-1}_{x_0,\alpha}\circ(\varphi_\alpha\circ\gamma))'(t_0), \tag{28}$$

which is coherent as $\pi_{x_0,\alpha}$ is even an isometry of $T_{x_0}X$ on E. So

$$(28_1) \qquad \gamma'(t_0) = \pi^{-1}_{x_0,\alpha}((\varphi_\alpha \circ \gamma)'(t_0)).$$

The definition is correct, if $(U_\beta, \varphi_\beta, E)$ is another local chart at x_0, then $\pi^{-1}_{x_0,\beta}((\varphi_\beta \circ \gamma)'(t_0)) = \pi^{-1}_{x_0,\alpha}((\varphi_\alpha \circ \gamma)'(t_0))$. Indeed, let $\xi := (\xi_\gamma)_{\gamma\in A_{x_0}}$ be with $\pi_{x_0,\alpha}(\xi) = (\varphi_\alpha \circ \gamma)'(t_0)$. Then $\pi^{-1}_{x_0,\beta}((\varphi_\beta \circ \gamma)'(t_0)) = \pi^{-1}_{x_0,\beta}((\varphi_\beta \circ \varphi_\alpha^{-1} \circ \varphi_\alpha \circ \gamma)'(t_0)) = \pi^{-1}_{x_0,\beta}((\varphi_\beta \circ \varphi_\alpha^{-1})'(\varphi_\alpha(x_0))((\varphi_\alpha \circ \gamma)'(t_0))) = \pi^{-1}_{x_0,\beta}((\varphi_\beta \circ \varphi_\alpha^{-1})'(\varphi_\alpha(x_0))(\xi_\alpha)) \overset{(16)}{=} \pi^{-1}_{x_0,\beta}(\xi_\beta) = \xi$ and hence the conclusion.

Retain

$$(29) \qquad \gamma'(t_0) \in T_{x_0}X,$$

$$\gamma \text{ differentiable at } t_0 \Rightarrow \gamma \text{ continuous at } t_0.$$

Remark. The definition (28) may be given for a differential C^1-manifold not necessary modelled.

Definitions. The path $\gamma: [a, b] \to X$ is *differentiable* if γ is differentiable on $[a, b]$. In this case, γ is of C^1 *class* (*curve of* C^1 *class*), $\gamma \in C^1([a, b]; X)$, if the application $t \to \gamma'(t)$ of $[a, b]$ in TX is continuous.

This definitions are extended in an obvious manner when X is C^p-manifold, $p \geq 1$ and one obtains the concept *curve of* C^p *class* respectively *curve of* C^∞ *class* (i.e. of C^p class $\forall\, p \geq 1$): TX is, when $p > 1$, C^{p-1}-manifold and $t \to \gamma'(t)$ is supposed differentiable on $[a, b]$ according to the definition (28), $T(TX)$ is C^{p-2}-manifold and $t \to \gamma''(t)$ is supposed differentiable on $[a, b]$ according to the definition (28) etc.

The definitions are extended in the following replacing the interval $[a, b]$ by an arbitrary interval J included in $\mathbf{R}$ not reduced at a point.

Moreover, $\gamma: J \to X$ is a *curve of* C^p *class*, $p \geq 1$ *by pieces* resp. *curve of* C^∞ *class by pieces* when γ is continuous and J has a finite partition with intervals on which γ is of C^p resp. C^∞ class.

16 *Let* X *be differential* C^1*-manifold* E*-modelled and* W *connected component of* X. *Two arbitrary points of* W *may be joint by a path of* C^1 *class*
(*R. S. Palais*).

Otherwise expressed

Every connected component of a C^1*-manifold* E*-modelled is connected by paths of* C^1 *class.*

■ Show firstly

(30) *Let* x, y, z *be from* W. *If* x, y *and also* y, z *may be joint in* W *by a path of* C^1 *class, then* x *and* z *have the same property.*

Let be $\gamma_1 \in C^1([0,\frac{1}{2}]; W)$, $\gamma_1(0) = x$, $\gamma_1(\frac{1}{2}) = y$ and $\gamma_2 \in C^1([\frac{1}{2}, 1]; W)$, $\gamma_2(\frac{1}{2}) = y$, $\gamma_2(1) = z$. Take $f: \mathbf{R} \to \mathbf{R}$ increasing of C^∞ class with $f(0) = 0$, $f(t) = \frac{1}{2}$ on $\left[\frac{1}{4}, \frac{3}{4}\right]$, $f(1) = 1$. Then $\gamma: [0, 1] \to W$, $\gamma(t) = \gamma_1(f(t))$ on $[0, \frac{3}{4}]$, $\gamma(t) = \gamma_2(f(t))$ on $[\frac{1}{4}, 1]$ is a curve of C^1 class in W with $\gamma(0) = x$ and $\gamma(1) = z$. We see that $\gamma(\frac{1}{2}) = y$.

Pass to prove the statement. X being locally connected (see above **8**), W is open, let x be any point in W and C_x the set of the points of W which may be joint with x by a C^1-path from W, obviously a nonempty set. If we show that C_x is open and closed, then $C_x = W$, whence the asserted property of the statement.

C_x *is open.* Let y be a point from C_x, γ a C^1-path in W with the extremities x, y, $(U_\alpha, \varphi_\alpha, E)$ local chart at y and S an open ball centred in $u := \varphi_\alpha(y)$ contained in $\varphi_\alpha(U_\alpha \cap W)$ (see **6**, 2°). We must show, to conclude C_x open, that a point z arbitrary from $U_\alpha \cap W$ may be joint to x by a C^1-path from W. But if $v := \varphi_\alpha(z)$, the segment $[u, v]$ ($= \{u + t(v - u): t \in [0, 1]\}$) is included in S and it remains only to consider $\sigma: [0,1] \to W$, $\sigma(t) = \varphi_\alpha^{-1}(u + t(v - u))$ and then to apply (30).

C_x *is closed.* An analogous justification as in the case above.

Remark. The analysis of the proof for **16** shows that in the Palais proposition the C^1 class may be replaced by the C^p class, $p \geq 1$.

Finsler structure. Let X be differential C^1-manifold E-modelled and $(U_\alpha, \varphi_\alpha, E)_{\alpha \in A}$ atlas of X. The functional

(31) $$\|\cdot\|: TX \to [0, +\infty)$$

is by definition **Finsler structure** *on* T X, the tangent sheaf, if:

1° For any x from X, the restriction $\|\cdot\|_x$ of $\|\cdot\|$ to $T_x X$ is norm on this having the property $v \to \|\pi_{x,\alpha}^{-1}(v)\|_x$ is equivalent to the norm $v \to |v|$ on E ($(U_\alpha, \varphi_\alpha, E)$ local chart at x, $\pi_{x,\alpha}$ the isomorphism $T_x X \to E$ of (17));

2° For any x_0 from X and $\lambda > 1$, $\exists\ U_\alpha$ at x_0 so that, $\forall\ x \in U_\alpha$,

(32) $$\frac{1}{\lambda}\left\|\pi_{x_0,\alpha}^{-1}(v)\right\|_{x_0} \le \|\pi_{x,\alpha}^{-1}(v)\|_x \le \lambda\left\|\pi_{x_0,\alpha}^{-1}(v)\right\|_{x_0}\quad \forall\ v \in E.$$

Definition. A differential C^1-manifold X with a Finsler structure on TX is named **Finsler manifold**.

Given a Finsler structure on TX, it can obtain a Finsler structure on T^*X, $|||\cdot|||: T^*X \to$ $\to [0, +\infty)$, by

(33) $$\xi \in T_x^* X \Rightarrow |||\xi|||_x = \sup\{|\xi(w)|: \|w\|_x \le 1\}.$$

17 *Every Finsler structure is continuous.*

■ We keep the notations above. Let w_1 be arbitrary fixed in TX, $w_1 \in T_{x_0}X$. For x_0 and $\lambda > 1$ $\exists\ U_\alpha$ at x_0 so that (32) is verified. Let be $w_1 := \pi_{x_0,\alpha}^{-1}(v_1)$ and $w_2 := \pi_{x,\alpha}^{-1}(v_2)$ with $x \in$ $\in U_\alpha$.

$$|\ \|w_2\|_x - \|w_1\|_{x_0}| \le |\ \|w_2\|_x - \left\|\pi_{x_0,\alpha}^{-1}(v_2)\right\|_{x_0}| + |\left\|\pi_{x_0,\alpha}^{-1}(v_2)\right\|_{x_0} - \left\|\pi_{x_0,\alpha}^{-1}(v_1)\right\|_{x_0}|.$$

But

$$|\left\|\pi_{x_0,\alpha}^{-1}(v_2)\right\|_{x_0} - \left\|\pi_{x_0,\alpha}^{-1}(v_1)\right\|_{x_0}| \le \left\|\pi_{x_0,\alpha}^{-1}(v_1 - v_2)\right\|_{x_0} \overset{1^0}{\le} \rho|v_1 - v_2|$$

and

$$|\ \|w_2\|_x - \left\|\pi_{x_0,\alpha}^{-1}(v_2)\right\|_{x_0}| \le (\lambda - 1)\left\|\pi_{x_0,\alpha}^{-1}(v_2)\right\|_{x_0}$$

(use (32), $2 - \lambda \le \frac{1}{\lambda}$), hence

(34) $$|\ \|w_2\|_x - \|w_1\|_{x_0}| \le \rho|v_1 - v_2| + (\lambda - 1)\left\|\pi_{x_0,\alpha}^{-1}(v_2)\right\|_{x_0}.$$

Let now be $\varepsilon > 0$. For v_2 sufficiently near to v_1 and λ sufficiently near to 1 the second member of (34) is majorized by ε and so it can find, *via* (34), a neighbourhood of w_1 in the afferent topology to the atlas on TX from (22) (see **6**) so that, on this, $|\ \|w_2\|_x -$ $- \|w_1\|_{x_0}| \le \varepsilon$. ■

Finsler manifold. Begin with an existence proposition.

18 *Every differential* C^1*-manifold* E*-modelled, which is also metric space, is Finsler manifold.*

■ Let X be a manifold as in the statement, $\mathcal{A} := (U_\alpha, \varphi_\alpha, E)_{\alpha\in A}$ atlas of X and $(f_\alpha)_{\alpha\in A}$ locally Lipschitz partition of the unit on X relative to $\mathcal{A}$ (see **2**). It obtains a Finsler structure as follows. For each x from X and for each w from $T_x X$ take ($|\cdot|$ is the norm on E)

(35) $$\|w\|_x = \sum_{\alpha\in A} f_\alpha(x)\, |\pi_{x,\alpha}(w)|,$$

$\pi_{x,\alpha}: T_x X \to E$ the isomorphism from (17). Let x_0 be a point from X and U_β at x_0. Because the family (supp $f_\alpha)_{\alpha\in A}$ is locally finite, there exists only a finite number of indices, let they be $\alpha_1, \alpha_2, \ldots, \alpha_m$, for which f_{α_i} are not identically equal to 0 on U_β (take the intersection of U_β with some neighbourhood of x_0). Then $\|w\|_{x_0} \overset{(25)}{=} \sum_{i=1}^{m} f_{\alpha_i}(x_0) \left|\pi_{x_0,\alpha_i}(w)\right|$, this clearly shows that $w \to \|w\|_{x_0}$ is a norm on $T_{x_0} X$ (the property 3° of the definition for **2** imposes $\exists f_{\alpha_1}(x_0) > 0$) and hence 1°. For 2°, use the relation

(36) $$\|\pi_{x,\beta}^{-1}(v)\|_x \overset{(35)}{=} \sum_{i=1}^{n} f_{\alpha_i}(x)\, |\, [\pi_{x,\alpha_i} \circ \pi_{x,\beta}^{-1} \circ \pi_{x_0,\beta} \circ \pi_{x_0,\alpha_i}^{-1}](\pi_{x_0,\alpha_i}(\pi_{x_0,\beta}^{-1}(v)))|.\ ■$$

Let X be Finsler manifold and $\gamma: [a, b] \to X$ a C^1-path (curve of C^1 class). The *length* $L(\gamma)$ of γ is

(37) $$L(\gamma) \overset{\text{def}}{=} \int_a^b \|\gamma'(t)\|_{\gamma(t)}\, dt.$$

The definition is correct as $\gamma'(t) \overset{(29)}{\in} T_{\gamma(t)} X$ and the Finsler structure is continuous (**17**).

Suppose now X *connected* Finsler manifold and let x, y be any points from X.

(38) $$d_F(x, y) \overset{\text{def}}{=} \inf\{L(\gamma): \gamma \in C^1([a, b]; X), \gamma(a) = x, \gamma(b) = y\}.$$

The definition is correct because, according to **16**, the set of (38) is nonempty.

19 *Let* X *be connected Finsler manifold. Then*

$$(x, y) \to d_F(x, y)$$

is a distance on X *and the afferent topology to this coincides with the topology of* X.

■ See [53]. ■

d_F is the **Finsler metric**.

20 *Corollary. Every connected Finsler manifold is paracompact and an* ANR.

■ Combine **19** with **9**. ■

We finish this section with

21 *Let* X *be Finsler manifold and* $\varphi : X \to \mathbf{R}$ *of* C^1 *class. For every point* x *not critical for* φ *there is* v(x) *vector from* $T_x X$ *so that*

1° $\|v(x)\| \le 2\|d\varphi(x)\|$,

2° $d\varphi(x)(v(x)) > \|d\varphi(x)\|^2$.

Attention ! The Finsler structures on *TX* and $T^* X$ have identical notation.

■ See [53]. ■

RIEMANN MANIFOLDS

We notify here a forced change of the notations : *M* instead of *X* differential manifold (*Mannigfaltigkeit*), *p*, *q*, *m* instead of *x*, *y*, *z* arbitrary points from *M* etc.

Riemann manifold. A differential C^∞-manifold separated connected *M* modelled on a Hilbert space *H*, with finite or infinite dimension, is by definition a *Riemann manifold*, if for every point *p* from *M* is defined on the tangent space $T_p M$ a symmetrical positive defined bilinear form $<\cdot, \cdot>_p$, which smoothly depends (C^∞ !) on *p* and in addition $\|\cdot\|_p$ is equivalent to the norm of *H*. In this case there exists on *M* a unique affine connexion $\nabla : (X, Y) \to \nabla_X Y$ having the properties:

(39) $$Z<X,Y> = <\nabla_Z X, Y> + <X, \nabla_Z Y>$$

(the parallel transport keeps the scalar product in the tangent spaces)

and

(40) $$\nabla_X Y - \nabla_Y X = XY - YX$$

(the torsion tensor is equal to 0),

$\forall X, Y, Z$ from $C^\infty(TM)$.

Define on each Riemann manifold *M* a *Riemann metric* in the following way. Let *p*, *q* be any points from *M*. For each $\gamma : [0, 1] \to M$ curve of C^∞ class by pieces (*Finsler manifolds*) with $\gamma(0) = p$, $\gamma(1) = q$ (γ *sets out* from *p* and *arrives* in *q*) one takes

(41) $$L(\gamma) \text{ (the } \textit{length} \text{ of } \gamma) \overset{\text{def}}{=} \int_0^1 \|\dot{\gamma}(t)\|_{\gamma(t)} dt .$$

Then

$$\text{(42)}\qquad \mathrm{d}(p, q) \overset{\text{def}}{=} \inf\{\mathrm{L}(\gamma) \mid \gamma : [0, 1] \to M \text{ of } \mathrm{C}^{\infty} \text{ class by pieces}, \gamma(0) = p, \gamma(1) = q\}.$$

The definition is correct, the set which appears in (42) being nonempty (see **16** with Remark). d is a distance on M. The topology defined by d coincides with the topology of M.

Notations

Let p be from M and $r > 0$.

$$S_{\mathrm{p}}^{\mathrm{r}} = \{\xi \in T_{\mathrm{p}} M : \|\xi\|_{\mathrm{p}} < r\},\ \sigma_{\mathrm{r}}^{\mathrm{p}} = \{\xi \in T_{\mathrm{p}} M : \|\xi\|_{\mathrm{p}} = r\},$$

$$\mathscr{S}_{\mathrm{p}}^{\mathrm{r}} = \{q \in M : \mathrm{d}(p, q) < r\},\ \Sigma_{\mathrm{p}}^{\mathrm{r}} = \{q \in M : \mathrm{d}(p, q) = r\}.$$

Geodesic. Let M be Riemann manifold and ∇ the afferent affine connexion. The solutions γ of the equation

$$\text{(43)}\qquad \nabla_{\dot{\gamma}(\mathrm{t})}\dot{\gamma}(t) = 0$$

are by definition the **geodesics** of M. If M is complete, every solution of (43) can be defined on **R**, that is in this case the geodesics of M can be indefinitely extended.

A geodesic γ which sets out from p and arrives in q, with the property $\mathrm{L}(\gamma) = \mathrm{d}(p, q)$ (see (41), (42)), is named *minimal geodesic*.

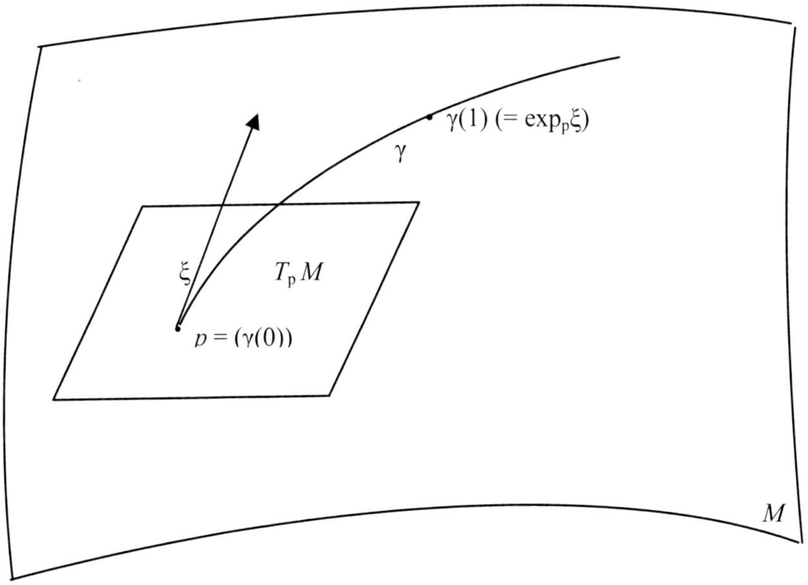

Fig. 6

Exponential map. Let M be complete Riemann manifold and p a point from M. For any ξ of $T_p M$ there exists $\gamma : [0, +\infty) \to M$ a unique solution of (43), hence a geodesic, which verifies the initial conditions

$$\gamma(0) = p, \gamma'(0) = \xi.$$

Set

$$\exp_p \xi := \gamma(1)$$

(see fig. 6).

So one finds a C^∞-map

$$\exp_p : T_p M \to M, \xi \to \exp_p \xi,$$

the *exponential map at the point* p. Retain

(44) $$\exp_p(0) = p.$$

It can assert that there is $\delta_0 > 0$ with the property:

$$\forall \delta \text{ from } (0, \delta_0)\ \exists\, \eta > 0 \text{ so that}$$

(45) $$m \in \mathcal{S}_p^\eta \Rightarrow \mathcal{S}_m^\delta \supset \mathcal{S}_p^\eta \text{ and } \exp_m : S_m^\delta \to \mathcal{S}_m^\delta \text{ is an isomorphism}$$

([30], 3).

For $m = p$ we find $\eta \leq \delta$. So, taking into account (44), one finds in particular

(46) $$d(m, \exp_m \xi) = \|\xi\|_m\ \forall \xi \in S_m^\delta .$$

Besides

(47) $$\forall q, m \in \mathcal{S}_p^\eta\ \exists\, \xi \text{ unique in } T_q M \text{ so that } m = \exp_q \xi$$

and the map $(m, q) \to \xi$ is of C^∞ class.

Otherwise expressed,

(48) Any two points of $\mathcal{S}_p^\eta$ may be joint by a minimal geodesic and this depends in a smooth manner of its extremities.

One designates by $\Delta(p)$ the supremum of the numbers η having the property expressed by (48).

So, according to the previous assertions, for any ρ, $0 < \rho < \Delta(p)$, $\exp_p$ is a C^∞-diffeomorphism of S_p^ρ on $\mathcal{S}_p^\rho$ and of σ_p^ρ on Σ_p^ρ and

(49) $$d(p, m) < \Delta(p) \Rightarrow \exists!\ \xi : m = \exp_p \xi \text{ and } \|\xi\|_p = d(p, m)$$

(see fig. 7).

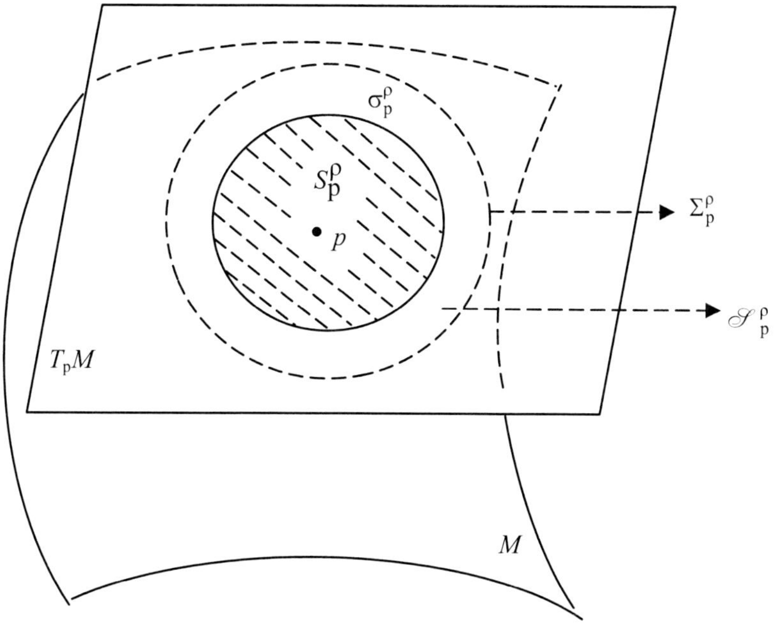

Fig. 7

BIBLIOGRAPHY

[1] *Ambrosetti A., Rabinowitz P. H., Dual variational methods in critical point theory and applications*, 1973, J. Functional Anal. **14**, 349-381.

[2] *Aubin J. P., Ekeland I., Applied nonlinear analysis*, 1984, John Wiley and Sons, New York Chichester Brisbane Toronto Singapore.

[3] *Bates P. W., Ekeland I., A saddle point theorem*, 1980, Differential equations, Academic Press, 123-126.

[3_1] *Benbrik A., Mbarki A., Lahrech S., Ouhab A., Ekeland's principle for vector-valued maps based on the characterization of uniform spaces via families of generalized quasi-metrics*, 2006, Lobachevskii Journal of Mathematics, vol. 21, 33-44.

[4] *Bishop E., Phelps R. R., The support functional of a convex set*, 1963, Convexity (Klee, ed.), Proc. Sympos. Pure Math., vol. 7, 27-35, Amer. Math. Soc., Providence, R.I.

[5] *Borwein J. M., Giles J. R., The proximal normal formula in Banach spaces*, 1987, Trans. Amer. Math. Soc. **302**, 371-381.

[6] *Borwein J. M., Preiss D., A smooth variational principle with applications to subdifferentiability and to differentiability of convex functions*, 1987, Transactions of the Amer. Math. Soc., vol. 303, No. 2, 517-527.

[7] *Borwein J. M., Strojwas H.M., Subderivatives and non-smooth analysis*, 1987, Canadian Journal of Math., **39**, 428-472.

[8] 1) *Brezis H., Analyse fonctionnelle*, 1992, Masson, Paris Milan Barcelone Bonn.

2) *Brezis H., The Ambrosetti-Rabinowitz MPL via Ekeland's minimization principle* (Manuscript).

[9] *Brezis H., Nirenberg L., Remarks on finding critical points*, 1991, Comm. Pure and Appl. Math. **44**, No. 8-9, 939-963.

[10] *Brezis H., Browder F., A general ordering principle in nonlinear functional analysis*, 1976, Advances in Math. **21**, 355-364.

[11] *Browder F. E., Normal solvability for nonlinear mappings into Banach spaces*, 1971, Bull. Amer. Soc. **77**, 73-77.

[12] *Cammaroto F., Chinni A., A complement to Ekeland's variational principle in Banach spaces*, 1996, Bull. of the Polish Academy of Sciences, vol. 44, No. 1.

[13] *Cammaroto F., Chinni A., Sturiale G., On an extension of Ekeland's principle for vector-valued functions*, 1998, Optimization, vol. 43, 19-28.

[14] *Caristi J., Fixed point theorems for mappings satisfying inwardness conditions*, 1976, Trans. Amer. Math. Soc., **215**, 241-251.

[15] *Caristi J., Kirk W. A., Geometric fixed point theory and inwardness conditions* in *The geometry of metric and linear spaces*, 1975, Lecture notes in Math., **490**, 74-83 Springer Verlag, Berlin Heidelberg New York.

[16] *Chang K. C., Variational methods for non-differentiable functionals and their applications to partial differential equations*, 1981, J. Math. Anal. Appl. **80**, 102-129.

[17] *Chang S. S., Cho I. J., Kim J. K., Ekeland's variational principle and Caristi's coincidence theorem for set-valued mappings in probabilistic metric spaces*, 1996, Periodica Mathematica Hungarica, vol. 33 (**2**), 83-92.

[18] *Chang S. S., Chen Y. Q., Guo J. L., Ekeland's variational principle and Caristi's fixed point theorem in probabilistic metric spaces*, 1991, Acta Math. Appl. : Sinica **3**, 217-230.

[19] *Chen G. Y., Huang X. X., Ekeland's variational principle for set-valued mappings*, 1993, Math. Meth. for Oper. Res., **48**, 181-186, Springer Verlag.

[20] 1) *Clarke F. H., Optimization and non-smooth analysis*, 1983, Canadian Math. Society.

2) *Clarke F. H., A new approach to Lagrange multipliers*, 1976, Math. Oper. Res. **1**, 165-174.

[21] *Constantin G., Istrătescu I., Elemente de analiză probabilistă şi aplicaţii*, 1981, Editura Academiei Române, Bucarest.

[22] *Costa D. G., Gonçalves J. V. A., Critical point theory for non-differentiable functionals and applications*, 1990, J. of Math. Anal. and Appl., **153**, 470-485.

[23] 1) *Crandall M. G., Lions P. L., Viscosity solutions of Hamilton-Jacobi equations*, 1983, Trans. Amer. Math. Soc. **277**, 1- 42.

2) *Crandall M. G., Lions P. L., Hamilton-Jacobi equations in infinite dimensions*, I + + II + III, J. Funct. Anal. **62** (1985), 379-398 + **65** (1986), 368-405 + **68** (1986), 214-247.

[24] *Dancs S., Hegedüs M., Medvegyev P., A general ordering and fixed point principle*

in complete metric space, 1983, Acta Sci. Mathem. (Szeged), vol. 46, 381-388.

[25] *Daneš J., A geometric theorem useful in nonlinear functional analysis*, 1972, Boll. Un. Math. Ital. **6**, 369-375.

[26] *Deimling K., Nonlinear functional analysis*, 1985, Springer Verlag, Berlin Heidelberg New York Tokyo.

[27] 1) *Deville R., Godefroy G., Zizler V., A smooth variational principle with applications to Hamilton-Jacobi equations in infinite dimensions*, 1993, J. of Func. Anal. **111**, 197-212.

2) *Deville R., Godefroy G., Zizler V., Smoothness and renormings in Banach spaces*, 1992, Monographs and Survey series, Longman, No. 64.

[28] *Diestel J., Geometry of Banach spaces*, 1975, Lecture Notes in Math., vol. 485, Springer Verlag.

[29] *Dugundji J., Topology*, 1973, Allyn and Bacon, Boston.

[30] 1) *Ekeland I., On the variational principle*, 1972, Cahiers de mathématique de la décision, No. 7217, Université Paris.

2) *Ekeland I., On the variational principle*, 1974, J. Math. Anal. and Appl. **47**, 324-353.

3) *Ekeland I., The Hopf-Rinow theorem in infinite dimension*, 1978, J. Differential Geometry **13**, 287-301.

4) *Ekeland I., Nonconvex minimization problems*, 1979, Bull. Amer. Math. Soc. **1**, 443-474.

5) *Ekeland I., Convexity methods in hamiltonian mechanics*, 1990, Springer Verlag.

6) *Ekeland I., Temam R., Analyse convexe et problèmes variationnels*, 1974, Dunod, Gauthier-Villars, Paris Bruxelles Montréal.

7) *Ekeland I., Lebourg G., Generic Fréchet differentiability and perturbed optimization problems in Banach spaces*, 1976, Trans. Amer. Math. Soc. **224**, 193-216.

[30_1] *Fang J.-X., The variational principle and fixed point theorem in certain topological spaces*, 1996, J. Math. Anal. Appl., 202: 398-412. MR **97d**: 54070.

[31] 1) *Fenchel W., On conjugate convex functions*, 1949, Canad. J. Math. **1**, 73-77.

2) *Fenchel W., Convex cones, sets and functions*, 1953, Notes de cours polycopiées, Princeton University.

[32] *de Figueiredo D. G., Lectures on the Ekeland variational principle with applications*

and detours, 1987, Springer Verlag, Berlin Heidelberg New York Tokyo.

[32_1] *Georgiev P. G., The strong Ekeland variational principle, the strong drop theorem and applications*, 1988, J. Math. Anal. Appl., 131, 1-21.

[33] *Gerth (Tammer) Chr., Weidner P., Nonconvex separation theorems and some applications in vector optimization*, 1990, J. of optimization theory and applications, vol. 67, No. 2, 297-320.

[34] *Ghoussoub N., Duality and perturbation methods in critical point theory*, 1993, Cambridge University Press.

[35] *Göpfert A., Tammer Chr., Zălinescu C., On the vectorial Ekeland 's variational principle and minimal points in product spaces*, 2000, Nonlinear Analysis **39**, 909-922.

[35_1] 1) *Hamel A., Phelp's lemma, Daneš drop theorem and Ekeland's principle in locally convex spaces*, 2003, Proc. Amer. Math. Soc., vol. 131, 3025-3038.

2) *Hamel A., Equivalents to Ekeland 's variational principle in uniform spaces*, 2005, Nonlinear Analysis 62, 913-924.

[35_2] *Hiriart-Urruty J. B., New concepts in nondifferentiable programming*, 1979, Bull. Soc. Math. France **60**, 57-85.

[36] *Hofer H., A geometric description of the neighbourhood of a critical point given by the mountain pass theorem*, 1985, J. Lond. Math. Soc. **31**, 566-570.

[37] *Isac G., The Ekeland principle and the Pareto ε-efficiency*, 1996, Lecture Notes in Economics and Mathematical Systems, vol. 432, 148-163, Springer Verlag, Berlin.

[38] *Ishii H., Perron's method for Hamilton-Jacobi equations*, 1987, Duke Math. J., **55**, 369-384.

[38_1] *Kada O., Suzuki T., Takahashi W., Nonconvex minimization theorems and fixed point theorems in complete metric spaces*, 1996, Math. Japonica **44**, 381-391.

[38_2] *Kelly J. L., General topology*, 1957, D. Van Nostrand Company, Inc. Princeton, New Jersey Toronto London New York.

[39] *Khank Ph. Q., On Caristi-Kirk's theorem and Ekeland 's variational principle for Pareto extrema*, 1989, Bull. of the Polish Acad. of Sc. Math., vol. 37, No. 1-6.

[39_1] *Köthe G., Topologische lineare Räume* I, 1966, Springer Verlag, Berlin.

[40] *Larman D. G., Phelps R. P.,Gâteaux differentiability of convex functions on Banach spaces*, 1979, J. London Math. Soc. **20**, 115-127.

[41] *Li Yongxin, Shi Shuzhong, A generalization of Ekeland 's variational principle and*

its Borwein-Preiss smooth variant, 2000, J. of Math. Anal. and Appl., **246**, 308-319.

[42] *Lindenstrauss J., Preiss D., On Fréchet differentiability of Lipschitz maps between Banach spaces*, 2003, Annals of Mathematics, **157**, 257-288.

[43] *Luc D. T., Theory of vector optimization*, 1989, Lecture Notes in Economics and Mathematical Systems **319**, Springer Verlag, Berlin Heidelberg New York.

[44] *Mawhin J., Problèmes de Dirichlet variationnels non linéaires*, 1987, Les Presses de L'Université de Montréal.

[45] *Mawhin J., Willem M., Critical points theory and Hamiltonian Systems*, 1989, Springer Verlag, New York Berlin Heidelberg London Paris Tokyo.

[46] *Meghea I., Minimax theorem, Mountain Pass theorem and Saddle Point theorem in β-differentiability*, 2003, Commun. Appl. Nonlinear Anal., No. 1, 55-66.

[47] *Meghea C., Meghea I., Differential Calculus and Integral Calculus*, vol. I, Editura Tehnică, 1997, Bucarest; vol. II, Editura Tehnică, 2000, Bucarest; vol. III, Editura Printech, 2002, Bucarest.

[48] *Menger K., Statistical metrics*, 1942, Proc. Nat. Acad. Sc. USA **28**, 535-537.

[49] *Milnor J., Morse theory*, 1969, Princeton University Press.

[50] *Moreau J. J., Fonctionnelles convexes*, 1966, Séminaire Équations aux dérivées partielles, Collège de France.

[51] 1) *Németh A. B., The nonconvex minimization principle in ordered regular Banach spaces*, 1981, Mathematica (Cluj), **23**, 43-46.

2) *Németh A. B., A nonconvex vector-minimization problem*, 1986, Nonlinear Analysis, Theory, Methods and Applications, **10**, No. 7, 669-678.

3) *Németh A. B., Ekeland variational principle in ordered abelian groups*, 2001, Nonlinear Anal. Forum 6(2), 299-312.

[52] *Oettli W., Théra M., Equivalents of Ekeland's variational principle*, 1993, Bull. Austral. Math. Soc., vol. 48, 385-392.

[53] *Palais R. S., Ljusternik-Schnirelmann theory on Banach manifolds*, 1966, Topology, **5**, 115-132.

[54] 1) *Park S., On extensions of the Caristi-Kirk fixed point theorem*, 1983, J. Koreean Math. Soc., vol. 19, No. 2.

2) *Park S., On generalizations of the Ekeland type variational principles*, 2000 Nonlinear Analysis, **39**, 881-889.

[55] *Penot J. P., The drop theorem, the petal theorem and Ekeland's variational*

principle, 1986, Nonlinear Analysis, Theory, Methods and Applications, vol. 10, No. 9, 813-822.

[56] 1) *Phelps R. R., Support cones in Banach spaces and their applications*, 1974, Advances in Mathematics, **13**, 1-19.

2) *Phelps R. R., Convex functions, monotone operators and differentiability*, 1993, Lecture Notes in Mathematics, Second edition, Springer Verlag, Berlin Heidelberg.

[57] *Preiss P., Differentiability of Lipschitz functions on Banach spaces*, 1990, J. Funct. Anal. **91**, 312-345.

[57_1] 1) *Qiu J.-H., Local completeness and drop theorem*, 2002, J. Math. Anal. Appl. **266**, 288-297.

2) *Qiu J.-H., Ekeland's variational principle in locally complete spaces*, 2003, Math. Nachr. **257**, 55-58.

3) *Qiu J.-H., Local completeness, drop theorem and Ekeland's variational principle,* 2005, J. Math. Anal. Appl. **311**, 23-39.

[58] 1) *Rabinowitz P. H., Some minimax theorems and applications to nonlinear partial differential equations*, 1978, Nonlinear Analysis, Academic Press, 161-177.

2) *Rabinowitz P. H., Some critical point theorems and applications to semilinear elliptic partial differential equations*, 1978, Ann. Scuola Norm. Sup. Pisa Ser. IV, 5, 161-177.

[59] 1) *Rockafellar R. T., Convex Analysis*, 1970, Princeton University Press.

2) *Rockafellar R. T., Generalized directional derivatives and subgradients of nonconvex functions*, 1980, Can. J. Math. **32**, 257-280.

3) *Rockafellar R. T., Directionally Lipschitzian functions and subdifferential calculus*, 1979, Proc. London Math.. Soc. (3) **39**, 331-355.

[60] *Schaefer H., Topological vector spaces*, 1971, Springer Verlag, third printing, Berlin Heidelberg New York.

[61] *Schweizer B., Sklar A., Statistical metric spaces*, 1960, Pacific J. Math. **10**, 323-334.

[62] *Schweizer B., Sklar A., Thorp E., The metrization of statistical metric spaces*, 1960, Pacific J. Math. **10**, 673-675.

[63] *Schweizer B., Sklar A., Probabilistic metric spaces*, 1983, North Holland.

[64] *Shen Jie Li, Hiao Qi Yang, Gung-Ya Chen, Vector Ekeland's variational principle*, 2000, Vector variat. ineq. and vector equil., 321-333, Kluiver Academic Publishers.

[65] *Shi Shuzhong, Ekeland's variational principle and the Mountain pass lemma*, 1985,

Acta mathematica Sinica, New Series, vol. I, No. 4, 348-355.

[66] *Siegel J., A new proof of Caristi's fixed point theorem*, 1977, Proc. Amer. Math. Soc. **66**, 54-56.

[67] *Sullivan F., A characterization of complete metric spaces*, 1981, Proc. of the Amer. Math. Soc., vol. 83, No. 2, October.

[68] *Szulkin A., Critical point theory of Ljusternik-Schnirelmann type and applications to partial differential equations*, 1989, Sem. Math. Sup., Presses Univ. Montréal.

[69] *Takahashi W., Existence theorems generalizing fixed point theorems for multivalued mappings* in *Fixed point theory and applications* (J. B. Baillon and M. Théra, Editors), 1991, Pitman Research Notes in Math. **252** (Longmann, Harlow), 397-406.

[70] *Tammer Chr., A generalization of Ekeland's variational principle*, 1992, Optimization, **25**, 129-141.

[70_1] *Truong Xuan Duc Ha, Variants of the Ekeland variational principle for a set-valued map involving the Clarke normal cone*, 2006, J. Math. Anal. Appl., **316**, 346-356.

[71] 1) *Zhong Cheng-Kui, On Ekeland's variational principle and a minimax theorem*, 1997, J. of Math. Anal. and Appl. **205**, 239-250.

2) *Zhong Cheng-Kui, A generalization of Ekeland's variational principle and application to the study of the relation between the weak P. S. condition and coercivity*, 1997, Nonlinear Anal., Theory, Methods and Applications, vol. 29, No. 12, 1421-1431, Elsevier Science Ltd.

Additional bibliography (post factum) with brief presentations

[1] *Chen G. Y., Huang X. X., Hou S. H., General Ekeland's variational principle for set-valued mappings*, 2000, J. of Optimization theory and applications, vol. 106, no.1, 151-164.

In the paper is presented a variant of the Ekeland principle for set-valued mappings in complete ordered metric spaces. Also a sufficient condition is given for the existence of approximate solutions for set-valued optimization problems. The obtained statements are generalizations of some propositions from [51], 2, [75], [37] and [19].

In the following

Y *is a separated real locally convex space and* C *pointed convex cone with vertex in* Y.

Some definitions

Let M be nonempty subset of Y and H nonempty subset of C. The point z_0 from M is an H *near to the minimum point of* M if

$$(z_0 - H - C) \cap M = \varnothing.$$

z_0 from M is *minimal point* or *efficient point* for M if

$$(z_0 - C) \cap M = \{z_0\}.$$

$\mathscr{E}(M)$ designates the set of minimal points of M.

M is *externally stable* if $\forall\, y$ from $M \,\exists\, y'$ in $\mathscr{E}(M)$ so that

$$y' \overset{C}{\le} y$$

($\overset{C}{\le}$ is the order relation from §8, 8.3).

For instance, any compact set is externally stable.

Let be $f: X \to 2^Y_*$ (the set of nonempty parts of Y), X a set. (x_0, y_0) from $X \times Y$ is an H *near to the minimum solution of* f *on* X if

$$y_0 \in \mathscr{E}(f(x_0)) \text{ and } (f(x) - y_0 + H) \cap -C = \varnothing \ \forall\, x \in X \setminus \{x_0\}.$$

Let $r: X \times X \to C$ be a C - metric on X (I, §8, *Minimal point theorem* IV) and $\mathscr{B}(0)$ a neighborhood base of 0 in Y. *Suppose* that the sets

$$\{x \in X: r(a, x) \in U\},$$

a goes over X and U over $\mathscr{B}(0)$, form a neighborhood basis for a separated topology on X.

A set-valued mapping $f: X \to 2^Y_*$ is said to be C-*submonotonous* at v_0 from X, if for any generalized sequence $(v_i)_{i \in I}$ from X, I totally ordered, with the properties

$$\lim_{i \in I} v_i = v_0,\ i \le j,\ y_i \in f(v_i) \text{ and } y_j \in f(v_j) \Rightarrow y_i \overset{C}{\le} y_j,$$

there exists y_0 in $f(v_0)$ so that

$$y_0 \overset{C}{\le} y_i \ \forall\, i \in I.$$

Assumption A1. *Let be* $f: X \to 2^Y_*$ and $\varepsilon > 0$. *For every* v_0 *from* X, y_0 *from* $f(v_0)$ *and* $(x_i, y_i)_{i \in I}$ *generalized sequence from* Graph f, I *totally ordered, with* $v_i \xrightarrow{i \in I} \bar{v}$ and

$$y_i - y_0 + \varepsilon r(v_i, v_0) \in -C \ \forall\, i \in I,$$

there is $\bar{y}$ *in* $f(\bar{v})$ *so that*

$$\bar{y} - y_0 + \varepsilon r(\bar{v}, v_0) \in -C.$$

Explanations. $r: X \times X \to C_0$ is a C_0 -metric on X, where C_0, convex cone with vertex in Y, is included in C.

$$\text{Graph } f = \{(x, y): x \in X, y \in f(x)\}.$$

Some statements .

Let be $f: X \to 2^Y_*$.

If C *is closed and* f *compact-valued and upper semicontinuous* (I, §9, section E), *then* f *is* C-*submonotonous*.

If C *is closed,* f *compact-valued upper semicontinuous and* r(· , a) *continuous* ∀ a ∈ ∈ X, *then* A1 *holds*.

If f *is* C-*submonotonous, then* f(v) *is externally stable* ∀ v ∈ X.

A sufficient condition for the existence of approximate optimal solutions of f on X is:

Let C_0 *be a complete* C-*bound regular convex cone and* U *from* $\mathscr{B}(0)$ *such that* $H := C_0 \setminus U \neq \varnothing$. *If* $f : X \to 2^Y_*$ *is* C-*lower bounded* (I, §9, section E) *and* $f(v)$ *externally stable* $\forall\, v \in X$, *then* f *has an* H *near to the minimum solution on* X.

Explanation. C_0, convex cone with vertex included in C, is C-bound regular if each C_0-increasing and C-bounded generalized sequence from C_0 converges to an element of C.

Main result

Let C_0 *be a complete* C-*bound regular*, (X, r) *a complete* C_0-*metric space and* $f : X \to 2^Y_*$ *satisfying*:

I f *is* C-*submonotonous with* C *closed*;

or

II C *is closed*, f(v) *is* C-*semicompact and* $f(v) \cap (y - C)$ *compact* $\forall\, v \in X$, $\forall\, y \in Y$ *and* A1 *holds*;

and

III $\exists\, (v_0, y_0)$ *in* Graph F *so that* f *is* C-*lower bounded on* $\{v \in X : (y_0 - C) \cap f(v) \neq \varnothing\}$.

Then $\forall\, \varepsilon > 0\ \exists\, (v^*, y^*)$ *in* Graph F *so that*;

IV $y_0 - y^* - \varepsilon r(v_0, v^*) \in C$;

V $(f(x) - y^* + \varepsilon r(x, v^*)) \cap -C = \varnothing\ \forall\, x \in X \setminus \{v^*\}$.

In addition, if $H := C_0 \setminus U \neq \varnothing$, U *from* $\mathscr{B}(0)$, *then*

VI $\exists\, (v_1, y_1)$ *in* Graph F *with* $y_1 \overset{C}{\leq} y_0$ *so that*

$$(f(x) - y_1 + \varepsilon H) \cap -C = \varnothing\ \forall\, x \in X \setminus \{v_1\}$$

and $\exists\, (v_\varepsilon, y_\varepsilon)$ *in* Graph F *so that* IV *and* V *hold with* (v_0, y_0) *and* (v^*, y^*) *replaced by* (v_1, y_1) *and* $(v_\varepsilon, y_\varepsilon)$ *respectively.* $(v_\varepsilon, y_\varepsilon)$ *is an* εH *near the minimum solution of* f *on* X *satisfying*

VII $r(v_1, v_\varepsilon) \in U$.

Explanation. The set-valued mapping $F : X \to 2^Y_*$ from III, IV, V, VI and VII is defined by

$$F(x) = \mathscr{E}(f(x))\ \forall\, x \in X.$$

[2] *Dincã G., Matei P., A variational method for a nonlinear Sturm-Liouville problem,* 1995, Applicable Analysis, Vol. 58, 101-121.

This paper studies the existence of a weak solution for the equation

$$(1) \qquad \sum_{i=1}^{m}(-1)^{i-1}\frac{d^{i-1}}{dx^{i-1}}\left[\alpha_i\left(\frac{d^i u}{dx^i}\right)\right]=f(x)$$

with the boundary conditions

$$(2) \qquad \begin{cases} u(a)=u'(a)=\ldots=u^{(m-1)}(a)=0 \\ u'(b)=u''(b)=\ldots=u^{(m-1)}(b)=0, \end{cases}$$

where $a, b \in \mathbf{R}$, $a < b$. The functions $\alpha_1, \ldots, \alpha_m : \mathbf{R} \to \mathbf{R}$ satisfy the condition

(H_1) *Each* α_i *is odd, strictly increasing, continuous and* $\lim_{t\to+\infty}\alpha_i(t) = +\infty$.

To solve the problem (1) + (2) Ekeland principle is used in the frame of Orlicz - Sobolev spaces. Firstly some preparations.

N-function (*M. A. Krasnoselski - Ia. B. Rutitki*). Let $\alpha : \mathbf{R}_+ \to \mathbf{R}_+$ be continuous strictly increasing with $\alpha(0) = 0$, $\lim_{t\to+\infty}\alpha(t) = +\infty$. The mapping $A : \mathbf{R} \to \mathbf{R}_+$

$$A(t) = \int_0^{|t|}\alpha(s)ds$$

is called N - *function*. The *N* - function

$$\widetilde{A} : \widetilde{A}(t) = \int_0^{|t|}\alpha^{-1}(s)ds, \ \alpha^{-1} \text{ the inverse function of } \alpha,$$

is by definition *complementary* of *A*. We have

$$(3) \qquad st \le A(s) + \widetilde{A}(t) \ \forall \ s, t \in \mathbf{R} \ (\textit{Young}).$$

Let *A*, *B* be *N* - functions. B *dominates* A *near infinity*, $A \prec B$, if $\exists \ k > 0$, $t_0 > 0$ so that

$$A(t) \le B(kt) \ \forall \ t \ge t_0 .$$

We have $A \prec B \Leftrightarrow \widetilde{B} \prec \widetilde{A}$ (complementary *N* - functions).

The *N* - function *A* satisfies the Δ_2 - *condition near infinity* when $\exists \ k > 0$, $t_0 > 0$ so that

$$A(2t) \le kA(t) \ \forall \ t \ge t_0 .$$

Orlicz spaces. Let Ω be open bounded set from $\mathbf{R}^n$ and *A* a *N* - function. The *Orlicz class* $K_A(\Omega)$ is the set of equivalence classes, modulo equality a.e. in Ω, of Lebesegue measurable real functions *u* on Ω satisfying

(4) $$\rho(A; u) := \int_{\Omega} A(u(x))dx < +\infty.$$

The *Orlicz space* $L_A(\Omega)$ is the smallest vector space, under pointwise addition and scalar multiplication, containing $K_A(\Omega)$ (the linear hull). We have $K_A(\Omega) = L_A(\Omega) \Leftrightarrow A$ satisfies the Δ_2 - condition near infinity, since $K_A(\Omega)$ is a vector space $\Leftrightarrow A$ satisfies the Δ_2–condition near infinity.

(5) $$\|u\|_A := \sup \{|< u, v >|: v \in K_{\tilde{A}}(\Omega), \rho(\tilde{A}; v) \le 1\},$$

where $< u, v > = \int_{\Omega} u(x)v(x)dx$, is a norm on $L_A(\Omega)$, the *Orlicz norm*, which makes it complete. Besides, according to (3),

(6) $$< u, v > \le \rho(A; u) + \rho(\tilde{A}; v).$$

Moreover,

(7) $$u \in L_A(\Omega),\ v \in L_{\tilde{A}}(\Omega) \Rightarrow uv \in L^1(\Omega) \text{ and } |< u, v >| \le \|u\|_A \|v\|_{\tilde{A}}.$$

Also, (8) the imbedding $L_B \to L_A$ holds $\Leftrightarrow A \prec B$.

On the vector space $L_A(\Omega)$ one sets also the norm

(9) $$\|u\|_{(A)} := \inf \{k > 0: \int_{\Omega} A\left(\frac{u(x)}{k}\right)dx \le 1\} \text{ (the \textit{Luxemburg norm})},$$

which makes it complete as well. We have

(10) $$\|u\|_{(A)} \le \|u\|_A \le 2\|u\|_{(A)}.$$

(10) shows that the property (8) is also kept in the Luxemburg norm case.

Moreover,

(11) $$|< u, v >| \le 2\|u\|_{(A)} \|v\|_{(\tilde{A})}.$$

Designate by $E_A(\Omega)$ the closure in $L_A(\Omega)$ of the vector space of bounded Lebesgue measurable functions on Ω. If A verifies the Δ_2 - condition near infinity, then

(12) $$E_A(\Omega) = K_A(\Omega) = L_A(\Omega).$$

In the contrary case,

(13) $$E_A(\Omega) \subset K_A(\Omega) \underset{\neq}{\subset} L_A(\Omega).$$

In both cases $E_A(\Omega)$ is maximal vector subspace of $K_A(\Omega)$. Besides we retain

(14) $E_A(\Omega) = L_A(\Omega) \Leftrightarrow A$ satisfies the Δ_2-condition near infinity.

The dual space $(E_A(\Omega))^*$ is isomorphic and homeomorphic to $L_{\tilde{A}}(\Omega)$. Also, every bounded sequence in $L_A(\Omega)$ contains an $E_{\tilde{A}}(\Omega)$ – weakly (or $\sigma(L_A(\Omega), E_{\tilde{A}}(\Omega))$) convergent subsequence ($(u_n)_{n\geq 1}$ from $L_A(\Omega)$ is $\sigma(L_A(\Omega), E_{\tilde{A}}(\Omega))$ convergent $\overset{def}{\Leftrightarrow}$

$\Leftrightarrow \left(\int_\Omega u_n(x)v(x)dx\right)_{n\geq 1}$ is convergent $\forall$ v from $E_{\tilde{A}}(\Omega)$). Finally, $L_{\tilde{A}}(\Omega)$ is reflexive $\Leftrightarrow$ A and $\tilde{A}$ satisfy the Δ_2 - condition near infinity, $E_A(\Omega)$ is separable, $L_A(\Omega)$ is separable $\Leftrightarrow$ A satisfies the Δ_2 - condition near infinity, hence $L_A(\Omega)$ is separable $\Leftrightarrow$ $L_A(\Omega) =$ $= E_A(\Omega)$ see (14).

Orlicz - Sobolev spaces. Let Ω be open bounded set in $\mathbf{R}^n$, A a N - function and m from $\mathbf{N}$. The *Orlicz - Sobolev space* $W^m L_A(\Omega)$ is the set of those equivalence classes in $L_A(\Omega)$ of functions u whose distributional derivatives $D^\alpha u$, $|\alpha| \leq m$, also belong to $L_A(\Omega)$. Similary define $W^m E_A(\Omega)$. $W^m L_A(\Omega)$ is a Banach space relative to the norm.

(15) $$\|u\|_{m,A} := \left(\sum_{0\leq|\alpha|\leq m} \|D^\alpha u\|_A^2\right)^{\frac{1}{2}}$$

and $W^m E_A(\Omega)$ is a closed subspace of this. We have (16) $W^m L_A(\Omega) = W^m E_A(\Omega) \Leftrightarrow A$ satisfies the Δ_2 - condition near infinity. $W^m E_A(\Omega)$ is separable. If A and $\tilde{A}$ satisfy the Δ_2 - condition near infinity, then $W^m L_A(\Omega) = W^m E_A(\Omega)$ and the Banach space is reflexive.

By definition $W_0^m L_A(\Omega)$ is taken to be the closure of $C_0^\infty(\Omega)$ (zero indicates compact support) in $W^m L_A(\Omega)$. In the definition, L_A may be replaced by E_A.

Pass to the subject of the paper ($\Omega = (a, b)$, $a, b \in \mathbf{R}$).

Consider the functions A_i, $i = \overline{1, m}$

$$A_i(t) = \int_0^t \alpha_i(s)ds ,$$

where α_i verify (H_1). These are N functions (α_i are odd). Moreover assume that

(H_2) $$A_i \succ A_{i+1} , i = \overline{i, m-1} .$$

Let be

$$W := \bigcap_{i=1}^{m} W_0^i E_{A_i}(a,b).$$

As $\mathscr{D}(a, b)$ (the set of testing functions on (a, b)) $= C_0^\infty(a, b)$, we have

$$\mathscr{D}(a, b) \subset W.$$

W is a Banach space for the norm

$$\|u\|_W := \left(\sum_{i=1}^{m} \left\| D^i u \right\|_{A_i}^2 \right)^{\frac{1}{2}}.$$

Let $\hat{W}$ be the closure of $\mathscr{D}(a, b)$ in W. The dual space $\hat{W}^*$ is isomorphic to the Banach space of those distributions T from $\mathscr{D}'(a, b)$ which verify, for some v_i from $L_{\widetilde{A}_i}(a, b)$, $i = \overline{1, m}$,

$$T(\varphi) = \sum_{i=1}^{m} \int_a^b v_i D^i \varphi dx \ \forall \ \varphi \in \mathscr{D}(a, b).$$

Consider the functional $F: W \to \mathbf{R}$,

(17) $$F(u) = \sum_{i=1}^{m} \int_a^b A_i(D^i u) dx - \sum_{i=1}^{m} \int_a^b f_i D^i u dx, \text{ where } f_i \in E_{\widetilde{A}_i}(a, b),\ i = \overline{1, m}.$$

F is lower bounded, continuous, convex and Gâteaux differentiable,. We have

(18) $$F'_W(u)(h) = \sum_{i=1}^{m} \int_a^b \alpha_i(D^i u) D^i h \, dx - \sum_{i=1}^{m} \int_a^b f_i D^i h \, dx \ \forall \ h \in W.$$

It follows the proposition which solves the existence of a weak solution for (1) + (2).

Assume verified (H_1), (H_2) *and* A_i, $\widetilde{A}_i, i = \overline{1, m}$ *satisfy the* Δ_2 *- condition near infinity. Let* f_i *be from* $E_{\widetilde{A}_i}(a, b), i = \overline{1, m}$. *Then*

a) *The element* u_0 *from* W *verifies the equation.*

$$(*) \ \sum_{i=1}^{m} \int_a^b \alpha_i(D^i u) D^i h \, dx - \sum_{i=1}^{m} \int_a^b f_i D^i h \, dx = 0 \ \forall h \in W,$$

iff u_0 *is a minimizer for the functional* (17) *with* f_i *from* (*).

b) *In* W *there is a unique element which verifies the equation* (*).

■ a) Combine (18) with I, § 5, corollary to 5.0 (F is convex).

b) *The existence*. Apply to F from (17) 3.2 from I, § 3, one obtains a sequence $(u_n)_{n\geq 1}$ from W with

$$F(u_n) \to \inf F(W),\ F'(u_n) \to 0.$$

This sequence has a subsequence $(u_{k_n})_{n\geq 1}$ with $u_{k_n} \xrightarrow{\sigma(E_{A_i}, E_{\tilde{A}_i})} u_0 \in E_{A_i}$,

$D^i u_{k_n} \xrightarrow{\sigma(E_{A_i}, E_{\tilde{A}_i})} D^i u_0$, $i = \overline{1, m}$. u_0 is in W and verifies (*).

The uniqueness. Let u_0, v_0 be solutions of (*). Then

$$\sum_{i=1}^{m} \int_a^b [\alpha_i(D^i u_0) - \alpha_i(D^i v_0)] D^i h\, dx = 0 \quad \forall\, h \in W. \tag{19}$$

Taking $h = u_0 - v_0$, it results $u_0 = v_0$, since if *par absurdum* $u_0 \neq v_0$, the functions α_i being strictly increasing one obtains that the left - hand side of (19) is > 0, contradiction. ■

P. Matei investigates in the following the more general case when the N - functions A_i and $\tilde{A}_i$, $i = \overline{1, m}$ does not necessarily verify the Δ_2 - condition near infinity (*An existence theorem for a nonlinear Sturm-Liouville problem*, Analele Universităţii Ovidiu Constanţa, vol. 3 (1), 1995, 148-156. Consider the normed spaces

$$H := \prod_{i=1}^{m} E_{A_i},\ \tilde{H} := \prod_{i=1}^{m} L_{\tilde{A}_i},\ \|u\|_H = \left(\sum_{i=1}^{m} \|u_i\|_{A_i}^2\right)^{\frac{1}{2}},\ \|v\|_{\tilde{H}} = \left(\sum_{i=1}^{m} \|v_i\|_{\tilde{A}_i}^2\right)^{\frac{1}{2}}.$$

$\tilde{H}$ may be identified with the dual H^*. The operator $\pi: \hat{W} \to H$, $\pi u = (D^1 u, \dots, D^m u)$ is linear and $\|\pi u\|_H = \|u\|_W$, hence π is an isometric isomorphism of $\hat{W}$ onto $\pi(\hat{W})$, hence $\pi(\hat{W})$ is a closed subspace of H. If X designates the $\sigma\left(\prod_{i=1}^{m} L_{A_i}, \prod_{i=1}^{m} E_{\tilde{A}_i}\right)$ - closure of $\pi(\mathcal{D}(a, b))$, it can state

Suppose (H_1), (H_2) *verified and let* f_i *be from* $E_{\tilde{A}_i}$, $i = \overline{1, m}$. *Then there exists a unique* u_0 *in* X *which satisfies the equation*

$$\sum_{i=1}^{m} \int_a^b \alpha_i(D^i u) D^i h\, dx - \sum_{i=1}^{m} \int_a^b f_i D^i h\, dx = 0 \quad \forall\, h \in X.$$

[3] *Kang B., Kim J., Applications of the Ekeland type variational principles*, 1999, Math. Sci. Res. Hot-line 3(3), 9 - 14.

This paper extends the main results of

[J] *J. S. Yung, Y. J. Cho, B. S. Lee, G. M. Lee, H. K. Pathak, Some nonconvex minimization theorems in complete metric spaces and applications*, 1997, Panamer. Math. J.**7**(3), 35-45.

This extension is made with simple proofs based on Th. 2 and Th. 4 from [54], 2 (see I, § 9, 9.67and 9.69). The considered spaces are quasi-metric [(X, d) is quasi-metric when d is not necessarily symmetric; the definition of Cauchy sequence remains the same; $x_n \to x \overset{\text{def}}{\Leftrightarrow} d(x_n, x) \to 0$ and $d(x, x_n) \to 0$, the limit is unique $(d(x, y) \le d(x, x_n) + d(x_n, y))$, every convergent sequence is Cauchy sequence $(d(x_m, x_n) \le d(x_m, x) + d(x, x_n))$, the definition of the completeness remains the same].

The following is an improved generalized version of Theorem 1 from [J].

Let X, Y *be complete quasi-metric spaces with* w - *distances* ω_X, ω_Y, $g: X \to Y$ *closed and* $h: g(X) \to \mathbf{R}$ *l.s.c. and lower bounded. Suppose that for any* u *from* X, *with* $\inf_{x\in X} h(g(x)) < h(g(u))$, *there exists* v *in* $X \setminus \{u\}$ *satisfying*

$$\max(\omega_X(u, v), \omega_Y(g(u), g(v))) \le h(g(u)) - h(g(v)).$$

Then there is x_0 *in* X *such that*

$$\inf_{x\in X} h(g(x)) = h(g(x_0)).$$

■ Take the function $F: g(X) \times g(X) \to \mathbf{R}$,

$$F(g(x), g(y)) = h(g(y)) - h(g(x).$$

F verifies (f1) and (f2) relative to $g(X)$, (I, § 9, J). Suppose *par absurdum*

$$\inf_{x\in X} h(g(x)) < h(g(y)) \quad \forall\, y \in X.$$

So for each y from X $\exists\, v$ in $X \setminus \{y\}$ so that

$$F(g(y), g(v)) + \max(\omega_X(y, v), \omega_Y(g(y), g(v))) \le 0.$$

So it can consider the function $f: y \to f(y) = v$. f has not any fixed point. But the last inequality tells that f verifies $\forall\, y$ from X

$$F(g(y), g(f(y))) + \max(\omega_X(y, f(y)), \omega_Y(g(y), g(f(y)))) \le 0$$

and 9.69 from § 9, J (take into account that h is lower bounded and takes values in **R**) imposes that f has a fixed point, contradiction and hence the conclusion. ■

With the proved statement it can improve Theorem 2 of [J], generalization of the Downing - Kirk theorem (see I, § 2, D).

Let X, Y *be complete quasi-metric spaces with* w - *distances* ω_X, ω_Y, $g: X \to Y$ *closed and* $h: g(X) \to \mathbf{R}$ *l.s.c. bounded from below. Suppose that* T: X → X *has the property*

$$\max(\omega_X(x, Tx), \omega_Y(g(x), g(Tx))) \le h(g(x)) - h(g(Tx)) \ \forall\, x \in X.$$

Then T *has a fixed point* x_0 *and*

$$\omega_X(x_0, x_0) = \omega_Y(g(x_0), g(x_0)) = 0.$$

■ Suppose that T has no fixed point. According to the hypothesis relative to T and with the proved statement one finds an $\bar{x}$ in X so that $\inf_{x\in X} h(g(x)) = h(g(\bar{x}))$. And then

$0 \le \omega_X(\bar{x}, T\bar{x}) \le \max(\omega_X(\bar{x}, T\bar{x}), \omega_Y(g(\bar{x}), g(T\bar{x}))) \le h(g(\bar{x})) - h(g(T\bar{x})) \le$ $\le h(g(T\bar{x})) - h(g(T\bar{x})) = 0$, which gives $\omega_X(\bar{x}, T\bar{x}) = 0$. Similarly, $\omega_X(\bar{x}, T^2\bar{x}) = 0$. Moreover, $\omega_X(\bar{x}, T^2\bar{x}) = 0$ (the triangle inequality). Then (W3) from I § 9, *J* gives $d(T\bar{x}, T^2\bar{x}) = 0$ (take $\varepsilon = \frac{1}{n}$, $n \in \in \mathbf{N}$), that is $T(\bar{x})$ is a fixed point of T, contradiction.

Let x_0 be a fixed point of T. Then

$0 \le \omega_X(x_0, x_0) = \omega_X(x_0, Tx_0) \le \max(\omega_X(x_0, Tx_0), \omega_Y(g(x_0), g(Tx_0))) \overset{\text{hypothesis}}{\le} h(g(x_0)) -$ $- h(g(Tx_0)) = 0$ and hence the conclusion. ■

[4] *Mc Linden L., An application of Ekeland's theorem to minimax problems*, 1982, Nonlinear Analysis, Theory, Methods and Applications, vol.6, no.2, 189-196.

Some definitions

Let X, Y be sets, $H: X \times Y \to [-\infty, +\infty]$, $\varnothing \ne A \times B \subset X \times Y$.

Minimax. (x^*, y^*) from $A \times B$ is a *minimax* for H on $A \times B$ if

(1) $$H(x^*, y) \le H(x^*, y^*) \le H(x, y^*) \ \forall\, (x, y) \in A \times B.$$

As

(2) $$\inf_{x\in A} H(x, y^*) \le \sup_B \inf_A H \le \inf_A \sup_B H \le \sup_{y\in B} H(x^*, y),$$

(1) is equivalent to

(3) $$\inf_{x\in A} H(x, y^*) = \sup_B \inf_A H = \inf_A \sup_B H = \sup_{y\in B} H(x^*, y).$$

The minimax (x^*, y^*) is *proper* when $H(x^*, y^*)$ is finite.

ε-minimax. Let ε be from $[0, +\infty)$. (x^*, y^*) from $A \times B$ is an *ε-minimax* for H on $A \times B$ if

(4) $$-\infty < \sup_{y\in B} H(x^*, y) \leq \varepsilon + \inf_{x\in A} H(x, y^*) < +\infty.$$

For ε = 0 one recovers the notion of proper minimax.

(α, β) - *local* ε - *minimax*. Let X, Y be real normed spaces, ε from $[0, +\infty)$ and α, β from $(0, +\infty]$. (x^*, y^*) from $A \times B$ is an (α, β) - *local* ε - *minimax* for H on $A \times B$ if

(5) $$-\infty < \sup_{y\in D} H(x^*, y) \leq \varepsilon + \inf_{x\in C} H(x, y^*) < +\infty,$$

where $C = A \cap S_\alpha(x^*)$, $D = B \cap S_\beta(y^*)$ closed balls).

Upper subderivative ([59)], 2, 3). Let X be real normed space, $f : X \to [-\infty, +\infty]$ and x_0, v_0 from X, $f(x_0)$ finite. *The upper subderivative of* f *at* x_0 *in the direction* v_0, $f^+(x_0; v_0)$, is

(6) $$f^+(x_0; v_0) = \sup_{t>0} \inf_{\varepsilon>0} \sup_{\substack{\|x-x_0\|<\varepsilon \\ f(x)<f(x_0)+\varepsilon \\ 0<t<\varepsilon}} \inf_{\|v-v_0\|<t} \frac{f(x+tv)-f(x)}{t}$$

(compare with the Clarke derivative, I, § 5, F, v becomes also variable).

The function $f^+(x_0, \cdot)$ is positively homogeneous convex l.s.c.

$\partial f(x_0)$ ([59], 2, 3). Let X be real normed space, $f : X \to [-\infty, +\infty]$ and x_0 from X with $f(x_0)$ finite. *The set of subgradients of* f *at* x_0, $\partial f(x_0)$, is

(7) $$\partial f(x_0) = \{\xi \in X^* : \xi(v) \leq f^+(x_0; v) \ \forall\, v \in X\}.$$

When $f(x_0)$ is infinite, one takes $\partial f(x_0) = \varnothing$. $\partial f(x_0)$ is a *-weak closed convex set.

Main results

X, Y are real Banach spaces, $\varnothing \neq A \times B \subset X \times Y$, $X \times Y$ has the norm $\|(x, y)\| = \|x\| + \|y\|$, $X^* \times Y^*$ has the norm $\|(\xi, \eta)\| = \max(\|\xi\|, \|\eta\|)$.

Associate with each (x^*, y^*), (α, β) - local ε - minimax for H on $A \times B$, the functions

(8) $$f(x) = \begin{cases} \sup_{y\in D} H(x, y), & x \in C \\ +\infty, & x \notin C \end{cases}, \quad g(y) = \begin{cases} \inf_{x\in C} H(x, y), & y \in D \\ -\infty, & y \notin D \end{cases},$$

where C and D are the sets from (5).

Theorem. Let (x^*, y^*) *be an* (α, β) - *local* ε - *minimax for* H *on* A × B *with* α, β *from* $(0, +\infty]$ *and* ε *from* $(\gamma, +\infty)$, *where* $\gamma = \inf_C \sup_D H - \sup_D \inf_C H$.

If f *is l.s.c. and* g *u.s.c.* (*see* (8)), *then* $\forall\ \lambda$ *from the interval* (0, min (α, β)) *there exists* (x_λ, y_λ) *in* $X \times Y$ *and* $(\xi_\lambda, \eta_\lambda)$ *in* $X^* \times Y^*$ *satisfying*

I $\|(x_\lambda, y_\lambda) - (x^*, y^*)\| \le \lambda$,

II $f(x_\lambda) - g(y_\lambda) \le f(x^*) - g(y^*)$,

III $f(x) - g(y) > f(x_\lambda) - g(y_\lambda) - \dfrac{\varepsilon - \gamma}{\lambda} \|(x, y) - (x_\lambda, y_\lambda)\|\ \forall\ (x, y) \in X \times Y \setminus \{(x_\lambda, y_\lambda)\}$,

IV $\|(\xi_\lambda, \eta_\lambda)\| \le \dfrac{\varepsilon - \gamma}{\lambda}$,

V $(\xi_\lambda, \eta_\lambda) \in \partial f(x_\lambda) \times \tilde{\partial} g(y_\lambda)$, *where* $\tilde{\partial} g = -\partial(-g)$.

■ Take the function $h : X \times Y \to [-\infty, +\infty]$, $h(x, y) = f(x) - g(y)$ $(+\infty - (-\infty) = +\infty)$. h is l.s.c. and $\inf\limits_{X\times Y} h = \gamma$ and $\gamma \in [0, +\infty)$ (use (5) and (2)). So $h(x^*, y^*) \le \varepsilon = (\varepsilon - \gamma) + \inf\limits_{X\times Y} h$. Apply Ekeland principle in the augmented form from Theorem 7, [59], 3 with $\varepsilon - \gamma$ and λ and it obtains (I), (II), and (III).

To prove (IV) and (V) one uses the functions $f^+(x_\lambda ; \cdot)$ and $-(-g)^+(y_\lambda ; \cdot)$. ■

Corollary 1. *Under the hypotheses of the theorem, we have for each* λ *from the interval* (0, min (α,β)) *the estimation*

$$\inf_C \sup_D H - \sup_D \inf_C H \le f(x^*) - g(y^*) - \lambda\mu,$$

where

$$\mu = \inf \{\|(\xi, \eta)\| : (\xi, \eta) \in \partial f(x) \times \tilde{\partial} g(y),\ \|(x, y) - (x^*, y^*)\| \le \lambda\} < +\infty.$$

Corollary 2. *Let be the sequences* $(\varepsilon_n)_{n\ge1}$, $(x_n^*, y_n^*)_{n\ge1}$ *so that* $\varepsilon_n \downarrow \varepsilon_\infty \in [0, +\infty)$ *and* (x_n^*, y_n^*) *is an* ε_n^2 *- minimax for* H *on* $A \times B$ *and the functions*

$$f(x) = \begin{cases} \sup\limits_{y\in B} H(x, y), & x \in A \\ +\infty, & x \notin A \end{cases}, \quad g(y) = \begin{cases} \inf\limits_{x\in A} H(x, y), & y \in B \\ -\infty, & y \notin B. \end{cases}$$

Then there are sequences $(x_n, y_n)_{n\ge1}$, $(\xi_n, y_n)_{n\ge1}$ *with the properties*

(I) $\|(x_n, y_n) - (x_n^*, y_n^*)\| \le \varepsilon_n$,

(II) $f(x_n) - g(y_n) \le f(x_n^*) - g(y_n^*)$,

(III) $f(x) - g(y) > f(x_n) - g(y_n) - \varepsilon_n \|(x, y) - (x_n, y_n)\|\ \forall\ (x, y) \in X \times Y \setminus \{(x_n, y_n)\}$,

(IV) $\|(\xi_n, \eta_n)\| \le \varepsilon_n$,

(V) $(\xi_n, \eta_n) \in \partial f(x_n) \times \tilde{\partial} g(y_n)$.

[5] *Pathak H. K., Mishra S. N., Some results related to Caristi's fixed point theorem and Ekeland's variational principle*, 2001, Demonstratio mathematica, vol. XXXIV, No.4.

This paper generalizes some results from

J. S. Jung, Y. J. Cho, J. K. Kim, Minimization theorems for fixed point theorems in fuzzy metric spaces and applications, 1994, Fuzzy Sets and Systems 61, 199-207

and

J. S. Jung, Y. J. Cho, S. M. Kang, S. S. Chang, Coincidence theorems for set-valued mappings and Ekeland's variational principle in fuzzy metric spaces, 1996, Fuzzy Sets and Systems 79, 239-250.

Definitions relative to the fuzzy metric spaces are found in [8].

The main result is the following proposition (very loaded statement!).

(*) *Let* (X_i, d_i, L, Max), $i = 1, 2$ *be two complete fuzzy metric spaces with* $\rho_{i\alpha}(x, y) < +\infty$ $\forall\, \alpha \in (0, 1]$, $\forall\, x, y \in X_i$, $E \subset X_1$ *nonempty*, $h : E \to X_1$, $g : h(E) \to X_1$ *surjective*, $f : X_1 \to X_2$ *a closed mapping*, $\theta : X_1 \to \mathbf{R}$, $\varphi : f(X_1) \to \mathbf{R}$ *bounded from below lower semicontinuous and* $(S_j)_{j\in I}$ *a family of set - valued mappings* $S_j : h(E) \to 2_*^{X_1}$. *Assume that*, $a, b, c > 0$ *and* $k : (0, 1) \to (0, +\infty)$ *decreasing with* $\sup_{r\in(0,1)} k(r) < +\infty$ *being given, if*

$x \in E \Rightarrow (g \circ h)(x) \notin \bigcap_{j\in I} S_j(h(x))$, *then* $\exists\, j_0$ *in* I *and* y *in* $S_{j_0}(h(x)) \setminus \{(g \circ h)(x)\}$ *so that*

$\max\{\rho_{1\alpha}((g \circ h)(x), y), c\rho_{2\alpha}(f((g \circ h)(x)), f(y))\} \le k(r) \min\{a[\theta((g \circ h)(x)) - \theta(y)], b[\varphi(f((g \circ h)(x))) - \varphi(f(y))]\}$ $\forall \alpha \in (0,1]$, $\forall r \in (0,1)$.

Then $\exists\, u$ *in* X_1 *so that*

$$(g \circ h)(u) \in \bigcap_{j\in I} S_j(h(u)).$$

Explanation. $[d_i(x, y)]_\alpha = [\lambda_{i\alpha}(x, y), \rho_{i\alpha}(x, y)]$, $2_*^{X_1}$ is the set of the nonempty parts of X_1. u is called *coincidence point*.

■ Define the order relation $\le$ in X_1, $x \le y \overset{\text{def}}{\Leftrightarrow} \max\{\rho_{1\alpha}(x, y), c\rho_{2\alpha}(f(x), f(y))\} \le k(r) \times \min\{a[\theta(x) - \theta(y)], b[\varphi(f(x)) - \varphi(f(y))]$ $\forall\, \alpha \in (0,1)$, $\forall\, r \in (0,1)$.

X_1 has a maximal element, let it be z. g being surjective, $\exists\, u$ in E so that $(g \circ h)(u) = z$. u is a coincidence point (*par absurdum*). ■

Corollary 1. Take in (*) only a set - valued mapping $S: h(E) \rightarrow 2_*^{X_1}$.

Corollary 2. One takes in (*) $E = X_1$, $h = g =$ the identity map and only a mapping S: $X_1 \rightarrow X_1$ ($S(x) = \{x\}$).

For $k = 1$, $b = 1$, $\theta = \varphi \circ f$ one obtains the Downing - Kirk fixed point theorem (I, § 2, 2.13) in a fuzzy metric space.

Corollary 3. Take in (*) $X_1 = X_2 = X$, $d_1 = d_2 = d$, $\theta = \varphi$, $a = b = c = 1$, f= the identity map.

Corollary 4. Take in Corollary 3 only a set-valued mapping $S: h(E) \rightarrow 2_*^{X}$.

Taking $E = X$ and $g = h =$ the identity map in Corollary 4, one obtains

Corollary 5. *Let be* (X, d, L, Max) *and* k *as above*, $\varphi: X \rightarrow \mathbf{R}$ *lower semcontinuous bounded from below and* $S: X \rightarrow X$ *with the property*

$$\rho_\alpha(x, S(x)) \leq k(r)[\varphi(x) - \varphi(S(x))] \ \forall \alpha \in (0,1], \forall r \in (0,1), \forall x \in X.$$

Then S *has a fixed point.*

Taking $k = 1$ in Corollary 5, one obtains the Caristi - Kirk fixed point theorem (I, §2, B) in a fuzzy metric space.

REMARK. In this paper the statement (3) from Theorem 3.1. has a false demonstration, the contrary affirmation of (3) being (the notation is this from original)

"$\exists\, w \neq x_0$ so that $\forall\, r$ from (0,1) $\exists\, \alpha_0$ in (0,1] for which

$$\rho_{\alpha_0}(f(x_0), f(w)) \leq k(r_0)[\Phi(f(x_0) - \Phi(f(w))]".$$

So being, "Ekeland's variational principle" must be taken out from the title.

[6] *Rozoveanu P., Ekeland's variational principle for vector - valued multifunctions*, 2000, Applied sciences, vol. 2, no.1, 56-60 (*complete presentation*).

Let X be a Banach space, Y a Banach lattice (the order relation is designated by $\leq$, exactly as those between the real numbers) with the Daniell property (every decreasing and bounded from below generalized sequence has infimum and tends to this) and $Y^+ =$ $= \{y \in Y, y \geq 0\}$.

Definition. $F: X \rightarrow 2_*^Y$ (the set of nonempty parts of Y) is *sequentially compact in* x_0, if for any $(x_i)_{i \in I}$ a generalized sequence from X with $x_i \xrightarrow{i \in I} x_0$, every sequence $(a_i)_{i \in I}$ from Y with $a_i \in F(x_i)$ has a convergent subsequence $(a_j)_{j \in J}$ (J cofinal part of I!) with

$\lim_{j\in J} a_j \in F(x_0)$.

Suppose that the multivalued function

$$(1) \qquad F: X \to 2_*^{Y^+}$$

has the properties

α) $F(x) + F(y) \subset F(x+y)\ \forall x, y \in X, F(0) = \{0\}$;

β) F *is sequentially compact in every point*;

γ) $\forall\, \varepsilon > 0\ \exists\, \delta > 0$ *so that*

$$\inf\{\|a\| : a \in F(x)\} \le \delta \Rightarrow \|x\| \le \varepsilon.$$

From γ) one deduces

$$(2) \qquad 0 \in F(x) \Rightarrow x = 0.$$

Indeed, let $\varepsilon > 0$ be any. $\exists\, \delta > 0$ given by γ), as $\inf\{\|a\| : a \in F(x)\} = 0 < \delta$, it results $\|x\| \le \varepsilon$.

Let be the set

$$\mathscr{K}(F) := \{(x, a) \in X \times Y : \exists\, b \in F(x) \text{ so that } a + b \le 0\}.$$

$\mathscr{K}(F)$ is nonempty because, according to α), $(0,0) \in \mathscr{K}(F)$. It can state:

(3) $\mathscr{K}(F) + \mathscr{K}(F) \subset \mathscr{K}(F)$;

(4) *The binary relation* $\preceq$ *in* X × Y $(x, a) \preceq (y, b) \overset{\text{def}}{\Leftrightarrow} (y, b) - (x, a) \in \mathscr{K}(F)$ *is an order relation*;

(5) *If the subset* M *of* $X \times Y^+$ *is nonempty closed, then* M *has a maximal element relative to* $\preceq$.

■ (3). Let $(x, a), (y, b)$ be from $\mathscr{K}(F)$, then $a + c_1 \le 0$ with $c_1 \in F(x)$, $b + c_2 \le 0$ with $c_2 \in F(y)$, so $(a + b) + (c_1 + c_2) \le 0$ and $c_1 + c_2 \overset{\alpha)}{\in} F(x + y)$, $(x, a) + (y, b) \in \mathscr{K}(F)$.

(4). *Antisymmetry* (*is not proved in the article*). Let be $(x, a) \preceq (x, b)$ and $(y, b) \preceq (x, a)$, that is $(y - x, b - a), (x - y, a - b \in \mathscr{K}(F)$, then $b - a + c_1 \le 0$, $c_1 \in F(y - x)$ and $a - b + c_2 \le 0$, $c_2 \in F(x - y)$, add, $c_1 + c_2 \le 0$, but $c_1 + c_2 \ge 0$, so $c_1 + c_2 = 0$, $c_1 = c_2 = 0$. It results on one hand $b - a = 0$, $b = a$ and on the other hand $0 \in F(y - x)$, hence $y = x$ (see (2)), $(x, a) = (y, b)$.

Transitivity. Let be $(x, a) \preceq (y, b)$ and $(y, b) \preceq (z, c)$, that is $(y, b) - (x, a) \in \mathscr{K}(F)$ and $(z, c) - (y, b) \in \mathscr{K}(F)$, add and take into account α).

(5). Apply in M the Zorn theorem. Let $(x_i, a_i)_{i\in I}$ be a totally ordered family of elements from M (we have bijective indexed). I is also totally ordered by

$$i \prec j \overset{\text{def}}{\Leftrightarrow} (x_i, a_i) \preceq (x_j, a_j).$$

So $(a_i)_{i\in I}$ is a generalized sequence. It is bounded from below by 0 and, in addition, decreasing: $i \prec j \overset{\text{def}}{\Rightarrow} (x_i, a_i) \preceq (x_j, a_j) \Rightarrow$ (6) $a_j - a_i + c_{ij} \le 0$, with $c_{ij} \in F(x_j - x_i)$, hence, as F takes positive values, $a_j \le a_i$. Y having the Daniell property, $\exists\, a^*$: $\inf_{i\in I} a_i = \lim_{i\in I} a_i$. From (6), $c_{ij} \le a_i - a_j$, hence $\|c_{ij}\| \le \|a_i - a_j\|$ and so, *via* γ), $(x_i)_{i\in I}$ is Cauchy sequence in X ($c_{ij} \in F(x_j - x_i)$). Let be $x^* = \lim_{i\in I} x_i$. It remains only to show that (x^*, a^*) bounds from above $(x_i, a_i)_{i\in I}$, because, M being closed, $(x^*, a^*) \in M$.

Return to (6) where we fix i. As $c_{ij} \in F(x_j - x_i)$ and $x_j - x_i \xrightarrow{j\in I} x^* - x_i$, one obtains by β) a convergent subsequence $(c_{ij})_{j\in J}$ of $(c_{ij})_{j\in I}$, $c_{ij} \xrightarrow[j\in J]{} c_i$ and $c_i \in F(x^* - x_i)$. From (6),

$$a^* - a_i + c_i \le 0,$$

that is, according to the definition, $(x_i, a_i) \preceq (x^*, a^*)$. ■

Now it can affirm

Let X *be a Banach space*, Y *Banach lattice with the Daniell property*, X_0 *nonempty subset of* X *and* $\varphi : X_0 \to 2_*^{Y^+}$ *multivalued with the property.*

$$\text{epi}\ \varphi := \{(x, y) \in X_0 \times Y^+ : y \in \varphi(x) + Y^+)$$

closed.

Assume that for $\varepsilon > 0$ *there exists* u_ε *in* X_0 *and* b_ε *in* $\varphi(u_\varepsilon)$ *so that*

$$b_\varepsilon \in \mu\left(\bigcup_{x\in X_0} (\varphi(x) + \varepsilon v_0)\right).$$

Then $\forall\, \lambda > 0\ \exists\, v_\lambda$ *in* X_0, a_λ *in* $\varphi(v_\lambda)$ *and* c_λ *in* $F(v_\lambda - u_\varepsilon)$ *so that*

$$(7)\quad c_\lambda \le \lambda v_0, \qquad (8)\quad a_\lambda \le b_\varepsilon - \frac{\varepsilon}{\lambda} c_\lambda,$$

$$(9)\qquad \text{epi}\ \varphi \cap ((v_\lambda, a_\lambda) + \mathscr{K}\left(\frac{\varepsilon}{\lambda} F\right)) = \{(v_\lambda, a_\lambda)\}.$$

Explanation. $\mu(A)$ is the set of the elements which bound from below the set A. F is the set-valued function from (1).

■ The set-valued mapping $\frac{\varepsilon}{\lambda}F$ has the properties from α) , β) and γ), hence may be taken into account the set $\mathscr{K}\left(\frac{\varepsilon}{\lambda}F\right)$, let $\preceq$ be the corresponding order relation. $M := $ $=$ epi φ being closed but also nonempty ($y + 0$, y from $\varphi(x)$ (nonempty set!), is in epi φ), it has a maximal element (v_λ, a_λ) (see (5)). The definition from (4) imposes (9).

Pass to (8). Because $(v_\lambda, a_\lambda) \in$ epi φ, the definition gives $a_\lambda \in b_\lambda + Y^+$ with $b_\lambda \in \varphi(v_\lambda)$. Then $0 \geq b_\lambda - a_\lambda = b_\lambda - a_\lambda + 0$ and $0 \overset{\alpha)}{\in} \frac{\varepsilon}{\lambda}F(v_\lambda - v_\lambda)$, i.e. $(v_\lambda, b_\lambda) \succeq (v_\lambda, a_\lambda)$. The second term being maximal, *a fortiori* $b_\lambda = a_\lambda$ and hence (10) $a_\lambda \in \varphi(v_\lambda)$. As $b_\varepsilon \in \varphi(u_\varepsilon)$, it results $(u_\varepsilon, b_\varepsilon) \in$ epi φ and hence

$$(u_\varepsilon, b_\varepsilon) \preceq (v_\lambda, a_\lambda),$$

that is (see (4)) $\exists\ c'_\lambda \in \frac{\varepsilon}{\lambda}F(v_\lambda - u_\varepsilon)$ so that $a_\lambda - b_\varepsilon + c'_\lambda \leq 0$, as $c'_\lambda = \frac{\varepsilon}{\lambda}c_\lambda$ with $c_\lambda \in F(v_\lambda -$ $- u_\varepsilon)$, (8) is proved.

Pass to (7). Because b_ε bounds from below $\bigcup_{x \in X_0}(\varphi(x) + \varepsilon v_0)$, we have $b_\varepsilon \leq b + \varepsilon v_0$ $\forall\ b \in \varphi(x)$, $\forall\ x \in X_0$ and combining with (8) one obtains

$$\frac{\varepsilon}{\lambda}c_\lambda \leq b_\varepsilon - a_\lambda \leq b + \varepsilon v_0 - a_\lambda \ \forall\ b \in \varphi(x),\ \forall\ x \in X_0 .$$

For $x = v_\lambda$ and $b = a_\lambda$ (see (10)) one finds $\frac{\varepsilon}{\lambda}c_\lambda \leq \varepsilon v_0$, that is (7). ■

Remarks relative to the article (in the author notations).

1° The relation $\preceq$ in $E \times F$ is proved to be only a preorder relation.

2° The set-valued functions δ and Φ (F and φ to us) must have as values only nonempty sets.

3° The definition of the multifunction *sequentially compact* must be given with generalized sequences.

4° In the statement of Th.1, Φ must take values in F_+ and epi Φ must be closed in $E \times$ $\times F_+$ for correct application of Proposition 1.

[7$_1$] *Suzuki T., Generalized distance and existence theorems in complete metric spaces*, 2001, J. Math, Anal. Appl., 253, 440-458.

{7_2} *Suzuki T., The strong Ekeland variational principle*, 2006, J. Math. Anal. Appl., 320, 787-794.

Definition ([7_1]). Let (X, d) be a metric space. The function $q: X \times X \to [0, +\infty)$ is *τ-distance on* X, if there exists $\eta : X \times [0, +\infty) \to [0, +\infty)$ which together with q verifies:

(τ1) $q(x, y) \le q(x, z) + q(z, y)\ \forall\ x, y, z \in X$;

(τ2) η is concave, continuous in its second variable and $\eta(x, 0) = 0$, $\eta(x, t) \ge t\ \forall\ x \in$ $\in X, \forall\ t \ge 0$;

(τ3) $\lim\limits_{n\to\infty} x_n = x_0$ and ∃ a sequence (z_n) so that $\lim\limits_{n\to\infty} \sup\{\eta(z_n, q(z_n, x_m)): m \ge n\} = 0 \Rightarrow$ $q(v, x_0) \le \varliminf\limits_{n\to\infty} q(v, x_n)\ \forall\ v \in X$;

(τ4) $\lim\limits_{n\to\infty} \sup\{q(x_n, y_m): m \ge n\} = 0$ and $\lim\limits_{n\to\infty} \eta(x_n, t_n) = 0 \Rightarrow \lim\limits_{n\to\infty} \eta(y_n, t_n) = 0$;

(τ5) $\lim\limits_{n\to\infty} \eta(z_n, q(z_n, x_n)) = 0$ and $\lim\limits_{n\to\infty} \eta(z_n, q(z_n, y_n)) = 0 \Rightarrow \lim\limits_{n\to\infty} d(x_n, y_n) = 0$.

If q is τ-distance on X, then $c\,q$, $c > 0$ has the same property.

For instance, d is τ-distance on X (take $\eta(x, t) = t$; for (τ3) the second condition is useless and for (τ4) – the first condition; for (τ5) use triangle inequality).

1. *τ-distance variant of Ekeland principle* ([7_1]). *Let* (X, d) *be a complete metric space*, q *τ-distance on* X *and* $\varphi : X \to (-\infty, +\infty]$ *lower bounded, lower semicontinuous and proper. Then*

1° ∀ u *from* X ∃ v *in* X *so that* $\varphi(v) \le \varphi(u)$ *and* $\varphi(v) < \varphi(w) + q(v, w)\ \forall\ w \in X \setminus \{v\}$;

2° ∀ u *from* X *with* $q(u, u) = 0$ *and* ∀ ε > 0 ∃ v_ε *in* X *so that*

$$\varphi(v_\varepsilon) + \varepsilon q(u, v_\varepsilon) \le \varphi(u),$$

$$\varphi(v_\varepsilon) < \varphi(w) + \varepsilon q(v_\varepsilon, w)\ \forall\ w \in X \setminus \{v_\varepsilon\}.$$

Definition ([7_1]). Let (X, d) be a metric space and q τ-distance on X. The sequence $(x_n)_{n\ge1}$ from X is q-*Cauchy sequence*, if for the function η which verifies (τ2) – (τ5) there exists a sequence $(z_n)_{n\ge1}$ which verifies the condition which appears in (τ3).

2. *If* ∃ z *in* X *so that* $\lim\limits_{n\to\infty} q(z, x_n) = 0$, *then* $(x_n)_{n\ge1}$ *is a* q-*Cauchy sequence* ([7_1]).

3. *Any* q-*Cauchy sequence is also Cauchy sequence relative to* d ([7_1]).

In [32_1] P.G. Georgiev proves the following *strong* Ekeland variational principle:

4. *Let* (X, d) *be complete metric space and* $\varphi : X \to (-\infty, +\infty]$ *lower bounded, lower* *semicontinuous and proper. For any* u *from* X *and* $\varepsilon > 0$, $\delta > 0$ $\exists$ $v_{\varepsilon,\delta}$ *in* X *so that*

1° $\varphi(v_{\varepsilon,\delta}) + \varepsilon\, d(u, v_{\varepsilon,\delta}) < \varphi(u) + \delta$;

2° $\varphi(v_{\varepsilon,\delta}) < \varphi(w) + \varepsilon\, d(v_{\varepsilon,\delta}, w)$ $\forall$ $w \in X \setminus \{v_{\varepsilon,\delta}\}$;

3° *If the sequence* $(x_n)_{n\geq 1}$ *from* X *verifies*

$$\lim_{n\to\infty} [\varphi(x_n) + \varepsilon\, d(v_{\varepsilon,\delta}, x_n)] = \varphi(v_{\varepsilon,\delta}), \textit{ then } \lim_{n\to\infty} x_n = v_{\varepsilon,\delta}.$$

The main result of [7_2] is the following generalization of **4**.

5. *τ-distance variant of the strong Ekeland variational principle. Let* X *be complete metric space,* q *τ-distance on* X *and* $\varphi: X \to (-\infty, +\infty]$ *lower bounded, lower semicontinuous and proper. For any* u *from* X *with* $q(u, u) = 0$ *and for each* $\varepsilon > 0$, $\delta > 0$, $\exists$ $v_{\varepsilon,\delta}$ *in* X *so that*

1° $\varphi(v_{\varepsilon,\delta}) \leq \varphi(u)$;

2° $\varphi(v_{\varepsilon,\delta}) + \varepsilon\, q(u, v_{\varepsilon,\delta}) < \varphi(u) + \delta$;

3° $\varphi(v_{\varepsilon,\delta}) < \varphi(w) + \varepsilon\, q(v_{\varepsilon,\delta}, w)$ $\forall$ $w \in X \setminus \{v_{\varepsilon,\delta}\}$;

4° *If the sequence* $(x_n)_{n\geq 1}$ *from* X *verifies*

$$\lim_{n\to\infty} [\varphi(x_n) + \varepsilon\, q(v_{\varepsilon,\delta}, x_n)] = \varphi(v),$$

then (x_n) *is a* q*-Cauchy sequence,* $\lim_{n\to\infty} x_n = v_{\varepsilon,\delta}$ and $q(v_{\varepsilon,\delta}, v_{\varepsilon,\delta}) = \lim_{n\to\infty} q(v_{\varepsilon,\delta}, x_n) =$ $= 0$.

■ Because 1° and 2° are obviously verified when $\varphi(u) = +\infty$ and in 3° and 4° u not appears, it can suppose $\varphi(u) < +\infty$.

Take ε' from $(0, \varepsilon)$ so that (1) $\dfrac{\varepsilon - \varepsilon'}{\varepsilon'}[\varphi(u) - \inf\varphi(X)] < \delta$. 2° from **1** gives $v_{\varepsilon,\delta}$, we note $v = v_{\varepsilon,\delta}$, so that

$$(2) \quad \varphi(v) \leq \varphi(u) - \varepsilon' q(u, v), \qquad (3) \quad \varphi(v) < \varphi(w) + \varepsilon' q(v, w) \ \forall\ w \in X \setminus \{v\}.$$

1° from the statement is obviously verified. In addition,

$$\varphi(v) = \left(1 + \frac{\varepsilon - \varepsilon'}{\varepsilon'}\right)\varphi(v) - \frac{\varepsilon - \varepsilon'}{\varepsilon'}\varphi(v) \overset{(2)}{\leq} \left(1 + \frac{\varepsilon - \varepsilon'}{\varepsilon'}\right)[\varphi(u) - \varepsilon' q(u, v)] - \frac{\varepsilon - \varepsilon'}{\varepsilon'}\varphi(v) = \varphi(u)$$

$$- \varepsilon q(u, v) + \frac{\varepsilon - \varepsilon'}{\varepsilon'}[\varphi(u) - \varphi(v)] \leq \varphi(u) - \varepsilon\, q(u, v) + \frac{\varepsilon - \varepsilon'}{\varepsilon'}[\varphi(u) - \inf\varphi(X)] \overset{(1)}{<} \varphi(u) -$$

$- \varepsilon\, q(u, v) + \delta$ and consequently 2° from the statement.

We have $\forall$ w from $X \setminus \{v\}$

$\varphi(w) \overset{(3)}{>} \varphi(v) - \varepsilon' q(v, w) > \varphi(v) - \varepsilon q(v, w)$ and consequently 3° from the statement.

Pass to 4° from the statement. Obviously (4) $\varphi(v) \overset{(3)}{\leq} \varphi(w) + \varepsilon' q(v, w)\ \forall\, w \in X$. Then $\overline{\lim_{n\to\infty}}\, q(v, x_n) = \overline{\lim_{n\to\infty}} \dfrac{\varepsilon q(v, x_n) - \varepsilon' q(v, x_n)}{\varepsilon - \varepsilon'} \overset{(4)}{\leq} \lim_{n\to\infty} \dfrac{\varepsilon q(v, x_n) + \varphi(x_n) - \varphi(v)}{\varepsilon - \varepsilon'} = 0$ (the hypothesis), hence $\lim_{n\to\infty} q(v, x_n) = 0$ and hence (x_n) is a q-Cauchy sequence (see **2**). But then (x_n) is also Cauchy sequence (see **3**), let so be $x_0 = \lim_{n\to\infty} x_n$. (τ3) and (τ2) give (5) $q(v, x_0) = 0$ (take $z_n = v$, $\lim_{m\to\infty} \eta(v, q(v, x_m)) = \eta(v, \lim_{m\to\infty} q(v, x_m)) = \eta(v, 0) = 0$). (6)$\varphi(x_0) \overset{\text{l.s.c.}}{\leq}$ $\leq \underline{\lim}_{n\to\infty}\, \varphi(x_n) \leq \lim_{n\to\infty} [\varphi(x_n) + \varepsilon q(v, x_n)] = \varphi(v)$ and if *par absurdum* $x_0 \neq v$ then $\varphi(x_0) \overset{(3)}{>}$ $> \varphi(v) - \varepsilon' q(v, x_0) \overset{(6)}{\geq} \varphi(v) \overset{(5)}{\geq} \varphi(x_0)$, consequently a contradiction and $v = x_0$, $q(v, v) = 0$. ■

[8] *Zhu J., Zhong Ch., Wang G., An extension of Ekeland's variational principle in fuzzy metric spaces and its applications*, 1999, Fuzzy Sets and Systems 108, 353-363.

Some preparations are necessary. For E preordered set and x an element from E, $E(x)$ designates the initial interval $\{y \in E \setminus \{x\} : y \leq x\}$.

Transfinite induction. Let E be a well-ordered set (every nonempty subset has a least element) and A a subset. If

$$\forall\, x \in E\colon E(x) \subset A \Rightarrow x \in A$$

then $A = E$.

■ Suppose *par absurdum* $E \setminus A \neq \varnothing$ and let x_0 be the least element of $E \setminus A$. This imposes $E(x_0) \subset A$ and hence $x_0 \in A$, a contradiction. ■

Ordinal numbers. A set α with the properties

(1) $$x \in \alpha \text{ and } y \in \alpha \Rightarrow \text{either } x \in y, \text{ or } y \in x, \text{ or } x = y$$

and

(2) $$x \in y \text{ and } y \in \alpha \Rightarrow x \in \alpha$$

is called *ordinal number.* The relation $\in$ is not an order relation as $x \in x$ is false. In each nonempty subset A of α there is a unique a_0 (the *first element* of A) so that, $\forall\, x$ from A, either $a_0 \in x$, or $a_0 = x$. The first element of α is $\varnothing$. If $z \in \alpha$, then z is also an ordinal number.

For any α, β ordinal numbers, we have

$$\text{either } \alpha \subset \beta, \text{ or } \beta \subset \alpha.$$

In addition,

$$\alpha \neq \beta \text{ and } \alpha \subset \beta \Leftrightarrow \alpha \in \beta .$$

Note by $\mathscr{X}$ the class of all ordinal numbers. $\mathscr{X}$ is well-ordered by the relation $\alpha \leq \beta \underset{\text{def}}{\Leftrightarrow} \alpha \subset \beta$. $\mathscr{X}$ is not a set. For any α from $\mathscr{X}$, all the elements of $\mathscr{X}\setminus\{\alpha\}$ which minorize α form a set $\mathscr{X}(\alpha)$ isomorphic to α (*isomorphism* ≡ order-preserving bijection). Each well-ordered set *E* is isomorphic to a suitable $\mathscr{X}(\alpha)$, α is called the *ordinal number of* E.

So the first few elements of $\mathscr{X}$ are

$$\varnothing;\ \{\varnothing\};\ \{\varnothing, \{\varnothing\}\};\ \{\varnothing, \{\varnothing\}, \{\varnothing, \{\varnothing\}\}\}.$$

∅, the least ordinal number, is denoted by 0, {∅} is denoted by 1, {0,1} is denoted by 2 etc.

If α is any ordinal number, $\alpha \cup \{\alpha\}$ is also, according to (1) and (2), an ordinal number, this is designated by α +1. This is by definition a *successor*. There are ordinal numbers ≠ 0 not of the form $\alpha \cup \{\alpha\}$, these are called *limit ordinal numbers* and have no immediate predecessor in $\mathscr{X}$.

An ordinal α is called *countable* if there is a bijection of α on $\mathbf{Z}_+$.

Ω designates the *first uncountable ordinal* (each equipotence class contains a least ordinal number).

Transfinite positive series. An expression of the from

$$\sum_{i \in \alpha} x_i ,$$

where $\alpha \in \Omega$ and x_i are real positive numbers, is called *transfinite positive series*. The *sum* of this is likewise designated and is defined by transfinite induction as follows. For *n* natural number, $\sum_{i \in n} x_i = x_0 + \ldots + x_{n-1}$ (see above). Suppose $\sum_{i \in \lambda} x_i$ defined, λ any from $\mathscr{X}(\alpha)$. Then if α = β + 1, $\sum_{i \in \alpha} x_i = \sum_{i \in \beta} x_i + x_\alpha$ and if α is a limit ordinal number, $\sum_{i \in \alpha} x_i =$ $\sup \left\{ \sum_{i \in \lambda} x_i : \lambda \in \alpha \right\}$. The series $\sum_{i \in \alpha} x_i$ is called *convergent* resp. *divergent* as its sum is finite resp. equal to $+\infty$. We have, $\forall\, k \geq 0$, $\sum_{i\in\alpha} kx_i = k \sum_{i\in\alpha} x_i$.

Fuzzy metric spaces (*O. Kaleva, S. Seikkala, On fuzzy metric spaces*, Fuzzy Sets and Systems (12) 1984, 215-229). The set of all upper semicontinuous convex fuzzy numbers is denoted by E, the set of all nonnegative fuzzy members in E by G and the additive and multiplicative identities of fuzzy numbers by $\bar{0}$ and $\bar{1}$ respectively.

Let x be from E and α from $\mathbf{R}_+$. The α-level set $[x]_\alpha$ of x is a closed interval $[a^\alpha, b^\alpha]$, where if $a^\alpha = -\infty$ resp. $b^\alpha = +\infty$ then $[a^\alpha, b^\alpha] = (-\infty, b^\alpha]$ resp. $[a^\alpha, b^\alpha] = [a^\alpha, +\infty)$. The sequence $(x_n)_{n\geq 1}$ from E is by definition α-*level convergent* to x from E if, $\forall\, \alpha$ from $(0,1]$, $\lim_{n\to\infty} a_n^\alpha = a^\alpha$ and $\lim_{n\to\infty} b_n^\alpha = b^\alpha$, where $[x_n]_\alpha = [a_n^\alpha, b_n^\alpha]$, $[x]_\alpha = [a^\alpha, b^\alpha]$.

Let X be nonempty set, $\mathrm{d} : X \times X \to G$ and $L : [0,1] \times [0,1] \to [0,1]$ symmetrical increasing in both arguments with $L(0,0) = 0$ and Max: $[0,1] \times [),1] \to [0,1]$. One notes, for x, y from X and α from $(0,1]$,

$$[\mathrm{d}(x, y)]_\alpha = [\lambda_\alpha(x, y), \rho_\alpha(x, y)].$$

The quadruple $(X, \mathrm{d}, L, \mathrm{Max})$ is by definition *fuzzy metric space* and d *fuzzy metric* if

(3) $\mathrm{d}(x, y) = \bar{0} \Leftrightarrow x = y$,

(4) $\mathrm{d}(x, y) = \mathrm{d}(y, x)\ \forall\, x, y \in X$,

(5) $\forall\, x, y, z$ from X we have

a) $\mathrm{d}(x, y)(s + t) \geq L(\mathrm{d}(x, z)(s), \mathrm{d}(z, y)(t))$,

whenever $s \leq \lambda_1(x, z)$, $t \leq \lambda_1(z, y)$, $s + t \leq \lambda_1(x, y)$;

b) $\rho_\alpha(x, y) \leq \rho_\alpha(x, z) + \rho_\alpha(z, y)\ \forall\, \alpha$ from $(0,1]$.

Let $(X, \mathrm{d}, L, \mathrm{Max})$ be a fuzzy metric space. The sets

(6) $$U_x(\varepsilon, \alpha) := \{y \in X : \rho_\alpha(x, y) < \varepsilon\},$$

where ε goes over $(0, +\infty)$, α over $(0,1]$ and x over X, form a basis for a metrisable topology. $\lim_{n\to\infty} x_n = x_0 \Leftrightarrow \lim_{n\to\infty} \mathrm{d}(x_n, x_0) = \bar{0}$. The limit is unique. Every convergent sequence in X is Cauchy sequence (it results from b) of (5)) and all its subsequences are convergent.

$(X, \mathrm{d}, L, \mathrm{Max})$ is by definition *complete* if every Cauchy sequence is convergent.

Pass to the subject.

The main result of this paper is the statement which follows, "fuzzy" variant of the theorem 1.1_1 from I, § 1.

(*) *Ekeland principle in fuzzy metric spaces. Let* (X, d, L, Max) *be a complete fuzzy metric space and* $\varphi : X \to (-\infty, +\infty]$ *bounded from below, lower semicontinuous and proper. For any* $\varepsilon > 0$ *and* u *from* X *with*

$$\text{(7)} \qquad \varphi(u) < \inf \varphi(X) + \varepsilon$$

and for each $\lambda > 0$, u_0 *from* X, $k : (0,1] \to (0, +\infty)$ *and* $h : [0, +\infty) \to [0, +\infty)$ *increasing with* $\int_0^{+\infty} \frac{dr}{1+h(r)} = +\infty$, *there exists* v_ε *in* X so *that*

(8) $\forall\, \alpha$ *from* (0,1] $\exists\, c(\alpha) > 0$ *satisfying*

$$\rho_\alpha(v_\varepsilon, u) \le \frac{\lambda}{\varepsilon}\, c(\alpha)\, k(\alpha)[\varphi(u) - \varphi(v_\varepsilon)];$$

$$\text{(9)} \qquad \rho_\alpha(v_\varepsilon, u) \le \lambda\, c(\alpha)\, k(\alpha),\ \rho_\alpha(u_0, v_\varepsilon) \le r_0(\alpha) + r_1(\alpha)\ \forall\, \alpha \text{ from } (0,1],$$

where $r_0(\alpha) = \rho_\alpha(u, u_0)$ *and* $r_1(\alpha)$ *verifies* $\displaystyle\int_{r_0(\alpha)}^{r_0(\alpha)+r_1(\alpha)} \frac{dr}{1+h(r)} \ge \lambda\, k(\alpha)$;

(10) $\forall\, w$ *from* $X \setminus \{v_\varepsilon\}$ $\exists\, \alpha_0$ *in* (0,1] *so that* $k(\alpha_0)[\varphi(v_\varepsilon) - \varphi(w)] < \dfrac{\varepsilon \rho_{\alpha_0}(v_\varepsilon, w)}{\lambda[1 + h(\rho_{\alpha_0}(u_0, v_\varepsilon))]}$.

■ Define the preorder relation on X $x \le y \Leftrightarrow \exists\, \nu \in \Omega$ and $\exists$ generalized sequence $(x_i)_{i \in \nu+1}$ with the properties:

(A_1) $x_0 = x$, $x_\nu = y$ and for any $\beta \in \nu + 1$ which is limit ordinal the generalized sequence $(x_i)_{i \in \beta}$ converges to x_β;

(A_2) $k(\alpha)\,\varphi(y) + \displaystyle\sum_{i \in \nu} \frac{\varepsilon \rho_\alpha(x_i, x_{i+1})}{\lambda\, [1 + h(\rho_\alpha(u_0, x_i))]} \le k(\alpha)\varphi(x)\ \forall\, \alpha$ from (0,1].

Consder the set

$$\pi(u) := \{x \in X : x \ge u\}$$

and let x be from $\pi(u)$. According to the definition $\exists\, \nu$ in Ω and $(x_i)_{i \in \nu+1}$ which verify (A_1) and (A_2). Then we have

$$\text{(11)} \qquad \rho_\alpha(u_0, x_i) < r_0(\alpha) + r_1(\alpha)\ \forall\, i \in \nu,\ \forall\, \alpha \in (0,1].$$

Moreover, every increasing sequence $(x_n)_{n \ge 1}$ from $\pi(u)$ has an upper bound. Then, as φ is decreasing with respect to $\le$ and lower bounded, it can apply 1.1_2 fron I, § 1 (where $v_1 \overset{B}{\le} v_2 \overset{def}{\Leftrightarrow} v_1 \ge v_2$), $\exists\, v_\varepsilon$ in $\pi(u)$ so that $\varphi(w) = \varphi(v_\varepsilon)\ \forall\, w \ge v_\varepsilon$. According to the definition, $\exists\, (x_i)_{i \in \nu+1}$ so that

(12) $$\sum_{i\in v}\frac{\varepsilon\rho_\alpha(x_i,x_{i+1})}{\lambda[1+h(\rho_\alpha(u_0,x_i))]}\le k(\alpha)[\varphi(u)-\varphi(v_\varepsilon)]\ \forall\ \alpha \text{ from } (0,1].$$

But

(13) $$\rho_\alpha(u,v_\varepsilon)\le c(\alpha)\sum_{i\in v}\frac{\rho_\alpha(x_i,x_{i+1})}{1+h(\rho_\alpha(u_0,x_i))}\quad \forall\alpha \text{ from } (0,1].$$

From (12) and (13) one obtains (8).

Combining with (7) and (8) one finds

$$\rho_\alpha(u,v_\varepsilon)\le\frac{\lambda}{\varepsilon}c(\alpha)k(\alpha)[\varphi(u)-\inf\varphi(X)]<\lambda c(\alpha)k(\alpha)\ \forall\alpha \text{ from } (0,1]$$

and from (11) results $\rho_\alpha(u_0,v_\varepsilon)\le r_0(\alpha)+r_1(\alpha)$ $\forall\alpha$ from (0,1].

Pass to (10) and suppose *par absurdum* that it is false, hence $\exists\ w\ne v_\varepsilon$ so that

(14) $$\frac{\varepsilon\rho_\alpha(v_\varepsilon,w)}{\lambda[1+h(\rho_\alpha(u_0,v_\varepsilon))]}\le k(\alpha)[\varphi(v_\varepsilon)-\varphi(w)]\ \forall\alpha \text{ from } (0,1],$$

i.e. $w\ge v_\varepsilon$, hence $\varphi(w)=\varphi(v_\varepsilon)$ and so, from (14), $\varphi_\alpha(v_\varepsilon, w)=0$ $\forall$ α from (0,1], that imposes $v_\varepsilon = w$, a contradiction. ■

The theorem (*) from above is applied to Menger PM - spaces (X, F, T) where $T =$ = Min (see I, § 10)), As results from a Kaleva - Seikkala statement (see above), defining $d: X\times X\to G$ by

$$d(x,y)(t)=\begin{cases}1-F_{xy}(t), & \text{if } t\ge t_{xy}\\ 0, & \text{if } t<t_{xy}\end{cases}$$

where $t_{xy}=\sup\{t\in\mathbf{R}: F_{xy}(t)=0\}$ and taking $L = 0$, one obtains (*X*, d, *L*, Max) fuzzy metric space. In this we have $\forall$ α from (0,1].

(15) $$\rho_\alpha(x,y)=\sup\{t\in\mathbf{R}: d(x,y)(t)\ge\alpha\}=\sup\{t\in\mathbf{R}: F_{xy}(t)\le 1-\alpha\},$$

that implies

$(x_n)_{n\ge1}$ Cauchy sequence in (*X*, *F*, Min) $\Leftrightarrow$ $(x_n)_{n\ge1}$ Cauchy sequence in (*X*, d, *L*, Max), consequently

(*X*, *F*, Min) is complete $\Leftrightarrow$ (*X*, d, *L*, Max) is complete.

So being, the teorem (*) from above becomes, *via* (15).

Let (X, F, Min) *be a complete Menger* PM - *space and* $\varphi: X\to(-\infty,+\infty]$ *lower bounded, lower semicontinuous and proper. For any* $\varepsilon>0$ *and* u *from* X *with* $\varphi(u) <$ $<\inf\varphi(X)+\varepsilon$ *and for each* λ, u_0, k *and* h *as to* (*), *there exists* v_ε *in* X *so that*

(16) $\forall\ \alpha$ *from* (0,1] $\exists\ c(\alpha) > 0$ *satisfying* $\sup\{t \in \mathbf{R}:\ F_{v_\varepsilon u}(t) \le 1 - \alpha\} \le \frac{\lambda}{\varepsilon}\, c(\alpha)k(\alpha)\times$

$\times[\varphi(u) - \varphi(v_\varepsilon)]$;

(17) $\sup\{t \in \mathbf{R}: F_{v_\varepsilon u}(t) \le 1 - \alpha\} \le \lambda\, c(\alpha)k(\alpha)$ *and* $\sup\{t \in \mathbf{R}: F_{v_\varepsilon u_0}(t) \le 1 - \varepsilon\} \le r_0(\alpha) +$

$+\, r_1(\alpha)\ \forall\ \alpha$ *from* (0,1],

where $r_0(\alpha) = \sup(\{t \in \mathbf{R}: F_{u_0 u}(t) \le 1 - \alpha\}$ *and* $r_1(\alpha)$ *is as in* (9);

(18) $\forall$ w *from* $X \setminus \{v_\varepsilon\}\ \exists\ \alpha_0$ *in* (0,1] *so that*

$$k(\alpha_0)[\varphi(v_\varepsilon) - \varphi(w)] < \frac{\varepsilon\ \sup\{t \in \mathbf{R}: F_{v_\varepsilon w}(t) \le 1 - \alpha_0\}}{\lambda\ [1 + h(\sup\{t \in \mathbf{R}: F_{v_\varepsilon u_0}(t) \le 1 - \alpha_0\})]}.$$

INDEX

U.

V.

W.